Golden® MATHEMATICS

CLASS X

Strictly as per Latest NCERT textbook and CBSE Guidelines
Including Questions as per latest
CCE
(Continuous and Comprehensive Evaluation)
Scheme of CBSE

MATHEMATICS

CLASS X

[A BOOK WITH A DIFFERENCE]

- Important Points to Remember
- NCERT Textbook Exercises with Solutions
- Additional Important Examples with Solutions
- Test Your Knowledge Questions for Practice
- Additional Multiple Choice Questions

Dr. Hari Kishan

LAXMI SCHOOL BOOKS DIVISION
NEW AGE INTERNATIONAL (P) LIMITED, PUBLISHERS

New Delhi • Bangalore • Chennai • Cochin • Guwahati • Hyderabad
Jalandhar • Kolkata • Lucknow • Mumbai • Ranchi
Visit us at www.newagepublishers.com

Copyright © 2010, New Age International (P) Ltd., Publishers
Published by New Age International (P) Ltd., Publishers
New Edition

Golden logo in English & Hindi is registered with Govt. of India, copyright office with Registration No. A-78534/2007 & A-78533/2007 respectively in name of Laxmi Publications (P) Ltd., and further New Age International (P) Ltd., is authorised to use the logo.

Rights are reserved. This logo may not be reproduced in any form, without the written permission of the copyright owner. Or else it should be an infringement of copyright which should be a criminal offence.

No part of this book may be reproduced in any form, by photostat, microfilm, xerography, or any other means, or incorporated into any information retrieval system, electronic or mechanical, without the written permission of the copyright owner.

Branches:

- 36, Malikarjuna Temple Street, Opp. ICWA, Basavanagudi, **Bangalore**-560 004. Tel.: (080) 26677815 Telefax: 26615247, E-mail: bangalore@newagepublishers.com
- 26, Damodaran Street, T. Nagar, **Chennai**-600 017. Tel.: (044) 24353401, Telefax: 24351463 E-mail: chennai@newagepublishers.com
- CC-39/1016, Carrier Station Road, Ernakulam South, **Cochin**-682 016. Tel.: (0484) 2377004, Telefax: 4051303 E-mail: cochin@newagepublishers.com
- Hemsen Complex, Mohd. Shah Road, Paltan Bazar, Near Starline Hotel, **Guwahati**-781 008. Tel.: (0361) 2513881 Telefax: 2543669, E-mail: guwahati@newagepublishers.com
- No. 105, 1st Floor, Madhiray Kaveri Tower, 3-2-19, Azam Jahi Road, Nimboliadda, **Hyderabad**-500 027. Tel.: (040) 24652456, Telefax: 24652457, E-mail: hyderabad@newagepublishers.com
- RDB Chambers (Formerly Lotus Cinema)106A, 1st Floor, S.N. Banerjee Road, **Kolkata**-700 014. Tel.: (033) 22273773, Telefax: 22275247, E-mail: kolkata@newagepublishers.com
- 16-A, Jopling Road, **Lucknow**-226 001. Tel.: (0522) 2209578, 4045297, Telefax: 2204098 E-mail: lucknow@newagepublishers.com
- 142C, Victor House, Ground Floor, N.M. Joshi Marg, Lower Parel, **Mumbai**-400 013. Tel.: (022) 24927869 Telefax: 24915415, E-mail: mumbai@newagepublishers.com
- 22, Golden House, Daryaganj, **New Delhi**-110 002. Tel.: (011) 23262370, 23262368, Telefax: 43551305 E-mail: sales@newagepublishers.com

ISBN : 978-81-224-2111-8

Rs. 270.00

C-10-01-4274

G10-4618-270-G. MATHEMATICS X E

Printed in India at Print 'O' Pack, Delhi.
Typeset at Goswami Associates, Delhi.

PUBLISHING FOR ONE WORLD
NEW AGE INTERNATIONAL (P) LIMITED, PUBLISHERS
4835/24, Ansari Road, Daryaganj, New Delhi-110002
Visit us at **www.newagepublishers.com**

PREFACE

The book **'Golden Mathematics'** has been written strictly as per the latest N.C.E.R.T. textbook and C.B.S.E. guidelines for the students associated with Class X.

Special features of the book are as follows:

- It has been designed primarily as a textbook for the students of Class X.
- It has been written in a lucid style and simple language.
- It includes the necessary material which provides a sound conceptual base in Mathematics.
- It contains a large number of well graded examples which will help the students to solve the exercise.
- It contains step by step working rule for solving the problems in words, supported by illustrations.
- It contains summary of each chapter at the end so as to take overview of the whole chapter.
- It gives due weightage to each topic according to the marks and periods allotted by the C.B.S.E.
- All exercises of the NCERT textbook and intext questions have been fully solved at proper places.
- Objective type questions based on CCE guidelines are given at the end of each chapter.
- Multiple Choice Questions to assess the practical skills of the students are given at the end.

I am confident that Golden Mathematics will be well received by all. Constructive suggestions for the improvement of the book shall be gracefully accepted.

AUTHOR

SYLLABUS

CLASS X–Theory

UNIT I : NUMBER SYSTEMS

1. Real Numbers   (15 Periods)

Euclid's division lemma, Fundamental Theorem of Arithmetic - statements after reviewing work done earlier and after illustrating and motivating through examples, Proofs of results - irrationality of $\sqrt{2}, \sqrt{3}, \sqrt{5}$, decimal expansions of rational numbers in terms of terminating/non-terminating recurring decimals.

UNIT II : ALGEBRA

1. Polynomials   (6 Periods)

Zeros of a polynomial. Relationship between zeros and coefficients of a polynomial with particular reference to quadratic polynomials. Statement and simple problems on division algorithm for polynomials with real coefficients.

2. Pair of Linear Equations in Two Variables   (15 Periods)

Pair of linear equations in two variables. Geometric representation of different possibilities of solutions/inconsistency.

Algebraic conditions for number of solutions. Solution of pair of linear equations in two variables algebraically, by substitution, by elimination and by cross multiplication. Simple situational problems must be included. Simple problems on equations reducible to linear equations may be included.

3. Quadratic Equations   (15 Periods)

Standard form of a quadratic equation $ax^2 + bx + c = 0, (a \neq 0)$. Solution of the quadratic equations (only real roots) by factorization and by completing the square, *i.e.*, by using quadratic formula. Relationship between discriminant and nature of roots.

Problems related to day to day activities to be incorporated.

4. Arithmetic Progressions   (8 Periods)

Motivation for studying AP. Derivation of standard results of finding the nth term and sum of first n terms.

UNIT III : TRIGONOMETRY

1. Trigonometric Ratios   (12 Periods)

Trigonometric ratios of an acute angle of a right-angled triangle. Proof of their existence (well defined); motivate the ratios, whichever are defined at 0° and 90°. Values (with proofs) of the trigonometric ratios of 30°, 45° and 60°. Relationships between the ratios.

2. **Trigonometric Identities** (16 Periods)

Proof and applications of the identity $\sin^2 A + \cos^2 A = 1$. Only simple identities to be given. Trigonometric ratios of complementary angles.

3. **Heights and Distances** (8 Periods)

Simple and believable problems on heights and distances. Problems should not involve more than two right triangles. Angles of elevation/depression should be only 30°, 45°, 60°.

UNIT IV : COORDINATE GEOMETRY

1. **Lines (In two-dimensions)** (15 Periods)

Review the concepts of coordinate geometry done earlier including graphs of linear equations. Awareness of geometrical representation of quadratic polynomials. Distance between two points and section formula (internal). Area of a triangle.

UNIT V : GEOMETRY

1. **Triangles** (15 Periods)

Definitions, examples, counter examples of similar triangles.

1. (Prove) If a line is drawn parallel to one side of a triangle to intersect the other two sides in distinct points, the other two sides are divided in the same ratio.
2. (Motivate) If a line divides two sides of a triangle in the same ratio, the line is parallel to the third side.
3. (Motivate) If in two triangles, the corresponding angles are equal, their corresponding sides are proportional and the triangles are similar.
4. (Motivate) If the corresponding sides of two triangles are proportional, their corresponding angles are equal and the two triangles are similar.
5. (Motivate) If one angle of a triangle is equal to one angle of another triangle and the sides including these angles are proportional, the two triangles are similar.
6. (Motivate) If a perpendicular is drawn from the vertex of the right angle of a right triangle to the hypotenuse, the triangles on each side of the perpendicular are similar to the whole triangle and to each other.
7. (Prove) The ratio of the areas of two similar triangles is equal to the ratio of the squares on their corresponding sides.
8. (Prove) In a right triangle, the square on the hypotenuse is equal to the sum of the squares on the other two sides.
9. (Prove) In a triangle, if the square on one side is equal to sum of the squares on the other two sides, the angles opposite to the first side is a right triangle.

2. **Circles** (8 Periods)

Tangents to a circle motivated by chords drawn from points coming closer and closer and closer to the point.

1. (Prove) The tangent at any point of a circle is perpendicular to the radius through the point of contact.
2. (Prove) The lengths of tangents drawn from an external point to circle are equal.

3. **Constructions** (8 Periods)
   1. Division of a line segment in a given ratio (internally).
   2. Tangent to a circle from a point outside it.
   3. Construction of a triangle similar to a given triangle.

UNIT VI : MENSURATION

1. **Areas of Plane Figures** (12 Periods)

   Motivate the area of a circle; area of sectors and segments of a circle. Problems based on areas and perimeter/circumference of the above said plane figures. (In calculating area of segment of a circle, problems should be restricted to central angle of 60°, 90° & 120° only. Plane figures involving triangles, simple quadrilaterals and circle should be taken.)

2. **Surface Areas and Volumes** (12 Periods)

   (i) Problems on finding surface areas and volumes of combinations of any two of the following : cubes, cuboids, spheres, hemispheres and right circular cylinders/cones. Frustum of a cone.

   (ii) Problems involving converting one type of metallic solid into another and other mixed problems. (Problems with combination of not more than two different solids be taken.)

UNIT VII : STATISTICS AND PROBABILITY

1. **Statistics** (15 Periods)

   Mean, median and mode of grouped data (bimodal situation to be avoided). Cumulative frequency graph.

2. **Probability** (10 Periods)

   Classical definition of probability. Connection with probability as given in Class IX. Simple problems on single events, not using set notation.

Scheme of Examination Reforms and Continuous and Comprehensive Evaluation (CCE)*

Class X– Academic Session 2010-11 onwards

1. **Senior Secondary Schools**

    In Senior Secondary Schools, there will be no Board examination at Class X since the students will be entering Class XI in the same school.

    These students will be assessed through the CCE internally by the school as per the strengthened CCE Scheme in Class X (*for two terms, the first term from April 2010 to September 2010 and the second from October–March 2011*).

    At the end of the academic year 2010-11, students will be issued the CCE certificate on the pre-printed stationery to be supplied by the Board. **These CCE certificates, once they are complete in all respects (for both Classes IX and X) will be required to be sent to the Regional Offices for the signatures of the Board official.**

    However the Board will provide **flexibility to the following students in Senior Secondary schools** also to appear in Board's external (pen and paper written/online) examination (described separately below):
    - The students wanting to terminate their studies in the school for admission in Pre-University, vocational course, etc.
    - The students wanting to shift to the other schools of other State Boards due to local reasons.

    Moreover, those students who wish to assess themselves vis-à-vis **their peers or for self-motivation** will be allowed to appear in an **On Demand** (pen and paper/online) **Proficiency test.**

2. **Secondary Schools:**

    In all schools upto secondary level there will be Board's external (*pen and paper written/on-line*) Examination at the end of Class X as detailed in *para 3 below* since the students will be moving out of these schools.

    Note: The students in Classes IX and X in Secondary Schools also will follow the CCE as described above. At the end of the Class X, students will be issued the CCE certificates on the pre-printed stationery supplied by the Board.

3. **External (pen and paper written/online) Examination**
    - These mainly application oriented external (*pen and paper written/online*) Examinations will be based on the same syllabi as detailed in the Curriculum Document 2011.
    - These will be certified by the CBSE.
    - **Concessions being given to the Differently Abled**

        All the relaxations such as use of scribe for visually challenged, choice of optional subjects, use of computers for visually challenged being provided the present Board Examinations of Class X to the differently-abled children need to be continued in the School Based Assessment also, at the formative as well as Summative level. Due consideration will also be given to these students in co-scholastic evaluation too.

    - **Aptitude Test**
        1. The Board will offer an Aptitude Test (*optional*) which along with other school records and CCE would help the students, parents and teachers in deciding the choice of the subjects at Class XI.
        2. The Board proposes to provide an opportunity to students to undertake the Aptitude Test at the end of Class IX and then at the end of Class X.
    - **Admission in Class XI (Academic Session 2011–12)**
        1. For the purpose of admission in Class XI the CCE certificate will be relied upon.
        2. It is also recommended that some amount of weightage be assigned to the co-scholastic aspects especially Life Skills and excellence in sports for allotting subjects in class XI. A multi-pronged approach for assigning subjects needs to be adopted. **Aptitude test, Scholastic Performance** and **Co-Scholastic Achievements, all need to be given weightage.**

---

* As per the CBSE Circular No. 39/20-09-2009

3. Students of the same school may be given preference over the students coming from any other school for admission in Class XI.

4.1 Evaluation of Scholastic areas:

Each term will have **two Formative assessments** and **one Summative assessment** for evaluation of Scholastic areas.

4.1.1 Formative Assessment:

Formative assessment is a tool used by the teacher to continuously monitor student progress in a non-threatening and supportive environment. If used effectively it can improve student performance tremendously while raising the self esteem of the child and reducing the work load of the teacher. Some of the main features of Formative assessment are that it is diagnostic and remedial, provides effective feedback to students, allows for the active involvement of students in their own learning, enables teachers to adjust teaching to take account of the results of assessment and recognizes the profound influence that assessment has on the motivation and self-esteem of students, both of which are crucial influences in learning.

It is highly recommended that the school should not restrict the Formative assessment to only a paper-pencil test. There are other means of testing such as through quizzes, conversations, interviews, oral testing, visual testing, projects, practicals and assignments.

Assessments done periodically will be shown to the students/parents so as to encourage continuous participatory improvement.

4.1.2 Summative Assessment:

The Summative assessment is the terminal assessment of performance at the end of instruction. Under the **end term Summative assessment**, the students will be tested internally based on the following criteria:

- Curriculum and Syllabus for Classes X will be the same as circulated by the Board earlier.
- The Summative assessment will be in the form of a pen-paper test conducted by the schools themselves. It will be conducted at the end of each term.
- In order to ensure standardisation, and to ensure uniformity, the Question Banks in different subjects to generate question papers will be forwarded by the Board to schools.
- In order to cater to difference in the pace of responding, the Schools will give flexible timing to the students during end term Summative assessment.
- Evaluation of answer scripts will be done by the school Teachers themselves on the basis of the Marking Scheme provided by the Board.
- There will be random verification of the assessments procedures carried on by schools by the Board officials/nominees appointed by the Board.

The Weightage of **Formative Assessment (FA)** and **Summative Assessment (SA)** shall be as follows:

Term	Type of Assessment	Percentage of weightage in academic session	Term-wise weightage	Total
FIRST TERM (April–Sept.)	Formative Assessment-1	10%	Formative Assessment-1 + 2 = 20%	Formative = 40%, Summative = 60% **Total 100%**
	Formative Assessment-2	10%		
	Summative Assessment-1	20%	Summative Assessment-1 = 20%	
SECOND TERM (Oct–March)	Formative Assessment-3	10%	Formative Assessment-3 + 4 = 20%	
	Formative Assessment-4	10%		
	Summative Assessment-2	40%	Summative Assessment-2 = 40%	

4.2 Evaluation of Co-Scholastic areas:

4.2.1 In addition to the Scholastic areas, co-scholastic areas like Life Skills; Attitudes & Values; Participation & Achievement in activities involving Literary & Creative Skills, Scientific Skills, Aesthetic Skills and Performing Arts & Clubs; and Health & Physical Education will also be evaluated. Most of the schools are already implementing activities involving these areas. The schools have been trained under Adolescence Education Programme (AEP), emphasising upon Life Skills; the schools are also aware about Comprehensive School Health Programme introduced in 2006 (*Circular No. 9/06/29/07, 27&48/08*). However, for ready reference and convenience of the schools, the activities under Co-Scholastic areas and evaluation thereof are also included in the comprehensive guidelines on various aspects of CCE (Refer Para 5 below).

5. Comprehensive guidelines on various aspects of CCE will be available in the *Teachers' Manual on School Based Assessment* shortly. This will also be hosted on the CBSE website (www.cbse.nic.in)

CONTENTS

Preface	...	(*v*)
Syllabus	...	(*vi*)

Chapter   **Pages**

1. Real Numbers ... 1
2. Polynomials ... 27
3. Pair of Linear Equations in Two Variables ... 68
4. Quadratic Equations ... 146
5. Arithmetic Progressions ... 180
6. Triangles ... 226
7. Coordinate Geometry ... 287
8. Inroduction to Trigonometry ... 316
9. Some Applications of Trigonometry ... 353
10. Circles ... 370
11. Constructions ... 384
12. Areas Related to Circles ... 398
13. Surface Areas and Volumes ... 425
14. Statistics ... 461
15. Probability ... 513
- Appendix A1: Proofs in Mathematics ... 535
- Appendix A2: Mathematical Modelling ... 554
- **Additional Multiple Choice Questions with Answers** ... **563–678**
- **Higher Order Thinking Skills (HOTS) Questions** ... **679–690**
- **Solutions to Higher Order Thinking Skills (HOTS) Questions** ... **691–728**
- **New Age Revision Test Paper 2011** ... **729–732**
- *Examination Papers* ... 733–757

1

Real Numbers

IMPORTANT POINTS

1. Properties of Positive Integers. There are two very important properties of positive integers. These are: (1) Euclid's Division Algorithm (2) The Fundamental Theorem of Arithmetic.

2. Euclid's Division Lemma. *Given positive integers a and b, there exist unique integers q and r satisfying $a = bq + r$, $0 \leq r < b$.*

Examples:

(i) Consider the pair of integers 17, 6. We can write the following relation for this pair.

$$17 = 6 \times 2 + 5$$

(6 goes into 17 twice and leaves a remainder of 5)

Here, $a = 17$
$b = 6$
$q = 2$
$r = 5$ $(0 \leq 5 < 6)$

(ii) Consider the pair of integers 10, 3. We can write the following relation for this pair.

$$10 = 3 \times 3 + 1$$

(3 goes into 10 thrice and leaves a remainder of 1)

Here, $a = 10$
$b = 3$
$q = 3$
$r = 1$ $(0 \leq 1 < 3)$

(iii) Consider the pair of integers 5, 12. We can write the following relation for this pair.

$$5 = 12 \times 0 + 5$$

(This relation holds since 12 is larger than 5)

Here, $a = 5$
$b = 12$
$q = 0$
$r = 5$ $(0 \leq 5 < 12)$

(iv) Consider the pair of integers 20, 4. We can write the following relation for this pair.

$$20 = 4 \times 5 + 0$$

(Here 4 goes into 20 five-times and leaves no remainder)

Here, $a = 20$
$b = 4$
$q = 5$
$r = 0$ $(0 \leq 0 < 4)$

(v) Consider the pair of integers 4, 19. We can write the following relation for this pair.

$$4 = 19 \times 0 + 4$$

(This relation holds since 19 is larger than 4)

Here, $a = 4$
$b = 19$
$q = 0$
$r = 0$ $(0 \leq 0 < 19)$

(vi) Consider the pair of integers 81, 3. We can write the following relation for this pair.

$$81 = 3 \times 27 + 0$$

(Here 3 goes into 81 twenty seven-times and leaves no remainder)

Here, $a = 81$
          $b = 3$
          $q = 27$
          $r = 0$ $(0 \leq 0 < 3)$

Note 1. From the above examples, we note that q and r can also be zero.

Note 2. From the above examples, we note that q and r are unique. These are the only integers satisfying the conditions $a = bq + r$, where $0 \leq r < b$.

Note 3. Euclid's Division Lemma is nothing but a restatement of the long division process with a, b, q and r as dividend, divisor, quotient and remainder respectively.

3. Euclid's Division Algorithm. This is based on Euclid's Division Lemma. Euclid's Division Algorithm is a technique to compute the Highest Common Factor (HCF) of two given positive integers. We know that the HCF of two positive integers a and b is the largest positive integer and that divides both a and b. According to this, the HCF of any two positive integers a and b, with $a > b$, is obtained as follows :

Step 1. Apply Euclid's division lemma to a and b to find whole numbers q and r such that $a = bq + r$, $0 \leq r < b$.

Step 2. If $r = 0$, b is the HCF of a and b. If $r \neq 0$, apply Euclid's division lemma to b and r.

Step 3. Continue the process that the remainder is zero. The divisor at this stage will be the required HCF. This algorithm works because HCF (a, b) = HCF (b, r) where the symbol HCF (a, b) denotes the HCF of a and b, etc.

Example:

Suppose we need to find the HCF of the integers 455 and 42. We start with the larger integer, that is, 455. Then we use Euclid's Division Lemma to get

$$455 = 42 \times 10 + 35$$

Now, consider the divisor 42 and the remainder 35 and apply Euclid's Division Lemma to get

$$42 = 35 \times 1 + 7$$

Now, consider the divisor 35 and the remainder 7 and apply Euclid's Division Lemma to get

$$35 = 7 \times 5 + 0$$

Muhammad ibn Musa al-Khawrizmi
(A.D. 780–850)

We notice that the remainder has become zero, and we can not proceed any further.

We *claim* that the *HCF of 455 and 42 is the divisor at this stage, i.e., 7.*

(We can verify this by listing all the factors of 455 and 42.)

We notice that 7 = HCF (35, 7) = HCF (42, 35) = HCF (455, 42). This is how this method works.

Remark: Euclid's division lemma and algorithm are so closely interlinked that people often call former as the division algorithm also.

ILLUSTRATIVE EXAMPLES

[NCERT Exercise 1.1]
(Page No. 7)

Example 1. *Use Euclid's division algorithm to find the HCF of :*
  (i) *135 and 225*   (ii) *196 and 38220*
  (iii) *867 and 255.*

Sol. (i) **135 and 225.** Start with the larger integer, that is, 225. Apply the division lemma to 225 and 135, to get
$$225 = 135 \times 1 + 90$$
Since the remainder $90 \neq 0$, we apply the division lemma to 135 and 90, to get
$$135 = 90 \times 1 + 45$$
We consider the new divisor 90 and the new remainder 45, and apply the division lemma to get
$$90 = 45 \times 2 + 0$$
The remainder has now become zero, so our procedure stops.

Since the divisor at this stage is 45, the HCF of 225 and 135 is 45.

(ii) **196 and 38220.** Start with the larger integer, that is, 38220. Apply the division lemma to 38220 and 196, to get
$$38220 = 196 \times 195 + 0$$
The remainder is zero, so our procedure stops.

Since the divisor is 196, the HCF of 38220 and 196 is 196.

(iii) **867 and 255.** Start with the larger integer, that is, 867. Apply the division lemma to get
$$867 = 255 \times 3 + 102$$
Since the remainder $102 \neq 0$, we apply the division lemma to 255 and 102, to get
$$255 = 102 \times 2 + 51$$
We, consider the new divisor 102 and the new remainder 51, and apply the division lemma to get
$$102 = 51 \times 2 + 0$$
The remainder has now become zero, so our procedure stops.

Since the divisor at this stage is 51, the HCF of 867 and 255 is 51.

Example 2. *Show that any positive odd integer is of the form $6q + 1$, or $6q + 3$ or $6q + 5$, where q is some integer.*

Sol. Let us start with taking a, where a is any positive odd integer. We apply the division algorithm, with a and $b = 6$. Since $0 \leq r < 6$, the possible remainders are 0, 1, 2, 3, 4, 5. That is, a can be $6q$, or $6q + 1$, or $6q + 2$, or $6q + 3$, or $6q + 4$, or $6q + 5$, where q is the quotient. However, since a is odd, we do not consider the cases $6q$, $6q + 2$ and $6q + 4$ (since all the three are divisible by 2).

Therefore, any positive odd integer is of the form $6q + 1$, or $6q + 3$, or $6q + 5$.

Example 3. *An army contingent of 616 members is to march behind an army band of 32 members in a parade. The two groups are to march in the same number of columns. What is the maximum number of columns in which they can march?*

Sol. By applying the Euclid's division lemma, we can find the maximum number of columns in which an army contingent of 616 members can march behind an army band of 32 members in a parade. HCF of 616 and 32 is equal to maximum number of columns in which 616 and 32 members can march.

Since $616 > 32$, we apply the division lemma to 616 and 32, to get
$$616 = 32 \times 19 + 8$$
Since the remainder $8 \neq 0$, we apply the division lemma to 32 and 8, to get
$$32 = 8 \times 4 + 0$$
The remainder has now become zero, so our procedure stops. Since the divisor at this stage is 8, the HCF of 616 and 32 is 8.

Therefore, the maximum number of columns in which an army contingent of 616 members can march behind an army band of 32 members in a parade is 8.

Example 4. *Use Euclid's division lemma to show that the square of any positive integer is either of the form $3m$ or $3m + 1$ for some integer m.*

[**Hint:** Let x be any positive integer, then it is of the form $3q$, $3q + 1$ or $3q + 2$. Now square each of these and show that they can be rewritten in the form $3m$ or $3m + 1$.]

Sol. Let a be any odd positive integer. We apply the division lemma with a and $b = 3$. Since $0 \le r < 3$, the possible remainders are 0, 1 and 2. That is, a can be $3q$, or $3q + 1$, or $3q + 2$, where q is the quotient.

Now, $(3q)^2 = 9q^2$

which can be written in the form $3m$, since 9 is divisible by 3.

Again, $(3q + 1)^2 = 9q^2 + 6q + 1$
$= 3(3q^2 + 2q) + 1$

which can be written in the form $3m + 1$ since $9q^2 + 6q$, i.e., $3(3q^2 + 2q)$ is divisible by 3.

Lastly,
$(3q + 2)^2 = 9q^2 + 12q + 4$
$= (9q^2 + 12q + 3) + 1$
$= 3(3q^2 + 4q + 1) + 1$

which can be written in the form $3m + 1$, since $9q^2 + 12q + 3$, i.e., $3(3q^2 + 4q + 1)$ is divisible by 3.

Therefore, the square of any positive integer is either of the form $3m$ or $3m + 1$ for some integer on.

Example 5. *Use Euclid's division lemma to show that the cube of any positive integer is of the form $9m$, $9m + 1$ or $9m + 8$.*

Sol. Let a be any positive integer. We apply the division lemma, with a and $b = 3$. Since $0 \le r < 3$, the posible remainders are 0, 1 and 2. That is, a can be $3q'$, or $3q' + 1$, or $3q' + 2$, where q' is the quotient.

Now, $(3q')^3 = 9q'^3$ which can be written in the from $9m$, since 9 is divisible by 9.

Again, $(3q' + 1)^3 = 27q'^3 + 27q'^2 + 9q' + 1$
$= 9(3q'^3 + 3q'^2 + q') + 1$

which can be written in the form

$9m + 1$, since $27q'^3 + 27q'^2 + 9q'$,
i.e., $9(3q'^3 + 3q'^2 + q')$ is divisible by 9.

Lastly,
$(3q' + 2)^3 = 27q'^3 + 54q'^2 + 36q' + 8$
$= 9(3q'^3 + 6q'^2 + 4q') + 8$

which can be written in the form $9m + 8$, since $27q'^3 + 54q'^2 + 36q'$, i.e., $9(3q'^3 + 6q'^2 + 4q')$ is divisible by 9.

Therefore, the cube of any positive integer is of the form $9m$, $9m + 1$ or $9m + 8$.

ADDITIONAL EXAMPLES

Example 6. *Use Euclid's division algorithm to find the HCF of 42237 and 75582.*

Sol. Start with the larger integer, that is, 75582. Apply the division lemma to 75582 and 42237 to get

$75582 = 42237 \times 1 + 33345$

Since the remainder $33345 \ne 0$, we apply the division lemma to 42237 and 33345, to get

$42237 = 33345 \times 1 + 8892$

We consider the new divisor 33345 and the new remainder 8892 and apply the division lemma to 33345 and 8892, to get

$33345 = 8892 \times 3 + 6669$

We consider the new divisor 8892 and the new remainder 6669 and apply the division lemma to 8892 and 6669, to get

$8892 = 6669 \times 1 + 2223$

We consider the new divisor 6669 and the new remainder 2223 and apply the division lemma to 6669 and 2223, to get

$6669 = 2223 \times 3 + 0$

The remainder has now become zero, so our procedure stops.

Since the divisor at this stage is 2223, the HCF of 75582 and 42237 is 2223.

REAL NUMBERS

Example 7. *Find the greatest weight which can be contained exactly in 6 kg 7 hg 4 dag 3g and 9 kg 9 dag 7g.*

Sol. 6 kg 7 hg 4 dag 3 g = 6743 g
9 kg 9 dag 7 g = 9097 g

The greatest weight required (in g) is the HCF of 6743 and 9097.

Start with the larger integer, that is, 9097. Apply the division lemma to 9097 and 6743, to get

$$9097 = 6743 \times 1 + 2354$$

Since the remainder $2354 \neq 0$, we apply the division lemma to 6743 and 2354, to get

$$6743 = 2354 \times 2 + 2035$$

We consider the new divisor 2354 and the new remainder 2035 and apply the division lemma to 2354 and 2035, to get

$$2354 = 2035 \times 1 + 319$$

We consider the new divisor 2035 and new the remainder 319 and apply the division lemma to 2035 and 319, to get

$$2035 = 319 \times 6 + 121$$

We consider the new divisor 319 and the new remainder 121 and apply the division lemma to 319 and 121, to get

$$319 = 121 \times 2 + 77$$

We consider the new divisor 121 and the new remainder 77 and apply the division lemma to 121 and 77, to get

$$121 = 77 \times 1 + 44$$

We consider the new divisor 77 and the new remainder 44 and apply the division lemma to 77 and 44, to get

$$77 = 44 \times 1 + 33$$

We consider the new divisor 44 and the new remainder 33 and apply the division lemma to 44 and 33, to get

$$44 = 33 \times 1 + 11$$

We consider the new divisor 33 and the new remainder 11 and apply the division lemma to 33 and 11, to get

$$33 = 11 \times 3 + 0$$

The remainder has now become zero, so our procedure stops.

Since the HCF at this stage is 11, HCF of 9097 and 6743 is 11.

Therefore, the required greatest weight is 11 g.

TEST YOUR KNOWLEDGE

1. Use Euclid's algorithm to find the HCF of 4052 and 12576.
   Using Euclid's algorithm:
2. Find the HCF 84 and 105, using Euclid's algorithm.
3. Find the HCF of 595 and 107, using Euclid's algorithm.
4. Find the HCF of 861 and 1353, using Euclid's algorithm.
5. Find the HCF of 616 and 1300, using Euclid's algorithm.
6. Show that every positive even integer is of the form $2q$, and that every positive odd integer is of the form $2q + 1$, where q is some integer.
7. Show that any positive odd integer is of the form $4q + 1$ or $4q + 3$, where q is some integer.
8. Show that one and only one out of n, $n + 2$ or $n + 4$ is divisible by 3, where n is any positive integer.
9. Find the greatest length which can be contained exactly in 10 m 5 dm 2 cm 4 mm and 12 m 7 dm 5 cm 2 mm.
10. Find the greatest measure which is exactly contained in 10 litres 857 millilitres and 15 litres 87 millilitres.

Answers

1. 4 2. 21
3. 119 4. 123
5. 4 9. 4 mm
10. 141 millilitres.

IMPORTANT POINTS

1. Natural numbers (Positive integers) as obtained by multiplying prime numbers. We know that a positive integer p ($p \neq 1$) is a prime if and the only positive divisor of p are 1 and p. For example : 2, 3, 5, 7, 11, 13, ... are the first few primes. Also, we know that any natural number can be written as a product of its prime factors. For instance, $2 = 2$, $4 = 2 \times 2$, $253 = 11 \times 23$, and so on. Now, we shall look at natural numbers from the other direction. That is, we shall try whether every natural number can be obtained by multiplying prime numbers.

Take any random collection of prime numbers, say : 2, 3, 7, 11 and 23. If we multiply some or all of these numbers, allowing them to repeat as many times as we wish, we can produce a large collection of positive integers from the above numbers (in fact, infinitely many). Few are listed below :

$$7 \times 11 \times 23 = 1771$$
$$2 \times 3 \times 7 \times 11 \times 23 = 10626$$
$$2^2 \times 3 \times 7 \times 11 \times 23 = 21252$$
$$3 \times 7 \times 11 \times 23 = 5313$$
$$2^3 \times 3 \times 7^3 = 8232$$

and so on.

Now, let us suppose that our collection of primes includes all the possible primes. Since, there are infinitely many primes, therefore, if we combine all these primes in all possible ways, we will get an infinite collection of numbers (natural numbers or positive integers), all the primes and all possible products of primes.

2. Composite numbers as products of primes. To decide whether all the composite numbers can be expressed as the products of the powers of primes, let us factorise positive integers using the factor tree. Let us take some large number, say, 32760 and factorise it as shown below:

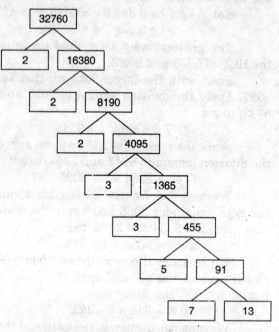

So we have factorised $32760 = 2 \times 2 \times 2 \times 3 \times 3 \times 5 \times 7 \times 13$ as a product of primes, i.e., $32760 = 2^3 \times 3^2 \times 5 \times 7 \times 13$ as a product of powers of primes. Let us try another number, say, 123456789. This can be written as $3^2 \times 3803 \times 3607$. Of course, this is true for all other natural numbers also.

This leads us to conjecture that every composite number can be written as the product of powers of primes. In fact, this statement is true and is called the **Fundamental theorem of Arithmetic** because of its basic crucial importance to the study of integers.

3. Fundamental Theorem of Arithmetic. *Every composite number can be expressed (factorised) as a product of primes, and this factorisation is unique, apart from the order in which the prime factors occur.*

Explanation:

The Fundamental Theorem of Arithmetic says that every composite number can be factorised as a product of primes. Actually it says more. It says that given any composite number it can be factorised as a product of prime numbers in a

REAL NUMBERS

'unique' way, except for the order in which the primes occur. That is, given any composite number there is one and only one way to write it as a product of primes, as long as we are not particular about the order in which the primes occur. So, for example, we regard $2 \times 3 \times 5 \times 7$ as the same as $3 \times 5 \times 7 \times 2$, or any other possible order in which these primes are written.

This fact is also stated in the following form:

The prime factorisation of a natural number is unique, except for the order of its factors.

Generalisation. In general, given a composite number x, we factorise it as $x = p_1 p_2 \ldots p_n$, where $p_1, p_2, \ldots, p_n$ are primes and written in ascending order, i.e., $p_1 \leq p_2 \leq \ldots \leq p_n$. If we combine the same primes, we will get powers of primes. For example,

$$32760 = 2 \times 2 \times 2 \times 3 \times 3 \times 5 \times 7 \times 13$$
$$= 2^3 \times 3^2 \times 5 \times 7 \times 13$$

Uniqueness of decomposition. Once we have decided that the order will be ascending, then the way the number is decomposed is unique.

4. The HCF of two or more numbers is the product of the smallest power of each common prime factor in the numbers, whereas the LCM of two or more numbers is the product of the greatest power of each prime factor, involved in the numbers.

Carl Friedrich Gauss
(1777–1855)

5. For any two positive integers **a** and **b**, HCF (a, b) × LCM (a, b) = a × b.

ILLUSTRATIVE EXAMPLES

[NCERT Exercise 1.2]
(Page No. 11)

Example 1. *Express each number as product of its prime factors:*

(i) 140   (ii) 156
(iii) 3825   (iv) 5005
(v) 7429.

Sol. (i) 140

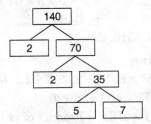

So, $140 = 2 \times 2 \times 5 \times 7 = 2^2 \times 5 \times 7$

(ii) 156

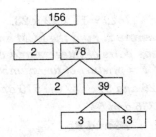

So, $156 = 2 \times 2 \times 3 \times 13 = 2^2 \times 3 \times 13$

(iii) **3825**

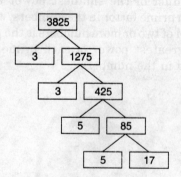

So, $3825 = 3 \times 3 \times 5 \times 5 \times 17$
$= 3^2 \times 5^2 \times 17$

(iv) **5005**

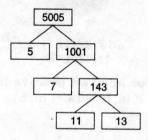

So, $5005 = 5 \times 7 \times 11 \times 13$.

(v) **7429**

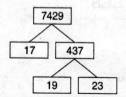

So, $7429 = 17 \times 19 \times 23$.

Example 2. *Find the LCM and HCF of the following pairs of integers and verify that LCM × HCF = product of two numbers.*

(i) 26 and 91   (ii) 510 and 92
(iii) 336 and 54

Sol. (i) **26 and 91**

So, $26 = 2 \times 13$

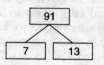

So, $91 = 7 \times 13$

Therefore,

LCM $(26, 91) = 2 \times 7 \times 13 = 182$

HCF $(26, 91) = 13$

Verification :

LCM × HCF $= 182 \times 13 = 2366$

and $26 \times 91 = 2366$

i.e., LCM × HCF = Product of two numbers.

(ii) **510 and 92**

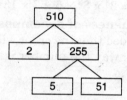

So, $510 = 2 \times 5 \times 51$

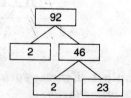

So, $92 = 2 \times 2 \times 23$

Therefore,

LCM $(510, 92)$
$= 2 \times 2 \times 5 \times 51 \times 23$
$= 23460$

HCF $(510, 92) = 2$

Verification :

LCM × HCF $= 23460 \times 2 = 46920$.

and $510 \times 92 = 46920$

i.e., LCM × HCF = Product of two numbers

(*iii*) **336 and 54**

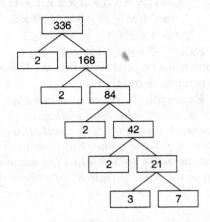

So, $336 = 2 \times 2 \times 2 \times 2 \times 21 = 2^4 \times 3 \times 7$

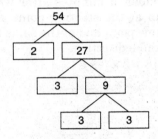

So, $54 = 2 \times 3 \times 3 \times 3 = 2 \times 3^3$
Therefore,
LCM (336, 52) = $2^4 \times 3^3 \times 7 = 3024$.
HCF (336, 52) = $2 \times 3 = 6$.

Verification:
    LCM × HCF = 3024 × 6 = 18144
and   336 × 54 = 18144
i.e.,   LCM × HCF = Product of two numbers.

Example 3. *Find the LCM and HCF of the following integers by applying the prime factorisation method.*
   (*i*) *12, 15 and 21*   (*ii*) *17, 23 and 29*
   (*iii*) *8, 9 and 25.*

Sol. (*i*) **12, 15 and 21**

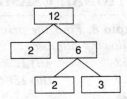

So, $12 = 2 \times 2 \times 3 = 2^2 \times 3$

So, $15 = 3 \times 5$

So, $21 = 3 \times 7$
Therefore,
   HCF (12, 15, 21) = 3
   LCM = (12, 15, 21)
          = $2^2 \times 3 \times 5 \times 7$
          = 420

(*ii*) **17, 23 and 29**
    17 = 17
    23 = 23
    29 = 29
Therefore,
   HCF (17, 23, 29) = 1
   LCM (17, 23, 29) = 17 × 23 × 29
                 = 11339

(*iii*) **8, 9 and 25**

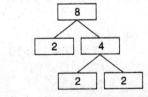

So, $8 = 2 \times 2 \times 2 = 2^3$

So, $9 = 3 \times 3 = 3^2$

So, $25 = 5 \times 5 = 5^2$

Therefore,

HCF $(8, 9, 25) = 1$

LCM $(8, 9, 25) = 2^3 \times 3^2 \times 5^2$
$= 1800.$

Example 4. *Given that HCF $(306, 657) = 9$, find LCM $(306, 657)$.*

Sol. LCM $(306, 657)$
$= \dfrac{306 \times 657}{\text{HCF }(306, 657)}$
$= \dfrac{306 \times 657}{9}$
$= 22338.$

Example 5. *Check whether 6^n can end with the digit 0 for any natural number n.*

Sol. If the number 6^n, for any natural number n, ends with digit 0, then it would be divisible by 5. That is, the prime factorisation of 6^n contain would the prime 5. This is not possible because $6^n = (2 \times 3)^n = 2^n \times 3^n$; so the only primes in the factorisation of 6^n are 2 and 3 and the uniqueness of the Fundamental Theorem of Arithmetic guarantees that there are no other primes in the factorisation of 6^n.

So, there is no natural number n for which 6^n ends with the digit zero.

Example 6. *Explain why $7 \times 11 \times 13 + 13$ and $7 \times 6 \times 5 \times 4 \times 3 \times 2 \times 1 + 5$ are composite numbers.*

Sol. (*i*) $7 \times 11 \times 13 + 13$
$= (7 \times 11 + 1) \times 13$
$= (77 + 1) \times 13$
$= 78 \times 13$
$= (2 \times 3 \times 13) \times 13$

So, $78 = 2 \times 3 \times 13$
$= 2 \times 3 \times 13^2$

Since, $7 \times 11 \times 13 + 13$ can be expressed as a product of primes, therefore, it is a composite number.

(*ii*) $7 \times 6 \times 5 \times 4 \times 3 \times 2 \times 1 + 5$
$= (7 \times 6 \times 4 \times 3 \times 2 \times 1 + 1) \times 5$
$= (1008 + 1) \times 5 = 1009 \times 5$
$= 5 \times 1009$

Since, $7 \times 6 \times 5 \times 4 \times 3 \times 2 \times 1 + 5$ can be expressed as a product of primes, therefore, it is a composite number.

Example 7. *There is a circular path around a sports field. Sonia takes 18 minutes to drive one round of the field, while Ravi takes 12 minutes for the same. Suppose they both start at the same point and at the same time, and go in the same direction. After how many minutes will they meet again at the starting point?*

Sol. By taking LCM of time taken (in minutes) by Sonia and Ravi, we can get the actual number of minutes after which they meet again at the starting point after both start at same point and at the same time, and go in the same direction.

$18 = 2 \times 3 \times 3$
$= 2 \times 3^2$

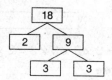

$12 = 2 \times 2 \times 3$
$= 2^2 \times 3$

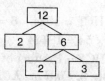

L.C.M $(18, 12) = 2^2 \times 3^2 = 36.$

Therefore, both Sonia and Ravi will meet again at the starting point after 36 minutes.

ADDITIONAL EXAMPLES

Example 8. *Find the prime factorisation of the following numbers :*

(*i*) 1152   (*ii*) 1296
(*iii*) 1440   (*iv*) 1584
(*v*) 1728.

Sol. *(i)* **1152**

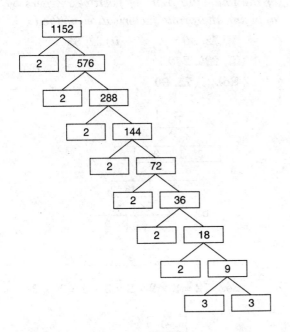

So, $1152 = 2 \times 2 \times 2 \times 2 \times 2 \times 2 \times 2 \times 3 \times 3$
$= 2^7 \times 3^2$

(ii) **1296**

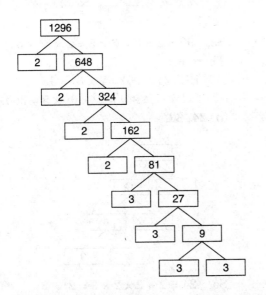

So, $1296 = 2 \times 2 \times 2 \times 2 \times 3 \times 3 \times 3 \times 3$
$= 2^4 \times 3^4$

(iii) **1440**

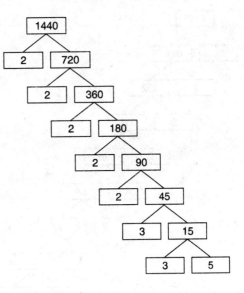

So, $1440 = 2 \times 2 \times 2 \times 2 \times 2 \times 3 \times 3 \times 5$
$= 2^5 \times 3^2 \times 5$

(iv) **1584**

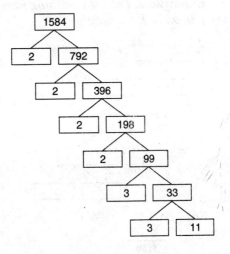

So, $1584 = 2 \times 2 \times 2 \times 2 \times 3 \times 3 \times 11$
$= 2^4 \times 3^2 \times 11$

(v) **1728**

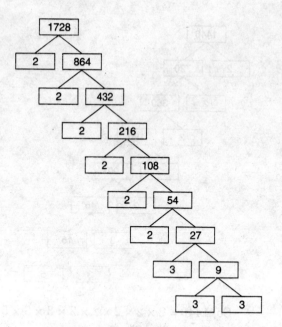

So, $1728 = 2 \times 2 \times 2 \times 2 \times 2 \times 2 \times 3 \times 3 \times 3$
$= 2^6 \times 3^3$.

Example 9. *Find the missing numbers in the following factorisation :*

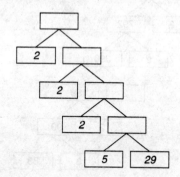

Sol.

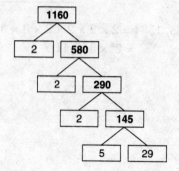

Example 10. *Find the HCF and LCM of the following pairs of positive integers by applying the prime factorisation method :*

(*i*) 72, 90   (*ii*) 24, 63

(*iii*) 225, 240.

Sol. (*i*) **72, 90**

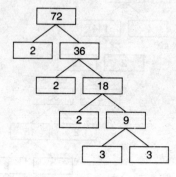

So, $72 = 2 \times 2 \times 2 \times 3 \times 3 = 2^3 \times 3^2$

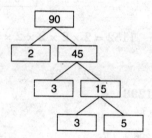

So, $90 = 2 \times 3 \times 3 \times 5 = 2 \times 3^2 \times 5$
Therefore,
     HCF (72, 90) = 2×3^2 = 18
     LCM (72, 90) = $2^3 \times 3^2 \times 5$ = 360

(*ii*) **24, 63**

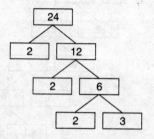

So, $24 = 2 \times 2 \times 2 \times 3 = 2^3 \times 3$

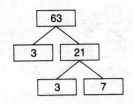

So, $63 = 3 \times 3 \times 7 = 3^2 \times 7$
Therefore,
 HCF (24, 63) = 3
 LCM (24, 63) = $2^3 \times 3^2 \times 7 = 504$

(iii) **225, 240**

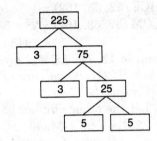

So, $225 = 3 \times 3 \times 5 \times 5 = 3^2 \times 5^2$

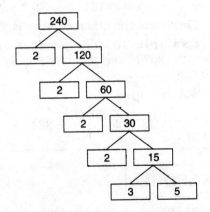

So, $240 = 2 \times 2 \times 2 \times 2 \times 3 \times 5$
$= 2^4 \times 3 \times 5$
Therefore,
 HCF (225, 240) = $3 \times 5 = 15$
 LCM (225, 240) = $2^4 \times 3^2 \times 5^2 = 3600.$

Example 11. *Find the LCM and HCF of the following positive integers by applying the prime factorisation method*
 (i) 35, 48, 56   (ii) 15, 55, 99
 (iii) 52, 63, 162.

Sol. (i) **35, 48, 56**

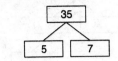

So, $35 = 5 \times 7$

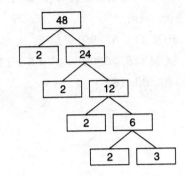

So, $48 = 2 \times 2 \times 2 \times 2 \times 3 = 2^4 \times 3$

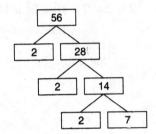

So, $56 = 2 \times 2 \times 2 \times 7 = 2^3 \times 7$
Therefore
 HCF (35, 48, 56) = 1
 LCM (35, 48, 56) = $2^4 \times 3 \times 5 \times 7$
      = 1680

(ii) **15, 55, 99**

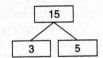

So, $15 = 3 \times 5$

So, $55 = 5 \times 11$

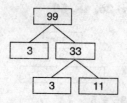

So, $99 = 3 \times 3 \times 11 = 3^2 \times 11$

Therefore,

HCF $(15, 55, 99) = 1$

LCM $(15, 55, 99) = 3^2 \times 5 \times 11 = 495$

(iii) **52, 63, 162**

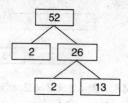

So, $52 = 2 \times 2 \times 13 = 2^2 \times 13$

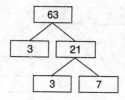

So, $63 = 3 \times 3 \times 7 = 3^2 \times 7$

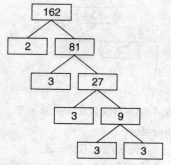

So, $162 = 2 \times 3 \times 3 \times 3 \times 3 = 2 \times 3^4$

Therefore,

HCF $(52, 63, 162) = 1$

LCM $(52, 63, 162) = 2^2 \times 3^4 \times 7 \times 13$
$= 29484$.

Example 12. *The LCM of two numbers is 64699, their HCF is 97 and one of the numbers is 2231. Find the other.*

Sol. Let the other number be x. Then,
$$2231\, x = 97 \times 64699$$
$$\Rightarrow \quad x = \frac{97 \times 64699}{2231}$$
$$\Rightarrow \quad x = 2813$$

Therefore, the other number is 2813.

Example 13. *The product of two numbers is 20736 and their HCF is 54. Find their LCM.*

Sol. Required
$$\text{LCM} = \frac{20736}{54} = 384.$$

TEST YOUR KNOWLEDGE

1. Consider the number 4^n, where n is a natural number. Check whether there is any value of $n \in N$ for which 4^n ends with the digit zero.
2. Find the LCM and HCF of 6 and 20 by the prime factorisation method.
3. Find the HCF of 12576 and 4052 by using the prime factorisation method.
4. Find the HCF and LCM of 6, 72 and 120 using the prime factorisation method.
5. Find the prime factorisation of the following numbers :
   (i) 1300   (ii) 1365
   (iii) 3456.
6. Find the LCM and HCF of 18, 24, 60, 150.
7. Find the HCF and LCM of 60, 32, 45, 80, 36, 120.
8. Show that the product of two numbers 60 and 84 is equal to the product of their HCF and LCM.
9. Split 4536 and 18711 into their prime factors and hence find their LCM and HCF.
10. The product of two numbers is 396×576 and their LCM is 6336. Find their HCF.

Answers

1. No
2. 60, 2
3. 4
4. 6, 360
5. (i) $2^2 \times 5^2 \times 13$ (ii) $3 \times 5 \times 7 \times 13$ (iii) $2^7 \times 3^3$
6. 1800, 6
7. 1, 1440
9. 149688, 567
10. 36.

IMPORTANT POINTS

1. Irrational Number. A number 's' is called irrational if it cannot be written in the form $\dfrac{p}{q}$, where p and q are integers and $q \neq 0$. For example :

$$\sqrt{2}, \sqrt{3}, \sqrt{15}, \pi, -\dfrac{\sqrt{2}}{\sqrt{3}},$$

0.10110111011110 ..... .

2. Theorem. Let p be a prime number. Show that if p divides a^2, then p divides a, where a is a positive integer.

Proof. Let the prime factorisation of a be as follows :

$$a = p_1 p_2 \ldots p_n,$$

where $p_1, p_2, \ldots, p_n$ are primes not necessarily distinct.

Therefore,

$$a^2 = (p_1 p_2 \ldots p_n)(p_1 p_2 \ldots p_n)$$
$$= p_1^2 p_2^2 \ldots p_n^2$$

Now, we are given that p divides a^2. Therefore, from the Fundamental Theorem of Arithmetic, it follows that p is one of prime factors of a^2. However, using the uniqueness part of the Fundamental Theorem of Arithmetic, we realise that the only prime factors of a^2 are $p_1, p_2, \ldots, p_n$. So, p is one of $p_1, p_2, \ldots, p_n$. Now, since $a = p_1 p_2 \ldots p_n$, p divides a.

3. Theorem. Prove that $\sqrt{2}$ is irrational.

The proof is based on a technique called 'proof by contradiction'. The idea is as follows :

- We want to show that $\sqrt{2}$ is irrational.
- We assume that it is not true, that is, $\sqrt{2}$ is rational.
- Based on this assumption, we make a series of statements which are logically correct.
- However, we end with a statement that contradicts our assumption that $\sqrt{2}$ is rational. Therefore, our assumption must be wrong i.e., $\sqrt{2}$ is irrational.

Proof. Let us assume, to the contrary, that $\sqrt{2}$ is rational.

So, we can find integers r and $s \,(\neq 0)$ such that $\sqrt{2} = \dfrac{r}{s}$.

Suppose r and s have a common factor other than one, then we divide by the common factor to get $\sqrt{2} = \dfrac{a}{b}$, where a and b are coprime.

So, $b\sqrt{2} = a$

Squaring on both sides, we get

$$2b^2 = a^2$$

Therefore, 2 divides a^2.

Therefore, 2 divides a,

by previous theorem

So, we can write

$a = 2c$ for some integer c.

Substituting for a, we get

$$2b^2 = 4c^2$$

that is, $\quad b^2 = 2c^2$

This means that 2 divides b^2, and so, by previous theorem, 2 divides b.

Therefore, a and b have at least 2 as a common factor.

But this contradicts the fact that a and b have no common factors other than 1.

This contradiction arose because of our incorrect assumption that $\sqrt{2}$ is rational.

So, we conclude that $\sqrt{2}$ is irrational.

ILLUSTRATIVE EXAMPLES

[NCERT Exercise 1.3]
(Page No. 14)

Example 1. *Prove that $\sqrt{5}$ is irrational.*

Sol. Let us assume, to the contrary, that $\sqrt{5}$ is rational.

So, we can find coprime integers a and b ($\neq 0$) such that

$$\sqrt{5} = \frac{a}{b}$$

$\Rightarrow \quad \sqrt{5}\, b = a$

Squaring on both sides, we get
$$5b^2 = a^2$$

Therefore, 5 divides a^2.
Therefore, 5 divides a
So, we can write
$$a = 5c \text{ for some integer } c.$$

Substituting for a, we get
$$5b^2 = 25c^2$$
$\Rightarrow \quad b^2 = 5c^2$

This means that 5 divides b^2, and so 5 divides b.

Therefore, a and b have at least 5 as a common factor.

But this contradicts the fact that a and b have no common factor other than 1.

This contradiction arose because of our incorrect assumption that $\sqrt{5}$ is rational.

So, we conclude that $\sqrt{5}$ is irrational.

Example 2. *Prove that $3 + 2\sqrt{5}$ is irrational.*

Sol. Let us assume, to the contrary, that $3 + 2\sqrt{5}$ is rational.

That is, we can find coprime integers a and b ($b \neq 0$) such that $3 + 2\sqrt{5} = \frac{a}{b}$

Therefore, $\frac{a}{b} - 3 = 2\sqrt{5}$

$\Rightarrow \quad \frac{a - 3b}{b} = 2\sqrt{5}$

$\Rightarrow \quad \frac{a - 3b}{2b} = \sqrt{5} \quad \Rightarrow \quad \frac{a}{2b} - \frac{3}{2}$

Since a and b are integers, we get $\frac{a}{2b} - \frac{3}{2}$ is rational, and so $\sqrt{5}$ is rational.

But this contradicts the fact that $\sqrt{5}$ is irrational.

This contradiction has arisen because of our incorrect assumption that $3 + 2\sqrt{5}$ is rational.

So, we conclude that $3 + 2\sqrt{5}$ is irrational.

Example 3. *Prove that the following are irrationals :*

(i) $\frac{1}{\sqrt{2}}$ (ii) $7\sqrt{5}$

(iii) $6 + \sqrt{2}$

Sol. (i) $\frac{1}{\sqrt{2}}$.

Let us assume, to the contrary, that $\frac{1}{\sqrt{2}}$ is rational.

So, we can find coprime integers a and b ($\neq 0$) such that

$$\frac{1}{\sqrt{2}} = \frac{a}{b}$$

$\Rightarrow \quad \sqrt{2} = \frac{a}{b}$

Since, a and b are integers, $\frac{a}{b}$ is rational, and so $\sqrt{2}$ is rational.

But this contradicts the fact that $\sqrt{2}$ is irrational.

So, we conclude that $\frac{1}{\sqrt{2}}$ is irrational.

(ii) **$7\sqrt{5}$.**

Let us assume to the contrary, that $7\sqrt{5}$ is rational.

So, we can find coprime integers a and $b \,(\neq 0)$ such that
$$7\sqrt{5} = \frac{a}{b}$$
$$\Rightarrow \quad \sqrt{5} = \frac{a}{7b}$$

Since, a and b are integers, $\frac{a}{7b}$ is rational, and so, $\sqrt{5}$ is rational.

But this contradicts the fact that $\sqrt{5}$ is irrational.

So, we conclude that $7\sqrt{5}$ is irrational.

(iii) **$6 + \sqrt{2}$.**

Let us assume to the contrary, that $\sqrt{2}$ is rational.

Then, $6 + \sqrt{2}$ is rational.

So, we can find coprime integers a and $b \,(\neq 0)$ such that
$$6 + \sqrt{2} = \frac{a}{b}$$
$$\Rightarrow \quad 6 - \frac{a}{b} = \sqrt{2}$$

Since, a and b are integers, we get $\frac{a}{b}$ is rational and so, $6 - \frac{a}{b}$ is rational and so, $\sqrt{2}$ is rational.

But this contradicts the fact that $\sqrt{2}$ is irrational.

So, we conclude that $6 + \sqrt{2}$ is irrational.

ADDITIONAL EXAMPLE

Example 4. *Prove that $\sqrt{3}$ is irrational.*

Sol. Let us assume, to the contrary, that $\sqrt{3}$ is irrational.

So, we can find coprime integers a and $b \,(\neq 0)$ such that
$$\sqrt{3} = \frac{a}{b} \quad \Rightarrow \quad \sqrt{3}b = a$$

Squaring on both sides, we get
$$3b^2 = a^2$$
Therefore, 3 divides a^2.
Therefore, 3 divides a.
So, we can write
$$a = 3c \text{ for some integer } c.$$
Substituting for a, we get
$$3b^2 = ac^2$$
$$\Rightarrow \quad b^2 = 3c^2$$

This means that 3 divides b^2, and so 3 divides b.

Therefore, a and b have at least 3 as a common factor.

But this contradicts the fact that a and b have no common factor other than 1.

This contradiction arose because of our incorrect assumption that $\sqrt{3}$ is rational.

So, we conclude that $\sqrt{3}$ is irrational.

TEST YOUR KNOWLEDGE

1. Prove that $\sqrt{5}$ is irrational.
2. Prove that $\sqrt{7}$ is irrational.
3. Prove that $\frac{1}{\sqrt{3}}$ is irrational.
4. Prove that $3\sqrt{5}$ is irrational.
5. Prove that $3 - \sqrt{3}$ is irrational.
6. Prove that $7 + \sqrt{2}$ is irrational.
7. Prove that $5 - \sqrt{3}$ is irrational.
8. Prove that $3\sqrt{2}$ is irrational.

IMPORTANT POINTS

1. Decimal Expansion of a rational number. We know that rational numbers have either a terminating decimal expansion or a non-terminating repeating decimal expansion. Now, we shall consider a rational number say $\frac{p}{q}$ ($q \neq 0$) and explore exactly when the decimal expansion of $\frac{p}{q}$ is terminating and when it is non-terminating repeating.

2. To explore when the decimal expansion of a rational number is terminating. Let us consider the following rational numbers:

(i) 0.375,   (ii) 0.104,
(iii) 0.0875,   (iv) 23.3408

Now,

(i) $0.375 = \dfrac{375}{1000} = \dfrac{375}{10^3}$

(ii) $0.104 = \dfrac{104}{1000} = \dfrac{104}{10^3}$

(iii) $0.0875 = \dfrac{875}{10000} = \dfrac{875}{10^4}$

(iv) $23.3408 = \dfrac{233408}{10000} = \dfrac{233408}{10^4}$

We notice that they can all be expressed as rational numbers whose denominators are powers of 10.

Let us try and cancel the common factors between the numerator and denominator and see what we get.

(i) $0.375 = \dfrac{375}{10^3} = \dfrac{3 \times 5^3}{2^3 \times 5^3} = \dfrac{3}{2^3}$

(ii) $0.104 = \dfrac{104}{10^3} = \dfrac{13 \times 2^3}{2^3 \times 5^3} = \dfrac{13}{5^3}$

(iii) $0.0875 = \dfrac{875}{10^4} = \dfrac{7 \times 5^3}{2^4 \times 5^4} = \dfrac{7}{2^4 \times 5}$

(iv) $23.3408 = \dfrac{233408}{10^4} = \dfrac{2^6 \times 7 \times 521}{2^4 \times 5^4}$

$= \dfrac{2^2 \times 7 \times 521}{5^4}$

We see a particular pattern. It looks like that we have converted a real number whose decimal expansion terminates to a rational number of the form $\frac{p}{q}$, where p and q are coprime, and the prime factorisation of each denominator (that is q) has only powers of 2 or powers of 5 or both. Obviously we should expect the denominator to look like this since powers of 10 can only have powers of 2 and 5 as factors.

The above inference is true for any real number. We can see that any real number which has a decimal expansion that terminates can be expressed as a rational number whose denominator is a power of 10. Cancelling out the common factors between the numerator and the denominator, we find that this real number is a rational number of the form $\frac{p}{q}$, where the prime factorization of q is of the form $2^n\, 5^m$, where n, m are some positive integers. Let us write our result formally:

Theorem. *Let x be a rational number whose decimal expansion terminates. Then, we can express x in the form $\frac{p}{q}$, where p and q are coprime, and the prime factorisation of q is of the form $2^n\, 5^m$, where n, m are non-negative integers.*

3. To explore whether a rational number of the form $\frac{p}{q}$, and the prime factorization of q is of the form $2^n\, 5^m$, where n, m are positive integers, has a terminating decimal expansion.

It is obvious that any rational number of the form $\frac{a}{b}$, where b is a power of 10, will have a terminating decimal expansion. So, it seems to make sense to convert a rational number of the form $\frac{p}{q}$, where q is of the form $2^n\ 5^m$, to an equivalent rational number of the form $\frac{a}{b}$, where b is a power of 10. Let us go back to above examples and work backwards.

(i) $\frac{3}{8} = \frac{3}{2^3} = \frac{3 \times 5^3}{2^3 \times 5^3} = \frac{375}{10^3} = 0.375$

(ii) $\frac{13}{125} = \frac{13}{5^3} = \frac{13 \times 2^3}{5^3 \times 2^3} = \frac{104}{10^3} = 0.104$

(iii) $\frac{7}{80} = \frac{7}{2^4 \times 5} = \frac{7 \times 5^3}{2^4 \times 5 \times 5^3} = \frac{875}{2^4 \times 5^4}$

$= \frac{875}{10^4} = 0.0875$

(iv) $\frac{14588}{625} = \frac{2^2 \times 7 \times 521}{5^4} = \frac{2^6 \times 7 \times 521}{2^4 \times 54}$

$= \frac{233408}{10^4} = 23.3408$

So, these examples show that we can convert a rational number of the form $\frac{p}{q}$, where q is of the form $2^n\ 5^m$, to an equivalent rational number of the form $\frac{a}{b}$, where b is a power of 10 and therefore, the decimal expansion of $\frac{p}{q}$ terminates. Let us write down our result formally :

Theorem. *Let $x = \frac{p}{q}$ be a rational number, such that the prime factorization of q is of the form $2^n\ 5^m$, where n, m are non-negative integers. Then, x has a decimal expansion which terminates.*

4. To explore when the decimal expansion of a rational number is recurring and non-terminating.

Consider the rational number $\frac{1}{7}$. Let us divide 1 by 7 by long division method.

```
        0.1428571
    ┌─────────────
  7 ) 10
      7
      ──
      ③0
      28
      ──
      ②0
      14
      ──
      ⑥0
      56
      ──
      ④0
      35
      ──
      ⑤0
      49
      ──
      ①0
       7
      ──
      ③0
```

Remainders : 3, 2, 6, 4, 5, 1, 3, 2, 6, 4, 5, 1, ...

Divisor : 7

We notice that the denominator, here 7, is not of the form $2^n\ 5^m$. Therefore, $\frac{1}{7}$ will not have a terminating decimal expansion (by theorem above). Therefore, 0 will not show up as a remainder and the remainders will start repeating after a certain stage. So, we will have a block of digits, namely, 142857, repeating in the quotient of $\frac{1}{7}$.

Let us put these ideas down as a theorem.

Theorem. *Let $x = \frac{p}{q}$ be a rational number, such that the prime factorization of q is not of the form $2^n\ 5^n$, where n, m are non-negative integers. Then, x has a decimal expansion which is non-terminating repeating recurring.*

5. An Important Inference. From the theorems given above we can conclude that the decimal expansion of every rational number is either terminating or non-terminating repeating.

ILLUSTRATIVE EXAMPLES

[NCERT Exercise 1.4]
(Page No. 17)

Example 1. *Without actually performing the long division, state whether the following rational numbers will have a terminating decimal expansion or a non-terminating repeating decimal expansion.*

(i) $\dfrac{13}{3125}$ (ii) $\dfrac{17}{8}$

(iii) $\dfrac{64}{455}$ (iv) $\dfrac{15}{1600}$

(v) $\dfrac{29}{343}$ (vi) $\dfrac{23}{2^3 \, 5^2}$

(vii) $\dfrac{129}{2^2 \, 5^7 \, 7^5}$ (viii) $\dfrac{6}{15}$

(ix) $\dfrac{35}{50}$ (x) $\dfrac{77}{210}$.

Sol. (i) $\dfrac{13}{3125}$

$$\dfrac{13}{3125} = \dfrac{13}{5^5}$$

Here, $q = 5^5$, which is of the form $2^n \, 5^m$ ($n = 0$, $m = 5$). So, the rational number $\dfrac{13}{3125}$ has a terminating decimal expansion.

(ii) $\dfrac{17}{8}$

$$\dfrac{17}{8} = \dfrac{17}{2^3}$$

Here, $q = 2^3$, which is of the form $2^n \, 5^m$ ($n = 3$, $m = 0$). So, the rational number $\dfrac{17}{8}$ has a terminating decimal expansion.

(iii) $\dfrac{64}{455}$

$$\dfrac{64}{455} = \dfrac{64}{5 \times 7 \times 13}$$

Here, $q = 5 \times 7 \times 13$, which is not of the form $2^n \, 5^m$. So, the rational number $\dfrac{64}{455}$ has a non-terminating repeating decimal expansion.

(iv) $\dfrac{15}{1600}$

$$\dfrac{15}{1600} = \dfrac{3}{320} = \dfrac{3}{2^6 \times 5^1}$$

Here, $q = 2^6 \times 5^1$, which is of the form $2^n \, 5^m$ ($n = 6$, $m = 1$). So, the rational number $\dfrac{15}{1600}$ has a terminating decimal expansion.

(v) $\dfrac{29}{343}$

$$\dfrac{29}{343} = \dfrac{29}{7^3}$$

Here, $q = 7^3$, which is not of the form $2^n \, 5^m$. So, the rational number $\dfrac{29}{343}$ has a non-terminating repeating decimal expansion.

(vi) $\dfrac{23}{2^3 \, 5^2}$

Here, $q = 2^3 \, 5^2$, which is of the from $2^n \, 5^m$ ($n = 3$, $m = 2$). So, the rational number $\dfrac{23}{2^3 \, 5^2}$ has a terminating decimal expansion.

(vii) $\dfrac{129}{2^2 \, 5^7 \, 7^5}$. Here, $q = 2^2 \, 5^7 \, 7^5$, which is not of the form $2^n \, 5^m$. So, the rational number $\dfrac{129}{2^2 \, 5^7 \, 7^5}$ has a non-terminating repeating decimal expansion.

(viii) $\dfrac{6}{15}$

$$\dfrac{6}{15} = \dfrac{2}{5}$$

Here, $q = 5$, which is of the form $2^n 5^m$ ($n = 0$, $m = 1$). So, the rational number $\dfrac{6}{15}$ has a terminating decimal expansion.

(ix) $\dfrac{35}{50}$

$$\dfrac{35}{50} = \dfrac{7}{10} = \dfrac{7}{2 \times 5}$$

Here $q = 2 \times 5$, which is of the form $2^n 5^m$ ($n = 1$, $m = 1$). So, the rational number $\dfrac{35}{50}$ has a terminating decimal expansion.

(x) $\dfrac{77}{210}$

$$\dfrac{77}{210} = \dfrac{11}{30} = \dfrac{11}{2 \times 3 \times 5}$$

Here, $q = 2 \times 3 \times 5$, which is not of the form $2^n 5^m$. So, the rational number $\dfrac{77}{210}$ has a non-terminating repeating decimal expansion.

Example 2. *Write down the decimal expansions of those rational numbers in Question 1 above which have terminating decimal expansions.*

Sol. (i) $\dfrac{13}{3125}$

$$= \dfrac{13}{5^5} = \dfrac{13 \times 2^5}{5^5 \times 2^5}$$

$$= \dfrac{416}{10^5} = .00416$$

(ii) $\dfrac{17}{8} = \dfrac{17}{2^3} = \dfrac{17 \times 5^3}{2^3 \times 5^3} = \dfrac{17 \times 5^3}{10^3}$

$$= \dfrac{2125}{10^3}$$

$$= 2.125$$

(iv) $\dfrac{15}{1600} = \dfrac{15}{2^4 \times 10^2}$

$$= \dfrac{15 \times 5^4}{2^4 \times 5^4 \times 10^2}$$

$$= \dfrac{9375}{10^6} = .009375$$

(vi) $\dfrac{23}{2^3 5^2} = \dfrac{23 \times 5^3 \times 2^2}{2^3 5^2 \times 5^3 \times 2^2}$

$$= \dfrac{11500}{10^5} = 0.115$$

(viii) $\dfrac{6}{15} = \dfrac{2}{5} = \dfrac{2 \times 2}{5 \times 2} = \dfrac{4}{10}$

$$= 0.4$$

(ix) $\dfrac{35}{50} = \dfrac{7}{10} = 0.7.$

Example 3. *The following real numbers have decimal expansions as given below. In each case, decide whether they are rational, or not. If they are rational, and of the form $\dfrac{p}{q}$, what can you say about the prime factors of q?*

(i) 43.123456789

(ii) 0.120 1200 12000 120000 ...

(iii) $43.\overline{123456789}$

Sol. (i) **43.123456789**

Since, the decimal expansion terminates, so the given real number is rational and therefore of the from $\dfrac{p}{q}$.

43.123456789

$$= \dfrac{43123456789}{1000\,000\,000}$$

$$= \dfrac{4312345679}{10^9}$$

$$= \dfrac{43123456789}{(2 \times 5)^9}$$

$$= \dfrac{4312345689}{2^9\, 5^9}$$

Here, $q = 2^9\, 5^9$

The prime factorization of q is of the form $2^n 5^m$, where $n = 9$, $m = 9$.

(ii) **0.120 1200 12000 120000 ...**

Since, the decimal expansion is neither terminating nor non-terminating repeating, therefore, the given real number is not rational.

(iii) $43.\overline{123456789}$

Since, the decimal expansion is non-terminating repeating, therefore, the given real number is rational and therefore of the form $\frac{p}{q}$.

Let $x = 43.\overline{123456789}$
$= 43.123456789...$ ...(1)

Multiplying both sides of (1) by 1000000000, we get

$1000000000\, x = 43123456789 . 123456789 ...$ ...(2)

Subtracting (1) from (2), we get
$999999999\, x = 43123456746$

$\Rightarrow \quad x = \dfrac{4323456746}{999999999}$

$= \dfrac{14374485583}{333333333}$

Here, $q = 333333333$ which is not of the form $2^n\, 5^m$, $n, m \in I$

ADDITIONAL EXAMPLE

Example 4. *Represent the following decimals in the form* $\dfrac{p}{q}$, *where* $p, q \in I$ *and* $q \neq 0$.

(i) $2.317\ 317\ 317...$
(ii) $0.2\overline{341}$.

Sol. (i) $2.317\ 317\ 317...$
Let $\quad x = 2.317\ 317\ 317...$ ...(1)
Multiplying both sides of (1) by 1000, we get

$1000\, x = 2317.\ 317\ 317\ 317...$ ...(2)

Subtracting (1) from (2), we get
$999\, x = 2315$

$\Rightarrow \quad x = \dfrac{2315}{999}$

Here $\quad p = 2315$
$\quad\quad q = 999$

(ii) $0.2\overline{341}$

Let $\quad x = 0.2\overline{341} = 0.2341\ 341\ 341...$ ...(1)

Multiplying both sides of (1) by 10, we get

$10\, x = 2.341\ 341\ 341...$ ...(2)

Multiplying both sides of (2) by 1000, we get

$10000\, x = 2341.341\ 341\ 341...$ ...(3)

Subtracting (2) from (3), we get
$9990\, x = 2339$

$\Rightarrow \quad x = \dfrac{2339}{9990}$

Here, $\quad p = 2339$
$\quad\quad q = 9990$.

TEST YOUR KNOWLEDGE

1. Without actually performing the long division, state whether the following rational numbers have a terminating decimal expansion or a non-terminating repeating decimal expansion:

   (i) $\dfrac{1}{7}$
   (ii) $\dfrac{1}{11}$
   (iii) $\dfrac{22}{7}$
   (iv) $\dfrac{3}{5}$
   (v) $\dfrac{7}{20}$
   (vi) $\dfrac{2}{13}$
   (vii) $\dfrac{27}{40}$
   (viii) $\dfrac{13}{125}$
   (ix) $\dfrac{23}{7}$
   (x) $\dfrac{42}{100}$

REAL NUMBERS

2. Write down the decimal expansions of the following rational numbers :

(i) $\dfrac{241}{2^3\,5^2}$ (ii) $\dfrac{19}{256}$

(iii) $\dfrac{25}{1600}$ (iv) $\dfrac{9}{30}$

(v) $\dfrac{133}{2^3\,5^4}$

3. Express the following in the form $\dfrac{p}{q}$:

(i) 0.6666...   (ii) 0.272727...
(iii) 3.7777...   (iv) 18.48 48 48...
(v) $0.\overline{3}$ (vi) $1.\overline{27}$
(vii) $0.2\overline{35}$ (viii) $0.\overline{47}$
(ix) $0.\overline{001}$ (x) 0.99 999... .

Answers

1. (i) non-terminating repeating
   (ii) non-terminating repeating
   (iii) non-terminating repeating
   (iv) terminating
   (v) terminating
   (vi) non-terminating repeating
   (vii) terminating
   (viii) terminating
   (ix) non-terminating repeating
   (x) terminating

2. (i) 1.205   (ii) 0.07421875
   (iii) .015625   (iv) 0.3
   (v) .0266

3. (i) $\dfrac{2}{3}$ (ii) $\dfrac{3}{11}$
   (iii) $\dfrac{34}{9}$ (iv) $\dfrac{610}{33}$
   (v) $\dfrac{1}{3}$ (vi) $\dfrac{14}{11}$
   (vii) $\dfrac{233}{990}$ (viii) $\dfrac{47}{99}$
   (ix) $\dfrac{1}{999}$ (x) 1.

MISCELLANEOUS EXERCISE

1. Use Euclid's division lemma to find the HCF of
   (i) 13281 and 15844
   (ii) 1128 and 1464
   (iii) 4059 and 2190
   (iv) 10524 and 12752
   (v) 10025 and 14035.

2. Show that 5309 and 3072 and prime to each other.

3. What is the greatest number by which 1037 and 1159 can both be divided exactly ?

4. Find the greatest number which both 2458090 and 867090 will contain an exact number of times.

5. Find the greatest height which can be contained exactly in 3 kg 7 hg 8 dag 1 g and 9 kg 1 hg 5 dag 4 g.

6. The HCF of two numbers in 119 and their LCM is 11781. If one of the numbers is 1071, find the other.

7. The LCM of two numbers is 2079 and their HCF is 27. If one of the numbers is 189, find the other.

8. Find the LCM of
   (i) 72, 90, 120   (ii) 24, 63, 70
   (iii) 455, 117, 338   (iv) 225, 240, 208
   (v) 2184, 2730, 3360.

   Apply prime factorization method.

9. Find the prime factorization of the following numbers :
   (i) 10000   (ii) 2160
   (iii) 396   (iv) 4725
   (v) 1188.

10. Find the missing numbers in the following factorisation :

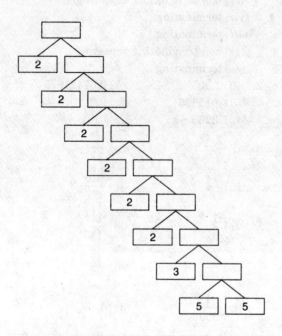

11. Prove that $\sqrt{13}$ is irrational.
12. Prove that $2\sqrt{2}$ in irrational.
13. Prove that $\dfrac{1}{\sqrt{5}}$ is irrational.
14. Prove that $7 + \sqrt{3}$ is irrational.
15. Prove that $8 - \sqrt{2}$ is irrational.
16. Without actually performing the long division, state whether the following rational numbers will have a terminating decimal expansion or a non-terminating repeating decimal expansion :

    (i) $\dfrac{11}{125}$ (ii) $\dfrac{19}{128}$

    (iii) $\dfrac{32}{405}$ (iv) $\dfrac{15}{3200}$

    (v) $\dfrac{29}{2401}$.

17. Write down the decimal expansions of the following rational numbers :

    (i) $\dfrac{5}{8}$ (ii) $\dfrac{12}{125}$

    (iii) $\dfrac{13}{625}$ (iv) $\dfrac{7}{64}$

    (v) $\dfrac{7}{8}$.

18. Convert each of the following real numbers in the form $\dfrac{p}{q}$; where $q \neq 0$, p and q are non-negative integers :

    (i) $15.7\overline{12}$ (ii) $0.00\overline{352}$

    (iii) 1.0001 (iv) $0.\overline{621}$

    (v) $125.\overline{3}$.

19. Chose the correct statements out of the following.

    (i) $\sqrt{2}$ is a rational number.

    (ii) π is an irrational number.

    (iii) A real number is either rational or irrational.

    (iv) The decimal expansion of every rational number is either terminating or repeating non-terminating.

    (v) $\dfrac{1}{7}$ will not have a terminating decimal expansion.

20. Let p be a prime number. Show that if p divides a^2, then p divides a, where a is a positive integer.

Answers

1. (i) 233   (ii) 24
   (iii) 3   (iv) 4
   (v) 2005.
3. 61   4. 10
5. 1 hg 9 dag 9 g   6. 1309
7. 297
8. (i) 360   (ii) 2520
   (iii) 106470   (iv) 46800
   (v) 43680.
9. (i) $2^4 \times 5^4$ (ii) $2^4 \times 3^3 \times 5$
   (iii) $2^2 \times 3^2 \times 11$ (iv) $3^3 \times 5^2 \times 7$
   (v) $2^2 \times 3^3 \times 11$.

10.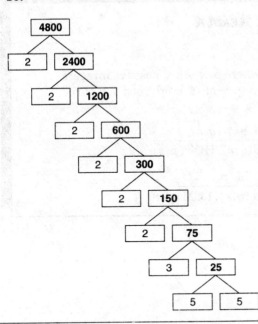

16. (i) terminating (ii) terminating
    (iii) non-terminating repeating
    (iv) terminating
    (v) non-terminating repeating

17. (i) 0.625 (ii) 0.96
    (iii) 0.0208 (iv) 0.109375
    (v) 0.875.

18. (i) $\dfrac{377}{66}$ (ii) $\dfrac{349}{99000}$
    (iii) $\dfrac{10001}{10000}$ (iv) $\dfrac{23}{39}$
    (v) $\dfrac{376}{3}$

19. (ii), (iii), (iv), (v).

SUMMARY

In this chapter, we have studied the following points:

1. **Euclid's division lemma:**
   Given position integers a and b, there exist whole numbers q and r satisfying $a = bq + r$, $0 \leq r < b$.

2. **Euclid's division algorithm** : This is based on Euclid's division lemma. According to this HCF of any two positive integers a and b, with $a > b$, is obtained as follows:
   Step 1: Apply the division lemma to find q and r where $a = bq + r$, $0 \leq r < b$.
   Step 2: If $r = 0$, the HCF is b. If $r \neq 0$, apply Euclid's lemma to b and r.
   Step 3: Continue the process till the remainder is zero. The divisor at this stage will be HCF (a, b). Also, HCF (a, b) = HCF (b, r).

3. **The Fundamental Theorem of Arithmetic:**
   Every composite number can be expressed (factorised) as a product of primes, and this factorisation is unique, apart from the order in which the prime factors occur.

4. If p is a prime and p divides a^2, then p divides q, where a is a positive integer.

5. To prove that $\sqrt{2}$, $\sqrt{3}$ are irrationals.

6. Let x be a rational number whose decimal expansion terminates. Then we can express x in the form $\dfrac{p}{q}$, where p and q are coprime, and the prime factorisation of q is of the form $2^n\,5^m$, where n, m are non-negative integers.

7. Let $x = \dfrac{p}{q}$ be a rational number, such that the prime factorisation of q is of the form $2^n\,5^m$, where n, m are non-negative integers. Then x has a decimal expansion which terminates.

8. Let $x = \dfrac{p}{q}$ be a rational number, such that the prime factorisation of q is not of the form $2^n\,5^m$, where n, m are non-negative integers. Then x has a decimal expansion which is non-terminating repeating (recurring).

A Note to the Reader

You have seen that :

HCF $(p, q, r) \times$ LCM $(p, q, r) \neq p \times q \times r$, where p, q, r are positive integers (See Example 8). However, the following results hold good for three numbers p, q and r :

$$\text{LCM}(p, q, r) = \frac{p \cdot q \cdot r \cdot \text{HCF}(p, q, r)}{\text{HCF}(p, q) \cdot \text{HCF}(q, r) \cdot \text{HCF}(p, r)}$$

$$\text{HCF}(p, q, r) = \frac{p \cdot q \cdot r \cdot \text{LCM}(p, q, r)}{\text{LCM}(p, q) \cdot \text{LCM}(q, r) \cdot \text{LCM}(p, r)}$$

2
Polynomials

IMPORTANT POINTS

1. Definition of a Polynomial. A polynomial in one variable x is an algebraic expression in x of the form
$$p(x) = a_n x^n + a_{n-1} x^{n-1} + a_{n-2} x^{n-2} + \ldots + a_2 x^2 + a_1 x + a_0$$
where $a_0, a_1, a_2, \ldots, a_n$ are constants and $a_n \neq 0$.

Note. n is essentially a whole number.

Examples :

$4x + 2$, $2y^2 - 3y + 4$, $5x^3 - 4x^2 + x - \sqrt{2}$,

$7u^6 - \dfrac{3}{2}u^4 + 4u^2 + u - 8$, etc.

2. Not a polynomial. The expression like $\dfrac{1}{x-1}, \sqrt{x}+2, \dfrac{1}{x^2+2x+3}$, etc. are not polynomials.

3. Degree of the Polynomial. The highest power of x in $p(x)$ is called the degree of the polynomial $p(x)$. For example, $4x + 2$ is a polynomial in the variable x of degree 1, $2y^2 - 3y + 4$ is a polynomial in the variable y of degree 2, $5x^3 - 4x^2 + x - \sqrt{2}$ is a polynomial in the variable x of degree 3 and $7u^6 - \dfrac{3}{2}u^4 + 4u^2 + u - 8$ is a polynomial in the variable u of degree 6.

4. Linear Polynomial. A polynomial of degree 1 is called a linear polynomial. For example, $2x - 3$, $\sqrt{3}x + 5$, $y + \sqrt{2}$, $x - \dfrac{2}{11}$, $3z + 4$, $\dfrac{2}{3}u + 1$, etc. are all linear polynomials. The most general form of a linear polynomial is $ax + b$, $a \neq 0$, a, b are reals.

Note. Polynomials such as $2x + 5 - x^2$, $x^3 + 1$, etc. are not linear polynomials.

5. Quadratic Polynomial. A polynomial of degree 2 is called a quadratic polynomial. The name 'quadratic' has been derived from 'quadrate', which means 'square'.

$2x^2 + 3x - \dfrac{2}{5}, y^2 - 2, 2 - x^2 + \sqrt{3}x, \dfrac{u}{3} - 2u^2 + 5$,

$\sqrt{5}v^2 - \dfrac{2}{3}v, 4z^2 + \dfrac{1}{7}$ are some examples of quadratic polynomials whose coefficients are real numbers. More generally, any quadratic polynomial in x with real coefficients is of the form $ax^2 + bx + c$, where a, b, c are real numbers and $a \neq 0$.

6. Cubic Polynomial. A polynomial of degree 3 is called a cubic polynomial. Some examples of a cubic polynomial are $2 - x^3$, x^3, $\sqrt{2}x^3$, $3 - x^2 + x^3$, $3x^3 - 2x^2 + x - 1$. In fact, the most general form of a cubic polynomial with coefficients as real numbers is $ax^3 + bx^2 + cx + d$, $a \neq 0$, a, b, c, d are reals.

7. Value of a Polynomial. If $p(x)$ is a polynomial in x, and if k is any real constant, then the real number obtained by replacing x by k in $p(x)$, is called the value of $p(x)$ at k, and is denoted by $p(k)$. For example, consider the polynomial $p(x) = x^2 - 3x - 4$. Then, putting

$x = 2$ in the polynomial, we get $p(2) = 2^2 - 3 \times 2 - 4 = -6$. The value -6, obtained by replacing x by 2 in $x^2 - 3x - 4$, is the value of $x^2 - 3x - 4$ at $x = 2$. Similarly, $p(0)$ is the value of $p(x)$ at $x = 0$, which is -4.

8. Zero of a Polynomial. A real number k is said to be a zero of a polynomial $p(x)$, if $p(k) = 0$. For example, consider the polynomial $p(x) = x^3 - 3x - 4$. Then,
$$p(-1) = (-1)^2 - \{3(-1)\} - 4 = 0$$
Also,
$$p(4) = (4)^2 - (3 \times 4) - 4 = 0$$
Here, -1 and 4 are called the zeroes of the quadratic polynomial $x^2 - 3x - 4$.

9. How to find the zero of a linear polynomial. In general, if k is a zero of $p(x) = ax + b$, then $p(k) = ak + b = 0$, i.e., $k = -\dfrac{b}{a}$. So, the zero of a linear polynomial $ax + b$ is $-\dfrac{b}{a} = -\dfrac{\text{(Constant term)}}{\text{Coefficient of } x}$. Thus, the zero of a linear polynomial is related to its coefficients. For example, consider the linear polynomial $p(x) = 2x + 3$. If k is a zero of $p(x)$, then $p(k) = 0$ gives us $2k + 3 = 0$, i.e., $k = -\dfrac{3}{2}$.

10. Geometrical Meaning of the zeroes of a polynomial. We know that a real number k is a zero of the polynomial $p(x)$ if $p(k) = 0$. But to understand the importance of finding the zeroes of a polynomial, first we shall see the geometrical representation of linear, quadratic and cubic polynomials and then geometrical meaning of their zeroes.

I. For Linear Polynomial. Consider first a linear polynomial $ax + b = 0$, $a \neq 0$. We know that the graph of $y = ax + b$ is a straight line. For example, the graph of $y = 2x + 3$ is a straight line passing through the points $(-2, -1)$ and $(2, 7)$.

Table

x	-2	2
$y = 2x + 3$	-1	7

From the figure, we can see that the graph of $y = 2x + 3$ intersects the x-axis mid-way between $x = -1$ and $x = -2$, that is, at $x = -\dfrac{3}{2}$. We also know that the zero of $2x + 3$ is $-\dfrac{3}{2}$. Thus, the zero of the polynomial $2x + 3$ is the x-coordinate of the point where the graph of $y = 2x + 3$ intersects the x-axis.

In general, for a linear polynomial $ax + b$, $a \neq 0$, the graph of $y = ax + b$ is a straight line which crosses the x-axis at exactly one point, namely $\left(-\dfrac{b}{a}, 0\right)$. Therefore, the linear polynomial $ax + b$, $a \neq 0$ has exactly one zero, namely, the x-coordinate of the point where the graph of $y = ax + b$ intersects the x-axis.

POLYNOMIALS

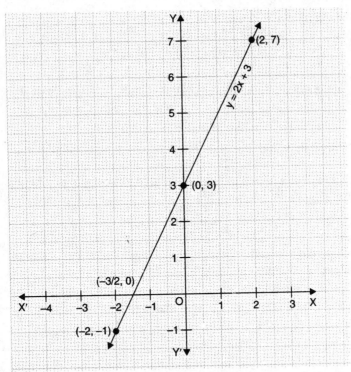

II. For Quadratic Polynomial. Consider the quadratic polynomial $x^2 - 3x - 4$. Let us see what the graph of $y = x^2 - 3x - 4$ looks like. Let us list a few values of $y = x^2 - 3x - 4$ corresponding to a few values for x as follows :

Table

x	-2	-1	0	1	2	3	4	5
$y = x^2 - 3x - 4$	6	0	-4	-6	-6	-4	0	6

If we locate the points above on a graph paper and draw the graph, it will actually look like the one given in the following figure.

(In fact for any quadratic polynomial $ax^2 + bx + c$, $a \neq 0$, the graph of the corresponding equation $y = ax^2 + bx + c$ has one of the two shapes either open upwards like $\cup$ or downwards like $\cap$ depending or whether $a > 0$ or $a < 0$. These curves are called parabolas)

We see from the above table that -1 and 4 are the zeroes of the quadratic polynomial. Also, we note that -1 and 4 are the x-coordinates of the points where the graph of $y = x^2 - 3x - 4$ intersects the x-axis. Thus, the zeroes of the quadratic polynomial $x^2 - 3x - 4$ are x-coordinates of the points where the graph of $y = x^2 - 3x - 4$ intersects the x-axis.

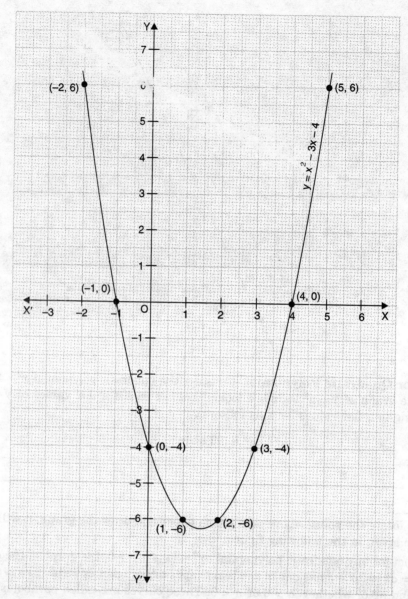

This fact is true for any quadratic polynomial, i.e., *the zeroes of a quadratic polynomial $ax^2 + bx + c = 0$, $a \neq 0$, are precisely the x-coordinates of the points where the parabola representing $y = ax^2 + bx + c$ intersects the x-axis.*

Shape of the graph of $y = ax^2 + bx + c$

Regarding the shape of the graph of $y = ax^2 + bx + c$, the following three cases can happen :

Case (i). Here, the graph cuts x-axis at two distinct points A and A'.

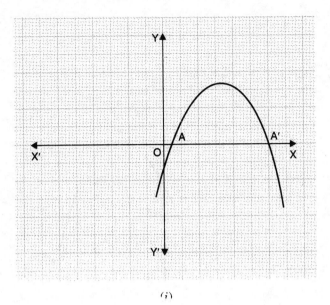

(i)

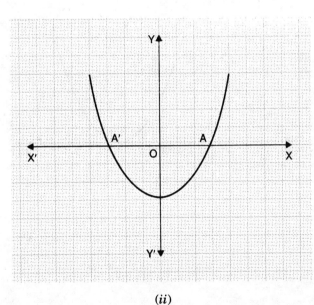

(ii)

The *x*-coordinates of A and A' are the **two zeroes** of the quadratic polynomial $ax^2 + bx + c$ in this case.

Case (ii). Here, the graph cuts the *x*-axis at exactly one point, *i.e.*, at two coincident points. So, the two points A and A' of case (*i*) coincide here to become one point A.

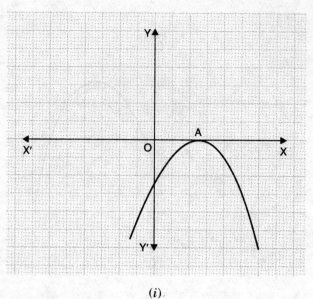

(i)

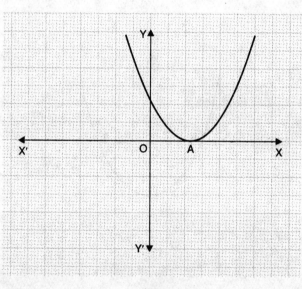

(ii)

The x-coordinate of A is the **only zero** for quadratic polynomial $ax^2 + bx + c$ in this case.

Case (iii). Here, the graph is either completely above the x-axis or completely below the x-axis. So, it does not cut the x-axis at any point.

So, the quadratic polynomial $ax^2 + bx + c$ has **no zero** in this case.

So, we can see geometrically that a quadratic polynomial can have either two distinct zeroes or two equal zeroes (i.e., one zero), or no zeroes. This also means that a polynomial of degree 2 has at most two zeroes.

POLYNOMIALS

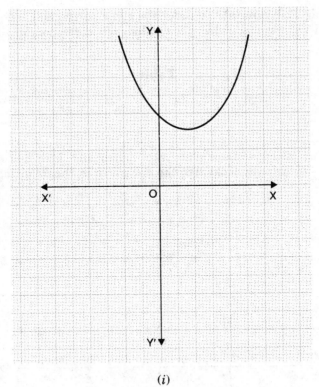

(i)

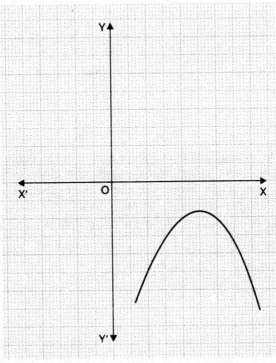

(ii)

III. For Cubic Polynomial. Consider the cubic polynomial $x^3 - 4x$. To see what the graph of $y = x^3 - 4x$ looks like, let us list a few values of y corresponding to a few values of x in the following table.

Table

x	-2	-1	0	1	2
$y = x^3 - 4x$	0	3	0	-3	0

Locating the points of the table on a graph paper and drawing the graph, we see that the graph of $y = x^3 - 4x$ actually looks like the one given in the following figure.

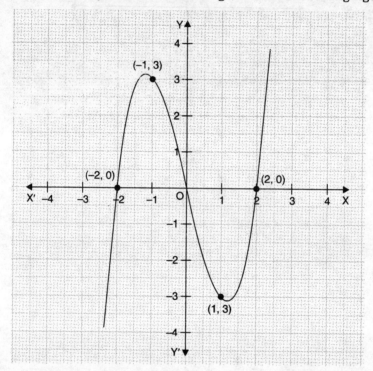

We see from the table that -2, 0 and 2 are zeroes of the cubic polynomial $x^3 - 4x$. Also, we observe that -2, 0 and 2 are, in fact, the x-coordinates of the only points where the graph of $y = x^3 - 4x$ intersects the x-axis. Since, the curve meets the x-axis in only these 3 points, their x-coordinates are the only zeroes of the polynomial.

Consider the cubic polynomial x^3. We draw the graph of $y = x^3$ as shown in the following figure.

POLYMOMIALS

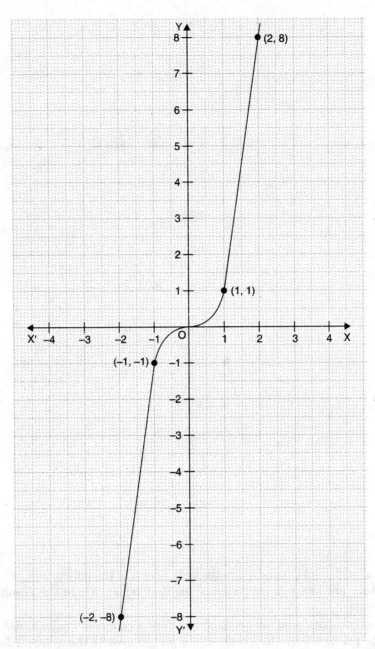

We note that 0 is the only zero of the polynomial x^3. Also, from figure, we can see that 0 is the x-coordinate of the only point, where the graph of $y = x^3$ intersects the x-axis.

Consider the cubic polynomial $x^3 - x^2$. We draw the graph of $y = x^3 - x^2$ as shown in the following figure.

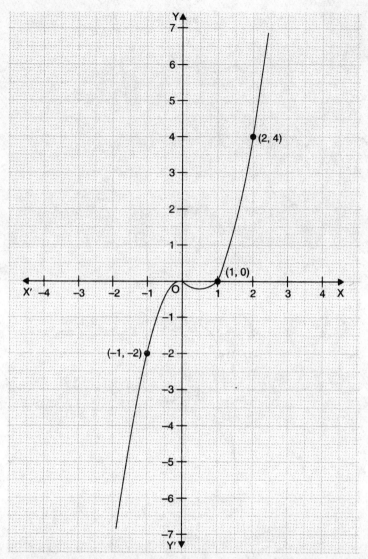

Since $x^3 - x^2 = x^2(x - 1)$, 0 and 1 are the only zeroes of the polynomial $x^3 - x^2$. Also, from figure, these values are the x-coordinates of the only points where the graph of $y = x^3 - x^2$ intersects the x-axis.

From the above examples, we see that the zeroes of a cubic polynomial $p(x)$ are the x-coordinates of the points where the graph of $y = p(x)$ intersects the x-axis. Also, there are at most 3 zeroes for the cubic polynomials given above. In fact, any polynomial of degree 3 can have at most three zeroes.

11. Generalisation. In general, given a polynomial $p(x)$ of degree n the graph of $y = p(x)$ crosses the x-axis at atmost n points. Therefore, a polynomial $p(x)$ of degree n has atmost n zeroes.

ILLUSTRATIVE EXAMPLES

[NCERT Exercise 2.1]
(Page No. 28)

Example 1. *The graphs of $y = p(x)$ are given in figure below, for some polynomials $p(x)$. Find the number of zeroes of $p(x)$, in each case.*

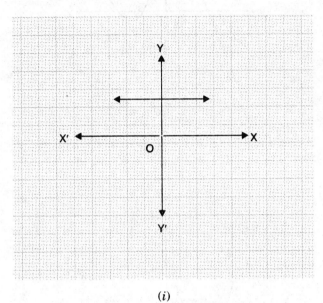

(i)

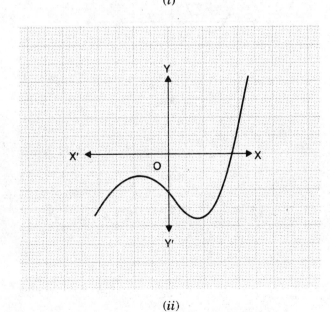

(ii)

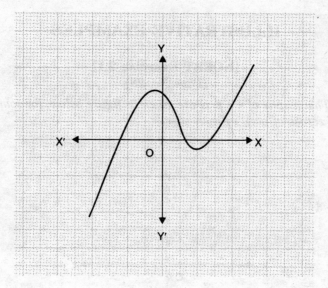

(iii)

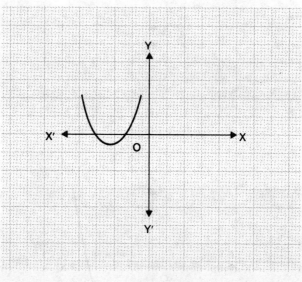

(iv)

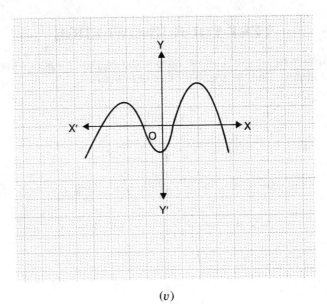

(v)

(vi)

Sol. (*i*) There is no zero as the graph does not intersect the *x*-axis at any point.

(*ii*) The number of zeroes is 1 as the graph intersects the *x*-axis at one point only.

(*iii*) The number of zeroes is 3 as the graph intersects the *x*-axis at three points.

(*iv*) The number of zeroes is 2 as the graph intersects the *x*-axis at two points.

(*v*) The number of zeroes is 4 as the graph intersects the *x*-axis at four points.

(*vi*) The number of zeroes is 3 as the graph intersects the *x*-axis at three points.

TEST YOUR KNOWLEDGE

1. Look at the graphs in figure given below. Each is the graph of $y = p(x)$, where $p(x)$ is a polynomial. For each of the graphs, find the number of zeroes of $p(x)$.

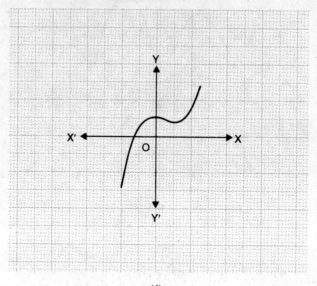

(i)

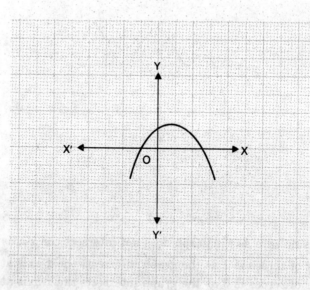

(ii)

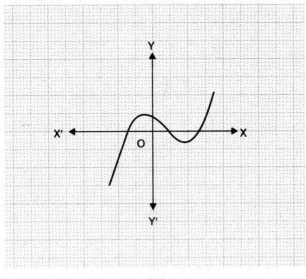

(iii)

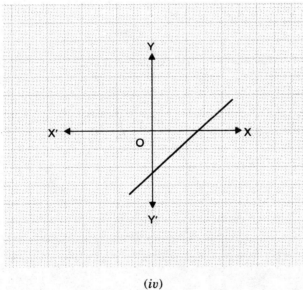

(iv)

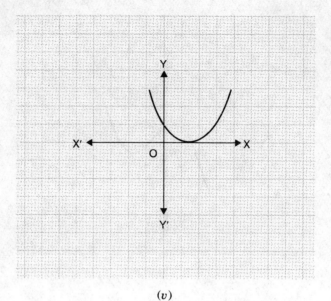

(v)

(vi)

Answers

1. (i) 1   (ii) 2   (iii) 3   (iv) 1
    (v) 1   (vi) 4

IMPORTANT POINTS

1. Relationship between the zeroes and the coefficients of a polynomial

I. For a quadratic polynomial. In general, if α and β are the zeroes of a quadratic polynomial $p(x) = ax^2 + bx + c$, $a \neq 0$, then we know that $x - \alpha$ and $x - \beta$ are the factors of $p(x)$. Therefore,

POLYNOMIALS

$$ax^2 + bx + c = k(x - \alpha)(x - \beta),$$
where k is a constant
$$= k[x^2 - (\alpha + \beta)x + \alpha\beta]$$
$$= kx^2 - k(\alpha + \beta)x + k\alpha\beta$$

Comparing the coefficients of x^2, x and constant terms on both the sides, we get
$$a = k$$
$$b = -k(\alpha + \beta)$$
and
$$c = k\alpha\beta$$

This gives,
$$\alpha + \beta = -\frac{b}{a} \quad \text{and} \quad \alpha\beta = \frac{c}{a}$$

Therefore,

Sum of the zeroes $= -\dfrac{b}{a}$

$$= -\frac{(\text{Coefficient of } x)}{\text{Coefficient of } x^2}$$

Product of zeroes $= \dfrac{c}{a} = \dfrac{\text{Constant term}}{\text{Coefficient of } x^2}$

Example 1. Consider a quadratic polynomial, say,
$$p(x) = 2x^2 - 8x + 6$$
Then, $p(x) = 2x^2 - 6x - 2x + 6$

| Splitting the middle term $-8x$ as a sum of two terms $-6x$ and $-2x$ whose product is $(-6x) \times (-2x) = 12x^2$

$$= 2x(x - 3) - 2(x - 3)$$
$$= (2x - 2)(x - 3)$$
$$= (2x - 1)(x - 3)$$

For zeroes of $p(x)$,
$$p(x) = 0$$
$$\Rightarrow \quad 2(x - 1)(x - 3) = 0$$
$$\Rightarrow \quad (x - 1)(x - 3) = 0$$
$$\Rightarrow \quad x - 1 = 0 \quad \text{or} \quad x - 3 = 0$$
$$\Rightarrow \quad x = 1 \quad \text{or} \quad x = 3$$
$$\Rightarrow \quad x = 1, 3$$

So, the zeroes of $p(x)$ are 1 and 3.
We observe that,
Sum of its zero
$$= 1 + 3 = 4$$

$$= -\frac{(-8)}{2}$$
$$= -\frac{(\text{Coefficient of } x)}{\text{Coefficient of } x^2}$$

Product of its zeroes $= 1 \times 3 = 3$
$$= \frac{6}{2} = \frac{\text{Constant term}}{\text{Coefficient of } x^2}.$$

Example 2. Consider a quadratic polynomial, say,
$$p(x) = 3x^2 + 5x - 2.$$
By the method of splitting the middle term,
$$3x^2 + 5x - 2 = 3x^2 + 6x - x - 2$$
$$= 3x(x + 2) - 1(x + 2)$$
$$= (3x - 1)(x + 2)$$

For zeroes of $p(x)$,
$$p(x) = 0$$
$$\Rightarrow \quad (3x - 1)(x + 2) = 0$$
$$\Rightarrow \quad 3x - 1 = 0 \quad \text{or} \quad x + 2 = 0$$
$$\Rightarrow \quad x = \frac{1}{3} \quad \text{or} \quad x = -2$$
$$\Rightarrow \quad x = \frac{1}{3}, -2$$

So, the zeroes of $p(x)$ are $\dfrac{1}{3}$ and -2.

We observe that,

Sum of its zeroes $= \dfrac{1}{3} + (-2) = -\dfrac{5}{3}$

$$= -\frac{(\text{Coefficient of } x)}{\text{Coefficient of } x^2}$$

Product of its zeroes
$$= \frac{1}{3} \times (-2) = \frac{-2}{3}$$
$$= \frac{\text{Constant term}}{\text{Coefficient of } x^2}$$

II. For a cubic polynomial. In general, if α, β, γ are the zeroes of a cubic polynomial $ax^3 + bx^2 + cx + d$, then
$$\alpha + \beta + \gamma = -\frac{b}{a}$$

$$= -\frac{\text{(Coefficient of } x^2)}{\text{Coefficient of } x^3}$$

$$\alpha\beta + \beta\gamma + \gamma\alpha = \frac{c}{a}$$

$$= \frac{\text{Coefficient of } x}{\text{Coefficient of } x^3}$$

$$\alpha\beta\gamma = -\frac{d}{a}$$

$$= -\frac{\text{Constant term}}{\text{Coefficient of } x^3}$$

Note. $\frac{b}{a}, \frac{c}{a}$ and $\frac{d}{a}$ are meaningful because $a \neq 0$.

Example 3. Consider a cubic polynomial, say,

$$p(x) = 2x^3 - 5x^2 - 14x + 8$$

For zeroes of $p(x)$

$$p(x) = 0$$
$$\Rightarrow \quad 2x^3 - 5x^2 - 14x + 8 = 0$$
$$\Rightarrow \quad 2x^2(x+2) - 9x(x+2) + 4(x+2) = 0$$
$$\Rightarrow \quad (x+2)(2x^2 - 9x + 4) = 0$$
$$\Rightarrow \quad (x+2)(2x^2 - 8x - x + 4) = 0$$
$$\Rightarrow \quad (x+2)\{2x(x-4) - 1(x-4)\} = 0$$
$$\Rightarrow \quad (x+2)(x-4)(2x-1) = 0$$
$$\Rightarrow \quad x = -2, 4, \frac{1}{2}$$

Since, $p(x)$ can have at most, three zeroes, these are the zeroes of
$$2x^3 - 5x^2 - 14x + 8.$$
Now,
Sum of the zeroes
$$= 4 + (-2) + \frac{1}{2}$$
$$= \frac{5}{2} = \frac{-(-5)}{2}$$
$$= \frac{-(\text{Coefficient of } x^2)}{\text{Coefficient of } x^3}$$

Product of the zeroes
$$= 4 \times (-2) \times \frac{1}{2} = -4$$
$$= \frac{-8}{2} = \frac{-\text{Constant term}}{\text{Coefficient of } x^3}$$

However, there is one more relationship here. Consider the sum of the products of the zeroes taken two at a time. This is

$$\{4 \times (-2)\} + \left\{(-2) \times \frac{1}{2}\right\} + \left\{\frac{1}{2} \times 4\right\}$$

$$= -8 - 1 + 2 = -7 = -\frac{14}{2}$$

$$= \frac{\text{Coefficient of } x}{\text{Coefficient of } x^3}.$$

ILLUSTRATIVE EXAMPLES

[NCERT Exercise 2.2]
(Page No. 33)

Example 1. *Find the zeroes of the following quadratic polynomials and verify the relationship between the zeroes and the coefficients.*

(i) $x^2 - 2x - 8$ (ii) $4s^2 - 4s + 1$
(iii) $6x^2 - 3 - 7x$ (iv) $4u^2 + 8u$
(v) $t^2 - 15$ (vi) $3x^2 - x - 4$

Sol. (i) $x^2 - 2x - 8$
Let $p(x) = x^2 - 2x - 8$

By the method of splitting the middle term,
$$x^2 - 2x - 8 = x^2 - 4x + 2x - 8$$
$$= x(x-4) + 2(x-4)$$
$$= (x-4)(x+2)$$
For zeroes of $p(x)$,
$$p(x) = 0$$
$$\Rightarrow \quad (x-4)(x+2) = 0$$

POLYNOMIALS

$\Rightarrow \quad x - 4 = 0 \text{ or } x + 2 = 0$
$\Rightarrow \quad x = 4 \text{ or } \quad x = -2$
$\Rightarrow \quad x = 4, -2$

So, the zeroes of $p(x)$ are 4 and -2.
We observe that,
Sum of its zeroes
$$= 4 + (-2) = 2$$
$$= \frac{-(-2)}{1}$$
$$= \frac{-(\text{Coefficient of } x)}{\text{Coefficient of } x^2}$$

Product of its zeroes
$$= 4 \times (-2) = -8 = \frac{-8}{1}$$
$$= \frac{\text{Constant term}}{\text{Coefficient of } x^2}.$$

(ii) $4s^2 - 4s + 1$

Let $p(s) = 4s^2 - 4s + 1$

By the method of splitting the middle term,
$$4s^2 - 4s + 1 = 4s^2 - 2s - 2s + 1$$
$$= 2s(2s - 1) - 1(2s - 1)$$
$$= (2s - 1)(2s - 1)$$

For zeroes of $p(s)$,
$p(s) = 0$
$\Rightarrow \quad (2s - 1)(2s - 1) = 0$
$\Rightarrow \quad 2s - 1 = 0 \text{ or } 2s - 1 = 0$
$\Rightarrow \quad s = \frac{1}{2} \text{ or } s = \frac{1}{2}$
$\Rightarrow \quad s = \frac{1}{2}, \frac{1}{2}$

So, the zeroes of $p(s)$ are $\frac{1}{2}$ and $\frac{1}{2}$.

We observe that,
Sum of its zeroes
$$= \frac{1}{2} + \frac{1}{2} = 1 = \frac{-(-4)}{4}$$
$$= \frac{-(\text{Coefficient of } s)}{\text{Coefficient of } s^2}$$

Product of its zeroes
$$= \left(\frac{1}{2}\right) \times \left(\frac{1}{2}\right) = \frac{1}{4}$$

$$= \frac{\text{Constant term}}{\text{Coefficient of } s^2}.$$

(iii) $6x^2 - 3 - 7x$

Let $p(x) = 6x^2 - 3 - 7x$
For zeroes of $p(x)$,
$$p(x) = 0$$
$\Rightarrow \quad 6x^2 - 3 - 7x = 0$
$\Rightarrow \quad 6x^2 - 7x - 3 = 0$
$\Rightarrow \quad 6x^2 - 9x + 2x - 3 = 0$
$\Rightarrow \quad 3x(2x - 3) + 1(2x - 3) = 0$
$\Rightarrow \quad (2x - 3)(3x + 1) = 0$
$\Rightarrow \quad 2x - 3 = 0 \text{ or } 3x + 1 = 0$
$\Rightarrow \quad x = \frac{3}{2} \text{ or } x = -\frac{1}{3}$
$\Rightarrow \quad x = \frac{3}{2}, -\frac{1}{3}$

So, the zeros of $p(x)$ are $\frac{3}{2}$ and $-\frac{1}{3}$.

We observe that
Sum of its zeroes
$$= \frac{3}{2} + \left(-\frac{1}{3}\right)$$
$$= \frac{3}{2} - \frac{1}{3}$$
$$= \frac{9-2}{6} = \frac{7}{6} = \frac{-(-7)}{6}$$
$$= -\frac{\text{Coefficient of } x}{\text{Coefficient of } x^2}$$

Product of its zeroes
$$= \left(\frac{3}{2}\right) \times \left(-\frac{1}{3}\right)$$
$$= -\frac{1}{2}$$
$$= -\frac{3}{6}$$
$$= \frac{\text{Constant term}}{\text{Coefficient of } x^2}.$$

(iv) **$4u^2 + 8u$**

Let $p(u) = 4u^2 + 8u$

For zeroes of $p(u)$,

$p(u) = 0$
$\Rightarrow \quad 4u^2 + 8u = 0$
$\Rightarrow \quad 4u(u + 2) = 0$
$\Rightarrow \quad u(u + 2) = 0$
$\Rightarrow \quad u = 0 \text{ or } u + 2 = 0$
$\Rightarrow \quad u = 0 \text{ or } u = -2$
$\Rightarrow \quad u = 0, -2$

So, the zeroes of $p(u)$ are 0 and -2.

We observe that,

Sum of its zeroes
$= (0) + (-2) = -2$
$= \dfrac{-(8)}{4}$
$= -\dfrac{\text{Coefficient of } u}{\text{Coefficient of } u^2}$

Product of its zeroes
$= (0)(-2) = 0 = \dfrac{0}{4}$
$= \dfrac{\text{Constant term}}{\text{Coefficient of } u^2}.$

(v) **$t^2 - 15$**

Let $p(t) = t^2 - 15$

For zeroes of $p(t)$,

$p(t) = 0$
$\Rightarrow \quad t^2 - 15 = 0$
$\Rightarrow \quad (t)^2 - (\sqrt{15})^2 = 0$
$\Rightarrow \quad (t - \sqrt{15})(t + \sqrt{15}) = 0$

| Using the identity $a^2 - b^2$
| $= (a - b)(a + b)$

$\Rightarrow \quad t - \sqrt{15} = 0 \text{ or } t + \sqrt{15} = 0$
$\Rightarrow \quad t = \sqrt{15} \text{ or } t = -\sqrt{15}$
$\Rightarrow \quad t = \sqrt{15}, -\sqrt{15}$

So, the zeroes of $p(t)$ are $\sqrt{15}$ and $-\sqrt{15}$.

We observe that,

Sum of its zeroes
$= (\sqrt{15}) + (-\sqrt{15}) = 0$
$= \dfrac{-0}{1} = \dfrac{-(\text{Coefficient of } t)}{\text{Coefficient of } t^2}$

Product of its zeroes
$= (\sqrt{15}) \times (-\sqrt{15})$
$= -15 = \dfrac{-15}{1}$
$= \dfrac{\text{Constant term}}{\text{Coefficient of } t^2}.$

(vi) **$3x^2 - x - 4$**

Let $p(x) = 3x^2 - x - 4$

By the method of splitting the middle term,

$3x^2 - x - 4$
$= 3x^2 - 4x + 3x - 4$
$= x(3x - 4) + 1(3x - 4)$
$= (3x - 4)(x + 1)$

For zeroes of $p(x)$,

$p(x) = 0$
$\Rightarrow \quad (3x - 4)(x + 1) = 0$
$\Rightarrow \quad 3x - 4 = 0 \text{ or } x + 1 = 0$
$\Rightarrow \quad 3x = 4 \text{ or } x = -1$
$\Rightarrow \quad x = \dfrac{4}{3} \text{ or } x = -1$
$\Rightarrow \quad x = \dfrac{4}{3}, -1$

So, the zeroes of $p(x)$ are $\dfrac{4}{3}$ and -1.

We observe that,

Sum of its zeroes
$= \dfrac{4}{3} + (-1) = \dfrac{4}{3} - 1$
$= \dfrac{4-3}{3} = \dfrac{1}{3} = \dfrac{-(-1)}{3}$
$= -\dfrac{(\text{Coefficient of } x)}{\text{Coefficient of } x^2}$

Product of its zeros
$= \left(\dfrac{4}{3}\right) \times (-1)$
$= -\dfrac{4}{3} = \dfrac{\text{Constant term}}{\text{Coefficient of } x^2}.$

Example 2. *Find a quadratic polynomial each with the given numbers as the sum and product of its zeroes respectively.*

(i) $\dfrac{1}{4}, -1$ (ii) $\sqrt{2}, \dfrac{1}{3}$

(iii) $0, \sqrt{5}$ (iv) $1, 1$

(v) $-\dfrac{1}{4}, \dfrac{1}{4}$ (vi) $4, 1$

Sol. (i) $\dfrac{1}{4}, -1$

Let the polynomial be $ax^2 + bx + c$, and its zeroes be α and β.

Then, $\alpha + \beta = \dfrac{1}{4} = -\dfrac{b}{a}$

and $\alpha\beta = -1 = \dfrac{c}{a}$

If $a = 4$, then $b = -1$ and $c = -4$

So, one quadratic polynomial which fits the given conditions is
$$4x^2 - x - 4.$$

(ii) $\sqrt{2}, \dfrac{1}{3}$

Let the polynomial be $ax^2 + bx + c$, and its zeroes be α and β.

Then, $\alpha + \beta = \sqrt{2} = -\dfrac{b}{a}$

and $\alpha\beta = \dfrac{1}{3} = \dfrac{c}{a}$

If $a = 3$, then $b = -3\sqrt{2}$ and $c = 1$.

So, one quadratic polynomial which fits the given conditions is $3x^2 - 3\sqrt{2}x + 1$.

(iii) $0, \sqrt{5}$

Let the polynomial be $ax^2 + bx + c$, and its zeroes be α and β.

Then, $\alpha + \beta = 0 = -\dfrac{b}{a}$

and $\alpha\beta = \sqrt{5} = \dfrac{c}{a}$

If $a = 1$, then $b = 0$ and $c = \sqrt{5}$.

So, one quadratic polynomial which fits the given conditions is $x^2 + \sqrt{5}$.

(iv) $1, 1$

Let the polynomial be $ax^2 + bx + c$ and its zeroes be α and β.

Then, $\alpha + \beta = 1 = -\dfrac{b}{a}$

and $\alpha\beta = 1 = \dfrac{c}{a}$

If $a = 1$, then $b = -1$ and $c = 1$.

So, one quadratic polynomial which fits the given conditions is $x^2 - x + 1$.

(v) $-\dfrac{1}{4}, \dfrac{1}{4}$

Let the polynomial be $ax^2 + bx + c$, and its zeroes be α and β.

Then, $\alpha + \beta = -\dfrac{1}{4} = -\dfrac{b}{a}$

and $\alpha\beta = \dfrac{1}{4} = \dfrac{c}{a}$

If $a = 4$, then $b = \dfrac{1}{4}$ and $c = 1$.

So, one quadratic polynomial which fits the given conditions is $4x^2 + x + 1$.

(vi) $4, 1$

Let the polynomial be $ax^2 + bx + c$, and is zeroes be α and β.

Then, $\alpha + \beta = 4 = -\dfrac{b}{a}$

and $\alpha\beta = 1 = \dfrac{c}{a}$

If $a = 1$, then $b = -4$ and $c = 1$.

So, one quadratic polynomial which fits the given conditions is $x^2 - 4x + 1$.

ADDITIONAL EXAMPLES

Example 3. *Verify that the numbers given alongside of the cubic polynomials below are their zeroes. Also verify the relationship between the zeroes and the coefficients in each case*

(i) $2x^3 + x^2 - 5x + 2$; $\dfrac{1}{2}, 1, -2$

(ii) $x^3 - 4x^2 + 5x - 2$; $2, 1, 1$.

Sol. (i) $2x^3 + x^2 - 5x + 2$; $\frac{1}{2}, 1, -2$

Comparing the given polynomial with $ax^3 + bx^2 + cx + d$, we get
$$a = 2, b = 1, c = -5, d = 2$$
Let $p(x) = 2x^3 + x^2 - 5x + 2$
Then,
$$p\left(\frac{1}{2}\right) = 2\left(\frac{1}{2}\right)^3 + \left(\frac{1}{2}\right)^2 - 5\left(\frac{1}{2}\right) + 2$$
$$= \frac{1}{4} + \frac{1}{4} - \frac{5}{2} + 2 = 0$$
$$p(1) = 2(1)^3 + (1)^2 - 5(1) + 2$$
$$= 2 + 1 - 5 + 2 = 0$$
$$p(-2) = 2(-2)^3 + (-2)^2 - 5(-2) + 2$$
$$= -16 + 4 + 10 + 2 = 0$$

Therefore, $\frac{1}{2}$, 1 and -2 are the zeroes of $2x^3 + x^2 - 5x + 2$.

So, $\alpha = \frac{1}{2}, \beta = 1$ and $\gamma = -2$
Therefore,
$$\alpha + \beta + \gamma = \frac{1}{2} + 1 + (-2)$$
$$= -\frac{1}{2} = -\frac{b}{a}$$
$$\alpha\beta + \beta\gamma + \gamma\alpha = \left(\frac{1}{2}\right) \times (1) + (1) \times (-2)$$
$$+ (-2) \times \left(\frac{1}{2}\right)$$
$$= \frac{1}{2} - 2 - 1 = -\frac{5}{2} = \frac{c}{a}$$
$$\alpha\beta\gamma = \left(\frac{1}{2}\right) \times (1) \times (-2) = -1$$
$$= \frac{-2}{2} = \frac{-d}{a}.$$

(ii) $x^3 - 4x^2 + 5x - 2$; 2, 1, 1

Comparing the given polynomial with $ax^3 + bx^2 + cx + d$, we get
$$a = 1, b = -4, c = 5, d = -2$$

Let $p(x) = x^3 - 4x^2 + 5x - 2$
Then, $p(2) = (2)^3 - 4(2)^2 + 5(2) - 2$
$$= 8 - 16 + 10 - 2 = 0$$
$$p(1) = (1)^3 - 4(1)^2 + 5(1) - 2$$
$$= 1 - 4 + 5 - 2 = 0$$

Therefore, 2, 1 and 1 are the zeroes of $x^3 - 4x^2 + 5x - 2$

So, $\alpha = 2, \beta = 1$ and $\gamma = 1$
Therefore,
$$\alpha + \beta + \gamma = 2 + 1 + 1 = 4 = -\frac{(-4)}{1} = -\frac{b}{a}$$
$$\alpha\beta + \beta\gamma + \gamma\alpha = (2)(1) + (1)(1) + (1)(2)$$
$$= 5 = \frac{5}{1} = \frac{c}{a}$$
$$\alpha\beta\gamma = (2)(1)(1) = 2$$
$$= \frac{-(-2)}{1} = \frac{-d}{a}.$$

Example 4. *Find a cubic polynomial with the sum, sum of the product of its zeroes taken two at a time, and product of its zeroes as 2, -7, -14, respectively.*

Sol. Let the cubic polynomial be $ax^3 + bx^2 + cx + d$ and its zeroes be α, β and γ.

Then, $\alpha + \beta + \gamma = 2 = \frac{-b}{a}$
$$\alpha\beta + \beta\gamma + \gamma\alpha = -7 = \frac{c}{a}$$
and $\alpha\beta\gamma = -14 = \frac{-d}{a}$

If $a = 1$, then $b = -2, c = -7, d = 14$

So, one cubic polynomial which fits the given conditions is $x^3 - 2x^2 - 7x + 14$.

Example 5. *Find the zeroes of the following quadratic polynomials and verify the relationship between the zeroes and their coefficients.*

(i) $x^2 + x - 12$ (ii) $x^2 - 121$
(iii) $x^2 - 10x + 25$ (iv) $x^2 - 6$
(v) $3x^2 + 15x$.

Sol. $x^2 + x - 12$
Let $p(x) = x^2 + x - 12$
For zeroes of $p(x)$, $p(x) = 0$
$\Rightarrow \quad x^2 + x - 12 = 0$

POLYNOMIALS

$\Rightarrow \quad x^2 + 4x - 3x - 12 = 0$
| By the method of splitting the middle term
$\Rightarrow \quad x(x + 4) - 3(x + 4) = 0$
$\Rightarrow \quad (x + 4)(x - 3) = 0$
$\Rightarrow \quad x + 4 = 0 \quad \text{or} \quad x - 3 = 0$
$\Rightarrow \quad x = -4 = 0 \quad \text{or} \quad x = 3$
$\Rightarrow \quad x = -4, 3$

So, the zeroes of $x^2 + x - 12$ are -4 and 3.

Sum of zeroes

$= (-4) + 3 = -1 = \dfrac{-1}{1}$

$= \dfrac{-\text{Coefficient of } x}{\text{Coefficient of } x^2}$

Product of zeroes

$= (-4) \times (3) = -12 = \dfrac{-12}{1}$

$= \dfrac{\text{Constant term}}{\text{Coefficient of } x^2}.$

(ii) $x^2 - 121$

Let $p(x) = x^2 - 121$

For zeroes of $p(x)$, $p(x) = 0$

$\Rightarrow \quad x^2 - 121 = 0$
$\Rightarrow \quad (x)^2 - (11)^2 = 0$
$\Rightarrow \quad (x - 11)(x + 11) = 0$
| Using the identity $a^2 - b^2 = (a - b)(a + b)$
$\Rightarrow \quad x - 11 = 0 \quad \text{or} \quad x + 11 = 0$
$\Rightarrow \quad x = 11 \quad \text{or} \quad x = -11$
$\Rightarrow \quad x = 11, -11$

So, the zeroes of $x^2 - 121$ are 11 and -11.

Sum of zeroes

$= (11) + (-11) = 0$

$= \dfrac{-0}{1} = \dfrac{-\text{Coefficient of } x}{\text{Coefficient of } x^2}$

Product of zeroes

$= (11) \times (-11)$

$= -121 = \dfrac{-121}{1}$

$= \dfrac{\text{Constant term}}{\text{Coefficient of } x^2}.$

(iii) $x^2 - 10x + 25$

Let $p(x) = x^2 - 10x + 25$

For zeroes of $p(x)$, $p(x) = 0$

$\Rightarrow \quad x^2 - 10x + 25 = 0$
$\Rightarrow \quad (x - 5)^2 = 0$
$\Rightarrow \quad x = 5, 5$

So, the zeroes of $x^2 - 10x + 25$ are 5 and 5.

Sum of zeroes

$= (5) + (5) = 10 = \dfrac{-(-10)}{1}$

$= \dfrac{-\text{Coefficient of } x}{\text{Coefficient of } x^2}$

Product of zeroes

$= (5)(5) = 25 = \dfrac{25}{1}$

$= \dfrac{\text{Constant term}}{\text{Coefficient of } x^2}$

(iv) $x^2 - 6$

Let $p(x) = x^2 - 6$

For zeroes of $p(x)$, $p(x) = 0$

$\Rightarrow \quad x^2 - 6 = 0$
$\Rightarrow \quad (x)^2 - (\sqrt{6})^2 = 0$
$\Rightarrow \quad (x - \sqrt{6})(x + \sqrt{6}) = 0$
| Using the identity $a^2 - b^2 = (a - b)(a + b)$
$\Rightarrow \quad x - \sqrt{6} = 0 \quad \text{or} \quad x + \sqrt{6} = 0$
$\Rightarrow \quad x = \sqrt{6} \quad \text{or} \quad x = -\sqrt{6}$
$\Rightarrow \quad x = \sqrt{6}, -\sqrt{6}$

So, the zeroes of $x^2 - 6$ and $\sqrt{6}$ and $-\sqrt{6}$.

Sum of zeroes

$= (\sqrt{6}) + (-\sqrt{6}) = 0 = \dfrac{-0}{1}$

$= \dfrac{-\text{Coefficient of } x}{\text{Coefficient of } x^2}$

Product of zeroes

$$= (\sqrt{6}) \times (-\sqrt{6}) = -6$$

$$= \frac{-6}{1} = \frac{\text{Constant term}}{\text{Coefficient of } x^2}.$$

(v) $3x^2 + 15x$

Let $p(x) = 3x^2 + 15x$

For zeroes of $p(x)$, $p(x) = 0$

$\Rightarrow \quad 3x^2 + 15x = 0$
$\Rightarrow \quad 3x(x + 5) = 0$
$\Rightarrow \quad x(x + 5) = 0$
$\Rightarrow \quad x = 0 \text{ or } x + 5 = 0$
$\Rightarrow \quad x = 0 \text{ or } x = -5$
$\Rightarrow \quad x = 0, -5$

So, the zeroes of $3x^2 + 15x$ are 0 and -5.

Sum of zeroes

$$= (0) + (-5) = -5 = \frac{-15}{3}$$

$$= \frac{-\text{Coefficient of } x}{\text{Coefficient of } x^2}$$

Product of zeroes

$$= (0) \times (-5) = 0 = \frac{0}{3}$$

$$= \frac{\text{Constant term}}{\text{Coefficient of } x^2}.$$

Example 6. *Find a quadratic polynomial each with the given numbers as the sum and product of its zeroes respectively.*

(i) $\frac{1}{3}, -1$ (ii) $\sqrt{3}, \frac{1}{2}$

(iii) $0, \sqrt{7}$.

Sol. (i) $\frac{1}{3}, -1$

Let the quadratic polynomial be $ax^2 + bx + c$ and its zeroes be α and β.

Then, $\alpha + \beta = \frac{1}{3} = \frac{-b}{a}$

and, $\alpha\beta = -1 = \frac{c}{a}$

If $a = 3$, then $b = -1 =$ and $c = -3$.

So, one quadratic polynomial which fits the given conditions is $3x^2 - x - 3$.

(ii) $\sqrt{3}, \frac{1}{2}$

Let the quadratic polynomial be $ax^2 + bx + c$ and its zeroes be α and β.

Then, $\alpha + \beta = \sqrt{3} = -\frac{b}{a}$

and, $\alpha\beta = \frac{1}{2} = \frac{c}{a}$

If $a = 2$, then $b = -2\sqrt{3}$ and $c = 1$. So, one quadratic polynomial which fits the given conditions is $2x^2 - 2\sqrt{3}x + 1$.

(iii) $0, \sqrt{7}$

Let the quadratic polynomial be $ax^2 + bx + c$ and its zeroes be α and β.

Then, $\alpha + \beta = 0 = -\frac{b}{a}$

and, $\alpha\beta = \sqrt{3} = \frac{c}{a}$

If $a = 1$, then $b = 0$ and $c = \sqrt{7}$

So, one quadratic polynomial which fits the given conditions is $x^2 + \sqrt{7}$.

Example 7. *Verify that the numbers alongside of the cubic polynomial below are their zeroes. Also verify the relationship between the zeroes and the coefficients in each case*

(i) $2x^3 - 5x^2 + x + 2 \; ; 1, -\frac{1}{2}, 2$

(ii) $x^3 + 6x^2 + 11x + 6 \; ; -1, -2, -3$.

Sol. (i) $2x^3 + 5x^2 + x + 2 \; ; 1, -\frac{1}{2}, 2$

Comparing the given polynomial with $ax^3 - bx^2 + cx + d$, we get

$a = 2, b = -5, c = 1, d = 2$

Let $p(x) = 2x^3 - 5x^2 + x + 2$

Then, $p(1) = 2(1)^3 - 5(1)^2 + (1) + 2$

$= 2 - 5 + 1 + 2 = 0$

$p\left(-\dfrac{1}{2}\right) = 2\left(-\dfrac{1}{2}\right)^3 - 5\left(-\dfrac{1}{2}\right)^2 + \left(-\dfrac{1}{2}\right) + 2$

$= -\dfrac{1}{4} - \dfrac{5}{4} - \dfrac{1}{2} + 2 = 0$

$p(2) = 2(2)^3 - 5(2)^2 + (2) + 2$

$= 16 - 20 + 2 + 2 = 0$

Therefore, $1, -\dfrac{1}{2}$ and 2 are the zeroes of $2x^3 - 5x^2 + x + 2$.

So, $\alpha = 1, \beta = -\dfrac{1}{2}$ and $\gamma = 2$

Therefore,

$\alpha + \beta + \gamma = 1 + \left(-\dfrac{1}{2}\right) + 2$

$= \dfrac{5}{2} = \dfrac{-(-5)}{2} = -\dfrac{b}{a}$

$\alpha\beta + \beta\gamma + \gamma\alpha = (1)\left(-\dfrac{1}{2}\right)$

$+ \left(-\dfrac{1}{2}\right)(2) + (2)(1)$

$= -\dfrac{1}{2} - 1 + 2$

$= \dfrac{1}{2} = \dfrac{c}{a}$

$\alpha\beta\gamma = 1\left(-\dfrac{1}{2}\right)2 = -1$

$= -\dfrac{2}{2} = \dfrac{d}{a}$

(ii) $x^3 + 6x^2 + 11x + 6, -1, -2, -3$

Comparing the given polynomial with $ax^3 + bx^2 + cx + d$, we get

$a = 1, b = 6, c = 11, d = 6$

Let $p(x) = x^3 + 6x^2 + 11x + 6$

Then,
$p(-1) = (-1)^3 + 6(-1)^2 + 11(-1) + 6$

$= -1 + 6 - 11 + 6 = 0$

$p(-2) = (-2)^3 + 6(-2)^2 + 11(-2) + 6$

$= -8 + 24 - 22 + 6 = 0$

$p(-3) = (-3)^3 + 6(-3)^2 + 11(-3) + 6$

$= -27 + 54 - 33 + 6 = 0$

Therefore, $-1, -2$ and -3 are the zeroes of $x^3 + 6x^2 + 11x + 6$.

So, $\alpha = -1, \beta = -2$ and $\gamma = -3$

Therefore,

$\alpha + \beta + \gamma = (-1) + (-2) + (-3)$

$= -6 = -\dfrac{6}{1} = -\dfrac{b}{a}$

$\alpha\beta + \beta\gamma + \gamma\alpha = (-1)(-2)$

$+ (-2)(-3) + (-3)(-1)$

$= 2 + 6 + 3$

$= 11 = \dfrac{11}{1} = \dfrac{c}{a}$

$\alpha\beta\gamma = (-1)(-2)(-3) = -6$

$= -\dfrac{6}{1} = -\dfrac{d}{a}$.

Example 8. *Find a cubic polynomial with the sum, sum of the products of its zeroes taken two at a time, and product of its zeroes as, 3, – 5, – 16, respectively.*

Sol. Let the cubic polynomial be $ax^3 + bx^2 + cx + d$ and its zeroes be α, β and γ.

Then,

$\alpha + \beta + \gamma = 3 = -\dfrac{b}{a}$,

$\alpha\beta + \beta\gamma + \gamma\alpha = -5 = \dfrac{c}{a}$,

and $\alpha\beta\gamma = -16 = -\dfrac{d}{a}$

If $a = 1$, then $b = -3, c = -5$ and $d = 16$.

So, one cubic polynomial which fits the given conditions is $x^3 - 3x^2 - 5x + 16$.

TEST YOUR KNOWLEDGE

1. Find the zeroes of the quadratic polynomial $x^2 + 7x + 10$, and verify the relationship between the zeroes and its coefficients.

2. Find the zeroes of $x^2 - 3$ and verify the relationship between the zeroes and its coefficients.

3. Find the zeroes of $64x^2 + 16x + 1$ and verify the relationship between the zeroes and its coefficients.

4. Find the zeroes of $x^2 - x + \frac{1}{4}$ and verify the relationship between the zeroes its coefficients.

5. Find the zeroes of $x^2 + x$ and verify the relationship between the zeroes and its coefficients.

6. Find a quadratic polynomial, the sum and product of whose zeroes are -1 and 1, respectively.

7. Find a quadratic polynomial, the sum and product of whose zeroes are $\frac{1}{4}$ and $-\frac{1}{8}$, respectively.

8. Find a quadratic polynomial, the sum and product of whose zeroes are 0 and 2, respectively.

9. Find a quadratic polynomial, the sum and product of whose zeroes are 3 and 0.

10. Find a quadratic polynomial, the sum and product of whose zeroes are 2 and 3.

11. Verify that $3, -1, -\frac{1}{3}$ are the zeroes of the cubic polynomial $p(x) = 3x^3 - 5x^2 - 11x - 3$, and then verity the relationship between the zeroes and its coefficients.

12. Verify that $-1, -1, 5$ are the zeroes of the cubic polynomial $p(x) = x^3 - 3x^2 - 9x - 5$, and then verify the relationship between the zeroes and its coefficient.

13. Verify that $1, 2, 3$ are the zeroes of the cubic polynomial $p(x) = x^3 - 6x^2 + 11x - 6$, and then verify the relationship between the zeroes and it coefficients.

14. Find the cubic polynomial with the sum, sum of the product of its zeroes taken two at a time and product of its zeroes as $1, 3, 6$, respectively.

15. Find a cubic polynomial with the sum, sum of the product of its zeroes taken two at a time, and product of its zeroes as $0, 0, 9$ respectively.

Answers

1. $-2, -5$
2. $\sqrt{3}, -\sqrt{3}$
3. $-\frac{1}{8}, -\frac{1}{8}$
4. $\frac{1}{2}, \frac{1}{2}$
5. $0, -1$
6. $x^2 + x + 1$
7. $x^2 - \frac{1}{4}x - \frac{1}{8}$
8. $x^2 + 2$
9. $x^2 - 3x$
10. $x^2 - 2x + 3$
14. $x^3 - x^2 + 3x - 6$
15. $x^3 - 9$.

IMPORTANT POINTS

1. How to find the other two zeroes of a cubic polynomial when its one zero is given.

Consider a polynomial $x^3 - 3x - x + 3$ one of whose zeroes is given to be 1. We are to find out its other two zeroes.

Clearly, $x - 1$ is a factor of $x^2 - 3x^2 - x + 3$.

So, we can divide $x^3 - 3x^2 - x + 3$ by $x - 1$ to get the quotient $x^2 - 2x - 3$.

$$x^2 - 2x - 3$$

$$\begin{array}{r}
x - 1 \overline{\smash{)}\, x^3 - 3x^2 - x + 3} \\
\underline{x^3 - x^2} \\
- + \\
\underline{-2x^2 - x + 3} \\
-2x^2 + 2x \\
+ - \\
\underline{-3x + 3} \\
-3x + 3 \\
+ - \\
\underline{\times}
\end{array}$$

Now, by splitting method
$x^2 - 2x - 3$
$= x^2 - 3x + x - 3$
| By splitting the middle term
$= x(x - 3) + 1(x - 3)$
$= (x - 3)(x + 1)$

So, the zeroes of $x^2 - 2x - 3$ are 3 and -1. These are nothing but the other two zeroes of $x^3 - 3x^2 - x + 3$. This gives us
$x^3 - 3x^2 - x + 3 = (x - 1)(x^2 - 2x - 3)$
$\qquad = (x - 1)(x - 3)(x + 1)$

Therefore, 1, 3 and -1 are the three zeroes of $x^3 - 3x^2 - x + 3$.

So, all the three zeroes of the given cubic polynomial are known to us.

2. Method of Dividing one polynomial by another

Steps/Algorithm

(i) We first arrange the terms of the dividend and the divisor in the decreasing order of their degrees. We know that arranging the terms in this order is called writing the polynomials in standard form :

(ii) To obtain the first term of the quotient, we divide the highest degree term of the dividend by the highest degree term of the divisor. Then, carry out the division process.

(iii) Now, to obtain the second term of the quotient, we divide the highest degree term of the new dividend by the highest degree term of the divisor. Again carry out the division process.

(iv) Continue the division process similarly further till either the remainder is zero or its degree is less than the degree of the divisor.

Note. This algorithm is similar to Euclid's divison algorithm.

3. Dividend = Quotient × Divisor
$\qquad$ + Remainder

4. Division algorithm for Polynomials

If $p(x)$ and $g(x)$ are any two polynomials with $g(x) \neq 0$, then we can find polynomials $q(x)$ and $r(x)$ such that
$p(x) = q(x) \times g(x) + r(x)$
where $r(x) = 0$ or degree of $r(x) <$ degree of $g(x)$.

This result is taken as **Division Algorithm** for polynomials.

ILLUSTRATIVE EXAMPLES

[NCERT Exercise 2.3]
(Page No. 36)

Example 1. *Divide the polynomial $p(x)$ by the polynomial $g(x)$ and find the quotient and remainder in each of the following :*
(i) $p(x) = x^3 - 3x^2 + 5x - 3$, $g(x) = x^2 - 2$
(ii) $p(x) = x^4 - 3x^2 + 4x + 5$,
$\qquad g(x) = x^2 + 1 - x$
(iii) $p(x) = x^4 - 5x + 6$, $g(x) = 2 - x^2$

Sol. (i) $p(x) = x^3 - 3x^2 + 5x - 3$,
$\qquad g(x) = x^2 - 2$

The given polynomial $p(x)$ and $g(x)$ are in standard from. Now, we apply the division algorithm to the given polynomials $p(x)$ and $g(x)$.

$$\begin{array}{r}
x - 3 \\
x^2 - 2 \overline{\smash{)} x^3 - 3x^2 + 5x - 3} \\
x^3 - 2x \\
\underline{-+} \\
-3x^2 + 7x - 3 \\
-3x^2 + 6 \\
\underline{+-} \\
7x - 9
\end{array}$$

We stop here since degree $(7x - 9) = 1 <$ degree $(x^2 - 2)$

So, quotient $= x - 3$,
$\qquad$ remainder $= 7x - 9$

Therefore,

Quotient × Divisor + Remainder
$$= (x - 3)(x^2 - 2) + (7x - 9)$$
$$= x^3 - 2x - 3x^2 + 6 + 7x - 9$$
$$= x^3 - 3x^2 + 5x - 3$$
$$= \text{Dividend}$$

Therefore, the division algorithm is verified.

(ii) $p(x) = x^4 - 3x^2 + 4x + 5$,
$g(x) = x^2 + 1 - x$

$p(x)$ is in standard form.

$g(x)$, in standard form, is $x^2 - x + 1$. Now, we apply the division algorithm to the given polynomial $p(x)$ and $g(x)$.

$$\begin{array}{r}
x^2 + x - 3 \\
x^2 - x + 1 \overline{)\, x^4 - 3x^2 + 4x + 5} \\
x^4 + x^2 - x^3 \\
\underline{- - +} \\
x^3 - 4x^2 + 4x + 5 \\
x^3 - x^2 + x \\
\underline{- + -} \\
-3x^2 + 3x + 5 \\
-3x^2 + 3x - 3 \\
\underline{+ - +} \\
8
\end{array}$$

We stop here since degree $(8) = 0 <$ degree $(x^2 - x + 1)$.

So, quotient $= x^2 + x - 3$,
remainder $= 8$

Therefore,

Quotient × Divisor + Remainder
$$= (x^2 + x - 3)(x^2 - x + 1) + 8$$
$$= x^4 - x^3 + x^2 + x^3 - x^2 + x - 3x^2 + 3x - 3 + 8$$
$$= x^4 - 3x^2 + 4x + 5$$
$$= \text{Dividend}$$

Therefore, the division algorithm is verified.

(iii) $p(x) = x^4 - 5x + 6$, $g(x) = 2 - x^2$

$p(x)$ is in standard form.

$g(x)$, in standard form, is $-x^2 + 2$. Now we apply the division algorithm is the given polynomial $p(x)$ and $g(x)$.

$$\begin{array}{r}
-x^2 - 2 \\
-x^2 + 2 \overline{)\, x^4 - 5x + 6.} \\
x^4 - 2x^2 \\
\underline{- +} \\
2x^2 - 5x + 6 \\
2x^2 - 4 \\
\underline{- +} \\
-5x + 10
\end{array}$$

We stop here since degree
$(-5x + 10) = 1 <$ degree $(-x^2 + 2)$

So, quotient $= -x^2 - 2$,
remainder $= -5x + 10$

Therefore,

Quotient × Divisor + Remainder
$$= (-x^2 - 2)(-x^2 + 2) + (-5x + 10)$$
$$= x^4 - 2x^2 + 2x^2 - 4 - 5x + 10$$
$$= x^4 - 5x + 6$$
$$= \text{Dividend}$$

Therefore, the division algorithm is verified.

Example 2. *Check whether the first polynomial is a factor of the second polynomial by dividing the second polynomial by the first polynomial :*

(i) $t^2 - 3$, $2t^4 + 3t^3 - 2t^2 - 9t - 12$
(ii) $x^2 + 3x + 1$, $3x^4 + 5x^3 - 7x^2 + 2x + 2$
(iii) $x^3 - 3x + 1$, $x^5 - 4x^3 + x^2 + 3x + 1$

Sol. (i) $t^2 - 3$, $2t^4 + 3t^3 - 2t^2 - 9t - 12$

The first and the second polynomials, both are in standard form. Now, we apply the division algorithm to the given polynomials.

POLYNOMIALS

$$\begin{array}{r}
2t^2 + 3t + 4 \\
t^2 - 3 \overline{\smash{)}\; 2t^4 + 3t^3 - 2t^2 - 9t - 12} \\
2t^4 \quad\quad\; - 6t^2 \\
\underline{-\quad\quad\quad\; +\quad\quad\quad\quad\quad} \\
3t^3 + 4t^2 - 9t - 12 \\
3t^3 \quad\quad\; - 9t \\
\underline{-\quad\quad\quad +\quad\quad\quad\quad} \\
4t^2 - 12 \\
4t^2 - 12 \\
\underline{-\quad\; +\quad\quad} \\
0
\end{array}$$

Since, the remainder is 0, therefore, the first polynomial is a factor of the second polynomial.

(ii) $x^2 + 3x + 1$, $3x^4 + 5x^3 - 7x^2 + 2x + 2$

The first and the second polynomials, both, are in standard form. Now, we apply division algorithm to the given polynomials.

$$\begin{array}{r}
3x^2 - 4x + 2 \\
x^2 + 3x + 1 \overline{\smash{)}\; 3x^4 + 5x^3 - 7x^2 + 2x + 2} \\
3x^4 + 9x^3 + 3x^2 \\
\underline{-\quad\; -\quad\; -\quad\quad\quad\quad} \\
-4x^3 - 10x^2 + 2x + 2 \\
-4x^3 - 12x^2 - 4x \\
\underline{+\quad\; +\quad\; +\quad\quad} \\
2x^2 + 6x + 2 \\
2x^2 + 6x + 2 \\
\underline{-\quad\; -\quad\; -\quad} \\
0
\end{array}$$

Since, the remainder is 0, therefore, the first polynomial is a factor of the second polynomial.

(iii) $x^3 - 3x + 1$, $x^5 - 4x^3 + x^2 + 3x + 1$

The first and the second polynomials, both, are in standard form. Now, we apply the division algorithm to the given polynomials.

$$\begin{array}{r}
x^2 - 1 \\
x^3 - 3x + 1 \overline{\smash{)}\; x^5 - 4x^3 + x^2 + 3x + 1} \\
x^5 - 3x^3 \quad\; + x^2 \\
\underline{-\quad\; +\quad\quad -\quad\quad\quad\quad} \\
-x^3 \quad\quad + 3x + 1 \\
-x^3 \quad\quad + 3x - 1 \\
\underline{+\quad\quad\quad -\quad\quad +} \\
2
\end{array}$$

Since, the remainder is 2 ($\neq 0$), therefore, the first polynomial is not a factor of the second polynomial.

Example 3. *Obtain all other zeroes of $3x^4 + 6x^3 - 2x^2 - 10x - 5$, if two of its zeroes are $\sqrt{\dfrac{5}{3}}$ and $-\sqrt{\dfrac{5}{3}}$.*

Sol. Since, two zeroes are $\sqrt{\dfrac{5}{3}}$ and $-\sqrt{\dfrac{5}{3}}$, therefore, $\left(x - \sqrt{\dfrac{5}{3}}\right)\left(x + \sqrt{\dfrac{5}{3}}\right) = x^2 - \dfrac{5}{3}$ is a factor of the given polynomial. Now, we apply the division algorithm to the given polynomial and $x^2 - \dfrac{5}{3}$.

$$\begin{array}{r}
3x^2 + 6x + 3 \\
x^2 - \dfrac{5}{3} \overline{\smash{)}\; 3x^4 + 6x^3 - 2x^2 - 10x - 5} \\
3x^4 \quad\quad\quad - 5x^2 \\
\underline{-\quad\quad\quad +\quad\quad\quad\quad\quad} \\
6x^3 + 3x^2 - 10x - 5 \\
6x^3 \quad\quad\; - 10x \\
\underline{-\quad\quad\quad +\quad\quad\quad} \\
3x^2 \quad\quad\; - 5 \\
3x^2 \quad\quad\; - 5 \\
\underline{-\quad\quad\; +\quad} \\
0
\end{array}$$

So, $3x^4 + 6x^3 - 2x^2 - 10x - 5$
$$= \left(x^2 - \frac{5}{3}\right)(3x^2 + 6x + 3)$$

Now, $3x^2 + 6x + 3$
$$= 3(x^2 + 2x + 1)$$
$$= 3(x + 1)^2$$

So, its zeroes are -1 and -1.

Therefore, the other zeroes of the given fourth degree polynomial are -1 and -1.

Example 4. *On dividing $x^3 - 3x^2 + x + 2$ by a polynomial $g(x)$, the quotient and remainder were $x - 2$ and $-2x + 4$, respectively. Find $g(x)$.*

Sol. Here,
$$p(x) = x^3 - 3x^2 + x + 2,$$
$$g(x) = x - 2,$$
and, $r(x) = -2x + 4$

By division algorithm for polynomials,
$$p(x) = q(x) \times g(x) + r(x)$$
$\Rightarrow x^3 - 3x^2 + x + 2$
$$= (x - 2)\ x) + (-2x + 4)$$
$\Rightarrow (x - 2)\, g(x)$
$$= x^3 - 3x^2 + x + 2 + 2x - 4$$
$$= x^3 - 3x^2 + 3x - 2$$
$\Rightarrow g(x) = \dfrac{x^3 - 3x^2 + 3x - 2}{x - 2}$

Let us apply division algorithm to $x^3 - 3x^2 + 3x - 2$ and $x - 2$.

```
              x² - x + 1
          ┌─────────────────
    x - 2 │ x³ - 3x² + 3x - 2
            x³ - 2x²
            -   +
            ─────────────
               -x² + 3x - 2
                x² + 2x
                +  -
            ─────────────
                     x - 2
                     x - 2
                     -  +
            ─────────────
                       0
```

Therefore, $g(x) = x^2 - x + 1$.

Example 5. *Give examples of polynomials $p(x)$, $g(x)$, $q(x)$ and $r(x)$, which satisfy the division algorithm and*
 (i) deg $p(x)$ = deg $q(x)$
 (ii) deg $q(x)$ = deg $r(x)$
 (iii) deg $r(x)$ = 0.

Sol. (i) **deg $p(x)$ = deg $q(x)$**
$$p(x) = 2x^2 - 2x + 14$$
$$g(x) = 2$$
$$q(x) = x^2 - x + 7$$
$$r(x) = 0$$
Clearly, $p(x) = q(x) \times g(x) + r(x)$.

(ii) **deg $q(x)$ = deg $r(x)$**
$$p(x) = x^3 + x^2 + x + 1$$
$$g(x) = x^2 - 1$$
$$q(x) = x + 1$$
$$r(x) = 2x + 2$$
Clearly, $p(x) = q(x) \times g(x) + r(x)$.

(iii) **deg $r(x)$ = 0**
$$p(x) = x^3 + 2x^2 - x - 2$$
$$g(x) = x^2 - 1$$
$$q(x) = x + 2$$
$$r(x) = 4.$$
Clearly, $p(x) = q(x) \times g(x) + r(x)$.

ADDITIONAL EXAMPLES

Example 6. *Apply the division algorithm to find the quotient and remainder an dividing $p(x)$ by $g(x)$ as given below.*

(i) $p(x) = x^5 + x^4 + x^3 + x^2 + 2x + 2$,
 $g(x) = x^3 + 1$

(ii) $p(x) = x^3 - 6x^2 + 11x - 6$,
 $g(x) = x^2 - 5x + 6$

(iii) $p(x) = x^4 + 2x^3 + 3x^2 + 2x + 20$,
 $g(x) = x^2 + 2x + 2$

(iv) $p(x) = x^6 + 5x^3 + 7x + 3$,
 $g(x) = x^2 + 2$

(v) $p(x) = x^4 + 1$, $g(x) = x + 1$.

Sol. (i) $p(x) = x^5 + x^4 + x^3 + x^2 + 2x + 2$,
$g(x) = x^3 + 1$

The given polynomials $p(x)$ and $g(x)$ are in standard form.

Now, we apply the division algorithm to the given polynomials $p(x)$ and $g(x)$.

$$\begin{array}{r}
x^2 + x + 1 \\
x^3 + 1 \overline{) x^5 + x^4 + x^3 + x^2 + 2x + 2}\\
\underline{x^5 + x^2 }\\
x^4 + x^3 + 2x + 2\\
\underline{x^4 + x }\\
x^3 + x + 2\\
\underline{x^3 + 1}\\
x + 1
\end{array}$$

We stop here since degree $(x + 1) = 1 <$ degree $(x^3 + 1)$

So, quotient $= x^2 + x + 1$,
remainder $= x + 1$

Therefore,
Quotient × divisor + Remainder
$= (x^2 + x + 1)(x^3 + 1) + (x + 1)$
$= x^5 + x^2 + x^4 + x + x^3 + 1 + x + 1$
$= x^5 + x^4 + x^3 + x^2 + 2x + 2$
$=$ Dividend

Therefore, the division algorithm is verified.

(ii) $p(x) = x^3 - 6x^2 + 11x - 6$,
$g(x) = x^2 - 5x + 6$

The given polynomials $p(x)$ and $g(x)$ are in standard form.

Now, we apply the division algorithm to the given polynomials $p(x)$ and $g(x)$.

$$\begin{array}{r}
x - 1 \\
x^2 - 5x + 6 \overline{) x^3 - 6x^2 + 11x - 6}\\
\underline{x^3 - 5x^2 + 6x }\\
-x^2 + 5x - 6\\
\underline{-x^2 + 5x - 6}\\
0
\end{array}$$

We stop here since the remainder is 0.

So, quotient $= x - 1$,
remainder $= 0$

Therefore,
Quotient × Divisor + Remainder
$= (x - 1)(x^2 - 5x + 6) + 0$
$= x^3 - 5x^2 + 6x - x^2 + 5x - 6$
$= x^3 - 6x^2 + 11x - 6$
$=$ Dividend

Therefore, the division algorithm is verified.

(iii) $p(x) = x^4 + 2x^3 + 3x^2 + 2x + 20$,
$g(x) = x^2 + 2x + 2$

The given polynomials $p(x)$ and $g(x)$ are in standard form.

Now, we apply the division algorithm to the given polynomials $p(x)$ and $g(x)$.

$$\begin{array}{r}
x^2 + 1 \\
x^2 + 2x + 2 \overline{) x^4 + 2x^3 + 3x^2 + 2x + 20}\\
\underline{x^4 + 2x^3 + 2x^2 }\\
x^2 + 2x + 20\\
\underline{x^2 + 2x + 2}\\
18
\end{array}$$

We stop here since degree
(18) = 0 < degree $(x^2 + 2x + 2)$
So, quotient $= x^2 + 1$,
    remainder $= 18$
Therefore,
Quotient × Divisor + Remainder
$= (x^2 + 1)(x^2 + 2x + 2) + 18$
$= x^4 + 2x^3 + 2x^2 + x^2 + 2x + 2 + 18$
$= x^4 + 2x^3 + 3x^2 + 2x + 20$
$=$ Dividend

Therefore, the division algorithm is verified.

(iv) $p(x) = x^6 + 5x^3 + 7x + 3$, $g(x) = x^2 + 2$

The given polynomials $p(x)$ and $g(x)$ are in standard form.

Now, we apply the division algorithm to the given polynomials $p(x)$ and $g(x)$.

$$\begin{array}{r}
x^4 - 2x^2 + 5x + 4 \\
x^2 + 2 \overline{\smash{)}\, x^6 + 5x^3 + 7x + 3} \\
\underline{x^6 + 2x^4} \\
-2x^4 + 5x^3 + 7x + 3 \\
\underline{-2x^4 - 4x^2} \\
5x^3 + 4x^2 + 7x + 3 \\
\underline{5x^3 + 10x} \\
4x^2 - 3x + 3 \\
\underline{4x^2 + 8} \\
-3x - 5
\end{array}$$

We stop here since degree
$(-3x - 5) = 1 <$ degree $(x^2 + 2)$
So, quotient $= x^4 - 2x^2 + 5x + 4$,
    remainder $= -3x - 5$
Therefore,
Quotient × Divisor + Remainder
$= (x^4 - 2x^2 + 5x + 4)(x^2 + 2) + (-3x - 5)$
$= x^6 + 2x^4 - 2x^4 - 4x^2 + 5x^3 + 10x + 4x^2$
$ + 8 - 3x - 5$
$= x^6 + 5x^3 + 7x + 3$
$=$ Dividend

Therefore, the division algorithm is verified.

(v) $p(x) = x^4 + 1$, $g(x) = x + 1$

The given polynomials $p(x)$ and $g(x)$ are in standard form.

Now, we apply the division algorithm to the given polynomials $p(x)$ and $g(x)$.

$$\begin{array}{r}
x^3 - x^2 + x - 1 \\
x + 1 \overline{\smash{)}\, x^4 + 1} \\
\underline{x^4 + x^3} \\
-x^3 + 1 \\
\underline{-x^3 - x^2} \\
x^2 + 1 \\
\underline{x^2 + x} \\
-x + 1 \\
\underline{-x - 1} \\
2
\end{array}$$

We stop here since degree (2) = 0 < degree $(x + 1)$
So, quotient $= x^3 - x^2 + x - 1$,
    remainder $= 2$
Therefore,
Quotient × Divisor + Remainder
$= (x^3 - x^2 + x - 1)(x + 1) + 2$
$= x^4 + x^3 - x^3 - x^2 + x^2 + x$
$ - x - 1 + 2$
$= x^4 + 1$
$=$ Dividend

Therefore, the division algorithm is verified.

Example 7. *Divide*
(i) $y^3 + y^2 - 2y + 1$ *by* $y - a$
(ii) $x^4 - a^4$ *by* $x - a$
(iii) $x^5 + a^5$ *by* $x + a$.

Sol. (i) $y^3 + y^2 - 2y + 1$ by $y - a$

$$
\begin{array}{r}
y^2 + (1+a)y + a^2 + a - 2 \\
y - a \overline{\smash{\big)}\, y^3 + y^2 - 2y + 1} \\
\underline{y^3 - ay^2} \\
-\quad + \\
(1+a)y^2 - 2y + 1 \\
\underline{(1+a)y^2 - (1+a)ay} \\
-\qquad\qquad + \\
(a^2 + a - 2)y + 1 \\
\underline{(a^2 + a - 2)y - a(a^2 + a - 2)} \\
-\qquad\qquad\qquad + \\
a^3 + a^2 - 2a + 1
\end{array}
$$

We stop here since degree $(a^3 + a^2 - 2a + 1) = 0 < $ degree $(y - a)$.

So, quotient $= y^2 + (1 + a)y + a^2 + a - 2$, remainder $= a^3 + a^2 - 2a + 1$

Therefore,
Quotient × Divisor + Remainder
$= \{y^2 + (1 + a)y + a^2 + a - 2\}(y - a)$
$\quad + (a^3 + a^2 - 2a + 1)$
$= y^3 - ay^2 + (1 + a)y^2 - a(1 + a)y$
$\quad + (a^2 + a - 2)y - a(a^2 + a - 2)$
$\quad + (a^3 + a^2 - 2a + 1)$
$= y^3 + y^2 - 2y + 1$
$=$ Dividend

Therefore, the division algorithm is verified.

(ii) $x^3 + a^4$ by $x - a$

$$
\begin{array}{r}
x^3 + ax^2 + a^2x + a^3 \\
x - a \overline{\smash{\big)}\, x^4 - a^4} \\
\underline{x^4 \qquad - ax^3} \\
- \quad + \\
ax^3 - a^4 \\
\underline{ax^3 \qquad - a^2x^2} \\
- \qquad + \\
a^2x^2 - a^4 \\
\underline{a^2x^2 \qquad - a^3x} \\
- \qquad + \\
a^3x - a^4 \\
\underline{a^3x - a^4} \\
- \quad + \\
0
\end{array}
$$

We stop here since the remainder is zero.

So, quotient $= x^3 + ax^2 + a^2x + a^3$, remainder $= 0$

Therefore,
Quotient × Divisor + Remainder
$= (x^3 + ax^2 + a^2x + a^3)(x - a) + 0$
$= x^4 - ax^3 + ax^3 - a^2x^2 + a^2x^2$
$\qquad - a^3x + a^3x - a^4$
$= x^4 - a^4$
$=$ Dividend

Therefore, the division algorithm is verified.

(iii) $x^5 + a^5$ by $x + a$

$$
\begin{array}{r}
x^4 - ax^3 + a^2x^2 - a^3x + a^4 \\
x + a \overline{\smash{\big)}\, x^5 + a^5} \\
\underline{x^5 \qquad + ax^4} \\
- \quad - \\
- ax^4 + a^5 \\
\underline{- ax^4 \qquad - a^2x^3} \\
+ \qquad + \\
a^2x^3 + a^5 \\
\underline{a^2x^3 \qquad + a^3x^2} \\
- \qquad - \\
- a^3x^2 + a^5 \\
\underline{- a^3x^2 \qquad - a^4x} \\
+ \qquad + \\
a^4x + a^5 \\
\underline{a^4x + a^5} \\
- \quad - \\
0
\end{array}
$$

We stop here since the remainder is zero.

So,
quotient $= x^4 - ax^3 + a^2x^2 - a^3x + a^4$, remainder $= 0$

Therefore,
Quotient × Divisor + Remainder
$= (x^4 - ax^3 + a^2x^2 - a^3x + a^4)(x + a) + 0$

$$= x^5 + ax^4 - ax^4 - a^2x^3 + a^2x^3$$
$$+ a^3x^2 - a^3x^2 - a^4x + a^4x + a^5$$
$$= x^5 + a^5$$
$$= \text{Dividend}$$

Therefore, the division algorithm is verified.

Example 8. *Obtain all the zeroes of $2x^4 - 7x^3 - 13x^2 + 63x - 45$, if two of its zeroes are 1 and 3.*

Sol. Since, two zeroes are 1 and 3, therefore, $(x-1)(x-3) = x^2 - 4x + 3$ is a factor of the given polynomial. Now, we apply the division algorithm to the given polynomial and $x^2 - 4x + 3$

$$\begin{array}{r}
2x^2 + x - 15 \\
x^2 - 4x + 3 \overline{\smash{)}\, 2x^4 - 7x^3 - 13x^2 + 63x - 45} \\
2x^4 - 8x^3 + 6x^2 \\
-\quad +\quad - \\
\hline
x^3 - 19x^2 + 63x - 45 \\
x^3 - 4x^2 + 3x \\
-\quad +\quad - \\
\hline
-15x^2 + 60x - 45 \\
-15x^2 + 60x - 45 \\
+\quad -\quad + \\
\hline
0
\end{array}$$

So, $2x^4 - 7x^3 - 13x^2 + 63x - 45$
$$= (x^2 - 4x + 3)(2x^2 + x - 15)$$
Now, $2x^2 + x - 15$
$$= 2x^2 + 6x - 5x - 15$$
| by splitting the middle term
$$= 2x(x+3) - 5(x+3)$$
$$= (x+3)(2x-5)$$

So, its zeroes are -3 and $\dfrac{5}{2}$.

Therefore, all the zeroes of the given fourth degree polynomial are $1, 3, -3$ and $\dfrac{5}{2}$.

Example 9. *Obtain all the zeroes of $3x^3 - x^2 - 3x + 1$ given that one of its zeroes is 1.*

Sol. Since, one zero of the given polynomial is 1, so $(x-1)$ is a factor of the given polynomial. Now, we apply the division algorithm to the given polynomial and $x-1$.

$$\begin{array}{r}
3x^2 + 2x - 1 \\
x - 1 \overline{\smash{)}\, 3x^3 - x^2 - 3x + 1} \\
3x^3 - 3x^2 \\
-\quad + \\
\hline
2x^2 - 3x + 1 \\
2x^2 - 2x \\
-\quad + \\
\hline
-x + 1 \\
-x + 1 \\
+\quad - \\
\hline
0
\end{array}$$

So,
$$3x^3 - x^2 - 3x + 1 = (x-1)(3x^2 + 2x - 1)$$
Now, $3x^2 + 2x - 1$
$$= 3x^2 + 3x - x - 1$$
| by splitting the middle term
$$= 3x(x+1) - 1(x+1)$$
$$= (x+1)(3x-1)$$

So, its zeroes are -1 and $\dfrac{1}{3}$.

Therefore, all the zeroes of the given third degree polynomial are $1, -1$ and $\dfrac{1}{3}$.

Example 10. *On dividing $p(x)$ by $x^2 + 2x + 2$, the quotient and the remainder are $x^2 + 1$ and 18 respectively. Find $p(x)$.*

Sol. Here
$$q(x) = x^2 + 1$$
$$r(x) = 18$$
$$g(x) = x^2 + 2x + 1$$
By divison algorithm,
Dividend = Divisor × Quotient
$\qquad\qquad\qquad\qquad$ + Remainder
$$\Rightarrow p(x) = q(x) \times g(x) + r(x)$$
$$\Rightarrow p(x) = (x^2 + 2x + 2)(x^2 + 1) + 18$$
$$= x^4 + x^2 + 2x^3 + 2x + 2x^2$$
$$\qquad\qquad\qquad\qquad + 2 + 18$$
$$= x^4 + 2x^3 + 3x^2 + 2x + 20.$$

Example 11. *Check whether the polynomial $x^2 - 10x + 16$ is a factor of the polynomial $5x^3 - 70x^2 + 153x - 342$ by applying the division algorithm.*

Sol. The given polynomials are in standard form. Now, we apply the division algorithm to the given polynomials

$$\begin{array}{r}
5x - 20 \\
x^2 - 10x + 16 \overline{)5x^3 - 70x^2 + 153x - 342}\\
5x^3 - 50x^2 + 80x \\
- + -\\
\overline{ -20x^2 + 73x - 342}\\
-20x^2 + 200x - 320\\
+ - +\\
\overline{ -127x - 22}
\end{array}$$

Since, the remainder is not 0, therefore, the polynomial $x^2 - 10x + 16$ is not a factor of the polynomial $5x^3 - 70x^2 + 153x - 342$.

Example 12. *Check whether the polynomial $x^2 - 4x + 3$ is a factor of the polynomial $x^3 - 3x^2 - x + 3$.*

Sol. The given polynomials are in standard form. Now, we apply the division algorithm to the given polynomials.

$$\begin{array}{r}
x + 1 \\
x^2 - 4x + 3 \overline{)x^3 - 3x^2 - x + 3}\\
x^3 - 4x^2 + 3x \\
- + -\\
\overline{ x^2 - 4x + 3}\\
x^2 - 4x + 3\\
- + -\\
\overline{ 0}
\end{array}$$

Since, the remainder is zero, therefore, the polynomial $x^2 - 4x + 3$ is a factor of the polynomial $x^3 - 3x^2 - x + 3$.

Example 13. *Find the value of b for which the polynomial $2x^3 + 9x^2 - x - b$ is divisible by $2x + 3$.*

Sol.

$$\begin{array}{r}
x^2 + 3x - 5 \\
2x + 3 \overline{)2x^3 + 9x^2 - x - b}\\
2x^3 + 3x^2 \\
- -\\
\overline{ 6x^2 - x - b}\\
6x^2 - 9x \\
- -\\
\overline{ -10x - b}\\
-10x - 15\\
+ +\\
\overline{ 15 - b}
\end{array}$$

If the polynomial $2x^3 + ax^2 - x - b$ is divisible by $2x + 3$, then the remainder must be zero.

So, $\quad 15 - b = 0$

$\Rightarrow \quad b = 15.$

Example 14. *What must be added to $6x^5 + 5x^4 + 11x^3 - 3x^2 + x + 1$, so that the polynomial so obtained is exactly divisible by $3x^2 - 2x + 4$?*

Sol.

$$\begin{array}{r}
2x^3 + 3x^2 + 3x - 3 \\
3x^2 - 2x + 4 \overline{)6x^5 + 5x^4 + 11x^3 - 3x^2 + x + 1}\\
6x^5 - 4x^4 + 8x^3 \\
- + -\\
\overline{ 9x^4 + 3x^3 - 3x^2 + x + 1}\\
9x^4 - 6x^3 + 12x^2 \\
- + -\\
\overline{ 9x^3 - 15x^2 + x + 1}\\
9x^3 - 6x^2 + 12x \\
- + -\\
\overline{ -9x^2 - 11x + 1}\\
-9x^2 + 6x - 12\\
+ - +\\
\overline{ -17x + 13}
\end{array}$$

Therefore, we must add $-(-17x + 13)$, *i.e.*, $17x - 13$.

Example 15. *What must be subtracted from* $2x^4 - 11x^3 + 29x^2 - 40x + 29$, *so that the resulting polynomial is exactly divisible by* $x^2 - bx + 4$?

Sol.

$$\begin{array}{r} 2x^2 - 5x + 6 \\ x^2 - 3x + 4 \overline{\smash{\big)}\, 2x^4 - 11x^3 + 29x^2 - 40x + 29} \\ 2x^4 - 6x^3 + 8x^2 \\ -\quad +\quad - \\ \hline -5x^3 + 21x^2 - 40x + 29 \\ -5x^3 + 15x^2 - 20x \\ +\quad -\quad + \\ \hline 6x^2 - 20x + 29 \\ 6x^2 - 18x + 24 \\ -\quad +\quad - \\ \hline -2x + 5 \end{array}$$

Therefore, we must subtract $(-2x + 5)$.

TEST YOUR KNOWLEDGE

1. Divide $2x^2 + 3x + 1$ by $x + 2$.
2. Divide $3x^3 + x^2 + 2x + 5$ by $1 + 2x + x^2$.
3. Find all the zeroes of $2x^4 - 3x^3 - 3x^2 + 6x - 2$ if you know that two of its zeroes are $\sqrt{2}$ and $-\sqrt{2}$.
4. Divide $3x^2 - x^3 - 3x + 5$ by $x - 1 - x^2$, and verify the division algorithm.
5. Divide $4x^4 - 3x^3 - 9x^2 + x + 2$ by $2x^2 - 1$.
6. Find whether or not the first polynomial is a factor of the second polynomial.
   (i) $3x + 2$, $3x^4 + 5x^3 - x^2 + 13x + 10$
   (ii) $x^2 + 1$, $x^4 - 3x^3 - 4x^2 + 3x + 2$.
7. Obtain all the zeroes of $x^4 + 10x^3 + 35x^2 + 50x + 24$, if two of its zeroes are -1 and -2.
8. Obtain all the zeroes of $2x^4 + x^3 - 14x^2 - 19x - 6$, if two of its zeroes are -1 and -2.
9. If one zero of the polynomial $x^3 - 23x^2 + 142x - 120$ is 1, then find the other two zeroes.
10. If one zero of the polynomial $y^3 - 7y + 6$ is -3, then find the other two zeroes.
11. What must be added to $x^4 + 2x^3 - 2x^2 + x - 1$ so that the result is exactly divisible by $x^2 + 2x - 3$?
12. What must be subtracted from $4x^4 - 2x^3 - 6x^2 + x - 5$ so that the result is exactly divisible by $2x^2 + x - 1$?
13. For what value of a is $2x^3 + ax^2 + 11x + a + 3$ exactly divisible by $(2x - 1)$?
14. Prove that $2x^4 - 6x^3 + 3x^2 + 3x - 2$ is exactly divisible by $x^2 - 3x + 2$.
15. Let $f(x) = x^2 + 2$ and $g(x) = x^3 + 7x$. Divide $f(x)$ by $g(x)$. Find the quotient and the remainder.
16. Divide $x^5 + x^4 - 2x^3 + 4x^2 - 4x - 15$ by $x^2 - x + 3$. Find the quotient and the remainder.
17. If the dividend is $x^3 - 14x^2 + 37x - 60$, quotient is $x^2 - 12x + 13$ and the remainder is -34, then find the divisor.
18. In each of the following questions, divide the polynomial p by g and find the quotient

and the remainder. Find in which case g is a factor of p.

(i) $p(u) = u^3 + 3u^2 - 12u + u$, $g(u) = u - 2$
(ii) $p(x) = x^3 - 14x^2 + 37x - 60$, $g(x) = x - 2$
(iii) $p(y) = y^6 + 3y^2 + 10$, $g(y) = y^3 + 1$
(iv) $p(x) = x^5 + 5x^3 + 3x^2 + 5x + 3$, $g(x) = x^2 + 4x + 2$
(v) $p(x) = 2x^2 - 3x + 5$, $g(x) = x - a$.

Answers

1. Quotient = $2x - 1$, Remainder = 3
2. Quotient = $3x - 5$, Remainder = $9x + 10$
3. $\sqrt{2}, -\sqrt{2}, \dfrac{1}{2}, 1$
4. Quotient = $x - 2$, Remainder = 3
5. Quotient = $2x^2 - \dfrac{3}{2}x - \dfrac{7}{2}$, Remainder = $-\dfrac{1}{2}x - \dfrac{3}{2}$.
6. (i) Quotient = $x^3 + x^2 - x + 5$, Remainder = 0, Yes.
(ii) Quotient = $x^2 - 3x - 5$, Remainder = $6x + 7$, No.
7. $-1, -2, -3, -4$
8. $-1, -2, 3, -\dfrac{1}{2}$
9. 10, 12
10. 1, 2
11. $x - 2$
12. -6
13. 7
15. Quotient = 0, Remainder = $x^2 + 2$
16. Quotient = $x^3 + 2x^2 - 3x - 5$, Remainder = 0.
17. $x - 2$
18. (i) Quotient = $u^2 + 5u - 2$, Remainder = 0, g is a factor of p.
(ii) Quotient = $x^2 - 12x + 13$, Remainder = -34, g is not a factor of p.
(iii) Quotient = $y^3 - 1$, Remainder = $3y^2 + 11$, g is not a factor of p.
(iv) Quotient = $x^3 - 4x^2 + 19x - 65$, Remainder = $223x + 133$, g is not a factor of p.
(v) Quotient = $2x + 2a - 3$, Remainder = $2a^2 - 3a + 5$, g is not a factor of p.

MISCELLANEOUS EXERCISE

1. Find the values of a and b so that $x^4 + x^3 + 8x^2 + ax + b$ is divisible by $4x^2 + 3x - 2$.
   [**Hint.** Remainder = $x(a - 1) + b - 7 = 0 = x \cdot 0 + 0 \Rightarrow a = 1, b = 7$]
2. Divide the polynomial $5x(x^2 - x + 1) - (9 + 4x^4)$ by $4x - 1$.
3. Divide the polynomial $3y^4 - y^3 + 12y^2 + 2$ by $3y^2 - 1$.
4. Obtain all the zeroes of $x^4 - 2x^3 - 7x^2 + 8x + 12$, if two of its zeroes are 2 and 3.
5. Obtain all the zeroes of $x^4 + x^3 - 7x^2 - x + 6$, if two of its zeroes are 1 and -1.
6. Obtain the other two zeroes of $x^3 - 6x^2 + 11x - 6$, if one of its zeroes is 1.
7. If one zero of $x^3 - 6x^2 + 3x + 10$ is 5, then find the other two zeroes.
8. Find the zeroes of the following quadratic polynomials and verify the relationship between the zeroes and their coefficients.
   (i) $x^2 - 10x + 9$
   (ii) $x^2 + 10x + 25$
   (iii) $x^2 - 49$
   (iv) $4p^2 - 17p - 21$
   (v) $24p^2 - 41p + 12$.
9. Find a quadratic polynomial each with the given numbers as the sum and product of its zeroes respectively.
   (i) $0, \sqrt{3}$
   (ii) $\sqrt{5}, \dfrac{1}{7}$
   (iii) $\dfrac{1}{9}, -2$.
10. Verify that the numbers given alongside of the cubic polynomials below are their zeroes. Also verify the relationship between the zeroes and the coefficients in each case.
    (i) $x^3 - 10x^2 - 53x - 42$; $1, 14, -3$
    (ii) $x^3 + 13x^2 + 31x - 45$; $1, -5, -9$
    (iii) $y^3 - 2y^2 - 29y - 42$; $-2, -3, 7$
    (iv) $2y^3 - 5y^2 - 19y + 42$; $-3, 2, \dfrac{7}{2}$

(v) $3u^3 - 4u^2 - 12u + 16$; $2, -2, \dfrac{4}{3}$.

11. Find a cubic polynomial with the sum, sum of the product of its zeroes taken two at a time, and product of its zeroes as $4, -9, -17$, respectively.

12. What must be added to or subtracted from $15x^5 + 19x^4 + 37x^3 - 19x^2 + 40x - 25$ so that it may be exactly divisible by $5x^2 - 2x - 4$?

13. What must be subtracted from or added to $8x^4 + 14x^3 - 2x^2 + 8x - 12$ so that it may be exactly divisible by $4x^2 + 3x - 2$?

14. For what value of c, the polynomial $x^3 + 4x^2 - cx + 8$ is exactly divisible by $x - 2$?

15. Find the value of c for which the polynomial $2x^3 - 7x^2 - x + c$ is exactly divisible by $2x + 3$?

16. What must be subtracted from $x^3 - 6x^2 - 15x + 80$ so that the result is exactly divisible by $x^2 + x - 12$?

17. What must be added to $3x^3 + x^2 - 22x + 9$ so that the result is exactly divisible by $3x^2 + 7x - 6$?

18. If $x^3 + ax^2 - bx + 10$ is divisible by $x^2 - 3x + 2$, find the values of a and b.

19. Find the values of m and n so that $x^4 + mx^3 + nx^2 - 3x + n$ is divisible by $x^2 - 1$.

20. Find, whether or not the first polynomial is a factor of the second:

    (i) $x + 1$, $2x^2 + 5x + 4$

    (ii) $3x - 1$, $6x^2 + x - 1$

    (iii) $4y + 1$, $8y^2 - 2y + 1$

    (iv) $2a - 3$, $10a^2 - 9a - 5$

    (v) $4 - z$, $3z^2 - 13z + 4$

    (vi) $4z^2 - 5$, $4z^4 + 7z^2 + 15$

    (vii) $y - 2$, $3y^3 + 5y^2 + 5y + 2$.

Answers

1. $a = 1$, $b = 7$
2. Quotient $= -x^3 + x^2 - x + 1$, Remainder $= -8$
3. Quotient $= y^2 - \dfrac{1}{3}y + \dfrac{13}{3}$.

   Remainder $= -\dfrac{1}{3}y + \dfrac{19}{3}$
4. $-1, -2, 2, 3$ 5. $1, -1, 2, -3$
6. $2, 3$ 7. $-1, 2$
8. (i) $1, 9$ (ii) $-5, -5$

   (iii) $7, -7$ (iv) $\dfrac{21}{4}, -1$

   (v) $\dfrac{3}{8}, \dfrac{4}{3}$

9. (i) $x^2 + \sqrt{3}$

   (ii) $x^2 - \sqrt{5}x + \dfrac{1}{7}$

   (iii) $x^2 - \dfrac{1}{9}x - 2$

11. $x^3 - 4x^2 - 9x + 17$
12. $5 - 2x$, $2x - 5$
13. $15x - 14$, $-15x + 14$
14. 16 15. 21
16. $4x - 4$ 17. $2x + 3$
18. $a = 2$, $b = 13$ 19. $m = 3$, $n = -3$
20. (i) No   (ii) Yes

    (iii) No   (iv) No

    (v) Yes   (vi) No

    (vii) No.

ILLUSTRATIVE EXAMPLES

[NCERT Exercise 2.4 (Optional)*]
(Page No. 36)

Example 1. *Verify that the numbers given alongside of the cubic polynomials below are their zeroes. Also verify the relationship between the zeroes and the coefficients in each case:*

(i) $2x^3 + x^2 - 5x + 2$; $\dfrac{1}{2}, 1, -2$

(ii) $x^3 - 4x^2 + 5x - 2$; $2, 1, 1$

Sol.

(i) Let $p(x) = 2x^3 + x^2 - 5x + 2$

---

*These exercises are not from the examination point of view.

Then, we have
$$p\left(\frac{1}{2}\right) = 2\left(\frac{1}{2}\right)^3 + \left(\frac{1}{2}\right)^2 - 5\left(\frac{1}{2}\right) + 2$$
$$= \frac{1}{4} + \frac{1}{4} - \frac{5}{2} + 2$$
$$= 0$$
$$p(1) = 2(1)^3 + (1)^2 - 5(1) + 2$$
$$= 2 + 1 - 5 + 2$$
$$= 0$$
and, $\quad p(-2) = 2(-2)^3 + (-2)^2 - 5(-2) + 2$
$$= -16 + 4 + 10 + 2$$
$$= 0$$

Therefore, $\frac{1}{2}$, 1, and -2 are the zeroes of $2x^3 + x^2 - 5x + 2$.

Comparing the given polynomial with $ax^3 + bx^2 + cx + d$, we get
$$a = 2$$
$$b = 1$$
$$c = -5$$
$$d = 2$$

Let $\quad \alpha = \frac{1}{2}$
$$\beta = 1$$
and $\quad \gamma = -2.$

Then, we have,
$$\alpha + \beta + \gamma = \frac{1}{2} + 1 + (-2) = \frac{-1}{2} = -\frac{b}{a},$$
$$\alpha\beta + \beta\gamma + \gamma\alpha = \left(\frac{1}{2}\right)(1) + (1)(-2) + (-2)\left(\frac{1}{2}\right)$$
$$= \frac{1}{2} - 2 - 1$$
$$= -\frac{5}{2} = \frac{c}{a}$$
$$\alpha\beta\gamma = \left(\frac{1}{2}\right)(1)(-2) = -1$$
$$= \frac{-2}{2} = \frac{-d}{a}$$

(ii) Let $p(x) = x^3 - 4x^2 + 5x - 2$

Then, we have
$$p(2) = (2)^3 - 4(2)^2 + 5(2) - 2$$
$$= 8 - 16 + 10 - 2$$
$$= 0$$
$$p(1) = (1)^3 - 4(1)^2 + 5(1) - 2$$
$$= 1 - 4 + 5 - 2$$
$$= 0$$

Therefore, 2 and 1 are the two zeroes of $x^3 - 4x^2 + 5x - 2$. Hence $(x-2)(x-1)$, i.e., $x^2 - 3x + 2$ is a factor of the given polynomial.

Now, we apply the division algorithm to the given polynomial and $x^2 - 3x + 2$.

$$\begin{array}{r}
x - 1 \\
x^2 - 3x + 2 \overline{\smash{)}x^3 - 4x^2 + 5x - 2} \\
x^3 - 3x^2 + 2x \\
\underline{-+-} \\
-x^2 + 3x - 2 \\
-x^2 + 3x - 2 \\
\underline{+-+} \\
0
\end{array}$$

So, $x^3 - 4x^2 + 5x - 2 = (x^2 - 3x + 2)(x - 1)$
$\Rightarrow x^3 - 4x^2 + 5x - 2$
$$= (x - 2)(x - 1)(x - 1)$$

Hence, 2, 1 and 1 are the zeroes of $x^3 - 4x^2 + 5x - 2$.

Comparing the given polynomial with $ax^3 + 4x^2 + cx + d$, we get
$$a = 1$$
$$b = -4$$
$$c = 5$$
$$d = -2$$

Let $\quad \alpha = 2$
$$\beta = 1$$
and $\quad \gamma = 1$

Then, we have
$$\alpha + \beta + \gamma = 2 + 1 + 1 = 4$$
$$= -\frac{(-4)}{1} = -\frac{b}{a}$$
$$\alpha\beta + \beta\gamma + \gamma\alpha = (2)(1) + (1)(1) + (1)(2)$$
$$= 5 = \frac{5}{1} = \frac{c}{a}$$

$$\alpha\beta\gamma = (2)(1)(1) = 2$$
$$= -\frac{2}{1} = -\frac{d}{a}$$

Example 2. *Find a cubic polynomial with the sum, sum of the product of its zeroes taken two at a time, and the product of its zeroes as 2, – 7, – 14 respectively.*

Sol. Let the cubic polynomial be $ax^3 + bx^2 + cx + d$ and its zeroes be α, β and γ. Then,

$$\alpha + \beta + \gamma = 2 = -\frac{b}{a}$$

$$\alpha\beta + \beta\gamma + \gamma\alpha = -7 = \frac{c}{a}$$

$$\alpha\beta\gamma = -14 = \frac{-d}{a}$$

If $\alpha = 1$, then $\beta = -2$, $c = -7$ and $d = 14$. Hence, one cubic polynomial which fits the given conditions is

$$x^3 - 2x^2 - 7x + 14.$$

Example 3. *If the zeroes of the polynomial $x^3 - 3x^2 + x + 1$ are $a - b$, a, $a + b$, find a and b.*

Sol. The given polynomial is
$$x^3 - 3x^2 + x + 1$$

Comparing with $Ax^3 + Bx^2 + Cx + D$, we get

$$A = 1$$
$$B = -3$$
$$C = 1$$
$$D = 1$$

Let
$$\alpha = a - b$$
$$\beta = a$$
$$\gamma = a + b$$

Then, we have

$$\alpha + \beta + \gamma = -\frac{B}{A} = -\frac{(-3)}{1} = 3$$
$$\Rightarrow a - b + a + a + b = 3$$
$$\Rightarrow 3a = 3$$
$$\Rightarrow a = 1$$

$$\alpha\beta\gamma = \frac{-D}{A} = -1$$

$$\Rightarrow (a - b) a (a + b) = -1$$
$$\Rightarrow (1 - b) 1 (1 + b) = -1$$
$$\Rightarrow 1 - b^2 = -1$$
$$\Rightarrow b^2 = 2$$
$$\Rightarrow b = \pm\sqrt{2}$$

Example 4. *If two zeroes of the polynomial $x^4 - 6x^3 - 26x^2 + 138x - 35$ are $2 \pm \sqrt{3}$, find other zeroes.*

Sol. Since two zeroes of the given polynomial $x^4 - 6x^3 - 26x^2 + 138x - 35$ are $2 \pm \sqrt{3}$, therefore,

$$\{x - (2 + \sqrt{3})\} \{x - (2 - \sqrt{3})\},$$

i.e., $\{(x - 2) - \sqrt{3}\} (x - 2) + \sqrt{3})\}$,

i.e., $(x - 2)^2 - (\sqrt{3})^2$, i.e., $x^2 - 4x + 1$ is a factor of the given polynomial. Now, we apply the division algorithm to the given polynomial and $x^2 - 4x + 1$.

$$\begin{array}{r}
x^2 - 2x - 35 \\
x^2 - 4x + 1 \overline{) x^4 - 6x^3 - 26x^2 + 138x - 35} \\
x^4 - 4x^3 + x^2 \\
-\quad +\quad - \\
\hline
-2x^3 - 27x^2 + 138x - 35 \\
-2x^3 + 8x^2 - 2x \\
+\quad -\quad + \\
\hline
-35x^2 + 140x - 35 \\
-35x^2 + 140x - 35 \\
+\quad -\quad + \\
\hline
0
\end{array}$$

So, $x^4 - 6x^3 - 26x^2 + 138x - 35$
$$= (x^2 - 4x + 1)(x^2 - 2x - 35)$$
$$= (x^2 - 4x + 1)(x - 7)(x + 5)$$

Hence, the other two zeroes are 7 and – 5.

Example 5. *If the polynomial $x^4 - 6x^3 + 16x^2 - 25x + 10$ is divided by another polynomial $x^2 - 2x + k$, the remainder comes out to be $x + a$, find k and a.*

Sol. Let us apply the division algorithm to the given polynomial $x^4 - 6x^3 + 16x^2 - 25x + 10$ and another polynomial $x^2 - 2x + k$.

POLYNOMIALS

$$\begin{array}{r}
x^2 - 4x + z(8-k) \\
x^2 - 2x + k \overline{\smash{\big)}\, x^4 - 6x^3 + 16x^2 - 25x + 10}\\
x^4 - 2x^3 + kx^2 \\
\underline{-+-}\\
-4x^3 + (16-k)x^2 - 25x + 10\\
-4x^3 + 8x^2 - 4kx \\
\underline{+-+}\\
(8-k)x^2 + (4k-25)x + 10\\
(8-k)x^2 - 2(8-k)x + k(8-k)\\
\underline{-+-}\\
(2k-9)x - k(8-k) + 10
\end{array}$$

∴ Remainder
 = $(2k-9)x - k(8-k) + 10$

But the remainder is given to be $x + a$.

Therefore, $2k - 9 = 1$
⇒ $\qquad 2k = 10 \Rightarrow k = 5$

and $\quad -k(8-k) + 10 = a$
⇒ $\quad -5(8-5) + 10 = a$
⇒ $\qquad\qquad -5 = a$
⇒ $\qquad\qquad a = -5$

SUMMARY

1. A polynomial of degree 1, 2 or 3 is called a linear polynomial, a quadratic polynomial or a cubic polynomial, respectively.

2. A quadratic polynomial in x with real coefficients is of the form $ax^2 + bx + c$ where a, b, c are real constants with $a \neq 0$.

3. The zeroes of a polynomial $p(x)$ are precisely the x-coordinates of the points where the graph of $y = p(x)$ intersects the x-axis.

4. A polynomial of degree n can have at most n zeroes. So, a quadratic polynomial can have at most 2 zeroes and a cubic polynomial can have at most 3 zeroes.

5. If α and β are the zeroes of a quadratic polynomial $ax^2 + bx + c$, then

$$\alpha + \beta = -\frac{b}{a} \quad \alpha\beta = \frac{c}{a}$$

6. If α, β, γ are the zeroes of a cubic polynomial $ax^3 + bx^2 + cx + d = 0$, then

$$\alpha + \beta + \gamma = \frac{-b}{a}$$

$$\alpha\beta + \beta\gamma + \gamma\alpha = \frac{c}{a}$$

$$\alpha\beta\gamma = \frac{-d}{a}$$

7. The division algorithm states that given any polynomial $p(x)$ and any non-zero and $r(x)$ polynomial $g(x)$, there are polynomials $q(x)$ such that

$$p(x) = g(x)\, q(x) + r(x),$$

where $r(x) = 0$ or degree $r(x)$ < degree $q(x)$.

3

Pair of Linear Equations in Two Variables

IMPORTANT POINTS

1. A situation. In everyday life, we come across situations like the one given below :

Akhila went to a fair in her village. She wanted to enjoy rides on the Giant Wheel and play Hoopla (a game in which one throws a ring on the items kept in the stall, and if the ring covers any object completely, then one gets it). The number of times she played Hoopla is half the number of rides she had on the Giant wheel. Each ride costs Rs. 3, and a game of Hoopla costs Rs. 4. She spent Rs. 20 which she had saved for this purpose. Now the question is to find out the number of rides she had and the number of times she played the Hoopla. For this either we may try it out case by case (whether she has one ride, two rides, and so on) or we may use the previous knowledge to represent such situations as linear equations in two variables. Let us try the latter approach.

Let us denote the number of rides that Akhila had by x, and the number of times she played Hoopla by y. Now the situation can be represented by the two equations

$x = 2y$ ...(1)

and $3x + 4y = 20$ ...(2)

There are several ways of finding the solutions of such a pair of equations. We shall study all these ways in this chapter.

2. Linear Equation in two variables. *An equation which can be put in the form $ax + by + c = 0$, where a, b and c are real numbers, and a and b are not both zero, is called a linear equation in two variables x and y.* [We often denote the condition a and b are not both zero by $a^2 + b^2 \neq 0$]. The following are examples of linear equations in two variables :

$2x + 3y = 5$

$x - 2y - 3 = 0$

and $x - 0y = 2$, *i.e.*, $x = 2$.

3. Solution. A solution of a linear equation in two variables x and y is a pair of values, one for x and the other for y, which makes the two sides of the equation equal. For example, in the equation $2x + 3y = 5$, let us substitute $x = 1$ and $y = 1$ in the left hand side (L.H.S.) of the equation. Then,

L.H.S. $= 2(1) + 3(1) = 2 + 3 = 5$,

which is equal to the right hand side (R.H.S.) of the equation.

Therefore, $x = 1$ and $y = 1$ is a solution of the equation $2x + 3y = 5$.

4. Not a solution. Now let us substitute $x = 1$, and $y = 7$ in the equation $2x + 3y = 5$. Then,

L.H.S. $= 2(1) + 3(7) = 2 + 21 = 23$,

which is not equal to the R.H.S.

Therefore, $x = 1$ and $y = 7$ is not a solution of the equation.

5. Geometrical Meaning. Geometrically, it means that the point (1, 1) lies on the line represented by the equation $2x + 3y = 5$, and the point (1, 7) does not lie on it. So, **every**

PAIR OF LINEAR EQUATIONS IN TWO VARIABLES

solution of the equation is a point on the line representing it.

6. Generalisation. In fact, this is true for any linear equation, that is, **each solution (x, y) of a linear equation in two variables, ax + by + c = 0, corresponds to a point on the line representing the equation, and vice versa.**

7. Linear Pair. Equations (1) and (2) given above, taken together, represent the information we have about Akhila at the fair.

These two linear equations are **in the same two variables x and y.** Equations like these are called *a pair of linear equations in two variables,* and sometimes we call them a **linear pair.**

8. Algebraic Look. The general form for a pair of linear equations in two variables x and y is

$$a_1x + b_1y + c_1 = 0$$

and $\quad a_2x + b_2y + c_2 = 0,$

where $a_1, b_1, c_1, a_2, b_2, c_2$ are all numbers and $a_1^2 + b_1^2 \neq 0, a_2^2 + b_2^2 \neq 0.$

Some examples of pairs of linear equations in two variables are :

(1) $2x + 3y - 7 = 0$ and $9x - 2y + 8 = 0$

(2) $5x = y$ and $-7x + 2y + 3 = 0$

(3) $x + y = 7$ and $17 = y.$

9. Geometrical Look. We know that the geometrical (*i.e.*, graphical) representation of a linear equation in two variables is a straight line. It suggests that a pair of linear equations in two variables will be two straight lines, both to be considered together, geometrically.

10. Three Possibilities. Given two lines in a plane, only one of the following three possibilities can happen :

(*i*) The two lines will intersect at one point.

(*ii*) The two lines will not intersect *i.e.*, they are parallel.

(*iii*) The two lines are coincident lines.

We show all these possibilities in the following figure :

In Fig. (*a*), they intersect ;

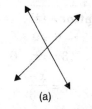

(a)

In Fig. (*b*), they are parallel ;

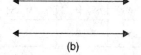

(b)

In Fig. (*c*), they are coinciding.

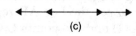

(c)

Note. Both ways of representing a linear pair go hand-in-hand–the algebraic and the geometric ways.

ILLUSTRATIVE EXAMPLES

[NCERT Exercise 3.1]
(*Page No. 44*)

Example 1. *Aftab tells his daughter, "Seven years ago, I was seven times as old as you were then. Also, three years from now, I shall be three times as old as you will be". (Isn't this interesting ?) Represent this situation algebraically and graphically.*

Sol. Let the present ages of Aftab and his daughter be x years and y years respectively. Then the algebraic representation is given by the following equations :

$$x - 7 = 7(y - 7)$$

and $\quad x + 3 = 3(y + 3)$

$$\Rightarrow \quad x - 7y + 42 = 0 \quad ...(1)$$
and $\quad x - 3y - 6 = 0 \quad ...(2)$

To represent these equations graphically, we find two solutions for each equation. These solutions are given below :

For Equation (1)
$$x - 7y + 42 = 0$$
$$\Rightarrow \quad 7y = x + 42$$
$$\Rightarrow \quad y = \frac{x + 42}{7}$$

Table 1 of solutions

x	0	7
y	6	7

For equations (2)
$$x - 3y - 6 = 0$$
$$\Rightarrow \quad 3y = x - 6$$
$$\Rightarrow \quad y = \frac{x - 6}{3}$$

Table 2 of solutions

x	0	6
y	-2	0

We plot the points A(0, 6) and B(7, 7) corresponding to the solutions in table 1 on a graph paper to get the line AB representing the equation (1) and the points C(0, $-$ 2) and D(6, 0) corresponding to the solutions in table 2 on the same graph paper to get the line CD representing the equation (2), as shown in the figure given below.

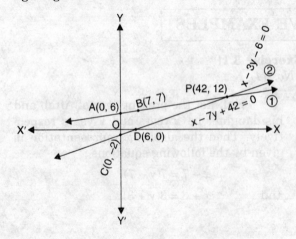

We observe in figure that the two lines representing the two equations are intersecting at the point P(42, 12).

Note : We know that are infinitely many solutions of each linear equation. So we can choose any two values, which may not be the ones we have chosen. For the sake of convenience, we choose $x = 0$ because when one of the coordinates is zero, the equation reduces to a linear equations in one variable, which can be solved easily. However, we prefer to find a pair of values of x and y in which x and y both are integral.

Example 2. *The coach of a cricket team buys 3 bats and 6 balls for Rs. 3900. Later, she buys another bat and 3 more balls of the same kind for Rs. 1300. Represent this situation algebraically and geometrically.*

Sol. Let us denote the cost of 1 bat by Rs. x and one ball by Rs. y. Then the algebraic representation is given by the following equations :

$$3x + 6y = 3900$$
and $\quad x + 3y = 1300$.
$$\Rightarrow \quad x + 2y = 1300 \quad ...(1)$$
and $\quad x + 3y = 1300 \quad ...(2)$

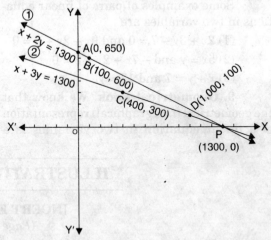

To represent these equations graphically, we find two solutions for each equation. These solutions are given below :

For equation (1)
$$x + 2y = 1300$$
$$\Rightarrow \quad 2y = 1300 - x$$

PAIR OF LINEAR EQUATIONS IN TWO VARIABLES

$$\Rightarrow \quad y = \frac{1300 - x}{2}$$

Table 1 of solutions

x	0	100
y	650	600

For equation (2)

$$x + 3y = 1300$$
$$\Rightarrow \quad 3y = 1300 - x$$
$$\Rightarrow \quad y = \frac{1300 - x}{3}$$

Table 2 of solutions

x	400	1,000
y	300	100

We plot the points A(0, 650) and B(100, 600) corresponding to the solutions in table 1 on a graph paper to get the line AB representing the equation (1) and the points C(400, 300) and D(1,000, 100) corresponding to the solutions in table 2 on the same graph paper to get the line CD representing the equation (2), as shown in the figure given below.

We observe in figure that the two lines representing the two equations are intersecting at the point P(1300, 0).

Example 3. *The cost of 2 kg of apples and 1 kg of grapes in a day was found to be Rs. 160. After a month, the cost of 4 kg of apples and 2 kg of grapes is Rs. 300. Represent the situation algebraically and geometrically.*

Sol. Let the cost of 1 kg of apples be Rs. x and of 1 kg of grapes be Rs. y. Then, the algebraic representation is given by the following equations :

$$2x + y = 160 \quad \ldots(1)$$
$$4x + 2y = 300$$
$$\Rightarrow \quad 2x + y = 150 \quad \ldots(2)$$

To represent these equations graphically, we find two solutions for each equation. These solutions are given below :

For equation (1)

$$2x + y = 160$$
$$\Rightarrow \quad y = 160 - 2x$$

Table 1 of solutions

x	50	40
y	60	80

For equation (2)

$$2x + y = 150$$
$$\Rightarrow \quad y = 150 - 2x$$

Table 2 of solutions

x	50	30
y	50	90

We plot the points A(50, 60) and B(40, 80) corresponding to the solutions in table 1 on a graph paper for get the line. AB representing the equation (1) and the points C(50, 50) and D(30, 90) corresponding to the solutions in table 2 on the same graph paper to get the line CD representing the equation (2) as shown in the figure given below :

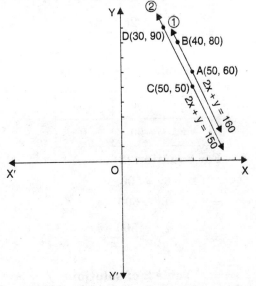

We observe in figure that, the two lines do not intersect anywhere *i.e.,* they are parallel.

ADDITIONAL EXAMPLES

Example 4. *Sangeeta went to a bookseller's shop and purchased 2 textbooks of IX Mathematics and 3 textbooks of X Mathematics for Rs. 250. Her friend Meenu also bought 4 textbooks of IX Mathematics and 6 textbooks of X Mathematics of the same kind for Rs. 500. Represent this situation algebraically and graphically.*

Sol. Let us denote the cost of 1 textbook of IX Mathematics by Rs. x and the cost of 1 textbook of X Mathematics by Rs. y. Then the algebraic representation is given by the following equations

$$2x + 3y = 250 \quad ...(1)$$
and $$4x + 6y = 500 \quad ...(2)$$

To represent these equations graphically, we find two solutions for each equation. These solutions are given below :

For equation (1)

$$2x + 3y = 250$$
$$\Rightarrow 3y = 250 - 2x$$
$$\Rightarrow y = \frac{250 - 2x}{3}$$

Table 1 of solutions

x	50	125
y	50	0

For equation (2)

$$4x + 6y = 500$$
$$\Rightarrow 6y = 500 - 4x$$
$$\Rightarrow y = \frac{500 - 4x}{6}$$

Table 2 of solutions

x	50	125
y	50	0

We plot these points on a graph paper. We find that both the lines coincide. This is so, because, both the equations are equivalent, i.e., one can be derived from the other.

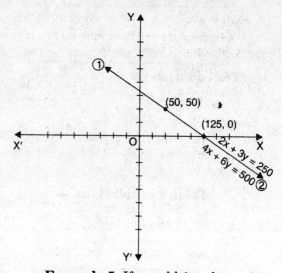

Example 5. *If we add 1 to the numerator of a fraction, it reduces to $\frac{1}{2}$. If we subtract 1 from the denominator, it reduces to $\frac{1}{3}$. Represent this situation algebraically and graphically.*

Sol. Let the numerator and the denominator of the fraction be x and y respectively. Then the algebraically representation is given by the following equations.

$$\frac{x+1}{y} = \frac{1}{2}$$
$$\Rightarrow 2(x+1) = y$$
$$\Rightarrow 2x + 2 = y$$
$$\Rightarrow 2x - y = -2 \quad ...(1)$$

and $$\frac{x}{y-1} = \frac{1}{3}$$
$$\Rightarrow 3x = y - 1$$
$$\Rightarrow 3x - y = -1 \quad ...(2)$$

To represent these equation graphically, we find two solutions for each equation.

These solution are given below :

For equation (1)

$$2x - y = -2$$
$$\Rightarrow y = 2x + 2$$

Table 1 of solutions

x	0	-1
y	2	0

For equation (2)

$$3x - y = -1$$
$$\Rightarrow y = 3x + 1$$

Table 2 of solutions

x	0	-1
y	1	-2

We plot the points A(0, 2) and B(−1, 0) corresponding to the solutions in Table 1 on a graph paper to get the line AB representing the equation (1) and the points C(0, 1) and D(−1, −2) corresponding to the solutions in table 2 on the same graph paper to get the line CD representing the equation (2) as shown in the figure given below.

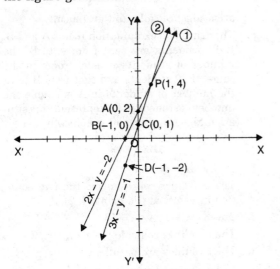

We observe in figure that the two lines representing the two equations are intersecting at the point P (1, 4).

Example 6. *Sangeeta has socks and handkerchiefs which are together 40 in number. If she has 5 less handkerchiefs and 5 more socks, the number of socks becomes four times the number of handkerchiefs. Represent this situation algebraically and graphically.*

Sol. Let Sangeeta have x socks and y handkerchiefs. Then the algebraic representation is given by

$$x + y = 40 \qquad ...(1)$$

and $\quad x + 5 = 4(y - 5)$

$$\Rightarrow x + 5 = 4y - 20$$
$$\Rightarrow x - 4y = -25 \qquad ...(2)$$

To represent these equations graphically, we find two solutions for each equation. These solutions are given below :

For equation (1)

$$x + y = 40$$
$$\Rightarrow y = 40 - x$$

Table 1 of solutions

x	20	40
y	20	0

For equation (2)

$$x - 4y = -25$$
$$\Rightarrow 4y = x + 25$$
$$\Rightarrow y = \frac{x + 25}{4}$$

Table 2 of solutions

x	15	55
y	10	20

We plot the points A(20, 20) and B(40, 0) corresponding to the solutions given in table 1 on a graph paper to get the line AB representing the equation (1) and the points C(15, 10) and D(55, 20) corresponding to the solutions in table 2 on the same graph paper to get the line CD representing the equation (2), as shown in the figure given below :

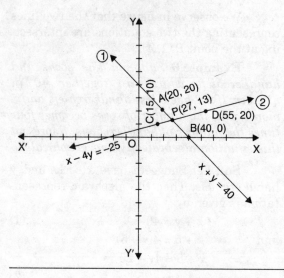

We observe in figure that the two lines representing the two equations are intersecting at the point P(27, 13).

TEST YOUR KNOWLEDGE

1. Akhila went to a fair in her village. She wanted to enjoy rides on the Giant Wheel and play Hoopla (a game in which you know a ring on the items kept in the stall, and if the ring covers any object completely you get it). The number of times she played Hoopla is half the number of sides she had on the Giant Wheel. Each ride costs Rs. 3, and a game of Hoopla costs Rs 4. She goes to the fair with Rs. 20. Represent this situation algebraically and graphically.

2. Romila went to a stationery stall and purchased 2 pencils and 3 erasers for Rs. 9. Her friend Sonali saw the new variety of pencils and erasers with Romila and she also bought 4 pencils and 6 erasers of the same kind for Rs. 18. Represent this situation algebraically and graphically.

3. The path of train A is given by the equation $x + 2y - 4 = 0$ and the path of another train B is given by the equation $2x + 4y - 12 = 0$. Represent this situation geometrically.

4. 4 kg of apples and 3 kg of guavas together cost Rs. 36.50 while 3 kg of apples and 2 kg of guavas cost Rs. 26.50. Represent this situation algebraically and graphically.

5. If three times the larger of the two numbers is divided by the smaller one, we get 4 as the quotient and 3 as the remainder. Also, if seven times the smaller number is divided by the larger one, we get 5 as quotient and 1 as remainder. Represent this situation algebraically and graphically.

6. The ages of two girls are in the ratio 5 : 7. Eight years ago, their ages were in the ratio 7 : 13. Represent this situation algebraically and graphically.

7. If the numerator of a fraction is multiplied by 2 and its denominator is increased by 2, it becomes $\frac{6}{7}$. If instead, we multiply the denominator by 2 and increase the numerator by 2, it reduces to $\frac{1}{2}$. Represent this situation algebraically and graphically.

8. There are two examination rooms A and B. If 10 students are sent from A to B, the number of students in each room is the same. If 20 students and sent from B to A, the number of students in A is double the number of student in B. Represent this situation algebraically and graphically,

Answers

1. $x = 2y$, $3x + 4y = 20$;
   The two lines representing the two equation intersect at the point (4, 2).

2. $2x + 3y = 9$, $4x + 6y = 18$;
   The two lines coincide

3. The two lines are parallel.

4. $4x + 3y = 36.50$, $3x + 2y = 26.50$;
   The two lines intersect at (6.5, 3.5)
5. $3x - 4y = 3$, $7y = 5x + 1$;
   The two lines intersect at (25, 18)
6. $7x - 5y = 0$; $13(x - 8) = 7(y - 8)$
   $\Rightarrow 13x - 7y = 48$;
   The two lines intersect at the point (15, 21).
7. $\dfrac{2x}{y+2} = \dfrac{6}{7}$, $\dfrac{x+2}{2y} = \dfrac{1}{2}$;
   The two lines intersect at the point (3, 5)
8. $x - 10 = y + 10$, $x + 20 = 2(y - 20)$;
   The two lines intersect at the point (100, 80).

GRAPHICAL METHOD OF SOLUTION OF A PAIR OF LINEAR EQUATIONS

IMPORTANT POINTS

1. Three Possibilities. The lines representing a pair of linear equations in two variables may either (i) intersect or (ii) be parallel or (iii) coincide.

2. Inconsistent pair. A pair of linear equations which has no solution, is called an inconsistent pair of linear equations.

3. Consistent pair. A pair of linear equations in two variables, which has a solutions (unique or infinitely many), is called a consistent pair of linear equations.

4. Dependent consistent pair. A pair of linear equations which are equivalent has infinitely many distinct solutions. Such a pair is called an dependent consistent pair of linear equations in two variables.

5. Summary. We can summarise the behaviour of lines representing a pair of linear equations in two variables and the existence of solutions as follows :

(i) the lines may intersect in a single point. In this case, the pair of equations has a unique solution (consistent pair of equations).

(ii) the lines may be parallel. In this case, the equations have no solution (inconsistent pair of equations).

(iii) the lines may be coincident. In this case, the equations have infinitely many solutions (consistent dependent pair of equatons).

6. Remark. (i) We must always verify the solution, we get.

(ii) Sometimes, as in the case $x + y = 0$, putting $x = 0$, or putting $y = 0$, gives us the same solution (0, 0). Then we can substitute another value of x to get another point on the graph.

7. Three conditions. If the lines $a_1 x + b_1 y + c_1 = 0$ and $a_2 x + b_2 y + c_2 = 0$ are

(i) intersecting then $\dfrac{a_1}{a_2} \neq \dfrac{b_1}{b_2}$

(ii) coincident then $\dfrac{a_1}{a_2} = \dfrac{b_1}{b_2} = \dfrac{c_1}{c_2}$

(iii) parallel then $\dfrac{a_1}{a_2} = \dfrac{b_1}{b_2} \neq \dfrac{c_1}{c_2}$

In fact, the converse is also true for any pair of lines.

ILLUSTRATIVE EXAMPLES

[NCERT Exercise 3.2]
(Page No. 49)

Example 1. *Form the pair of linear equations in the following problems, and find their solutions graphically.*

(i) *10 students of Class X took part in a Mathematics quiz. If the number of girls is 4 more than the number of boys, find the number of boys and girls who took part in the quiz.*

(ii) 5 pencils and 7 pens together cost Rs. 50, whereas 7 pencils and 5 pens together cost Rs. 46. Find the cost of one pencil and that of one pen.

Sol. (i) Let the number of boys and girls who took part in the quiz be x and y respectively. Then the pair of linear equations formed is

$$x + y = 10 \quad ...(1)$$
and
$$y = x + 4 \quad ...(2)$$

Let us draw the graphs of equations (1) and (2) by finding two solutions for each of the equations. These two solutions of the equations (1) and (2) are given below in table 1 and table 2 respectively.

For equation (1)

$$x + y = 10$$
$$\Rightarrow \quad y = 10 - x$$

Table 1 of solutions

x	6	4
y	4	6

For equation (2)

$$y = x + y$$

Table 2 of solutions

x	0	1
y	4	5

We plot the points A(6, 4) and B(4, 6) on a graph paper and join these points to form the line AB representing the equation (1) as shown in the paper. Also, we plot the points C(0, 4) and D(1, 5) on the same graph paper and join these points to form the line CD representing the equation (2) as shown in the same figure.

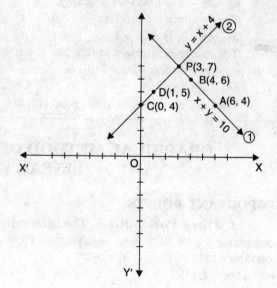

In the figure, we observe that the two lines intersect at the point P(3, 7). So $x = 3$ and $y = 7$ is the required solution of the pair of linear equations formed, i.e., the number of boys and girls who took part in the quiz is 3 and 7 respectively.

Verification. Substituting $x = 3$ and $y = 7$ in (1) and (2), we find that both the equations are satisfied as shown below :

$$x + y = 3 + 7 = 10$$
$$x + y = 3 + 4 = 7 = y$$

This verifies the method.

(ii) Let the cost of one pencil and a pen be Rs. x and Rs. y respectively. Then the pair of linear equations formed is

$$5x + 7y = 50 \quad ...(1)$$
and
$$7x + 5y = 46 \quad ...(2)$$

Let us draw the graphs of equations (1) and (2) by finding two solutions for each of the equations. These two solutions of the equations (1) and (2) are given below in table 1 and table 2 respectively.

For equation (1)

$$5x + 7y = 50$$
$$\Rightarrow \quad 7y = 50 - 5x$$
$$\Rightarrow \quad y = \frac{50 - 5x}{7}$$

Table 1 of solutions

x	3	-4
y	5	10

For equation (2)

$$7x + 5y = 46$$
$$\Rightarrow \quad 5y = 46 - 7x$$
$$\Rightarrow \quad y = \frac{46 - 7x}{5}$$

Table 2 of solutions

x	-2	8
y	12	-2

We plot the points A (3, 5) and B(– 4, 10) on a graph paper and join these points to form the line AB representing the equation (1) as shown in the figure. Also, we plot the points C(– 2, 12) and D(8, – 2) on the same graph paper and join these points to form the line CD representing the equation (2) as shown in the same figure.

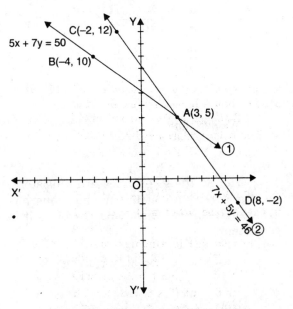

In the figure we observe that the two lines intersect at the point A(3, 5). So $x = 3$

and $y = 5$ is the required solution of the pair of linear equations for used, *i.e.,* the cost of one pencil and a pen is Rs. 3 and Rs. 5 respectively.

Verification. Substituting $x = 3$ and $y = 5$ in (1) and (2), we find that both the equations are satisfied as shown below :

$$5x + 7y = 5(3) + 7(5) = 15 + 35 = 50$$
$$7x + 5y = 7(3) + 5(5) = 21 + 25 = 46$$

This verifies the solution.

Example 2. *On comparing the ratios $\frac{a_1}{a_2}, \frac{b_1}{b_2}$ and $\frac{c_1}{c_2}$, find out whether the lines representing the following pairs of linear equations intersect at a point, are parallel or coincident:*

(i) $5x - 4y + 8 = 0$
   $7x + 6y - 9 = 0$

(ii) $9x + 3y + 12 = 0$
    $18x + 6y + 24 = 0$

(iii) $6x - 3y + 10 = 0$
     $2x - y + 9 = 0$

Sol. (i) $5x - 4y + 8 = 0$
   $7x + 6y - 9 = 0$

Here, $a_1 = 5, b_1 = -4, c_1 = 8$
      $a_2 = 7, b_2 = 6, c_2 = -9$

We see that $\frac{a_1}{a_2} \neq \frac{b_1}{b_2}$

Hence, the lines representing the given pair of linear equations intersect at a point.

(ii) $9x + 3y + 12 = 0$
    $18x + 6y + 24 = 0$

Here, $a_1 = 9, b_1 = 3, c_1 = 12$
      $a_2 = 18, b_2 = 6, c_2 = 24$

We see that

$$\frac{a_1}{a_2} = \frac{b_1}{b_2} = \frac{c_1}{c_2}$$

Hence, the lines represented the given pair of linear equations are coincident.

(iii) $6x - 3y + 10 = 0$
     $2x - y + 9 = 0$

Here, $a_1 = 6, b_1 = -3, c_1 = 10$
$a_2 = 2, b_2 = -1, c_2 = 9$

We see that $\dfrac{a_1}{a_2} = \dfrac{b_1}{b_2} \neq \dfrac{c_1}{c_2}$

Hence, the lines representing the given pair of linear equations are parallel.

Example 3. *On comparing the ratios* $\dfrac{a_1}{a_2}, \dfrac{b_1}{b_2}$ *and* $\dfrac{c_1}{c_2}$, *find out whether the following pair of linear equations are consistent, or inconsistent.*

(i) $3x + 2y = 5 \; ; \; 2x - 3y = 7$

(ii) $2x - 3y = 8 \; ; \; 4x - 6y = 9$

(iii) $\dfrac{3}{2}x + \dfrac{5}{3}y = 7 \; ; \; 9x - 10y = 14$

(iv) $5x - 3y = 11 \; ; \; -10x + 6y = -22$

(v) $\dfrac{4}{3}x + 2y = 8 \; ; \; 2x + 3y = 12$

Sol. (i) $3x + 2y = 5 \; ; \; 2x - 3y = 7$
Here, $a_1 = 3, b_1 = 2, c_1 = 5$
$a_2 = 2, b_2 = -3, c_2 = 7$

$\therefore \dfrac{a_1}{a_2} = \dfrac{3}{2}, \dfrac{b_1}{b_2} = \dfrac{-2}{3}, \dfrac{c_1}{c_2} = \dfrac{5}{7}$

We see that $\dfrac{a_1}{a_2} \neq \dfrac{b_1}{b_2}$

Hence, the given lines are intersecting. So, the given pair of linear equations has exactly one solution and therefore it is consistent.

(ii) $2x - 3y = 8 \; ; \; 4x - 6y = 9$
Here, $a_1 = 2, b_1 = -3, c_1 = 8$
$a_2 = 4, b_2 = -6, c_2 = 9$

$\therefore \dfrac{a_1}{a_2} = \dfrac{2}{4} = \dfrac{1}{2}, \dfrac{b_1}{b_2} = \dfrac{-3}{-6} = \dfrac{1}{2}, \dfrac{c_1}{c_2} = \dfrac{8}{9}$

We see that $\dfrac{a_1}{a_2} = \dfrac{b_1}{b_2} \neq \dfrac{c_1}{c_2}$

Hence, the given lines are parallel. So, the given pair of linear equation it has no solution and therefore it is inconsistent.

(iii) $\dfrac{3}{2}x + \dfrac{5}{3}y = 7 \; ; \; 9x - 10y = 14$

Here, $a_1 = \dfrac{3}{2}, b_1 = \dfrac{5}{3}, c_2 = 7$
$a_2 = 9, b_2 = -10, c_2 = 14$

$\therefore \dfrac{a_1}{a_2} = \dfrac{\frac{3}{2}}{9} = \dfrac{1}{6}$

$\dfrac{b_1}{b_2} = \dfrac{\frac{5}{3}}{-10} = -\dfrac{1}{6}$

$\dfrac{c_1}{c_2} = \dfrac{7}{14} = \dfrac{1}{2}$

We see that $\dfrac{a_1}{a_2} \neq \dfrac{b_1}{b_2}$

Hence, the given lines are intersecting. So the given pair of linear equations has exactly one solution and therefore it is consistent.

(iv) $5x - 3y = 11 \; ; \; -10x + 6y = -22$
Here, $a_1 = 5, b_1 = -3, c_1 = 11$
$a_2 = -10, b_2 = 6, c_2 = -22$

$\therefore \dfrac{a_1}{a_2} = \dfrac{5}{-10} = -\dfrac{1}{2}$

$\dfrac{b_1}{b_2} = -\dfrac{3}{6} = -\dfrac{1}{2}$

$\dfrac{c_1}{c_2} = \dfrac{11}{-22} = -\dfrac{1}{2}$

We see that

$\dfrac{a_1}{a_2} = \dfrac{b_1}{b_2} = \dfrac{c_1}{c_2}$

Hence, the given lines are consistent. So, the given pair of linear equations has infinitely many solutions and therefore it is consistent.

(v) $\dfrac{4}{3}x + 2y = 8 \; ; \; 2x + 3y = 12$

Here, $a_1 = \dfrac{4}{3}, b_1 = 2, c_1 = 8$

$a_2 = 2, b_2 = 3, c_2 = 12$

PAIR OF LINEAR EQUATIONS IN TWO VARIABLES

$$\therefore \quad \frac{a_1}{a_2} = \frac{\frac{4}{3}}{2} = \frac{2}{3}$$

$$\frac{b_1}{b_2} = \frac{2}{3}$$

$$\frac{c_1}{c_2} = \frac{8}{12} = \frac{2}{3}$$

We see that

$$\frac{a_1}{a_2} = \frac{b_1}{b_2} = \frac{c_1}{c_2}$$

Hence, the given lines are consistent. So the given pair of linear equations has infinitely many solutions and therefore it is consistent.

Example 4. *Which of the following pairs of linear equations are consistent/inconsistent ? If consistent, obtain the solution graphically:*

(i) $x + y = 5$, $2x + 2y = 10$
(ii) $x - y = 8$, $3x - 3y = 16$
(iii) $2x + y - 6 = 0$, $4x - 2y - 4 = 0$
(iv) $2x - 2y - 2 = 0$, $4x - 4y - 5 = 0$.

Sol. (i) $x + y = 5$ ...(1)
    $2x + 2y = 10$ ...(2)

Here, $a_1 = 1$, $b_1 = 1$, $c_1 = -5$
      $a_2 = 2$, $b_2 = 2$, $c_2 = -10$

We see that

$$\frac{a_1}{a_2} = \frac{b_1}{b_2} = \frac{c_1}{c_2}$$

Hence, the lines represented by the equations (1) and (2) are coincident.

Therefore, equations (1) and (2) have infinitely many common solutions, *i.e.*, the given pair of linear equations is consistent.

Graphical Representation. We draw the graphs of the equations (1) and (2) by finding two solutions for each of the equations. These two solutions of the equations (1) and (2) are given below in table 1 and table 2 respectively.

For equation (1)

$x + y = 5$
$\Rightarrow \quad y = 5 - x$

Table 1 of solutions

x	0	5
y	5	0

For equation (2)

$2x + 2y = 10$
$\Rightarrow \quad 2y = 10 - 2x$
$\Rightarrow \quad y = \dfrac{10 - 2x}{2}$
$\Rightarrow \quad y = 5 - x$

Table 2 of solutions

x	1	2
y	4	3

We plot the points A(0, 5) and B(5, 0) on a graph paper and join these points to form the line AB representing the equation (1) as shown in the figure. Also, we plot the points C(1, 4) and D(2, 3) on the same graph paper and join these points to form the line CD representing the equation (2) as shown in the same figure.

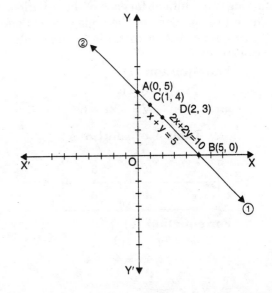

In the figure we observe that the two lines AB and CD coincide.

(ii) $x - y = 8$ ...(1)
          $3x - 3y = 16$ ...(2)

Here, $a_1 = 1, b_1 = -1, c_1 = -8$
         $a_2 = 3, b_2 = -3, c_2 = -16$

We see that
$$\frac{a_1}{a_2} = \frac{b_1}{b_2} \neq \frac{c_1}{c_2}$$

Hence, the lines represented by the equations (1) and (2) are parallel. Therefore, equations (1) and (2) have no solution, *i.e.*, the given pair of linear equation is inconsistent.

(iii) $2x + y - 6 = 0$ ...(1)
       $4x - 2y - 4 = 0$ ...(2)

Here, $a_1 = 2, b_1 = 1, c_1 = -6$
         $a_2 = 4, b_2 = -2, c_2 = -4$

We see that
$$\frac{a_1}{a_2} \neq \frac{b_1}{b_2}$$

Hence, the lines represented by the equations (1) and (2) are intersecting. Therefore, equations (1) and (2) have exactly one (unique) solution *i.e.*, the given pair of linear equation is consistent.

Graphical Representation. We draw the graphs of the equations (1) and (2) by finding two solutions for each of the equations. These two solutions of the equations (1) and (2) are given below in table 1 and table 2 respectively.

For equation (1)
$2x + y - 6 = 0$
$\Rightarrow y = -2x + 6$

Table 1 of solutions

x	0	3
y	6	0

For equation (2)
$4x - 2y - 4 = 0$
$\Rightarrow 2y = 4x - 4$
$\Rightarrow y = \dfrac{4x - 4}{2}$
$\Rightarrow y = 2x - 2$

Table 2 of solutions

x	0	1
y	-2	0

We plot the points A $(0, 6)$ and B$(3, 0)$ on a graph paper and join these points to form the line AB representing the equation (1) as shown in the figure. Also, we plot the points C$(0, -2)$ and D$(1, 0)$ on the same graph paper and join these points to form the line CD representing the equation (2) as shown in the same figure.

In the figure we observe that the two lines intersect at the point P$(2, 1)$. So $x = 2$ and $y = 1$ is the required unique solution of the pair of linear equations formed.

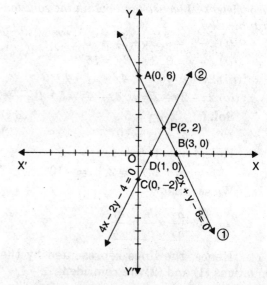

Verification. Substituting $x = 2$ and $y = 1$ in (1) and (2), we find that both the equations are satisfied as shown below :

$2x + y - 6 = 2(2) + 2 = 6$
$4x - 2y - 4 = 4(2) - 2(2) - 4 = 0$

This verifies the solution

(iv) $2x - 2x - 2 = 0$ ...(1)
      $4x - 4y - 5 = 0$ ...(2)

Here, $a_1 = 2, b_1 = -2, c_1 = -2$
         $a_2 = 4, b_2 = -4, c_2 = -5$

We see that

$$\frac{a_1}{a_2} = \frac{b_1}{b_2} \neq \frac{c_1}{c_2}$$

Hence the lines represented by the equations (1) and (2) are parallel. Therefore, equations (1) and (2) have no solution, *i.e.*, the given pair of linear equation is inconsistent.

Example 5. *Half the perimeter of a rectangular garden, whose length is 4 m more than its width, is 36 m. Find the dimensions of the garden.*

Sol. Let the dimensions (*i.e.*, length and width) of the garden be x m and y m respectively. Then,

$$x = y + 4$$

and $\quad \frac{1}{2}(2x + 2y) = 36$

$\Rightarrow \quad x - y = 4 \quad \ldots(1)$

$\quad x + y = 36 \quad \ldots(2)$

Let us draw the graphs of equations (1) and (2) by finding two solutions for each of the equations. These two solutions of the equations (1) and (2) are given below in table 1 and table 2 respectively.

For equation (1)

$x - y = 4$

$\Rightarrow y = x - 4$

Table 1 of solutions

x	4	2
y	0	−2

For equation (2)

$x + y = 36$

$\Rightarrow \quad y = 36 - x$

Table 2 of solutions

x	20	16
y	16	20

We plot the points A (4, 0) and B(2, −2) on a graph paper and join these points to form the line AB representing. The equation (1) as shown in the figure. Also, we plot the points C (20, 16) and D(16, 20) on the same graph paper and join these points to form the line CD representing the equation (2) as shown in the same figure.

In the figure we observe that the two lines intersect at the point C(20, 16). So $x = 20$, $y = 16$ is the required solution of the pair of linear equations formed, *i.e.*, the dimensions of the garden are 20 m and 16 m.

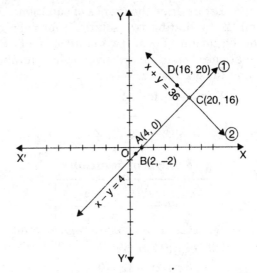

Verification. Substituting $x = 20$ and $y = 16$ in (1) and (2), we find that both the equations are satisfied as shown below :

$$20 - 16 = 4$$
$$20 + 16 = 36$$

This verifies the solution.

Example 6. *Given the linear equation $2x + 3y - 8 = 0$, write another linear equation in two variables such that the geometrical representation of the pair so formed is :*

(*i*) *intersecting lines*

(*ii*) *parallel lines*

(*iii*) *coincident lines*

Sol. (*i*) $2x + 3y - 8 = 0$

$\quad 3x + 2y - 7 = 0$

(*ii*) $2x + 3y - 8 = 0$

$\quad 2x + 3y - 12 = 0$

(iii) $2x + 3y - 8 = 0$
$4x + 6y - 16 = 0$

Example 7. *Draw the graphs of the equations $x - y + 1 = 0$ and $3x + 2y - 12 = 0$. Determine the coordinates of the vertices of the triangle formed by these lines and the x-axis, and shade the triangular region.*

Sol. The given equations are

$x - y + 1 = 0$ ...(1)

$3x + 2y - 12 = 0$ ...(2)

Let us draw the graphs of equations (1) and (2) by finding two solutions for each of the equation. These two solutions of the equations (1) and (2) are given below in table 1 and table 2 respectively.

For equation (1)

$x - y + 1 = 0$

$\Rightarrow \quad y = x + 1$

Table 1 of solutions

x	0	−1
y	1	0

For equation (2)

$3x + 2y - 12 = 0$

$\Rightarrow \quad 2y = 12 - 3x$

$\Rightarrow \quad y = \dfrac{12 - 3x}{2}$

Table 2 of solutions

x	4	0
y	0	6

We plot the points A(0, 1) and B(− 1, 0) on a graph paper and join these points to form the line AB representing the equation (1) as shown in the figure. Also, we plot the points C(4, 0) and D (0, 6) on the same graph paper and join these points to form the line CD representing the equation (2) as shown in the same figure.

In the figure we observe that the coordinates of the vertices of the triangle formed by these given lines and the x-axis are E (2, 3), B (− 1, 0) and C (4, 0).

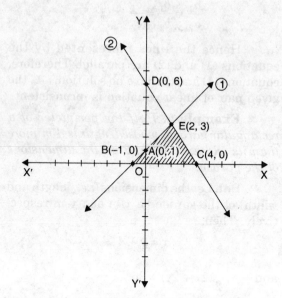

The triangular region EBC has been shaded.

ADDITIONAL EXAMPLES

Example 8. *Determine whether the following system of linear equations is inconsistent or not :*

$3x - 5y = 20$

$6x - 10y = -40.$

Sol. The given system of linear equations is

$3x - 5y = 20$ ...(1)

$6x - 10y = -40$ ...(2)

Here, $a_1 = 3, b_1 = -5, c_1 = 20$

$a_2 = 6, b_2 = -10, c_2 = -40$

We see that

$\dfrac{a_1}{a_2} = \dfrac{b_1}{b_2} \neq \dfrac{c_1}{c_2}$

Hence, the lines representing the given pair of linear equations are parallel. Therefore, equations (1) and (2) have no common solution, *i.e.,* the given pair of linear equations is inconsistent.

PAIR OF LINEAR EQUATIONS IN TWO VARIABLES

Example 9. *Examine whether the solution set of the system of equations*
$$3x - 4y = -7$$
$$3x - 4y = -9$$
is consistent or inconsistent.

Sol. The given system of equations is
$$3x - 4y = -7 \quad ...(1)$$
$$3x - 4y = -9 \quad ...(2)$$
Here, $a_1 = 3, b_1 = -4, c_1 = 7$
$a_2 = 3, b_2 = -4, c_2 = 9$
We see that
$$\frac{a_1}{a_2} = \frac{b_1}{b_2} \neq \frac{c_1}{c_2}$$

Hence, the lines represented by the given pair of linear equations are parallel. Therefore, equations (1) and (2) have no common solution, *i.e.*, the solution set of the given system equations is inconsistent.

Example 10. *Obtain the condition for the following system of linear equations to have a unique solution :*
$$ax + by = c$$
$$lx + my = n.$$

Sol. The given system of linear equations is
$$ax + by = c \quad ...(1)$$
$$lx + my = n \quad ...(2)$$
Here, $a_1 = a, b_1 = b, c_1 = -c$
$a_2 = l, b_2 = m, c_2 = -n$

If the given system of linear equations has a unique solution, then
$$\frac{a_1}{a_2} \neq \frac{b_1}{b_2}$$
$$\Rightarrow \quad \frac{a}{l} \neq \frac{b}{m}$$
$$\Rightarrow \quad am \neq bl$$

This is the required condition.

Example 11. *Determine whether the following system of linear equation has a unique solution :*
$$x - 2y = 8$$
$$5x - 10y = 10.$$

Sol. The given system of linear equations is
$$x - 2y = 8 \quad ...(1)$$
$$5x - 10y = 10 \quad ...(2)$$
Here, $a_1 = 1, b_1 = -2, c_1 = 8$
$a_2 = 5, b_2 = -10, c_3 = 10$
We see that
$$\frac{a_1}{a_2} = \frac{b_1}{b_2}$$

Hence, the given system of linear equations does not have a unique solution.

Example 12. *Determine whether the following system of linear equations has infinitely many solutions.*
$$3x - 5y = 20$$
$$6x - 10y = 40.$$

Sol. The given system of linear equations is
$$3x - 5y = 20 \quad ...(1)$$
$$6x - 10y = 40 \quad ...(2)$$
Here, $a_1 = 3, b_1 = -5, c_1 = 20$
$a_2 = 6, b_2 = -10, c_2 = 40$
We see that
$$\frac{a_1}{a_2} = \frac{b_1}{b_2} = \frac{c_1}{c_2}$$

Hence, the given system of linear equation has similarly many solutions.

Example 13. *5 books and 7 pens together cost Rs. 79 whereas 7 books and 5 pens together cost Rs. 77. Find the total cost of 1 book and 2 pens.*

Sol. Let the cost of 1 book be Rs. x and that of 1 pen be Rs. y.

Then, according to the question,
$$5x + 7y = 79 \quad ...(1)$$
and $\quad 7x + 5y = 77 \quad ...(2)$

Let us draw the graphs of equations (1) and (2) by finding two solutions for each of the equations. These two solutions of the equations (1) and (2) are given below in table 1 and table 2 respectively.

For equation (1)

$5x + 7y = 79$

$\Rightarrow \quad 7y = 79 - 5x$

$\Rightarrow \quad y = \dfrac{79 - 5x}{7}$

Table 1 of solutions

x	6	−8
y	7	17

For equation (2)

$7x + 5y = 77$

$\Rightarrow \quad 5y = 77 - 7x$

$\Rightarrow \quad y = \dfrac{77 - 7x}{5}$

Table 2 of solutions

x	1	−4
y	14	21

We plot the points A (6, 7) and B(− 8, 13) on a graph paper and join these points to form the line AB representing the equation (1) as shown in the figure. Also, we plot the points C (1, 14) and D(− 4, 21) on the same graph paper and join these points to form the line CD representing the equation (2) as shown in the same figure.

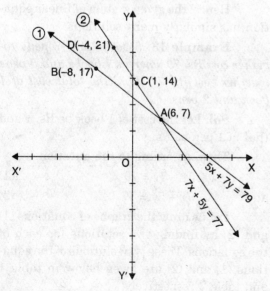

In the figure we observe that the two lines intersect at the point A (6, 7). So $x = 6$ and $y = 7$ is the required solution of the pair of linear equation formed, *i.e.,* the cost of 1 book is Rs. 6 and of 1 pen is Rs. 7.

Therefore, cost of 1 book and 2 pens
$= 6 + 2 \times 7 =$ Rs. 20.

Example 14. *Solve the following system of linear equations graphically.*

$2x - y - 5 = 0$

$x - y - 3 = 0$

Find the points where the line meets the y-axis.

Sol. The given system of linear equations is

$2x - y - 5 = 0$ ...(1)

$x - y - 3 = 0$ ...(2)

Let us draw the graphs of equations (1) and (2) by finding two solutions for each of the equations. These two solutions of the equations (1) and (2) are given below in table 1 and table 2 respectively.

For equation (1)

$2x - y - 5 = 0$

$\Rightarrow \quad y = 2x - 5$

Table 1 of solutions

x	3	2
y	1	−1

For equation (2)

$x - y - 3 = 0$

$\Rightarrow \quad y = x - 3$

Table 2 of solutions

x	3	4
y	0	1

We plot the points A (3, 1) and B (2, − 1) on the graph paper and join these points to form the line AB representing the equation (1) as shown in the figure. Also, we plot the points C(3, 0) and D(4, 1) on the same graph paper and join these points to form the line CD representing the equation (2) as shown in the same figure.

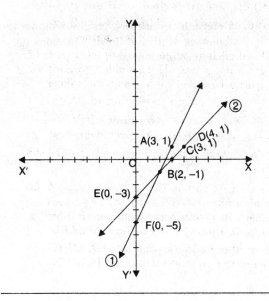

In the figure we observe that the two lines intersect at the point B(2, −1). So $x = 2$, $y = -1$ is the required solution of the given pair of linear equations.

Also we observe that the lines (1) and (2) meet the y-axis in the points E(0, −3) and F(0, −5) respectively.

TEST YOUR KNOWLEDGE

1. Akhila went to a fair in her village. She wanted to enjoy rides on the Giant Wheel and play Hoopla (a game in which you throw a ring on the items kept in the stall, and if the ring covers any object completely you get it). The number of times she played Hoopla is half the number of rides she had on the Giant Wheel. Each side costs Rs. 3 and a game of Hoopla upto Rs. 4. She goes to the fair with Rs. 20 and wants to have rides on the Giant wheel and play Hoopla. Find out how many rides on the Giant wheel Akhila had, and how many times she played Hoopla.

2. Romila went to a stationery stall and purchased 2 pencils and 3 erasers for Rs 9. Her friend Sonali saw the new variety of pencils and erasers with Romila and she also bought 4 pencils and 6 erasers of the same kind for Rs. 18. Find the cost 4 of each pencil and each eraser.

3. The parts of train A is given by the equation $x + 2y - 4 = 0$ and the parts of another train B is given by the equation $2x + 4y - 12 = 0$. Can the parts of the two trains cross?

4. Check whether the pair of equations
$$x + 3y = 6 \qquad \ldots(1)$$
$$2x - 3y = 12 \qquad \ldots(2)$$
is consistent. If so, solve then graphically.

5. Graphically, find whether the following pair of equations has two solution, a unique solution or infinitely many solutions
$$5x - 8y + 1 = 0 \qquad \ldots(1)$$
$$3x - \frac{24}{5}y + \frac{3}{5} = 0 \qquad \ldots(2)$$

6. Champa went to a 'Sale' to purchase some pants and skirts. When her friend asked her how many of each she had bought, she answered, "The number of skirts is two less than twice the number of pants purchased. Also, the number of skirts is four less than four times the number of pants purchased." Help her friends to find how many pants and skirts champa bought.

7. Solve the following system of linear equations graphically:
$$2x - y - 4 = 0$$
$$x + y + 1 = 0$$
Find the points where the lines meet y-axis.

8. Draw the graphs of $x - y + 1 = 0$ and $3x + 2y - 12 = 0$. Calculate the area bounded by these lines and x-axis.

9. Solve the following system of linear equations graphically :
$$x + 2y = 1$$
$$x - 2y = -7$$
Also, read the points from the graphs where the lines meet the axis of x.

10. Solve the following system of simultaneous linear equations graphically :
$$x + 3y + 6 = 0$$
$$2x + 6y + 12 = 0$$
Determine whether the given system is consistent or inconsistent.

11. Determine graphically whether the following system of equations
$$2x - 3y = 5$$
$$6y - 4x = 3$$
is consistent or inconsistent.

12. Determine by drawing graphs, whether the following system of linear equations has a unique solution or not :
$$x - 3y = 3$$
$$3x - 9y = 2.$$

Answers

1. 4, 2
2. Rs. 3, Re. 1 ; Rs. 3.75, Rs. 0.50, and so on.
3. No.
4. Yes ; $x = 6, y = 0$
5. Infinitely many solution
6. 1, 0
7. 1, – 2 ; (0, – 4), (0, – 1)
8. 7.5 square units
9. (– 3, 2) ; (1, 0), (– 7, 0)
10. infinitely many solutions, consistent
11. inconsistent
12. No.

ALGEBRAIC METHODS OF SOLVING A PAIR OF LINEAR EQUATIONS

IMPORTANT POINTS

1. Limitations of Graphical Methods. The graphical method is not convenient in cases when the point representing the solution of the linear equations has non-integral coordinates like $(\sqrt{3}, 2\sqrt{7})$, $(-1.75, 3.3)$, $\left(\dfrac{y}{13}, \dfrac{1}{19}\right)$ etc. There is every permeability of making mistakes while reading such coordinates. To overcome this difficulty, we prefer to use algebraic methods of solving a pair of linear equations given below.

2. Substitution Method. It is an algebraic method of solving a pair of linear equations. The stepwise procedure of this method is given below :

Step 1 : Find the value of one variable, say y in terms of the other variable, *i.e.*, x from either equation, whichever is convenient.

Step 2 : Substitute this value of y in the other equation, and reduce it to an equation in one variable, *i.e.*, in terms of x, which can be solved by simplifying. Sometimes, we can get equations with no variable. If this equation is true you can conclude that the linear pair has infinitely many solutions. If the equation in Step 2 is false, then the linear pair is inconsistent.

Step 3 : Substitute the value of x (or y) obtained in Step 2 in the equation used in Step 1 to obtain the value of the other variable.

Remark : We have substituted the value of one variable by expressing it in terms of the other variable to solve the pair of linear equations. This is why the method is known as the *substitution method*.

PAIR OF LINEAR EQUATIONS IN TWO VARIABLES

ILLUSTRATIVE EXAMPLES

[NCERT Exercise 3.3]
(Page No. 53)

Example 1. *Solve the following pair of linear equations by the substitution method.*

(i) $x + y = 14$
$x - y = 4$

(ii) $s - t = 3$
$\dfrac{s}{3} + \dfrac{t}{2} = 6$

(iii) $3x - y = 3$
$9x - 3y = 9$

(iv) $0.2x + 0.3y = 1.3$
$0.4x + 0.5y = 2.3$

(v) $\sqrt{2}x + \sqrt{3}y = 0$
$\sqrt{3}x - \sqrt{8}y = 0$

(vi) $\dfrac{3x}{2} - \dfrac{5y}{3} = -2$
$\dfrac{x}{3} + \dfrac{y}{2} = \dfrac{13}{6}$

Sol. (i) $x + y = 14$
$x - y = 4$

The given pair of linear equations is
$x + y = 14$ ...(1)
$x - y = 4$ ...(2)

From equation (1),
$y = 14 - x$ ...(3)

Substitute this value of y in equation (2), we get
$x - (14 - x) = 4$
$\Rightarrow x - 14 + x = 4$
$\Rightarrow 2x - 14 = 4$
$\Rightarrow 2x = 4 + 14$
$\Rightarrow 2x = 18$
$\Rightarrow x = \dfrac{18}{2} = 9$

Substituting this value of x in equation (3), we get
$y = 14 - 9 = 5$

Therefore the solution is
$x = 9, \quad y = 5$

Verification. Substituting $x = 9$ and $y = 5$, we find that both the equations (1) and (2) are satisfied as shown below :
$x + y = 9 + 5 = 14$
$x - y = 9 - 5 = 4$

This verifies the solution.

(ii) $s - t = 3$
$\dfrac{s}{3} + \dfrac{t}{2} = 6$

The given pair of linear equations is
$s - t = 3$ ...(1)
$\dfrac{s}{3} + \dfrac{t}{2} = 6$ ...(2)

From equation (1),
$s = t + 3$ ...(3)

Substitute this value of s in equation (2), we get
$\dfrac{t + 3}{3} + \dfrac{t}{2} = 6$
$\Rightarrow \dfrac{2(t + 3) + 3t}{6} = 6$
$\Rightarrow 2(t + 3) + 3t = 36$
$\Rightarrow 2t + 6 + 3t = 36$
$\Rightarrow 5t + 6 = 36$
$\Rightarrow 5t = 36 - 6$
$\Rightarrow 5t = 30$
$\Rightarrow t = \dfrac{30}{5} = 6$

Substituting this value of t in equation (3), we get
$s = 6 + 3 = 9$

Therefore the solution is
$s = 9, t = 6$

Verification. Substituting $s = 9$ and $t = 6$, we find that both equation (1) and (2) are satisfied as shown below :
$$s - t = 9 - 6 = 3$$
$$\frac{s}{3} + \frac{t}{2} = \frac{9}{3} + \frac{6}{2} = 3 + 3 = 6$$

This verifies the solution.

(iii) $3x - y = 3$
$9x - 3y = 9$

The given pair of linear equations is
$$3x - y = 3 \qquad ...(1)$$
$$9x - 3y = 9 \qquad ...(2)$$

From equation (1),
$$y = 3x - 3 \qquad ...(3)$$

Substitute this value of y in equation (2), we get
$$9x - 3(3x - 3) = 9$$
$$\Rightarrow \quad 9x - 9x + 9 = 9$$
$$\Rightarrow \quad 9 = 9$$

which is true. Therefore, equations (1) and (2) have infinitely many solutions.

(iv) $0.2x + 0.3y = 1.3$
$0.4x + 0.5y = 2.3$

The given system of linear equations is
$$0.2x + 0.3y = 1.3 \qquad ...(1)$$
$$0.4x + 0.5y = 2.3 \qquad ...(2)$$

From equation (1),
$$0.3y = 1.3 - 0.2x$$
$$\Rightarrow \quad y = \frac{1.3 - 0.2x}{0.3} \qquad ...(3)$$

Substituting this value of y in equation (2), we get
$$0.4x + 0.5\left(\frac{1.3 - 0.2x}{0.3}\right) = 2.3$$
$$\Rightarrow \quad 0.21x + 0.65 - 0.1x = 0.69$$
$$\Rightarrow \quad 0.12x - 0.1x = 0.69 - 0.65$$
$$\Rightarrow \quad 0.02x = 0.04$$
$$\Rightarrow \quad x = \frac{0.04}{0.02} = 2$$

Substituting this value of x in equation (3), we get

$$y = \frac{1.3 - 0.2(2)}{0.3} = \frac{1.3 - 0.4}{0.3}$$
$$= \frac{0.9}{0.3} = 3$$

Therefore the solutions is $x = 2$, $y = 3$.

Verification. Substituting $x = 2$ and $y = 3$, we find that both the equations (1) and (2) are satisfied as shown below :
$$0.2x + 0.3y = (0.2)(2) + (0.3)(3)$$
$$= 0.4 + 0.9 = 1.3$$
$$0.4x + 0.5y = (0.4)(2) + (0.5)(3)$$
$$= 0.8 + 1.5 = 2.23$$

This verifies the solution.

(v) $\sqrt{2}x + \sqrt{3}y = 0$
$\sqrt{3}x - \sqrt{8}y = 0$

The given pair of linear equations is
$$\sqrt{2}x + \sqrt{3}y = 0 \qquad ...(1)$$
$$\sqrt{3}x - \sqrt{8}y = 0 \qquad ...(2)$$

From equation (2),
$$\sqrt{3}x = \sqrt{8}y$$
$$\Rightarrow \quad x = \frac{\sqrt{8}}{\sqrt{3}}y \qquad ...(3)$$

Substituting this value of x in equation (1), we get
$$\sqrt{2} \cdot \frac{\sqrt{8}}{\sqrt{3}}y + \sqrt{3}y = 0$$
$$\Rightarrow \quad \frac{4}{\sqrt{3}}y + \sqrt{3}y = 0$$
$$\Rightarrow \quad \left(\frac{4}{\sqrt{3}} + \sqrt{3}\right)y = 0$$
$$\Rightarrow \quad y = 0 \quad \left[\because \frac{4}{\sqrt{3}} + \sqrt{3} \neq 0\right]$$

Substituting this value of y in equation (3), we get
$$x = \frac{\sqrt{8}}{\sqrt{3}}(0) = 0$$

Therefore, the solution is
$$x = 0, y = 0.$$
Verification. Substituting $x = 0$ and $y = 0$, we find that both the equations (1) and (2) are satisfied as shown below :
$$\sqrt{2}x + \sqrt{3}y = \sqrt{2}(0) + \sqrt{3}(0) = 0$$
$$\sqrt{3}x - \sqrt{8}y = \sqrt{3}(0) - \sqrt{8}(0) = 0$$
This verifies the solution.

(vi) $\dfrac{3x}{2} - \dfrac{5y}{3} = -2$

$\dfrac{x}{3} + \dfrac{y}{2} = \dfrac{13}{6}$

The given system of linear equations is
$$\dfrac{3x}{2} - \dfrac{5y}{3} = -2 \qquad \ldots(1)$$
$$\dfrac{x}{3} + \dfrac{y}{2} = \dfrac{13}{6} \qquad \ldots(2)$$
$$\Rightarrow \quad 9x - 10y = -12 \qquad \ldots(3)$$
$$2x + 3y = 13 \qquad \ldots(4)$$

From equation (3)
$$9x - 10y = -12$$
$$9x = 10y - 12$$
$$x = \dfrac{10y - 12}{9}$$

Substituting the value of y in equation (4), we get
$$2\left(\dfrac{10y - 12}{9}\right) + 3y = 13$$
$$20y - 24 + 27y = 117$$
$$47y = 117 + 24$$
$$y = \dfrac{141}{47}$$
$$y = 3$$

Substituting the value of y in equation (4), we get
$$2x + 3 \times 3 = 13$$
$$2x + 9 = 13$$
$$2x = 13 - 9$$
$$x = \dfrac{4}{2} = 2$$

Therefore, the solution is
$$x = 2, y = 3$$

Verification. Substituting $x = 2$ and $y = 3$, we find that both the equations (1) and (2) are satisfied as shown below :
$$\dfrac{3}{2}x - \dfrac{5y}{3} = \dfrac{3}{2}(2) - \dfrac{5}{3}(3) = 3 - 5 = -2$$
$$\dfrac{x}{3} + \dfrac{y}{2} = \dfrac{2}{3} + \dfrac{3}{2} = \dfrac{13}{6}$$
This verifies the solution.

Example 2. *Solve $2x + 3y = 11$ and $2x - 4y = -24$ and hence find the value of 'm' for which $y = mx + 3$.*

Sol. The given pair of linear equations is
$$2x + 3y = 11 \qquad \ldots(1)$$
$$2x - 4y = -24 \qquad \ldots(2)$$
From equation (1),
$$3y = 11 - 2x$$
$$\Rightarrow \quad y = \dfrac{11 - 2x}{3} \qquad \ldots(3)$$

Substituting this value of y in equation (2), we get
$$2x - 4\left(\dfrac{11 - 2x}{3}\right) = -24$$
$$\Rightarrow \quad 6x - 44 + 8x = -72$$
$$\Rightarrow \quad 14x - 44 = -72$$
$$\Rightarrow \quad 14x = 44 - 72$$
$$\Rightarrow \quad 14x = -28$$
$$\Rightarrow \quad x = -\dfrac{28}{14} = -2$$

Substituting this value of x in equation (3), we get
$$y = \dfrac{11 - 2(-2)}{3} = \dfrac{11 + 4}{3} = \dfrac{15}{3} = 5$$

Verification. Substituting $x = -2$ and $y = 5$, we find that both the equations (1) and (2) are satisfied as shown below :
$$2x + 3y = 2(-2) + 3(5) = -4 + 15 = 11$$
$$2x - 4y = 2(-2) - 4(5) = -4 - 20 = -24$$
This verifies the solution.

Now, $y = mx + 3$
$\Rightarrow \quad 5 = m(-2) + 3$
$\Rightarrow \quad -2m = 5 - 3$
$\Rightarrow \quad -2m = 2$
$\Rightarrow \quad m = \dfrac{2}{-2} = -1.$

Example 3. *Form the pair of linear equations in the following problems and find their solution by substitution method.*

(i) *The difference between two numbers is 26 and one number is three times the other. Find them.*

(ii) *The larger of two supplementary angles exceeds the smaller by 18 degrees. Find them.*

(iii) *The coach of a cricket team buys 7 bats and 6 balls for Rs. 3800. Later, she buys 3 bats and 5 balls for Rs. 1750. Find the cost of each bat and each ball.*

(iv) *The taxi charges in a city consist of a fixed charge together with the charge for the distance covered. For a distance of 10 km, the charge paid is Rs 105 and for a journey of 15 km, the charge paid is Rs. 155. What are the fixed charges and the charge per kilometer? How much does a person have to pay for travelling a distance of 25 km?*

(v) *A fraction becomes $\dfrac{9}{11}$, if 2 is added to both the numerator and the denominator. If, 3 is added to both the numerator and the denominator it becomes $\dfrac{5}{6}$. Find the fraction.*

(vi) *Five years hence, the age of Jacob will be three times that of his son. Five years ago, Jacob's age was seven times that of his son. What are their present ages?*

Sol. (*i*) Let the two numbers be x and y ($x > y$) then, according to the question, the pair of linear equations formed is

$x - y = 26$ ...(1)
$x = 3y$ ...(2)

Substitute the value of x from equation (2) in equation (1). We get

$3y - y = 26$

$\Rightarrow \quad 2y = 26$
$\Rightarrow \quad y = \dfrac{26}{2}$
$\Rightarrow \quad y = 13$

Substituting this value of y in equation (2), we get

$x = 3(13) = 39$

Hence, the required numbers are 39 and 13.

Verification. Substituting $x = 39$ and $y = 13$, we find that both the equation (1) and (2) are satisfied as shown below.

$x - y = 39 - 13 = 26$
$3y = 3(13) = 39 = x$

This verifies the solution.

(*ii*) Let the larger and the smaller of two supplementary angles be $x°$ and $y°$ respectively.

Then, according to the question.

The pair of linear equations formed is

$x° = y° + 18°$ ...(1)
$x° + y° = 180°$ ...(2)

| $\because$ The two angles are supplementary

Substitute the value of $x°$ from equation (1) in equation (2), we get

$y° + 18° + y° = 180°$
$\Rightarrow \quad 2y° + 18° = 180°$
$\Rightarrow \quad 2y° = 180° - 18°$
$\Rightarrow \quad 2y° = 162°$
$\Rightarrow \quad y° = \dfrac{162°}{2} = 81°$

Substituting this value of $y°$ in equation (1), we get

$x° = 81° + 18° = 99°$

Hence, the larger and the smaller of the two supplementary angles are 99° and 81° respectively.

Verification. Substituting $x° = 99°$ and $y° = 81°$, we find that both the equations (1) and (2) are satisfied as shown below :

$y° + 18° = 81° + 18° = 99° = x°$
$x° + y° = 99° + 81° = 180°$

This verifies the solution.

PAIR OF LINEAR EQUATIONS IN TWO VARIABLES

(iii) Let the cost of each bat and each ball be Rs. x and Rs. y respectively.

Then, according to the question,
The pair of linear equations formed is
$$7x + 6y = 3800 \quad ...(1)$$
$$3x + 5y = 1750 \quad ...(2)$$

From equation (2),
$$5y = 1750 - 3x$$
$$y = \frac{1750 - 3x}{5} \quad ...(3)$$

Substitute this value of y in equation (1), we get
$$7x + 6\left(\frac{1750 - 3x}{5}\right) = 3800$$
$$\Rightarrow 35x + 10500 - 18x = 19000$$
$$\Rightarrow 17x + 10500 = 19000$$
$$\Rightarrow 17x = 19000 - 10500$$
$$\Rightarrow 17x = 8500$$
$$\Rightarrow x = \frac{8500}{17} = 500$$

Substituting this value of x in equation (3), we get
$$y = \frac{1750 - 3(500)}{5}$$
$$= \frac{1750 - 1500}{5} = \frac{250}{5} = 50$$

Hence, the cost of each bat and each ball is Rs. 500 and Rs. 50 respectively.

Verification. Substituting $x = 500$ and $y = 50$, we find that both the equations (1) and (2) are satisfied as shown below :
$$7x + 6y = 7(500) + 6(50)$$
$$= 3500 + 300 = 3800$$
$$3x + 5y = 3(500) + 5(50)$$
$$= 1500 + 250 = 1750$$

This verifies the solution.

(iv) Let the fixed charges be Rs. x and the charge per kilometre be Rs. y.

Then, according to the question
The pair of linear equations formed is
$$x + 10y = 105 \quad ...(1)$$
$$x + 15y = 155 \quad ...(2)$$

From equation (1),
$$x = 105 - 10y \quad ...(3)$$

Substitute this value of x in equation (2), we get
$$105 - 10y + 15y = 155$$
$$\Rightarrow 105 + 5y = 155$$
$$\Rightarrow 5y = 155 - 105$$
$$\Rightarrow 5y = 50$$
$$\Rightarrow y = \frac{50}{5} = 10$$

Substituting this value of y in equation (3), we get
$$x = 105 - 10(10)$$
$$= 105 - 100 = 5$$

Hence, the fixed charges are Rs. 5 and the charge per kilometre is Rs. 10.

Verification. Substituting $x = 5$ and $y = 10$, we find that both the equations (1) and (2) are satisfied as shown below :
$$x + 10y = 5 + 10(10) = 5 + 100 = 105$$
$$x + 15y = 5 + 15(10) = 5 + 150 = 155$$

This verifies the solution.

Again, for travelling a distance of 25 km, a person will have to pay $= 5 + 10(25)$
$$= 5 + 250$$
$$= \text{Rs. } 255$$

(v) Let the fraction be $\frac{x}{y}$

Then, according to the question, the pair of linear equations formed is
$$\frac{x+2}{y+2} = \frac{9}{11}$$
$$11(x + 2) = 9(y + 2)$$
$$\Rightarrow 11x + 22 = 9y + 18$$
$$\Rightarrow 11x = 9y + 18 - 22$$
$$\Rightarrow 11x - 9y + 4 = 0 \quad ...(1)$$

and
$$\frac{x+3}{y+3} = \frac{5}{6}$$
$$\Rightarrow 6(x + 3) = 5(y + 3)$$
$$\Rightarrow 6x + 18 = 5y + 15$$
$$\Rightarrow 6x - 5y = -3 \quad ...(2)$$

From equation (1),

$$\Rightarrow x = \frac{9y-4}{11} \quad \ldots(3)$$

Substitute this value of x in equation (2), we get

$$\Rightarrow 6\left(\frac{9y-4}{11}\right) - 5y = -3$$
$$\Rightarrow 6(9y-4) - 55y = -33$$
$$\Rightarrow 54y - 24 - 55y = -33$$
$$\Rightarrow -y + 9 = 0$$
$$\Rightarrow y = 9$$

Substituting this value of y in equation (3), we get

$$x = \frac{9(9)-4}{11} = \frac{81-4}{11} = \frac{77}{11} = 7$$

Hence, the required fraction is $\frac{7}{9}$.

Verification. Substituting $x = 7$ and $y = 9$, we find that both the equations (1) and (2) are satisfied as shown below :

$$\frac{x+2}{y+2} = \frac{7+2}{9+2} = \frac{9}{11}$$

$$\frac{x+3}{y+3} = \frac{7+3}{9+3} = \frac{10}{12} = \frac{5}{6}.$$

This verifies the solution.

(vi) Let the present ages of Jacob and his son be x years and y years respectively.

Then, according to the question, the pair of linear equation formed is

$$x + 5 = 3(y + 5)$$
$$x - 5 = 7(y - 5)$$
$$\Rightarrow x - 3y = 10 \quad \ldots(1)$$
$$x - 7y = -30 \quad \ldots(2)$$

From equation (1),

$$x = 3y + 10 \quad \ldots(3)$$

Substitute this value of x in equation (2), we get

$$(3y + 10) - 7y = -30$$
$$\Rightarrow 3y - 7y = -30 - 10$$
$$\Rightarrow -4y = -40$$
$$\Rightarrow y = \frac{-40}{-4} = 10$$

Substituting this value of y in equation (3), we get

$$x = 3(10) + 10 = 30 + 10 = 40$$

Hence, the present ages of Jacob and his son are 40 years and 10 years respectively.

Verification. Substituting $x = 40$ and $y = 10$, we find that both the equations (1) and (2) are satisfied as shown below :

$$x - 3y = 40 - 3(10) = 40 - 30 = 10$$
$$x - 7y = 40 - 7(10) = 40 - 70 = -30$$

This verifies the solution.

ADDITIONAL EXAMPLES

Example 4. *Solve the following system of equations by substitution method.*

$$x + y = 7$$
$$2x - 3y = 11.$$

Sol. The given system of equations is

$$x + y = 7 \quad \ldots(1)$$
$$2x - 3y = 11 \quad \ldots(2)$$

From equation (1),

$$y = 7 - x \quad \ldots(3)$$

Substitute this value of y in equation (2), we get

$$2x - 3(7 - x) = 11$$
$$2x - 21 + 3x = 11$$
$$\Rightarrow 5x - 21 = 11$$
$$\Rightarrow 5x - 21 = 11$$
$$\Rightarrow 5x = 11 + 21$$
$$\Rightarrow 5x = 32$$
$$\Rightarrow x = \frac{32}{5}$$

Substituting this value of x in equation (3), we get

$$y = 7 - \frac{32}{5} = \frac{35-32}{5} = \frac{3}{5}$$

PAIR OF LINEAR EQUATIONS IN TWO VARIABLES

Therefore the solution is
$$x = \frac{32}{5},\ y = \frac{3}{5}.$$

Verification. Substituting $x = \frac{32}{5}$ and $y = \frac{3}{5}$, we find that both the equations (1) and (2) are satisfied as shown below:

$$x + y = \frac{32}{5} + \frac{3}{5} = \frac{32+3}{5} = \frac{35}{5} = 7$$

$$2x - 3y = 2\left(\frac{32}{5}\right) - 3\left(\frac{3}{5}\right)$$

$$= \frac{64}{5} - \frac{9}{5}$$

$$= \frac{64-9}{5} = \frac{55}{5} = 11$$

This verifies the solution.

Example 5. *Solve the following system of equation by substitution method.*

$$2x - 7y = 1$$
$$4x + 3y = 15.$$

Sol. The given system of equation is

$$2x - 7y = 1 \qquad \ldots(1)$$
$$4x + 3y = 15 \qquad \ldots(2)$$

From equation (1),
$$7y = 2x - 1$$
$$\Rightarrow \quad y = \frac{2x-1}{7} \qquad \ldots(3)$$

Substitute this value of y in equation (2), we get

$$4x + 3\left(\frac{2x-1}{7}\right) = 15$$

$$\Rightarrow \quad 28x + 6x - 3 = 105$$
$$\Rightarrow \quad 34x - 3 = 105$$
$$\Rightarrow \quad 34x = 105 + 3$$
$$\Rightarrow \quad x = \frac{108}{34}$$
$$\Rightarrow \quad x = \frac{54}{17}$$

Substituting this value of x in equation (3), we get

$$y = \frac{2\left(\frac{54}{17}\right) - 1}{7} = \frac{\frac{108}{17} - 1}{7} = \frac{108 - 17}{119}$$

$$= \frac{91}{119} = \frac{13}{17}$$

Therefore the solution is
$$x = \frac{54}{17},\ y = \frac{13}{17}.$$

Verification. Substituting $x = \frac{54}{17}$ and $y = \frac{13}{17}$, we find that both the equations (1) and (2) are satisfied as shown below:

$$2x - 7y = 2\left(\frac{54}{17}\right) - 7\left(\frac{13}{17}\right) = \frac{108}{17} - \frac{21}{17}$$

$$= \frac{108 - 91}{17}$$

$$= \frac{17}{17} = 1$$

$$4x + 3y = 4\left(\frac{54}{17}\right) + 3\left(\frac{13}{17}\right)$$

$$= \frac{216}{17} + \frac{39}{17}$$

$$= \frac{216 + 39}{17} = \frac{255}{17} = 15$$

This verifies the solution.

Example 6. *Solve the following system of equations by substitution method:*

$$4x - 3y - 8 = 0$$
$$6x - y - \frac{29}{3} = 0.$$

Sol. The given system of equations is

$$4x - 3y - 8 = 0 \qquad \ldots(1)$$

$$6x - y - \frac{29}{3} = 0 \qquad \ldots(2)$$

From equation (2),

$$y = 6x - \frac{29}{3} \qquad \ldots(3)$$

Substitute this value of y in equation (1), we get

$$4x - 3\left(6x - \frac{29}{3}\right) - 8 = 0$$
$$\Rightarrow \quad 4x - 18x + 29 - 8 = 0$$
$$\Rightarrow \quad -14x + 21 = 0$$
$$\Rightarrow \quad 14x = 21$$
$$\Rightarrow \quad x = \frac{21}{14} = \frac{3}{2}$$

Substituting this value of x in equation (3), we get

$$y = 6\left(\frac{3}{2}\right) - \frac{29}{3} = 9 - \frac{29}{3}$$
$$= \frac{27 - 29}{3} = \frac{-2}{3}$$

Therefore the solution is

$$x = \frac{3}{2}, y = -\frac{2}{3}.$$

Verification. Substituting $x = \frac{3}{2}$ and $y = -\frac{2}{3}$, we find that both the equation (1) and (2) are satisfied as shown below :

$$4x - 3y - 8 = 4\left(\frac{3}{2}\right) - 3\left(\frac{-2}{3}\right) - 8$$
$$= 6 + 2 - 8 = 0$$
$$6x - y - \frac{29}{3} = 6\left(\frac{3}{2}\right) - \left(\frac{-2}{3}\right) - \frac{29}{3}$$
$$= 9 + \frac{2}{3} - \frac{29}{3} = 0$$

This verifies the solution.

Example 7. *Solve the following system of linear equations by substitution method :*

$$3x + 5y = 7$$
$$11x - 13y = 9.$$

Sol. The given system of equation is
$$3x + 5y = 7 \qquad ...(1)$$
$$11x - 13y = 9 \qquad ...(2)$$
From equation (1),
$$5y = 7 - 3x$$

$$\Rightarrow \quad y = \frac{7 - 3x}{5} \qquad ...(3)$$

Substitute this value of y in equation (2), we get

$$11x - 13\left(\frac{7 - 3x}{5}\right) = 9$$
$$\Rightarrow \quad 55x - 91 + 39x = 45$$
$$\Rightarrow \quad 94x = 45 + 91$$
$$\Rightarrow \quad 94x = 136$$
$$\Rightarrow \quad 47x = 68$$
$$\Rightarrow \quad x = \frac{68}{47}.$$

Substituting this value of x in equation (3), we get

$$y = \frac{7 - 3\left(\frac{68}{47}\right)}{5}$$
$$= \frac{329 - 204}{235} = \frac{125}{235} = \frac{25}{47}$$

Therefore the solution is

$$x = \frac{68}{47}, y = \frac{25}{47}.$$

Verification. Substituting $x = \frac{68}{47}$, $y = \frac{25}{47}$, we find that both the equations (1) and (2) are satisfied as shown below :

$$3x + 5y = 3\left(\frac{68}{47}\right) + 5\left(\frac{25}{47}\right)$$
$$= \frac{204}{47} + \frac{125}{47} = \frac{329}{47} = 7$$
$$11x - 13y = 11\left(\frac{68}{47}\right) - 13\left(\frac{25}{47}\right)$$
$$= \frac{748}{47} - \frac{325}{47} = \frac{748 - 325}{47}$$
$$= \frac{423}{47} = 9$$

This verifies the solution.

PAIR OF LINEAR EQUATIONS IN TWO VARIABLES

Example 8. *Solve the following system of linear equations by substitution method :*
$$2x - y = 11$$
$$5x + 4y = 1.$$

Sol. The given system of equations is
$$2x - y = 11 \quad \ldots(1)$$
$$5x + 4y = 1 \quad \ldots(2)$$

From equation (1),
$$y = 2x - 11 \quad \ldots(3)$$

Substitute this value of y in equation (2), we get
$$5x + 4(2x - 11) = 1$$
$$\Rightarrow \quad 5x + 8x - 44 = 1$$
$$\Rightarrow \quad 13x - 44 = 1$$
$$\Rightarrow \quad 13x = 44 + 1$$
$$\Rightarrow \quad 13x = 45$$
$$\Rightarrow \quad x = \frac{45}{13}$$

Substituting this value of x in equation (3), we get
$$y = 2\left(\frac{45}{13}\right) - 11$$
$$= \frac{90}{13} - 11 = \frac{90 - 143}{13} = -\frac{53}{13}$$

Therefore the solution is
$$x = \frac{45}{13}, y = -\frac{53}{13}.$$

Verification. Substituting $x = \frac{45}{13}$, $y = -\frac{53}{13}$, we find that both the equations (1) and (2) are satisfied as shown below :

$$2x - y = 2\left(\frac{45}{13}\right) - \left(-\frac{53}{13}\right)$$
$$= \frac{90}{13} + \frac{53}{13} = \frac{143}{13} = 11$$

$$5x + 4y = 5\left(\frac{45}{13}\right) + 4\left(-\frac{53}{13}\right)$$
$$= \frac{225}{13} - \frac{212}{13}$$
$$= \frac{225 - 212}{13} = \frac{13}{13} = 1$$

This verifies the solution.

Example 9. *Solve the following system of linear equations :*
$$4x + 7y = 20$$
$$21x - 13y = 21.$$

Sol. The given system of linear equations is
$$4x + 7y = 20 \quad \ldots(1)$$
$$21x - 13y = 21 \quad \ldots(2)$$

From equation (2),
$$13y = 21x - 21$$
$$\Rightarrow \quad y = \frac{21x - 21}{13} \quad \ldots(3)$$

Substitute this value of y in equation (1), we get
$$4x + 7\left(\frac{21x - 21}{13}\right) = 20$$
$$\Rightarrow \quad 52x + 147x - 147 = 260$$
$$\Rightarrow \quad 199x = 147 + 260$$
$$\Rightarrow \quad 199x = 407$$
$$\Rightarrow \quad x = \frac{407}{199}$$

Substituting this value of x in equation (3), we get
$$y = \frac{21\left(\frac{407}{199}\right) - 21}{13}$$
$$= \frac{8547 - 4179}{2587} = \frac{4368}{2587} = \frac{336}{199}$$

Therefore the solution is
$$x = \frac{407}{199}, y = \frac{336}{199}.$$

Verification. Substituting $x = \frac{407}{199}$, $y = \frac{336}{199}$, we find that both the equations (1) and (2) are satisfied as shown below :

$$4x + 7y = 4\left(\frac{407}{199}\right) + 7\left(\frac{336}{199}\right)$$

$$= \frac{1628 + 2352}{199}$$

$$= \frac{3980}{199} = 20$$

$$21x - 13y = 21\left(\frac{407}{199}\right) - 13\left(\frac{336}{199}\right)$$

$$= \frac{8547 - 4368}{199}$$

$$= \frac{4179}{199} = 21$$

This verifies the solution.

Example 10. *If $2x + y = 35$ and $3x + 4y = 65$, find the value of $\frac{x}{y}$.*

Sol. The given system of equations is

$2x + y = 35$ ...(1)
$3x + 4y = 65$ ...(2)

From equation (1),

$y = 35 - 2x$ ...(3)

Substitute this value of y in equation (2), we get

$3x + 4(35 - 2x) = 65$
$\Rightarrow \quad 3x + 140 - 8x = 65$
$\Rightarrow \quad -5x + 140 = 65$
$\Rightarrow \quad -5x = 65 - 140$
$\Rightarrow \quad -5x = -75$
$\Rightarrow \quad x = \frac{-75}{-5} = 15$

Substituting this value of x in equation (3), we get

$y = 35 - 2(15) = 35 - 30 = 5$

Therefore

$$\frac{x}{y} = \frac{15}{5} = 3.$$

Verification. Substituting $x = 15$, $y = 5$, we find that both the equations (1) and (2) are satisfied as shown below.

$2x + y = 2(15) + 5 = 30 + 5 = 35$
$3x + 4y = 3(15) + 4(5) = 45 + 20 = 65$

This verifies the solution.

Example 11. *A and B each have certain number of oranges, A say to B, "If you give me 10 of your oranges, I will have twice the number of oranges left with you". B replies "If you give me 10 of your oranges. I will have the same number of oranges as left with you". Find the number of oranges with A and B respectively.*

Sol. Let the number of oranges with A and B be x and y respectively. Then, according to the question,

$x + 10 = 2(y - 10)$
$x - 10 = y + 10$
$\Rightarrow \quad x - 2y = -30$ ...(1)
$\quad\quad x - y = 20$ ...(2)

From equation (2),

$y = x - 20$ ...(3)

Substitute this value of y in equation (1), we get

$x - 2(x - 20) = -30$
$\Rightarrow \quad x - 2x + 40 = -30$
$\Rightarrow \quad -x + 40 = -30$
$\Rightarrow \quad -x = -30 - 40$
$\Rightarrow \quad -x = -70$
$\Rightarrow \quad x = 70$

Substituting this value of x in equation (3), we get

$y = 70 - 20 = 50$

Hence, the number of oranges with A and B are 70 and 50 respectively.

Verification. Substituting $x = 70$, $y = 50$, we find that both the equations (1) and (2) are satisfied as shown below :

$x - 2y = 70 - 2(50) = 70 - 100 = -30$
$x - y = 70 - 50 = 20$

The verifies the solution.

Example 12. *If we buy 2 tickets from station A to station B, and 3 from station A to station C, we have to pay Rs. 795. But 3 tickets from station A to B and 5 tickets from A to C cost a total of Rs. 1300. What is the fare*

PAIR OF LINEAR EQUATIONS IN TWO VARIABLES

from station A to B and that from station A to C ?

Sol. Let the fare from station A to B be Rs. x and that from station A to C be Rs. y.

Then, according to the question,

$2x + 3y = 795$ ...(1)

$3x + 5y = 1300$ ...(2)

From equation (1),

$3y = 795 - 2x$

$\Rightarrow \quad y = \dfrac{795 - 2x}{3}$ ...(3)

Substitute this value of y in equation (2), we get

$3x + 5\left(\dfrac{795 - 2x}{3}\right) = 1300$

$\Rightarrow \quad 9x + 3975 - 10x = 3900$

$\Rightarrow \quad -x + 3975 = 3900$

$\Rightarrow \quad -x = 3900 - 3975$

$\Rightarrow \quad -x = -75$

$\Rightarrow \quad x = 75$

Substituting the value of x in equation (3), we get

$y = \dfrac{795 - 2(75)}{3} = \dfrac{795 - 150}{3}$

$= \dfrac{645}{3} = 215$

Hence, the fare from station A to B is Rs. 75 and that from station A to C is Rs. 215.

Verification. Substituting $x = 75$, $y = 215$, we find that both the equations (1) and (2) are satisfied as shown below :

$2x + 3y = 2(75) + 3(215)$

$= 150 + 645 = 795$

$3x + 5y = 3(75) + 5(215)$

$= 225 + 1075 = 1300$

This verifies the solution.

TEST YOUR KNOWLEDGE

1. Solve the following pair of equations by substitution method :

    $7x - 15y = 2$

    $x + 2y = 3$

2. Aftab tells his daughter, "seven years ago, I was seven times as old as you were then. Also, three years from now, I shall be three times as old as you will be". Find the present ages of Aftab and his daughter using the method of substitution.

3. The cost of 2 pencils and 3 erasers is Rs. 9 and the cost of 4 pencils and 6 erasers is Rs. 18. Find the cost of each pencil and each eraser.

4. The path of a train A is given by the equation $x + 2y - 4 = 0$ and the path of another train B is given by the equation $2x + 4y - 12 = 0$. Will the paths cross ? Solve by the method of substitution.

5. Solve the following pair of equation by substitution method :

    $2x + 3y - 10 = 0$

    $y = 5 - \dfrac{3}{2}x.$

6. Solve the following system of equations by substitution method :

    $5y - 9x = -24$

    $5y = 11x.$

7. Solve the following system of equations by substitution method :

    $y = 2x$

    $7x - 2y = 35.$

8. Solve the following system of equations by substitution method :

    $x = 4$

    $2x + 3y = 14.$

9. Solve the following system of equations by substitution method :

    $3x - y = 2$

    $x + 2y = 3.$

10. Solve the following system of equations by substitution method.

    $3x + 4y = 10$

    $5x - 2y = 8.$

11. The sum of the numerator and denominator of a fraction is 3 less than twice the denominator. If the numerator

and denominator are decreased by 1, the numerator becomes half the denominator. Determine the fractions.

12. If twice the son's age in years is added to the age of his father, the sum is 90. If twice the father's age in years is added to the age of the son, the sum is 120. Find their ages.

13. 5 pens and 6 pencils together cost Rs. 9 and 3 pens and 2 pencils cost Rs. 5. Find the cost of 1 pen and 1 pencils separately.

14. The sum of the digits of a two-digit number is 15. The number is decreased by 27, if the digits are reversed. Find the number.

15. A part of monthly hostel charges in a college are fixed and the remaining depends upon the number of days one has taken food in the mess. When a student A takes food for 20 days, he has to pay Rs. 1000 as hostel charges whereas a student B, who takes food for 26 days, pay Rs. 1180 as hostel charges. Find the fixed charge and the cost of food per day.

Answers

1. $x = \dfrac{49}{29}, y = \dfrac{19}{29}$ 2. 42 years, 12 years
3. infinitely many solutions.
4. No. 5. $x = 2, y = 2$
6. $x = -12, y = -\dfrac{132}{5}$
7. $x = 7, y = 14$ 8. $x = 4, y = 2$
9. $x = 1, y = 1$ 10. $x = 2, y = 1$
11. $\dfrac{4}{7}$
12. Son → 20 years, Father → 50 years
13. Rs. 1.50, Re. 0.25
14. 96
15. Rs. 400, Rs. 30

ELIMINATION METHOD

IMPORTANT POINTS

1. Elimination Method : This is another method of eliminating, *i.e.*, removing one variable. This is sometimes more convenient than the substitution method.

2. Steps in elimination method :

Step 1 : First multiply both the equations by suitable non-zero constants to make the coefficient of a variable (either x or y) numerically equal.

Step 2 : Then add or subtract one equation from the other so that one variable gets eliminated. If you get an equation in one variable, go to Step 3.

If, in Step 2, we obtain a true equation involving no variable, then the original pair of equations has infinitely many solutions.

If, in Step 2, we obtain a false equation involving no variable, then the original pair of equations has no solution, *i.e.*, it is inconsistent.

Step 3 : Solve the equation in one variable (x or y) so obtained to obtain get its value.

Step 4 : Substitute this value of x (or y) in either of the original equations to get the value of the other variable.

3. Reason for the Name : This method is called **elimination method** because in the method we eliminate one variable first, to get a linear equation in one variable.

ILLUSTRATIVE EXAMPLES

[NCERT Exercise 3.4]
(Page No. 56)

Example 1. *Solve the following pair of linear equations by the elimination method and the substitution method:*

(i) $x + y = 5$ and $2x - 3y = 4$
(ii) $3x + 4y = 10$ and $2x - 2y = 2$

(iii) $3x - 5y - 4 = 0$ and $9x = 2y + 7$

(iv) $\dfrac{x}{2} + \dfrac{2y}{3} = -1$ and $x - \dfrac{y}{3} = 3$.

Sol. (i) $x + y = 5$ and $2x - 3y = 4$

(I) **By elimination method**

The given system of equations is
$$x + y = 5 \quad \ldots(1)$$
$$2x - 3y = 4 \quad \ldots(2)$$

Multiplying equation (1) by 3, we get
$$3x + 3y = 15 \quad \ldots(3)$$

Adding equation (2) and equation (3), we get
$$5x = 19$$
$$x = \dfrac{19}{5}$$

Substituting this value of x in equation (1), we get
$$\dfrac{19}{5} + y = 5$$
$$\Rightarrow y = 5 - \dfrac{19}{5}$$
$$\Rightarrow y = \dfrac{6}{5}$$

So, the solution of the given system of equations is
$$x = \dfrac{19}{5}, y = \dfrac{6}{5}$$

(II) **By Substitution Method**

The given systems of equations is
$$x + y = 5 \quad \ldots(1)$$
$$2x - 3y = 4 \quad \ldots(2)$$

From equation (1),
$$y = 5 - x \quad \ldots(3)$$

Substitute this value of y in equation (2), we get
$$2x - 3(5 - x) = 4$$
$$\Rightarrow 2x - 15 + 3x = 4$$
$$\Rightarrow 5x - 15 = 4$$
$$\Rightarrow 5x = 15 + 4$$
$$\Rightarrow 5x = 19$$
$$\Rightarrow x = \dfrac{19}{5}$$

Substituting this value of x in equation (3), we get
$$y = 5 - \dfrac{19}{5} = \dfrac{6}{5}$$

So, the solution of the given system of equations is
$$x = \dfrac{19}{5}, y = \dfrac{6}{5}.$$

Verification. Substituting $x = \dfrac{19}{5}$, $y = \dfrac{6}{5}$, we find that both the equation (1) and (2) are satisfied shown below:
$$x + y = \dfrac{19}{5} + \dfrac{6}{5} = \dfrac{19 + 6}{5} = \dfrac{25}{5} = 5$$
$$2x - 3y = 2\left(\dfrac{19}{5}\right) - 3\left(\dfrac{6}{5}\right)$$
$$= \dfrac{38}{5} - \dfrac{18}{5} = \dfrac{38 - 18}{5} = \dfrac{20}{5} = 4$$

Hence the solution is correct.

(ii) $3x + 4y = 10$ and
$2x - 2y = 2$

(I) **By elimination method**

The given system of equation is
$$3x + 4y = 10 \quad \ldots(1)$$
$$2x - 2y = 2 \quad \ldots(2)$$

Multiplying equation (2), by 2, we get
$$4x - 4y = 4 \quad \ldots(3)$$

Adding equation (1) and equation (3), we get
$$7x = 14$$
$$\Rightarrow x = \dfrac{14}{7} = 2$$

Substituting this value of x in equation (2), we get
$$2(2) - 2y = 2$$
$$\Rightarrow 4 - 2y = 2$$
$$\Rightarrow 2y = 4 - 2$$
$$\Rightarrow 2y = 2$$
$$\Rightarrow y = \dfrac{2}{2} = 1$$

So, the solutions of the given system of equations is $x = 2, y = 1$.

(II) By substitution method

The given system of equations is

$3x + 4y = 10$ ...(1)
$2x - 2y = 2$ ...(2)

From equation (1),

$2y = 2x - 2$

$\Rightarrow y = \dfrac{2x - 2}{2}$

$\Rightarrow y = x - 1$ ...(3)

Substituting this value of y in equation (1), we get

$3x + 4(x - 1) = 10$
$\Rightarrow 3x + 4x - 4 = 10$
$\Rightarrow 7x - 4 = 10$
$\Rightarrow 7x = 10 + 4$
$\Rightarrow 7x = 14$
$\Rightarrow x = \dfrac{14}{7} = 2$

Substituting this value of x in equation (3), we get

$y = 2 - 1$
$\Rightarrow y = 1.$

So, the solution of the given system of equation is

$x = 2, y = 1$

Verification. Substituting $x = 2$, $y = 1$, we find that both the equations (1) and (2) are satisfied as shown below.

$3x + 4y = 3(2) + 4(1) = 6 + 4 = 0$
$2x + 2y = 2(2) - 2(1) = 4 - 2 = 2$

Hence, the solution is correct.

(iii) $3x - 5y - 4 = 0$ and $9x = 2y + 7$

(I) By elimination method

The given system of equations is

$3x - 5y - 4 = 0$ ...(1)
$9x = 2y + 7$
$\Rightarrow 9x - 2y - 7 = 0$ ...(2)

Multiplying equation (1) by 3, we get

$9x - 15y - 12 = 0$ ...(3)

Subtracting equation (3) from equation (2), we get

$13y + 5 = 0$
$\Rightarrow 13y = -5$
$\Rightarrow y = \dfrac{-5}{13}$

Substituting this value of y in equation (1), we get

$3x - 5\left(\dfrac{-5}{13}\right) - 4 = 0$

$\Rightarrow 3x + \dfrac{25}{13} - 4 = 0$

$\Rightarrow 3x - \dfrac{27}{13} = 0$

$\Rightarrow 3x = \dfrac{27}{13}$

$\Rightarrow x = \dfrac{9}{13}$

So, the solution of the given system of equations is $x = \dfrac{9}{13}, y = \dfrac{-5}{13}$.

(II) By substitution method

The given system of equations is

$3x - 5y - 4 = 0$ ...(1)
$9x = 2y + 7$ ...(2)

From equation (2),

$x = \dfrac{2y + 7}{9}$ ...(3)

Substitute this value of x in equation (1), we get

$3\left(\dfrac{2y + 7}{9}\right) - 5y - 4 = 0$

$\Rightarrow \dfrac{2y + 7}{3} - 5y - 4 = 0$

$\Rightarrow 2y + 7 - 15y - 12 = 0$

$\Rightarrow -13y - 5 = 0$

$\Rightarrow 13y = -5$

$\Rightarrow y = \dfrac{-5}{13}$

Substituting this value of y in equation (3), we get

$x = \dfrac{2\left(\dfrac{-5}{13}\right) + 7}{9}$

$= \dfrac{\dfrac{-10}{13} + 7}{9} = \dfrac{-10 + 91}{117}$

$$= \frac{81}{117} = \frac{9}{13}$$

So, the solution of the given system of equations is

$$x = \frac{9}{13}, y = \frac{-5}{13}$$

Verification. Substituting $x = \frac{9}{13}$, $y = \frac{-5}{13}$, we find that both the equations (1) and (2) are satisfied as shown below :

$$3x - 5y - 4 = 3\left(\frac{9}{13}\right) - 5\left(\frac{-5}{13}\right) - 4$$

$$= \frac{27}{13} + \frac{25}{13} - 4$$

$$= \frac{52}{13} - 4$$

$$= 4 - 4 = 0$$

$$2y + 7 = 2\left(\frac{-5}{13}\right) + 7$$

$$= \frac{-10}{13} + 7 = \frac{81}{13}$$

$$= 9\left(\frac{9}{13}\right) = 9y$$

Hence, the solution is correct.

(iv) $\frac{x}{2} + \frac{2y}{3} = -1$ and $x - \frac{y}{3} = 3$

(I) **By elimination method**

The given system of equations is

$$\frac{x}{2} + \frac{2y}{3} = -1 \qquad \ldots(1)$$

$$x - \frac{y}{3} = 3 \qquad \ldots(2)$$

Multiplying equation (2) by 2, we get

$$2x - \frac{2y}{3} = 6 \qquad \ldots(3)$$

Adding equation (1) and equation (2), we get

$$\frac{5}{2} x = 5$$

$$\Rightarrow \quad x = \frac{5 \times 2}{5}$$

$$\Rightarrow \quad x = 2$$

Substituting this value of x in equation (2), we get

$$2 - \frac{y}{3} = 3$$

$$\Rightarrow \quad \frac{y}{3} = 2 - 3 = -1$$

$$\Rightarrow \quad y = -3$$

So, the solution of the given system of equations is $x = 2, y = -3$.

(II) **By substitution method**

The given system of equations is

$$\frac{x}{2} + \frac{2y}{3} = -1 \qquad \ldots(1)$$

$$x - \frac{y}{3} = 3 \qquad \ldots(2)$$

From equation (2),

$$x = \frac{y}{3} + 3 \qquad \ldots(3)$$

Substitute this value of x in equation (1), we get

$$\frac{1}{2}\left(\frac{y}{3} + 3\right) + \frac{2y}{3} = -1$$

$$\Rightarrow \quad \frac{y}{6} + \frac{3}{2} + \frac{2y}{3} = -1$$

$$\Rightarrow \quad \frac{5y}{6} = -1 - \frac{3}{2}$$

$$\Rightarrow \quad \frac{5y}{6} = -\frac{5}{2}$$

$$\Rightarrow \quad y = -3$$

Substituting this value of y in equation (3), we get

$$x = -\frac{3}{3} + 3 = -1 + 3 = 2$$

So, the solution of the given system of equations

$$x = 2, y = -3.$$

Verification. Substituting $x = 2, y = -3$, we find that both the equations (1) and (2) are satisfied as shown below :

$$\frac{x}{2} + \frac{2y}{3} - 1 = \frac{2}{2} + \frac{2(-3)}{3} - 1$$
$$= 1 - 1 - 1 = -1$$
$$x - \frac{y}{3} = 2 - \frac{(-3)}{3} = 2 + 1 = 3$$

Hence, the solution is correct.

The substitution method was must efficient in this case.

Example 2. *Form the pair of linear equations in the following problems, and find their solutions (if they exist) by the elimination method:*

(i) If we add 1 to the numerator and subtract 1 from the denominator, a fraction reduces to 1. It becomes $\frac{1}{2}$ if we only add 1 to the denominator. What is the fraction?

(ii) Five years ago, Nuri was thrice as old as Sonu. Ten years later, Nuri will be twice as old as Sonu. How old are Nuri and Sonu?

(iii) The sum of the digits of a two-digit number is 9. Also, nine times this number is twice the number obtained by reversing the order of the number. Find the number.

(iv) Meena went to a bank to withdraw Rs. 2000. She asked the cashier to give her Rs. 50 and Rs. 100 notes only. Meena got 25 notes in all. Find how many notes of Rs. 50 and Rs. 100 she received.

(v) A lending library has a fixed charge for the first three days and an additional charge for each day thereafter. Saritha paid Rs. 27 for a book kept for seven days. While Susy paid Rs. 21 for the book she kept five days. Find the fixed charge and the charge for each extra day.

Sol. *(i)* Let the fraction be $\frac{x}{y}$.

Then, according to the question,

$$\frac{x+1}{y-1} = 1 \qquad \ldots(1)$$

$$\frac{x}{y+1} = \frac{1}{2} \qquad \ldots(2)$$
$$\Rightarrow \quad x + 1 = y - 1 \qquad \ldots(3)$$
$$2x = y + 1 \qquad \ldots(4)$$
$$\Rightarrow \quad x - y = -2 \qquad \ldots(5)$$
$$2x - y = 1 \qquad \ldots(6)$$

Substituting equation (5) from equation (6), we get

$$x = 3$$

Substituting this value of x in equation (5), we get

$$3 - y = -2$$
$$\Rightarrow \quad y = 3 + 2$$
$$\Rightarrow \quad y = 5$$

Hence, the required fraction is $\frac{3}{5}$.

Verification. Substituting the values of $x = 3$ and $y = 5$, we find that both the equations (1) and (2) are satisfied as shown below :

$$\frac{x+1}{y-1} = \frac{3+1}{5-1} = \frac{4}{4} = 1$$

$$\frac{x}{y+1} = \frac{3}{5+1} = \frac{3}{6} = \frac{1}{2}$$

Hence, the solution is correct.

(ii) Let Nuri and Sonu be x years and y years old respectively at present.

Then, according to the question,

$$x - 5 = 3(y - 5)$$
$$x + 10 = 2(y + 10)$$
$$\Rightarrow \quad x - 5 = 3y - 15$$
$$x + 10 = 2y + 20$$
$$\Rightarrow \quad x - 3y = -10 \qquad \ldots(1)$$
$$x - 2y = 10 \qquad \ldots(2)$$

Subtracting equation (2) from equation (1), we get

$$-y = -20$$
$$\Rightarrow \quad y = 20$$

Substituting this value of y in equation (2), we get

$$x - 2(20) = 10$$
$$\Rightarrow \quad x - 40 = 10$$
$$\Rightarrow \quad x = 40 + 10$$
$$\Rightarrow \quad x = 50$$

PAIR OF LINEAR EQUATIONS IN TWO VARIABLES

Hence, Nuri and Sonu are 50 years and 20 years old respectively at present.

Verification. Substituting the values of $x = 50$ and $y = 20$, we find that both the equations (1) and (2) are satisfied as shown below :
$$x - 3y = 50 - 3(20) = 50 - 60 = -10$$
$$x - 2y = 50 - 2(20) = 50 - 40 = 10$$

Hence, the solution is correct.

(iii) Let the unit's digit and the ten's digit in the two-digit number be x and y respectively.

Then, the number $= 10y + x$

Also, the number obtained by reversing the order of the digits $= 10x + y$

According to the question,
$$x + y = 9 \qquad ...(1)$$
$$9(10y + x) = 2(10x + y)$$
$$\Rightarrow \quad 90y + 9x = 20x + 2y$$
$$\Rightarrow \quad 11x - 88y = 0$$
$$\Rightarrow \quad x - 8y = 0 \qquad ...(2)$$

Subtracting equation (2) from equation (1), we get
$$9y = 9$$
$$\Rightarrow \quad y = \frac{9}{9} = 1$$

Substituting this value of y in equation (1), we get
$$x + 1 = 9$$
$$\Rightarrow \quad x = 9 - 1 = 8$$

Hence, the required number is 18.

Verification. Substituting $x = 8$ and $y = 1$, we find that both the equations (1) and (2) are satisfied as shown below.
$$x + y = 8 + 1 = 9$$
$$x - 8y = 8 - 8(1) = 0$$

Hence, the solution is correct.

(iv) Suppose that Meena received x notes of Rs. 50 and y notes Rs. 100.

Then, according to the question,
$$x + y = 25 \qquad ...(1)$$
$$50x + 100y = 2000$$
$$\Rightarrow \quad x + 2y = 40 \qquad ...(2)$$

Subtracting equation (1) from equation (2), we get
$$y = 15$$

Substituting this value of y in equation (1), we get
$$x + 15 = 25$$
$$\Rightarrow \quad x = 25 - 15 = 10$$

Hence, Meena received 10 notes of Rs. 50 and 15 notes of Rs. 100.

Verification. Substituting $x = 10$ and $y = 15$, we find that both the equations (1) and (2) are satisfied as shown below :
$$x + y = 10 + 15 = 25$$
$$x + 2y = 10 + 2(15)$$
$$= 10 + 30 = 40$$

Hence, the solution is correct.

(v) Let the fixed charge be Rs. a and the charge for each extra day be Rs. b.

Then, according to the question,
$$a + 4b = 27 \qquad ...(1)$$
| Extra days $= 7 - 3 = 4$
$$a + 2b = 21 \qquad ...(2)$$
| Extra days $= 5 - 3 = 2$

Subtracting equation (2) from equation (1), we get
$$2b = 6$$
$$\Rightarrow \quad b = \frac{6}{2} = 3$$

Substituting this value of b in equation (2), we get
$$a + 2(3) = 21$$
$$\Rightarrow \quad a + 6 = 21$$
$$\Rightarrow \quad a = 21 - 6 = 15$$

Hence, the fixed charges one Rs. 15 and the charge for each extra day is Rs. 3.

Verification. Substituting $a = 15$ and $b = 3$ we, find that both the equation. (1) and (2) are satisfied as shown below.
$$a + 4b = 15 + 4(3) = 15 + 12 = 27$$
$$a + 2b = 15 + 2(3) = 15 + 6 = 21$$

Hence, the solution is correct.

ADDITIONAL EXAMPLES

Example 3. *Solve* $4x + \dfrac{6}{y} = 15$ *and* $6x - \dfrac{8}{y} = 14$ *by eliminating method and hence find 'p' if* $y = px - 2$.

Sol. The given system of equations is

$$4x + \frac{6}{y} = 15 \quad \ldots(1)$$

$$6x - \frac{8}{y} = 14 \quad \ldots(2)$$

Multiplying equation (1) by 3 and equation (2) by 2, we get

$$12x + \frac{18}{y} = 45 \quad \ldots(3)$$

$$12x - \frac{16}{y} = 28 \quad \ldots(4)$$

Subtracting equation (4) from equation (3), we get

$$\frac{34}{y} = 17$$

$$\Rightarrow \quad y = \frac{34}{17} = 2$$

Substituting this value of y in equation (1), we get

$$4x + \frac{6}{2} = 15$$

$$\Rightarrow \quad 4x + 3 = 15$$

$$\Rightarrow \quad 4x = 15 - 3$$

$$\Rightarrow \quad 4x = 12$$

$$\Rightarrow \quad x = \frac{12}{4}$$

$$\Rightarrow \quad x = 3$$

So, the solution of the given system of equations is $x = 3, y = 2$.

Verification. Substituting $x = 3, y = 2$, we find that both the equations (1) and (2) are satisfied as shown below :

$$4x + \frac{6}{y} = 4(3) + \frac{6}{2} = 12 + 3 = 15$$

$$6x - \frac{8}{y} = 6(3) - \frac{8}{2} = 18 - 4 = 14$$

Hence, the solution is correct.

Now, $y = px - 2$

$\Rightarrow \quad 2 = 3p - 2$

$\Rightarrow \quad 3p = 4$

$\Rightarrow \quad p = \dfrac{4}{3}$.

Example 4. Solve the following system of equation by elimination method

$5x + 3y = 70$

$3x - 7y = 60$.

Sol. The given system of equations is

$5x + 3y = 70 \quad \ldots(1)$

$3x - 7y = 60 \quad \ldots(2)$

Multiplying equation (1) by 3 and equation (2) by 5, we get

$15x + 9y = 210 \quad \ldots(3)$

$15x - 35y = 300 \quad \ldots(4)$

Subtracting equation (4) from equation (3), we get

$44y = -90$

$\Rightarrow \quad y = \dfrac{-90}{44} = \dfrac{-45}{22}$

Substituting this value of y in equation (1), we get

$$5x + 3\left(\frac{-45}{22}\right) = 70$$

$$\Rightarrow \quad 5x - \frac{135}{22} = 70$$

$$\Rightarrow \quad 5x = 70 + \frac{135}{22}$$

$$\Rightarrow \quad 5x = \frac{1540 + 135}{22} = \frac{1675}{22}$$

$$\Rightarrow \quad x = \frac{335}{22}$$

So, the solution of the given system of equations is

$$x = \frac{335}{22}, y = \frac{-45}{22}.$$

Verification. Substituting $x = \dfrac{335}{22}$, $y = \dfrac{-45}{22}$, we find that both the equations (1) and (2) are satisfied as shown below.

$$5x + 3y = 5\left(\dfrac{335}{22}\right) + 3\left(\dfrac{-45}{22}\right)$$

$$= \dfrac{1675}{22} - \dfrac{135}{22}$$

$$= \dfrac{1675 - 135}{22} = \dfrac{1540}{22} = 70$$

$$3x - 7y = 3\left(\dfrac{335}{22}\right) - 7\left(\dfrac{-45}{22}\right)$$

$$= \dfrac{1005}{22} + \dfrac{315}{22}$$

$$= \dfrac{1005 + 315}{22} = \dfrac{1320}{20} = 60.$$

Hence, the solution is correct.

Example 5. *Solve the following system of equations by elimination method :*

$$\dfrac{x}{2} - \dfrac{y}{5} = 4$$

$$\dfrac{x}{7} + \dfrac{y}{15} = 3.$$

Sol. The given system of equations is

$$\dfrac{x}{2} - \dfrac{y}{5} = 4 \qquad \ldots(1)$$

$$\dfrac{x}{7} + \dfrac{y}{15} = 3 \qquad \ldots(2)$$

Multiplying equation (2) by 3, we get

$$\dfrac{3x}{7} + \dfrac{y}{5} = 9 \qquad \ldots(3)$$

Adding equation (1) and equation (3), we get

$$\dfrac{x}{2} + \dfrac{3x}{7} = 13$$

$$\Rightarrow \qquad \dfrac{13}{14} x = 13$$

$$\Rightarrow \qquad x = \dfrac{13 \times 14}{13} = 14$$

Substituting this value of x in equation (2), we get

$$\dfrac{14}{7} + \dfrac{y}{15} = 3$$

$$\Rightarrow \qquad 2 + \dfrac{y}{15} = 3 \Rightarrow \dfrac{y}{15} = 3 - 2$$

$$\Rightarrow \qquad \dfrac{y}{15} = 1$$

$$\Rightarrow \qquad y = 15$$

So, the solution of the given system of equations is

$$x = 14, y = 15$$

Verification. Substituting $x = 14$, $y = 15$, we find that both the equations (1) and (2) are satisfied as shown below :

$$\dfrac{x}{2} - \dfrac{y}{5} = \dfrac{14}{2} - \dfrac{15}{5} = 7 - 3 = 4$$

$$\dfrac{x}{7} + \dfrac{y}{15} = \dfrac{14}{7} + \dfrac{15}{15} = 2 + 1 = 3$$

Hence, the solution is correct.

Example 6. *90% and 97% pure acid solutions and mixed to obtain 21 litres of 95% pure acid solution. Find the amount of each type of acid to be mixed to form the mixture.*

Sol. Let x litres of 90% pure acid solution and y litres of 97% pure acid solution be mixed. Then, total volume of the mixture

$$= (x + y) \text{ litres}$$

According to the question,

$$x + y = 21 \qquad \ldots(1)$$

and, 90% of x + 97% of y = 95% of 21

$$\Rightarrow \qquad \dfrac{90}{100} x + \dfrac{97}{100} y = \dfrac{95}{100} \times 21$$

$$\Rightarrow \qquad 90x + 97y = 1995 \qquad \ldots(2)$$

Multiplying equation (1) by 90, we get

$$90x + 90y = 1890 \qquad \ldots(3)$$

Subtracting equation (3) from equation (2), we get

$$7y = 105$$

$$\Rightarrow \qquad y = \dfrac{105}{7} = 15$$

Substituting this value of y in equation (1), we get

$$x + 15 = 21$$
$$\Rightarrow \quad x = 21 - 15 = 6$$
$\therefore$ Amount of 90% pure acid solution
$$= 6 \text{ litres}$$
and, amount of 97% pure acid solution
$$= 15 \text{ litres}.$$

Verification. Substituting $x = 6, y = 15$, we find that both the equations (1) and (2) are satisfied as shown below :

$$x + y = 6 + 15 = 21$$
$$90x + 97y = (90)(6) + (97)(15)$$
$$= 540 + 1455$$
$$= 1995$$

Hence, the solution is correct.

TEST YOUR KNOWLEDGE

1. The ratio of income of two persons is 9 : 7 and the ratio of their expenditure is 4 : 3. If each of them on equals to save Rs. 2000 per month, find their monthly income.

2. Use elimination method to find all possible solutions of the following linear pair :
   $$2x + 3y = 8$$
   $$4x + 6y = 7.$$

3. The sum of a two-digit number and the number obtained by reversing the digits is 66. If the digits of the number differ by 2, find the number. How many such numbers are there ?

4. Solve by the method of elimination, the system of equations
   $$3x - 4y = 1$$
   $$4x - 3y = 6.$$

5. Solve the following system of equations,
   $$2x + 3y = 0$$
   $$3x + 4y = 5.$$

6. Solve the following system of equations using the elimination method by equating co-efficients :
   $$11x - 5y + 61 = 0$$
   $$3x - 20y - 2 = 0.$$

7. Solve the following system of linear equations using elimination method :
   $$3x + 2y = 14$$
   $$-x + 4y = 7.$$

8. Solve the following system of equations by elimination method :
   $$-6x + 5y = 2$$
   $$-5x + 6y = 9.$$

9. Solve the following system of equations using elimination method :
   $$3x - 5y = 19$$
   $$3y - 3x = -1.$$

10. Solve the following system of equations using elimination method :
    $$7x - 2y = 1$$
    $$3x + 4y = 15.$$

11. In a two-digit number, the sum of the digits is 9. If the digits are reversed, the number is increased by 9. Find the number.

12. 2 tables and 3 chairs together cost Rs. 2000 whereas 3 tables and 2 chairs together cost Rs. 2500. Find the total cost of 1 table and 5 chairs.

13. A man has only 20 paise coins and 25 paise coins in his pure. If he has 50 coins in all totalling Rs. 11.25, how many coins of each type does he have ?

14. 10 years ago father was 12 times as old as his son and 10 years hence he will he twice as old as his son. Find their present ages.

15. Two numbers are in the ratio 2 : 3. If 5 is added to each number, the ratio becomes 5 : 7. Find the numbers.

16. The sum of the numerator and denominator of a fraction is 8 of 3 is added to both the numerator and the denominator, the fraction becomes $\frac{3}{4}$. Find the fraction.

    (*CBSE 2003*)

17. Scooter charges consist of fixed charges and remaining depending upon the distance

travelled in kilometres. If a person travels 12 km, he pays Rs. 45 and for travelling 20 km, he pay Rs. 13. Express the above statements in the form of simultaneous equations and hence pure the fixed charges and the rate per km. (CBSE 2000)

Answers

1. Rs. 18000 and Rs. 14000
2. The pair of equations has no solution
3. 42 or 24
4. $x = 3, y = 2$
5. $x = 15, y = -10$
6. $x = -6, y = -1$
7. $x = 3, y = \dfrac{5}{2}$
8. $x = 3, y = 4$
9. $x = 1, y = -5$
10. $x = 1, y = 3$
11. 45
12. Rs. 1700
13. 25, 25
14. 34 years, 12 years
15. 20, 30
16. $\dfrac{3}{5}$
17. $x + 12y = 45, x + 20y = 73$; Rs. 3.00; Rs. 3.50.

CROSS MULTIPLICATION METHOD

IMPORTANT POINTS

1. Cross-multiplication method : It is another algebraic method to solve a pair of linear equations. For many reasons, it is a very useful method of solving these equations.

2. Working of cross-multiplication method : Consider any pair of linear equations in two variables of the form

$$a_1 x + b_1 y + c_1 = 0 \quad \ldots(1)$$
and
$$a_2 x + b_2 y + c_2 = 0 \quad \ldots(2)$$

To obtain the values of x and y as shown above we follow the following steps :

Step 1 : Multiply Equation (1) by b_2 and Equation (2) by b_1, to get

i.e.,
$$b_2 a_1 x + b_2 b_1 y + b_2 c_1 = 0 \quad \ldots(3)$$
$$b_1 a_2 x + b_1 b_2 y + b_1 c_2 = 0 \quad \ldots(4)$$

Subtracting Equation (4) from (3), we get :

$$(b_2 a_1 - b_1 a_2) x + (b_2 b_1 - b_1 b_2) y + (b_2 c_1 - b_1 c_2) = 0$$

i.e., $(b_2 a_1 - b_1 a_2) x = b_1 c_2 - b_2 c_1$

So, $$x = \dfrac{b_1 c_2 - b_2 c_1}{a_1 b_2 - a_2 b_1}$$

Substituting this value of x in (1) or (2), we get

$$y = \dfrac{c_1 a_2 - c_2 a_1}{a_1 b_2 - a_2 b_1}$$

Now two possibilities arise for the value of $a_1 b_2 - a_2 b_1$.

Case 1 : $a_1 b_2 - a_2 b_1 \neq 0$. In this case $\dfrac{a_1}{a_2} \neq \dfrac{b_1}{b_2}$. Then the pair of linear equations has a unique solution.

Case 2 : $a_1 b_2 - a_2 b_1 = 0$. If we write $\dfrac{a_1}{a_2} = \dfrac{b_1}{b_2} = k$, then $a_1 = k\, a_2$, $b_1 = k b_2$.

Substituting the values of a_1 and b_1 in the equation $a_1 x + b_1 y + c_1 = 0$, we get

$$k(a_2 x + b_2 y) + c_1 = 0.$$

It can be observed that the equations $k(a_2 x + b_2 y) + c_1 = 0$ and $a_2 x + b_2 y + c_2 = 0$ can both be satisfied only if $c_1 = k c_2$.

If $c_1 = k c_2$ any solution of $a_2 x + b_2 y + c_2 = 0$ will satisfy the equation $a_1 x + b_1 y + c_1 = 0$, and vice versa. So, if $\dfrac{a_1}{a_2} = \dfrac{b_1}{b_2} = \dfrac{c_1}{c_2} = k$, then there are infinitely many solutions to the pairs of linear equations given by (1) and (2).

If $c_1 \neq k c_2$, then the pair has no solution. So, if $\dfrac{a_1}{a_2} = \dfrac{b_1}{b_2} = k$, and $c_1 \neq k c_2$, then the pair of linear equations has no solution.

3. Summary : We can summarize the consistency of the pair of linear equations given by (1) and (2) as follows :

(i) When $\dfrac{a_1}{a_2} \neq \dfrac{b_1}{b_2}$, we get a unique solution (consistent).

(ii) When $\dfrac{a_1}{a_2} = \dfrac{b_1}{b_2} = \dfrac{c_1}{c_2}$, there are infinitely many solutions (consistent).

(iii) When $\dfrac{a_1}{a_2} = \dfrac{b_1}{b_2} \neq \dfrac{c_1}{c_2}$, there is no solution (inconsistent).

4. An aid to memory. The following diagram help us in remembering, and writing the solution in Case 1 given by Equations (5) and (6) in the following form :

$$\dfrac{x}{b_1 c_2 - b_2 c_1} = \dfrac{y}{c_1 a_2 - c_2 a_1} = \dfrac{1}{a_1 b_2 - a_2 b_1}$$

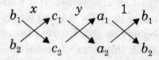

The arrows between the two numbers indicate that they are to be multiplied and the second product is to be subtracted from the first.

This also gives us an indication why this method is called the cross-multiplication method.

ILLUSTRATIVE EXAMPLES

[NCERT Exercise 3.5]
(Page No. 62)

Example 1. *Which of the following pairs of linear equations has unique solution, no solution, or infinitely many solutions. In case there is a unique solution, find it by using cross multiplication method.*

(i) $x - 3y - 3 = 0$
$3x - 9y - 2 = 0$
(ii) $2x + y = 5$
$3x + 2y = 8$
(iii) $3x - 5y = 20$
$6x - 10y = 40$
(iv) $x - 3y - 7 = 0$
$3x - 3y - 15 = 0$

Sol.
(i) $\mathbf{x - 3y - 3 = 0}$
$\mathbf{3x - 9y - 2 = 0}$

The given pair of linear equations is
$x - 3y - 3 = 0$
$3x - 9y - 2 = 0$
Here, $a_1 = 1, b_1 = -3, c_1 = -3$
$a_2 = 3, b_2 = -9, c_2 = -2$
We see that
$$\dfrac{a_1}{a_2} = \dfrac{b_1}{b_2} \neq \dfrac{c_1}{c_2}$$

Hence, the given pair of linear equations has no solution.

(ii) $\mathbf{2x + y = 5}$
$\mathbf{3x + 2y = 8}$

The given pair of linear equations is
$2x + y = 5$
$3x + 2y = 8$
$\Rightarrow 2x + y - 5 = 0$
$3x + 2y - 8 = 0$
Here, $a_1 = 2, b_1 = 1, c_1 = -5$
$a_2 = 3, b_2 = 2, c_2 = -8$
We see that
$$\dfrac{a_1}{a_2} \neq \dfrac{b_1}{b_2}$$

Hence, the given pair of linear equations has a unique solution.

To solve the given equation by cross multiplication method, we draw the diagram below :

PAIR OF LINEAR EQUATIONS IN TWO VARIABLES

Then,
$$\frac{x}{(1)(-8)-(2)(-5)} = \frac{y}{(-5)(3)-(-8)(2)}$$
$$= \frac{1}{(2)(2)-(3)(1)}$$

$\Rightarrow \quad \dfrac{x}{-8+10} = \dfrac{y}{-15+16} = \dfrac{1}{4-3}$

$\Rightarrow \quad \dfrac{x}{2} = \dfrac{y}{1} = \dfrac{1}{1}$

$\Rightarrow \quad x = 2, y = 1$

Hence, the required solution of the given pair of linear equation is $x = 2, y = 1$.

Verification. Substituting $x = 2, y = 1$, we find that both the equations (1) and (2) are satisfied as shown below :

$2x + y = 2(2) + 1 = 4 + 1 = 5$

$3x + 2y = 3(2) + 2(1) = 6 + 2 = 8$

Hence, the solution is correct.

(iii) **3x – 5y = 20**
   6x – 10y = 40

The given pair of linear equations is
$3x - 5y = 20$
$6x - 10y = 40$

$\Rightarrow \quad 3x - 5y - 20 = 0$
$\quad 6x - 10y - 40 = 0$

Here, $a_1 = 3, b_1 = -5, c_1 = -20$
$a_2 = 6, b_2 = -10, c_2 = -40$

We see that
$$\frac{a_1}{a_2} = \frac{b_1}{b_2} = \frac{c_1}{c_2}$$

Hence, the given pair of linear equations has infinitely many solution.

(iv) **x – 3y – 7 = 0**
   3x – 3y – 15 = 0

The given pair of linear equations is
$x - 3y - 7 = 0$
$3x - 3y - 15 = 0$

Here, $a_1 = 1, b_1 = -3, c_1 = -7$
$a_2 = 3, b_2 = -3, c_2 = -15$

We see that
$$\frac{a_1}{a_2} \ne \frac{b_1}{b_2}$$

Hence, the given pair of linear equations has a unique solution.

To solve the given equations by cross multiplication method, we draw the diagram below :

$$\begin{matrix} & x & & y & & 1 & \\ -3 & & -7 & & 1 & & -3 \\ -3 & & -15 & & 3 & & -3 \end{matrix}$$

Then,
$$\frac{x}{(-3)(-15)-(-3)(-3)}$$
$$= \frac{y}{(-3)(3)-(-15)(1)}$$
$$= \frac{1}{(1)(-3)-(3)(-3)}$$

$\Rightarrow \quad \dfrac{x}{45-21} = \dfrac{y}{-21+15} = \dfrac{1}{-3+9}$

$\Rightarrow \quad \dfrac{x}{24} = \dfrac{y}{-6} = \dfrac{1}{6}$

$\Rightarrow \quad x = \dfrac{24}{6} = 4, y = -\dfrac{6}{6} = -1$

Hence, the required method of the given pair of linear equation is $x = 4, y = -1$.

Verification. Substituting $x = 4, y = -1$, we find that both the equations (1) and (2) are satisfied as shown below :

$x - 3y - 7 = 4 - 3(-1) - 7 = 0$

$3x - 3y - 15 = 3(4) - 3(-1) - 15 = 0$

Hence, the solution is correct.

Example 2. (i) *For which values of a and b does the following pair of linear equations have an infinite number of solutions ?*

$2x + 3y = 7$

$(a - b)x + (a + b)y = 3a + b - 2$

(ii) *For which value of k will the following pair of linear equations have no solution ?*

$3x + y = 1$

$(2k - 1)x + (k - 1)y = 2k + 1$

Sol. (i) The given pair of linear equations is

$2x + 3y = 7$

$(a - b)x + (a + b)y = 3a + b - 2$

Here, $a_1 = 2, b_1 = 3, c_1 = 7$
$a_2 = a - b, b_2 = a + b,$
$c_2 = 3a + b - 2$

For having an infinite number of solutions, we must have

$$\frac{a_1}{a_2} = \frac{b_1}{b_2} = \frac{c_1}{c_2}$$

$\Rightarrow \quad \dfrac{2}{a-b} = \dfrac{3}{a+b} = \dfrac{7}{3a+b-2}$

From first two,

$\dfrac{2}{a-b} = \dfrac{3}{a+b}$

$\Rightarrow \quad 2(a+b) = 3(a-b)$
$\Rightarrow \quad 2a + 2b = 3a - 3b$
$\Rightarrow \quad a - 5b = 0 \qquad \qquad ...(1)$

For last two,

$\dfrac{3}{a+b} = \dfrac{7}{3a+b-2}$

$\Rightarrow \quad 3(3a+b-2) = 7(a+b)$
$\Rightarrow \quad 9a + 3b - 6 = 7a + 7b$
$\Rightarrow \quad 2a - 4b - 6 = 0$
$\Rightarrow \quad a - 2b - 3 = 0$
| dividing throughout by 2 ...(2)

To solve the equations (1) and (2) by cross-multiplication method, we draw the diagram below :

$$\begin{array}{cccc} -5 & a & 0 & b & 1 & 1 & -5 \\ -2 & & -3 & & 1 & & -2 \end{array}$$

Then,

$\dfrac{a}{(-5)(-3)-(-2)(0)} = \dfrac{b}{(0)(1)-(-3)(1)}$

$= \dfrac{1}{(1)(-2)-(1)(-5)}$

$\Rightarrow \quad \dfrac{a}{15} = \dfrac{b}{3} = \dfrac{1}{3}$

$\Rightarrow \quad a = \dfrac{15}{3} = 5$

and $\quad b = \dfrac{3}{3} = 1$

Hence, the required values of a and b are 5 and 1 respectively.

(ii) The given pair of linear equations is

$3x + y = 1$
$(2k-1)x + (k-1)y = 2k+1$
$\Rightarrow \quad 3x + y - 1 = 0$

Here, $(2k-1)x + (k-1)y - (2k+1) = 0$
$a_1 = 3, b_1 = 1, c_1 = -1$
$a_2 = 2k-1, b_2 = k-1,$
$c_2 = -(2k+1)$

For having no solution, we must have

$$\frac{a_1}{a_2} = \frac{b_1}{b_2} \neq \frac{c_1}{c_2}$$

$\Rightarrow \quad \dfrac{3}{2k-1} = \dfrac{1}{k-1} \neq \dfrac{-1}{-(2k+1)}$

From above we have

$\dfrac{3}{2k-1} = \dfrac{1}{k-1}$

$\Rightarrow \quad 3(k-1) = 2k - 1$
$\Rightarrow \quad 3k - 3 = 2k - 1$
$\Rightarrow \quad 3k - 2k = 3 - 1$
$\Rightarrow \quad k = 2$

Hence, the required value of k is 2.

Example 3. *Solve the following pair of linear equations by the substitution and cross-multiplication methods.*

$8x + 5y = 9$
$3x + 2y = 4.$

Sol. The given pair of linear equations is

$8x + 5y = 9 \qquad \qquad ...(1)$
$3x + 2y = 4 \qquad \qquad ...(2)$

(I) By substitution method

From equation (2),

$2y = 4 - 3x$

$\Rightarrow \quad y = \dfrac{4-3x}{2} \qquad \qquad ...(3)$

Substitute this value of y in equation (1), we get

$8x + 5\left(\dfrac{4-3x}{2}\right) = 9$

PAIR OF LINEAR EQUATIONS IN TWO VARIABLES

$\Rightarrow \quad 16x + 20 - 15x = 18$

$\Rightarrow \quad x + 20 = 18$

$\Rightarrow \quad x = 18 - 20$

$\Rightarrow \quad x = -2$

Substituting this value of x in equation (3), we get

$$y = \frac{4 - 3(-2)}{2} = \frac{4+6}{2} = \frac{10}{2} = 5$$

So the solution of the given pair of linear equations is $x = -2, y = 5$.

(II) By cross-multiplication method

Let us write the given pair of linear equations is

$8x + 5y - 9 = 0$ ...(1)

$3x + 2y - 4 = 0$ ...(2)

To solve the equations (1) and (2) by cross multiplication method, we draw the diagram below :

$$\frac{x}{(5)(-4) - (2)(-9)} = \frac{y}{(-9)(3) - (-4)(8)}$$

$$= \frac{1}{(8)(2) - (3)(5)}$$

$$= \frac{x}{-20 + 18} = \frac{y}{-27 + 32} = \frac{1}{16 - 15}$$

$\Rightarrow \quad \dfrac{x}{-2} = \dfrac{y}{5} = \dfrac{1}{1}$

$\Rightarrow \quad x = -2$ and $y = 5$

Hence, the required solution of the given pair of linear equations is

$x = -2, y = 5$.

Verification. Substituting $x = -2$, $y = 5$, we find that both the equations (1) and (2) are satisfied as shown below :

$8x + 5y = 8(-2) + 5(5) = -16 + 25 = 9$

$3x + 2y = 3(-2) + 2(5) = -6 + 10 = 4$

Hence, the solution is correct.

Example 4. *Form the pair of linear equations in the following problems and find their solutions (if they exist) by any algebraic method:*

(i) A part of monthly hostel charges is fixed and the remaining depends on the number of days one has taken food in the mess. When a student A takes food for 20 days she has to pay Rs 1000 as hostel charges whereas a student B, who takes food for 26 days, pay's Rs 1180 as hostel charges. Find the fixed charge and the cost of food per day.

(ii) A fraction becomes $\dfrac{1}{3}$ when 1 is subtracted from the numerator and it becomes $\dfrac{1}{4}$ when 8 is added to its denominator. Find the fraction.

(iii) Yash scored 40 marks in a test, getting 3 marks for each right answer and losing 1 mark for each wrong answer. Had 4 marks been awarded for each correct answer and 2 marks been deducted for each incorrect answer, then yash would have scored 50 marks. How many questions were there in the test ?

(iv) Places A and B are 100 km apart on a highway. One car starts from A and another from B at the same time. If the cars travel in the same direction at a different speeds, they meet in 5 hours. If they travel towards each other, they meet in 1 hour. What are the speeds of the two cars ?

(v) The area of a rectangle gets reduced by 9 square units if its length is reduced by 5 units and breadth is increased by 3 units. If we increase the length by 3 units and the breadth by 2 units, the area increases by 67 square units. Find the dimensions of the rectangle.

Sol. (*i*) Let the fixed charge be Rs x and the cost of food per day be Rs y.

Then, according to the question,

$x + 20y = 1000$ ...(1)

$x + 26y = 1180$ ...(2)

$\Rightarrow \quad x + 20y - 1000 = 0$ ...(3)

$x + 26y - 1180 = 0$ ...(4)

To solve the equations (3) and (4) by cross-multiplication method, we draw the diagonal below :

$$\begin{array}{cccccc} 20 & x & -1000 & y & 1 & 1 & 20 \\ 26 & & -1180 & & 1 & & 26 \end{array}$$

Then,

$$\frac{x}{(20)(-1180)-(26)(-1000)}$$

$$= \frac{y}{(-1000)(1)-(-1180)(1)}$$

$$= \frac{1}{(1)(26)-(1)(20)}$$

$$= \frac{x}{-23600+26000}$$

$$= \frac{y}{-1000+1180} = \frac{1}{26-20}$$

$$\Rightarrow \frac{x}{2400} = \frac{y}{180} = \frac{1}{6}$$

$$\Rightarrow x = \frac{2400}{6} = 400$$

and $y = \frac{180}{6} = 30$

Hence, the fixed charges and Rs 400 and the cost of food per day is Rs 30.

Verification. Substituting $x = 400$, $y = 30$, we find that both the equations (1) and (2) are satisfied as shown below :

$x + 20y = 400 + 20(30)$
$= 400 + 600 = 1000$
$x + 26y = 400 + 26(30)$
$= 400 + 780 = 1180$

Hence, the solution we have got is correct.

(ii) Let the fraction be $\frac{x}{y}$.

Then, according to the question,

$$\frac{x-1}{y} = \frac{1}{3} \qquad ...(1)$$

$$\frac{x}{y+8} = \frac{1}{4} \qquad ...(2)$$

$\Rightarrow 3(x-1) = y \qquad ...(3)$
$4x = y + 8 \qquad ...(4)$
$\Rightarrow 3x - y - 3 = 0 \qquad ...(5)$
$4x - y - 8 = 0 \qquad ...(6)$

To solve the equations (5) and (6) by cross-multiplication method, we draw the diagram below :

$$\begin{array}{cccccc} -1 & x & -3 & y & 3 & 1 & -1 \\ -1 & & -8 & & 4 & & -1 \end{array}$$

Then,

$$\frac{x}{(-1)(-8)-(-1)(-3)}$$

$$= \frac{y}{(-3)(4)-(-8)(3)}$$

$$= \frac{1}{(3)(-1)-(4)(-1)}$$

$$\Rightarrow \frac{x}{8-3} = \frac{y}{-12+24} = \frac{1}{-3+4}$$

$$\Rightarrow \frac{x}{5} = \frac{y}{12} = \frac{1}{1}$$

$$\Rightarrow x = 5 \text{ and } y = 12$$

Hence, the required fraction is $\frac{5}{12}$.

Verification. Substituting $x = 5, y = 12$, we find that both the equations (1) and (2) are satisfied as shown below :

$$\frac{x-1}{y} = \frac{5-1}{12} = \frac{4}{12} = \frac{1}{3}$$

$$\frac{x}{y+8} = \frac{5}{12+8} = \frac{5}{20} = \frac{1}{4}$$

Hence, the solution we have got is correct.

(iii) Suppose that yash gave right answers to x questions and wrong answers to y questions. Then, the total number of questions in the test $= x + y$. Also, according to the question.

PAIR OF LINEAR EQUATIONS IN TWO VARIABLES

$$3x - y = 40 \qquad \ldots(1)$$
$$4x - 2y = 50 \qquad \ldots(2)$$
$$\Rightarrow \quad 3x - y - 40 = 0 \qquad \ldots(3)$$
$$4x - 2y - 50 = 0 \qquad \ldots(4)$$

To solve the equations (3) and (4) by cross-multiplication method, we draw the diagram below :

$$\begin{array}{ccccc} -1 & x & -40 & y & 3 & 1 & -1 \\ & \times & & \times & & \times & \\ -2 & & -50 & & 4 & & -2 \end{array}$$

Then,

$$\Rightarrow \quad \frac{x}{(-1)(-50)-(-2)(-40)}$$

$$= \frac{y}{(-40)(4)-(-50)(3)}$$

$$= \frac{1}{(3)(-2)-(4)(-1)}$$

$$\Rightarrow \quad \frac{x}{50-80} = \frac{y}{-160+150} = \frac{1}{-6+4}$$

$$\Rightarrow \quad \frac{x}{-30} = \frac{y}{-10} = \frac{1}{-2}$$

$$\Rightarrow \quad x = \frac{-30}{-2} = 15$$

and $\quad y = \dfrac{-10}{-2} = 5$

Hence, yash gave right answers to 15 questions and wrong answers to 5 questions.

Therefore, total number of questions in the test $= x + y = 15 + 5 = 20$.

Verification. Substituting $x = 15$, $y = 5$, we find that both the equations (1) and (2) are satisfied as shown below :

$$3x - y = 3(15) - 5 = 45 - 5 = 40$$
$$4x - 2y = 4(15) - 5$$
$$= 60 - 10 = 50$$

Hence, the solution we have got is correct.

(iv) Let the speeds of two cars be x km/hour and y km/hour respectively.

Case I. When the cars travel in the same direction

Let the meet at P.

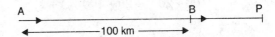

Distance travelled by the car starting from A in 5 hours = AP = $5x$ km

| Dividing = speed × time

Distance travelled by the car starting from B in 5 hours = BP = $5y$ km

| Distance = speed × time

Now, AP − BP = 100
$$\Rightarrow \quad 5x - 5y = 100$$
$$\Rightarrow \quad x - y = 20 \qquad \ldots(1)$$

| Dividing throughout by 5

Case II. When the cars travel towards each other

Let then meet at Q.

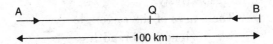

Distance travelled by the car starting from A in 1 hour = AQ = x km

Distance travelled by the car starting from B in 1 hour

$$= BQ = y \text{ km}$$

Now, AQ + BQ = 100
$$\Rightarrow \quad x + y = 100 \qquad \ldots(2)$$

Equation (1) and (2) can be re-written as
$$x - y - 20 = 0 \qquad \ldots(3)$$
$$x + y - 100 = 0 \qquad \ldots(4)$$

To solve the equations (3) and (4) by cross-multiplication method, we draw the diagram below :

$$\begin{array}{ccccc} -1 & x & -20 & y & 1 & 1 & -1 \\ & \times & & \times & & \times & \\ 1 & & -100 & & 1 & & 1 \end{array}$$

Then,

$$\frac{x}{(-1)(-100)-(1)(-20)}$$

$$= \frac{y}{(-20)(1)-(-100)(1)}$$

$$= \frac{1}{(1)(1)-(1)(-1)}$$

$$\Rightarrow \frac{x}{100+20} = \frac{y}{-20+100} = \frac{1}{1+1}$$

$$= \frac{x}{120} = \frac{y}{80} = \frac{1}{2}$$

$$\Rightarrow x = \frac{120}{2} = 60 \text{ and } y = \frac{80}{2} = 40$$

Hence, the speeds of the two cars are 60 km/hour and 40 km/hour respectively.

Verification. Substituting $x = 60$, $y = 40$, we find that both the equations (1) and (2) are satisfied as shown below :

$$x - y = 60 - 40 = 20$$
$$x + y = 60 + 40 = 100$$

Hence, the solution we have got is correct.

(v) Let the dimensions (i.e., the length and the breath) of the rectangle be x units and y units respectively.

Then, area of the rectangle
= length × breadth
= xy square units

According to the question,

$$xy - 9 = (x - 5)(y + 3)$$
$$\Rightarrow xy - 9 = xy + 3x - 5y - 15$$
$$\Rightarrow 3x - 5y - 6 = 0 \qquad ...(1)$$

and, $xy + 67 = (x + 3)(y + 2)$
$$\Rightarrow xy + 67 = xy + 2x + 3y + 6$$
$$\Rightarrow 2x + 3y - 61 = 0 \qquad ...(2)$$

To solve the equations (1) and (2) by cross-multiplication method, we draw the diagram below :

$$\begin{array}{ccccc} x & & y & & 1 \\ -5 & \diagdown & -6 & \diagdown & 3 & \diagdown & -5 \\ 3 & \diagup & -61 & \diagup & 2 & \diagup & 3 \end{array}$$

Then,
$$\frac{x}{(-5)(-61)-(3)(-6)}$$
$$= \frac{y}{(-6)(2)-(-61)(3)}$$
$$= \frac{1}{(3)(3)-(2)(-5)}$$

$$\Rightarrow \frac{x}{305+18} = \frac{y}{-12+183} = \frac{1}{9+10}$$

$$= \frac{x}{323} = \frac{y}{179} = \frac{1}{19}$$

$$\Rightarrow x = \frac{323}{19} = 17 \text{ and } y = \frac{171}{19} = 9$$

Hence, the dimensions (i.e., the length and the breadth) of the rectangle are 17 units and 9 units respectively.

Verification. Substituting $x = 13$, $y = 9$, we find that both the equations (1) and (2) are satisfied as shown below :

$$3x - 5y - 6 = 3(17) - 5(9) - 6$$
$$= 51 - 45 - 6 = 0$$
$$2x + 3y - 61 = 2(17) + 3(9) - 61$$
$$= 34 + 27 - 61 = 0$$

Hence, the solution we have got is correct.

ADDITIONAL EXAMPLES

Example 5. *For what value of k, will the system of equations*

$$x + 2y = 5$$
$$3x + ky + 15 = 0$$

have

(i) *a unique solution*
(ii) *no solution.* (CBSE 2001)

Sol. The given system of equations is
$$x + 2y = 5$$
$$3x + ky + 15 = 0$$
$$\Rightarrow x + 2y - 5 = 0$$
$$3x + ky + 15 = 0$$

Here, $a_1 = 1, b_1 = 2, c_1 = -5$
$a_2 = 3, b_2 = k, c_2 = 15$

(i) For having a unique solution, we must have

$$\frac{a_1}{a_2} \neq \frac{b_1}{b_2}$$

$$\Rightarrow \frac{1}{3} \neq \frac{2}{k}$$

$$\Rightarrow k \neq 6$$

(ii) For having no solution, we must have

$$\frac{a_1}{a_2} = \frac{b_1}{b_2} \neq \frac{c_1}{c_2}$$

$$\Rightarrow \frac{1}{3} = \frac{2}{k} \neq \frac{-5}{15}$$

$$\Rightarrow k = 6$$

Example 6. *Find the values of a and b for which the following system of equations has infinitely many solutions :*

$$(2a - 1)x - 3y = 5$$
$$3x + (b - 2)y = 3. \quad \text{(CBSE 2002)}$$

Sol. The given system of equations is
$$(2a - 1)x - 3y = 5$$
$$3x + (b - 2)y = 3$$
$$\Rightarrow (2a - 1)x - 3y - 5 = 0$$
$$3x + (b - 2)y - 3 = 0$$

Here, $a_1 = 2a - 1, b_1 = -3, c_1 = -5$
$a_2 = 3, b_2 = b - 2, c_2 = -3$

For having infinitely many solutions, we must have

$$\frac{a_1}{a_2} = \frac{b_1}{b_2} = \frac{c_1}{c_2}$$

$$\Rightarrow \frac{2a - 1}{3} = \frac{-3}{b - 2} = \frac{-5}{-3}$$

First and last give

$$\frac{2a - 1}{3} = \frac{-5}{-3}$$

$$\Rightarrow \frac{2a - 1}{3} = \frac{-5}{-3}$$

$$\Rightarrow 2a - 1 = 5$$

$$\Rightarrow 2a = 5 + 1 = 6$$

$$\Rightarrow a = \frac{6}{2} = 3$$

Second and last give

$$\Rightarrow \frac{-3}{b - 2} = \frac{5}{3}$$

$$\Rightarrow \frac{-3}{b - 2} = \frac{5}{3}$$

$$\Rightarrow 5(b - 2) = -9$$

$$\Rightarrow 5b - 10 = -9$$

$$\Rightarrow 5b = 10 - 9 = 1$$

$$\Rightarrow b = \frac{1}{5}.$$

Example 7. *Solve the following system of equations by using cross-multiplication method :*

$$11x + 15y + 23 = 0$$
$$7x - 2y - 20 = 0.$$

Sol. The given system of equations is
$$11x + 15y + 23 = 0 \quad \ldots(1)$$
$$7x - 2y - 20 = 0 \quad \ldots(2)$$

To solve the equations (1) and (2) by cross-multiplication method, we draw the diagram below :

Then,

$$\frac{x}{(15)(-20) - (-2)(23)}$$
$$= \frac{y}{(23)(7) - (-20)(11)}$$
$$= \frac{1}{(11)(-2) - (7)(15)}$$

$$\Rightarrow \frac{x}{-300 + 46} = \frac{y}{161 + 220}$$
$$= \frac{1}{-22 - 105}$$

$$\Rightarrow \frac{x}{-254} = \frac{y}{381} = \frac{1}{-127}$$

$$\Rightarrow x = \frac{-254}{-127} = 2$$

and $\qquad y = \frac{381}{-127} = -3$

Hence, the required solution of the given pair of equations is $x = 2, y = -3$.

Verification. Substituting $x = 2, y = -3$, we find that both the equations (1) and (2) are satisfied as shown below :

$$11x + 15y + 23 = 11(2) + 15(-3) + 23$$
$$= 22 - 45 + 23 = 0$$
$$7x - 2y - 20 = 7(2) - 2(-3) - 20$$
$$= 14 + 6 - 20 = 0$$

Hence, the solution we have got is correct.

Example 8. *The sum of a two digit number and the number formed by interchanging its digits is 110. If 10 is subtracted from the first number. The new number is 4 more than 5 times the sum of the digits in the first number. Find the first number.* (CBSE 2002)

Sol. Let in the first two digit number,
Unit's digit $= x$
and, ten's digit $= y$
Then, the number $= 10y + x$
On interchanging the digits, in the new number unit's digit $= y$
and, ten's digit $= x$
∴ The new number $= 10x + y$
According to the question,
$(10y + x) + (10x + y) = 110$
$\Rightarrow \quad 11x + 11y = 110$
$\Rightarrow \quad x + y = 10$
| Dividing throughout by 11
$\Rightarrow \quad x + y - 10 = 0$ ...(1)
and $\quad (10y + x) - 10 = 5(x + y) + 4$
$\Rightarrow \quad 10y + x - 10 = 5x + 5y + 4$
$\Rightarrow \quad 4x - 5y = -14$
$\Rightarrow \quad 4x - 5y + 14 = 0$ ...(2)

To solve the equations (1) and (2) by cross-multiplication method, we draw the diagram below :

$$\begin{array}{ccccccc} 1 & x & -10 & y & 1 & & 1 \\ & \times & & \times & & \times & \\ -5 & & 14 & & 4 & & -5 \end{array}$$

Then, $\dfrac{x}{(1)(14) - (-5)(-10)}$

$= \dfrac{y}{(-10)(4) - (14)(1)} = \dfrac{1}{(1)(-5) - (4)(1)}$

$\Rightarrow \quad \dfrac{x}{14 - 50} = \dfrac{y}{-40 - 14} = \dfrac{1}{-5 - 4}$

$\Rightarrow \quad \dfrac{x}{-36} = \dfrac{y}{-54} = \dfrac{1}{-9}$

$\Rightarrow \quad x = \dfrac{-36}{-9} = 4 \text{ and } y = \dfrac{-54}{-9} = 6$

Hence, the first two digit number
$= 10 \times 6 + 4 = 60 + 4 = 64.$

Verification. Substituting $x = 4, y = 6$, we find that both the equations (1) and (2) are satisfied as shown below :
$x + y - 10 = 4 + 6 - 10 = 0$
$4x - 5y + 14 = 4(4) - 5(6) + 14$
$16 - 30 + 14 = 0$
Hence, the solution we have got is correct.

Example 9. *Father's age is three times the sum of the ages of his two children. After 5 years, his age will be twice the sum of the ages of two children. Find the age of father.*
(CBSE 2003)

Sol. Let the present age of the father be x years and the sum of the present ages of his two children be y years. Then, according to the question,
$\quad\quad x = 3y$
$\Rightarrow \quad x - 3y = 0$ ...(1)
and, $\quad x + 5 = 2(y + 5 + 5)$
$\Rightarrow \quad x + 5 = 2(y + 10)$
$\Rightarrow \quad x + 5 = 2y + 20$
$\Rightarrow \quad x - 2y - 15 = 0$ ...(2)

To solve the equations (1) and (2) by cross-multiplication method, we draw the diagram below :

$$\begin{array}{ccccccc} -3 & x & 0 & y & 1 & & -3 \\ & \times & & \times & & \times & \\ -2 & & -15 & & 1 & & -2 \end{array}$$

Then, $\dfrac{x}{(-3)(-15) - (-2)(0)}$

$= \dfrac{y}{(0)(1) - (-15)(1)}$

$= \dfrac{1}{(1)(-2) - (1)(-3)}$

$\Rightarrow \quad \dfrac{x}{45} = \dfrac{y}{15} = \dfrac{1}{1}$

$\Rightarrow \quad x = 45 \text{ and } y = 15$

Hence, the age of father is 45 years.

Verification. Substituting $x = 45$, $y = 15$, we find that both the equations (1) and (2) are satisfied as shown in below :

$x - 3y = 45 - 3(15) = 45 - 45 = 0$
$x - 2y - 15 = 45 - 2(15) - 15$
$= 45 - 30 - 15 = 0$

Hence, the solution we have got is correct.

TEST YOUR KNOWLEDGE

1. The cost of 5 oranges and 3 apples is Rs. 35 and the cost of 2 oranges and 4 apples is Rs. 28. Find the cost of an orange and an apple.

2. From Bangalore bus stand, if we buy 2 tickets to Malleswarm and 3 tickets to Yeshwanthpur, the total cost is Rs. 46; but if we buy 3 tickets to Malleswarm and 5 tickets to Yeshwanthpur, the total cost is Rs. 74. Find the fare from Bangalore to Malleswarm and to Yeshwanthpur.

3. For which value of p does the pair of equations given below has unique solution?
   $4x + py + 8 = 0$
   $2x + 2y + 2 = 0.$

4. For what values of k will the following pair of linear equations have infinitely many solutions?
   $kx + 3y - (k - 3) = 0$
   $12x + 10y - k = 0$

5. Solve the following system of equations by cross-multiplication method:
   (i) $4y - 3x - 23 = 0$
   $3y + 4x - 11 = 0$
   (ii) $3x + 4y = 12$
   $4y + 7x = 12$
   (iii) $7x + 4y = 12$
   $3x + 4y = -4$
   (iv) $3x - 5y = -1$
   $x - y = -1.$

6. 3 chairs and 4 tables cost Rs. 2250 and 4 chairs and 3 tables cost Rs. 1950. Find the total cost of 2 chairs and 1 table.

7. The denominator of a fraction is 4 more than twice the numerator. When both the numerator and denominator and decreased by 6, then the denominator becomes 12 times the numerator. Determine the fraction. *(CBSE 2001)*

8. 10 years ago father was 12 times as old as his son and 10 years hence he will be twice as old as his son. Find their present ages. *(CBSE 1999)*

9. The area of a rectangle gets reduced by 80 square units. If its length is reduced by 5 units and the breadth is increased by 2 units. If the increase the length by 10 units and decrease the breadth by 5 units, the area is increased by 50 square units. Find the length and breadth of the rectangle.
   (CBSE 1998)

10. Places A and B are 100 km apart from each other on a highway. A car starts from A and another from B at the same time. If they move in the same direction, they meet in 10 hours and if they move in opposite directions, they meet in 1 hour and 40 minutes. Find the speeds of the cars.
    (CBSE 2002)

11. Determine the value of k so that the following linear equations have no solution.
    $(3k - 1)x + 3y - 2 = 0$
    $(k^2 + 1)x + (k - 2)y - 5 = 0.$
    (CBSE 2001)

12. Find the values of a and b for which the following system of equations has infinitely many solutions.
    $2x - (2a + 5)y = 5$
    $(2b + 1)x - 9y = 15.$ *(CBSE 2002)*

Answers

1. Rs. 4, Rs. 5
2. Rs. 8, Rs. 10
3. $p \neq 4$
4. $k = 6$
5. (i) $x = -1, y = 5$ (ii) $x = 0, y = 3$
   (iii) $x = 4, y = -4$ (iv) $x = -2, y = -1$
6. Rs. 750
7. $\dfrac{7}{18}$
8. 34 years, 12 years
9. 40 units, 30 units.
10. 35 km/hour, 25 km/hour
11. $k = -1$
12. $a = 1, b = \dfrac{5}{2}.$

IMPORTANT POINTS

1. Equations reducible to a linear pair in two variables

Sometimes, we find equations like $\frac{1}{x} + \frac{7}{y} = 10$, which is not a linear equation. But, by making some substitution, we can reduce it to a linear equation, which we know how to solve. Then, using that solution, we can solve the unique equation.

ILLUSTRATIVE EXAMPLES

[NCERT Exercise 3.6]
(Page No. 67)

Example 1. *Solve the following pairs of equations by reducing them to a pair of linear equations:*

(i) $\frac{1}{2x} + \frac{1}{3y} = 2$

$\frac{1}{3x} + \frac{1}{2y} = \frac{13}{6}$

(ii) $\frac{2}{\sqrt{x}} + \frac{3}{\sqrt{y}} = 2$

$\frac{4}{\sqrt{x}} - \frac{9}{\sqrt{y}} = -1$

(iii) $\frac{4}{x} + 3y = 14$

$\frac{3}{x} - 4y = 23$

(iv) $\frac{5}{x-1} + \frac{1}{y-2} = 2$

$\frac{6}{x-1} - \frac{3}{y-2} = 1$

(v) $\frac{7x - 2y}{xy} = 5$

$\frac{8x + 7y}{xy} = 15.$

(vi) $6x + 3y = 6xy$

$2x + 4y = 5xy$

(vii) $\frac{10}{x+y} + \frac{2}{x-y} = 4$

$\frac{15}{x+y} - \frac{5}{x-y} = -2$

(viii) $\frac{1}{3x+y} + \frac{1}{3x-y} = \frac{3}{4}$

$\frac{1}{2(3x+y)} - \frac{1}{2(3x-y)} = \frac{-1}{8}$

Sol.

(i) $\frac{1}{2x} + \frac{1}{3y} = 2$

$\frac{1}{3x} + \frac{1}{2y} = \frac{13}{6}$

The given pair of equations is

$\frac{1}{2x} + \frac{1}{3y} = 2$ ...(1)

$\frac{1}{3x} + \frac{1}{2y} = \frac{13}{6}$ ...(2)

Put $\frac{1}{x} = X$ ...(3)

and $\frac{1}{y} = Y$ ...(4)

Then the equations (1) and (2) can be rewritten as

$\frac{1}{2}X + \frac{1}{3}Y = 2$ ...(5)

$\frac{1}{3}X + \frac{1}{2}Y = \frac{13}{6}$ ...(6)

$\Rightarrow$ 3X + 2Y = 12 ...(7)

2X + 3Y = 13 ...(8)

Multiplying equation (7) by 3 and equation (8) by 2, we get

$$9X + 6Y = 36 \qquad ...(9)$$
$$4X + 6Y = 26 \qquad ...(10)$$

Subtracting equation (10) from equation (9), we get

$$5X = 10$$
$$\Rightarrow \quad X = \frac{10}{5} = 2 \qquad ...(11)$$

Substituting this value of X in equation (9), we get

$$9(2) + 6Y = 36$$
$$\Rightarrow \quad 18 + 6Y = 36$$
$$\Rightarrow \quad 6Y = 36 - 18 = 18$$
$$\Rightarrow \quad Y = \frac{18}{6} = 3 \qquad ...(12)$$

From equation (3) and equation (11), we get

$$\frac{1}{x} = 2 \Rightarrow x = \frac{1}{2}$$

From equation (4) and equation (12), we get

$$\frac{1}{y} = 3 \Rightarrow y = \frac{1}{3}$$

Hence, the solution of the given pair of equations is

$$x = \frac{1}{2}, y = \frac{1}{3}.$$

Verification. Substituting $x = \frac{1}{2}$, $y = \frac{1}{3}$, we find that both the equations (1) and (2) are satisfied as shown below :

$$\frac{1}{2x} + \frac{1}{3y} = \frac{1}{2 \cdot \frac{1}{2}} + \frac{1}{3 \cdot \frac{1}{3}} = 1 + 1 = 2$$

$$\frac{1}{3x} + \frac{1}{2y} = \frac{1}{3 \cdot \frac{1}{2}} + \frac{1}{2 \cdot \frac{1}{3}} = \frac{2}{3} + \frac{3}{2} = \frac{13}{6}$$

Hence, the solution we have got is correct.

(ii) $\dfrac{2}{\sqrt{x}} + \dfrac{3}{\sqrt{y}} = 2$

$\dfrac{4}{\sqrt{x}} - \dfrac{9}{\sqrt{y}} = -1$

The given pair of equations is

$$\frac{2}{\sqrt{x}} + \frac{3}{\sqrt{y}} = 2 \qquad ...(1)$$

$$\frac{4}{\sqrt{x}} - \frac{9}{\sqrt{y}} = -1 \qquad ...(2)$$

Put $\dfrac{1}{\sqrt{x}} = u \qquad ...(3)$

and $\dfrac{1}{\sqrt{y}} = v \qquad ...(4)$

Then equations (1) and (2) can be rewritten as

$$2u + 3v = 2 \qquad ...(5)$$
$$4u - 9v = -1 \qquad ...(6)$$

Multiplying equation (5) by 3, we get

$$6u + 9v = 6 \qquad ...(7)$$

Adding equation (6) and equation (7), we get

$$10u = 5$$
$$\Rightarrow \quad u = \frac{5}{10} = \frac{1}{2} \qquad ...(8)$$

Substituting the value of x in equation (5), we get

$$2\left(\frac{1}{2}\right) + 3v = 2$$
$$\Rightarrow \quad 1 + 3v = 2$$
$$\Rightarrow \quad 3v = 2 - 1 = 1$$
$$\Rightarrow \quad v = \frac{1}{3} \qquad ...(9)$$

From equation (3) and equation (8), we get

$$\frac{1}{\sqrt{x}} = \frac{1}{2}$$
$$\Rightarrow \quad \sqrt{x} = 2$$
$$\Rightarrow \quad x = 4 \qquad \text{| squaring}$$

From equation (4) and equation (9), we get

$$\frac{1}{\sqrt{y}} = \frac{1}{3}$$

$\Rightarrow \quad \sqrt{y} = 3$

$\Rightarrow \quad y = 9$ | squaring

Hence, the solution if the given pair of equations is

$$x = 4, y = 9.$$

Verification. Substituting $x = 4, y = 9$, we find that both the equations (1) and (2) are satisfied as shown below.

$$\frac{2}{\sqrt{x}} + \frac{3}{\sqrt{y}} = \frac{2}{\sqrt{4}} + \frac{3}{\sqrt{9}}$$

$$= \frac{2}{2} + \frac{3}{3} = 1 + 1 = 2$$

$$\frac{4}{\sqrt{x}} - \frac{9}{\sqrt{y}} = \frac{4}{\sqrt{9}} = \frac{9}{\sqrt{9}} = \frac{4}{2} - \frac{9}{3}$$

$$= 2 - 3 = -1$$

Hence, the solution we have got is correct.

(iii) $\dfrac{4}{x} + 3y = 14$

$\dfrac{3}{x} - 4y = 23$

The given pair of equations is

$\dfrac{4}{x} + 3y = 14$ ...(1)

$\dfrac{3}{x} - 4y = 23$ ...(2)

Put $\dfrac{1}{x} = X$ ...(3)

Then equations (1) and (2) can be re-written as

$4X + 3y = 14$ ...(4)

$3X - 4y = 23$ ...(5)

From equation (5),

$4y = 3X - 23$

$\Rightarrow \quad y = \dfrac{3X - 23}{4}$ ...(6)

Substitute this value of y in equation (4), we get

$$4X + 3\left(\frac{3X - 23}{4}\right) = 14$$

$\Rightarrow \quad 16X + 9X - 69 = 125$

$\Rightarrow \quad 25X = 56 + 69 = 125$

$\Rightarrow \quad X = \dfrac{125}{25} = 5$ ...(7)

Substituting this value of X in equation (6), we get

$$y = \frac{3(5) - 23}{4} = \frac{15 - 23}{4}$$

$$= \frac{-8}{4} = -2$$ ...(8)

From equation (3) and equation (7), we get

$$\frac{1}{x} = 5$$

$\Rightarrow \quad x = \dfrac{1}{5}$ ...(9)

Hence, the solution of the given pair of equation is $x = \dfrac{1}{5}, y = -2$.

Verification. Substituting $x = \dfrac{1}{5}, y = -2$, we find that both the equations (1) and (2) are satisfied as shown below :

$$\frac{4}{x} + 3y = \frac{4}{\left(\frac{1}{5}\right)} + 3(-2) = 20 - 6 = 14$$

$$\frac{3}{x} - 4y = \frac{3}{\left(\frac{1}{5}\right)} - 4(-2) = 15 + 8 = 23$$

Hence, the solution is correct.

(iv) $\dfrac{5}{x-1} + \dfrac{1}{y-2} = 2$

$\dfrac{6}{x-1} - \dfrac{3}{y-2} = 1$

The given pair of equations is

$\dfrac{5}{x-1} + \dfrac{1}{y-2} = 2$ ...(1)

$$\frac{6}{x-1} - \frac{3}{y-2} = 1 \quad \ldots(2)$$

Put $\dfrac{1}{x-1} = u$ ...(3)

and $\dfrac{1}{y-2} = v$ ...(4)

Then equations (1) and (2) can be re-written as

$5u + v = 2$ ...(5)
$6u - 3v = 1$ ...(6)

Multiplying equation (5) by 3, we get

$15u + 3v = 6$ ...(7)

Adding equation (6) and equation (7), we get

$21u = 7$

$\Rightarrow u = \dfrac{7}{21} = \dfrac{1}{3}$ ...(8)

Substituting this value of u in equation (5), we get

$5\left(\dfrac{1}{3}\right) + v = 2$

$\Rightarrow \dfrac{5}{3} + v = 2$

$\Rightarrow v = 2 - \dfrac{5}{3} = \dfrac{1}{3}$ ...(9)

From equation (3) and equation (8), we get

$\dfrac{1}{x-1} = \dfrac{1}{3}$

$\Rightarrow x - 1 = 3$
$\Rightarrow x = 1 + 3 = 4$

From equation (4) and equation (9), we get

$\dfrac{1}{y-2} = \dfrac{1}{3}$

$\Rightarrow y - 2 = 3$
$\Rightarrow y = 3 + 2 = 5$

Hence, the solution of the given pair of equations is

$x = 4, y = 5$.

Verification. Substituting $x = 4, y = 5$, we find that both the equations (1) and (2) are satisfied as shown below :

$$\frac{5}{x-1} + \frac{1}{y-2} = \frac{5}{4-1} + \frac{1}{5-2}$$

$$= \frac{5}{3} + \frac{1}{3} = 2$$

$$\frac{6}{x-1} - \frac{3}{y-2} = \frac{6}{4-1} - \frac{3}{5-2}$$

$$= 2 - 1 = 1$$

Hence, the solution we have got is correct.

(v) $\dfrac{7x - 2y}{xy} = 5$

$\dfrac{8x + 7y}{xy} = 15$

The given pair of equations is

$\dfrac{7x - 2y}{xy} = 5$ ...(1)

$\dfrac{8x + 7y}{xy} = 15$ ...(2)

$\Rightarrow \dfrac{7x}{xy} - \dfrac{2y}{xy} = 5$ ...(3)

$\dfrac{8x}{xy} + \dfrac{7y}{xy} = 15$ ...(4)

$\Rightarrow \dfrac{7}{y} - \dfrac{2}{x} = 5$ ...(5)

$\dfrac{8}{y} + \dfrac{7}{x} = 15$ ...(6)

Put $\dfrac{1}{x} = u$ ...(7)

and $\dfrac{1}{y} = v$ ...(8)

Then the equations (5) and (6) can be rewritten as

$7v - 2u = 5$ ...(9)
$8v + 7u = 15$ ...(10)

$\Rightarrow$ $\quad 7v - 2u - 5 = 0$ ...(11)
$\quad 8v + 7u - 15 = 0$ ...(12)

To solve the equations by the cross-multiplication method, we draw the diagram below :

$$\begin{array}{ccccc} -2 & v & -5 & u & 1 & -2 \\ & \times & & \times & & \times \\ 7 & & -15 & & 8 & & 7 \end{array}$$

Then,

$$\frac{v}{(-2)(-15) - (7)(-5)}$$

$$= \frac{u}{(-5)(8) - (-15)(7)}$$

$$= \frac{1}{(7)(7) - (8)(-2)}$$

$\Rightarrow \quad \dfrac{v}{30 + 35} = \dfrac{u}{-40 + 105} = \dfrac{1}{49 + 16}$

$\Rightarrow \quad \dfrac{v}{65} = \dfrac{u}{65} = \dfrac{1}{65}$

$\Rightarrow \quad v = \dfrac{65}{65} = 1$ ...(13)

and $\quad u = \dfrac{65}{65} = 1$ ...(14)

From equation (7) and equation (14), we get

$\dfrac{1}{x} = 1 \implies x = 1$

From equation (8) and equation (13), we get

$\dfrac{1}{y} = 1 \implies y = 1$

Hence, the solution of the given pair of equations is

$x = 1, y = 1$.

Verification. Substituting $x = 1, y = 1$ we find that both the equations (1) and (2) are satisfied as shown below :

$\dfrac{7x - 2y}{xy} = \dfrac{7(1) - 2(1)}{(1)(1)} = 5$

$\dfrac{8x + 7y}{xy} = \dfrac{8(1) + 7(1)}{(1)(1)} = 15$

Hence, the solution we have got is correct.

(vi) **6x + 3y = 6xy**
2x + 4y = 5xy

The given pair of equations is

$6x + 3y = 6xy$

$\Rightarrow \quad \dfrac{6x}{xy} + \dfrac{3y}{xy} = \dfrac{6xy}{xy}$

| Dividing throughout by xy

$\Rightarrow \quad \dfrac{6}{y} + \dfrac{3}{x} = 6$ ...(1)

and, $2x + 4y = 5xy$

$\Rightarrow \quad \dfrac{2x}{xy} + \dfrac{4y}{xy} = \dfrac{5xy}{xy}$

| Dividing throughout by xy

$\Rightarrow \quad \dfrac{2}{y} + \dfrac{4}{x} = 5$ ...(2)

Put $\dfrac{1}{x} = u$ ...(3)

and $\dfrac{1}{y} = v$ ...(4)

Then equations (1) and (2) can be re-written as

$6v + 3u = 6$ ...(5)
$2v + 4u = 5$ ...(6)

Multiplying equation (6) by 3, we get

$6v + 12u = 15$ ...(7)

Subtracting equation (5) from equation (7), we get

$9u = 9$

$\Rightarrow \quad u = \dfrac{9}{9} = 1$

$\Rightarrow \quad \dfrac{1}{x} = 1$ | using (3)

$\Rightarrow \quad x = 1$

Substituting this value of u in equation (5), we get

$6v + 3 \times 1 = 6$

$\Rightarrow \qquad 6v + 3 = 6$

$\Rightarrow \qquad 6v = 6 - 3 = 3$

$\Rightarrow \qquad v = \dfrac{3}{6} = \dfrac{1}{2}$

$\Rightarrow \qquad \dfrac{1}{y} = \dfrac{1}{2}$ | Using (4)

$\Rightarrow \qquad y = 2$

Hence, the solution of the given pair. If equation is $x = 1, y = 2$.

Verification. Substituting $x = 1, y = 2$, we find that both the equations (1) and (2) are satisfied as shown below :

$$\dfrac{6}{y} + \dfrac{3}{x} = \dfrac{6}{2} + \dfrac{3}{1} = 3 + 3 = 6$$

$$\dfrac{2}{y} + \dfrac{4}{x} = \dfrac{2}{2} + \dfrac{4}{1} = 1 + 4 = 5$$

Hence, the solution is correct.

(vii) $\dfrac{10}{x+y} + \dfrac{2}{x-y} = 4$

$\dfrac{15}{x+y} - \dfrac{5}{x-y} = -2$

The given pair of equations is

$$\dfrac{10}{x+y} + \dfrac{2}{x-y} = 4 \qquad \ldots(1)$$

$$\dfrac{15}{x+y} - \dfrac{5}{x-y} = -2 \qquad \ldots(2)$$

Put $\qquad \dfrac{1}{x+y} = u \qquad \ldots(3)$

and $\qquad \dfrac{1}{x-y} = v \qquad \ldots(4)$

Then equations (1) and (2) can be re-written as

$10u + 2v = 4 \qquad \ldots(3)$

$15u - 5v = -2 \qquad \ldots(4)$

(3) gives

$5u + v = 2 \qquad \ldots(5)$

| Dividing throughout by 2

Multiplying equation (5) by 3, we get

$15u + 3v = 6 \qquad \ldots(6)$

Subtracting equation (4) from equation (6), we get

$8v = 8$

$\Rightarrow \qquad v = \dfrac{8}{8} = 1 \qquad \ldots(7)$

Substituting this value of v in equation (5), we get

$5u + 1 = 2$

$\Rightarrow \qquad 5u = 2 - 1 = 1$

$\Rightarrow \qquad u = \dfrac{1}{5} \qquad \ldots(8)$

From equation (3) and equation (8), we get

$\dfrac{1}{x+y} = \dfrac{1}{5}$

$\Rightarrow \qquad x + y = 5 \qquad \ldots(9)$

From equation (4) and equation (7), we get

$\dfrac{1}{x-y} = 1$

$\Rightarrow \qquad x - y = 1 \qquad \ldots(10)$

Adding equation (9) and equation (10), we get

$2x = 6$

$\Rightarrow \qquad x = \dfrac{6}{2} = 3$

Substitute this value of x in equation (9), we get

$3 + y = 5$

$\Rightarrow \qquad y = 5 - 3 = 2$

Hence, the solution of the given pair of equations is

$x = 3, y = 2$

Verification. Substituting $x = 3, y = 2$, we find that both the equations (1) and (2) are satisfied as shown below :

$$\dfrac{10}{x+y} + \dfrac{2}{x-y} = \dfrac{10}{3+2} + \dfrac{2}{3-2}$$

$$= 2 + 2 = 4$$

$$\dfrac{15}{x+y} - \dfrac{5}{x-y} = \dfrac{15}{3+2} - \dfrac{5}{3-2}$$

$$= 3 - 5 = -2$$

This verifies the solution.

(viii) $\dfrac{1}{(3x+y)} + \dfrac{1}{(3x-y)} = \dfrac{3}{4}$

$\dfrac{1}{2(3x+y)} - \dfrac{1}{2(3x-y)} = \dfrac{-1}{8}$

The given pair of equations is

$\dfrac{1}{3x+y} + \dfrac{1}{3x-y} = \dfrac{3}{4}$ ...(1)

$\dfrac{1}{2(3x+y)} - \dfrac{1}{2(3x-y)} = \dfrac{-1}{8}$ ...(2)

Put $\dfrac{1}{3x+y} = u$ ...(3)

and $\dfrac{1}{3x-y} = v$ ...(4)

Then equations (1) and (2) can be re-written as

$u + v = \dfrac{3}{4}$ ...(5)

$\dfrac{1}{2}u + \dfrac{1}{2}v = -\dfrac{1}{8}$ ...(6)

(6) gives

$u - v = -\dfrac{1}{4}$ ...(7)

| Multiplying both sides by 2

Adding equation (5) and equation (7), we get

$2u = \dfrac{3}{4} - \dfrac{1}{4} = \dfrac{1}{2}$

$\Rightarrow u = \dfrac{1}{4}$ ...(8)

Subtracting equation (7) from equation (5), we get

$2v = \dfrac{3}{4} + \dfrac{1}{4} = 1$

$\Rightarrow v = \dfrac{1}{2}$ ...(9)

From equation (3) and equation (8), we get

$\dfrac{1}{3x+y} = \dfrac{1}{4}$

$\Rightarrow 3x + y = 4$ ...(10)

From equation (4) and equation (9), we get

$\dfrac{1}{3x-y} = \dfrac{1}{2}$

$\Rightarrow 3x - y = 2$ ...(11)

Adding equation (10) and equation (11), we get

$6x = 6$

$\Rightarrow x = \dfrac{6}{6} = 1$

Substituting this value of x in equation (10), we get

$3(1) + y = 4$

$\Rightarrow 3 + y = 4$

$\Rightarrow y = 4 - 3 = 1$

Hence, the solution of the given pair of equations is

$x = 1, y = 1$.

Verification. Substituting $x = 1, y = 1$, we find that both the equations (1) and (2) are satisfied as shown below:

$\dfrac{1}{3x+y} + \dfrac{1}{3x-y} = \dfrac{1}{3(1)+1} + \dfrac{1}{3(1)-1}$

$= \dfrac{1}{4} + \dfrac{1}{2} = \dfrac{3}{4}$

$\dfrac{1}{2(3x+y)} - \dfrac{1}{2(3x-y)}$

$= \dfrac{1}{2(3.1+1)} - \dfrac{1}{2(3.1-1)}$

$= \dfrac{1}{8} - \dfrac{1}{4} = -\dfrac{1}{8}$

This verifies the solution.

Example 2. *Formulate the following problems as a pair of equations, and hence find their solutions:*

(i) Ritu can row downstream 20 km in 2 hours, and upstream 4 km in 2 hours. Find her speed of rowing in still water and the speed of the current.

(ii) 2 women and 5 men can together finish an embroidery work in 4 days, while 3 women and 6 men can finish it in 3 days. Find

the time taken by 1 woman alone to finish the work, and also that taken by 1 man alone.

(*iii*) *Roohi travels 300 km to her home partly by train and partly by bus. She takes 4 hours if she travels 60 km by train and the remaining by bus. If she travels 100 km by train and the remaining by bus, she takes 10 minutes longer. Find the speed of the train and the bus separately.*

Sol. (*i*) Let her speed of rowing in still water be x km/hour and the speed of the current by y km/hour. Then, her speed of rowing downstream $= (x + y)$ km/hour

and, her speed of rowing upstream $= (x - y)$ km/hour

Also, time $= \dfrac{\text{distance}}{\text{speed}}$

In the first case, when she goes 20 km downstream, then the time taken is 2 hours.

$\therefore \quad \dfrac{20}{x + y} = 2$

$\Rightarrow \quad x + y = 10 \quad \quad \quad \ldots(1)$

In the second case, when she goes 4 km upstream, then the time taken is 2 hours.

$\therefore \quad \dfrac{4}{x - y} = 2$

$\Rightarrow \quad x - y = 2 \quad \quad \quad \ldots(2)$

Adding equation (1) and equation (2), we get

$2x = 12$

$\Rightarrow \quad x = \dfrac{12}{2} = 6$

Substituting this value of x in equation (1), we get

$6 + y = 10$

$\Rightarrow \quad y = 10 - 6 = 4$

Hence, the speed of her rowing in still water is 6 km/hour and the speed of the current is 4 km/hour.

Verification. Substituting $x = 6, y = 4$, we find that both the equations (1) and (2) are satisfied as shown below :

$x + y = 6 + 4 = 10$
$x - y = 6 - 4 = 2$

Hence, the solution we have got is correct.

(*ii*) Let the taken by 1 women alone to finish the embroidery be x days and the time taken by 1 man alone to finish the embroidery be y days. Then

1 woman's 1 day's work $= \dfrac{1}{x}$

and   1 man's 1 day's work $= \dfrac{1}{y}$

$\therefore$ 2 women's 1 day's work $= \dfrac{2}{x}$

and   5 men's 1 day's work $= \dfrac{5}{y}$

$\because$ 2 women and 5 men can together finish a piece of embroidery in 4 days.

$\therefore \quad 4\left(\dfrac{2}{x} + \dfrac{5}{y}\right) = 1$

$\Rightarrow \quad \dfrac{2}{x} + \dfrac{5}{y} = \dfrac{1}{4} \quad \quad \ldots(1)$

Again, 3 women's 1 day's work $= \dfrac{3}{x}$

and   6 men's 1 day's work $= \dfrac{6}{y}$

$\because$ 3 women and 6 men can together finish a piece of embroidery in 3 days

$\therefore \quad 3\left(\dfrac{3}{x} + \dfrac{6}{y}\right) = 1$

$\Rightarrow \quad \dfrac{3}{x} + \dfrac{6}{y} = \dfrac{1}{3} \quad \quad \ldots(2)$

Put $\dfrac{1}{x} = u \quad \quad \ldots(3)$

and $\dfrac{1}{y} = v \quad \quad \ldots(4)$

Then equations (1) and (2) can be re-written as

$2u + 5v = \dfrac{1}{4} \quad \quad \ldots(5)$

$$3u + 6v = \frac{1}{3} \qquad \ldots(6)$$

Multiplying equation (5) by 3 and equation (6) by 2, we get

$$6u + 15v = \frac{3}{4} \qquad \ldots(7)$$

$$6u + 12v = \frac{2}{3} \qquad \ldots(8)$$

Subtracting equation (8) from equation (7), we get

$$3v = \frac{3}{4} - \frac{2}{3} = \frac{1}{12}$$

$$\Rightarrow \quad v = \frac{1}{36} \qquad \ldots(9)$$

Substituting this value of v in equation (5), we get

$$2u + 5\left(\frac{1}{36}\right) = \frac{1}{4}$$

$$\Rightarrow \quad 2u + \frac{5}{36} = \frac{1}{4}$$

$$\Rightarrow \quad 2u = \frac{1}{4} - \frac{5}{36}$$

$$\Rightarrow \quad 2u = \frac{9}{36} - \frac{5}{36}$$

$$\Rightarrow \quad 2u = \frac{4}{36} = \frac{1}{9}$$

$$\Rightarrow \quad u = \frac{1}{18} \qquad \ldots(10)$$

From equation (3) and equation (10), we get

$$\frac{1}{x} = \frac{1}{18}$$

$$\Rightarrow \quad x = 18$$

From equation (4) and equation (9), we get

$$\frac{1}{y} = \frac{1}{36}$$

$$\Rightarrow \quad y = 36$$

Hence, the time taken by 1 women alone to finish the embroidery is 18 days and the time taken by 1 man alone to finish the embroidery is 36 days.

Verification. Substituting $x = 18$, $y = 36$, we find that both the equations (1) and (2) are satisfied as shown below :

$$\frac{2}{x} + \frac{5}{y} = \frac{2}{18} + \frac{5}{36} = \frac{1}{9} + \frac{5}{36} = \frac{1}{4}$$

$$\frac{3}{x} + \frac{6}{y} = \frac{3}{18} + \frac{6}{36} = \frac{1}{6} + \frac{1}{6} = \frac{1}{3}$$

This verifies the solution.

(*iii*) Let the speed of the train and the bus be x km/hour and y km/hour respectively.

Case I. When she travels 60 km by train and the remaining (300 − 60) km, *i.e.*, 240 km by bus, the time taken is 4 hours

$$\therefore \quad \frac{60}{x} + \frac{240}{y} = 4 \;\Big|\; \because \text{ time} = \frac{\text{Distance}}{\text{Speed}}$$

$$\Rightarrow \quad \frac{1}{x} + \frac{4}{y} = \frac{1}{15}$$

... (1) | Dividing by 60

Case II. When she travels 100 km by train and the remaining (300 − 100) km, *i.e.*, 200 km by bus, the time taken is 4 hours 10 minutes, *i.e.*, $\frac{25}{6}$ hours.

$$\therefore \quad \frac{100}{x} + \frac{200}{y} = \frac{25}{6}$$

$$\Rightarrow \quad \frac{4}{x} + \frac{8}{y} = \frac{1}{6} \qquad \ldots(2)$$

| Dividing by 25

Multiplying equation (1) by 2, we get

$$\frac{2}{x} + \frac{8}{y} = \frac{2}{15} \qquad \ldots(3)$$

Subtracting equation (3) from equation (2), we get

$$\frac{2}{x} = \frac{1}{6} - \frac{2}{15} = \frac{1}{30}$$

$$\Rightarrow \quad x = 60$$

Substituting this value of x in equation (3), we get

$$\frac{2}{60} + \frac{8}{y} = \frac{2}{15}$$

$\Rightarrow \quad \dfrac{1}{30} - \dfrac{8}{y} = \dfrac{2}{15}$

$\Rightarrow \quad \dfrac{8}{y} = \dfrac{2}{15} - \dfrac{1}{30} = \dfrac{1}{10}$

$\Rightarrow \quad y = 80$

So, the solution of the equations (1) and (2) is $x = 60$ and $y = 80$.

Hence, the speed of the train is 60 km/hour and the speed of the bus is 80 km/hour.

Verification. Substituting $x = 60$, $y = 80$, we find that both the equations (1) and (2) and satisfied as shown below :

$\dfrac{1}{x} + \dfrac{1}{y} = \dfrac{1}{60} + \dfrac{4}{80} = \dfrac{1}{60} + \dfrac{1}{20} = \dfrac{1}{15}$

$\dfrac{4}{x} + \dfrac{8}{y} = \dfrac{4}{60} + \dfrac{8}{80} = \dfrac{1}{15} + \dfrac{1}{10} = \dfrac{1}{6}$

This verifies the solution.

ADDITIONAL EXAMPLES

Example 3. *Solve the following system of linear equations :*

$(a - b) x + (a + b) y = a^2 - 2ab - b^2.$

$(a + b) (x + y) = a^2 + b^2.$

(CBSE 2003)

Sol. The given system of linear equations is

$(a - b) x + (a + b) y = a^2 - 2ab - b^2$ ...(1)

$(a + b) (x + y) = a^2 + b^2$ ...(2)

Equation (2) can be rewritten as

$(a + b) x + (a + b) y = a^2 + b^2$ ...(3)

Subtracting equation (3) from equation (1), we get

$-2bx = -2ab - 2b^2$

$\Rightarrow \quad -2bx = -2b(a + b)$

$\Rightarrow \quad x = a + b$

Substituting this value of x in equation (1), we get

$(a - b)(a + b) + (c + b) y = a^2 - 2ab - b^2$

$\Rightarrow \quad a^2 - b^2 + (a + b) y = a^2 - 2ab - b^2$

$\Rightarrow \quad (a + b) y = -2ab$

$\Rightarrow \quad y = -\dfrac{2ab}{a + b}$

Hence, the solution of the given pair of linear equations is

$x = a + b, y = \dfrac{-2ab}{a + b}$

Verification. Substituting $x = a + b$, $y = \dfrac{-2ab}{a + b}$, we find that both the equations (1) and (2) are satisfied as shown below :

$(a - b) x + (a + b) y$

$= (a - b)(a + b) + (a + b) \left\{ \dfrac{-2ab}{a + b} \right\}$

$= a^2 - b^2 - 2ab$

$(a + b)(x + y) = (a + b) \left\{ a + b - \dfrac{2ab}{a + b} \right\}$

$= (a + b)^2 - 2ab$

$= a^2 + b^2$

Hence, the solution we have got is correct.

Example 4. *Solve the following system of equations :*

$35x + 23y = 209$

$23x + 35y = 197.$

Sol. The given system of equations is

$35x + 23y = 209$ ...(1)

$23x + 35y = 197$ ...(2)

Adding equation (1) and equation (2), we get

$58x + 58y = 406$

$\Rightarrow \quad x + y = 7$ ...(3)

| Dividing throughout by 58

Subtracting equation (2) from equation (1), we get

$12x - 12y = 12$

$\Rightarrow \quad x - y = 1$ ...(4)

| Dividing throughout by 12

Adding equation (3) and equation (4), we get

$2x = 8$

$$\Rightarrow \qquad x = \frac{8}{2} = 4$$

Subtracting equation (4) from equation (3), we get

$$2y = 6$$

$$\Rightarrow \qquad y = \frac{6}{2} = 3$$

Hence, the solution of the given pair of equations is $x = 4, y = 3$.

Verification. Substituting $x = 4, y = 3$ we find that both the equations (1) and (2) are satisfied as shown below :

$$35x + 23y = 35(4) + 23(3)$$
$$= 140 + 69 = 209$$
$$23x + 35y = 23(4) + 35(3)$$
$$= 92 + 105 = 197$$

Hence, the solution is correct.

Example 5. *Solve for x and y :*

$$\frac{4}{x} + 5y = 7$$

$$\frac{3}{x} + 4y = 5. \qquad \text{(CBSE 2003)}$$

Sol. The given system of equations is

$$\frac{4}{x} + 5y = 7 \qquad ...(1)$$

$$\frac{3}{x} + 4y = 5 \qquad ...(2)$$

Put $\quad \frac{1}{x} = X \qquad ...(3)$

Then equations (1) and (2) can be re-written as

$$4X + 5y = 7 \qquad ...(4)$$
$$3X + 4y = 5 \qquad ...(5)$$
$$\Rightarrow 4X + 5y - 7 = 0 \qquad ...(6)$$
$$3X + 4y - 5 = 0 \qquad ...(7)$$

To solve the equations by the cross-multiplication method, we draw the diagram below :

$$\begin{array}{ccccc} & X & & y & & 1 \\ 5 & \diagdown & -7 & \diagdown & 4 & \diagdown & 5 \\ 4 & \diagup & -5 & \diagup & 3 & \diagup & 4 \end{array}$$

Then, $\dfrac{X}{(5)(-5)-(-4)(-7)}$

$$= \frac{y}{(-7)(3)-(-5)(4)}$$

$$= \frac{1}{(4)(4)-(3)(5)}$$

$$\Rightarrow \quad \frac{X}{-25+8} = \frac{y}{-2+20} = \frac{1}{16-15}$$

$$\Rightarrow \quad \frac{X}{+3} = \frac{y}{4} = \frac{1}{1}$$

$$\Rightarrow \quad X = +3 \quad \text{and} \quad y = -1$$

$$\Rightarrow \quad \frac{1}{x} = +3 \quad \text{and} \quad y = -1$$

| using (3)

$$\Rightarrow \quad x = +\frac{1}{3} \quad \text{and} \quad y = -1$$

Hence, the solution of the given system of equations is $x = +\dfrac{1}{3}, y = -1$

Verification. Substituting $x = +\dfrac{1}{3}$, $y = -1$, we find that both the equations (1) and (2) are satisfied as shown below :

$$\frac{4}{x} + 5y = \frac{4}{\left(+\frac{1}{3}\right)} + 5(-1) = 12 - 5 = 7$$

$$\frac{3}{x} + 4y = \frac{3}{\left(\frac{1}{3}\right)} + 4(-1) = 9 - 4 = 5$$

This verifies the solution.

Example 6. *A motor boat takes 6 hours to cover 100 km downstream and 30 km upstream. If the boat goes 75 km downstream and returns back to the starting point in 8 hours. Find the speed of the boat in still water and the speed of the stream.* (CBSE 2002)

Sol. Let the speed of the motor boat in still water be x km/hour and the speed of the stream by y km/hour. Then,

The speed of the motor boat downstream
$$= (x + y) \text{ km/hour}$$
and, the speed of the motor boat upstream
$$= (x - y) \text{ km/hour}$$

Also, time = $\dfrac{\text{distance}}{\text{speed}}$

In the first case, the motor boat takes 6 hours to cover 100 km downstream and 30 km upstream.

$$\therefore \quad \frac{100}{x+y} + \frac{30}{x-y} = 6 \quad \ldots(1)$$

In the second case, the motor boat goes 75 km downstream and returns back to the starting point in 8 hours.

$$\therefore \quad \frac{75}{x+y} + \frac{75}{x-y} = 8 \quad \ldots(2)$$

Put $\dfrac{1}{x+y} = u$ ...(3)

and, $\dfrac{1}{x-y} = v$ ...(4)

Then equations (1) and (2) can be re-written as
$$100u + 30v = 6 \quad \ldots(5)$$
$$75u + 75v = 8 \quad \ldots(6)$$

From equation (5)
$$50u + 15v = 3 \quad \ldots(7)$$
| Dividing by 2

Multiplying equation (7) by 5, we get
$$250u + 75u = 15 \quad \ldots(8)$$

Subtracting equation (6) from equation (8), we get
$$175u = 7$$
$$\Rightarrow \quad u = \frac{7}{175} = \frac{1}{25} \quad \ldots(9)$$

Substituting this values of u in equation (7), we get
$$50\left(\frac{1}{25}\right) + 15v = 3$$
$$\Rightarrow \quad 2 + 15v = 3$$

$$\Rightarrow \quad 15v = 1$$
$$\Rightarrow \quad v = \frac{1}{15} \quad \ldots(10)$$

From equation (3) and equation (9), we get
$$\frac{1}{x+y} = \frac{1}{25}$$
$$\Rightarrow \quad x + y = 25 \quad \ldots(11)$$

From equation (4) and equation (10), we get
$$\frac{1}{x-y} = \frac{1}{15}$$
$$\Rightarrow \quad x - y = 15 \quad \ldots(12)$$

Adding equation (11) and equation (12), we get
$$2x = 40$$
$$\Rightarrow \quad x = \frac{40}{2} = 20$$

Substituting this value of x in equation (11), we get
$$20 + y = 25$$
$$\Rightarrow \quad y = 5$$

Hence, the speed of the boat in still water is 20 km/hour and the speed of the stream is 5 km/hour.

Verification. Substituting $x = 20$, $y = 5$, we find that both the equations (1) and (2) are satisfied as shown below :

$$\frac{100}{x+y} + \frac{30}{x-y} = \frac{100}{20+5} + \frac{30}{20-5}$$
$$= 4 + 2 = 6$$
$$\frac{75}{x+y} + \frac{75}{x-y} = \frac{75}{20+5} + \frac{75}{20-5}$$
$$= 3 + 5 = 8$$

Hence, the solution we have got is correct.

TEST YOUR KNOWLEDGE

1. Solve the pair of linear equations:
$$\frac{2}{x} + \frac{3}{y} = 13$$
$$\frac{5}{x} - \frac{4}{y} = -2.$$

2. Solve the following pair of equations by reducing them to a pair of linear equations:
$$\frac{5}{x-1} + \frac{1}{y-2} = 2$$
$$\frac{6}{x-1} - \frac{3}{y-2} = 1.$$

3. Solve the following system of equations:
$$\frac{1}{2(2x+3y)} + \frac{12}{7(3x-2y)} = \frac{1}{2}$$
$$\frac{7}{2x+3y} + \frac{4}{3x-2y} = 2;$$
$2x + 3y \neq 0, 3x - 2y \neq 0.$

4. Solve the following system of equations:
$$\frac{6}{x+y} = \frac{7}{x-y} + 3$$
$$\frac{1}{2(x+y)} = \frac{1}{3(x-y)}; x+y \neq 0, x-y \neq 0.$$

5. Solve the system of equations
$$\frac{3}{x} - \frac{1}{y} = -9$$
$$\frac{2}{x} + \frac{3}{y} = 5.$$

6. Solve the following system of equations
$$\frac{15}{u} + \frac{2}{v} = 17$$
$$\frac{1}{u} + \frac{1}{v} = \frac{36}{5}.$$

7. Solve the following system of equations:
$3(2u + v) = 7uv$
$3(u + 3v) = 11uv.$

8. Solve the following system of equations:
$$\frac{1}{7x} + \frac{1}{6y} = 3$$
$$\frac{2}{2x} - \frac{1}{3y} = 5; x \neq 0, y \neq 0.$$

9. Solve the following system of equations:
$9x + 5y = 37xy$
$7x - 4y = 13xy; x \neq 0, y \neq 0.$

10. 2 men and 7 boys can do a piece of work in 4 days. The same work is done in 3 days by 4 men and 4 boys. How long respectively would it take for one man and one boy is do it?

Answers

1. $x = \frac{1}{2}, y = \frac{1}{3}$
2. $x = 4, y = 5$
3. $x = 2, y = 1$
4. $x = -\frac{5}{4}, y = -\frac{1}{4}$
5. $x = -\frac{1}{2}, y = \frac{1}{3}$
6. $u = 5, v = \frac{1}{7}$
7. $u = 1, v = \frac{3}{2}$
8. $x = \frac{1}{14}, y = \frac{1}{6}$
9. $x = \frac{1}{2}, y = \frac{1}{3}$
10. 15 days, 60 days.

ILLUSTRATIVE EXAMPLES

[NCERT Exercise 3.7 (Optional)*]
(Page No. 68)

Example 1. *The ages of two friends Ani and Biju differ by 3 years. Ani's father Dharam is twice as old as Ani and Biju is twice as old as his sister Cathy. The ages of Cathy and Dharam differ by 30 years. Find the ages of Ani and Biju.*

Sol. Let the ages of Ani and Biju be x years and y years respectively. Then, according to the question,
$$x - y = \pm 3 \qquad \ldots(1)$$
Age of Ani's father Dharam = $2x$ years

*These exercises are not from examinations point of view.

Age of Biju's sister = $\frac{y}{2}$ years.

According to the question,

$$2x - \frac{y}{2} = 30$$

$$\Rightarrow 4x - y = 60 \qquad ...(2)$$

Case I. When $x - y = 3$

Then, we have

$$x - y = 3 \qquad ...(1)$$
$$4x - y = 60 \qquad ...(2)$$

Subtracting equation (1) from eq. (2)

$$3x = 57 \qquad ...(4)$$

$$x = \frac{57}{3} = 19 \text{ years}$$

Substituting the value of x in eqn. (1)

$$19 - y = 3$$
$$y = 19 - 3$$
$$= 16$$

Ani's age = 19 years

Biju's age = 16 years

Verification

$$x - y = 19 - 16 = 3$$

$$4x - \frac{y}{4} = 4 \times 19 - 16 = 76 - 16$$

$$= 60$$

This verifies the solution.

Case II. When $x - y = -3$

Then, we have

$$x - y = -3 \qquad ...(1)$$
$$4x - y = 60 \qquad ...(2)$$

Subtracting equation (1) from equation (2), we get,

$$3x = 63$$

$$\Rightarrow x = \frac{63}{3} = 21$$

Substituting the value of x the equation (1), we get

$$21 - y = -3$$

$$\Rightarrow y = 24$$

Anil age = 21 years

Biju age = 24 years

$$x - y = 21 - 24 = -3$$
$$4x - y = 4(21) - 24$$
$$= 84 - 24 = 60$$

This satisfied the solution.

Example 2. *One says, "Give me a hundred, friend! I shall then become twice as rich as you". The other replies, "If you give me ten, I shall be six times as rich as you". Tell me what is the amount of their (respective) capital? [From the Bijaganita of Bhaskara II]*

[**Hint :** $x + 100 = 2(y - 100)$, $y + 10 = 6(x - 10)$.]

Sol. Let the amounts of their respective capitals be Rs x and Rs y respectively.

Then, according to the question,

$$x + 100 = 2(y - 100)$$

$$\Rightarrow x - 2y = -300 \qquad ...(1)$$

and $\quad 6(x - 10) = y + 10$

$$\Rightarrow 6x - y = 70 \qquad ...(2)$$

From equation (1),

$$x = 2y - 300 \qquad ...(3)$$

Substitute the value of x in equation (2), we get

$$6(2y - 300) - y = 70$$
$$\Rightarrow 12y - 1800 - y = 70$$
$$\Rightarrow 11y = 1870$$
$$\Rightarrow y = \frac{1870}{11} = 170$$

Substituting this value of y in equation (3), we get

$$x = 2(170) - 300$$
$$= 340 - 300 = 40$$

So, the solution of the equations (1) and (2) is $x = 40$ and $y = 170$. Hence, the amounts of their respective capitals are Rs. 40 and Rs. 170 respectively.

Verification. Substituting $x = 40$, $y = 170$, we find that both the equations (1) and (2) are satisfied as shown below :

$$x - 2y = 40 - 2(170)$$
$$= 40 - 340 = -300$$
$$6x - y = 6(40) - 170 = 240 - 170$$
$$= 70$$

Hence, the solution we have got is correct.

Example 3. *A train covered a certain distance at a uniform speed. If the train would have been 10 km/h faster, it would have taken 2 hours less than the scheduled time. And, if the train were slower by 10 km/h ; it would have taken 3 hours more than the scheduled time. Find the distance covered by the train.*

Sol. Let the actual speed of the train be x km/hour and the actual time taken by y hours. Then,

$$\text{distance} = (xy) \text{ km} \qquad ...(1)$$

| Distance = speed × time

According to the equation,

$$xy = (x + 10)(y - 2)$$
$$\Rightarrow \quad xy = xy - 2x + 10y - 20$$
$$\Rightarrow \quad 2x - 10y + 20 = 0$$
$$\Rightarrow \quad x - 5y + 10 = 0 \qquad ...(1)$$

| Dividing throughout by 2

and, $\quad xy = (x - 10)(y + 3)$
$$\Rightarrow \quad xy = xy + 3x - 10y - 30$$
$$\Rightarrow \quad 3x - 10y - 30 = 0 \qquad ...(2)$$

To solve the equations (1) and (2) by the cross-multiplication method, we draw the diagram below :

```
    x       y       1
  -5   10   -5   1    -5
     ╳       ╳       ╳
 -10   -30   3       -10
```

Then, $\dfrac{x}{(-5)(-30) - (-10)(10)}$

$= \dfrac{y}{(10)(3) - (-30)(1)}$

$= \dfrac{1}{(1)(-10) - (3)(-5)}$

$$\Rightarrow \quad \dfrac{x}{150 + 100} = \dfrac{y}{30 + 30} = \dfrac{1}{-10 + 15}$$

$$\Rightarrow \quad \dfrac{x}{250} = \dfrac{y}{60} = \dfrac{1}{5}$$

$$\Rightarrow \quad x = \dfrac{250}{5} = 50$$

and $\quad y = \dfrac{60}{5} = 12$

So, the solution of the equations (1) and (2) is $x = 50$ and $y = 12$. Hence, the distance covered by the train is $50 \times 12 = 600$ km.

Verification. Substituting $x = 50$, $y = 12$, we find that both the equations (1) and (2) are satisfied as shown below. Hence, the solution we have got is correct.

Example 4. *The students of a class are made to stand in rows. If 3 students are extra in a row, there would be 1 row less. If 3 students are less in a row, there would be 2 rows more. Find the number of students in the class.*

Sol. Let the number of students in the class be x and the number of rows be y. Then, number of students in each now = $\dfrac{x}{y}$. According to the equation.

If 3 students are extra in row. Then there would be 1 row less, *i.e.*, when each row has $\left(\dfrac{x}{y} + 3\right)$ students. Then the number of rows is $(y - 1)$.

∴ Total number of students = number of row × Number of students in each row

$$\Rightarrow \quad x = \left(\dfrac{x}{y} + 3\right)(y - 1)$$

$$\Rightarrow \quad x = x - \dfrac{x}{y} + 3y - 3$$

$$\Rightarrow \quad \dfrac{x}{y} - 3y + 3 = 0 \qquad ...(1)$$

and, if 3 students are less in a row, then there would be 2 rows more, *i.e.*, when each row has $\left(\dfrac{x}{y} + 3\right)$ students, then the number of rows is $(y + 2)$.

∴ Total number of students = Number of rows × Number of students in each row

$$\Rightarrow \quad x = \left(\dfrac{x}{y} - 3\right)(y + 2)$$

PAIR OF LINEAR EQUATIONS IN TWO VARIABLES

$\Rightarrow \quad x = x + \dfrac{2x}{y} - 3y - 6$

$\Rightarrow \quad \dfrac{2x}{y} - 3y - 6 = 0 \qquad \ldots(2)$

Put $\dfrac{x}{y} = u \qquad \ldots(3)$

Then equations (1) and (2) can be re-written as

$u - 3y + 3 = 0 \qquad \ldots(4)$
$2u - 3y - 6 = 0 \qquad \ldots(5)$

Subtracting equation (4) from equation (5), we get

$u - 9 = 0$
$\Rightarrow \quad u = 9 \qquad \ldots(6)$

Substituting this value of x in equation (4), we get

$9 - 3y + 3 = 0$
$\Rightarrow \quad -3y + 12 = 0$
$\Rightarrow \quad 3y = 12$
$\Rightarrow \quad y = \dfrac{12}{3} = 4 \qquad \ldots(7)$

From equation (3) and equation (6), we get

$\dfrac{x}{y} = 9$

$\Rightarrow \quad \dfrac{x}{4} = 9 \qquad$ | using (7)

$\Rightarrow \quad x = 36 \qquad \ldots(8)$

So, the solution of the equations (1) and (2) is $x = 36$ and $y = 4$. Hence, the number of students in the class $= xy = (36)(4) = 144$.

Verification. Substituting $x = 36$, $y = 4$, we find that both the equations (1) and (2) are satisfied as shown below :

$\dfrac{x}{y} - 3y + 3 = \dfrac{36}{4} - 3(4) + 3$

$\qquad = 9 - 12 + 3 = 0$

$\dfrac{2x}{y} - 3y - 6 = \dfrac{2(36)}{4} - 3(4) - 6$

$\qquad = 18 - 12 - 6 = 0$

Hence, the solution is correct.

Example 5. *In a $\triangle ABC$, $\angle C = 3$, $\angle B = 2(\angle A + \angle B)$. Find the three angles.*

Sol. We have

$\angle C = 3 \angle B = 2(\angle A + \angle B) \qquad \ldots(1)$

We know that the sum of the measures of three angles of a triangle is 180°.

$\therefore \quad \angle A + \angle B + \angle C = 180° \qquad \ldots(2)$

(1) and (2) give

$\angle A + \angle B + 2(\angle A + \angle B) = 180°$
$\Rightarrow \quad 3\angle A + 3\angle B = 180°$
$\Rightarrow \quad \angle A + \angle B = 60° \qquad \ldots(3)$

| Dividing throughout by 3

and, $\angle A + \angle B + 3\angle B = 180°$
$\Rightarrow \quad \angle A + 4\angle B = 180° \qquad \ldots(4)$

Subtracting equation (3) from equation (4), we get

$3\angle B = 120°$

$\Rightarrow \quad \angle B = \dfrac{120°}{3} = 40°$

Substituting this value $\angle B = 40°$ in equation (3), we get

$\angle A + 40° = 60°$
$\Rightarrow \quad \angle A = 60° - 40°$
$\Rightarrow \quad \angle A = 20°$

Again, from (1)

$\angle C = 3 \angle B = 3(40°) = 120°$

Hence, the three angles of the $\triangle ABC$ are given by $\angle A = 20°$, $\angle B = 40°$ and $\angle C = 120°$.

Verification. Substituting $\angle A = 20°$, $\angle B = 40°$ and $\angle C = 120°$, we find that (1) is satisfied as shown below :

$\angle C = 3 \angle B = 2(\angle A + \angle B)$
$\Rightarrow \quad 120° = 3(40°) = 2(20° + 40°)$

This verifies the solution.

Example 6. *Draw the graphs of the equations $5x - y = 5$ and $3x - y = 3$. Determine the co-ordinate of the vertices of the triangle formed by these lines and the y axis.*

Sol. The given equations are

$5x - y = 5 \qquad \ldots(1)$
$3x - y = 3 \qquad \ldots(2)$

Let us draw the graphs of equations (1) and (2) by finding two solutions of each of the equations. Those two solutions of each of the equations (1) and (2) are given below in table 1 and table 2 respectively.

For equation (1)

$y = 5x - 5$

Table 1 of solutions

x	1	2
y	0	5

For equation (2)

$y = 3x - 3$

Table 2 of solutions

x	2	3
y	3	6

We plot the points A(1, 0) and B(2, 5) on a graph paper and join there points to

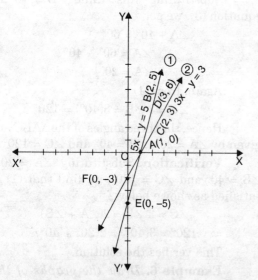

form the line AB representing the equation (1) as shown in the figure. Also, we plot the points C(2, 3) and D(3, 6) on the same graph paper and join these points to form the line CD representing the equation (2) as shown in the same figure.

In the figure we observe that the coordinates of the vertices of the triangle AEF and A(1, 0), B(0, −5) and F(0, −3).

Example 7. *Solve the following pair of linear equations :*

(i) $px + qy = p - q$
$qx - py = p + q$

(ii) $ax + by = c$
$bx + ay = 1 + c$

(iii) $\dfrac{x}{a} - \dfrac{y}{b} = 0$
$ax + by = a^2 + b^2$

(iv) $(a - b)x + (a + b)y = a^2 - 2ab - b^2$
$(a + b)(x + y) = a^2 + b^2$

(v) $152x - 378y = -74$
$-378x + 152y = -604$

Sol. (i) $px + qy = p - q$
$qx - py = p + q$

The given pair of linear equations is
$px + qy = p - q$ ...(1)
$qx - py = p + q$ ...(2)

Multiplying equation (1) by p and equation (2) by q, we get
$p^2x + pqy = p^2 - pq$ ...(3)
$q^2x - pqy = pq + q^2$ ...(4)

Adding equation (3) and equation (4), we get
$(p^2 + q^2)x = p^2 + q^2$
$\Rightarrow \quad x = \dfrac{p^2 + q^2}{p^2 + q^2} = 1$

Substituting this value of x in equation (1), we get
$p(1) + qy = p - q$
$\Rightarrow \quad qy = -q$
$\Rightarrow \quad y = \dfrac{-q}{q} = -1$

So, the solution of the given pair of linear equations is $x = +1, y = -1$.

Verification. Substituting $x = 1, y = -1$, we find that both the equations (1) and (2) are satisfied as shown below :
$px + qy = p(1) + q(-1) = p - q$
$qx - py = q(1) - p(-1) = q + p = p + q$

This verifies the solution.

PAIR OF LINEAR EQUATIONS IN TWO VARIABLES

(ii) **ax + by = c**
    bx + ay = 1 + c

The given pair of linear equations is

$$ax + by = c \quad \ldots(1)$$
$$bx + ay = 1 + c \quad \ldots(2)$$
$$\Rightarrow \quad ax + by - c = 0 \quad \ldots(3)$$
$$bx + ay - (1 + c) = 0 \quad \ldots(4)$$

To solve the equations by the cross-multiplication method, we draw the diagram below :

$$\begin{array}{cccc} b & x & -c & y & a & 1 & b \\ & \times & & \times & & \times & \\ a & & -(1+c) & & b & & a \end{array}$$

Then, $\dfrac{x}{(b)\{-(1+c)\} - (a)(-c)}$

$= \dfrac{y}{(-c)(b) - \{-(1+c)\}(a)}$

$= \dfrac{1}{(a)(a) - (b^2)(b)}$

$= \dfrac{x}{-b - bc + ac} = \dfrac{y}{-bc + a + ac}$

$= \dfrac{1}{a^2 - b^2}$

$\Rightarrow \quad x = \dfrac{-b - bc + ac}{a^2 - b^2}$

$y = \dfrac{-bc + a + ac}{a^2 - b^2}$

Hence, the solution of the given pair of linear equations is

$$x = \dfrac{-b - bc + ac}{a^2 - b^2}, \; y = \dfrac{-bc + a + ac}{a^2 - b^2}$$

Verification. Substituting

$$x = \dfrac{-b - bc + ac}{a^2 - b^2},$$

$$y = \dfrac{-bc + a + ac}{a^2 - b^2},$$

we find that both the equations (1) and (2) are satisfied as shown below :

$$ax + by = a\left(\dfrac{-b - bc + ac}{a^2 - b^2}\right)$$

$$+ b\left(\dfrac{-bc + a + ac}{a^2 - b^2}\right)$$

$$= \dfrac{-ab - abc + a^2c - b^2c + ab + abc}{a^2 - b^2} = c$$

$$bx + ay = b\left(\dfrac{-b - bc + ac}{a^2 - b^2}\right)$$

$$+ a\left(\dfrac{-bc + a + ac}{a^2 - b^2}\right)$$

$$= \dfrac{-b^2 - b^2c + abc - abc + a^2 + a^2c}{a^2 - b^2}$$

$$= \dfrac{a^2 - b^2 + (a^2 - b^2)c}{a^2 - b^2} = 1 + c$$

This verifies the solution.

(iii) $\dfrac{\mathbf{x}}{\mathbf{a}} - \dfrac{\mathbf{y}}{\mathbf{b}} = 0$
    $\mathbf{ax + by = a^2 + b^2}$

The given pair of linear equations is

$$\dfrac{x}{a} - \dfrac{y}{b} = 0 \quad \ldots(1)$$
$$ax + by = a^2 + b^2 \quad \ldots(2)$$

From equation (1),

$$\dfrac{y}{b} = \dfrac{x}{a}$$

$\Rightarrow \quad y = \dfrac{b}{a}x \quad \ldots(3)$

Substituting the value of y in equation (2), we get

$$ax + b\left(\dfrac{b}{a}x\right) = a^2 + b^2$$

$$= \dfrac{a^2 + b^2}{a}x = a^2 + b^2$$

$\Rightarrow \quad x = \dfrac{a(a^2 + b^2)}{a^2 + b^2} = a$

Substituting this value of x in equation (3), we get

$$y = \frac{b}{a}(a) = b$$

Hence, the solution of the given pair of linear equations is $x = a, y = b$.

Verification. Substituting $x = a, y = b$, we find that both the equations (1) and (2) are satisfied as shown below :

$$\frac{x}{a} - \frac{y}{b} = \frac{a}{a} - \frac{b}{b} = 1 - 1 = 0$$

$$ax + by = a(a) + b(b) = a^2 + b^2$$

This verifies the solution.

(iv) **(a − b) x + (a + b) y = a² − 2ab − b²**
(a + b) (x + y) = a² + b²

The given pair of linear equations is

$$(a - b)x + (a + b)y = a^2 - 2ab - b^2 \quad ...(1)$$

$$(a + b)(x + y) = a^2 + b^2$$

$$\Rightarrow (a + b)x + (a + b)y = a^2 + b^2 \quad ...(2)$$

Subtracting equation (2) from equation (1), we get

$$-2bx = -2ab - 2b^2$$

$$\Rightarrow x = \frac{-2ab - 2b^2}{-2b}$$

$$= a \frac{-2b(a + b)}{-2b} = a + b$$

Substituting this value of x in equation (1), we get

$$(a - b)(a + b) + (a + b)y = a^2 - 2ab - b^2$$

$$\Rightarrow a^2 - b^2 + (a + b)y = a^2 - 2ab - b^2$$

$$\Rightarrow (a + b)y = -2ab$$

$$\Rightarrow y = \frac{-2ab}{a + b}$$

Hence, the solution of the given pair of linear equations is

$$x = a + b, y = \frac{-2ab}{a + b}$$

Verification. Substituting $x = a + b$, $y = \frac{-2ab}{a+b}$, we find that both the equation (1) and (2) are satisfied as shown below :

$(a - b)x + (a + b)y$

$$= (a - b)(a + b) + (a + b)\left(\frac{-2ab}{a+b}\right)$$

$$= a^2 - b^2 - 2ab$$

$$(a + b)(x + y) = (a + b)\left\{a + b - \frac{2ab}{a+b}\right\}$$

$$= (a + b)^2 - 2ab = a^2 + b^2$$

This verifies the solution.

(v) **152x − 378y = − 74**
− 378x + 152y = − 604

The given pair of linear equations is

$$152x - 378y = -74 \quad ...(1)$$

$$-378x + 152y = -604 \quad ...(2)$$

Adding equation (1) and equation (2), we get

$$-226x - 226y = -678$$

$$\Rightarrow x + y = 3 \quad ...(3)$$

| Dividing throughout by − 226

Subtracting equation (2) from equation (1), we get

$$530x - 530y = 530$$

$$\Rightarrow x - y = 1 \quad ...(4)$$

| Dividing throughout by 530

Adding equation (3) and equation (4), we get

$$2x = 4$$

$$\Rightarrow x = \frac{4}{2} = 2$$

Subtracting equation (4) from equation (3), we get

$$2y = 2$$

$$\Rightarrow y = \frac{2}{2} = 1$$

Hence, the solution of the given pair of linear equations is $x = 2, y = 1$.

PAIR OF LINEAR EQUATIONS IN TWO VARIABLES

Verification. Substituting $x = 2$, $y = 1$, we find that both the equations (1) and (2) are satisfied as shown below

$$152x - 378y = (152)(2) - (378)(1)$$
$$= 304 - 378 = -74$$
$$-378x + 152y = (-378)(2) + (152)(1)$$
$$= -756 + 152 = -604$$

This verifies the solution.

Example 8. *ABCD is a cyclic quadrilateral. Find the angles of the cyclic quadrilateral*

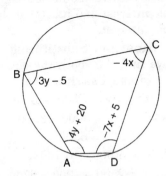

Sol. We know that the opposite angles of a cyclic quadrilateral are supplementary, therefore,

$$A + C = 180°$$
$$\Rightarrow 4y + 20° + (-4x) = 180°$$
$$\Rightarrow 4y - 4x = 160°$$
$$\Rightarrow y - x = 40° \qquad ...(1)$$

| dividing throughout by 4

and,
$$B + D = 180°$$
$$\Rightarrow 3y - 5 + (-7x + 5) = 180°$$
$$\Rightarrow 3y - 5 - 7x + 5 = 180°$$
$$\Rightarrow 3y - 7x = 180° \qquad ...(2)$$

From equation (1), $y = 40° + x \qquad ...(3)$

Substituting this value of y in equation (2), we get

$$3(40° + x) - 7x = 180°$$
$$120° + 3x - 7x = 180°$$
$$-4x = 60°$$
$$x = \frac{60}{-40} = -15°$$

Substituting $x = -15°$ in equation (3), we get,
$$y = 40° - 15° = 25°$$

$$\therefore A = 4(25) + 20 = 120°$$
$$B = 3(25) - 5 = 70°$$
$$C = -4 \times (-15) = 60°$$
$$D = -7 \times (-15) + 5$$
$$= 105 + 5 = 110°.$$

Hence, the angles of the cyclic quadrilateral are

$$\angle A = 120°, \angle B = 70°, \angle C = 60°$$
and $\angle D = 110°$.

Verification. Substituting $x = 15$, $y = 25$. We find that both the equations (1) and (2) are satisfied as shown below :

$$x + y = 15 + 25 = 40°$$
$$7x + 3y = 7(15) + 3(25)$$
$$= 105° + 75° = 180°$$

This verifies the solution.

ADDITIONAL EXAMPLES

Example 9. *Seven times a two-digit number is equal to four times the number obtained by reversing the digits. If the difference between the digits is 3. Find the number.*

Sol. Let in the two-digit number,
unit's digit = x
and, ten's digit = y
$\therefore$ The two-digit number = $10y + x$
$\therefore$ The number obtained by reversing the digit = $10x + y$

According to the question,
$$7(10y + x) = 4(10x + y)$$
$$\Rightarrow 70y + 7x = 40x + 4y$$
$$\Rightarrow 33x - 66y = 0$$
$$\Rightarrow x - 2y = 0$$
...(1) | Dividing by 33

Also, we are given that the difference between the digits is 3.
$$\therefore x - y = 3$$
...(2) | $\because$ From (1), $x > y$

Subtracting equation (2) from equation (1), we get
$$-y = -3$$
$$\Rightarrow y = 3$$

Substituting this value of y in equation (2), we get
$$x - 3 = 3$$
$$\Rightarrow \quad x = 3 + 3 = 6$$
Hence, the required number is 36.

Verification. Substituting $x = 6$, $y = 3$, we find that both the equations (1) and (2) are satisfied as shown below:
$$x - 2y = 6 - 2(3) = 0$$
$$x - y = 6 - 3 = 3$$
Hence, the solution we have got is correct.

Example 10. *A person invested some amount the rate of 12% simple interest and the remaining at 10%. He received yearly interest of Rs 130 but if he had interchanged the amounts invested, he would have received Rs 4 more interest. How much money did he invest at different rates?*

Sol. Suppose that he invested Rs x at the rate of 12% simple interest and Rs y at the rate of 10% simple interest.

Then, according to the question,
$$\frac{y . 12 . 1}{100} + \frac{x . 10 . 1}{100} = 130 + 4$$
$$\Rightarrow \quad 12x + 10y = 13000$$
| Dividing throughout by 2
$$\Rightarrow \quad 6x + 5y = 6500 \qquad ...(1)$$
and,
$$\frac{y . 12 . 1}{100} + \frac{x . 10 . 1}{100} = 130 + 4$$
$$\Rightarrow \quad 12y + 10x = 13400$$
$$\Rightarrow \quad 6y + 5x = 6700$$
| Dividing throughout by 2
$$\Rightarrow \quad 5x + 6y = 6700 \qquad ...(2)$$

Multiplying equation (1) by 6 and equation (2) by 5, we get
$$36x + 30y = 39000 \qquad ...(3)$$
$$25x + 30y = 33500 \qquad ...(4)$$
Subtracting equation (4) from equation (3), we get
$$11x = 5500$$
$$\Rightarrow \quad x = \frac{5500}{11} = 500$$

Substituting this value of x in equation (1), we get
$$6(500) + 5y = 6500$$
$$\Rightarrow \quad 3000 + 5y = 6500$$
$$\Rightarrow \quad 5y = 6500 - 3000$$
$$\Rightarrow \quad 5y = 3500$$
$$\Rightarrow \quad y = \frac{3500}{5} = 700$$

So, the solution of the equations (1) and (2) is $x = 500$ and $y = 700$.

Hence, he invested Rs 500 at the rate of 12% simple interest and Rs 700 at the rate of 10% simple interest.

Verification. Substituting $x = 500$, $y = 700$, we find that both the equations (1) and (2) are satisfied as shown below:
$$6x + 5y = 6(500) + 5(700)$$
$$= 3000 + 3500$$
$$= 6500$$
$$5x + 6y = 5(500) + 6(700)$$
$$= 2500 + 4200 = 6700$$
This verified the solution.

Example 11. *A and B have certain number of oranges A says to B "If you give me 10 of yours oranges I will have twice the number of oranges left with you". B replies," If you give me 10 of your oranges I will have the same number of oranges as left with you". Find the number of oranges with A and B separately.*

Sol. Let the number of oranges with A and B separately be x and y respectively.

Then, according to the question,
$$x + 10 = 2(y - 10)$$
$$\Rightarrow \quad x + 10 = 2y - 20$$
$$\Rightarrow \quad x - 2y = -20 - 10$$
$$\Rightarrow \quad x - 2y = -30 \qquad ...(1)$$
$$x - 10 = y + 10$$
$$\Rightarrow \quad x - y = 10 + 10$$
$$\Rightarrow \quad x - y = 20 \qquad ...(2)$$

Subtracting equation (1) from equations (2), we get
$$y = 50$$

Substituting this value of y in equation (2), we get
$$x - 50 = 20$$
$$\Rightarrow \quad x = 50 + 20$$

PAIR OF LINEAR EQUATIONS IN TWO VARIABLES

$\Rightarrow \qquad x = 70$

So the solution of the equations (1) and (2) is $x = 70$ and $y = 50$. Hence, the number of mangoes with A and B separately and 70 and 50 respectively.

Verification. Substituting $x = 70$, $y = 50$, we find that both the equations (1) and (2) are satisfied as shown below :

$x - 2y = 70 - 2(50) = 70 - 100 = -30$

$x - y = 70 - 50 = 20$

Hence, the solution is correct.

Example 12. *A takes 3 hours more than B to walk a distance of 30 km. But, if A doubles his pace (speed) he is ahead of B by $1\frac{1}{2}$ hours, find their speeds of walking.*

Sol. Let the speeds of walking of A and B be x km/hour and y km/hour respectively.

Time taken by A to walk a distance of 30 km = $\dfrac{30}{x}$ hours

$\left|\text{Time} = \dfrac{\text{Distance}}{\text{Speed}}\right.$

Time taken by B to walk a distance of 30 km = $\dfrac{30}{y}$ hours

$\left|\text{Time} = \dfrac{\text{Distance}}{\text{Speed}}\right.$

According to the question,

$$\dfrac{30}{x} = \dfrac{30}{y} + 3$$

$\Rightarrow \quad \dfrac{30}{x} - \dfrac{30}{y} = 3$

$\Rightarrow \quad \dfrac{1}{x} - \dfrac{1}{y} = \dfrac{3}{30}$

| Dividing throughout by 30

$\Rightarrow \quad \dfrac{1}{x} - \dfrac{1}{y} = \dfrac{1}{10}$ ...(1)

When A doubles his pace (speed), then speed of A = $2x$ km/hour. Now,

Time taken by A to walk a distance of 30 km = $\dfrac{30}{2x}$

$\left|\text{Time} = \dfrac{\text{Distance}}{\text{Speed}}\right.$

$= \dfrac{15}{x}$ hours

According to the question,

$$\dfrac{15}{x} + 1\dfrac{1}{2} = \dfrac{30}{y}$$

$\Rightarrow \quad \dfrac{15}{x} = \dfrac{30}{y} - \dfrac{3}{2}$

$\Rightarrow \quad \dfrac{15}{x} - \dfrac{30}{y} = -\dfrac{3}{2}$

$\Rightarrow \quad \dfrac{1}{x} - \dfrac{2}{y} = -\dfrac{1}{10}$ ...(2)

| Dividing throughout by 15

Subtracting equation (2) from equation (1), we get

$\dfrac{1}{y} = \dfrac{1}{5}$

$\Rightarrow \quad y = 5$

Substituting $y = 5$ in equation (1), we get

$\dfrac{1}{x} - \dfrac{1}{5} = \dfrac{1}{10}$

$\Rightarrow \quad \dfrac{1}{x} = \dfrac{1}{5} + \dfrac{1}{10}$

$\Rightarrow \quad \dfrac{1}{x} = \dfrac{3}{10}$

$\Rightarrow \quad x = \dfrac{10}{3}$

So, the solution of the equations (1) and (2) is $x = \dfrac{1}{3}$ and $y = 5$. Hence, their speeds of walking and $\dfrac{10}{3}$ km/hour and 5 km/hour respectively.

Verification. Substituting $x = \dfrac{10}{3}$, $y = 5$, we find that both the equations (1) and (2) are satisfied as shown below :

$$\frac{1}{x} - \frac{1}{y} = \frac{1}{\left(\frac{10}{3}\right)} - \frac{1}{5} = \frac{3}{10} - \frac{1}{5} = \frac{1}{10}$$

$$\frac{1}{x} - \frac{2}{y} = \frac{1}{\left(\frac{10}{3}\right)} - \frac{2}{5} = \frac{3}{10} - \frac{2}{5} = -\frac{1}{10}$$

This verifies the solution.

Example 13. *There are two class rooms A and B. If 10 students are sent from A to B, the number of students in each room becomes the same. If 20 students are sent from B to A, the number of students in A becomes double the number of students in B. Find the number of students in each class room.* $(A = 100, B = 80)$.

Sol. Let the number of students in class rooms A and B be x and y respectively.

Then, according to the question,

$$x - 10 = y + 10$$
$$\Rightarrow \quad x - y - 20 = 0 \qquad \ldots(1)$$

and $\quad x + 20 = 2(y - 20)$
$$\Rightarrow \quad x - 2y + 60 = 0 \qquad \ldots(2)$$

To solve the equations (1) and (2) by the cross-multiplication method, we draw the diagram below:

$$\begin{array}{ccccc}
-1 & x & -20 & y & 1 & & -1 \\
-2 & & 60 & & 1 & & -2
\end{array}$$

Then, $\dfrac{x}{(-1)(60) - (-2)(-20)}$

$$= \dfrac{y}{(-20)(1) - (60)(1)}$$

$$= \dfrac{1}{(1)(-2) - (1)(-1)}$$

$$= \dfrac{x}{-60 - 40} = \dfrac{y}{-20 - 60} = \dfrac{1}{-2 + 1}$$

$$\Rightarrow \quad \dfrac{x}{-100} = \dfrac{y}{-80} = \dfrac{1}{-1}$$

$$\Rightarrow \quad x = \dfrac{-100}{-1} = 100$$

and $\quad y = \dfrac{-80}{-1} = 80$

So, the solution of the equations (1) and (2) is $x = 100$ and $y = 80$.

Hence, the number of students in class rooms A and B are 100 and 80 respectively.

Verification. Substituting $x = 100$, $y = 80$, we find that both the equations (1) and (2) are satisfied as shown below.

$$x - y - 20 = 100 - 80 - 20 = 0$$
$$x - 2y + 60 = 100 - 2(80) + 60 = 0$$

Hence, the solution we have got is correct.

Example 14. *On selling a T.V. at 5% gain and a fridge at 10% gain. A shopkeeper gains Rs 2000. But if he sells the T.V. at 10% gain and the fridge at 5% loss. He gains Rs 1500 on the transaction. Find the actual price of the T.V. and the fridge.*

Hint: $\left[\begin{array}{l} \dfrac{5}{100}x + \dfrac{10}{100}y = 200 \\ \dfrac{10}{100}x + \dfrac{5}{100}y = 1500. \end{array}\right]$

Sol. Let the actual price of the T.V. and the fridge be Rs x and Rs y respectively.

Then, according to the question,

$$\left(\dfrac{105}{100}x - x\right) + \left(\dfrac{110}{100}y - y\right) = 2000$$

$$\Rightarrow \quad \dfrac{x}{20} + \dfrac{y}{10} = 2000$$

$$\Rightarrow \quad x + 2y = 40000 \qquad \ldots(1)$$

and, $\left(\dfrac{110}{100}x - x\right) - \left(y - \dfrac{95}{100}y\right) = 1500$

$$\Rightarrow \quad \dfrac{x}{10} - \dfrac{y}{20} = 1500$$

$$\Rightarrow \quad 2x - y = 30000 \qquad \ldots(2)$$

Multiplying equation (2) by 2, we get

$$4x - 2y = 60000 \qquad \ldots(3)$$

Adding equation (1) from equation (3), we get

$$5x = 100000$$

$$\Rightarrow \quad x = \dfrac{100000}{5} = 20000$$

Substituting $x = 20000$ in equation (1), we get

PAIR OF LINEAR EQUATIONS IN TWO VARIABLES

$20000 + 2y = 4000$

$\Rightarrow \quad 2y = 40000 - 20000 = 20000$

$\Rightarrow \quad y = \dfrac{20000}{2} = 10000$

So, the solution of the given equations is $x = 20000$ and $y = 10000$. Hence, the actual price of the T.V. and the fridge are Rs. 20000 and Rs. 10000 respectively.

Verification. Substituting $x = 20000$ and $y = 10000$, we find that both the equations (1) and (2) are satisfied as shown below.

$x + 2y = 20000 + 2(10000) = 40000$

$2x - y = 2(20000) - 10000 = 30000$

Hence, the solution we have got is correct.

Example 15. *A wizard having powers of mystic in candations and magical medicines seeing a cock, fight going on, spoke privately to both the owners of the cocks. To one he said, 'If your bird coins, then you give me your stake-money, but if you do not win, I shall give you two-thirds of that'. Going to the other, he promised in the same way to give three fourths. From both of them his gain would be only 12 gold pieces. Tell me O Ornament of the first rate mathematicians, the stake money of each of the cock-owners example from mahanvira.*

[Hint : $x - \dfrac{3}{4}y = 12$, $y - \dfrac{2}{3}x = 12$]

Sol. Let the stake money of each of the cock-owners be Rs. x and Rs. y respectively.

Then, according to the question,

$x - \dfrac{3}{4}y = 12$

$\Rightarrow \quad 4x - 3y = 48 \qquad \ldots(1)$

| Multiplying both sides by 4

and, $y - \dfrac{2}{3}x = 12$

$\Rightarrow \quad 3y - 2x = 36$

| Multiplying both sides by 3

$\Rightarrow \quad -2x + 3y = 36 \qquad \ldots(2)$

Adding equation (1) and equation (2), we get

$2x = 84$

$\Rightarrow \quad x = \dfrac{84}{2} = 42$

Substituting this value of x in equation (2), we get

$-2(42) + 3y = 36$

$\Rightarrow \quad -84 + 3y = 36$

$\Rightarrow \quad 3y = 84 + 36$

$\Rightarrow \quad 3y = 120$

$\Rightarrow \quad y = \dfrac{120}{3} = 40$

So, the solution of the equations (1) and (2) is $x = 42$ and $y = 40$. Hence, the stake money of each of the cock-owners is Rs. 42 and Rs. 40 respectively.

Verification. Substituting $x = 42$, $y = 40$, we find that both the equations (1) and (2) are satisfied as shown below :

$4x - 3y = 4(42) - 3(40)$

$\qquad = 168 - 120 = 48$

$-2x + 3y = -2(42) + 3(40)$

$\qquad = -84 + 120 = 36$

This verifies the solution.

MISCELLANEOUS EXERCISE

1. Mala purchased 5 chairs and 2 tables for Rs. 162.5 Reshma purchased 2 chairs and 1 table for Rs. 750. Represent this situation algebraically and graphically.

2. The taxi charges in a city comprise of a fixed charge together with the charge for the distance covered. For a journey of 10 km, the charges paid is Rs. 75 and for a journey of 15 km, the charges paid is Rs. 110. Represent this situation algebraically and graphically.

3. The path of a train A is given by the equation $2x + 7y = 11$ and the path of another train B is given by the equation $5x + \dfrac{35}{2}$

$y = 25$. Represent this situation geometrically.

4. The cost of 3 pens and 4 pencils is Rs. 20. The cost of 6 pens and 8 pencils is Rs. 40. Represent this situation algebraically and graphically.

5. Apala is walking along the path joining $(-3, -1)$ and $(0, -2)$ while Meenu is walking alone the path joining $(-3, -3)$ and $(0, -4)$. Represent this situation graphically.

6. Draw the graph of $2y - x = 7$ and determine from the graph whether $x = 3$, $y = 2$ is a solution or not.

7. Draw the graphs of $2y = 4x - 6$ and $3x = y + 3$ and determine whether this system of linear equations has unique solution or not.

8. Determine graphically whether the following system of equations :
$$3x - 4y = 1$$
$$8y - 6x = 4$$
is consistent or inconsistent.

9. Determine whether the graphs of the following equations are coincident or not.
$$2x + 3y = 5$$
$$6x + 9y = 15.$$

10. Solve graphically the following system of linear equations :
$$2x - y = 2$$
$$4x - y = 8$$
Also, find the coordinaters of the points where the lines meet the axis of x.

11. Solve graphically the following system of linear equations :
$$2x + y = 0$$
$$x - 2y = -2$$
Also, find the coordinates of the points where the lines meet the axes of x.

12. Solve each of the following system of equations by drawing the graphs. Determine whether these have
    (i) only one solution
    (ii) infinite number of solutions
    (iii) no solution
    In case of only one solution, find the solution.
    (i) $5x + y = 4$; $5x - y = 0$
    (ii) $2x + y - 4 = 0$; $-2x + y + 3 = 0$
    (iii) $3x - 4y + 5 = 0$; $6x - 8y + 10 = 0$
    (iv) $3x - 4y + 5 = 0$; $6x + 8y + 9 = 0$
    (v) $4x + 6y - 10 = 0$; $6x + 9y - 14 = 0$
    (vi) $4x + 6y - 10 = 0$; $6x + 9y - 15 = 0$
    (vii) $0.3x + 0.4y = 3.2$; $0.6x + 0.8y = 2.4$
    (viii) $2x + 3y = 0$; $2x - 3y = 0$
    (ix) $2x + 3y = 0$; $4x + 6y = 0$
    (x) $\dfrac{3}{2}x - \dfrac{5}{4}y = 6$; $6x - 5y = 20$.

13. Solve the following system of equations by substitution method :
$$2x - 3y = 1.3$$
$$y - x = -0.5.$$

14. Solve the following system of equations by substitution method :
$$3x + y - 13 = 0$$
$$x - 3y + 9 = 0.$$

15. Solve the following system of equations by substitution method :
$$2x - y = 9$$
$$5x - 2y = 27.$$

16. Solve the following system of equations by substitution method :
$$x + \dfrac{y}{2} = 4$$
$$\dfrac{x}{3} + 2y = 5.$$

17. A person invested some amount at the rate of 8% p.a. simple interest and some other amount at the rate of 6% p.a. simple interest. He received yearly interest of Rs. 76. But if he had interchanged the amounts invested, he would have received Rs. 2 more as interest. How much amounts did he invest at different rates ?

18. The car rental charges in a city comprise of a fixed charge together with the charge for the distance covered. For a journey of 13 km, the charge paid is Rs. 96 and for a journey of 18 km, the charge paid is Rs. 131. What will a person have to pay for travelling a distance of 25 km ?

19. A part of monthly hostel charges in a college are fixed and the remaining depend on the number of days one has taken food in the mess. When a student A takes food for 22 days, he has to pay Rs. 1380 as hostel charges whereas a student B, who takes

PAIR OF LINEAR EQUATIONS IN TWO VARIABLES

food for 28 days pahy Rs. 1680 as hostel charges. Find the fixed charge and the cost of food per day.

20. 7 audio cassettes and 3 video cassettes cost Rs. 1110 and 5 audio cassettes and 4 video cassettes cost Rs. 1350. Find the cost of an audio cassette and a video cassette.

21. A and B are friends and A is elder to B by 2 years. A's father D is twice as old as A and B is twice as old as his sister C. If the ages of D and C differ by 40 years, find the ages of A and B.

22. Solve the following system of equations by elimination method :
$$\frac{x}{2} + \frac{y}{4} = 3$$
$$2x - y = 4.$$

23. Solve the following system of equations by elimination method :
$$x + 2y = -1$$
$$2x - 3y = 12.$$

24. Solve the following system of equations by elimination method :
$$x + 2y = \frac{3}{2}$$
$$2x + y = \frac{3}{2}.$$

25. Solve the following system of equation by elimination method :
$$3x + 2y = -25$$
$$-2x - y = 10.$$

26. A part of monthly expends of a family in constant and the remaining varies with the price of wheat. When the rate of wheat is Rs. 250 a quantal, the total monthly expenses of the family are Rs. 1000 and when it is Rs. 240 a quantal, the total monthly expenses are Rs. 980. Find the total monthly expenses of the family when the cost of wheat is Rs. 350 a quantal.

27. Two numbers are in the ratio 5 : 6. If 5 is subtracted from each number, the ratio becomes 4 : 5. Find the numbers.

28. Solve the following system of equations by using cross-multiplication method :
    (i) $2x + 3y = 7$
    $6x - 5y = 11$

    (ii) $3x - 5y = 20$
    $7x + 2y = 17$

    (iii) $7x - 2y = 3$
    $11x - \frac{3}{2}y = 8$

    (iv) $6x + 5y = 11$
    $9x + 10y = 21$

    (v) $4x + 7y = 10$
    $10x - \frac{35}{2}y = 25$

    (vi) $4x + \frac{2}{3}y - 1 = 0$
    $6x - y + 2 = 0$

    (vii) $\frac{5}{3}x + \frac{3}{5}y - 1 = 0$
    $\frac{3}{5}x - \frac{5}{3}y + 2 = 0.$

29. Find the values of for which the following system of equations have exactly one solution :
    (i) $px + 2y = 5$
    $3x + y = 1$
    (ii) $px + 3y = 7$
    $2x - y = 6$
    (iii) $ax + py - 1 = 0$
    $3x + 4y - 2 = 0$
    (iv) $7x - 5y - 4 = 0$
    $14x + py + 4 = 0.$

30. For what value of k, the following system of equations
$$4x + 5y = 3$$
$$kx + 15y = 9$$
has infinitely many solutions ?

31. Solve the following system of equations :
$$\frac{15}{u} + \frac{2}{v} = 17$$
$$\frac{1}{u} + \frac{1}{v} = \frac{36}{5}$$

32. Solve the following system of equations :
$$\frac{11}{v} - \frac{7}{u} = 1$$
$$\frac{9}{v} - \frac{4}{u} = 6.$$

33. Find the solution such that $u \neq 0$, $v \neq 0$ of the following :

$$2u + v = \frac{7}{3} uv$$

$$u + 3v = \frac{11}{3} uv.$$

34. Solve the following system of equations :

$$2x - \frac{3}{y} = 12$$

$$5x + \frac{7}{y} = 1; y \neq 0.$$

35. A person can row downstream 20 km in 2 hours and upstream 4 km in 2 hours. Find man's speed of rowing in still water and the speed of the current.

36. Umesh travels 600 km to his home partly by train and partly by car. He takes 8 hours if he travels 120 km by train and the rest by car. He takes 20 minutes longer if he travels 200 km by train and the rest by car. Find the speed of the train and the car.

Answers

1. $5x + 2y = 1625$,
   $2x + y = 750$; the two lines intersect at the point (125, 500).
2. $x + 10y = 75$
   $x + 15y = 110$; the two lines intersect at the point (5, 7)
3. The two lines are parallel
4. $3x + 4y = 20$,
   $6x + 8y = 40$; the two lines coincide.
5. The two paths are parallel.
6. No   7. No
8. Inconsistent   9. Yes
10. (3, 4) ; (1, 0), (2, 0)
11. (2, 2) ; (3, 0), (– 2, 0)
12. (i) only one solution ; $x = \frac{2}{5}, y = 2$

    (ii) only one solution ; $x = \frac{7}{4}, y = \frac{1}{2}$

    (iii) infinite number of solutions
    (iv) no solution
    (v) no solution
    (vi) infinite number of solutions
    (vii) no solution

(viii) only one solutions ; $x = 0, y = 0$
(ix) infinite number of solutions
(x) no solution

13. $x = 0.2, y = -0.3$ 14. $x = 3, y = 4$
15. $x = 5, y = 1$ 16. $x = 3, y = 2$
17. Rs. 500 at 8% p.a., Rs. 600 at 6% p.a.
18. Rs. 180.   19. Rs. 280, Rs. 50.
20. Rs. 30, Rs. 300   21. 26 years, 24 years.
22. $x = 4, y = 4$ 23. $x = 3, y = -2$
24. $x = \frac{1}{2}, y = \frac{1}{2}$ 25. $x = 5, y = -20$
26. Rs. 1200   27. 25, 30
28. (i) $x = \frac{17}{7}, y = \frac{5}{7}$

    (ii) $x = \frac{125}{41}, y = -\frac{89}{41}$

    (iii) $x = 1, y = 2$

    (iv) $x = \frac{1}{3}, y = \frac{9}{5}$

    (v) $x = \frac{5}{2}, y = 0$

    (vi) $x = -\frac{1}{24}, y = \frac{7}{4}$

    (vii) $x = \frac{105}{706}, y = \frac{885}{706}$

29. (i) $p \neq 6$ (ii) $p \neq -6$
    (iii) $p \neq 12$ (iv) $p \neq -10$

30. 12   31. $u = 5, v = \frac{1}{7}$

32. $u = \frac{1}{3}, v = \frac{1}{2}$ 33. $u = 1, v = \frac{3}{2}$

34. $x = 3, y = -\frac{1}{2}$

35. 6 km/hour, 4 km/hour
36. 60 km/hour, 80 km/hour.

SUMMARY

In this chapter, you have studied the following points :

1. Two linear equations in the same two variables are called a pair of linear equations in two variables, or briefly, a linear pair.

The most general form of a linear pair is
$$a_1x + b_1y + c_1 = 0$$
$$a_2x + b_2y + c_2 = 0$$
where $a_1, a_2, b_1, b_2, c_1, c_2$ are real numbers, such that $a_1^2 + b_1^2 \neq 0$, $a_2^2 + b_2^2 \neq 0$.

2. A pair of linear equations in two variables can be represented, and solved, by the :
   (i) graphical method
   (ii) algebraic method

3. Graphical Method :
   The graph of the pair of linear equations in two variables is represented by a pair of lines.
   (i) If the pairs intersect at a point, then that point is the unique solution of the two equations. In this case, the pair of equations is **consistent.**
   (ii) If the lines coincide, then it has infinitely many solutions—each point on the line being a solution. In this case, the pair of equation is **consistent (dependent).**
   (iii) If the two lines are parallel, then the pair has no solution, and is called **inconsistent.**

4. Algebraic Method : We have discussed the following methods for finding the solution(s) of a pair of linear equations.
   (i) Substitution Method
   (ii) Elimination Method
   (iii) Cross-multiplication Method

5. If $a_1x + b_1y + c_1 = 0$ and $a_2x + b_2y + c_2 = 0$, then the following situations can arise.

   (i) $\dfrac{a_1}{a_2} \neq \dfrac{b_1}{b_1}$: In this case the pair of linear equations is inconsistent.

   (ii) $\dfrac{a_1}{a_2} = \dfrac{b_1}{b_1} \neq \dfrac{c_1}{c_2}$: In this case the pair of linear equations is inconsistent.

   (iii) $\dfrac{a_1}{a_2} = \dfrac{b_1}{b_2} = \dfrac{c_1}{c_2}$: In this case, the pair o linear equation is dependent and consistent.

6. There are several situations which can be mathematically represented by two equations that are not linear to start with. But we alter them so that they are reduced to a linear pair.

4
Quadratic Equations

IMPORTANT POINTS

1. Quadratic Equation. We know many kinds of polynomials like linear, quadratic, cubic, biquadratic, etc. Particularly, a quadratic polynomial is of the form $ax^2 + bx + c$, $a \neq 0$. When the equate this polynomial to zero, we get a quadratic equation. Thus, a quadratic equation is of the form $ax^2 + bx + c = 0$, $a \neq 0$.

2. Quadratic equations in real life situations. Quadratic equations come up when the deal with many real-life situations. For instance, suppose a charity trust decides to build a prayer hall having a carpet area of 300 square metres with its length one metre more than twice its breadth. Suppose the breadth of the hall is x metres. Then its length should be $(2x + 1)$ metres. We can depict this information pictorially as shown in the figure given below :

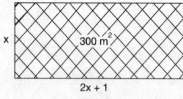

Fig.

Now, area of the hall
$= (2x + 1) x = (2x^2 + x) \, m^2$.
But, $2x^2 + x = 300$ | Given
Therefore, $2x^2 + x - 300 = 0$

So, the breadth of the hall should satisfy the equation $2x^2 + x - 300 = 0$. So, to find the dimensions of the hall to be built, we need to solve the above quadratic equation.

3. Historical Facts. Many people believe that Babylonians were the first to solve quadratic equations. For instance, they knew how to find two positive numbers with a given positive sum and a given positive product, and this problem is equivalent to solving a quadratic equation of the form $x^2 - px + q = 0$. Greek mathematician Euclid developed a geometrical approach for finding out lengths which, in our present day terminology, are solutions of quadratic equations. The solution of quadratic equations, in general form, is often credited to ancient Indian mathematicians. In fact, Brahmagupta (598–665 A.D.) gave an explicit formula to solve a quadratic equation of the form $ax^2 + bx = c$. Later, Sridharacharya (1025 A.D.) gave the method of derivation of this formula now known as the quadratic formula (as quoted by Bhaskara II) Arab mathematician Al-Khwarizmi (about 800 A.D.) also studied quadratic equations of different types. Abraham bar Hiyya Ha-Nasi, in his famed book 'Liber embadorum' published in Europe in 1145 A.D.) gave complete solutions of quadratic equations.

4. Standard Form of a Quadratic Equation. We know that a quadratic equation in the variable x is an equation of the form $ax^2 + bx + c = 0$, where a, b, c are real numbers, $a \neq 0$. For example, $2x^2 + x - 300 = 0$ is a quadratic equation. Similarly, $2x^2 - 3x + 1 = 0$, $4x - 3x^2 + 2 = 0$ and $1 - x^2 + 300 = 0$ are quadratic equations. Consider the quadratic equation $2x^2 - 3x + 1 = 0$. By rearranging the terms we can write this as $1 - 3x + 2x^2 = 0$, $-3x + 2x^2 + 1 = 0$, or $-3x + 1 + 2x^2 = 0$, $1 + 2x^2 - 3x = 0$.

QUADRATIC EQUATIONS

All these equations are also quadratic equations. In fact, any equation of the form $p(x) = 0$, where $p(x)$ is polynomial of degree 2, is a quadratic equation. But when we write the terms of $p(x)$ in descending order of their degrees, then we get the standard form of the equation. That is, $ax^2 + bx + c = 0$, $a \neq 0$ is called the **standard form of a quadratic equation.**

5. A check. Sometimes the given equation appears to be quadratic equation, but it is not a quadratic equation whereas sometimes the given equation does not appears to be a quadratic equation but it is a quadratic equation. So we must simplify the given equation before deciding whether it is quadratic or not.

If after simplification, the given equation is of the form $ax^2 + bx + c = 0$, $a \neq 0$, then it is a quadratic equation otherwise not.

ILLUSTRATIVE EXAMPLES

[NCERT Exercise 4.1]
(Page No. 73)

Example 1. *Check whether the following are quadratic equations :*
   (i) $(x + 1)^2 = 2(x - 3)$
   (ii) $x^2 - 2x = (-2)(3 - x)$
   (iii) $(x - 2)(x + 1) = (x - 1)(x + 3)$
   (iv) $(x - 3)(2x + 1) = x(x + 5)$
   (v) $(2x - 1)(x - 3) = (x + 5)(x - 1)$
   (vi) $x^2 + 3x + 1 = (x - 2)^2$
   (vii) $(x + 2)^3 = 2x(x^2 - 1)$
   (viii) $x^3 - 4x^2 - x + 1 = (x - 2)^3$

Sol. (i) **$(x + 1)^2 = 2(x - 3)$**

The given equation is
$$(x + 1)^2 = 2(x - 3)$$
$$\Rightarrow x^2 + 2x + 1 = 2x - 6$$
$$\Rightarrow x^2 + 7 = 0$$

It is of the form $ax^2 + bx + c = 0$, $a \neq 0$.

Therefore, the given equation is a quadratic equation.

(ii) **$x^2 - 2x = (-2)(3 - x)$**

The given equation is
$$x^2 - 2x = (-2)(3 - x)$$
$$\Rightarrow x^2 - 2x = -6 + 2x$$
$$\Rightarrow x^2 - 4x + 6 = 0$$

It is of the form $ax^2 + bx + c = 0$, $a \neq 0$.

Therefore, the given equation is a quadratic equation.

(iii) **$(x - 2)(x + 1) = (x - 1)(x + 3)$**

The given equation is
$$(x - 2)(x + 1) = (x - 1)(x + 3)$$
$$\Rightarrow x^2 + x - 2x - 2 = x^2 + 3x - x - 3$$
$$\Rightarrow x^2 - x - 2 = x^2 + 2x - 3$$
$$\Rightarrow 3x - 1 = 0.$$

It is not of the form $ax^2 + bx + c = 0$, $a \neq 0$. Therefore, the given equation is not a quadratic equation.

Note. Actually it is a linear equation.

(iv) **$(x - 3)(2x + 1) = x(x + 5)$**

The given equation is
$$(x - 3)(2x + 1) = x(x + 5)$$
$$\Rightarrow 2x^2 + x - 6x - 3 = x^2 + 5x$$
$$\Rightarrow 2x^2 - 5x - 3 = x^2 + 5x$$
$$\Rightarrow x^2 - 10x - 3 = 0$$

It is of the form $ax^2 + bx + c = 0$, $a \neq 0$. Therefore, the given equation is a quadratic equation.

(v) **$(2x - 1)(x - 3) = (x + 5)(x - 1)$**

The given equation is
$$(2x - 1)(x - 3) = (x + 5)(x - 1)$$
$$\Rightarrow 2x^2 - 6x - x + 3 = x^2 - x + 5x - 5$$
$$\Rightarrow 2x^2 - 7x + 3 = x^2 + 4x - 5$$
$$\Rightarrow x^2 - 11x + 8 = 0$$

It is of the form $ax^2 + bx + c = 0$, $a \neq 0$. Therefore, the given equation is a quadratic equation.

(vi) **$x^2 + 3x + 1 = (x - 2)^2$**

The given equation is
$$x^2 + 3x + 1 = (x - 2)^2$$
$$\Rightarrow x^2 + 3x + 1 = x^2 - 4x + 4$$
$$\Rightarrow 7x - 3 = 0$$

It is not of the form $ax^2 + bx + c = 0$, $a \neq 0$. Therefore, the given equation is not a quadratic equation.

Note. Actually it is a linear equation.

(vii) $(x + 2)^3 = 2x(x^2 - 1)$

The given equation is
$$(x + 2)^3 = 2x(x^2 - 1)$$
$$\Rightarrow x^3 + 6x^2 + 12x + 8 = 2x^3 - 2x$$
$$\Rightarrow x^3 - 6x^2 - 14x - 8 = 0$$

It is not of the form
$$ax^2 + bx + c = 0, a \neq 0$$

Therefore, the given equation is not a quadratic equation.

Note. Actually it is a cubic equation.

(viii) $x^3 - 4x^2 - x + 1 = (x - 2)^3$

The given equation is
$$x^3 - 4x^2 - x + 1 = (x - 2)^3$$
$$\Rightarrow x^3 - 4x^2 - x + 1 = x^3 - 8 - 6x^2 + 12x$$
$$\Rightarrow 2x^2 - 13x + 9 = 0$$

It is of the from $ax^2 + bx + c = 0, a \neq 0$.

Therefore, the given equation is a quadratic equation.

Example 2. *Represent the following situation in the form of quadratic equations:*

(i) The area of rectangular plot is 528 m^2. The length of the plot (in metres) is one more than twice its breadth. We need to find the length and breadth of the plot.

(ii) The product of two consecutive positive integers is 306. We need to find the integers.

(iii) Rohan's mother is 26 years older than him. The product of their ages (in years) 3 years from now will be 360. We would like to find Rohan's present age.

(iv) A train travels a distance of 480 km at a uniform speed. If the speed had been 8 km/h less, then it would have taken 3 hours more to cover the same distance. We need to find the speed of the train.

Sol. (i) Let the breadth of the rectangular plot be x metres.

Then the length of the rectangular plot = $(2x + 1)$ metres.

Therefore, the area of the rectangular plot = Length × Breadth = $(2x + 1) x$ m^2

According to the problem situation,
$$(2x + 1) x = 528$$
$$\Rightarrow 2x^2 + x = 528$$
$$\Rightarrow 2x^2 + x - 528 = 0$$

Therefore, the breadth of the rectangular plot satisfies the quadratic equation
$$2x^2 + x - 528 = 0.$$

(ii) Let the two consecutive positive integers be x and $x + 1$.

Then, their product = $x(x + 1)$

According to the problem situation.
$$x(x + 1) = 306$$
$$\Rightarrow x^2 + x = 306$$
$$\Rightarrow x^2 + x - 306 = 0$$

Therefore, the smaller positive integer satisfies the quadratic equation.
$$x^2 + x - 306 = 0$$

(iii) Let Rohan's present age be x years.

Then present age of Rohan's mother
= $(x + 26)$ years

3 years from now, age of Rohan = $(x + 3)$ years

age of Rohan's mother = $(x + 26 + 3)$ years

Therefore, the product of their ages 3 years from now = $(x + 3) (x + 29)$

According to the problem situation,
$$(x + 3) (x + 29) = 360$$
$$\Rightarrow x^2 + 29x + 3x + 87 = 360$$
$$\Rightarrow x^2 + 32x - 273 = 0$$

Therefore, Rohan's present age satisfies the quadratic equation $x^2 + 32x - 273 = 0$

(iv) Let the uniform speed of the train be x km/hour.

Then, time taken by the train to cover a distance of 480 km = $\dfrac{480}{x}$ hours

$$\left| \text{Time} = \dfrac{\text{Distance}}{\text{Speed}} \right.$$

If the speed had been 8 km/h less, then the speed of the train = $(x - 8)$ km/h.

Therefore, time taken by the train to cover a distance of 480 km now = $\dfrac{480}{x - 8}$ hours

$$\left| \text{Time} = \dfrac{\text{Distance}}{\text{Speed}} \right.$$

QUADRATIC EQUATIONS

According to the problem situation,

$$\frac{480}{x-8} = \frac{480}{x} + 3$$

$$\Rightarrow \frac{480}{x-8} - \frac{480}{x} = 3$$

$$\Rightarrow 480\left(\frac{1}{x-8} - \frac{1}{x}\right) = 3$$

$$\Rightarrow \frac{1}{x-8} - \frac{1}{x} = \frac{3}{480}$$

$$\Rightarrow \frac{x-(x-8)}{(x-8)x} = \frac{1}{160}$$

$$\Rightarrow \frac{x-x+8}{x^2-8x} = \frac{1}{160}$$

$$\Rightarrow \frac{8}{x^2-8x} = \frac{1}{160}$$

$$\Rightarrow x^2 - 8x = (8)(160)$$
$$\Rightarrow x^2 - 8x = 1280$$
$$\Rightarrow x^2 - 8x - 1280 = 0$$

Therefore, the uniform speed of the train satisfies the quadratic equation
$$x^2 - 8x - 1280 = 0.$$

ADDITIONAL EXAMPLES

Example 3. *Which of the following are quadratic equations ?*
(i) $x^2 - 6x - 4 = 0$
(ii) $3x^2 - 7x - 2 = 0$
(iii) $x^3 - 6x^2 + 2x - 1 = 0$
(iv) $7x = 2x^2$
(v) $x^2 + \frac{1}{x} = 2 \ (x \neq 0)$
(vi) $x + \frac{1}{x} = x^2 \ (x \neq 0).$

Sol. (i) $x^2 - 6x - 4 = 0$
The given equation is
$$x^2 - 6x - 4 = 0$$
It is of the form $ax^2 + bx + c = 0, a \neq 0.$
Therefore, the given equation is a quadratic equation.

(ii) $3x^2 - 7x - 2 = 0$
The given equation is
$$3x^2 - 7x - 2 = 0$$
It is of the form $ax^2 + bx + c = 0, a \neq 0$
Therefore, the given equation is a quadratic equation.

(iii) $x^3 - 6x^2 + 2x - 1 = 0$
The given equation is
$$x^3 - 6x^2 + 2x - 1 = 0$$
It is not of the form $ax^2 + bx + c = 0,$
$$a \neq 0.$$
Therefore, the given equation is not a quadratic equation.

Note : Actually it is a cubic equation

(iv) $7x = 2x^2$
The given equation is
$$7x = 2x^2$$
$$\Rightarrow 2x^2 - 7x = 0$$
It is of the form $ax^2 + bx + c = 0,$
$$a \neq 0.$$
Therefore, the given equation is a quadratic equation.

(v) $x^2 + \frac{1}{x^2} = 2 (x \neq 0)$
The given equation is
$$x^2 + \frac{1}{x^2} = 2 \ (x \neq 2)$$

$$\Rightarrow \frac{x^4+1}{x^2} = 2$$

$$\Rightarrow x^4 + 1 = 2x^2$$
$$\Rightarrow x^4 - 2x^2 + 1 = 0$$
It is not of the form $ax^2 + bx + c = 0,$
$$a \neq 0.$$
Therefore, the given equation is not a quadratic equation.

Note. Actually it is a bi quadratic equation.

(vi) $x + \frac{1}{x} = x^2 \ (x \neq 0)$
The given equation is
$$\Rightarrow x + \frac{1}{x} = x^2 \ (x \neq 0)$$

$$\Rightarrow \quad \frac{x^2+1}{x} = x^2$$
$$\Rightarrow \quad x^2 + 1 = x^3$$
$$\Rightarrow \quad x^3 - x^2 - 1 = 0$$

It is not of the form
$$ax^2 + bx + c = 0, a \neq 0.$$

Therefore, the given equation is not a quadratic equation.

Note. Actually it is a cubic equation.

Example 2. *Represent the following situations mathematically :*

(i) *The sum of two numbers is 40. We need to find the numbers, if the sum of their reciprocals is $\frac{2}{5}$.*

(ii) *We need to find two consecutive odd integers, the sum of whose squares is 202.*

Sol. (i) Let one of the two numbers be x.

Then, the other number = $40 - x$

| ∵ The sum of two numbers is 40.

Their reciprocal are $\frac{1}{x}$ and $\frac{1}{40-x}$.

Therefore, the sum of their reciprocals

$$= \frac{1}{x} + \frac{1}{40-x}$$

According to the problem situation.

$$\frac{1}{x} + \frac{1}{40-x} = \frac{2}{5}$$

$$\Rightarrow \quad \frac{40 - x + x}{x(40-x)} = \frac{2}{5}$$

$$\Rightarrow \quad \frac{40}{x(40-x)} = \frac{2}{5}$$

$$\frac{20}{x(40-x)} = \frac{1}{5}$$

$$\Rightarrow \quad x(40-x) = (20)(5)$$
$$\Rightarrow \quad 40x - x^2 = 100$$
$$\Rightarrow \quad x^2 - 40x + 100 = 0$$

Therefore, one of the two numbers satisfies the quadratic equation.

$$x^2 - 40x + 100 = 0$$

(ii) Let the smaller odd integer be x.

Then the consecutive greater odd integer = $x + 2$

| ∵ Two consecutive odd integers differ by 2

Therefore, the sum of their squares
$$= x^2 + (x+2)^2$$

According to the problem situation,
$$x^2 + (x+2)^2 = 202$$
$$\Rightarrow \quad x^2 + x^2 + 4x + 4 = 202$$
$$\Rightarrow \quad 2x^2 + 4x - 198 = 0$$
$$\Rightarrow \quad x^2 + 2x - 99 = 0$$

| Dividing throughout by 2

Therefore, the smaller of the two consecutive odd integers satisfies the quadratic equation $x^2 + 2x - 99 = 0$.

TEST YOUR KNOWLEDGE

1. Represent the following situations mathematically :

   (i) John and Jivanti together have 45 marbles. Both of them lost 5 marbles each, and the product of the number of marbles they now have is 124. We would like to find out how many marbles they had to start with.

   (ii) A cottage industry produces a certain number of toys in a day. The cost of production of each toy (in rupees) was found to be 55 minus the number of articles produced in a day. On a particular day, the total cost of production was Rs. 750. We would like to find out the number of toys produced on that day.

2. Check whether the following are quadratic equation :

   (i) $(x-2)^2 + 1 = 2x - 3$
   (ii) $x(x+1) + 8 = (x+2)(x-2)$
   (iii) $x(2x+3) = x^2 + 1$
   (iv) $(x+2)^3 = x^3 - 4.$

3. Check whether the following are quadratic equations :
   (i) $3x^2 - 4 = 0$
   (ii) $(x + 1)(x + 3) = 0$
   (iii) $(2x + 1)(3x + 2) = 6(x - 1)(x - 2)$
   (iv) $16x^2 - 3 = (2x + 5)(5x - 3)$.

4. Represent the following situations mathematically :
   (i) There are three consecutive positive integers such that the sum of the square of the first and the product of the other two is 154. We would like to find out the smallest positive integer.
   (ii) Vikram wishes to fit three rods together in the shape of a right triangle. The hypotenuse is to be 2 cm longer than the base and 4 cm longer than the altitude. We would like to find out the length of the smallest rod.

Answers

1. (i) $x^2 - 45x + 324 = 0$
   (ii) $x^2 - 55x + 750 = 0$
2. (i) Yes   (ii) No
   (iii) Yes   (iv) Yes
3. (i) Yes   (ii) Yes
   (iii) No   (iv) Yes
4. (i) $2x^2 + 3x - 152 = 0$
   (ii) $x^2 - 12x + 20 = 0$.

IMPORTANT POINTS

Finding the Roots of Quadratic Equations by the Factorisation Method

1. Root of a quadratic equation. Consider the quadratic equation $2x^2 - 3x + 1 = 0$. If we replace x by 1 on the LHS of this equation, we get $(2 \times 1^2) - (3 \times 1) + 1 = 0$ = RHS of the equation. We say that 1 is a root of the quadratic equation $2x^2 - 3x + 1 = 0$.

Note : Since $(2 \times 1^2) - (3 \times 1) + 1 = 0$, this also means that 1 is a zero of the qudratic polynomial $2x^2 - 3x + 1$.

2. Condition for a real number to be a root. In general, a real number α is called a root of the quadratic equation $ax^2 + bx + c = 0$, $a \neq 0$ if $a\alpha^2 + b\alpha + c = 0$. We also say that α **is a solution of the equation,** or that α **satisfies the quadratic equation.**

Note. The zeroes of the quadratic polynomial $ax^2 + bx + c$ and the roots of the quadratic equation $ax^2 + bx + c = 0$ are the same.

3. Number of roots of a quadratic equation. We know that a quadratic polynomial can have at most two zeroes. So, any quadratic can have at most two roots.

4. Determination of roots by Factorisation Method. We know how to factorize a quadratic polynomial into two linear factors by splitting its middle term. We shall use this knowledge for finding the roots of a quadratic equation. We shall equate each factor to zero and find out the roots of the corresponding quadratic equation.

ILLUSTRATIVE EXAMPLES

[NCERT Exercise 4.2]
(Page No. 76)

Example 1. *Find the roots of the following quadratic equations by factorisation:*
(i) $x^2 - 3x - 10 = 0$
(ii) $2x^2 + x - 6 = 0$
(iii) $\sqrt{2}x^2 + 7x + 5\sqrt{2} = 0$
(iv) $2x^2 - x + \dfrac{1}{8} = 0$
(v) $100x^2 - 20x + 1 = 0$

Sol. (i) $x^2 - 3x - 10 = 0$
The given quadratic equation is
$$x^2 - 3x - 10 = 0$$

$\Rightarrow \quad x^2 - 5x + 2x - 10 = 0$
$\Rightarrow \quad x(x - 5) + 2(x - 5) = 0$
$\Rightarrow \quad (x - 5)(x + 2) = 0$
$\Rightarrow \quad x - 5 = 0 \text{ or } x + 2 = 0$
$\Rightarrow \quad x = 5 \text{ or } x = -2$
$\Rightarrow \quad x = 5, -2$

Therefore, 5 and -2 are the roots of the quadratic equation $x^2 - 3x - 10 = 0$

Verification. (i) **For root $x = 5$**

Put $x = 5$ in the quadratic equation $x^2 - 3x - 10 = 0$. We see that
$$x^2 - 3x - 10$$
$$= (5)^2 - 3(5) - 10$$
$$= 25 - 15 - 10$$
$$= 0, \text{ which is correct.}$$
Hence, the equation is satisfied.
Therefore, $x = 5$ is a root of the quadratic equation $x^2 - 3x - 10 = 0$

(ii) **For root x = −2**

Put $x = -2$ in the quadratic equation $x^2 - 3x - 10 = 0$. We see that
$$x^2 - 3x - 10$$
$$= (-2)^2 - 3(-2) - 10$$
$$= 4 + 6 - 10$$
$$= 0, \text{ which is correct.}$$
Hence, the equation is satisfied.
Therefore, $x = -2$ is a root of the quadratic equation $x^2 - 3x - 10 = 0$.

(ii) **$2x^2 + x - 6 = 0$**

The given quadratic equation is
$$2x^2 + x - 6 = 0$$
$\Rightarrow \quad 2x^2 + 4x - 3x - 6 = 0$
$\Rightarrow \quad 2x(x + 2) - 3(x + 2) = 0$
$\Rightarrow \quad (x + 2)(2x - 3) = 0$
$\Rightarrow \quad x + 2 = 0 \text{ or } 2x - 3 = 0$
$\Rightarrow \quad x = -2 \text{ or } x = \dfrac{3}{2}$
$\Rightarrow \quad x = -2, \dfrac{3}{2}$

Therefore, -2 and $\dfrac{3}{2}$ are the roots of the quadratic equation $2x^2 + x - 6 = 0$

Verification. (i) **For root $x = -2$**

Put $x = -2$ in the quadratic equation $2x^2 + x - 6 = 0$. We see that
$$2x^2 + x - 6 = 2(-2)^2 + (-2) - 6$$
$$= 8 - 2 - 6$$
$$= 0, \text{ which is correct.}$$
Hence, the equation is satisfied. Therefore, $x = -2$ is a root of the quadratic equation $2x^2 + x - 6 = 0$

(ii) **For root x = $\dfrac{3}{2}$**

Put $x = -\dfrac{3}{2}$ in the quadratic equation.
We see that
$$2x^2 + x - 6$$
$$= 2\left(\dfrac{3}{2}\right)^2 + \left(\dfrac{3}{2}\right) - 6$$
$$= \dfrac{9}{2} + \dfrac{3}{2} - 6$$
$$= 0, \text{ which is correct}$$
Hence, the equation is satisfied.

Therefore, $x = \dfrac{3}{2}$ is a root of the quadratic equation $2x^2 + x - 6 = 0$

(iii) **$\sqrt{2}\,x^2 + 7x + 5\sqrt{2} = 0$**

The given quadratic equation is
$$\sqrt{2}\,x^2 + 7x + 5\sqrt{2} = 0$$
$\Rightarrow \quad \sqrt{2}\,x^2 + 2x + 5x + 5\sqrt{2} = 0$
$\Rightarrow \quad \sqrt{2}\,x(x + \sqrt{2}) + 5(x + \sqrt{2}) = 0$
$\Rightarrow \quad (x + \sqrt{2})(\sqrt{2}\,x + 5) = 0$
$\Rightarrow \quad x + \sqrt{2} = 0 \text{ or } \sqrt{2}\,x + 5 = 0$
$\Rightarrow \quad x = -\sqrt{2} \text{ or } x = \dfrac{-5}{\sqrt{2}}$
$\Rightarrow \quad x = -\sqrt{2}, \dfrac{-5}{\sqrt{2}}$

Therefore, $-\sqrt{2}$ and $\dfrac{-5}{\sqrt{2}}$ are the roots of the quadratic equation $\sqrt{2}\,x^2 + 7x + 5\sqrt{2} = 0$.

QUADRATIC EQUATIONS

Verification (i) For root $x = -\sqrt{2}$

Put $x = -\sqrt{2}$ in the quadratic equation
$$\sqrt{2}x^2 + 7x + 5\sqrt{2} = 0$$
We see that
$$\sqrt{2}x^2 + 7x + 5\sqrt{2}$$
$$= \sqrt{2}(-\sqrt{2})^2 + 7(-\sqrt{2}) + 5\sqrt{2}$$
$$= 2\sqrt{2} - 7\sqrt{2} + 5\sqrt{2} = 0$$
which is correct.
Hence, the equation in satisfied.
Therefore, $x = -\sqrt{2}$ is a root of the quadratic equation $\sqrt{2}x^2 + 7x + 5\sqrt{2} = 0$

(ii) **For root** $x = \dfrac{-5}{\sqrt{2}}$

Put $x = \dfrac{-5}{\sqrt{2}}$ in the quadratic equation
$$\sqrt{2}x^2 + 7x + 5\sqrt{2} = 0.$$
We see that $\sqrt{2}x^2 + 7x^2 + 5\sqrt{2}$
$$= \sqrt{2}\left(\dfrac{-5}{\sqrt{2}}\right)^2 + 7\left(\dfrac{-5}{\sqrt{2}}\right) + 5\sqrt{2}$$
$$= \dfrac{25}{\sqrt{2}} - \dfrac{35}{\sqrt{2}} + 5\sqrt{2}$$
$$= 0, \text{ which is correct.}$$
Hence, the equation is satisfied.

Therefore, $x = \dfrac{-5}{\sqrt{2}}$ is a root of the quadratic equation $\sqrt{2}x^2 + 7x + 5\sqrt{2} = 0$

(iv) $2x^2 - x + \dfrac{1}{8} = 0$

The given quadratic equation is
$$2x^2 - x + \dfrac{1}{8} = 0$$
$$\Rightarrow \quad 2x^2 - \dfrac{1}{2}x - \dfrac{1}{2}x + \dfrac{1}{8} = 0$$
$$\Rightarrow \quad x\left(2x - \dfrac{1}{2}\right) - \dfrac{1}{4}\left(2x - \dfrac{1}{2}\right) = 0$$

$$\Rightarrow \quad \left(2x - \dfrac{1}{2}\right)\left(x - \dfrac{1}{4}\right) = 0$$
$$\Rightarrow \quad 2x - \dfrac{1}{2} = 0 \text{ or } x - \dfrac{1}{4} = 0$$
$$\Rightarrow \quad x = \dfrac{1}{4} \quad \text{or} \quad x = \dfrac{1}{4}$$
$$\Rightarrow \quad x = \dfrac{1}{4}, \dfrac{1}{4}$$

So this root is repeated twice.
Therefore, both the roots of the quadrilateral equation $2x^2 - x + \dfrac{1}{8} = 0$ are $\dfrac{1}{4}$.

Verification. For repeated root $x = \dfrac{1}{4}$.

Put $x = \dfrac{1}{4}$ in the quadratic equation
$2x^2 - x + \dfrac{1}{8} = 0$. We see that $2x^2 - x + \dfrac{1}{8}$
$$= 2\left(\dfrac{1}{4}\right)^2 - \left(\dfrac{1}{4}\right) + \dfrac{1}{8}$$
$$= \dfrac{1}{8} - \dfrac{1}{4} + \dfrac{1}{8}$$
$$= 0, \text{ which is correct.}$$
Hence, the equation is satisfied.

Therefore, $x = \dfrac{1}{4}$ is a root (repeated twice) of the quadratic equation $2x^2 - x + \dfrac{1}{8} = 0$

(v) $100x^2 - 20x + 1 = 0$

The given quadratic equation is
$$100x^2 - 20x + 1 = 0$$
$$\Rightarrow \quad 100x^2 - 10x - 10x + 1 = 0$$
$$\Rightarrow \quad 10x(10x - 1) - 1(10x - 1) = 0$$
$$\Rightarrow \quad (10x - 1)(10x - 1) = 0$$
$$\Rightarrow \quad 10x - 1 = 0 \text{ or } 10x - 1 = 0$$
$$\Rightarrow \quad x = \dfrac{1}{10} \quad \text{or} \quad x = \dfrac{1}{10}$$
$$\Rightarrow \quad x = \dfrac{1}{10}, \dfrac{1}{10}$$

So, this root is repeated twice, once for each repeated factor $10x - 1$.

Therefore, both of roots of the quadratic equation $100x^2 - 20x + 1 = 0$ are $\dfrac{1}{10}$.

Verification. Put $x = \dfrac{1}{10}$ in the quadratic equatuion $100x^2 - 20x + 1 = 0$. We see that

$100x^2 - 20x + 1$

$= 100\left(\dfrac{1}{10}\right)^2 - 20\left(\dfrac{1}{10}\right) + 1$

$= 1 - 2 + 1$

$= 0$, which is correct.

Hence, the equation is satisfied.

Therefore, $x = \dfrac{1}{10}$ is a root (repeated twice) of the quadratic equation $100x^2 - 20x + 1 = 0$.

Example 2. *Solve the problems given in Example 1 of NCERT Book page 76.*

Sol. (*i*) Let the number of marbles John had be x.

Then the number of marbles Jivanti had $= 45 - x$

| ∵ John and Jivanti together have 45 marbles

The number of marbles left with John when he lost 5 marbles $= x - 5$

The number of marbles left with Jivanti when she lost when 5 marbles $= 45 - x - 5$

$= 40 - x$

Therefore, their product

$= (x - 5)(40 - x)$

$= 40x - x^2 - 200 + 5x$

$= -x^2 + 45x - 200$

Now, $-x^2 + 45x - 200 = 124$ (Given)

i.e., $-x^2 + 45x - 324 = 0$

i.e., $-x^2 + 45x - 324 = 0$

Therefore, the number of marbles John had, satisfies the quadratic equation

$x^2 - 45x + 324 = 0$

$\Rightarrow \quad x^2 - 9x - 36x + 324 = 0$

$\Rightarrow \quad x(x - 9) - 36(x - 9) = 0$

$\Rightarrow \quad (x - 9)(x - 36) = 0$

$\Rightarrow \quad x - 9 = 0 \quad \text{or} \quad x - 36 = 0$

$\Rightarrow \quad x = 9 \quad \text{or} \quad x = 36$

$\Rightarrow \quad x = 9, 36.$

$\Rightarrow \quad 45 - x = 36, 9$

So, John and Jivanti had 9 and 36 or 36 and 9 marbles respectively to start with.

(*ii*) Let the number of toys produced on that day be x.

Therefore, the cost of production (in rupees) of each toy that day $= 55 - x$

So, the total cost of production (in rupees) that day $= x(55 - x)$

Therefore, $x(55 - x) = 750$

i.e., $55x - x^2 = 750$

i.e., $-x^2 + 55x - 750 = 0$

i.e., $x^2 - 55x + 750 = 0$

Therefore, the number of toys produced that day satisfies the quadratic equation

$x^2 - 55x + 750 = 0$

$\Rightarrow \quad x^2 - 25x - 30x + 750 = 0$

$\Rightarrow \quad x(x - 25) - 30(x - 25) = 0$

$\Rightarrow \quad (x - 25)(x - 30) = 0$

$\Rightarrow \quad x - 25 = 0 \quad \text{or} \quad x - 30 = 0$

$\Rightarrow \quad x = 25 \quad \text{or} \quad x = 30$

$\Rightarrow \quad x = 25, 30$

So, the number of toys produced on that day was 25 or 30.

Example 3. *Find two numbers whose sum is 27, and product is 182.*

Sol. Let one number be x.

Then the other number $= 27 - x$

| ∵ Sum of two numbers is 27.

Therefore, their product $= x(27 - x)$

According to the question,

$x(27 - x) = 182$

$\Rightarrow \quad 27x - x^2 = 182$

$\Rightarrow \quad x^2 - 27x + 182 = 0$

$\Rightarrow \quad x^2 - 13x - 14x + 182 = 0$

$\Rightarrow \quad x(x - 13) - 14(x - 13) = 0$

$\Rightarrow \quad (x - 13)(x - 14) = 0$

$\Rightarrow \quad x - 13 = 0 \quad \text{or} \quad x - 14 = 0$

$\Rightarrow \quad x = 13 \quad \text{or} \quad x = 14$

$\Rightarrow \quad x = 13, 14$

$\Rightarrow \quad 27x = 14, 13$

So, the required two numbers are 13 and 14.

QUADRATIC EQUATIONS

Example 4. *Find two consecutive positive integers, sum of whose squares is 365.*

Sol. Let the consecutive positive integers be x and $x + 1$.

| ∵ Two consecutive positive integers differ by 1

Then, the sum of their squares
$$= x^2 + (x + 1)^2$$
$$= x^2 + x^2 + 2x + 1$$
$$= 2x^2 + 2x + 1$$

According to the question,
$$2x^2 + 2x + 1 = 365$$
$$\Rightarrow 2x^2 + 2x - 364 = 0$$
$$\Rightarrow x^2 + x - 182 = 0$$

| Dividing throughout by 2

$$\Rightarrow x^2 + 14x - 13x - 192 = 0$$
$$\Rightarrow x(x + 14) - 13(x + 14) = 0$$
$$\Rightarrow (x + 14)(x - 13) = 0$$
$$\Rightarrow x + 14 = 0 \text{ or } x - 13 = 0$$
$$\Rightarrow x = -14 \text{ or } x = 13$$
$$\Rightarrow x = -14, 13$$

∵ x is a positive integer

∴ $x = -14$ is unadmissible

So, $x = 13$
$$\Rightarrow x + 1 = 13 + 1 = 14$$

Hence, the required two consecutive positive integers are 13 and 14.

Example 5. *The altitude of a right triangle is 7 cm less than its base. If the hypotenuse is 13 cm, find the other two sides.*

Sol. Let the base of the right triangle be x cm.

Then the height of the right triangle $= (x - 7)$ cm

By Pythagoras theorem,
$(\text{Base})^2 + (\text{Height})^2 = (\text{Hypotenuse})^2$
$$\Rightarrow x^2 + (x - 7)^2 = 13^2$$
$$\Rightarrow x^2 + x^2 - 14x + 49 = 169$$
$$\Rightarrow 2x^2 - 14x - 120 = 0$$
$$\Rightarrow x^2 - 7x - 60 = 0$$

| Dividing throughout by 2

$$\Rightarrow x^2 - 12x + 5x - 60 = 0$$
$$\Rightarrow x(x - 12) + 5(x - 12) = 0$$
$$\Rightarrow (x - 12)(x + 5) = 0$$
$$\Rightarrow x - 12 = 0 \text{ or } x + 5 = 0$$
$$\Rightarrow x = 12 \text{ or } x = -5$$
$$\Rightarrow x = 12, -5$$

$x = -5$ is inadmissible

| ∵ x is the length of the base of the right triangle and length can not be negative

∴ $x = 12$
$$\Rightarrow x - 7 = 12 - 7 = 5$$

Therefore, the lengths of the other two sides are 5 cm and 12 cm.

Example 6. *A cottage industry produces a certain number of pottery articles in a day. It was observed on a particular day that the cost of production of each article (in rupees) was 3 more than twice the number of articles produced on that day. If the total cost of production on that day was Rs 90, find the number of articles produced and the cost of each article.*

Sol. Let the number of articles produced on that day be x.

Then, the cost of production of each article on that day.
$$= \text{Rs } (2x + 3)$$

Therefore, total cost of production on that day = Number of articles produced on that day × cost of produced of each article on that day
$$= \text{Rs } x(x + 3)$$

According to the question,
$$x(2x + 3) = 90$$
$$\Rightarrow 2x^2 + 3x - 90 = 0$$
$$\Rightarrow 2x^2 + 15x - 12x - 90 = 0$$
$$\Rightarrow x(2x + 15) - 6(2x + 15) = 0$$
$$\Rightarrow (2x + 15)(x - 6) = 0$$
$$\Rightarrow 2x + 15 = 0 \text{ or } x - 6 = 0$$
$$\Rightarrow x = -\frac{15}{2} \text{ or } x = 6$$
$$\Rightarrow x = -\frac{15}{2}, 6$$

Since, x is the number of articles, it cannot be negative. So, the number of articles produced on that day = 6

The cost of each article = $2x + 3$
$$= 2 \times 6 + 3$$
$$= \text{Rs } 15$$

ADDITIONAL EXAMPLES

Example 6. *Solve the following quadratic equation by factorisation :*

$$x^2 + \left(\frac{a}{a+b} + \frac{a+b}{a}\right)x + 1 = 0.$$

Sol. The given quadratic equation is

$$x^2 + \left(\frac{a}{a+b} + \frac{a+b}{a}\right)x + 1 = 0$$

$$\Rightarrow x^2 + \frac{a}{a+b}x + \frac{a+b}{a}x + 1 = 0$$

$$\Rightarrow x\left(x + \frac{a}{a+b}\right) + \frac{a+b}{a}\left(x + \frac{a}{a+b}\right) = 0$$

$$\Rightarrow \left(x + \frac{a}{a+b}\right)\left(x + \frac{a+b}{a}\right) = 0$$

$$\Rightarrow x + \frac{a}{a+b} = 0 \text{ or } x + \frac{a+b}{a} = 0$$

$$\Rightarrow x = -\frac{a}{a+b} \text{ or } x = -\frac{a+b}{a}$$

$$\Rightarrow x = -\frac{a}{a+b}, -\frac{a+b}{a}$$

Therefore, the roots of $x^2 + \left(\frac{a}{a+b} + \frac{a+b}{a}\right)x + 1 = 0$ are $\frac{-a}{a+b}$ and $-\frac{a+b}{a}$.

Example 7. *A person on tour has Rs. 360 for his daily expenses. If the exceeds his tour programme by 4 days, he must cut down his daily expenses by Rs 3 per day. Find the number of days of his tour programme.*

Sol. Let the number of days of his tour programme be x.

Total amount for daily expenses
= Rs 360

Therefore, his daily expenditure
= Rs $\dfrac{360}{x}$

If he exceeds his tour programme for 4 days. Then the number of days of his tour programme now = $x + 4$

Therefore, his daily expenditure now
= Rs. $\dfrac{360}{x+4}$

According to the question,

$$\frac{360}{x+4} = \frac{360}{x} - 3$$

$$\Rightarrow \frac{360}{x} - \frac{360}{x+4} = 3$$

$$\Rightarrow 360\left(\frac{1}{x} - \frac{1}{x+4}\right) = 3$$

$$\Rightarrow \frac{1}{x} - \frac{1}{x+4} = \frac{3}{360}$$

$$\Rightarrow \frac{1}{x} - \frac{1}{x+4} = \frac{1}{120}$$

$$\Rightarrow \frac{(x+4) - x}{x(x+4)} = \frac{1}{120}$$

$$\Rightarrow \frac{4}{x(x+4)} = \frac{1}{120}$$

$$\Rightarrow x(x+4) = 480$$

$$\Rightarrow x^2 + 4x = 480$$

$$\Rightarrow x^2 + 4x - 480 = 0$$

$$\Rightarrow x^2 + 24x - 20x - 480 = 0$$

$$\Rightarrow x(x+24) - 20(x+24) = 0$$

$$\Rightarrow (x+24)(x-20) = 0$$

$$\Rightarrow x + 24 = 0 \quad \text{or} \quad x - 20 = 0$$

$$\Rightarrow x = -24 \quad \text{or} \quad x = 20$$

$$\Rightarrow x = -24, 20$$

Since x is the number of days, it cannot be negative.

So, the number of days of his tour programme is 20.

QUADRATIC EQUATIONS

Example 8. *In a game of cards, Neeraj scored 3 points more than twice the number of points Pankaj scored. If the product of their scores was 65, how many points did each score?*

Sol. Let Pankaj score x points.
Then Neeraj scored $(2x + 3)$ points.
Therefore, product of their scores
$$= x(2x + 3)$$
According to the question,
$$x(2x + 3) = 65$$
$$\Rightarrow 2x^2 + 3x = 65$$
$$\Rightarrow 2x^2 + 3x - 65 = 0$$
$$\Rightarrow 2x^2 + 13x - 10x - 65 = 0$$
$$\Rightarrow x(2x + 13) - 5(2x + 13) = 0$$
$$\Rightarrow (2x + 13)(x - 5) = 0$$
$$\Rightarrow 2x + 13 = 0 \text{ or } x - 5 = 0$$
$$\Rightarrow x = -\frac{13}{2} \text{ or } x = 5$$
$$\Rightarrow x = -\frac{13}{2}, 5$$

Since, x is the number of points, it cannot be negative.

So, the Pankaj scored 5 points and Neeraj scored $2x + 3 = 2(5) + 3 = 10 + 3 = 13$ points.

TEST YOUR KNOWLEDGE

1. Use the factorisation method to find the roots of the equation $2x^2 - 5x + 3 = 0$.
2. Find the roots of the quadratic equation $6x^2 - x - 2 = 0$ by the factorisation method.
3. Find the roots of the quadratic equation
$$3x^2 - 2\sqrt{6}x + 2 = 0.$$
4. A charity decides to build a prayer hall having a carpet area 300 square metres with its length one metre more than twice its breadth. Find the dimensions of the prayer hall.
5. There are three consecutive integers such that the square of the first increased by the product of the other two gives 154. What are the intergers?
6. A takes 6 day less than the time taken by B to finish a piece of work. If both A and B together can finish it in 4 days, find the time taken by B to finish the work.
7. Solve the following quadratic equation by factorisation method.
$$abx^2 + (b^2 - ac)x - bc = 0$$
8. Two circles touch externally. The sum of their areas is 130π square centimeters and the distance between their centres is 14 cm. Find the radii of the circles.

Answers

1. $1, \dfrac{3}{2}$
2. $\dfrac{2}{3}, -\dfrac{1}{2}$
3. $\sqrt{\dfrac{2}{3}}, \sqrt{\dfrac{2}{3}}$
4. $25\ m, 12\ n$
5. 8, 9, 10
6. 12 days
7. $\dfrac{c}{b}, -\dfrac{b}{a}$
8. 3 cm, 11 cm

METHOD OF COMPLETING THE SQUARES

IMPORTANT POINTS

1. Quadratic Formula. Consider the quadratic equation
$$ax^2 + bx + c = 0, a \neq 0$$
Dividing throughout by a, we get
$$x^2 + \frac{b}{a}x + \frac{c}{a} = 0$$
$$\Rightarrow \left(x + \frac{b}{2a}\right)^2 - \left(\frac{b}{2a}\right)^2 + \frac{c}{a} = 0$$
$$\Rightarrow \left(x + \frac{b}{2a}\right)^2 - \frac{b^2 - 4ac}{4a^2} = 0$$

$$\Rightarrow \quad \left(x+\frac{b}{2a}\right)^2 = \frac{b^2-4ac}{4a^2} \quad \ldots(1)$$

If $b^2 - 4ac \geq 0$, then by taking the square roots in (1), we get

$$x + \frac{b}{2a} = \pm \frac{\sqrt{b^2-4ac}}{2a}$$

$$\Rightarrow \quad x = -\frac{b}{2a} \pm \frac{\sqrt{b^2-4ac}}{2a}$$

$$\Rightarrow \quad x = \frac{-b \pm \sqrt{b^2-4ac}}{2a}$$

So, the roots of $ax^2 + bx + c = 0$ are

$$\frac{-b+\sqrt{b^2-4ac}}{2a} \text{ and } \frac{-b-\sqrt{b^2-4ac}}{2a}$$

if $b^2 - 4ac \geq 0$.

Therefore, if $b^2 - 4ac \geq 0$, then the roots of the quadratic equation $ax^2 + bx + c = 0$ are given by $\dfrac{-b \pm \sqrt{b^2-4ac}}{2a}$.

This formula for finding the roots of a quadratic equation is known as the **quadratic formula**.

ILLUSTRATIVE EXAMPLES

[NCERT Exercise 4.3]
(Page No. 87)

Example 1. *Find the roots of the following quadratic equations, if they exist, by the method of completing the square:*

(i) $2x^2 - 7x + 3 = 0$
(ii) $2x^2 + x - 4 = 0$
(iii) $4x^2 + 4\sqrt{3}x + 3 = 0$
(iv) $2x^2 + x + 4 = 0$.

Sol. (i) $2x^2 - 7x + 3 = 0$

The given quadratic equation is
$$2x^2 - 7x + 3 = 0$$

$$\Rightarrow \quad x^2 - \frac{7}{2}x + \frac{3}{2} = 0$$

| dividing throughout by 2

$$\Rightarrow \quad \left\{x - \frac{1}{2}\left(\frac{7}{2}\right)\right\}^2 - \left\{\frac{1}{2}\left(\frac{7}{2}\right)\right\}^2 + \frac{3}{2} = 0$$

$$\Rightarrow \quad \left(x - \frac{7}{4}\right)^2 - \frac{25}{16} = 0$$

$$\Rightarrow \quad \left(x - \frac{7}{4}\right)^2 = \frac{25}{16}$$

$$\Rightarrow \quad x - \frac{7}{4} = \pm\sqrt{\frac{25}{16}}$$

$$\Rightarrow \quad x - \frac{7}{4} = \pm \frac{5}{4}$$

$$\Rightarrow \quad x = \frac{7}{4} \pm \frac{5}{4}$$

$$\Rightarrow \quad x = \frac{7}{4} + \frac{5}{4} \text{ and } \frac{7}{4} - \frac{5}{4}.$$

$$\Rightarrow \quad x = 3 \text{ and } \frac{1}{2}$$

Hence, the roots of the quadratic equation $2x^2 - 7x + 3 = 0$ are 3 and $\dfrac{1}{2}$.

Verification (*i*) **For root x = 3**

Put $x = 3$ in $2x^2 - 7x + 3 = 0$. We see that

$2x^2 - 7x + 3$
$= 2(3)^2 - 7(3) + 3$
$= 18 - 21 + 3$
$= 0$, which is correct.

So, $x = 3$ is a root of quadratic equation $2x^2 - 7x + 3 = 0$

(*ii*) **For root x = $\dfrac{1}{2}$**

Put $x = \dfrac{1}{2}$ in $2x^2 - 7x + 3 = 0$. We see that

QUADRATIC EQUATIONS

$2x^2 - 7x + 3$

$= 2\left(\dfrac{1}{2}\right)^2 - 7\left(\dfrac{1}{2}\right) + 3$

$= \dfrac{1}{2} - \dfrac{7}{2} + 3$

$= 0$, which is correct.

So, $x = \dfrac{1}{2}$ is a root of the quadratic equation $2x^2 - 7x + 3 = 0$

(ii) **$2x^2 + x - 4 = 0$**

The given quadratic equation is
$2x^2 + x - 4 = 0$

$\Rightarrow 4x^2 + 2x - 8 = 0$

| Multiplying throughout by 2

$\Rightarrow (2x^2) + 2(2x)\left(\dfrac{1}{2}\right) - \left(\dfrac{1}{2}\right)^2 - \left(\dfrac{1}{2}\right)^2 - 8$
$= 0$

$\Rightarrow \left(2x + \dfrac{1}{2}\right)^2 - \dfrac{33}{4} = 0$

$\Rightarrow \left(2x + \dfrac{1}{2}\right)^2 = \dfrac{33}{4}$

$\Rightarrow 2x + \dfrac{1}{2} = \pm\sqrt{\dfrac{33}{4}}$

$\Rightarrow 2x + \dfrac{1}{2} = \pm\dfrac{\sqrt{33}}{2}$

$\Rightarrow 2x = -\dfrac{1}{2} \pm \dfrac{\sqrt{33}}{2}$

$\Rightarrow 2x = \dfrac{-1 \pm \sqrt{33}}{2}$

$\Rightarrow 2x = \dfrac{-1+\sqrt{33}}{2}$ and $\dfrac{-1-\sqrt{33}}{2}$

$\Rightarrow x = \dfrac{-1+\sqrt{33}}{4}$ and $\dfrac{-1-\sqrt{33}}{4}$

Hence, the roots of the quadratic equation $2x^2 + x - 4 = 0$ are $\dfrac{-1+\sqrt{33}}{4}$ and $\dfrac{-1-\sqrt{33}}{4}$.

Verification

(i) **For root $x = \dfrac{-1+\sqrt{33}}{4}$**

Put $x = \dfrac{-1+\sqrt{33}}{4}$ in $2x^2 + x - 4 = 0$.

We see that $2x^2 + x - 4$

$= 2\left(\dfrac{-1+\sqrt{33}}{4}\right)^2 + \left(\dfrac{-1+\sqrt{33}}{4}\right) - 4$

$= \dfrac{2(1 + 33 - 2\sqrt{33})}{16} + \left(\dfrac{-1+\sqrt{33}}{4}\right) - 4$

$= \dfrac{34 - 2\sqrt{33}}{8} + \left(\dfrac{-1+\sqrt{33}}{4}\right) - 4$

$= \dfrac{34 - 2\sqrt{33} + (-2 + 2\sqrt{33}) - 32}{8}$

$= 0$, which is correct.

So, $x = \dfrac{-1+\sqrt{33}}{4}$ is root of the quadratic equation $2x^2 + x - 4 = 0$

(ii) **For root $x = \dfrac{-1-\sqrt{33}}{4}$**

Put $x = \dfrac{-1-\sqrt{33}}{4}$ in $2x^2 + x - 4 = 0$, we see that $2x^2 + x - 4$

$= 2\left(\dfrac{-1-\sqrt{33}}{4}\right)^2 + \left(\dfrac{-1-\sqrt{33}}{4}\right) - 4$

$= 2\left(\dfrac{1 + 33 + 2\sqrt{33}}{16}\right) + \left(\dfrac{-1-\sqrt{33}}{4}\right) - 4$

$= \dfrac{34 + 2\sqrt{33} - 2 - 2\sqrt{33} - 32}{8}$

$= 0$, Which is correct.

So, $x = \dfrac{-1-\sqrt{33}}{4}$ is a root of the quadratic equation $2x^2 + x - 4 = 0$

(iii) $4x^2 + 4\sqrt{3}x + 3 = 0$

The given quadratic equation is
$$4x^2 + 4\sqrt{3}x + 3 = 0$$
$\Rightarrow (2x)^2 + 2(2x)(\sqrt{3}) + (\sqrt{3})^2 - (\sqrt{3})^2 + 3 = 0$
$\Rightarrow (2x + \sqrt{3})^2 = 0$
$\Rightarrow 2x + \sqrt{3} = 0, 2x + \sqrt{3} = 0$
$\Rightarrow x = -\dfrac{\sqrt{3}}{2}, x = -\dfrac{\sqrt{3}}{2}$
$\Rightarrow x = -\dfrac{\sqrt{3}}{2}$ and $-\dfrac{\sqrt{3}}{2}$

Hence, the roots of the quadratic equation $4x^2 + 4\sqrt{3}x + 3 = 0$ are $-\dfrac{\sqrt{3}}{2}$ and $-\dfrac{\sqrt{3}}{2}$.

Verification. For root $x = -\dfrac{\sqrt{3}}{2}$

Put $x = -\dfrac{\sqrt{3}}{2}$ in $4x^2 + 4\sqrt{3}x + 3 = 0$.
We see that
$4x^2 + 4\sqrt{3}x + 3$
$= 4\left(-\dfrac{\sqrt{3}}{2}\right)^2 + 4\sqrt{3}\left(-\dfrac{\sqrt{3}}{2}\right) + 3$
$= 3 - 6 + 3$
$= 0$, which is correct.

So, $x = -\dfrac{\sqrt{3}}{2}$ is a root (repeated twice) of the quadratic equation $4x^2 + 4\sqrt{3}x + 3 = 0$

(iv) $2x^2 + x + 4 = 0$

The given quadratic equation is
$$2x^2 + x + 4 = 0$$
$\Rightarrow x^2 + \dfrac{1}{2}x + 2 = 0$
| Dividing throughout by 2
$\Rightarrow \left\{x + \dfrac{1}{2}\left(\dfrac{1}{2}\right)\right\}^2 - \left\{\dfrac{1}{2}\left(\dfrac{1}{2}\right)\right\}^2 + 2 = 0$

$\Rightarrow \left(x + \dfrac{1}{4}\right)^2 - \dfrac{1}{16} + 2 = 0$

$\Rightarrow \left(x + \dfrac{1}{4}\right)^2 + \dfrac{31}{16} = 0$

$\Rightarrow \left(x + \dfrac{1}{4}\right)^2 = -\dfrac{31}{16} \; (<0)$

But $\left(x + \dfrac{1}{4}\right)^2$ cannot be negative for any real value of x because the square of real number cannot be negative. So there is the real value of x satisfying the given equation. Therefore, the given equation has no real roots, *i.e.*, the solution of the given equation does not exist.

Example 2. *Find the roots of the quadratic equations given in Q. 1 above by applying the quadratic formula.*

Sol.

(i) $2x^2 - 7x + 3 = 0$

The given quadratic equation is
$$2x^2 - 7x + 3 = 0$$
Here, $a = 2$
$b = -7$
$c = 3$
So, $b^2 - 4ac = (-7)^2 - 4(2)(3)$
$= 49 - 24$
$= 25 \geq 0$
Therefore,
$$x = \dfrac{-b \pm \sqrt{b^2 - 4ac}}{2a}$$
$= -\dfrac{(-7) \pm \sqrt{25}}{2(2)}$
$= \dfrac{7 \pm 5}{4}$
$= \dfrac{7+5}{4}, \dfrac{7-5}{4}$
$= 3, \dfrac{1}{2}$

QUADRATIC EQUATIONS

So, the roots are 3 and $\dfrac{1}{2}$.

(ii) $2x^2 + x - 4 = 0$

The given quadratic equation is
$$2x^2 + x - 4 = 0$$
Here, $\quad a = 2$
$\quad\quad\quad b = 1$
$\quad\quad\quad c = -4$

So, the $b^2 - 4ac = (1)^2 - 4(2)(-4)$
$\quad\quad\quad\quad\quad = 1 + 32$
$\quad\quad\quad\quad\quad = 33 \geq 0$

Therefore,
$$x = \dfrac{-b \pm \sqrt{b^2 - 4ac}}{2a}$$
$$= \dfrac{-1 \pm \sqrt{33}}{2(2)}$$
$$= \dfrac{-1 \pm \sqrt{33}}{4}$$
$$= \dfrac{-1 + \sqrt{33}}{4}, \dfrac{-1 - \sqrt{33}}{2(2)}$$

So, the roots are $\dfrac{-1 + \sqrt{33}}{4}$ and $\dfrac{-1 - \sqrt{33}}{4}$.

(iii) $4x^2 + 4\sqrt{3}\,x + 3 = 0$

The given quadratic equation is $4x + 4\sqrt{3}\,x + 3 = 0$
Here, $\quad a = 4$
$\quad\quad\quad b = 4\sqrt{3}$
$\quad\quad\quad c = 3$

So, $b^2 - 4ac = (4\sqrt{3})^2 - 4(4)(3)$
$\quad\quad\quad\quad = 0$

Therefore,
$$x = \dfrac{-b \pm \sqrt{b^2 - 4ac}}{2a}$$
$$= \dfrac{-4\sqrt{3} \pm \sqrt{0}}{2(4)}$$
$$= -\dfrac{\sqrt{3}}{2}, -\dfrac{\sqrt{3}}{2}$$

So, both the roots are $-\dfrac{\sqrt{3}}{2}$.

(iv) $2x^2 + x + 4 = 0$

The given quadratic equation is
$$2x^2 + x + 4 = 0$$
Here, $\quad a = 2$
$\quad\quad\quad b = 1$
$\quad\quad\quad c = 4$

So, $b^2 - 4ac = (1)^2 - 4(2)(4)$
$\quad\quad\quad\quad = -31 < 0$

But $\quad x = \dfrac{-b \pm \sqrt{b^2 - 4ac}}{2a}$

Since, the square of a real number cannot be negative, therefore x will now have any real value.

So, there are no real roots for the given equation, i.e. the solution of the given equation does not exist.

Example 3. *Find the roots of the following equations:*

(i) $x - \dfrac{1}{x} = 3,\ x \neq 0$

(ii) $\dfrac{1}{x+4} - \dfrac{1}{x-7} = \dfrac{11}{30},\ x \neq -4, 7$

Sol. (i) $x - \dfrac{1}{x} = 3,\ x \neq 0$

The given equation is
$$x - \dfrac{1}{x} = 3$$
$$\Rightarrow \dfrac{x^2 - 1}{x} = 3$$
$$\Rightarrow x^2 - 1 = 3x$$
$$\Rightarrow x^2 - 3x - 1 = 0$$

which is the quadratic equation in x.

Here,
$a = 1, b = -3, c = -1$

So, $b^2 - 4ac = (-3)^2 - 4(1)(-1)$
$\quad\quad\quad\quad = 9 + 4$
$\quad\quad\quad\quad = 13 \geq 0$

Therefore,

$$x = \frac{-b \pm \sqrt{b^2 - 4ac}}{2a}.$$

Using the quadratic formula.

$$= \frac{-(-3) \pm \sqrt{13}}{2(1)}$$

$$= \frac{3 \pm \sqrt{13}}{2}$$

$$= \frac{3 + \sqrt{13}}{2}, \frac{3 - \sqrt{13}}{2}$$

So, the roots are $\frac{3 + \sqrt{13}}{2}$ and $\frac{3 - \sqrt{13}}{2}$.

Verification. (*i*) For root $x = \dfrac{3 + \sqrt{13}}{2}$

Put $x = \dfrac{3 + \sqrt{13}}{2}$ in $x^2 - 3x - 1 = 0$. We see that

$$x^2 - 3x - 1$$

$$= \left(\frac{3 + \sqrt{13}}{2}\right)^2 - 3\left(\frac{3 + \sqrt{13}}{2}\right) - 1$$

$$= \frac{9 + 13 + 6\sqrt{13}}{4} - \left(\frac{9 + 3\sqrt{13}}{2}\right) - 1$$

$$= \frac{22 + 6\sqrt{13}}{4} - \frac{9 + 3\sqrt{13}}{2} - 1$$

$$= \frac{22 + 6\sqrt{13} - 18 - 6\sqrt{13} - 4}{4}$$

$= 0$, which is correct.

So, $x = \dfrac{3 + \sqrt{13}}{2}$ is a root of the quadratic equation $x^2 - 3x - 1 = 0$.

(*ii*) For root $x = \dfrac{3 - \sqrt{13}}{2}$

Put $x = \dfrac{3 - \sqrt{13}}{2}$ in $x^2 - 3x - 1 = 0$. We get that

$$x^2 - 3x - 1$$

$$= \left(\frac{3 - \sqrt{13}}{2}\right)^2 - 3\left(\frac{3 - \sqrt{13}}{2}\right) - 1$$

$$= \frac{9 + 13 - 6\sqrt{13}}{4} - \frac{9 - 3\sqrt{13}}{2} - 1$$

$$= \frac{9 + 13 - 6\sqrt{13} - 18 + 6\sqrt{13} - 4}{4}$$

$= 0$, which is correct.

So, $x = \dfrac{3 - \sqrt{13}}{2}$ is a root of the quadratic equation $x^2 - 3x - 1 = 0$

(*ii*) $\dfrac{1}{x + 4} - \dfrac{1}{x - 7} = \dfrac{11}{30}$, $x \neq -4, 7$

The given equation is

$$\frac{1}{x+4} - \frac{1}{x-7} = \frac{11}{30}$$

$$\Rightarrow \frac{(x-7) - (x+4)}{(x+4)(x-7)} = \frac{11}{30}$$

$$\Rightarrow \frac{x - 7 - x - 4}{(x+4)(x-7)} = \frac{11}{30}$$

$$\Rightarrow \frac{-11}{(x+4)(x-7)} = \frac{11}{30}$$

$$\Rightarrow -(x+4)(x-7) = 30$$

$$\Rightarrow -(x^2 - 7x + 4x - 28) = 30$$

$$\Rightarrow -(x^2 - 3x - 28) = 30$$

$$\Rightarrow -x^2 + 3x + 28 = 30$$

$$\Rightarrow x^2 - 3x + 2 = 0$$

which is a quadratic equation in x.

Here, $a = 1$
$b = -3$
$c = 2$

So, $b^2 - 4ac = (-3)^2 - 4(1)(2) = 9 - 8$

$\Rightarrow \quad 1 \geq 0$

QUDRATIC EQUATIONS

Therefore,
$$x = \frac{-b \pm \sqrt{b^2 - 4ac}}{2a}$$

| Using the quadratic formula

$$\Rightarrow x = \frac{-(-3) \pm \sqrt{1}}{2(1)}$$

$$\Rightarrow x = \frac{3 \pm 1}{2}$$

$$\Rightarrow x = \frac{3+1}{2}, \frac{3-1}{2}$$

$$\Rightarrow x = 2, 1$$

So, the roots are 2 and 1.

Verification in (*i*) **For root x = 1**

Put $x = 1$ and $x^2 - 3x + 2 = 0$. We see that
$x^2 - 3x + 2$
$= (1)^2 - 3(1) + 2$
$= 1 - 3 + 2$
$= 0$, which is correct.

So, $x = 1$ is a root of the quadratic equation
$x^2 - 3x + 2 = 0$

(*ii*) **For root x = 2**

Put $x = 2$ in $x^2 - 3x + 2 = 0$. We see that
$x^2 - 3x + 2$
$= (2)^2 - 3(2) + 2$
$= 4 - 6 + 2$
$= 0$, which is correct.

So, $x = 2$ is a root of the quadratic equation $x^2 - 3x + 2 = 0$.

Example 4. *The sum of the reciprocals of Rehman's ages, (in years) 3 years ago and 5 years from now is $\frac{1}{3}$. Find his present age.*

Sol. Let the present age of Rehman be x years.

Then,
Rehman's age 3 years ago = $(x - 3)$ years
and, Rehman's age 5 years from now
= $(x + 5)$ years

According to the question,

$$\frac{1}{x-3} + \frac{1}{x+5} = \frac{1}{3}$$

$$\Rightarrow \frac{(x+5) + (x-3)}{(x-3)(x+5)} = \frac{1}{3}$$

$$\Rightarrow \frac{2x+2}{(x-3)(x+5)} = \frac{1}{3}$$

$$\Rightarrow (x-3)(x+5) = 3(2x+2)$$
$$\Rightarrow x^2 + 5x - 3x - 15 = 6x + 6$$
$$\Rightarrow x^2 + 2x - 15 = 6x + 6$$
$$\Rightarrow x^2 - 4x - 21 = 0$$

which is a quadratic equation in x.
Here, $a = 1, b = -4, c = -21$
Using the quadratic formula, we get

$$x = \frac{-b \pm \sqrt{b^2 - 4ac}}{2a}$$

$$= \frac{-(-4) \pm \sqrt{(-4)^2 - 4(1)(-21)}}{2(1)}$$

$$= \frac{4 \pm \sqrt{100}}{2}$$

$$= \frac{4 \pm 10}{2}$$

$$= \frac{4+10}{2}, \frac{4-10}{2}$$

$$= 7, -3$$

Since, x is the age, it cannot be negative. So, we ignore the root
$x = -3$.

Therefore, $x = 7$ gives the present age of Rehmans 7 years.

Example 5. *In a class test, the sum of Shefali's marks in Mathematics and English is 30. Had she got 2 marks more in Mathematics and 3 marks less in English, the product would have been 210. Find here marks in the two subjects.*

Sol. Let Shefali's marks in Mathematics be x.

Then, Shefali's marks in English = $(30 - x)$
| ∵ The sum of Shefali's marks in Mathematics and English is 30

According to the question,
$$(x + 2)\{(30 - x) - 3\} = 210$$
$$\Rightarrow (x + 2)(27 - x) = 210$$
$$\Rightarrow 27x - x^2 + 54 - 2x = 210$$
$$\Rightarrow x^2 - 25x + 156 = 0$$
which is a quadratic equation in x.
Here, $a = 1, b = -25, c = 156$
Using the quadratic formula, we get
$$x = \frac{-b \pm \sqrt{b^2 - 4ac}}{2a}$$
$$= \frac{-|-25| \pm \sqrt{(-25)^2 - 4(1)(156)}}{2(1)}$$
$$= \frac{25 \pm \sqrt{625 - 624}}{2}$$
$$= \frac{25 \pm 1}{2}$$
$$= \frac{25+1}{2}, \frac{25-1}{2} = 13, 12$$
$$\Rightarrow 30 - x = 30 - 13 = 30 - 12$$
$$\Rightarrow 30 - x = 17, 18$$

Therefore, either Shefali's marks in Mathematics are 13 and in English are 17 or her marks in Mathematics are 12 and in English are 18.

Example 6. *The diagonal of a rectangular field is 60 metres more than the shorter side. If the longer side is 30 metres more than the shorter side, find the sides of the field.*

Sol. Let the shorter side of the rectangular field be x metres.

Then, the longer side of the rectangular field $= (x + 30)$ metres

Therefore, diagonal of the rectangular field
$$= \sqrt{(\text{Length of the shorter side})^2 + (\text{Length of the longer side})^2}$$
| By Pythagoras Theorem
$$= \sqrt{x^2 + (x + 30)^2} \text{ metres}$$
According to the question,
$$\sqrt{x^2 + (x + 30)^2} = x + 60$$

Squaring both sides, we get
$$x^2 + (x + 30)^2 = (x + 60)^2$$
$$\Rightarrow x^2 + x^2 + 60x + 900 = x^2 + 120x + 3600$$
$$\Rightarrow x^2 - 60x - 2700 = 0$$
which is a quadratic equation in x.
Here, $a = 1, b = -25, c = 156$
Using the quadratic formula, we get
$$x = \frac{-b \pm \sqrt{b^2 - 4ac}}{2a}$$
$$= \frac{-(-60) \pm \sqrt{(-60)^2 - 4(1)(-2700)}}{2(1)}$$
$$= \frac{60 \pm \sqrt{3600 + 10800}}{2}$$
$$= \frac{60 \pm \sqrt{14400}}{2}$$
$$= \frac{60 \pm 120}{2}$$
$$= \frac{60 + 120}{2}, \frac{60 - 120}{2}$$
$$\Rightarrow x = 90, -30$$

Since x cannot be negative, being a dimension, the length of the shorter side of the rectangular field is 90 metres. The length of the longer side $= x + 30$
$$= 90 + 30$$
$$= 120 \text{ metres.}$$

Example 7. *The difference of squares of two numbers is 180. The square of the smaller number is 8 times the larger number. Find the two numbers.*

Sol. Let the larger number be x.

Then, (smaller number)^2 = 8(larger number)
$$= 8x$$
$$\Rightarrow \text{smaller number} = \sqrt{8x}$$
According to the question,
$$x^2 - 8x = 180$$
$$\Rightarrow x^2 - 8x - 180 = 0$$
Using the quadratic formula, we get
$$x = \frac{-b \pm \sqrt{b^2 - 4ac}}{2a}$$

QUDRATIC EQUATIONS

$$= \frac{-(-8) \pm \sqrt{(-8)^2 - 4(1)(-180)}}{2(1)}$$

$$= \frac{8 \pm \sqrt{64 + 720}}{2}$$

$$= \frac{8 \pm \sqrt{784}}{2}$$

$$= \frac{8 \pm 28}{2}$$

$$= \frac{8 + 28}{2}, \frac{8 - 28}{2}$$

$\Rightarrow \quad x = 18, -10$

$x = -10$ is inadmissible as then smaller number $= \sqrt{8(-10)} = \sqrt{-80}$ which does not exist.

$\therefore \quad x = 18$

$\therefore \quad \sqrt{8x} = \sqrt{8 \times 18} = \sqrt{144} = \pm 12$

Hence, the two numbers are 18, 12 or 18, –12.

Example 8. *A train travels 360 km at a uniform speed. If the speed had been 5 km/h more, it would have taken 1 hour less for the same journey. Find the speed of the train.*

Sol. Let the uniform speed of the train be x km/h.

Then, the time taken by the train to travel 360/km $= \dfrac{360}{x}$ | Time $= \dfrac{\text{distance}}{\text{speed}}$

If the speed had been 5 km/h more, than the time taken by the train to travel 360 km

$= \dfrac{360}{x+5}$ | Time $= \dfrac{\text{distance}}{\text{speed}}$

According to the question,

$$\frac{360}{x+5} = \frac{360}{x} - 1$$

$\Rightarrow \quad \dfrac{360}{x} - \dfrac{360}{x+5} = 1$

$\Rightarrow \quad 360 \left(\dfrac{1}{x} - \dfrac{1}{x+5} \right) = 1$

$\Rightarrow \quad \dfrac{1}{x} - \dfrac{1}{x+5} = \dfrac{1}{360}$

$\Rightarrow \quad \dfrac{x+5-x}{x(x+5)} = \dfrac{1}{360}$

$\Rightarrow \quad \dfrac{5}{x(x+5)} = \dfrac{1}{360}$

$\Rightarrow \quad x(x+5) = (5)(360)$

$\Rightarrow \quad x(x+5) = 1800$

$\Rightarrow \quad x^2 + 5x - 1800 = 0$

Using quadratic formula, we get

$$x = \frac{-b \pm \sqrt{b^2 - 4ac}}{2a}$$

$$= \frac{-5 \pm \sqrt{(5)^2 - 4(1)(-1800)}}{2(1)}$$

$$= \frac{-5 \pm \sqrt{25 + 7200}}{2}$$

$$= \frac{-5 \pm \sqrt{7225}}{2}$$

$$= \frac{-5 \pm 85}{2}$$

$$= \frac{-5 + 85}{2}, \frac{-5 - 85}{2}$$

$\Rightarrow \quad x = 40, -45.$

Since, x is the speed of the train, it cannot be negative. So, we ignore the root $x = -45$.

Therefore, $x = 40$ gives the speed of the train as 40 km/h.

Example 9. *Two water taps together can fill a tank in $9\dfrac{3}{8}$ hours. The tap of larger diameter takes 10 hours less than the smaller one to fill the tank separately. Find the time in which each tap can separately fill the tank.*

Sol. Let the smaller tap fill the tank in x hours.

Then the larger tap will fill the tank in $(x - 10)$ hours.

∴ Tank filled by the smaller tap in 1 hours = $\dfrac{1}{x}$

and, Tank filled by the larger tap in 1 hour = $\dfrac{1}{x-10}$.

∴ Tank filled by the two taps together in 1 hour = $\left(\dfrac{1}{x} + \dfrac{1}{x-10}\right)$

∴ Tank filled by the two taps together in $9\dfrac{3}{8}$ hours = $\dfrac{75}{8}\left(\dfrac{1}{x} + \dfrac{1}{x-10}\right)$

$\left| \because 9\dfrac{3}{8} = \dfrac{75}{8} \right.$

According to the question,

$$\dfrac{75}{8}\left(\dfrac{1}{x} + \dfrac{1}{x-10}\right) = 1$$

$\Rightarrow \dfrac{1}{x} + \dfrac{1}{x-10} = \dfrac{8}{75}$

$\Rightarrow \dfrac{x-10+x}{x(x-10)} = \dfrac{8}{75}$

$\Rightarrow \dfrac{2x-10}{x(x-10)} = \dfrac{8}{75}$

$\Rightarrow 8x(x-10) = 75(2x-10)$

$\Rightarrow 8x^2 - 80x = 150x - 750$

$\Rightarrow 8x^2 - 230x + 750 = 0$

Comparing with $ax^2 + bx + c = 0$, we get

$a = 8$
$b = -230$
$c = 750$

Using the quadratic formula, we get

$$x = \dfrac{-b \pm \sqrt{b^2 - 4ac}}{2a}$$

$$x = \dfrac{-(-230) \pm \sqrt{(-230)^2 - 4(8)(750)}}{2(8)}$$

$= \dfrac{230 \pm \sqrt{52900 - 24000}}{16}$

$= \dfrac{230 \pm \sqrt{28900}}{16}$

$= \dfrac{230 \pm 170}{16}$

$= \dfrac{230+170}{16}, \dfrac{230-170}{16}$

$= \dfrac{400}{16}, \dfrac{60}{16}$

$= 25, \dfrac{15}{4}$

$x = \dfrac{15}{4}$ is in admissible is then

$x - 10 = \dfrac{15}{4} - 10 = -\dfrac{25}{4}$ which is negative and the time cannot be negative.

∴ $x = 25$

$\Rightarrow x - 10 = 25 - 10 = 15$.

Hence, the times in which the smaller tap and the larger tap fill the tank separately are 25 hours and 15 hours respectively.

Example 10. *An express train takes 1 hour less than a passenger train to travel 132 km between Mysore and Banglore (without taking into consideration the time they stop at intermediate stations). If the average speed of the express train is 11 km/h more than that of the passenger train, find the average speed of the two trains.*

Sol. Let the average speed of the passenger train be x km/h.

Then, the average speed of the express train = $(x + 11)$ km/h.

Time taken by the passenger train to travel 132 km between Mysore and Banglore

$= \dfrac{132}{x}$ h $\left|\text{ Time} = \dfrac{\text{Distance}}{\text{Speed}}\right.$

Time taken by the express train to travel 132 km between Mysore and Banglore

$= \dfrac{132}{x+11}$ h $\left|\text{ Time} = \dfrac{\text{Distance}}{\text{Speed}}\right.$

QUDRATIC EQUATIONS

According to the question,

$$\frac{132}{x+11} = \frac{132}{x} - 1$$

$$\Rightarrow \quad \frac{132}{x} - \frac{132}{x+11} = 1$$

$$\Rightarrow \quad 132\left(\frac{1}{x} - \frac{1}{x+11}\right) = 1$$

$$\Rightarrow \quad \frac{1}{x} - \frac{1}{x+11} = \frac{1}{132}$$

$$\Rightarrow \quad \frac{(x+11) - x}{x(x+11)} = \frac{1}{132}$$

$$\Rightarrow \quad \frac{11}{x(x+11)} = \frac{1}{132}$$

$$\Rightarrow \quad x(x+11) = (11)(132)$$

$$\Rightarrow \quad x(x+11) = 1452$$

$$\Rightarrow \quad x^2 + 11x - 1452 = 0$$

This is of the form $ax^2 + bx + c = 0$ where $a = 1, b = 11, c = -1452$.

So, the quadratic formula gives as

$$x = \frac{-b \pm \sqrt{b^2 - 4ac}}{2a}$$

$$= \frac{-11 \pm \sqrt{(11)^2 - 4(1)(-1452)}}{2(1)}$$

$$= \frac{-11 \pm \sqrt{121 + 5808}}{2}$$

$$= \frac{-11 \pm \sqrt{5929}}{2}$$

$$= \frac{-11 \pm 77}{2}$$

$$= \frac{-11 + 77}{2}, \frac{-11 - 77}{2}$$

$$= \frac{66}{2}, \frac{-88}{2}$$

$$= 33, -44$$

Since x is the average speed of the passenger train, it cannot be negative. So, we ignore the root $x = -44$.

Therefore, $x = 33$ gives the average speed of the passenger train as 33 km/h.

Average speed of the express train

$$= x + 11$$
$$= 33 + 11$$
$$= 44 \text{ km/h.}$$

Example 11. *Sum of the areas of two squares is 468 m^2. If the difference of their perimeters is 24 m, find the sides of the two squares.*

Sol. Let the side of the smaller square be x m.

Then, perimeter of the smaller square
$$= 4x \text{ m}$$

Therefore, perimeter of the larger square
$$= (4x + 24) \text{ m}$$

Therefore, side of the larger square

$$= \frac{4x + 24}{4} \text{ m} = \frac{4(x + 6)}{4} \text{ cm}$$

$$= (x + 6) \text{ m}$$

Again,

Area of the smallar square = x^2 cm^2

Area of the larger square = $(x + 6)^2$ cm^2

According to the question,

$$x^2 + (x + 6)^2 = 468$$

$$\Rightarrow \quad x^2 + x^2 + 12x + 36 = 468$$

$$\Rightarrow \quad 2x^2 + 12x - 432 = 0$$

$$\Rightarrow \quad x^2 + 6x - 216 = 0$$

| dividing throughout by 2

which is a quadratic equation in x.

Comparing with $ax^2 + bx + c = 0$, we get

$$a = 1$$
$$b = 6$$
$$c = -216$$

Therefore, $b^2 - 4ac = (6)^2 - 4(1)(-216)$
$$= 36 + 864$$
$$= 900 \geq 0$$

So, the given equation can be solved for x.

Using the quadratic formula, we get

$$x = \frac{-b \pm \sqrt{b^2 - 4ac}}{2a}$$

$$= \frac{-6 \pm \sqrt{900}}{2(1)}$$

$$= \frac{-6 \pm 30}{2}$$

$$= \frac{-6 + 30}{2}, \frac{-6 - 30}{2}$$

$$= 12, -18$$

Since, x cannot be negative, being the length of side of the smaller square, the length of the side of the smaller square is 12 m.

The length of the side of the larger square $= x + 6$
$= 12 + 6$
$= 18$ m.

ADDITIONAL EXAMPLES

Example 12. S*olve for* x
$12abx^2 - (9a^2 - 8b^2)x - 6ab = 0$.
(CBSE 2006)

Sol. The given quadratic equation is
$12abx^2 - (9a^2 - 8b^2)x - 6ab = 0$
Comparing with $Ax^2 + Bx + C = 0$, we get
$A = 12\,ab$
$B = -(9a^2 - 8b^2)$
$C = -6ab$

Using the quadratic formula, we get

$$x = \frac{-B \pm \sqrt{B^2 - 4AC}}{2A}$$

$\Rightarrow \quad x = -\{-(9a^2 - 8b^2)\}$

$$\pm \frac{\sqrt{\{-(9a^2 - 8b^2)\}^2 - 4(12ab)(-6ab)}}{2(12ab)}$$

$$= \frac{9a^2 - 8b^2 \pm \sqrt{\begin{array}{c}81a^4 + 64b^4 \\ -144a^2b^2 + 288a^2b^2\end{array}}}{24\,ab}$$

$$= \frac{9a^2 - 8b^2 \pm \sqrt{81a^4 + 64b^4 + 144a^2b^2}}{24\,ab}$$

$$= \frac{9a^2 - 8b^2 \pm \sqrt{(9a^2 + 8b^2)^2}}{24\,ab}$$

$$= \frac{9a^2 - 8b^2 \pm (9a^2 + 8b^2)}{24\,ab}$$

$$= \frac{9a^2 - 8b^2 + 9a^2 + 8b^2}{24\,ab},$$

$$\frac{9a^2 - 8b^2 - 9a^2 - 8b^2}{24\,ab}$$

$$= \frac{18a^2}{24\,ab}, \frac{-16b^2}{24\,ab}$$

$$= \frac{3a}{4b}, \frac{-2b}{3a}$$

Therefore, the solutions of the given quadratic equation are $\frac{3a}{4b}$ and $\frac{-2b}{3a}$.

Example 13. *Using the quadratic formula, solve the equation :*
$a^2b^2x^2 - (4b^4 - 3a^4)x - 12a^2b^2 = 0$.
(A.I. CBSE 2006)

Sol. The given quadratic equation is
$a^2b^2x^2 - (4b^4 - 3a^4)x - 12a^2b^2 = 0$
Comparing with $Ax^2 + Bx + C = 0$, we get
$A = a^2b^2$
$B = -(4b^4 - 3a^4)$
$C = -12a^2b^2$

Using the quadratic formula, we get

$$x = \frac{-B \pm \sqrt{B^2 - 4AC}}{2A}$$

QUDRATIC EQUATIONS

$$= -\frac{\{-(4b^4-3a^4)\} \pm \sqrt{\{-(4b^4-3a^4)\}^2 - 4(a^2b^2)(-12a^2b^2)}}{2a^2b^2}$$

$$= \frac{4b^4 - 3a^4 \pm \sqrt{16b^8 + 9a^8 - 24a^4b^4 + 48a^4b^4}}{2a^2b^2}$$

$$= \frac{4b^4 - 3a^4 \pm \sqrt{16b^8 + 9a^8 + 24a^4b^4}}{2a^2b^2}$$

$$= \frac{4b^4 - 3a^4 \pm \sqrt{(4b^4 + 3a^4)^2}}{2a^2b^2}$$

$$= \frac{4b^4 - 3a^4 \pm (4b^4 + 3a^4)}{2a^2b^2}$$

$$= \frac{4b^4 - 3a^4 + 4b^4 + 3a^4}{2a^2b^2},$$

$$\frac{4b^4 - 3a^4 - 4b^4 + 3a^4}{2a^2b^2}$$

$$= \frac{8b^4}{2a^2b^2}, \frac{-6a^4}{2a^2b^2}$$

$$= \frac{4b^2}{a^2}, -\frac{3a^2}{b^2}$$

Therefore the solutions of the given quadratic equation one $\frac{4b^2}{a^2}$ and $\frac{-3a^2}{b^2}$.

Example 14. *Solve for x :*

$$\frac{x-1}{x-2} + \frac{x-3}{x-4} = 3\frac{1}{3} \ (x \neq 2, 4).$$

(A.I. CBSE 2005)

Sol. The given equation is

$$\frac{x-1}{x-2} + \frac{x-3}{x-4} = 3\frac{1}{3} \ (x \neq 2, 4)$$

$$\Rightarrow \frac{(x-1)(x-4)+(x-3)(x-2)}{(x-2)(x-4)} = \frac{10}{3}$$

$$\Rightarrow \frac{x^2-4x-x+4+x^2-2x-3x+6}{x^2-4x-2x+8} = \frac{10}{3}$$

$$\Rightarrow \frac{2x^2-10x+10}{x^2-6x+8} = \frac{10}{3}$$

$$\Rightarrow 3(2x^2-10x+10) = 10(x^2-6x+8)$$

$$\Rightarrow 6x^2-30x+30 = 10x^2-60x+80$$

$$\Rightarrow 4x^2-30x+50 = 0$$

$$\Rightarrow (2x)^2 - 2(2x)\left(\frac{15}{2}\right) + \left(\frac{15}{2}\right)^2$$

$$- \left(\frac{15}{2}\right)^2 + 50 = 0$$

$$\Rightarrow \left(2x-\frac{15}{2}\right)^2 - \frac{225}{4} + 50 = 0$$

$$\Rightarrow \left(2x-\frac{15}{2}\right)^2 - \frac{25}{4} = 0$$

$$\Rightarrow \left(2x-\frac{15}{2}\right)^2 = \frac{25}{4}$$

$$\Rightarrow 2x - \frac{15}{2} = \pm\frac{25}{4}$$

$$\Rightarrow 2x - \frac{15}{2} = \pm\frac{5}{2}$$

$$\Rightarrow 2x = \frac{15}{2} \pm \frac{5}{2}$$

$$\Rightarrow 2x = \frac{15}{2} + \frac{5}{2}, \frac{15}{2} - \frac{5}{2}$$

$$\Rightarrow 2x = 10, 5$$

$$\Rightarrow x = 5, \frac{5}{2}$$

Hence, the solutions of the given equation and 5 and $\frac{5}{2}$.

Example 15. *The speed of a boat in still water is 11 km/hour. It can go 12 km upstream and return downstream to the engine point in*

2 hours 45 minutes. Find the speed of the stream.   (A.I. CBSE 2006)

Sol. Let the speed of the stream be x km/hour speed of the boat in still water

$= 11$ km/hr.

So, Speed of the boat with upstream

$= (11 - x)$ km/hour.

and, speed of the boat with downstream

$= (11 + x)$ km/hour

Time taken by the boat in going 12 km

in Upstream $= \dfrac{12}{11-x}$ hours

| Time $= \dfrac{\text{Distance}}{\text{Speed}}$

Time taken by the boat in returning downstream to the engine point.

$= \dfrac{12}{11+x}$ hours

| Time $= \dfrac{\text{Distance}}{\text{Speed}}$

According to the question,

$\dfrac{12}{11-x} + \dfrac{12}{11+x} = 2\dfrac{3}{4}$

∵ 2 hours 45 minutes
$= 2$ hours $+ 45$ minutes
$= 2$ hours $+ \dfrac{45}{60}$ hours
$= 2$ hours $+ \dfrac{3}{4}$ hour
$= 2\dfrac{3}{4}$ hours

$\Rightarrow 12\left(\dfrac{1}{11-x} + \dfrac{1}{11+x}\right) = \dfrac{11}{4}$

$\Rightarrow \dfrac{1}{11-x} + \dfrac{1}{11+x} = \dfrac{11}{48}$

$\Rightarrow \dfrac{11+x+11-x}{(11-x)(11+x)} = \dfrac{11}{48}$

$\Rightarrow \dfrac{22}{121-x^2} = \dfrac{11}{48}$

$\Rightarrow \dfrac{2}{121-x^2} = \dfrac{1}{48}$

$\Rightarrow 121 - x^2 = 96$

$\Rightarrow x^2 = 121 - 96$

$\Rightarrow x^2 = 25$

$\Rightarrow x = \pm \sqrt{25}$

$\Rightarrow x = \pm 5$

$\Rightarrow x = 5, -5$

Since, x is the speed of the stream, it cannot be negative. So, we ignore the root $x = -5$

Therefore, $x = 5$ gives the speed of the stream as 5 km/hour.

Example 16. *The sum of first 2 even natural numbers is given by the relation* $s = n(n + 1)$. *Find n, if the sum is 420.*

Sol. $n(n + 1) = 420$ | Given

$\Rightarrow n^2 + n = 420$

$\Rightarrow n^2 + n - 420 = 0$

Comparing with $An^2 + Bn + C = 0$, we get

$A = 1$
$B = 1$
$C = -420$

Using the quadratic formula, we get

$n = \dfrac{-B \pm \sqrt{B^2 - 4AC}}{2A}$

$\Rightarrow \dfrac{-1 \pm \sqrt{(1)^2 - 4(1)(-420)}}{2(1)}$

$\Rightarrow \dfrac{-1 \pm \sqrt{1 + 1680}}{2} = \dfrac{-1 \pm \sqrt{1681}}{2}$

$= \dfrac{-1 \pm 41}{2} = \dfrac{-1 + 41}{2}, \dfrac{-1 - 41}{2}$

$= 20, -21$

$n = -21$ is in admissible as n is the number of terms.

∴ $n = 20$

Hence, the required value of n is 20.

QUDRATIC EQUATIONS

Example 17. *The denominator of a fraction is more than twice the numerator. If the sum of the fraction and its reciprocal is $2\frac{16}{21}$, find the fraction.*

Sol. Let the numerator of the fraction be x.

Then, the denominator of the fraction
$$= 2x + 1$$

$\therefore$ Fraction $= \dfrac{x}{2x+1}$

$\therefore$ Reciprocal of the fraction $= \dfrac{2x+1}{x}$

According to the question,

$$\dfrac{x}{2x+1} + \dfrac{2x+1}{x} = 2\dfrac{16}{21}$$

$\Rightarrow \dfrac{x^2 + (2x+1)^2}{(2x+1)\,x} = \dfrac{58}{21}$

$\Rightarrow \dfrac{x^2 + 4x^2 + 4x + 1}{2x^2 + x} = \dfrac{58}{21}$

$\Rightarrow \dfrac{5x^2 + 4x + 1}{2x^2 + x} = \dfrac{58}{21}$

$\Rightarrow 21(5x^2 + 4x + 1) = 58(2x^2 + x)$
$\Rightarrow 105x^2 + 84x + 21 = 116x^2 + 58x$
$\Rightarrow 11x^2 - 26x - 21 = 0$

Comparing with $ax^2 + bx + c = 0$, we get
$$a = 11$$
$$b = -26$$
$$c = -21$$

Using the quadratic formula, we get

$$x = \dfrac{-b \pm \sqrt{b^2 - 4ac}}{2a}$$

$$= \dfrac{-(-26) \pm \sqrt{(-26)^2 - 4(11)(-21)}}{2(11)}$$

$$= \dfrac{26 \pm \sqrt{676 + 924}}{22}$$

$$= \dfrac{26 \pm \sqrt{1600}}{22}$$

$$= \dfrac{26 \pm 40}{22}$$

$$= \dfrac{26 + 40}{22}, \dfrac{26 - 40}{22}$$

$$= 3, -\dfrac{7}{11}$$

$x = -\dfrac{7}{11}$ is in admissible as x, being the numerator of a fraction, is a natural number.

$\therefore x = 3$
$\therefore 2x + 1 = 2(3) + 1 = 7$

$\therefore$ Required fraction $= \dfrac{3}{7}$.

Example 18. *A man travels a distance of 300 km at a uniform speed. If the speed of the train is increased by 5 km an hour, the joining would have taken two hours less. Find the original speed of the train.* (CBSE 2006)

Sol. Let the original speed of the train be x km an hour.

Then, the total taken by the train to travel a distance of 300 km at a uniform speed of x km an hour

$$= \dfrac{300}{x} \text{ hours} \qquad \left| \; \because \text{Time} = \dfrac{\text{Distance}}{\text{Speed}} \right.$$

Increased speed of the train $= (x + 5)$ km an hour time taken by the train to travel a distance of 300 km at the increased speed

$$= \dfrac{300}{x+5} \text{ hours} \qquad \left| \; \because \text{Time} = \dfrac{\text{Distance}}{\text{Speed}} \right.$$

According to the equation,

$$\dfrac{300}{x} - 2 = \dfrac{300}{x+5}$$

$$\Rightarrow \frac{300}{x} - \frac{300}{x+5} = 2$$

$$\Rightarrow 300\left(\frac{1}{x} - \frac{1}{x+5}\right) = 2$$

$$\Rightarrow \frac{1}{x} - \frac{1}{x+5} = \frac{2}{300}$$

$$\Rightarrow \frac{1}{x} - \frac{1}{x+5} = \frac{1}{150}$$

$$\Rightarrow \frac{x+5-x}{x(x+5)} = \frac{1}{150}$$

$$\Rightarrow \frac{5}{x^2+5x} = \frac{1}{150}$$

$$\Rightarrow \quad x^2 + 5x = 750$$

$$\Rightarrow \quad x^2 + 5x - 750 = 0$$

Comparing with $ax^2 + bx + c = 0$, we get
$a = 1, b = 5, c = -750$

Using the quadratic formula, we get

$$x = \frac{-b \pm \sqrt{b^2 - 4ac}}{2a}$$

$$= \frac{-5 \pm \sqrt{(5)^2 - 4(1)(-750)}}{2(1)}$$

$$= \frac{-5 \pm \sqrt{25 + 3000}}{2}$$

$$= \frac{-5 \pm \sqrt{3025}}{2} = \frac{-5 \pm 55}{2}$$

$$= \frac{-5 + 55}{2}, \frac{-5 - 55}{2}$$

$$= 25, -30$$

$x = -30$ is inadmissible as x is the speed of the train and speed cannot be negative.

$\therefore \quad x = 25$

Hence, the original speed of the train is 25 km an hour.

TEST YOUR KNOWLEDGE

1. The product of Sunita's age (in years) two years ago and her age four years from now is one more than twice her present age. What is her present age ?
2. Solve $x^2 + 4x - 5 = 0$ by the method of completing the square.
3. Solve $9x^2 - 15x + 6 = 0$ by the method of completing the square.
4. Solve $3x^2 - 5x + 2 = 0$ by the method of completing the square.
5. Solve the equation $2x^2 - 5x + 3 = 0$ by the method of completing the square.
6. Find the roots of the equation $5x^2 - 6x - 2 = 0$ by the method of completing the square.
7. Find the roots of $4x^2 + 3x + 5 = 0$ by the method of completing the square.
8. The area of a rectangular plot is 528 m². The length of the plot (in metres) is one more than twice its breadth. Find the length and breadth of the plot.
9. Find two consecutive odd positive integers sum of whose squares is 290.
10. I want to design a rectangular park whose breadth is 3 m less than its length. Its area is to be 4 square metres more than the area of a park that has already been made in the shape of a isosceles triangle with its base as the breadth of the rectangular park and altitude 12 m. Is it possible to have such a rectangular park. If so, find its length and breadth.
11. Find the roots of the following quadratic equations of they exist, using the quadratic formula :
    (i) $3x^2 - 5x + 2 = 0$
    (ii) $x^2 + 4x + 5 = 0$
    (iii) $2x^2 - 2\sqrt{2}\,x + 1 = 0$
12. Find the roots of the following equations :
    (i) $x + \frac{1}{x} = 3, x \neq 0$
    (ii) $\frac{1}{x} - \frac{1}{x-2} = 3, x \neq 0, 2$.

QUDRATIC EQUATIONS

13. A motor boat whose speed is 18 km/h in still water takes 1 hour more to go 24 kms upstream than to return downstream to the same spot. Find the speed of the stream.

14. The sum of two natural number is 8. Determine the numbers if the sum of their reciprocals is $\dfrac{8}{15}$.   (A.I. CBSE 2006)

15. If the price of a book is reduced by Rs. 5, a person can buy 5 more books for Rs. 300. Find the original list price of the book.   (CBSE 2003)

16. A dealer sells an article for Rs. 75 and gains as much percent as the cost price of the article. What is the cost price of the article?

17. In a society for children, each child gives a gift to every of the child. If the number of gifts is 132, find the number of children.

18. The perimeter of a rectangle is 82 m and its area is 400 m². Find the breadth of the rectangle.

Answers

1. 3 years
2. 1, –5
3. $1, \dfrac{2}{3}$
4. $1, \dfrac{2}{3}$
5. $\dfrac{3}{2}, 1$
6. $\dfrac{3+\sqrt{19}}{5}, \dfrac{3-\sqrt{19}}{5}$
7. no real roots
8. 33 m, 16 m
9. 11, 13
10. Yes 7 m, 4 m
11. (i) $\dfrac{2}{3}, 1$ (ii) no real roots
    (iii) $\dfrac{1}{\sqrt{2}}, \dfrac{1}{\sqrt{2}}$
12. (i) $\dfrac{3+\sqrt{5}}{2}, \dfrac{3-\sqrt{5}}{2}$
    (ii) $\dfrac{3+\sqrt{3}}{3}, \dfrac{3-\sqrt{3}}{3}$
13. 6 km/h
14. 3, 5
15. Rs. 20
16. Rs. 50
17. 12
18. 16 m or 25 m.

NATURE OF ROOTS

IMPORTANT POINTS

1. Roots of the quadratic equation $ax^2 + bx + c = 0$. We know that the roots of the equation $ax^2 + bx + c = 0$ are the values of x satisfying

$$\left(x + \dfrac{b}{2a}\right)^2 = \dfrac{b^2 - 4ac}{4a^2} \quad \ldots(1)$$

Now these arise three cases :

Case I. $b^2 - 4ac < 0$

If $b^2 - 4ac < 0$, then $\dfrac{b^2 - 4ac}{4a^2} < 0$. But there is no real number whose square is negative.

So, there is no real value of x satisfying (1). Therefore there is no real roots for the given quadratic equation in this case.

Case II. $b^2 - 4ac > 0$

If $b^2 - 4ac > 0$, then $\dfrac{b^2 - 4ac}{4a^2} > 0$. By taking square roots on both sides in (1), we get

$$x + \dfrac{b}{2a} = \dfrac{\pm\sqrt{b^2 - 4ac}}{2a}$$

i.e.,

$$x = -\dfrac{b}{2a} \pm \sqrt{\dfrac{b^2 - 4ac}{2a}}$$

So, if $b^2 - 4ac > 0$, we get two distinct roots $-\dfrac{b}{2a} + \sqrt{\dfrac{b^2 - 4ac}{2a}}$ and $-\dfrac{b}{2a} - \sqrt{\dfrac{b^2 - 4ac}{2a}}$

Case III. $b^2 - 4ac = 0$

If $b^2 - 4ac = 0$,

then $\quad x = \dfrac{b}{2a} \pm \sqrt{\dfrac{0}{2a}} = -\dfrac{b}{2a} \pm 0$

i.e., $\quad x = -\dfrac{b}{2a}$ or $-\dfrac{b}{2a}$.

So, the roots of the equation

$ax^2 + bx + c = 0$ are both $\dfrac{-b}{2a}$.

Therefore, we say that the quadratic equation $ax^2 + bx + c = 0$ has two equal real roots in this case.

2. Discriminant. Since, $b^2 - 4ac$ determines whether the quadratic equation $ax^2 + bx + c = 0$ has real roots or not, $b^2 - 4ac$ is called the **discriminant** of this quadratic equation.

3. Inference. A quadratic equation
$$ax^2 + bx + c = 0$$
(i) has no real roots if $b^2 - 4ac < 0$.
(ii) has two equal real roots if
$$b^2 - 4ac = 0.$$
(iii) has two distinct real roots if
$$b^2 - 4ac > 0.$$

ILLUSTRATIVE EXAMPLES

[NCERT Exercise 4.4]
(Page No. 91)

Example 1. *Find the nature of the roots of the following quadratic equations. If the real roots exist, find them:*

(i) $2x^2 - 3x + 5 = 0$
(ii) $3x^2 - 4\sqrt{3}x + 4 = 0$
(iii) $2x^2 - 6x + 3 = 0$.

Sol. (i) $\mathbf{2x^2 - 3x + 5 = 0}$

The given quadratic equation is
$2x^2 - 3x + 5 = 0$

Here, $\quad a = 2$
$\quad\quad\quad b = -3$
$\quad\quad\quad c = 5$

Therefore, discriminant $= b^2 - 4ac$
$= (-3)^2 - 4(2)(5)$
$= 9 - 40$
$= -31 < 0$

So, the given quadratic equation has no real roots.

(ii) $\mathbf{3x^2 - 4\sqrt{3}x + 4 = 0}$

The given quadratic equation is
$3x^2 - 4\sqrt{3}x + 4 = 0$

Here, $\quad a = 3$
$\quad\quad\quad b = -4\sqrt{3}$
$\quad\quad\quad c = 4$

Therefore, discriminant $= b^2 - 4ac$
$= (-4\sqrt{3})^2 - 4(3)(4)$
$= 48 - 48$
$= 0$

Hence, the given quadratic equation has two equal real roots.

The roots are $-\dfrac{b}{2a}, -\dfrac{b}{2a}$,

i.e., $-\dfrac{(-4\sqrt{3})}{2 \times 3}, -\dfrac{(-4\sqrt{3})}{2 \times 3}$, i.e., $\dfrac{2}{\sqrt{3}}, \dfrac{2}{\sqrt{3}}$.

(iii) $\mathbf{2x^2 - 6x + 3 = 0}$

The given quadratic equation is
$2x^2 - 6x + 3 = 0$

Here, $a = 2$
$\quad\quad b = -6$
$\quad\quad c = 3$

Therefore, discriminant $= b^2 - 4ac$
$= (-6)^2 - 4(2)(3)$
$= 36 - 24$
$= 12 > 0$

So, the given quadratic equation has two distinct real roots.

Solving the quadratic equation $2x^2 - 6x + 3 = 0$, by the quadratic formula, we get

$$x = \frac{-b \pm \sqrt{b^2 - 4ac}}{2a}$$

$$= \frac{(-6) \pm \sqrt{12}}{2(2)}$$

$$= \frac{6 \pm 2\sqrt{3}}{4}$$

$$= \frac{3 \pm \sqrt{3}}{2}$$

Therefore, the roots are $\frac{3 \pm \sqrt{3}}{2}$, i.e., $\frac{3 + \sqrt{3}}{2}$ and $\frac{3 - \sqrt{3}}{2}$.

Example 2. *Find the values of k for each of the following quadratic equations, so that they have two real equal roots.*

(i) $2x^2 + kx + 3 = 0$ (ii) $kx(x - 2) + 6 = 0$

Sol. (i) $2x^2 + kx + 3 = 0$

The given quadratic equation is
$$2x^2 + kx + 3 = 0$$

Here, $a = 2$
$b = k$
$c = 3$

Therefore, discriminant $= b^2 - 4ac$
$$= (k)^2 - 4(2)(3)$$
$$= k^2 - 24$$

If the given quadratic equation has two real equal roots, then
$$b^2 - 4ac = 0$$
$\Rightarrow \quad k^2 - 24 = 0$
$\Rightarrow \quad k^2 = 24$
$\Rightarrow \quad k = \pm \sqrt{24}$
$\Rightarrow \quad k \pm 2\sqrt{6}$

Hence, the required values of k are $\pm 2\sqrt{6}$, i.e., $2\sqrt{6}$ and $-2\sqrt{6}$.

(ii) **kx (x − 2) + 6 = 0**

The given quadratic equation is
$$kx(x - 2) + 6 = 0$$

$\Rightarrow \quad kx^2 - 2kx + 6 = 0$

Here, $a = k$
$b = -2k$
$c = 6$

Therefore, discriminant
$$= b^2 - 4ac$$
$$= (-2k)^2 - 4(k)(6)$$
$$= 4k^2 - 24k$$

If the given quadratic equation has two real equal roots, then
$$b^2 - 4ac = 0$$
$\Rightarrow \quad 4k^2 - 24k = 0$
$\Rightarrow \quad 4k(k - 6) = 0$
$\Rightarrow \quad k(k - 6) = 0 \quad | \because \ 4 \neq 0$
$\Rightarrow \quad k = 0, 6$

$k = 0$ is in admissible as then the given quadratic equation reduces to $b = 0$, which is not true. [See that for $k = 0$, the given equation no longer no remains a quadratic equation]

$\therefore \quad k = 6$

Hence, the required value of k is 6.

Example 3. *Is it possible to design a rectangular mango grove whose length is twice its breadth, and area is 800 m^2 ? If so, find its length and breadth.*

Sol. Let the breadth of the rectangle mango grove be x m.

Then, the length of the rectangular mango grove $= 2x$ m

Therefore,

Area of the rectangular mango grove
$$= \text{Length} \times \text{Breadth}$$
$$= 2x \times x \ \text{m}^2$$
$$= 2x^2 \ \text{m}^2$$

According to the question,
$$2x^2 = 800$$
$\Rightarrow \quad x^2 = \frac{800}{2} = 400$
$\Rightarrow \quad x = \pm \sqrt{400}$
$\Rightarrow \quad x = \pm 20$, i.e., 20, -20

$x = -20$

is in admissible as x is a dimension.

Therefore, $x = 20$

$\Rightarrow \quad 2x = 2(20) = 40$

Hence, it is possible to design the rectangular mango grove and its length and breadth and 40 m and 20 m, respectively.

Example 4. *Is the following situation possible? If so, determine their present ages. The sum of the ages of two friends is 20 years. Four years ago, the product of their ages in years was 48.*

Sol. Let the age of first friend be x years.

Then the age of second friend = $(20 - x)$ years.

| $\because$ The sum of the ages of the two friends is 20 years

Four years ago

Age of first friend = $(x - 4)$ years

Age of second friend = $(20 - x - 4)$ years
$\qquad = (16 - x)$ years

Therefore, product of their ages
$\qquad = (x - 4)(16 - x)$

According to the question,
$\qquad (x - 4)(16 - x) = 48$
$\Rightarrow \quad 16x - x^2 - 64 + 4x = 48$
$\Rightarrow \quad x^2 - 20x + 112 = 0 \qquad \ldots(1)$

Here, $\quad a = 1$
$\qquad b = -20$
$\qquad c = 112$

Therefore, discriminant = $b^2 - 4ac$
$\qquad = (-20)^2 - 4(1)(112)$
$\qquad = 400 - 448$
$\qquad = -48 < 0$

Hence, the quadratic equation (1) has no real roots.

So, the given situation is not possible.

Example 5. *Is it possible to design a rectangular park of perimeter 80 m and area 400 m^2? If so, find its length and breadth.*

Sol. Let the breadth of the rectangular park be x m.

Perimeter of the rectangular park
$\qquad = 80$ m $\qquad$ | Given

$\Rightarrow \quad$ 2(Length + breadth) = 80

$\Rightarrow \quad$ Length + Breadth = $\dfrac{80}{2} = 40$

$\Rightarrow \quad$ Length + x = 40

$\Rightarrow \quad$ Length = $(40 - x)$ m

Therefore, area of the rectangular park
$\qquad$ = Length × Breadth
$\qquad = (40 - x) x$ m^2

According to the question,
$\qquad (40 - x) x = 400$
$\Rightarrow \quad 40x - x^2 = 400$
$\Rightarrow \quad x^2 - 40x + 400 = 0 \qquad \ldots(1)$

Here, $a = 1$
$\qquad b = -40$
$\qquad c = 400$

Therefore, discriminant = $b^2 - 4ac$
$\qquad = (-40)^2 - 4(1)(400)$
$\qquad = 1600 - 1600$
$\qquad = 0$

So, the quadratic equation (1) has equal real roots and it is possible to design the rectangular park.

Each root = $-\dfrac{b}{2a} = \dfrac{-(-40)}{2(1)} = 20$

Hence, the length and the breadth of the rectangular park are 20 m and 20 m respectively.

ADDITIONAL EXAMPLES

Example 6. *If the equation $(1 + m^2) x^2 + 2mcx + (c^2 - a^2) = 0$ has equal roots, prove that*
$\qquad c^2 = a^2 (1 + m^2).$

Sol. The given quadratic equation is
$(1 + m^2) x^2 + 2mcx + (c^2 - a^2) = 0$

Here $\qquad A = 1 + m^2$
$\qquad B = 2mc$
$\qquad C = c^2 - a^2$

Therefore, discriminant = $B^2 - 4AC$
$\qquad = (2mc)^2 - 4(1 + m^2)(c^2 - a)^2$
$\qquad = 4m^2c^2 - 4(c^2 - a^2 + m^2c^2 - m^2a^2)$

QUDRATIC EQUATIONS

$= 4m^2c^2 - 4c^2 + 4a^2 - 4m^2c^2 + 4m^2a^2$
$= -4c^2 + 4a^2 + 4m^2a^2$

If the given quadratic equation has equal roots, then

$$B^2 - 4AC = 0$$
$\Rightarrow \quad -4c^2 + 4a^2 + 4m^2a^2 = 0$
$\Rightarrow \quad 4c^2 = 4a^2 + 4m^2a^2$
$\Rightarrow \quad 4c^2 = 4a^2(1 + m^2)$
$\Rightarrow \quad c^2 = a^2(1 + m^2)$

| Dividing throughout by 4

Example 7. *Determine the positive values of k for which the equations $x^2 + kx + 64 = 0$ and $x^2 - 8x + k = 0$ will both have real roots.*

Sol. First quadratic equation is
$$x^2 + kx + 64 = 0$$
Here, $a = 1$
$b = k$
$c = 64$

Therefore, discriminant $(D_1) = b^2 - 4ac$
$= k^2 - 4(1)(64)$
$= k^2 - 256$

If the first quadratic equation has real roots, then
$$D_1 \geq 0$$
$\Rightarrow \quad k^2 - 256 \geq 0$
$\Rightarrow \quad k^2 \geq 256$
$\Rightarrow \quad k \geq 16 \quad ...(1) \mid \because \; k > 0$

Second quadratic equation is
$$x^2 - 8x + k = 0$$
Here, $a = 1$
$b = -8$
$c = k$

Therefore, discriminant $(D_2) = b^2 - 4ac$
$= (-8)^2 - 4(1)(k)$
$= 64 - 4k$

If the second quadratic equation has real roots, then
$$D_2 \geq 0$$
$\Rightarrow \quad 64 - 4k \geq 0$
$\Rightarrow \quad 4k \leq 64$
$\Rightarrow \quad k \leq 16 \quad ...(2)$

We see that $k = 16$ satisfies (1) and (2) both.

Hence, both the equations will have real roots for $k = 16$.

Example 8. *Determine p so that the equation $x^2 + 5px + 16 = 0$ has no real roots.*

Sol. The given quadratic equation is
$$x^2 + 5px + 16 = 0$$
Here, $a = -1$
$b = 5p$
$c = 16$

Therefore, discriminant $= b^2 - 4ac$
$= (5p)^2 - 4(1)(16)$
$= 25p^2 - 64$

If the given quadratic equation has no real roots, then
$$b^2 - 4ac < 0$$
$\Rightarrow \quad 25p^2 - 64 < 0$
$\Rightarrow \quad 25p^2 < 64$
$\Rightarrow \quad p^2 < \dfrac{64}{25}$
$\Rightarrow \quad p^2 - \dfrac{64}{25} < 0$
$\Rightarrow \quad p^2 - \left(\dfrac{8}{5}\right)^2 < 0$
$\Rightarrow \quad \left(p - \dfrac{8}{5}\right)\left(p + \dfrac{8}{5}\right) < 0$
$\Rightarrow$ **Either** $p - \dfrac{8}{5} > 0$ and $p + \dfrac{8}{5} < 0$,

i.e., $p > \dfrac{8}{5}$ and $p < -\dfrac{8}{5}$ which is not possible

or $p - \dfrac{8}{5} < 0$ and $p + \dfrac{8}{5} > 0$,

i.e., $p < \dfrac{8}{5}$

and $p > -\dfrac{8}{5}$,

i.e., $-\dfrac{8}{5} < p < \dfrac{8}{5}$.

TEST YOUR KNOWLEDGE

1. Find the discriminant of the quadratic equation $2x^2 - 4x + 3 = 0$ and hence find the nature of its roots.

2. A pole has to be erected at a point on the boundary of a circular park of diameter 13 metres in such a way that the differences of its distances from two diametrically opposite fixed gates A and B on the boundary in 7 metres. Is it possible to do so ? If yes, at what distances from the two gates should the pole be erected ?

3. Find the discriminant of the equation $3x^2 - 2x + \dfrac{1}{3} = 0$ and hence, find the nature of its roots. Find them, if they are real.

4. Find the value of k for which the quadratic equation $(k + 4)x^2 + (k + 1) x + 1 = 0$ has equal roots. *(CBSE 2000)*

5. For what values of k, does the quadratic equation $9x^2 + 8kx + 16 = 0$ have equal roots ? *(CBSE 2001)*

6. Determine k so that the equation $x^2 - 4x + k = 0$ has
    (i) two real roots
    (ii) two distinct roots
    (iii) coincident roots
    (iv) no real roots

7. Without finding the roots, comment on the nature of the roots of the following quadratic equation :
$9a^2b^2x^2 - 48abcdx + 64c^2d^2 = 0, a \neq 0, b \neq 0$.

8. The hypotenuse of a right triangle is $3\sqrt{10}$ cm. If the smaller leg is tripled and the longer leg doubled, new hypotenuse will be $9\sqrt{5}$ cm. Is it possible to draw such a right triangle ? If so, how long are the legs of the triangle ?

Answers

1. -8, no real roots.
2. Yes, 5 m
3. 0 ; $\dfrac{1}{3}$ and $\dfrac{1}{3}$.
4. ± 7
5. ± 3
6. (i) $k \leq 4$
   (ii) $k < 4$
   (iii) $k = 4$
   (iv) $k > 4$
7. Real and equal
8. Yes, 3 cm and 9 cm

MISCELLANEOUS EXERCISE

1. Which of the following are quadratic equations ?
    (i) $x^3 + 1 = 0$
    (ii) $x^3 = 1$
    (iii) $3x^2 - 48 = 0$
    (iv) $8x^2 + 6x = 24$
    (v) $4x^2 + 5x + 5 = 5x + 30$
    (vi) $5x^2 - 64 = 4x^2$
    (vii) $(x + 2)(x + 3) = 5x + 7$
    (viii) $\dfrac{1}{x+1} + \dfrac{1}{x} = 2$
    (ix) $(2x + 5)(x - 4) = (x + 1)(x - 4)$
    (x) $x^4 + a^4 = 0$.

2. Find the solution set of the equation $10x^2 + 3bx + a^2 - 7ax - b^2 = 0$.

3. Find the roots of the equation
   $\sqrt{7} x^2 - 6x - 13\sqrt{7} = 0$.

4. Prove that the equation
   $x^2(a^2 + b^2) + 2x(ac + bd) + (c^2 + d^2) = 0$ has no real roots, if $ad \neq bc$.

5. Prove that the roots of equation
   $2x^2 + 3(b - a) x - 4a^2 - 3ab + b^2 = 0$ are real and distinct.

6. If the roots of the equation
   $(c^2 - ab)x^2 + 2(bc - a^2)x + (b^2 - ac) = 0$ are equal, then prove that either $a = 0$ or $a^3 + b^3 + c^3 = 3abc$.

7. Solve the following equations, if possible.
    (i) $x^2 - 8x + 15 = 0$
    (ii) $x^2 + 2x - 5 = 0$
    (iii) $5z^2 + 12z + 10 = 0$

(iv) $5x^2 = x + 1$
(v) $x^2 - 4x + 7 = 0$.

8. Using factorisation method, solve each of the following equations :

(i) $x^2 - \frac{11}{4}x + \frac{15}{8} = 0$

(ii) $x + \frac{4}{x} = -4$ $(x \neq 0)$

(iii) $\frac{x+1}{x-1} + \frac{x-2}{x+2} = 3$ $(x \neq 1, x \neq -2)$

(iv) $\frac{1}{x-3} - \frac{1}{x+9} = \frac{1}{6}$ $(x \neq 3, x \neq -5)$

9. Using quadratic formula, solve the following equation for x :
$abx^2 + (b^2 - ac)x - bc = 0$.

10. Determine the least value of k so that the equation $2x^2 - kx + 1 = 0$ has equal roots.

11. Determine λ so that the equation $x^2 + \lambda x + \lambda + 1.25 = 0$ has
(i) two distinct roots.
(ii) two coincident roots.

12. Determine the value of $k(k > 0)$ such that the equations $x^2 + kx + 64 = 0$ and $x^2 - 8x + k = 0$ both will have real roots.

13. A farmer wishes to grow a 100 m² rectangular vegetable garden. Since he has with him, only 30 m barbed wire, he fences three sides of the rectangular garden letting compound wall of his house act as the fourth side-fence. Is it possible to do so ? If yes, find the dimensions of his garden.

14. The sum of n successive odd natural numbers starting from 3 is given by the relation $S = n(n + 2)$. Determine n, if the sum is 168.

15. The product of two successive integral multiples of 5 is 300. Determine the multiples.

16. Divide 16 into two parts such that twice the square of the larger part exceeds the square of the smaller part by 164.

17. B is a point in the line segment AC such that it has between A and C. If AB = 9 cm and AB × AC = 12 (BC)², find BC.

18. A segment AB of 2 m length is divided at C into two parts such that $AC^2 = AB \cdot CB$. Find the length of the parts CB.

19. An aeroplane takes one hour less for a journey of 1200 km, if its speed is increased by 100 km/h from its usual speed. Find its usual speed.

20. The angry Arjun carried some arrows for fighting with Bheshm. With half of the arrows he cut down the arrows thrown by Bheshm on him and with six other arrows he killed the Rath driver of Bheshm. With one arrow each he knocked down respectively the Rath, flag and bow of Bheshum. Finally, with one more than four times the square root of arrows, he laid Bheshm unconscious on an arrow bed. Find the total number of arrows Arjun had. (*CBSE 1996*)

Answers

1. (*iii*), (*iv*), (*v*), (*vii*), (*viii*), (*ix*)

2. $\frac{a-b}{2}, \frac{a+b}{5}$ 3. $\frac{13\sqrt{7}}{7}, -\sqrt{7}$

7. (*i*) 3, 5   (*ii*) $1, -\frac{5}{3}$

(*iii*) No solution   (*iv*) $\frac{1}{2}, -\frac{1}{3}$

(*v*) No solution

8. (*i*) $\frac{3}{2}, \frac{5}{4}$ (*ii*) $-2, -2$

(*iii*) 2, -5 (*iv*) $-9, 7$

9. $\frac{c}{b}, -\frac{b}{a}$ 10. $-2\sqrt{2}$

11. (*i*) $\lambda < -1$ or $\lambda < 5$
(*ii*) $\lambda = 5$ or -1

12. 16

13. 10 m, 10 m or 20 m, 5 m

14. 12

15. 15, 20

16. 6, 10

17. 3 cm

18. $(3 - \sqrt{5})$ m

19. 300 km/h

20. 100 arrows

5

Arithmetic Progressions

IMPORTANT POINTS

1. Patterns in nature. We observe that in nature, many things follow a certain pattern, such as the petals of a sunflower, the holes of a honeycomb, the grains on maize cob, the spirals on a pineapple and on a pine cone etc.

2. Patterns in day-to-day life. Let us look for some patterns which occur in our day-to-day life. Some such examples are :

(i) Reena applied for a job and got selected. She has been offered a job with a starting monthly salary of Rs 8000, with an annual increment of Rs 500 in her salary. Her salary (in Rs) for the 1st, 2nd, 3rd,... years will be, respectively

8000, 8500, 9000, ...

(ii) The lengths of the rungs of a ladder decrease uniformly by 2 cm from bottom to top (see figure below). The bottom rung is 45 cm in length. The lengths (in cm) of the 1st, 2nd, 3rd, ..., 8th rung from the bottom to the top are, respectively

45, 43, 41, 39, 37, 35, 33, 31

(iii) In saving scheme, the amount becomes $\frac{5}{4}$ times of itself after every 3 years.

The maturity amount (in Rs) of an investment of Rs 8000 after 3, 6, 9 and 12 years will be, respectively

10000, 12500, 15625, 19531.25

(iv) The number of unit squares in squares with side 1, 2, 3, ... units (see figure below) are, respectively

$1^2, 2^2, 3^2, \ldots$

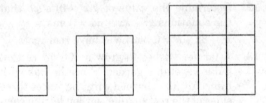

(v) Shakila put Rs 100 into her daughter's money box when she was one year old and increased the amount by Rs 50 every year. The amounts of money (in Rs) in the box on the 1st, 2nd, 3rd, 4th, ... birthday were

100, 150, 200, 250, ... respectively.

(vi) A pair of rabbits are too young to produce in their first month. In the second, and every subsequent month, they produce a new pair. Each new pair of rabbits produce a new pair in their second month and in every subsequent month (see figure on next page). Assuming no rabbit dies, the number of pairs of rabbits at the start of the 1st, 2nd, 3rd, ..., 6th month, respectively are :

1, 1, 2, 3, 5, 8

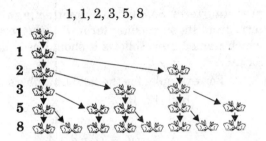

In the examples above, we observe some patterns. In some, we find that the succeeding terms are obtained by adding a fixed number, in other by multiplying with a fixed number, in another we find that they are squares of consecutive numbers, and so on.

Here, in this chapter, we shall confine our selves to the study of those patterns in which succeeding terms are obtained by adding a fixed number to the preceding terms. We shall also learn how to find their nth terms and the sum of n consecutive terms, and use this knowledge in solving some daily life problems.

3. Arithmetic Progressions. Consider the following lists of numbers :

(i) 1, 2, 3, 4, ...

(ii) 100, 70, 40, 10, ...

(iii) $-3, -2, -1, 0$, ...

(iv) 3, 3, 3, 3, ...

(v) $-1.0, -1.5, -2.0, -2.5$, ...

Each of the numbers in the list is called a **term**.

Given a term, we can write the next term in each of the lists above by following a pattern or rule. Let us observe and write the rule for each list given above.

In (i), each term is 1 more than the term preceding it.

In (ii), each term is 30 less than the term preceding it.

In (iii), each term is obtained by adding 1 to the term preceding it.

In (iv), all the terms in the list are 3, i.e., each term is obtained by adding (or subtracting) 0 to the term preceding it.

In (v), each term is obtained by adding -0.5 to (i.e., subtracting 0.5 from) the term preceding it.

Thus, in all the lists above, we see that successive terms are obtained by adding a fixed number to the preceding terms. Such list of numbers is said to form an **Arithmetic Progression (AP)**.

4. Definition of an arithmetic progression. So, an arithmetic progression is a list of numbers in which each term is obtained by adding a fixed number to the preceding term except the first term. It is abbreviated as an A.P.

This fixed number is called the **common difference** of the A.P. Remember that **it can be positive, negative or zreo.**

5. Symbolical form. Let us denote the first term of an AP by a_1, second term by a_2, ... nth term by a_n and the common difference by d. Then the AP becomes $a_1, a_2, a_3, ..., a_n$.

So, $a_2 - a_1 = a_3 - a_2 = ... = a_n - a_{n-1} = d$.

6. More Examples. Some more examples of AP are :

(a) The heights (in cm) of some students of a school standing in a queue in the morning assembly are 147. 148, 149, ..., 157.

(b) The minimum temperatures (in degree celsius) recorded for a week in the month of January in a city, arranged in ascending order are

$-3.1, -3.0, -2.9, -2.8, -2.7, -2.6, -2.5$

(c) The balance money (in Rs) after paying 5% of the total loan of Rs 1000 every month is 950, 900, 850, 800, ..., 50.

(d) The cash prizes (in Rs) given by a school to the toppers of Classes I to XII are, respectively, 200, 250, 300, 350, ..., 750.

(e) The total savings (in Rs) after every month for 10 months when Rs 50 are saved each month are 50, 100, 150, 200, 250, 300, 350, 400, 450, 500.

7. General Form of an AP. The list of numbers

$$a, a + d, a + 2d, a + 3d, ...$$

represents an arithmetic progression where a is the first term and d the common difference. This is called the **general form of an AP.**

8. Finite AP. An AP in which there are only a finite number of terms is called a finite AP. It may be noted that each such AP has a last term.

The examples (a) to (e) given above are examples of a finite AP.

9. Infinite AP. An AP in which the number of terms is not finite is called infinite AP. It is note worthy that such APs donot have a last term. The example (i) to (iv) given one examples of an infinite A.P.

10. Minimum information needed. To know about an AP, the minimum information we need to know is to know both—the first term a and the common difference d.

For instance if the first term a is 6 and the common difference d is 3, then AP is

$$6, 9, 12, 15, \ldots$$

and if a is 6 and d is -3, then the AP is

$$6, 3, 0, -3, \ldots$$

Similarly, when

$a = -7, d = -2,$ the AP is $-7, -9, -11, -13, \ldots$

$a = 1.0, d = 0.1,$ the AP is $1.0, 1.1, 1.2, 1.3, \ldots$

$a = 0, d = 1\frac{1}{2},$ the AP is $0, 1\frac{1}{2}, 3, 4\frac{1}{2}, 6, \ldots$

$a = 2, d = 0,$ the AP is $2, 2, 2, 2, \ldots$

So, if we know what a and d are we can list the AP.

11. Criterion for a list of numbers to be an AP. Consider a list of numbers given to us. We are to find out whether this list of numbers is an AP or not and, if so, then to find a and d. Since, a is the first term, it can easily be written. We know that in an AP, every succeeding term is obtained by adding d to the preceding term. So, d found by subtracting any term from its succeeding term, i.e., the term which immediately follows it should be same for an AP.

For example, for the list of numbers :
$$6, 9, 12, 15, \ldots ,$$

We have $a_2 - a_1 = 9 - 6 = 3,$
$a_3 - a_2 = 12 - 9 = 3,$
$a_4 - a_3 = 15 - 12 = 3$

Here, the difference of any two consecutive terms in each case is 3. So, the given list is an AP whose first term a is 6 and common difference d is 3.

For the list of numbers : $6, 3, 0, -3, \ldots,$

$a_2 - a_1 = 3 - 6 = -3$
$a_3 - a_2 = 0 - 3 = -3$
$a_4 - a_3 = -3 - 0 = -3$

Similarly this is also an AP whose first term is 6 and the common difference is -3.

Generalisation. In general, for an AP $a_1, a_2, \ldots, a_n$, we have $d = a_{k+1} - a_k$

where a_{k+1} and a_k are the $(k+1)$th and the kth terms respectively.

An Advice. To obtain d in a given AP, we need not find all of $a_2 - a_1, a_3 - a_2, a_4 - a_3, \ldots$

It is enough to find only one of them. We can find d using any two consecutive terms, once we know that the numbers are in AP.

12. Not an AP. Consider the list of numbers 1, 1, 2, 3, 5, ... By looking at it, you can tell that the difference between any two succeeding terms is not the same. So, this is not an AP.

13. Negative d. To find d in the AP 6, 3, 0, $-3, \ldots$, we should subtracted 6 from 3 and not 3 from 6, i.e., we should subtract the kth term from the $(k+1)$th term even if the $(k+1)$th term is smaller. In this case,

$$a = 3 - 6 = -3 (<0).$$

ARITHMETIC PROGRESSIONS

ILLUSTRATIVE EXAMPLES

[NCERT Exercise 5.1]
(Page No. 99)

Example 1. *In which of the following situation, does the list of numbers involved make an arithmetic progression, and why?*

(i) The taxi fare after each km when the fare is Rs 15 for the first km and Rs 8 for each additional km.

(ii) The amount of air present in a cylinder when a vacuum pump removes $\frac{1}{4}$ of the air remaining in the cylinder at a time.

(iii) The cost of digging a well after every metre of digging, when it costs Rs 150 for the first metre and rises by Rs 50 for each subsequent metre.

(iv) The amount of money in the account every year, when Rs 10000 is deposited at compound interest at 8% per annum.

Sol. (i) Taxi fare for 1 km = Rs 15 = a_1
Taxi fare for 2 kms
= Rs 15 + Rs 8 = Rs 23 = a_2
Taxi fare for 3 kms
= Rs 23 + Rs 8 = Rs 31 = a_3
Taxi fare for 4 kms
= Rs 31 + Rs 8 = Rs 39 = a_4
and so on.

$a_2 - a_1$ = Rs 23 − Rs 15 = Rs 8
$a_3 - a_2$ = Rs 31 − Rs 23 = Rs 8
$a_4 - a_3$ = Rs 39 − Rs 31 = Rs 8

i.e., $a_{k+1} - a_k$ is the same everytime.

So, this list of numbers form an arithmetic progression with the first term a = Rs 15 and the common difference d = Rs 8.

(ii) Amount of air present in the cylinder = x units (say) = a_1

Amount of air present in the cylinder after one time removal of air by the vacuum pump = $x - \frac{x}{4} = \frac{3x}{4}$ units = a_2

Amount of air present in the cylinder after two times removal of air by the vacuum pump = $\frac{3x}{4} - \frac{1}{4}\left(\frac{3x}{4}\right) = \frac{3x}{4} - \frac{3x}{16} = \frac{9x}{16}$ units

= $\left(\frac{3}{4}\right)^2 x$ units = a_3

Amount of air present in the cylinder after three times removal of air by the vacuum pump

= $\left(\frac{3}{4}\right)^2 x - \frac{1}{4}\left(\frac{3}{4}\right)^2 x$

= $\left(1 - \frac{1}{4}\right)\left(\frac{3}{4}\right)^2 x$

= $\left(\frac{3}{4}\right)\left(\frac{3}{4}\right)^2 x$

= $\left(\frac{3}{4}\right)^3 x$ units = a_4

and so on.

$a_2 - a_1 = \frac{3x}{4} - x = \frac{x}{4}$ units

$a_3 - a_2 = \left(\frac{3}{4}\right)^2 x - \frac{3}{4} x = \frac{3x}{4}$ units

As $a_2 - a_1 \neq a_3 - a_2$, this list of numbers does no form an AP.

(iii) Cost of digging the well after 1 metre of digging = Rs 150 = a_1

Cost digging the well after 2 metres of digging
= Rs 150 + Rs 50 = Rs 200 = a_2

Cost of digging the well after 3 metres of digging
= Rs 150 + Rs 50 = Rs 200 = a_3

Cost of digging the well after 4 metres of digging
= Rs 200 + Rs 50 = Rs 250 = a_4

and so on.

183

$a_2 - a_4$ = Rs 200 – Rs 150 = Rs 50
$a_3 - a_2$ = Rs 250 – Rs 200 = Rs 50
$a_4 - a_3$ = Rs 350 – Rs 250 = Rs 50

i.e., $a_{k+1} - a_k$ is the same everytime. So this list of numbers forms an AP with the first term a = Rs 150 and the common difference d = Rs 50.

(iv) Amount of money after 1 year

$$= \text{Rs } 10000 \left(1 + \frac{8}{100}\right) = a_1$$

Amount of money after I years

$$= \text{Rs } 10000 \left(1 + \frac{8}{100}\right)^2 = a_2$$

Amount of money after 3 years

$$= \text{Rs } 10000 \left(1 + \frac{8}{100}\right)^3 = a_3$$

Amount of money after 4 years

$$= \text{Rs } 10000 \left(1 + \frac{8}{100}\right)^4 = a_4$$

$$a_2 - a_1 = \text{Rs } 10000 \left(1 + \frac{8}{100}\right)^2$$

$$- \text{Rs } 10000 \left(1 + \frac{8}{100}\right)$$

$$= \text{Rs } 10000 \left(1 + \frac{8}{100}\right)\left(1 + \frac{8}{100} - 1\right)$$

$$= \text{Rs } 10000 \left(1 + \frac{8}{100}\right)\left(\frac{8}{100}\right)$$

$$a_3 - a_2 = \text{Rs } 10000 \left(1 + \frac{8}{100}\right)^3$$

$$- \text{Rs } 10000 \left(1 + \frac{8}{100}\right)^2$$

$$= \text{Rs } 10000 \left(1 + \frac{8}{100}\right)^2 \left(1 + \frac{8}{100} - 1\right)$$

$$= \text{Rs } 10000 \left(1 + \frac{8}{100}\right)^2 \left(\frac{8}{100}\right)$$

As $a_2 - a_1 \ne a_3 - a_2$, this list of numbers does not form in AP.

Example 2. *Write first four terms of the AP, when the first term a and the common difference d are given as follows :*

(i) $a = 10, d = 10$

(ii) $a = -2, d = 0$

(iii) $a = 4, d = -3$

(iv) $a = -1, d = \frac{1}{2}$

(v) $a = -1.25, d = -0.25$.

Sol. (i) **a = 10, d = 10**

First term = a = 10

Second term = $18 + d$ = 10 + 10 = 20

Third term = $20 + d$ = 20 + 10 = 30

Fourth term = $30 + d$ = 30 + 10 = 40

Hence, first four terms of the given AP are

10, 20, 30, 40.

(ii) **a = – 2, d = 0**

First term = a = – 2

Second term = $-2 + d = -2 + 0 = -2$

Third term = $-2 + d = -2 + 0 = -2$

Fourth term = $-2 + d = -2 + 0 = -2$

Hence, first four terms of the given AP are – 2, – 2, – 2, – 2 respectively.

(iii) **a = 4, d = – 3**

First term = a = 4

Second term = $4 + d = 4 + (-3) = 1$

Third term = $1 + d = 1 + (-3) = -2$

Fourth term = $-2 + d = -2 + (-3) = -5$

Hence, four first terms of the given AP are 4, 1, – 2, – 5.

(iv) **a = – 1, d = $\frac{1}{2}$**

First term = a = – 1

Second term = $-1 + d = -1 + \frac{1}{2} = -\frac{1}{2}$

Third term = $-\frac{1}{2} + d = -\frac{1}{2} + \frac{1}{2} = 0$

Fourth term = $0 + d = 0 + \frac{1}{2} = \frac{1}{2}$

ARITHMETIC PROGRESSIONS

Hence, first four terms of the given AP are $-1, -\frac{1}{2}, 0, \frac{1}{2}$.

(v) a = – 1.25, d = – 0.25

First term = a = – 1.25

Second term = – 1.25 + d
= – 1.25 + (– 0.25)
= – 1.50

Third term = – 1.50 + d
= – 1.50 + (– 0.25)
= – 1.75

Fourth term = – 1.75 + d
= – 1.75 + (– 0.25)
= – 2.00

Hence, first four terms of the given AP are – 1.25, – 1.50, – 1.75, – 2.00.

Example 3. *For the following APs, write the first term and the common difference :*

(i) 3, 1, – 1, – 3, ...
(ii) – 5, – 1, 3, 7, ...
(iii) $\frac{1}{3}, \frac{5}{3}, \frac{9}{3}, \frac{13}{3}, ...$
(iv) 0.6, 1.7, 2.8, 3.9, ...

Sol. (i) 3, 1, – 1, – 3, ...

First term (a) = 3

Common difference (d) = 1 – 3 = – 2

(ii) – 5, – 1, 3, 7, ...

First term (a) = – 5

Common difference (d) = – 1 – (– 5) = – 1 + 5 = 4

(iii) $\frac{1}{3}, \frac{5}{3}, \frac{2}{3}, \frac{13}{3}, ...$

First term (a) = $\frac{1}{3}$

Common difference (d) = $\frac{5}{3} - \frac{1}{3} = \frac{4}{3}$

(iv) 0.6, 1.7, 2.8, 3.9, ...

First term (a) = 0.6

Common difference (d) = 1.7 – 0.6 = 1.1.

Example 4. *Which of the following are APs ? If they form an AP, find the common difference d and write three more terms.*

(i) 2, 4, 8, 16, ...
(ii) 2, $\frac{5}{2}$, 3, $\frac{7}{2}$, ...
(iii) – 1.2, – 3.2, – 5.2, – 7.2, ...
(iv) – 10, – 6, – 2, 2, ...
(v) 3, 3 + $\sqrt{2}$, 3 + 2$\sqrt{2}$, 3 + 3$\sqrt{2}$, ...
(vi) 0.2, 0.22, 0.222, 0.2222, ...
(vii) 0, – 4, – 8, – 12, ...
(viii) $-\frac{1}{2}, -\frac{1}{2}, -\frac{1}{2}, -\frac{1}{2}, ...$
(ix) 1, 3, 9, 27, ...
(x) $a, 2a, 3a, 4a, ...$
(xi) $a, a^2, a^3, a^4, ...$
(xii) $\sqrt{2}, \sqrt{8}, \sqrt{18}, \sqrt{32}, ...$
(xiii) $\sqrt{3}, \sqrt{6}, \sqrt{9}, \sqrt{12}, ...$
(xiv) $1^2, 3^2, 5^2, 7^2, ...$
(xv) $1^2, 5^2, 7^2, 73, ...$

Sol. (i) **2, 4, 8, 16, ...**

$a_2 - a_1 = 4 - 2 = 2$
$a_3 - a_2 = 8 - 4 = 4$
$a_4 - a_3 = 16 - 8 = 8$

Here, $a_2 - a_1 \ne a_3 - a_2$

So, the given list of numbers does not form an AP.

(ii) **2, $\frac{5}{2}$, 3, $\frac{7}{2}$, ...**

$a_2 - a_1 = \frac{5}{2} - 2 = \frac{1}{2}$

$a_3 - a_2 = 3 - \frac{5}{2} = \frac{1}{2}$

$a_4 - a_3 = \frac{7}{2} - 3 = \frac{1}{2}$

i.e., $a_{k+1} - a_k$ is the same every time so, the given list of numbers forms an A.P. with the common difference $d = \frac{1}{2}$

The next three terms are :

$\frac{7}{2} + \frac{1}{2} = 4$, $4 + \frac{1}{2} = \frac{9}{2}$

and $\quad \dfrac{9}{2} + \dfrac{1}{2} = 5$

(iii) $-1.2, -3.2, -5.2, -7.2, \ldots$

$a_2 - a_1 = -3.2 - (-1.2)$
$\qquad = -3.2 + 1.2$
$\qquad = -2.0$
$a_3 - a_2 = -5.2 - (-3.2)$
$\qquad = -5.2 + 3.2$
$\qquad = -2.0$
$a_4 - a_3 = -7.2 - (-5.2)$
$\qquad = -7.2 + 5.2$
$\qquad = -2.0$

i.e., $a_{k+1} - a_k$ is the same everytime. So, the given list of numbers forms an AP with the common difference $d = -2.0$.

The next three terms are :
$\qquad -7.2 + (-2.0) = -9.2,$
$\qquad -9.2 + (-2.0) = -11.2$
and $\quad -11.2 + (-2.0) = -13.2$

(iv) $-10, -6, -2, 2, \ldots$

$a_2 - a_1 = -6 - (-10)$
$\qquad = -6 + 10 = 4$
$a_3 - a_2 = -2 - (-6)$
$\qquad = -2 + 6 = 4$
$a_4 - a_3 = 2 - (-2)$
$\qquad = 2 + 2 = 4$

i.e., $a_{k+1} - a_k$ is the same every time. So, the given list of numbers forms an AP with the common difference $d = 4$.

The next three terms are :
$\qquad 2 + 4 = 6, 6 + 4 = 10$
and $\quad 10 + 4 = 14.$

(v) $3, 3 + \sqrt{2}, 3 + 2\sqrt{2}, 3 + 3\sqrt{2}, \ldots$

$a_2 - a_1 = (3 + \sqrt{2}) - 3 = \sqrt{2}$
$a_3 - a_2 = (3 + 2\sqrt{2}) - (3 + \sqrt{2}) = \sqrt{2}$
$a_4 - a_3 = (3 + 3\sqrt{2}) - (3 + 2\sqrt{2}) = \sqrt{2}$

i.e., $a_{k+1} - a_k$ is the same every time. So, the given list of numbers forms an AP with the common difference $d = \sqrt{2}$.

The next three terms are :
$\qquad (3 + 3\sqrt{2}) + \sqrt{2} = 3 + 4\sqrt{2},$
$\qquad (3 + 4\sqrt{2}) + \sqrt{2} = 3 + 5\sqrt{2}$
and $\quad (3 + 5\sqrt{2}) + \sqrt{2} = 3 + 6\sqrt{2}$

(vi) $0.2, 0.22, 0.222, 0.2222, \ldots$

$a_2 - a_1 = 0.22 - 0.2 = 0.02$
$a_3 - a_2 = 0.222 - 0.22 = 0.002$

As $a_2 - a_1 \neq a_3 - a_2$, the given list of numbers does not form an AP.

(vii) $0, -4, -8, -12, \ldots$

$a_2 - a_1 = -4 - 0 = -4$
$a_3 - a_2 = -8 - (-4) = -8 + 4 = -4$
$a_4 - a_3 = -12 - (-8) = -12 + 8 = -4$

i.e., $a_{k+1} - a_k$ is the same everytime. So, the given list of numbers forms an AP with the common difference $d = -4$.

The next three terms are :
$\qquad -12 + (-4) = -12 - 4 = -16,$
$\qquad -16 + (-4) = -16 - 4 = -20$
and $\quad -20 + (-4) = -20 - 4 = -24$

(viii) $-\dfrac{1}{2}, -\dfrac{1}{2}, -\dfrac{1}{2}, -\dfrac{1}{2}, \ldots$

$a_2 - a_1 = -\dfrac{1}{2} - \left(-\dfrac{1}{2}\right) = -\dfrac{1}{2} + \dfrac{1}{2} = 0$

$a_3 - a_2 = -\dfrac{1}{2} - \left(-\dfrac{1}{2}\right) = -\dfrac{1}{2} + \dfrac{1}{2} = 0$

$a_4 - a_3 = -\dfrac{1}{2} - \left(-\dfrac{1}{2}\right) = -\dfrac{1}{2} + \dfrac{1}{2} = 0$

i.e., $a_{k+1} - a_k$ is the same everytime. So, the given list of numbers forms an AP with the common difference $d = 0$.

The next three terms are :
$\qquad -\dfrac{1}{2} + 0 = -\dfrac{1}{2}, -\dfrac{1}{2} + 0 = -\dfrac{1}{2}$
and $\quad -\dfrac{1}{2} + 0 = -\dfrac{1}{2}$

(ix) $1, 3, 9, 27, \ldots$

$a_2 - a_1 = 3 - 1 = 2$
$a_3 - a_2 = 9 - 3 = 6$

ARITHMETIC PROGRESSIONS

As $a_2 - a_1 \neq a_3 - a_2$, the given list of numbers does not form as AP.

(x) **a, 2a, 3a, 4a, ...**
$a_2 - a_1 = 2a - a = a$
$a_3 - a_2 = 3a - 2a = a$
$a_4 - a_3 = 4a - 3a = a$

i.e., $a_{k+1} - a_k$ is the same everytime. So, the given list of numbers from an AP with the common difference $d = a$.

The next three terms are :
$4a + a = 5a, 5a + a = 6a$
and $6a + a = 7a$.

(xi) **a, a², a³, a⁴, ...**
$a_2 - a_1 = a^2 - a = a(a-1)$
$a_3 - a_2 = a^3 - a^2 = a^2(a-1)$

As $a_2 - a_1 \neq a_3 - a_2$, the given list of numbers does not form an AP.

(xii) $\sqrt{2}, \sqrt{8}, \sqrt{18}, \sqrt{32}, ...$
$a_2 - a_1 = \sqrt{8} - \sqrt{2}$
$= 2\sqrt{2} - \sqrt{2} = \sqrt{2}$
$a_3 - a_2 = \sqrt{18} - \sqrt{8}$
$= 3\sqrt{2} - 2\sqrt{2} = \sqrt{2}$
$a_4 - a_3 = \sqrt{32} - \sqrt{18} = 4\sqrt{2} - 3\sqrt{2} = \sqrt{2}$

i.e., $a_{k+1} - a_k$ is the same everytime. So, the given list of numbers forms an AP with the common difference $d = \sqrt{2}$.

The next three terms are :
$\sqrt{32} + \sqrt{2} = 4\sqrt{2} + \sqrt{2} = 5\sqrt{2} = \sqrt{50}$,
$5\sqrt{2} + \sqrt{2} = 6\sqrt{2} = \sqrt{72}$
and $6\sqrt{2} + \sqrt{2} = 7\sqrt{2} = \sqrt{98}$.

(xiii) $\sqrt{3}, \sqrt{6}, \sqrt{9}, \sqrt{12}, ...$
$a_2 - a_1 = \sqrt{6} - \sqrt{3} = \sqrt{3}(\sqrt{2} - 1)$
$a_3 - a_2 = \sqrt{9} - \sqrt{6} = \sqrt{3}(\sqrt{3} - \sqrt{2})$

As $a_2 - a_1 \neq a_3 - a_2$, the given list of numbers does not form an AP.

(xiv) **1², 3², 5², 7², ...**
$a_2 - a_1 = 3^2 - 1^2 = 9 - 1 = 8$
$a_3 - a_2 = 5^2 - 3^2 = 25 - 9 = 16$

As $a_2 - a_1 \neq a_3 - a_2$, the given list of numbers does not form an AP.

(xv) **1², 5², 7², 73, ...**
$a_2 - a_1 = 5^2 - 1^2 = (5-1)(5+1)$
$= (4)(6) = 24$
$a_3 - a_2 = 7^2 - 5^2 = (7-5)(7+5)$
$= (2)(12) = 24$
$a_4 - a_3 = 73 - 7^2 = 73 - 49 = 24$

i.e., $a_{k+1} - a_k$ is the same everytime. So, the given list of numbers forms an AP with the common difference $d = 24$.

The next three terms are :
$73 + 24 = 97, 97 + 24 = 121$
and $121 + 24 = 145$.

ADDITIONAL EXAMPLES

Example 5. *Find out the value of k such that $\dfrac{2}{3}, k, \dfrac{5}{8}$ are the three consecutive terms of an AP.*

Sol. $a_2 - a_1 = k - \dfrac{2}{3}$

$a_3 - a_2 = \dfrac{5}{8} - k$

If $\dfrac{2}{3}, k, \dfrac{5}{8}$ and the three consecutive terms of an AP, then
$a_2 - a_1 = a_3 - a_2$

$\Rightarrow k - \dfrac{2}{3} = \dfrac{5}{8} - k$

$\Rightarrow k + k = \dfrac{5}{8} + \dfrac{2}{3}$

$\Rightarrow 2k = \dfrac{31}{24}$

$\Rightarrow k = \dfrac{31}{48}$

Hence, the required value of k is $\dfrac{31}{48}$.

Example 6. *A sequence is given by $a_n = n^2 - 1, n \in N$, Prove that it is not an AP.*

Sol. $a_n = n^2 - 1$ | Given

Put $n = 1, 2, 3, 4, \ldots$

we get,

$a_1 = 1^2 - 1 = 1 - 1 = 0$
$a_2 = 2^2 - 1 = 4 - 1 = 3$
$a_3 = 3^2 - 1 = 9 - 1 = 8$
$a_4 = 4^2 - 1 = 16 - 1 = 15$, and so on.

Hence, the given sequence is

$0, 3, 8, 15, \ldots$

$a_2 - a_1 = 3 - 0 = 3$
$a_3 - a_2 = 8 - 3 = 5$

As $a_2 - a_1 \neq a_3 - a_2$, the given sequence is not an AP.

Example 7. *Find the common difference and write the next four terms of the following arithmetic progression* :

$$-1, \frac{1}{4}, \frac{3}{2}, \ldots$$

Sol. The common difference d

$$= a_2 - a_1 = \frac{1}{4} - (4) = \frac{1}{4} + 1 = \frac{5}{4}$$

The next four terms are :

$$\frac{3}{2} + \frac{5}{4} = \frac{11}{4}, \frac{11}{4} + \frac{5}{4} = 4,$$

$$4 + \frac{5}{4} = \frac{21}{4}$$

and $\quad \dfrac{21}{4} + \dfrac{5}{4} = \dfrac{13}{2}$.

TEST YOUR KNOWLEDGE

1. For the AP : $\dfrac{3}{2}, \dfrac{1}{2}, -\dfrac{1}{2}, -\dfrac{3}{2}, \ldots$, write the first term a and the common difference d.

2. Which of the following list of numbers does form an AP ? If they form an AP, write the next two terms :

    (i) $4, 10, 16, 22, \ldots$

    (ii) $1, -1, -3, -5, \ldots$

    (iii) $-2, 2, -2, 2, -2, \ldots$

    (iv) $1, 1, 1, 2, 2, 3, 3, \ldots$

3. Determine k so that $8k + 4$, $6k - 2$ and $2k - 7$ are the three consecutive terms of an AP.

4. Find out a and d for each of the following arithmetic progressions :

    (i) $1, 4, 7, 10, 13, 16, \ldots$

    (ii) $11, 16, 21, 26, 31, 36, \ldots$

    (iii) $-4, 0, 4, 8, 12, 16, \ldots$

    (iv) $-12, -2, 8, 18, 28, 38, \ldots$

    (v) $16, 8, 0, -8, -16, -24, \ldots$

    (vi) $21, 6, -9, -24, -39, -54, \ldots$

    (vii) $1.0, 1.5, 2.0, 2.5, 3.0, 3.5, \ldots$

5. Show that $a - b$, a and $a + b$ form consecutive terms of an AP.

Answers

1. $a = \dfrac{3}{2}, d = -1$

2. (i) Yes ; 28, 34
   (ii) Yes ; $-7, -9$
   (iii) No
   (iv) No

3. $k = \dfrac{1}{2}$

4. (i) $a = 1, d = 3$,   (ii) $a = 11, d = 5$
   (iii) $a = -4, d = 4$, (iv) $a = -12, d = 10$
   (v) $a = 16, d = -8$ (vi) $a = 21, d = -15$
   (vii) $a = 1.0, d = 0.5$

n^{th} TERM OF AN AP

IMPORTANT POINTS

1. A situation versus an example. Let us consider the following situation :

Reena applied for a job and got selected. She has been offered the job with a starting monthly salary of Rs 8000, with an annual increment of Rs 500. Then,

Her monthly salary for the second year
$= $ Rs $(8000 + 500) = $ Rs 8500

ARITHMETIC PROGRESSIONS

In the same way, we can find the monthly salary for the 3rd, 4th and 5th year by adding Rs 500 to the salary of the previous year.

So, the salary for the 3rd year
= Rs (8500 + 500)
= Rs (8000 + 500 + 500)
= Rs (8000 + 2 × 500)
= Rs [8000 + (**3 – 1**) × 500]
                (for the **3rd year**)
= Rs 9000

Salary for the 4th year
= Rs (9000 + 500)
= Rs (8000 + 500 + 500 + 500)
= Rs (8000 + 3 × 500)
= Rs [8000 + (**4 – 1**) × 500]
                (for the **4th year**)
= Rs 9500

Salary for the 5th year
= Rs (9500 + 500)
= Rs (8000 + 500 + 500 + 500 + 500)
= Rs (8000 + 4 × 500)
= Rs [8000 + (**5 – 1**) × 500]
                (for the **5th year**)
= Rs 10000

We observe that we are setting a list of numbers.

8000, 8500, 9000, 9500, 10000, ... clearly more numbers one in AP because each number is obtained by adding a fixed-number (500 here) to the preceding number except the first number.

Now, looking at the pattern formed above, we can find her monthly salary for the 6th year, the 15th year, the 25th year assuming that she will be still working in the job. We would calculated this by adding Rs 500 each time to the salary of the previous year to give the answer. However, we can make the process shorter as we have already got some idea from the way we have obtained the salaries above.

Salary for the 15th year
= Salary for the 14th year + Rs 500

= Rs $\left[8000 + \underbrace{500 + 500 + 500 + ... + 500}_{13 \text{ times}} \right]$ + Rs 500

= Rs [8000 + 14 × 500]
= Rs [8000 (**15 – 1**) × 500]
= Rs 15000

i.e., First salary + (15 – 1) × Annual increment.

In the same way, her monthly salary for the 25th year would be

Rs [8000 + (**25 – 1**) × 500] = Rs 20000
              = First salary + (**25 – 1**)
              × Annual increment

This example gives us some idea about how to write the 15th term or the 25th term, and more generally, the nth term of the AP.

2. nth term of an AP. Let $a_1, a_2, a_3,$ ...... be an AP whose first term a_1 is a and the common differences is d.

Then, the **second term** $a_2 = a + d = a + $ (**2 – 1**) d

the **third** term
$$a_3 = a_2 + d = (a + d) + d$$
$$= a + 2d = a + (\mathbf{3 - 1})\,d$$

the **fourth** term
$$a_4 = a_3 + d = (a + 2d) + d$$
$$= a + 3d = a + (\mathbf{4 - 1})\,d$$

....
....

Looking at the pattern, we can say that the **n**th term $a_n = a + (n - 1)\,d$

So, **the nth a_n of the AP with first term a and common difference d is given $a_n = a + (n - 1)\,d$.**

a_n is also called the **general term of the AP.** If there are m terms in the AP, then a_m represents the **last term which is sometimes also denoted by** l.

ILLUSTRATIVE EXAMPLES

[NCERT Exercise 5.2]
(Page No. 105)

Example 1. *Fill in the blanks in the following table, given that a is the first term, d the common difference and a_n the nth term of the AP :*

	a	d	n	a_n
(i)	7	3	8	...
(ii)	−18	...	10	0
(iii)	...	−3	18	−5
(iv)	−18.9	2.5	...	3.6
(v)	3.5	0	105	...

Sol. (i) $a_n = a + (n-1)d$
$\Rightarrow a_n = 7 + (8-1)3$
$\Rightarrow a_n = 7 + (7)3$
$\Rightarrow a_n = 7 + 21$
$\Rightarrow a_n = 28$

(ii) $a_n = a + (n-1)d$
$\Rightarrow 0 = -18 + (10-1)d$
$\Rightarrow 18 = 9d$
$\Rightarrow d = \dfrac{18}{9} = 2$

(iii) $a_n = a + (n-1)d$
$\Rightarrow -5 = a + (18-1)(-3)$
$\Rightarrow -5 = a - 51$
$\Rightarrow a = 51 - 5$
$\Rightarrow a = 46$

(iv) $a_n = a + (n-1)d$
$\Rightarrow 3.6 = -18.9 + (n-1)(2.5)$
$\Rightarrow 3.6 + 18.9 = (n-1)(2.5)$
$\Rightarrow 22.5 = (n-1)(2.5)$
$\Rightarrow n - 1 = \dfrac{22.5}{2.5}$
$\Rightarrow n - 1 = 9$
$\Rightarrow n - 1 = 9 + 1$
$\Rightarrow n = 10$

(v) $a_n = a + (n-1)d$
$\Rightarrow a_n = 3.5 + (105-1)0$
$\Rightarrow a_n = 3.5.$

Example 2. *Choose the correct choice in the following and justify :*

(i) 30th term of the AP : 10, 7, 4, ... is
(A) 97 (B) 77 (C) −77
(D) −87

(ii) 11th term of the AP : $-3, -\dfrac{1}{2}, 2, ...$ is
(A) 28 (B) 22 (C) −38
(D) $-48\dfrac{1}{2}$.

Sol. The given A.P. is
10, 7, 4, ...
Here, $a = 10$,
$d = 7 - 10 = -3$
and $n = 30$
We have $a_n = a + (n-1)d$
So, $a_{30} = 10 + (30-1)(-3)$
$\Rightarrow a_{30} = 10 - 87$
$\Rightarrow a_{30} = -77$
Hence, the correct choice is **(C) −77.**

(ii) The given AP is
$-3, \dfrac{1}{2}, 2, ...$
Here, $a = -3$,
$d = -\dfrac{1}{2} - (-3) = -\dfrac{1}{2} + 3 = \dfrac{5}{2}$
and $n = 11$
We have
$a_n = a + (n-1)d$
So, $a_{11} = -3 + (11-1)\left(\dfrac{5}{2}\right)$
$\Rightarrow a_{11} = -3 + 25 \Rightarrow a_{11} = 22$
Hence, the correct choice is **(B) 22.**

Example 3. *In the following APs, find the missing terms in the boxes :*
(i) 2, ☐, 26
(ii) ☐, 13, ☐, 3

(iii) 5, ☐, ☐, $9\frac{1}{2}$

(iv) −4, ☐, ☐, ☐, ☐, 6

(v) ☐, 38, ☐, ☐, ☐, −22

Sol. (i) Let the common difference of the given AP be d.

Then,

Third term $= 2 + d + d = 2 + 2d$

According to the question,

$2 + 2d = 26$

$\Rightarrow \quad 2d = 26 - 2$

$\Rightarrow \quad 2d = 24$

$\Rightarrow \quad d = \frac{24}{2} = 12$

So, second term $= 2 + d = 2 + 12 = 14$

Hence, the missing termed in the box is $\boxed{14}$.

(ii) Let the first term and the common difference of the given AP be a and d respectively.

Second term $= 13$

$\Rightarrow \quad a + (2 - 1)d = 13$

$\Rightarrow \quad a + d = 13$ ...(1)

Fourth term $= 3$

$\Rightarrow \quad a + (4 - 1)d = 3$

$\Rightarrow \quad a + 3d = 3$ ...(2)

Solving (1) and (2), we get

$a = 18$

$d = -5$

Therefore,

Third term $= a + (3 - 1)d$

$= a + 2d$

$= 18 + 2(-5)$

$= 18 - 10 = 8$

Hence, the missing terms in the boxes are $\boxed{18}$ and $\boxed{8}$.

(iii) Let the common difference of the given AP be d.

$a = 5$

4th term $= 9\frac{1}{2}$

$\Rightarrow 5 + (4 - 1)d = \frac{19}{2}$

$[\because a_n = a + (n - 1)d]$

$\Rightarrow 3d = \frac{19}{2} - 5$

$\Rightarrow 3d = \frac{9}{2}$

$\Rightarrow d = \frac{3}{2}$

Therefore,

Second term $= 5 + \frac{3}{2} = \frac{13}{2} = 6\frac{1}{2}$

and, Third term $= \frac{13}{2} + \frac{3}{2} = 8$

Hence, the missing terms in the boxes are

$\boxed{6\frac{1}{2}}$ and $\boxed{8}$.

(iv) Let the common difference of the given AP be d.

$a = -4$

6th term $= 6$

$\Rightarrow -4 + (6 - 1)d = 6$

$[\because a_n = a + |n - 1|d$

$\Rightarrow -4 + 5d = 6$

$\Rightarrow 5d = 6 + 4$

$\Rightarrow 5d = 10$

$\Rightarrow d = \frac{10}{5}$

$\Rightarrow d = 2$

Therefore,

Second term $= -4 + 2 = -2$

Third term $= -2 + 2 = 0$

Fourth term $= 0 + 2 = 2$

Fifth term $= 2 + 2 = 4$

Hence, the missing terms in the boxes are

$\boxed{-2}$, $\boxed{0}$, $\boxed{2}$, $\boxed{4}$

(v) Let the first terms and the common difference of the given AP be a and d respectively.

Second term $= 38$

$\Rightarrow a + (2 - 1)d = 38$

$[\because a_n = a + (n - 1)d$

$\Rightarrow a + d = 38$ ...(1)

Sixth term = -22
$\Rightarrow a + (6-1)d = -22$
$\Rightarrow \quad a + 5d = -22$ ...(2)
Solving (1) and (2), we get
$\quad a = 53$
$\quad d = -15$
Therefore,
Third term = $53 + (3-1)(-15)$
$\quad |\because a_n = a + (n-1)d$
$= 53 - 30$
$= 23$
Fourth term = $53 + (4-1)(-15)$
$\quad |\because a_n = a + (n-1)d$
$= 8$
Fifth four = $53 + (5-1)(-15)$
$\quad |\because a_n = a + (n-1)d$
$= -7$
Hence, the missing terms in the boxes are

$\boxed{53}, \boxed{23}, \boxed{8}, \boxed{-7}$.

Example 4. *Which term of the AP :*
3, 8, 13, 18, ... is 78.
Sol. The given AP is
$\quad 3, \quad 8, \quad 13, \quad 18, ...$
Here, $a = 3$
$\quad d = 8 - 3 = 5$
Let the nth term of the AP be 78.
Then, $a_n = a + (n-1)d$
$\Rightarrow \quad 78 = 3 + (n-1)(5)$
$\Rightarrow \quad 5(n-1) = 78 - 3$
$\Rightarrow \quad 5(n-1) = 75$
$\Rightarrow \quad n - 1 = \dfrac{75}{5}$
$\Rightarrow \quad n - 1 = 15$
$\Rightarrow \quad n = 15 + 1$
$\Rightarrow \quad n = 16$
Hence, 16th term of the AP is 78.

Example 5. *Find the number of terms in each of the following APs :*
(i) 7, 13, 19, ..., 205
(ii) 18, $15\dfrac{1}{2}$, 13, ... , -47.

Sol. (i) 7, 13, 19, ..., 205
Here, $a = 7$
$\quad d = 13 - 7 = 6$
$\quad a_n = 205$
Let the number of terms be n.
Then, $\quad a_n = 205$
$\Rightarrow \quad a + (n-1)d = 205$
$\Rightarrow \quad 7 + (n-1)6 = 205$
$\Rightarrow \quad 6(n-1) = 205 - 7$
$\Rightarrow \quad 6(n-1) = 198$
$\Rightarrow \quad n - 1 = \dfrac{198}{6}$
$\Rightarrow \quad n - 1 = 33$
$\Rightarrow \quad n = 33 + 1$
$\Rightarrow \quad n = 34$
Hence, the number of terms of the given AP is 34.

(ii) 18, $15\dfrac{1}{2}$, 13, ..., -47
Here, $a = 18$
$\quad d = 15\dfrac{1}{2} - 18 = \dfrac{31}{2} - 18 = -\dfrac{5}{2}$
$\quad a_n = -47$
Let the number of terms be n.
Then, $\quad a_n = -47$
$\Rightarrow \quad a + (n-1)d = -47$
$\Rightarrow \quad 18 + (n-1)\left(-\dfrac{5}{2}\right) = -47$
$\Rightarrow \quad -\dfrac{5}{2}(n-1) = -47 - 18$
$\Rightarrow \quad -\dfrac{5}{2}(n-1) = -65$
$\Rightarrow \quad \dfrac{5}{2}(n-1) = 65$
$\Rightarrow \quad n - 1 = \dfrac{65 \times 2}{5}$
$\Rightarrow \quad n - 1 = 26$
$\Rightarrow \quad n = 26 + 1$
$\Rightarrow \quad n = 27$
Hence, the number of terms of the given AP is 27.

ARITHMETIC PROGRESSIONS

Example 6. *Check whether* -150 *is a term of the AP : 11, 8, 5, 2, ...*

Sol. The given list of numbers is
$$11, 8, 5, 2, ...$$
$$a_2 - a_1 = 8 - 11 = -3$$
$$a_3 - a_2 = 5 - 8 = -3$$
$$a_4 - a_3 = 2 - 5 = -3$$

i.e., $a_{k+1} - a_k$ is the same every time. So, the given list of numbers forms an AP with first term $a = 11$ and the common difference $d = -3$.

Let -150 be the nth term of the given AP.

Then, $a_n = -150$
$$\Rightarrow \quad a + (n-1)d = -150$$
$$\Rightarrow \quad 11 + (n-1)(-3) = -150$$
$$\Rightarrow \quad -3(n-1) = -150 - 11$$
$$\Rightarrow \quad -3(n-1) = -161$$
$$\Rightarrow \quad 3(n-1) = 161$$
$$\Rightarrow \quad n - 1 = \frac{161}{3}$$
$$\Rightarrow \quad n = \frac{161}{3} + 1$$
$$\Rightarrow \quad n = \frac{164}{3}$$

But n should be a positive integer. So -150 is not a term of
$$11, 8, 5, 2, ...$$

Example 7. *Find the 31st term of an AP whose 11th term is 38 and the 16th term is 73.*

Sol. Let the first term and the common difference of the AP be a and d respectively. Then,

11th term = 38 | Given
$$\Rightarrow \quad a + (11-1)d = 38$$
$$\quad\quad | \because a_n = a + (n-1)d$$
$$\Rightarrow \quad a + 10d = 38 \quad ...(1)$$
and,   16th term = 73
$$\Rightarrow \quad a + (16-1)d = 73$$
$$\quad\quad | \because a_n = a + (n-1)d$$
$$\Rightarrow \quad a + 15d = 73 \quad ...(2)$$

Solving (1) and (2), we get
$$a = -32$$

$$d = 7$$
Therefore, 31st term
$$= a + (31-1)d$$
$$= a + 30d$$
$$= -32 + (30)(7)$$
$$= -32 + 210 = 178$$

Hence, the 31st term of the AP is 178.

Example 8. *An AP consists of 50 terms of which 3rd term is 12 and the last term is 106. Find the 29th term.*

Sol. Let the first term and the common difference of the AP be a and d respectively.

3rd term = 12 | Given
$$\Rightarrow \quad a + (3-1)d = 12$$
$$\quad\quad | \because a_n = a + (n-1)d$$
$$\Rightarrow \quad a + 2d = 12 \quad ...(1)$$

Last term = 106 | Given
$$\Rightarrow \quad \text{So the term} = 106$$
$$\quad | \because \text{The AP consists of 50 terms}$$
$$\Rightarrow \quad a + (50-1)d = 106$$
$$\Rightarrow \quad a + 49d = 106 \quad ...(2)$$

Solving (1) and (2), we get
$$a = 8$$
$$d = 2$$

Therefore,
29th term of the AP
$$= 9 + (29-1)d$$
$$\quad\quad | \because a_n = a + (n-1)d$$
$$= 9 + 28d$$
$$= 8 + (28)(2)$$
$$= 8 + 56$$
$$= 64.$$

Example 9. *If the 3rd and the 9th terms of an AP are 4 and -8 respectively, which term of this AP is zero ?*

Sol. Let the first term and the common difference of the AP be a and d respectively.

3rd term = 4 | Given
$$a + (3-1)d = 4$$
$$\quad\quad | \because a_n = a + (n-1)d$$
$$\Rightarrow \quad a + 2d = 4 \quad\quad (1)$$

9th term = -8 | Given

$\Rightarrow \quad a + (9-1)d = -8$
$\Rightarrow \quad a + 8d = -8$ ...(2)

Solving (1) and (2), we get
$a = 8$
$d = -2$

Let the nth term of the AP be zero. Then,
$9 + (n-1)d = 0$
$\qquad |\because\ a_n = a + (n-1)d$
$\Rightarrow \quad 8 + (n-1)(-2) = 0$
$\Rightarrow \quad -2(n-1) = -8$
$\Rightarrow \quad 2(n-1) = 8$
$\Rightarrow \quad n - 1 = \dfrac{8}{2} = 4$
$\Rightarrow \quad n = n + 1 = 5$

Hence, the fifth term of AP is zero.

Example 10. *The 17th term of an AP exceeds its 10th term by 7. Find the common difference.*

Sol. Let the first term and the common difference of the AP be a and d respectively.

According to the question,
$a_{17} = a_{10+7}$
$\Rightarrow \quad a + (17-1)d = a + (10-1)d + 7$
$\qquad |\because\ a_n = a + (n-1)d$
$\Rightarrow \quad a + 16d = a + 9d + 7$
$\Rightarrow \quad 16d - 9d = 7$
$\Rightarrow \quad 7d = 7$
$\Rightarrow \quad d = \dfrac{7}{7} = 1$

Hence, the common differences is 1.

Example 11. *Which term of the AP : 3, 15, 27, 39, ... will be 132 more than its 54th term ?*

Sol. The given AP is
3, 15, 27, 39, ...

Here, $a = 3$
$d = 15 - 3 = 12$

Let the nth term be 132 more than 54th term.

Then, $a_n = a_{54} + 132$
$= a + (n-1)\,12$
$= a + (54-1)\,12 + 132$

$\Rightarrow \quad (n-1)\,12 = (53)\,(12) + 132$
$\Rightarrow \quad (n-54)\,12 = 132$
$\Rightarrow \quad (n-54) = \dfrac{132}{12} = 11$
$\Rightarrow \quad n = 54 + 11 = 65$

Hence, the 65th term will be 132 more than the 54th term.

Example 12. *Two APs have the same common differences. The difference between their 100th terms is 100, what is the difference between their 1000th terms ?*

Sol. Let the first terms of two APs be a_1 and $a_2\ (a_1 > a_2)$ respectively. Let d be the same common difference of the two APs. Then,

100th term of the first AP
$= a_1 + (100-1)d$
$\qquad |\because\ a_n = a + (n-1)d$
$= a_1 + 99d$.

100th term of the second AP
$= a_2 + (100-1)d$
$\qquad |\because\ a_n = a + (n-1)d$
$= a_2 + 99d$

According to the question,
$(a_1 + 99d) - (a_2 + 99d) = 100$
$\Rightarrow \quad a_1 - a_2 = 100$ ...(1)

Now, 1000th term of the first AP
$= a_1 + (1000-1)d$
$\qquad |\because\ a_n = a + (n-1)d$
$= a_1 + 999d$

1000th term of the second AP
$= a_2 + (1000-1)d$
$\qquad |\because\ a_n = a + (n-1)d$
$= a_2 + 999d$

Therefore, difference between their 1000th terms
$= (a_1 + 999d) - (a_2 + 999d)$
$= a_1 - a_2$
$= 100$ | From (1)

Hence, the difference between their 1000th terms is 100

Example 13. *How many three-digit numbers are divisible by 7 ?*

Sol. The three digit numbers divisible by 7 are :

105, 112, 119, 126, ..., 994
$$a_2 - a_1 = 112 - 105 = 7$$
$$a_3 - a_2 = 119 - 112 = 7$$
$$a_4 - a_3 = 126 - 119 = 7$$

i.e., $a_{k+1} - a_k$ is the same every time. So, the above list of numbers forms an AP with the first term $a = 105$ and the common difference $d = 7$.

Last term $(l) = 994$

Let their be n terms in this AP.

Then, nth term $= l$
$$\Rightarrow a + (n - 1)d = 994$$
$$\Rightarrow 105 + (n - 1)7 = 994$$
$$\Rightarrow (n - 1)7 = 994 - 105$$
$$\Rightarrow (n - 1)7 = 889$$
$$\Rightarrow n - 1 = \frac{889}{7}$$
$$\Rightarrow n - 1 = 127$$
$$\Rightarrow n = 127 + 1$$
$$\Rightarrow n = 128$$

Hence, these are 128 three-digit numbers divisible by 7.

Example 14. *How many multiples of 4 lie between 10 and 250 ?*

Sol. The multiples of 4 that lie between 10 and 250 are :
12, 16, 20, 24, ..., 248
$$a_2 - a_1 = 16 - 12 = 4$$
$$a_3 - a_2 = 20 - 16 = 4$$
$$a_4 - a_3 = 24 - 20 = 4$$

As $a_{k+1} - a_k$ is the same for $k = 1, 2, 3,$ etc.,

The above list of numbers forms and AP with the first term $a = 12$ and the common difference $d = 4$.

Last term $(l) = 248$

Let there be n terms in this AP. Then, nth term $= l$
$$\Rightarrow a + (n - 1)d = 248$$
$$\Rightarrow 12 + (n - 1)4 = 248$$
$$\Rightarrow (n - 1)4 = 248 - 12$$
$$\Rightarrow (n - 1)4 = 236$$
$$\Rightarrow n - 1 = \frac{236}{4}$$
$$\Rightarrow n - 1 = 59$$
$$\Rightarrow n = 59 + 1$$
$$\Rightarrow n = 60$$

Hence, 60 multiples of 4 lie between 10 and 250.

Example 15. *For what value of n, are the nth terms of two APs : 63, 65, 67, ... and 3, 10, 17, ... equal ?*

Sol. First APs
63, 65, 67, ...

Here, $a = 63$
$$d = 65 - 63 = 2$$
$\therefore$ **nth term** $= 63 + (n - 1)2$
| $\because a_n = a + (n - 1)d$

Second APs
3, 10, 17, ...

Here, $a = 3$
$$d = 10 - 3 = 7$$
$\therefore$ nth term $= 3 + (n - 1)7$
| $\because a_n = a + (n - 1)d$

If the nth terms of the two APs are equal, then
$$63 + (n - 1)2 = 3 + (n - 7)7$$
$$\Rightarrow (n - 1)2 - (n - 1)7 = 3 - 63$$
$$\Rightarrow (n - 1)(2 - 7) = -60$$
$$\Rightarrow (n - 1)(-5) = -60$$
$$\Rightarrow n - 1 = \frac{-60}{-5}$$
$$\Rightarrow n - 1 = 12$$
$$\Rightarrow n = 12 + 1$$
$$\Rightarrow n = 13$$

Hence, for $n = 13$, the nth terms of the two APs are equal.

Example 16. *Determine the AP whose third term is 16 and the 7th term exceeds the 5th term by 12.*

Sol. Let the first term and the common difference of the AP be a and d respectively. Then,

Third term $= 16$
$$\Rightarrow a + (3 - 1)d = 16$$
| $\because a_n = a + (n - 1)d$

and,
$$\Rightarrow \quad a + 2d = 16 \quad \ldots(1)$$
7th term = 5th term + 12
$$\Rightarrow \quad a + (7-1)d = a + (5-1)d + 12$$
$$|\because a_n = a + (n-1)d$$
$$\Rightarrow \quad a + 6d = a + 4d + 12$$
$$\Rightarrow \quad 6d - 4d = 12$$
$$\Rightarrow \quad 2d = 12$$
$$\Rightarrow \quad d = \frac{12}{2} = 6$$

Put $d = 6$ in (1), we get
$$a + 2(6) = 16$$
$$\Rightarrow \quad a + 12 = 16$$
$$\Rightarrow \quad a = 16 - 12$$
$$\Rightarrow \quad a = 4$$

Hence, the required APs also
$4, 4 + 6, 4 + 6 + 6, 4 + 6 + 6 + 6, \ldots$
i.e., $4, 10, 16, 22, \ldots$

Example 17. *Find the 20th term from the last term of the AP : 3, 8, 13, ..., 253.*

Sol. The given AP is
$3, 8, 13, \ldots, 253$
Here, $a = 3$
$d = 8 - 3 = 5$
$l = 253$

Let the number of terms of the AP be n.
Term, nth term = l
$$\Rightarrow \quad 3 + (n-1)5 = 253$$
$$|\because a_n = a + (n-1)d$$
$$\Rightarrow \quad (n-1)5 = 253 - 3$$
$$\Rightarrow \quad (n-1)5 = 250$$
$$\Rightarrow \quad n - 1 = \frac{250}{5}$$
$$\Rightarrow \quad n - 1 = 50$$
$$\Rightarrow \quad n = 50 + 1$$
$$\Rightarrow \quad n = 51$$

So, there are 51 terms in the given AP.
Now, 20th term from the last term
= (51 − 20 + 1)th term from the beginning
= 32th term from the beginning
= 3 + (32 − 1) 5
$$|\because a_n = a + (n-1)d$$
= 3 + 155
= 158

Hence, the 20th term from the last term of the given AP is 158.

Aliter. Let us write the given AP in the reverse order. Then the AP becomes
$253, 248, 243, \ldots, 3$
Here, $a = 253$
$d = 248 - 253 = -5$
Therefore, Required term
= 20th term of the AP
= $253 + (20 - 1)(-5)$
$$|\because a_n = a + (n-1)d$$
= 253 − 95
= 158

Hence, the 20th term from the last term of the given AP is 158.

Example 18. *The sum of the 4th and 8th terms of an AP is 24 and the sum of the 6th and 10th term is 44. Find the first three terms of the AP.*

Sol. Let the first term and the common difference of the AP be a and d respectively. Then
4th term = $a + (4-1)d = a + 3d$
$$|\because a_n = a + (n-1)d$$
8th term = $a + (8-1)d = a + 7d$
$$|\because a_n = a + 1(n-1)d$$
6th = $a + (6-1)d = a + 5d$
$$|\because a_n = a + (n-1)d$$
and 10th term = $a + (10-1)d = a + 9d$
$$|\because a_n = a + (n-1)d$$

According to the question,
4th term + 8th term = 24
$$\Rightarrow \quad (a + 3d) + (a + 7d) = 24$$
$$\Rightarrow \quad 2a + 10d = 24$$
$$\Rightarrow \quad a + 5d = 12 \quad \ldots(1)$$
| Dividing throughout by 2

and, 6th term + 10th term = 44
$$\Rightarrow \quad (a + 5d) + (a + 9d) = 44$$
$$\Rightarrow \quad 2a + 14d = 44$$
$$\Rightarrow \quad a + 7d = 22 \quad \ldots(2)$$
| Dividing throughout by 2

ARITHMETIC PROGRESSIONS

Solving (1) and (2), we get

$a = -13$

$d = 5$

So, First term $= -13$

Second term $= -13 + 5 = -8$

Third term $= -8 + 5 = -3$

Hence, the first three terms of the given AP are $-13, -8$ and -3.

Example 19. *Subba Rao started work in 1995 at an annual salary of Rs 5000 and received an increment of Rs 200 each year. In which year did his income reach Rs 7000 ?*

Sol. Here,

$a = $ Rs 5000

$d = $ Rs 200

$l = $ Rs 7000

Suppose that his income reached Rs 7000 after n years.

Then, $l = a + (n-1)d$

$\Rightarrow 7000 = 5000 + (n-1)\,200$

$\Rightarrow (n-1)\,200 = 7000 - 5000$

$\Rightarrow (n-1)\,200 = 2000$

$\Rightarrow n - 1 = \dfrac{2000}{200}$

$\Rightarrow n - 1 = 10$

$\Rightarrow n = 11$

Hence, his income reached Rs 7000 in 11th year.

Example 20. *Ramkali saved Rs 5 in the first week of a year and then increased her weekly savings by Rs 1.75. If in the nth week, her weekly savings become Rs 20.75, find n.*

Sol. Here,

$a = $ Rs 5

$d = $ Rs 1.75

$a_n = $ Rs 20.75

We know that

$a_n = a + (n-1)d$

$\Rightarrow 20.75 = 5 + (n-1)(1.75)$

$\Rightarrow (n-1)(1.75) = 20.75 - 5$

$\Rightarrow (n-1)(1.75) = 15.75$

$\Rightarrow n - 1 = \dfrac{15.75}{1.75}$

$\Rightarrow n - 1 = 9$

$\Rightarrow n = 10$

Hence, the required value of n is 10.

ADDITIONAL EXAMPLES

Example 21. *The 6th term of an Arithmetic progression (AP) is -10 and the 10th term is -26. Determine the 15th term of the AP.*

(CBSE 2006)

Sol. Let the first term and the common difference of the AP be a and d respectively.

6th term $= -10$ | Given

$\Rightarrow a + (6-1)d = -10$

| $\because a_n = a + (n-1)d$

$\Rightarrow a + 5d = -10$ ...(1)

10th term $= -26$ | Given

$\Rightarrow a + (10-1)d = -26$

$\Rightarrow a + 9d = -26$ ...(2)

Solving (1) and (2), we get

$a = 10$

$d = -4$

Therefore, 15th term of the AP

$= a + (15-1)d$

| $\because a_n = a + (n-1)d$

$= a + 14d$

$= 10 - 14(-4)$

$= 10 - 56 = -46$

Hence, the 15th term of the AP is -46.

Example 22. *Find the 6th term from end of the AP 17, 14, 11, ..., -40.*

(CBSE 2005)

Sol. The given APs

$17, 14, 11, \ldots, -40$

Here,

$a = 7$

$d = 14 - 17 = -3$

$l = -40$

Let there between in the given A.P.

Then, nth term $= -40$

$\Rightarrow a + (n-1)d = -40$

| $\because a_n = a + (n-1)d$

$\Rightarrow \quad 17 + (n-1)(-3) = -40$

$\Rightarrow \quad (n-1)(-3) = -40 - 17$

$\Rightarrow \quad (n-1)(-3) = -57$

$\Rightarrow \quad n - 1 = \dfrac{-57}{-3}$

$\Rightarrow \quad n - 1 = 19$

$\Rightarrow \quad n = 19 + 1$

$\Rightarrow \quad n = 20$

Hence, there are 20 terms in the given AP.

Now, 6th term from the end
$= (20 - 6 + 1)$th term from the beginning
$= 15^{th}$ term from the beginning
$= a + (15 - 1)d$
  | $\because a_n = a + (n-1)d$
$= a + 14d$
$= 17 + 14(-3)$
$= 17 - 42$
$= -25$

Hence, the 6th term from the end of the given AP is -25.

Example 23. *The 8th term of an Arithmetic progression is zero. Prove that its 38th term is triple of its 18th term.*

(A.I. CBSE 2005)

Sol. Let the first and the common difference of the AP be a and d respectively.

8th term = 0 | Given

$\Rightarrow \quad a + (8-1)d$
  | $\because a_n = a + (n-1)d$

$\Rightarrow \quad a + 7d = 0$ ...(1)

Again,

38th term $= a + (38-1)d$
  | $\because a_n = a + (n-1)d$
$= a + 37d$

18th term, $a + (18-1)d$
$= a + 17d$
  | $\because a_n = a + (n-1)d$

If 38th term is triple of 18th term, then

$a + 37d = 3(a + 17d)$

$\Rightarrow \quad a + 37d = 3a + 51d$

$\Rightarrow \quad 2a + 14d = 0$

$\Rightarrow \quad 2(a + 7d) = 0$

$\Rightarrow \quad a + 7d = 0$

which is true by (1).

Hence, 38th term is triple of 18th term.

Example 24. *For A.P. $a_1, a_2, a_3, \ldots$, if $\dfrac{a_4}{a_7} = \dfrac{2}{3}$, find $\dfrac{a_6}{a_8}$.*

Sol. Let the first term and the common difference of the AP be a and d respectively.

Then, $a_4 = a + (4-1)d = a + 3d$

$a_6 = a + (6-1)d = a + 5d$

$a_7 = a + (7-1)d = a + 6d$

$a_8 = a + (8-1)d = a + 7d$
  | $\because a_n = a + (n-1)d$

$\dfrac{a_4}{a_7} = \dfrac{2}{3}$

$\Rightarrow \quad \dfrac{a + 3d}{a + 6d} = \dfrac{2}{3}$

$\Rightarrow \quad 3(a + 3d) = 2(a + 6d)$

$\Rightarrow \quad 3a + 9d = 2a + 12d$

$\Rightarrow \quad a = 3d$ ...(1)

Now, $\dfrac{a_6}{a_8} = \dfrac{a + 5d}{a + 7d}$

$= \dfrac{3d + 5d}{3d + 7d}$ | From (1)

$= \dfrac{8d}{10d} = \dfrac{4}{5}$.

Example 25. *Is 200 any term of the sequence*

$3, 7, 11, 15, \ldots$.

Sol. The given sequence is

$3, 7, 11, 15, \ldots$

$a_2 - a_1 = 7 - 3 = 4$

$a_3 - a_2 = 11 - 7 = 4$

$a_4 - a_3 = 15 - 11 = 4$

As $a_{k+1} - a_k$ is the same for $k = 1, 2, 3$, etc., the given sequence form an AP.

Here, $a = 3$

$d = 4$

ARITHMETIC PROGRESSIONS

Let 200 be nth term of the given sequence. Then,
$$a_n = 200$$
$$\Rightarrow a + (n-1)d = 200$$
$$\Rightarrow 3 + (n-1)4 = 200$$
$$\Rightarrow (n-1)4 = 200 - 3 = 197$$
$$\Rightarrow n - 1 = \frac{197}{4} \Rightarrow n = \frac{197}{4} + 1$$
$$\Rightarrow n = \frac{201}{4}.$$

But n should be a positive integer. So, 200 is not any term of the given sequence.

Example 26. *Which term of the sequence* $20, 19\frac{1}{4}, 18\frac{1}{2}, 17\frac{3}{4}, \ldots$ *is the first negative term ?*

Sol. The given sequence is
$$20, 19\frac{1}{4}, 18\frac{1}{2}, 17\frac{3}{4}, \ldots$$
$$a_2 - a_1 = 19\frac{1}{4} - 20 = -\frac{3}{4}$$
$$a_3 - a_2 = 18\frac{1}{2} - 19\frac{1}{4} = -\frac{3}{4}$$
$$a_4 - a_3 = 17\frac{3}{4} - 18\frac{1}{2} = -\frac{3}{4}$$

i.e., The given $a_{k+1} - a_k$ is the same everytime.
So, the given sequence form an AP.
Here, $a = 20$
$$d = -\frac{3}{4}$$

Let the nth term be the first negative term.

Then, $a_n < 0$
$$\Rightarrow a + (n-1)d < 0$$
$$\Rightarrow 20 + (n-1)\left(-\frac{3}{4}\right) < 0$$
$$\Rightarrow 20 < \frac{3}{4}(n-1)$$
$$\Rightarrow 80 < 3(n-1)$$
$$\Rightarrow 3(n-1) > 80$$
$$\Rightarrow n - 1 > \frac{80}{3}$$
$$\Rightarrow n > \frac{80}{3} + 1$$
$$\Rightarrow n > \frac{83}{3}$$
$$\Rightarrow n > 27\frac{2}{3}$$

∴ Least positive integral value of $n = 28$

Hence, 28th term of the given sequence is the first negative term.

Example 27. *If the nth term of a sequence is an expression of first degree in n, show that it is an AP.*

Sol. Let the nth term of the sequence be given by
$$a_n = a_n + b$$
Put $n = 1, 2, 3, 4, \ldots$, we get
$$a_1 = a + b$$
$$a_2 = 2a + b$$
$$a_3 = 3a + b$$
$$a_4 = 4a + b$$
$$\vdots \qquad \vdots$$
$$a_2 - a_1 = (2a + b) - (a + b) = a$$
$$a_3 - a_2 = (3a + b) - (2a + b) = a$$
$$a_4 - a_3 = (4a + b) - (3a + b) = a$$

i.e., $a_{k+1} - a_k$ is the same everytime.

Hence, the given segment forms an AP.

Example 28. *If the pth term of an AP is q and the qth term is p, then prove that its nth term is $(p + 1 - n)$ and hence prove that its $(p + q)$ th term is zero.*

Sol. Let the first term and the common difference of the AP be a and d respectively.

pth term = a | Given
$$\Rightarrow a + (p-1)d = q \qquad \ldots(1)$$
qth term = p | Given
$$\Rightarrow a + (q-1)d = p \qquad \ldots(2)$$
Subtracting equation (2) from equation (1), we get
$$(p, q)d = q - p$$
$$\Rightarrow d = -1$$

substituting $d = -1$ in equation (1), we get
$$a + (p-1)(-1) = q$$
$$\Rightarrow a = p + q - 1$$
Now, pth term of the AP
$$= a + (p-1)d$$
$$= (p+q-1) + (p-1)(-1)$$
$$= p + q - n$$
Hence,
$(p+q)$th term of the AP
$$= p + q - (p+q)$$
$$= 0.$$

Example 29. *If a, b, c are the pth, qth and rth terms of an AP, then prove that*
$$a(q-r) + b(r-p) + c(p-q) = 0.$$

Sol. Let the first term and the common difference of the AP be A and D respectively.

pth term $= a$ | Given
$$= A + (p-1)D = a \quad ...(1)$$
qth term $= D$ | Given
$$\Rightarrow A + (q-1)D = b \quad ...(2)$$
rth term $= c$ | Given
$$\Rightarrow A + (r-1)D = c \quad ...(3)$$

Multiplying equations (1), (2) and (3) by $q-r, r-p$ and $p-q$ respectively, we get
$$a(q-r) + b(r-p) + c(p-q)$$
$$= [A + (p-1)D](q-r)$$
$$\quad + [A + (q-1)D](r-p)$$
$$\quad + [A + (r-1)D](p-q)]$$
$$= A[q - r + r - p + p - q] +$$
$$D[(p-q)(q-r) + (q-1)(r-p) + (r-1)(p-q)]$$
$$= A(0) + D(0)$$
$$= 0$$

Example 30. *Two AP's have the same common difference. The difference between their 100th terms is 111222333. What is the difference between their millionth terms?*

[**Hint :** 1 million = 10,00,000]

Sol. For First AP
$$a_1 = a(\text{say}), d_1 = d(\text{say})$$
For Second AP
$$A_1 = A(\text{say}), D_1 = d$$
100th term of the first AP
$$= a_1 + (100-1)d_1$$
$$\quad | \because a_n = a + (n-1)d$$
$$= a_1 + 99d_1 = a + 99d$$
100th term of the second AP
$$= A_1 + (100-1)D_1$$
$$\quad | \because a_n = a + (n-1)d$$
$$= A_1 + 99D_1$$
$$= A + 99d$$
Their differences $= 111222333$
$$\Rightarrow (A + 99d) = (a + 99d) = 111223333$$
$$\Rightarrow A - a = 111222333 \quad ...(1)$$
Now, millionth term of the first AP
$$= a_4 + (1000000 - 1)d_1$$
$$= a_4 + 999999 d_1$$
$$= a + 999999d$$
millionth term of the second AP
$$= A_1 + (1000000 - 1)D_1$$
$$= A_1 + 999999d$$
$$= A + 999999d$$
Their difference $= (A + 999999d) - (9 + 999999d)$
$$= A - 1$$
$$= 111222333 \quad | \text{From (1)}$$
Hence, the difference between their millionth terms is 111222333.

TEST YOUR KNOWLEDGE

1. Find the 10th term of the AP :
   2, 7, 12, ...
2. Which term of the AP : 21, 18, 15, ... is -81? Also, is any term 0? Give reason for your answer.
3. Determine the AP whose 3rd term is 5 and the 7th term is 9.
4. Is 301 a term of the list of numbers 5, 11, 17, 23, ...? why?
5. How many two-digit numbers are divisible by 37?

6. Find the 11th term from the last term (towards the first term) of the AP :
   10, 7, 4, ..., – 62.
7. A sum of Rs 1000 is invested at 8% simple interest per year. Calculate the interest at the end of each year. Do these interests form an AP ? If so, find the interest as the end of 30 years making use of this fact.
8. In a flower bed, there are 23 rose plants in the first row, 21 in the second, 19 in the third and so on. There are 5 rose plants in the last row. How many roses are there in the flower bed ?
9. The 5th term of an Arithmetic progression (AP) is 26 and the 10th term is 51. Determine the 15th term of the AP.
   (A.I. CBSE 2006)
10. Find the 8th term from the end of the AP : 7, 10, 13, ..., 1814.   (CBSE 2005)
11. The 4th term of an arithmetic progression is zero. Prove that its 25th term is triple its 11th term.   (A.I. CBSE 2005)
12. The 6th term of an arithmetic progression is zero. Prove that its 21st term is triple its 11th term.   (A.I. CBSE 2005)
13. The 7th term of AP is 32 and its 13th term is 62. Find the AP.   (CBSE 2004)
14. The 7th term of an AP is 20 and its 13th term is 32. Find the AP.   (CBSE 2004)
15. The 8th term of an Arithmetic progression (AP) is 37 and its 12th term is 57. Find the AP.   (A.I. CBSE 2004)
16. The 8th term of an Arithmetic progression (AP) is – 23 and its 12th term is – 39. Find the AP.   (A.I. CBSE 2004)
17. The 8th term of an Arithmetic Progression (AP) is 32 and its 12th term is 52. Find the AP.   (A.I. CBSE 2004)
18. Which term of the arithmetic progression 27, 24, 21, 18, ... is zero ?
19. The last term of the sequence 8, 15, 22, ... is 218. Find the number of terms.
20. The fourth term of an AP is equal to 3 times the first term and the seventh term exceeds twice the third term by 1. Find the first term and the common differences.
21. The 5th term of the AP is 11 and 9th term is 7. Find the 16th term of the progression.
22. A sum of Rs 1000 is invested at 8% simple interest per annum. Calculated the interest at the end of 1, 2, 3, ... years. Is the sequence of interests an AP's. Find the integer at the end of 30 years.

Answers

1. 47
2. 35th ; Yes, 8th
3. 3, 4, 5, 6, 7, ...
4. No.
5. 30
6. – 32
7. Yes, Rs 2400
8. 10
9. 76
10. 163
13. 2, 7, 12, 17, ...
14. 8, 10, 12, 14, ...
15. 2, 7, 12, 17, ...
16. 5, 1, – 3, – 7, ...
17. – 3, 2, 7, 12, ...
18. 10th
19. 31.
20. 3, 2
21. 0
22. Yes, Rs 2400

SUM OF FIRST *n* TERMS OF AN AP

IMPORTANT POINTS

1. Need. Let us consider a situation given below :

Shakila puts Rs 100 into her daughter's money box when she was one year. Old and increasing the amount by Rs 50 every year, puts Rs 150 on her second birthday, Rs 200 on her third birthday and will continue in the same way. How much money will be collected in the money box by the time her daughter is 21 years old ?

Here, the amount of money (in Rs) put in the money box on her first, second, third, fourth ...... birthday were respectively 100, 150, 200, 250, ... till her 21st birthday. To find the total amount in the money box on her 21st birthday, we have to write each of the 21 numbers in the list above and then add them up. But it would be a tedious and time consuming process. This process can be made shorter by finding a method getting this sum.

2. Gauss Problem. We consider the problem to Gauss (about whom you read in Chapter 1), to solve when he was just 10 years old. He was asked to find the sum of the positive integers from 1 to 100. He immediately replied that the sum is 5050. He found the sum as follows :

$$S = 1 + 2 + 3 + \ldots + 99 + 100$$

And then, reversed the numbers to write

$$S = 100 + 99 + \ldots + 3 + 2 + 1$$

Adding these two, he got

$$2S = (100 + 1) + (99 + 2) + \ldots$$
$$+ (3 + 98) + (2 + 99) + (1 + 100)$$
$$= 101 + 101 + \ldots + 101 +$$
$$101 \ (100 \text{ times})$$

So, $S = \dfrac{100 \times 101}{2} = 5050.$

3. Sum of the first n terms of an AP using Gauss's technique. We will now use Gauss's technique to find the sum of the first n terms of an AP :

$a, a + d, a + 2d, \ldots$

The nth term of this AP is $a + (n-1)d$. Let S denote the sum of the first n terms of the AP. We have,

$$S = a + (a + d) + (a + 2d) + \ldots$$
$$+ [a + (n-1)d] \quad \ldots(1)$$

Rewriting the terms in reverse order, we have

$$S = [a + (n-1)d] + [a + (n-2)d]$$
$$+ \ldots + (a + d) + a \quad \ldots(2)$$

On adding (1) and (2), term-wise, we get

$$2S = \dfrac{[2a+(n-1)d] + [2a+(n-1)d] + \ldots + [2a+(n-1)d] + [2a+(n-1)d]}{n \text{ times}}$$

or $\quad 2S = n[2a + (n-1)d]$

(Since, there are n terms)

or, $\quad S = \dfrac{n}{2}[2a + (n-1)d]$

So, **the sum of the first n terms of an AP is given by**

$$S = \dfrac{n}{2}[2a + (n-1)d]$$

We can also write this as

$$S = \dfrac{n}{2}[a + a + (n-1)d]$$

i.e., $\quad S = \dfrac{n}{2}(a + a_n) \quad \ldots(3)$

Now, if there are only n terms in an AP, then $a_n = l$, the last term.

From (3), we see that

$$S = \dfrac{n}{2}(a + l) \quad \ldots(4)$$

This form of the result is useful when the first and the last terms of an AP are given and the common difference is not given.

4. Solution to situation given in point 1 above. The amount of money (in Rs) in the money box of Shakila's daughter on 1st, 2nd, 3rd, 4th birthday, ..., were 100, 150, 200, 250, respectively.

This is an AP. We have to find the total money collected on her 21st birthday, i.e., the sum of the first 21 terms of this AP.

Here, $a = 100$, $d = 50$ and $n = 21$. Using the formula :

$$S = \dfrac{n}{2}[2a + (n-1)d],$$

we have

$$S = \dfrac{21}{2}[2 \times 100 + (21-1) \times 50]$$
$$= \dfrac{21}{2}[200 + 1000]$$
$$= \dfrac{21}{2} \times 1200 = 12600$$

So, the amount of money collected on her 21st birthday is Rs 12600.

We see that the use of formula made it easier to solve the problem.

5. Notations. We use S_n also in place of S to denote the sum of first n terms of the AP. We write S_{20} to denote the sum of the first 20 terms of an AP. The formula for the sum of

ARITHMETIC PROGRESSIONS

the first n terms involves four quantities S, a, d and n. If we know any three of them, we can find the fourth.

Remark : The nth term of an AP is the difference of the sum to n terms and the sum to $(n-1)$ terms of it, i.e., $a_n = S_n - S_{n-1}$.

ILLUSTRATIVE EXAMPLES

[NCERT Exercise 5.3]
(Page No. 112)

Example 1. *Find the sum of the following APs :*

(i) 2, 7, 12, ..., to 10 terms.
(ii) – 37, – 33, – 29, ..., to 12 terms.
(iii) 0.6, 1.7, 2.8, ..., to 100 terms.
(iv) $\dfrac{1}{15}, \dfrac{1}{12}, \dfrac{1}{10}$, ..., to 11 terms.

Sol. (i) **2, 7, 12, ..., to 10 terms**
Here, $a = 2$
$d = 7 - 2 = 5$
$n = 10$

We know that
$$S_n = \dfrac{n}{2}[2a + (n-1)d]$$

$\Rightarrow \quad S_{10} = \dfrac{10}{2}[2(2) + (10-1)5]$
$\Rightarrow \quad S_{10} = 5[4 + 45]$
$\Rightarrow \quad S_{10} = 245$

So, the sum of the first 10 terms of the given AP is 245.

(ii) **– 37, – 33, – 29, ... to 12 terms**
Here, $a = -37$
$d = -33 - (-37)$
$= -33 + 37 = 4$
$n = 12$

We know that
$$S_n = \dfrac{n}{2}[2a + (n-1)d]$$

$\Rightarrow \quad S_{12} = \dfrac{12}{2}[2(-37) + (12-1)4]$
$\Rightarrow \quad S_{12} = 6[-74 + 44]$
$\Rightarrow \quad S_{12} = 6[-30]$
$\Rightarrow \quad S_{12} = -180$

So, the sum of the first 12 terms of the given AP is – 180.

(iii) **0.6, 1.7, 2.8, ... to 100 terms**
Here, $a = 0.6$
$d = 1.7 - 0.6 = 1.1$
$n = 100$

We know that
$$S_n = \dfrac{n}{2}[2a + (n-1)d]$$

$\Rightarrow \quad S_{100} = \dfrac{100}{2}[2(0.6) + (100-1)(1.1)]$
$\Rightarrow \quad S_{100} = 50[1.2 + 108.9]$
$\Rightarrow \quad S_{100} = 50[110.1]$
$\Rightarrow \quad S_{100} = 5505$

So, the sum of the first 100 terms of the given AP is 5505.

(iv) $\dfrac{1}{15}, \dfrac{1}{12}, \dfrac{1}{10}$, ... **to 11 terms**

Here, $a = \dfrac{1}{15}$

$d = \dfrac{1}{12} - \dfrac{1}{15} = \dfrac{1}{60}$

$n = 11$

We know that
$\Rightarrow \quad S_n = \dfrac{n}{2}[2a + (n-1)d]$

$\Rightarrow \quad S_{11} = \dfrac{11}{2}\left[2\left(\dfrac{1}{15}\right) + (11-1)\left(\dfrac{1}{60}\right)\right]$

$\Rightarrow \quad S_{11} = \dfrac{11}{2}\left[\dfrac{2}{15} + \dfrac{1}{6}\right]$

$\Rightarrow \quad S_{11} = \dfrac{11}{2}\left[\dfrac{3}{10}\right]$

$\Rightarrow S_{11} = \dfrac{33}{20}$

So, the sum of the first 11 terms of the given AP is $\dfrac{33}{20}$.

Example 2. *Find the sums given below:*

(i) $7 + 10\dfrac{1}{2} + 14 + \ldots + 84$

(ii) $34 + 32 + 30 + \ldots + 10$

(iii) $-5 + (-8) + (-11) + \ldots + (-230)$

Sol. (i) $7 + 10\dfrac{1}{2} + 14 + \ldots + 84$

This is an AP.
Here, $a = 7$

$$d = 10\dfrac{1}{2} - 7 = 3\dfrac{1}{2} = \dfrac{7}{2}$$

Let the number of terms of the AP be n.
We know that
$$l = a + (n-1)d$$
$\Rightarrow 84 = 7 + (n-1)\dfrac{7}{2}$

$\Rightarrow (n-1)\dfrac{7}{2} = 84 - 7$

$\Rightarrow (n-1)\dfrac{7}{2} = 77$

$\Rightarrow (n-1) = 22$

$\Rightarrow n = 22 + 1$

$\Rightarrow n = 23$

Again, we know that
$$S_n = \dfrac{n}{2}(a + l)$$

$\Rightarrow S_{23} = \dfrac{23}{2}(7 + 84)$

$\Rightarrow S_{23} = \dfrac{2093}{2}$

$\Rightarrow S_{23} = 1046\dfrac{1}{2}$

Hence, the required sum is $1096\dfrac{1}{2}$.

(ii) $34 + 32 + 30 + \ldots + 10$
This is an AP.
Here, $a = 34$
$d = 32 - 34 = -2$
$l = 10$

Let the number of terms of the AP be n.
We know that
$$l = a + (n-1)d$$
$\Rightarrow 10 = 34 + (n-1)(-2)$

$\Rightarrow (n-1)(-2) = -12$

$\Rightarrow n - 1 = \dfrac{-24}{-2} = 12$

$\Rightarrow n = 13$

Again, we know that
$$S_n = \dfrac{n}{2}(a + l)$$

$\Rightarrow S_{13} = \dfrac{13}{2}(34 + 10)$

$\Rightarrow S_{13} = 286$

Hence, the required sum is 286.

(iii) $-5 + (-8) + (-11) + \ldots + (-230)$

This is an AP.
Here, $a = -5$
$d = -8 - (-5) = -8 + 50 = -3$
$l = -230$

Let the number of terms of the AP be n.
We know that
$$l = a + (n-1)d$$
$\Rightarrow -230 = -5 + (n-1)(-3)$

$\Rightarrow (n-1)(-3) = -230 + 5$

$\Rightarrow (n-1)(-3) = -225$

$\Rightarrow n - 1 = \dfrac{-225}{-3} = 75$

$\Rightarrow n = 75 + 1$

$\Rightarrow n = 76$

Again, we know that
$$S_n = \dfrac{n}{2}(a + l)$$

$\Rightarrow S_{76} = \dfrac{76}{2}[(-5) + (-230)]$

$\Rightarrow S_{76} = 38(-235)$

$\Rightarrow S_{76} = -8930$

Hence, the required sum is -8930.

ARITHMETIC PROGRESSIONS

Example 3. *In an AP:*

(i) *given* $a = 5$, $d = 3$, $a_n = 50$, *find* n *and* S_n.

(ii) *given* $a = 7$, $a_{13} = 35$, *find* d *and* S_{13}.

(iii) *given* $a_{12} = 37$, $d = 3$, *find* a *and* S_{12}.

(iv) *given* $a_3 = 15$, $S_{10} = 125$, *find* d *and* a_{10}.

(v) *given* $d = 5$, $S_9 = 75$, *find* a *and* a_9.

(vi) *given* $a = 2$, $d = 8$, $S_n = 90$, *find* n *and* a_n.

(vii) *given* $a = 8$, $a_n = 62$, $S_n = 210$, *find* n *and* d.

(viii) *given* $a_n = 4$, $d = 2$, $S_n = -14$, *find* n *and* a.

(ix) *given* $a = 3$, $n = 8$, $S = 192$, *find* d.

(x) *given* $l = 28$, $S = 144$, *and there are total 9 terms. Find* a.

Sol. (i) Here,
$$a = 5$$
$$d = 3$$
$$a_n = 50$$
We know that
$$a_n = a + (n-1)d$$
$\Rightarrow \quad 50 = 5 + (n-1)3$
$\Rightarrow \quad (n-1)3 = 50 - 5$
$\Rightarrow \quad (n-1)3 = 45$
$\Rightarrow \quad n - 1 = \dfrac{45}{3}$
$\Rightarrow \quad n - 1 = 15$
$\Rightarrow \quad n = 15 + 1$
$\Rightarrow \quad n = 16$

Again, we know that
$$S_n = \dfrac{n}{2}[2a + (n-1)d]$$
$\Rightarrow \quad S_n = \dfrac{16}{2}[2(5) + (16-1)3]$
$\Rightarrow \quad S_n = 8[10 + 45]$
$\Rightarrow \quad S_n = 8(55)$
$\Rightarrow \quad S_n = 440.$

(ii) Here,
$$a = 7$$
$$a_{13} = 35$$

We know that
$$a_n = a + (n-1)d$$
$\Rightarrow \quad a_{13} = a + (13-1)d$
$\Rightarrow \quad a_{13} = a + 12d$
$\Rightarrow \quad 35 = 7 + 12d$
$\Rightarrow \quad 12d = 35 - 7$
$\Rightarrow \quad 12d = 28$
$\Rightarrow \quad d = \dfrac{28}{12}$
$\Rightarrow \quad d = \dfrac{7}{3}$

Again, we know that
$$S_n = \dfrac{n}{2}[2a + (n-1)d]$$
$\Rightarrow \quad S_{13} = \dfrac{13}{2}[2a + (13-1)d]$
$\Rightarrow \quad S_{13} = \dfrac{13}{2}[2a + 12d]$
$\Rightarrow \quad S_{13} = \dfrac{13}{2}\left[2(7) + 12\left(\dfrac{7}{3}\right)\right]$
$\Rightarrow \quad S_{13} = \dfrac{13}{2}(14 + 28)$
$\Rightarrow \quad S_{13} = \dfrac{13}{2}(42)$
$\Rightarrow \quad S_{13} = (13)(21)$
$\Rightarrow \quad S_{13} = 273.$

(iii) Here,
$$a_{12} = 37$$
$$d = 3$$
We know that
$$a_n = a + (n-1)d$$
$\Rightarrow \quad a_{12} = a + (12-1)d$
$\Rightarrow \quad a_{12} = a + 11d$
$\Rightarrow \quad 37 = a + 11 \times 3$
$\Rightarrow \quad 37 = a + 33$
$\Rightarrow \quad a = 37 - 33 = 4$

Again, we know that
$$S_n = \dfrac{n}{2}[2a + (n-1)d]$$

$\Rightarrow\ S_{12} = \dfrac{12}{2}[2a + (12-1)d]$

$\Rightarrow\ S_{12} = 6[2a + 11d]$

$\Rightarrow\ S_{12} = 6[2 \times 4 + 11 \times 3]$

$\Rightarrow\ S_{12} = 6[8 + 33]$

$\Rightarrow\ S_{12} = 6 \times 41$

$\Rightarrow\ S_{12} = 246.$

(iv) Here,
$$a_3 = 15$$
$$S_{10} = 125$$
We know that
$$a_n = a + (n-1)d$$
$\Rightarrow\ a_3 = a + (3-1)d$

$\Rightarrow\ a_3 = a + 2d$

$\Rightarrow\ 15 = a + 2d$

$\Rightarrow\ a + 2d = 15$ ...(1)

Again, we know that
$$S_n = \dfrac{n}{2}[2a + (n-1)d]$$
$\Rightarrow\ S_{10} = \dfrac{10}{2}[2a + (10-1)d]$

$\Rightarrow\ S_{10} = 5(2a + 9d)$

$\Rightarrow\ 125 = 5(2a + 9d)$

$\Rightarrow\ 25 = 2a + 9d$

$\Rightarrow\ 2a + 9d = 25$ ...(2)

Solving equation (1) and equation (2), we get
$$a = 17$$
$$d = -1$$
Now, $a_n = a + (n-1)d$

$\Rightarrow\ a_{10} = a + (10-1)d$

$\Rightarrow\ a_{10} = a + 9d$

$\Rightarrow\ a_{10} = 17 + 9(-1)$

$\Rightarrow\ a_{10} = 17 - 9$

$\Rightarrow\ a_{10} = 8.$

(v) Here,
$$d = 5$$
$$S_9 = 75$$
We know that
$$S_n = \dfrac{n}{2}[2a + (n-1)d]$$
$\Rightarrow\ S_9 = \dfrac{9}{2}[2a + (9-1)d]$

$\Rightarrow\ S_9 = \dfrac{9}{2}[2a + 8d]$

$\Rightarrow\ S_9 = 9[a + 4d]$

$\Rightarrow\ S_9 = 9[a + 4 \times 5]$

$\Rightarrow\ S_9 = 9[a + 20]$

$\Rightarrow\ 75 = 9a + 180$

$\Rightarrow\ 9a = 75 - 180$

$\Rightarrow\ 9a = -105$

$\Rightarrow\ a = -\dfrac{105}{9}$

$\Rightarrow\ a = -\dfrac{35}{3}$

Again, we know that
$$a_n = a + (n-1)d$$
$\Rightarrow\ a_9 = a + (9-1)d$

$\Rightarrow\ a_9 = a + 8d$

$\Rightarrow\ a_9 = -\dfrac{35}{3} + 8(5)$

$\Rightarrow\ a_9 = -\dfrac{35}{3} + 40$

$\Rightarrow\ a_9 = \dfrac{-35 + 120}{3}$

$\Rightarrow\ a_9 = \dfrac{85}{3}.$

(vi) Here,
$$a = 2$$
$$d = 8$$
$$S_n = 90$$
We know that
$$S_n = \dfrac{n}{2}[2a + (n-1)d]$$
$\Rightarrow\ 90 = \dfrac{n}{2}[2(2) + (n-1)8]$

$\Rightarrow\ 90 = n[2 + (n-1)4]$

$\Rightarrow\ 90 = n[2 + 4n - 4]$

$\Rightarrow\ 90 = n[4n - 2]$

$\Rightarrow\ 90 = 2n[2n - 1]$

$\Rightarrow\ 45 = n[2n - 1]$

$\Rightarrow\ 45 = 2n^2 - n$

$\Rightarrow \qquad 2n^2 - n - 45 = 0$

$\Rightarrow \qquad 2n^2 - 10n + 9n - 45 = 0$

$\Rightarrow \qquad 2n(n-5) + 9(n-5) = 0$

$\Rightarrow \qquad (n-5)(2n+9) = 0$

$\Rightarrow \qquad n - 5 = 0 \text{ or } 2n + 9 = 0$

$\Rightarrow \qquad n = 5 \text{ or } n = -\dfrac{9}{2}.$

$n = -\dfrac{9}{2}$

is inadmissible as n, being the number of terms, is a natural number.

$\therefore \qquad n = 5$

Again, we know that

$a_n = a + (n-1)d$

$\Rightarrow \qquad a_n = 2 + (5-1)8$

$\Rightarrow \qquad a_n = 2 + (4)8$

$\Rightarrow \qquad a_n = 2 + 32$

$\Rightarrow \qquad a_n = 34.$

(*vii*) Here,

$a = 8$

$a_n = 62$

$S_n = 210$

We know that

$a_n = a + (n-1)d$

$\Rightarrow \qquad 62 = 8 + (n-1)d$

$\Rightarrow \qquad 62 - 8 = (n-1)d$

$\Rightarrow \qquad 54 = (n-1)d$

$\Rightarrow \qquad (n-1)d = 54 \qquad \ldots(1)$

Again, we know that

$S_n = \dfrac{n}{2}[2n + (n-1)d]$

$\Rightarrow \qquad 210 = \dfrac{n}{2}[2(8) + (n-1)d]$

$\Rightarrow \qquad 210 = \dfrac{n}{2}[16 + (n-1)d]$

$\Rightarrow \qquad 210 = \dfrac{n}{2}[16 + 54] \qquad \text{| Using (1)}$

$\Rightarrow \qquad 210 = \dfrac{n}{2}(70)$

$\Rightarrow \qquad 210 = 35n$

$\Rightarrow \qquad n = \dfrac{210}{35}$

$\Rightarrow \qquad n = 6$

Putting $n = 6$ in equation (1), we get

$(6-1)d = 54$

$\Rightarrow \qquad 5d = 54$

$\Rightarrow \qquad d = \dfrac{54}{5}$

(*viii*) Here,

$a_n = 4$

$d = 2$

$S_n = -14$

We know that

$a_n = a + (n-1)d$

$\Rightarrow \qquad 4 = a + (n-1)2$

$\Rightarrow \qquad 4 = a + 2n - 2$

$\Rightarrow \qquad 4 + 2 = a + 2n$

$\Rightarrow \qquad 6 = a + 2n$

$\Rightarrow \qquad a + 2n = 6 \qquad \ldots(1)$

Again, we know that

$S_n = \dfrac{n}{2}[2a + (n-1)d]$

$\Rightarrow \qquad -14 = \dfrac{n}{2}[2a + (n-1)2]$

$\Rightarrow \qquad -14 = n[a + (n-1)]$

$\Rightarrow \qquad -14 = n(a + n - 1)$

$\Rightarrow \qquad -14 = n(6 - n - 1)$

| From (1), $a + 2n = 6 \Rightarrow a + n = 6 - n$

$\Rightarrow \qquad -14 = n(-n + 5)$

$\Rightarrow \qquad -14 = -n^2 + 5n$

$\Rightarrow \qquad n^2 - 5n - 14 = 0$

$\Rightarrow \qquad n^2 - 7n + 2n - 14 = 0$

$\Rightarrow \qquad n(n-7) + 2(n-7) = 0$

$\Rightarrow \qquad (n-7)(n+2) = 0$

$\Rightarrow \qquad n - 7 = 0 \text{ on } n + 2 = 0$

$\Rightarrow \qquad n = 7 \text{ or } n = -2$

$n = -2$

is inadmissible as n, being the number of terms, is a natural number.

$\therefore \qquad n = 7$

Putting $n = 7$ in equation (1), we get

$a + 2(7) = 6$

$\Rightarrow \quad a + 14 = 6$

$\Rightarrow \quad a = 6 - 14$

$\Rightarrow \quad a = -8.$

(ix) Here,

$a = 3$

$n = 8$

$S = 192$

We know that

$$S = \frac{n}{2}[2a + (n-1)d]$$

$\Rightarrow \quad 192 = \frac{8}{2}[2(3) + (8-1)d]$

$\Rightarrow \quad 192 = 4[6 + 7d]$

$\Rightarrow \quad \frac{192}{4} = 6 + 7d$

$\Rightarrow \quad 48 = 6 + 7d$

$\Rightarrow \quad 48 - 6 = 7d$

$\Rightarrow \quad 42 = 7d$

$\Rightarrow \quad 7d = 42$

$\Rightarrow \quad d = \frac{42}{7}$

$\Rightarrow \quad d = 6.$

(x) Here,

$l = 28$

$S = 144$

$n = 9$

We know that

$$S = \frac{n}{2}(a + l)$$

$\Rightarrow \quad 144 = \frac{9}{2}(a + 28)$

$\Rightarrow \quad \frac{(144)(2)}{9} = a + 28$

$\Rightarrow \quad 32 = a + 28$

$\Rightarrow \quad a + 28 = 32$

$\Rightarrow \quad a = 32 - 28$

$\Rightarrow \quad a = 4.$

Example 4. *How many terms of the AP : 9, 17, 25, ... must be taken to give a sum of 636 ?*

Sol. The given AP is

9, 17, 25, ...

Here, $a = 9$

$d = 17 - 9 = 8$

Let n terms of the AP must be taken.

Then, $S_n = 636$

$\Rightarrow \quad \frac{n}{2}[2a + (n-1)d] = 636$

$\Rightarrow \quad \frac{n}{2}[2(9) + (n-1)8] = 636$

$\Rightarrow \quad n[9 + (n-1)4] = 636$

$\Rightarrow \quad n[9 + 4n - 4] = 636$

$\Rightarrow \quad n[(4n + 5] = 636$

$\Rightarrow \quad 4n^2 + 5n = 636$

$\Rightarrow \quad 4n^2 + 5n - 636 = 0$

$\Rightarrow \quad 4n^2 + 53n - 48n - 636 = 0$

$\Rightarrow \quad n(4n + 53) - 12(4n + 53) = 0$

$\Rightarrow \quad (4n + 53)(n - 12) = 0$

$\Rightarrow \quad 4n + 53 = 0 \text{ or } n - 12 = 0$

$\Rightarrow \quad n = -\frac{53}{4} \text{ or } n = 12$

$n = -\frac{53}{4}$

is inadmissible as n, being the number of terms, is a natural number.

$\therefore \quad n = 12$

Hence, 12 terms of the AP must be taken.

Example 5. *The first term of an AP is 5, the last term is 45 and the sum is 400. Find the number of terms and the common difference.*

Sol. Here,

$a = 5$

$l = 45$

$S = 400$

We know that

$$S = \frac{n}{2}(a + l)$$

ARITHMETIC PROGRESSIONS

$\Rightarrow \qquad 400 = \dfrac{n}{2}(5 + 45)$

$\Rightarrow \qquad 400 = \dfrac{n}{2}(50)$

$\Rightarrow \qquad 400 = 25n$

$\Rightarrow \qquad n = \dfrac{400}{25}$

$\Rightarrow \qquad n = 16.$

Hence, the number of terms is 16.
Again, we know that
$$l = a + (n-1)d$$
$\Rightarrow \qquad 45 = 5 + (16-1)d$
$\Rightarrow \qquad 45 = 5 + 15d$
$\Rightarrow \qquad 45 - 5 = 15d$
$\Rightarrow \qquad 40 = 15d$
$\Rightarrow \qquad 15d = 40$
$\Rightarrow \qquad d = \dfrac{40}{15}$
$\Rightarrow \qquad d = \dfrac{8}{3}.$

Hence, the common difference is $\dfrac{8}{3}$.

Example 6. *The first and the last terms of an AP are 17 and 350 respectively. If the common difference is 9, how many terms are there and what is their sum?*

Sol. Here,
$\qquad a = 17$
$\qquad l = 350$
$\qquad d = 9$
We know that
$$l = a + (n-1)d$$
$\Rightarrow \qquad 350 = 17 + (n-1)9$
$\Rightarrow \qquad 350 - 17 = (n-1)9$
$\Rightarrow \qquad 333 = (n-1)9$
$\Rightarrow \qquad (n-1)9 = 333$
$\Rightarrow \qquad n - 1 = \dfrac{333}{9}$
$\Rightarrow \qquad n - 1 = 37$
$\Rightarrow \qquad n = 37 + 1$
$\Rightarrow \qquad n = 38$

Hence, there are 38 terms.

Again, we know that
$$S_n = \dfrac{n}{2}(a + l)$$
$\Rightarrow \qquad S_{38} = \dfrac{38}{2}(17 + 350)$
$\Rightarrow \qquad S_{38} = (19)(367)$
$\Rightarrow \qquad S_{38} = 6973$

Hence, their sum is 6973.

Example 7. *Find the sum of first 22 terms of an AP in which d = 7 and 22nd term is 149.*

Sol. Here,
$\qquad d = 7$
$\qquad a_{22} = 149$
Let the first term of the AP be a.
We know that
$\qquad a_n = a + (n-1)d$
$\Rightarrow \qquad a_{22} = a + (22-1)d$
$\Rightarrow \qquad a_{22} = a + 21d$
$\Rightarrow \qquad 149 = a + (21)(7)$
$\Rightarrow \qquad 149 = a + 147$
$\Rightarrow \qquad a + 147 = 149$
$\Rightarrow \qquad a = 149 - 147$
$\Rightarrow \qquad a = 2$

Again, we know that
$$S_n = \dfrac{n}{2}[2a + (n-1)d]$$
$\Rightarrow \qquad S_{22} = \dfrac{22}{2}[2(2) + (22-1)7]$
$\Rightarrow \qquad S_{22} = (11)[4 + 147]$
$\Rightarrow \qquad S_{22} = (11)(151)$
$\Rightarrow \qquad S_{22} = 1661$

Hence, the sum of first 22 terms of the AP is 1661.

Example 8. *Find the sum of first 51 terms of an AP whose second and third terms are 14 and 18 respectively.*

Sol. Let the first term and the common difference of the AP be a and d respectively.

Second term = 14 | Given
$\Rightarrow \qquad a + (2-1)d = 14$
$\qquad\qquad$ | $\because a_n = a + (n-1)d$

$\Rightarrow \quad a + d = 14$ ...(1)

Third term = 18 | Given

$\Rightarrow \quad a + (3-1)d = 18$

| $\because a_n = a + (n-1)d$

$\Rightarrow \quad a + 2d = 18$ ...(2)

Solving equation (1) and equation (2), we get

$a = 10$
$d = 4$

Now, sum of first 51 terms of the AP

$= S_{51}$

$= \dfrac{51}{2}[2a + (51-1)d]$

| $\because S_n = \dfrac{n}{2}[2a + (n-1)d]$

$= \dfrac{51}{2}[2a + 50d]$

$= 51(a + 25d)$

$= (51)(10 + 25 \times 4)$

$= (51)(10 + 100)$

$= (51)(110)$

$= 5610$.

Example 9. *If the sum of first 7 terms of an AP is 49 and that of 17 terms is 289, find the sum of first n terms.*

Sol. Let the first term and the common difference of the AP be a and d respectively.

Sum of first 7 terms = 49 | Given

$\Rightarrow \quad S_7 = 49$

$\Rightarrow \quad \dfrac{7}{2}[2a + (7-1)d] = 49$

| $\because S_n = \dfrac{n}{2}[2a + (n-1)d]$

$\Rightarrow \quad \dfrac{7}{2}[2a + 6d] = 49$

$\Rightarrow \quad 7(a + 3d) = 49$

$\Rightarrow \quad a + 3d = \dfrac{49}{7}$

$\Rightarrow \quad a + 3d = 7$ ...(1)

Sum of first 17 terms = 289 | Given

$\Rightarrow \quad S_{17} = 289$

$\Rightarrow \quad \dfrac{17}{2}[2a + (17-1)d] = 289$

| $\because S_n = \dfrac{n}{2}[2a + (n-1)d]$

$\Rightarrow \quad \dfrac{17}{2}[2a + 16d] = 289$

$\Rightarrow \quad 17(a + 8d) = 289$

$\Rightarrow \quad a + 8d = \dfrac{289}{17}$

$\Rightarrow \quad a + 8d = 17$ ...(2)

Solving equation (1) and equation (2), we get

$a = 1$
$d = 2$

Now, sum of first in terms

$= S_n$

$= \dfrac{n}{2}[2a + (n-1)d]$

$= \dfrac{n}{2}[2 \times 1 + (n-1)2]$

$= n[1 + n - 1]$

$= n(n)$

$= n^2$.

Example 10. *Show that $a_1, a_2, ..., a_n, ...$ form an AP where a_n is defined as below.*

(i) $a_n = 3 + 4n$ (ii) $a_n = 9 - 5n$.

Also find the sum of the first 15 terms in each case.

Sol. (i) $a_n = 3 + 4n$

We have

$a_n = 3 + 4n$

Put $n = 1, 2, 3, 4, ...$ in succession, we get

$a_1 = 3 + 4(1) = 3 + 4 = 7$
$a_2 = 3 + 4(2) = 3 + 8 = 11$
$a_3 = 3 + 4(3) = 3 + 12 = 15$
$a_4 = 3 + 4(4) = 3 + 16 = 19$
$\vdots \qquad \vdots \qquad \vdots \qquad \vdots$

$\therefore \quad a_2 - a_1 = 11 - 7 = 4$

$a_3 - a_2 = 15 - 11 = 4$

$a_4 - a_3 = 19 - 15 = 4$

i.e., $a_{k+1} - a_k$ is the same every time.

So, $a_1, a_2, ..., a_n, ...$ form an AP.
Here, $a = a_1 = 7$
$d = a_2 - a_1 = 4$
∴ Sum of the first 15 terms
$= S_{15}$
$= \dfrac{15}{2}[2a + (15-1)d]$

$\left| \because S_n = \dfrac{n}{2}[2a + (n-1)d] \right.$

$= \dfrac{15}{2}[2a + 14d]$
$= 15(a + 7d)$
$= (15)(7 + 7 \times 4)$
$= (15)(7 + 28)$
$= (15)(35)$
$= 525.$

(ii) $\mathbf{a_n = a - 5n}$
We have
$a_n = 9 - 5n$
Put $n = 1, 2, 3, 4, ...$ in succession, we get
$a_1 = 9 - 5(1) = 9 - 5 = 4$
$a_2 = 9 - 5(2) = 9 - 10 = -1$
$a_3 = 9 - 5(3) = 9 - 15 = -6$
$a_4 = 9 - 5(4) = 9 - 20 = -11$
$\vdots \qquad \vdots \qquad \vdots$
∴ $a_2 - a_1 = -1 - 4 = -5$
$a_3 - a_2 = -6 - (-1) = -6 + 1 = -5$
$a_4 - a_3 = -11 - (-6) = -11 + 6 = -5$
i.e., $a_{k+1} - a_k$ is the same everytime.
So, $a_1, a_2, ..., a_n, ...$ form an AP.
Here, $a = a_1 = 4$
$a = a_2 - a_1 = -5$
∴ Sum of the first 15 terms
$= S_{15}$
$= \dfrac{15}{2}[2a + (n-1)d]$

$\left| \because S_n = \dfrac{n}{2}[2a + (n-1)d] \right.$

$= \dfrac{15}{2}[2 \times a + (15-1)d]$

$= \dfrac{15}{2}[2a + 14d]$
$= 15(a + 7d)$
$= (15)[4 + 7(-5)]$
$= (15)(4 - 35)$
$= (15)(-31)$
$= -465.$

Example 11. *If the sum of the first n terms of an AP is $4n - n^2$, what is the first term (that is S_1) ? What is the sum of first two terms ? What is the second term ? Similarly, find the 3rd, the 10th and the nth terms.*

Sol. We have
Sum of the first n terms $= 4n - n^2$
$\Rightarrow \quad S_n = 4n - n^2$
Put, $n = 1$
$S_1 = 4(1) - (1)^2 = 4 - 1 = 3$
$\Rightarrow \quad a_1 = 3$
Hence, the first term is 3.
Put $n = 2$
$S_2 = 4(2) - (2)^2 = 8 - 4 = 4$
Hence, the sum of two terms is 4.
Second term
$= S_2 - S_1$
$= 4 - 3 = 1$
Put $n = 3$
$S_3 = 4(3) - (3)^2 = 12 - 9 = 3$
∴ 3rd term $= a_3 = S_3 - S_2$
$= 3 - 4$
$= -1$
Put $n = 9, 10$
$S_9 = 4(9) - (9)^2$
$= 36 - 81$
$= -45$
$S_{10} = 4(10) - (10)^2$
$= 40 - 100$
$= -60$
∴ 10th term $= a_{10} = S_{10} - S_9$
$= -60 - (-45)$
$= -60 + 45$
$= -15$

$S_{n-1} = 4(n-1) - (n-1)^2$
$= 4n - 4 - (n^2 - 2n + 1)$
$= 4n - 4 - n^2 + 2n - 1$
$= 6n - n^2 - 5$

∴ nth term $= a_n$
$= S_n - S_{n-1}$
$= (4n - n^2) - (6n - n^2 - 5)$
$= 5 - 2n.$

Example 12. *Find the sum of the first 40 positive integers divisible by 6.*

Sol. The first 40 positive integers divisible by 6 are

$$6, 12, 18, 24, \ldots\ldots$$

Here,
$a_2 - a_1 = 12 - 6 = 6$
$a_3 - a_2 = 18 - 12 = 6$
$a_4 - a_3 = 24 - 18 = 6$

i.e., $a_{k+1} - a_k$ is the same everytime.

So, the above list of numbers form an AP.

Here,
$a = 6$
$d = 6$
$n = 40$

∴ Sum of the first 40 positive integers
$= S_{40}$
$= \dfrac{40}{2}[2a + (40-1)d]$

$\left| \because S_n = \dfrac{n}{2}[2a + (n-1)d] \right.$

$= 20[2a + 39d]$
$= (20)[2 \times 6 + 39 \times 6]$
$= (20)(12 + 234)$
$= (20)(246)$
$= 4920.$

Example 13. *Find the sum of the first 15 multiples of 8.*

Sol. The first 15 multiples of 8 are

$$8, 16, 24, 32, \ldots\ldots$$

Here,
$a_2 - a_4 = 16 - 8 = 8$
$a_3 - a_2 = 24 - 16 = 8$

$a_4 - a_3 = 32 - 24 = 8$

i.e., $a_{k-1} - a_k$ is the same everytime.

So, the above list of numbers form an AP.

Here,
$a = 8$
$d = 8$
$n = 15$

∴ Sum of first 15 multiples of 8
$= S_{15}$
$= \dfrac{15}{2}[2a + (15-1)d]$

$\left| \because S_n = \dfrac{n}{2}[2a + (n-1)d] \right.$

$= \dfrac{15}{2}[2a + 14d]$
$= 15(a + 7d)$
$= (15)(8 + 7 \times 8)$
$= (15)(8 + 56)$
$= (15)(64)$
$= 960.$

Example 14. *Find the sum of the odd numbers between 0 and 50.*

Sol. The odd numbers between 0 and 50 are $1, 3, 5, 7, \ldots\ldots, 49$.

Here,
$a_2 - a_4 = 3 - 1 = 2$
$a_3 - a_2 = 5 - 3 = 2$
$a_4 - a_3 = 7 - 5 = 2$

i.e., $a_{k+1} - a_k$ is the same everytime.

So, the above list of numbers forms an AP.

Here,
$a = 1$
$d = 2$
$l = 49$

Let the number of terms of the AP be n.
Then, $l = a + (n-1)d$
$\Rightarrow \quad 49 = 1 + (n-1)2$
$\Rightarrow \quad 49 - 1 = (n-1)2$
$\Rightarrow \quad 48 = (n-1)2$
$\Rightarrow \quad (n-1)2 = 48$

ARITHMETIC PROGRESSIONS

$\Rightarrow \qquad n - 1 = \dfrac{48}{2}$

$\Rightarrow \qquad n - 1 = 24$

$\Rightarrow \qquad n = 24 + 1$

$\Rightarrow \qquad n = 25$

Hence, the number of terms of the AP be 25.

$\therefore$ Sum of the odd numbers between 0 and 50

$= S_{25}$

$= \dfrac{25}{2}(a + l) \qquad \left| \because S_n = \dfrac{n}{2}(a + l) \right.$

$= \left(\dfrac{25}{2}\right)(1 + 49)$

$= \left(\dfrac{25}{2}\right)(50)$

$= (25)(25)$

$= 625.$

Example 15. *A contract on construction job specifies a penalty for delay of completion beyond a certain date as follows : Rs 200 for the first day, Rs 250 for the second day, Rs 300 for the third day, etc., the penalty for each succeeding day being Rs 50 more than for the preceding day. How much money the contractor has to pay as penalty, if he has delayed the work by 30 days ?*

Sol. Since the penalty for each succeeding day is Rs 50 more than the preceding day, therefore, the penalties for the first day, the second day, the third day, etc. will form an AP.

Let us denote the penalty for the nth day by a_n. Then,

$a_1 = $ Rs 200

$a_2 = $ Rs 250

$a_3 = $ Rs 300

$\vdots \qquad \vdots$

Here,

$a = $ Rs 200

$d = a_2 - a_1$

$\quad = $ Rs 250 – Rs 200 = Rs 50

$n = 30$

$\therefore$ The money the contractor has to pay penalty, if he has delayed the work by 30 days

$= \dfrac{30}{2}[2a + (30 - 1)d]$

$\left| \because S_n = \dfrac{n}{2}[2a + (n - 1)d] \right.$

$= 15[2a + 29d]$

$= $ Rs $(15)\,[2 \times 200 + 29 \times 50]$

$= $ Rs $(15)\,(400 + 1450)$

$= $ Rs $(15)\,(1850)$

$= $ Rs 27750.

Example 16. *A sum of Rs 700 is to be used to give seven cash prizes to students of a school for their overall academic performance. If each prize is Rs 20 less than its preceding prize, find the value of each of the prizes.*

Sol. Since each prize is Rs 20 less than its preceding prize, therefore, the values of the seven successive cash prizes will form an AP.

Let the first prize be Rs a.

Then the winner prizes, in succession, will be Rs $(a - 20)$, Rs $(a - 40)$, Rs $(a - 60)$, etc.

Here,

$A = a$

$d = (a - 20) - a = -20$

$n = 7$

$S_n = 700$

We know that

$S_n = \dfrac{n}{2}[2A + (n - 1)d\,]$

$\Rightarrow \quad 700 = \dfrac{7}{2}[2a + (7 - 1)(-20)]$

$\Rightarrow \quad 700 = \dfrac{7}{2}[2a - 120]$

$\Rightarrow \quad 700 = 7(a - 60)$

$\Rightarrow \quad a - 60 = \dfrac{700}{7}$

$\Rightarrow \quad a - 60 = 100$

$\Rightarrow \qquad a = 100 + 60$

$\Rightarrow \qquad a = 160$

$\Rightarrow \quad$ Value of first prize = Rs 160

∴ Value of second prize
= Rs 160 – Rs 20
= Rs 140

Value of third prize
= Rs 140 – Rs 20
= Rs 120

Value of fourth prize
= Rs 120 – Rs 20
= Rs 100

Value of fifth prize
= Rs 100 – Rs 20
= Rs 80

Value of sixth prize
= Rs 80 – Rs 20
= Rs 60

Value of seventh prize
= Rs 60 – Rs 20
= Rs 40.

Example 17. *In a school, students thought of planting trees in and around the school to reduce air pollution. It was decided that the number of trees, that each section of each class will plant, will be the same as the class, in which they are studying, e.g., a section of Class I will plant 1 tree, a section of Class II will plant 2 trees and so on till Class XII. There are three sections of each class. How many trees will be planted by the students ?*

Sol. Number of trees planted by the students of class I
= 1 + 1 + 1 = 3

Number of trees planted by the students of class II
= 2 + 2 + 2 = 6

Number of trees planted by the students of class III
= 3 + 3 + 3 = 9

Number of trees planted by the students of class IV
= 4 + 4 + 4 = 12
⋮ ⋮

Number of trees planted by the students of class XII
= 12 + 12 + 12 = 36

Therefore, the numbers of trees form the sequence
3, 6, 9, 12, ..., 36
∴ $a_2 - a_1 = 6 - 3 = 3$
$a_3 - a_2 = 9 - 6 = 3$
$a_4 - a_3 = 12 - 9 = 3$
i.e., $a_{k+1} - a_k$ is the same everytime.

So, the above sequence forms are AP.
Here,
$a = 3$
$d = 6 - 3 = 3$
$l = 36$
$n = 12$

∴ Number of trees planted by the students
$= S_{12}$
$= \frac{12}{2}[a + l]$ $\quad \left| \because S_n = \frac{n}{2}(a + l) \right.$
$= (6)(3 + 36)$
$= (6)(39)$
$= 234.$

Example 18. *A spiral is made up of successive semicircles, with centres alternately at A and B, starting with centre at A, of radii 0.5 cm, 1.0 cm, 1.5 cm, 2.0 cm, ... as shown in figure. What is the total length of such a spiral made up of thirteen consecutive semicircles ?*
(Take $\pi = \frac{22}{7}$)

[**Hint.** Length of successive semicircles is $l_1, l_2, l_3, l_4, ...$ with centres at A, B, A, B, ..., respectively.]

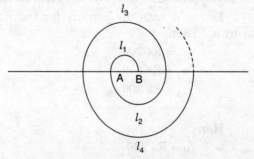

Sol. Lengths (in cm) of successive semi circles with centres at A, B, A, B, ... are respectively.

ARITHMETIC PROGRESSIONS

$\pi(0.5), \pi(1.0), \pi(1.5), \pi(2.0), \ldots\ldots$

$\therefore \quad a_2 - a_1 = \pi(1.0) - \pi(0.5) = \pi(0.5)$
$\quad a_3 - a_2 = \pi(1.5) - \pi(1.0) = \pi(0.5)$
$\quad a_4 - a_3 = \pi(2.0) - \pi(1.5) = \pi(0.5)$

i.e., $a_{k+1} - a_k$ is the same everytime.

So, the above list of numbers form an AP.

Here,
$$a = \pi(0.5)$$
$$d = \pi(0.5)$$
$$n = 13$$

$\therefore$ Total length of the spiral
$= S_{13}$
$= \dfrac{13}{2}[2a + (13-1)d]$

$\left| \because S_n = \dfrac{n}{2}[2a + (n-1)d] \right.$

$= \dfrac{13}{2}[2a + 12d]$
$= 13(a + 6d)$
$= (13)[\pi(0.5) + 6\pi(0.5)]$ cm
$= (13)[7\pi(0.5)]$ cm
$= (13)\left[7 \times \dfrac{22}{7} \times 0.5\right]$ cm
$= (13)(11)$ cm
$= 143$ cm.

Example 19. *200 logs are stacked in the following manner : 20 logs in the bottom row, 19 in the next row, 18 in the row next to it and so on (see figure). In how may rows are the 200 logs placed and how many logs are in the top row ?*

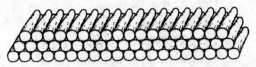

Sol. The numbers of logs in the bottom row, next row, row next to it and so on form the sequence

$20, 19, 18, 17, \ldots\ldots$

$\therefore \quad a_2 - a_1 = 19 - 20 = -1$
$\quad a_3 - a_2 = 18 - 19 = -1$
$\quad a_4 - a_3 = 17 - 18 = -1$

i.e., $a_{k+1} - a_k$ is the same everytime.

So, the above sequence forms an AP.
Here, $a = 20$
$d = -1$
$S_n = 200$
We know that
$$S_n = \dfrac{n}{2}[2a + (n-1)d]$$

$\Rightarrow \quad 200 = \dfrac{n}{2}[2(20) + (n-1)(-1)]$

$\Rightarrow \quad 200 = \dfrac{n}{2}[40 - n + 1]$

$\Rightarrow \quad 200 = \dfrac{n}{2}(41 - n)$

$\Rightarrow \quad 400 = n(41 - n)$

$\Rightarrow \quad n(41 - n) = 400$

$\Rightarrow \quad 41n - n^2 = 400$

$\Rightarrow \quad n^2 - 41n + 400 = 0$

$\Rightarrow \quad n^2 - 25n - 16n + 400 = 0$

$\Rightarrow \quad n(n - 25) - 16(n - 25) = 0$

$\Rightarrow \quad (n - 25)(n - 16) = 0$

$\Rightarrow \quad n - 25 = 0 \quad$ or $\quad n - 16 = 0$

$\Rightarrow \quad n = 25 \quad$ or $\quad n = 16$

$\Rightarrow \quad n = 25, 16$

Hence, the number of rows is either 25 or 16.

Now, Number of logs in row
$=$ Number of logs in 25th row
$= a_{25}$
$= a + (25 - 1)d$

$\quad | \because a_n = a + (n-1)d$

$= a + 24d$
$= 20 + 24(-1)$
$= 20 - 24$
$= -4$

which is not possible.

Therefore, $n = 16$.

and,
Number of log in top row
$=$ Number of logs in 16th row
$= a_{16}$
$= a + (16 - 1)d$

$\quad | \because a_n = a + (n-1)d$

$= a + 15d$

= 20 + 15(– 1)
= 20 – 15
= 5.

Example 20. *In a potato race, a bucket is placed at the starting point, which is 5m from the first potato, and the other potatoes are placed 3 m apart in a straight line. There are ten potatoes in the line (see figure).*

Each competitor starts from the bucket, picks up the nearest potato, runs back with it, drops it in the bucket, runs back to pick up the next potato, runs to the bucket to drop it in, and she continues in the same way until all the potatoes are in the bucket. What is the total distance the competitor has to run ?

[**Hint.** To pick up the first potato and the second potato, the total distance run (in metres) run by a competitor is $2 \times 5 + 2 \times (5 + 3)$].

Sol. To drop the first potato in the bucket, the distance run $= 2 \times 5$ m

To drop the second potato in the bucket, the distance run $= 2 \times (5 + 3)$ m,

To drop the third potato in the bucket, the distance run $= 2 \times (5 + 3 + 3)$ m, and so on.

$\therefore \quad a_2 - a_1 = 2 \times (5 + 3) \text{ m} - 2 \times 5 \text{ m}$
$\qquad = 2 \times 3 \text{ m} = 6 \text{ m}$
$a_3 - a_2 = 2 \times (5 + 3 + 3) \text{ m} - 2 \times (5 + 3) \text{ m}$
$\qquad = 2 \times 3 \text{ m}$
$\qquad = 6 \text{ m}$

i.e., $a_{k+1} - a_k$ is the same every time.

So, the above distances (in m) form an AP.

Here, $a = 2 \times 5 = 10$ m
$d = 6$ m
$n = 10$

$\therefore$ The total distance the competitor has to sum
$\qquad = S_{10}$

$= \dfrac{10}{2}[2a + (10 - 1)d]$

$\left| \because S_n = \dfrac{n}{2}[2a + (n - 1)d] \right.$

$= 5[2a + 9d]$
$= 5[2 \times 10 + 9 \times 6]$ m
$= 5(20 + 54)$ m
$= 370$ m.

ADDITIONAL EXAMPLES

Example 21. *Find the sum of all the two-digit natural numbers which are divisible by 4.* (CBSE 2006)

Sol. All the two-digit natural numbers divisible by 4 are

12, 16, 20, 24, ..., 96

Here, $a_1 = 12$
$a_2 = 16$
$a_3 = 20$
$a_4 = 24$
$\vdots \qquad \vdots$

$\therefore \quad a_2 - a_1 = 16 - 12 = 4$
$a_3 - a_2 = 20 - 16 = 4$
$a_4 - a_3 = 24 - 20 = 4$
$\vdots \qquad \vdots$

$\therefore \quad a_2 - a_1 = a_3 - a_2 = a_4 - a_3 = $ ......
(= 4 each)

$\therefore$ This sequence is an arithmetic progression whose common difference is 4.

Here, $a = 12$
$d = 4$
$l = 96$

Let the number of terms be n.
Then, $l = a + (n - 1)d$
$\Rightarrow \quad 96 = 12 + (n - 1)4$
$\Rightarrow \quad 96 - 12 = (n - 1)4$
$\Rightarrow \quad 84 = (n - 1)4$
$\Rightarrow \quad (n - 1)4 = 84$
$\Rightarrow \quad n - 1 = \dfrac{84}{4}$
$\Rightarrow \quad n - 1 = 21$

ARITHMETIC PROGRESSIONS

$\Rightarrow \quad n = 21 + 1$

$\Rightarrow \quad n = 22$

$\therefore \quad S_n = \dfrac{n}{2}(a + l) = \dfrac{22}{2}(12 + 96)$

$\qquad = (11)(108)$

$\qquad = 1188.$

Example 22. *Find the sum of all natural numbers 100 and 200 which are divisible by 4.* (CBSE 2006)

Sol. All natural numbers between 100 and 200 which are divisible by 4 are

$\qquad 104, 108, 112, 116, \ldots\ldots, 196$

Here, $a_1 = 104$

$\qquad a_2 = 108$

$\qquad a_3 = 112$

$\qquad a_4 = 116$

$\qquad \vdots \quad \vdots$

$\therefore \quad a_2 - a_1 = 108 - 104 = 4$

$\qquad a_3 - a_2 = 112 - 108 = 4$

$\qquad a_4 - a_3 = 116 - 112 = 4$

$\qquad \vdots \quad \vdots \quad \vdots$

$\because \quad a_2 - a_1 = a_3 - a_2 = a_4 - a_3 = \ldots\ldots$

$\qquad\qquad\qquad\qquad\qquad (= 4 \text{ each})$

$\therefore$ This sequence is an arithmetic progression whose common difference is 4.

Here, $a = 104$

$\qquad d = 4$

$\qquad l = 196$

Let the number of terms be n. Then

$\qquad l = a + (n - 1)d$

$\Rightarrow \quad 196 = 104 + (n - 1)4$

$\Rightarrow \quad 196 - 104 = (n - 1)4$

$\Rightarrow \qquad\quad 92 = (n - 1)4$

$\Rightarrow \qquad (n - 1)4 = 92$

$\Rightarrow \qquad n - 1 = \dfrac{92}{4}$

$\Rightarrow \qquad n - 1 = 23$

$\Rightarrow \qquad n = 23 + 1$

$\Rightarrow \qquad n = 24$

$\therefore \quad S_n = \dfrac{n}{2}(a + l)$

$\qquad = \left(\dfrac{24}{2}\right)(104 + 196)$

$\qquad = (12)(300)$

$\qquad = 3600.$

Example 23. *Find the sum of all natural numbers less than 200 which are divisible by 5.* (CBSE 2006)

Sol. All natural numbers less than 200 which are divisible by 5 are

$\qquad 5, 10, 15, 20, \ldots\ldots, 195$

Here, $a_1 = 5$

$\qquad a_2 = 10$

$\qquad a_3 = 15$

$\qquad a_4 = 20$

$\qquad \vdots \quad \vdots$

$\therefore \quad a_2 - a_1 = 10 - 5 = 5$

$\qquad a_3 - a_2 = 15 - 10 = 5$

$\qquad a_4 - a_3 = 20 - 15 = 5$

$\qquad \vdots \quad \vdots$

$\because \quad a_2 - a_1 = a_3 - a_2 = a_4 - a_3 = \ldots\ldots$

$\qquad\qquad\qquad\qquad\qquad (= 5 \text{ each})$

$\therefore$ This sequence is an arithmetic progression whose common difference is 5.

Here, $a = 5$

$\qquad d = 5$

$\qquad l = 195$

Let the numbers of terms be n. Then,

$\qquad l = a + (n - 1)d$

$\Rightarrow \quad 195 = 5 + (n - 1)5$

$\Rightarrow \quad 195 - 5 = (n - 1)5$

$\Rightarrow \qquad 190 = (n - 1)5$

$\Rightarrow \quad (n - 1)5 = 190$

$\Rightarrow \qquad n - 1 = \dfrac{190}{5}$

$\Rightarrow \qquad n - 1 = 38$

$\Rightarrow \qquad n = 38 + 1$

$\Rightarrow \qquad n = 39$

$\therefore \quad S_n = \dfrac{n}{2}(a + l)$

$\qquad = \left(\dfrac{39}{2}\right)(5 + 195)$

$$= \left(\frac{39}{2}\right)(200)$$
$$= (39)(100)$$
$$= 3900.$$

Example 24. *Find the sum of all the natural numbers less than 100 which are divisible by 6.* (AI CBSE 2006)

Sol. All the natural numbers less than 100 which are divisible by 6 are

$$6, 12, 18, 24, \ldots, 96$$

Here, $a_1 = 6$

$a_2 = 12$

$a_3 = 18$

$a_4 = 24$

$\vdots \quad \vdots$

$\therefore \quad a_2 - a_1 = 12 - 6 = 6$

$a_3 - a_2 = 18 - 12 = 6$

$a_4 - a_3 = 24 - 18 = 6$

$\vdots \quad \vdots \quad \vdots$

$\because \quad a_2 - a_1 = a_3 - a_2 = a_4 - a_3 = 6 \ldots$

(= 6 each)

$\therefore$ This sequence is an arithmetic progression whose difference is 6.

Here, $a = 6$

$d = 6$

$l = 96$

Let the number of terms be n. Then,

$l = a + (n - 1)d$

$\Rightarrow \quad 96 = 6 + (n - 1)6$

$\Rightarrow \quad 96 - 6 = (n - 1)6$

$\Rightarrow \quad 90 = (n - 1)6$

$\Rightarrow \quad (n - 1)6 = 90$

$\Rightarrow \quad n - 1 = \dfrac{90}{6}$

$\Rightarrow \quad n - 1 = 15$

$\Rightarrow \quad n = 15 + 1$

$\Rightarrow \quad n = 16$

$\therefore \quad S_n = \dfrac{n}{2}(a + l)$

$$= \left(\frac{16}{2}\right)(6 + 96)$$

$$= (8)(102)$$
$$= 816.$$

Example 25. *Find the sum of all natural numbers between 100 and 500 which are divisible by 8.* (AI CBSE 2006)

Sol. All the numbers between 100 and 500 which are divisible by 8 are

$$104, 112, 120, 128, \ldots, 496$$

Here, $a_1 = 104$

$a_2 = 112$

$a_3 = 120$

$a_4 = 128$

$\vdots \quad \vdots$

$\therefore \quad a_2 - a_1 = 112 - 104 = 8$

$a_3 - a_2 = 120 - 112 = 8$

$a_4 - a_3 = 128 - 120 = 8$

$\vdots \quad \vdots$

$\because \quad a_2 - a_1 = a_3 - a_2 = a_4 - a_3 = \ldots$

(= 8 each)

$\therefore$ This sequence is an arithmetic progression whose common difference is 8.

Here, $a = 104$

$d = 8$

$l = 496$

Let the number of terms be n. Then,

$l = a + (n - 1)d$

$\Rightarrow \quad 496 = 104 + (n - 1)8$

$\Rightarrow \quad 496 - 104 = (n - 1)8$

$\Rightarrow \quad 392 = (n - 1)8$

$\Rightarrow \quad (n - 1)8 = 392$

$\Rightarrow \quad n - 1 = \dfrac{392}{8}$

$\Rightarrow \quad n - 1 = 49$

$\Rightarrow \quad n = 49 + 1$

$\Rightarrow \quad n = 50$

$\therefore \quad S_n = \dfrac{n}{2}(a + l)$

$$= \left(\frac{50}{2}\right)(104 + 496)$$

ARITHMETIC PROGRESSIONS

$= (25)(600)$
$= 15000.$

Hence, the sum of all the natural numbers between 100 and 500 which are divisible by 8 is 15000.

Example 26. *Find the sum of all the natural numbers between 200 and 300 which are divisible by 4.* (AI CBSE 2006)

Sol. All the natural numbers between 200 and 300 which are divisible by 4 are
$$204, 208, 212, 216, \ldots, 196$$

Here, $a_1 = 204$
$a_2 = 208$
$a_3 = 212$
$a_4 = 216$
$\vdots \quad \vdots$

$\therefore \quad a_2 - a_1 = 208 - 204 = 4$
$a_3 - a_2 = 212 - 208 = 4$
$a_4 - a_3 = 216 - 212 = 4$
$\vdots \quad \vdots \quad \vdots$

$\because \quad a_2 - a_1 = a_3 - a_2 = a_4 - a_3 = \ldots$
(= 4 each)

$\therefore$ This sequence is an arithmetic progression whose common difference is 4.

Here, $a = 204$
$d = 4$
$l = 296.$

Let the number of terms be n. Then,
$l = a + (n-1)d$
$\Rightarrow \quad 296 = 204 + (n-1)4$
$\Rightarrow \quad 296 - 204 = (n-1)4$
$\Rightarrow \quad 92 = (n-1)4$
$\Rightarrow \quad (n-1)4 = 92$
$\Rightarrow \quad n - 1 = \dfrac{92}{4}$
$\Rightarrow \quad n - 1 = 23$
$\Rightarrow \quad n = 23 + 1$
$\Rightarrow \quad n = 24$

$\therefore \quad S_n = \dfrac{n}{2}(a + l)$

$= \left(\dfrac{24}{2}\right)(204 + 296)$

$= (12)(500)$
$= 6000$

Hence, the sum of all the natural numbers between 200 and 300 which are divisible by 4 is 6000.

Example 27. *Find the number of terms of the AP 54, 51, 48, ..., so that their sum is 513.* (CBSE 2005)

Sol. The given AP is 54, 51, 48, ...
Here, $a = 54$
$d = 51 - 54 = -3$

Let the sum of n terms of this AP be 513.

We know that
$$S_n = \dfrac{n}{2}[2a + (n-1)d]$$

$\Rightarrow \quad 513 = \dfrac{n}{2}[2(54) + (n-1)(-3)]$

$\Rightarrow \quad 513 = \dfrac{n}{2}[108 - 3n + 3]$

$\Rightarrow \quad 513 = \dfrac{n}{2}[111 - 3n]$

$\Rightarrow \quad 1026 = n[111 - 3n]$
$\Rightarrow \quad 1026 = 111n - 3n^2$
$\Rightarrow \quad 3n^2 - 111n + 1026 = 0$
$\Rightarrow \quad 3n^2 - 37n + 342 = 0$
| Dividing throughout by 3
$\Rightarrow \quad n^2 - 18n - 19n + 342 = 0$
$\Rightarrow \quad n(n - 18) - 19(n - 18) = 0$
$\Rightarrow \quad (n - 18)(n - 19) = 0$
$\Rightarrow \quad n - 18 = 0 \quad \text{or} \quad n - 19 = 0$
$\Rightarrow \quad n = 18 \quad \text{or} \quad n = 19$
$\Rightarrow \quad n = 18, 19$

Hence, the sum of 18 terms or 19 terms of the given AP is 513.

Note. Actually 19th term
$= a_{19}$
$= a + (19 - 1)d$
$\quad | \because a_n = a + (n-1)d$
$= a + 18d$
$= 54 + 18(-3)$
$= 54 - 54$
$= 0.$

Example 28. *If the nth term of an AP is $(2n + 1)$, find the sum of first n terms of the AP.* (CBSE 2005)

Sol. Here,
$$a_n = 2n + 1$$
Put $n = 1, 2$. Then,
$$a_1 = 2(1) + 1 = 3$$
$$a_2 = 2(2) + 1 = 5$$
$$\therefore \quad a = 3$$
$$d = a_2 - a_1 = 5 - 3 = 2$$
$\therefore$ Sum of first n-terms of the AP
$$= S_n$$
$$= \frac{n}{2}[2a + (n-1)d]$$
$$= \frac{n}{2}[2(3) + (n-1)2]$$
$$= n(3 + n - 1)$$
$$= n(n + 2).$$

Example 29. *Find the AP whose sum to n terms is $2n^2 + 2$.*

Sol. Here, $S_n = 2n^2 + 2$ | Given

Put $n = 1, 2, 3, 4, ...$, in succession, we get
$$S_1 = 2(1)^2 + 1 = 2 + 1 = 3$$
$$S_2 = 2(2)^2 + 2 = 8 + 2 = 10$$
$$S_3 = 2(3)^2 + 3 = 18 + 3 = 21$$
$$S_4 = 2(4)^2 + 4 = 32 + 4 = 36$$
and so on.
$$\therefore \quad a_1 = S_1 = 3$$
$$a_2 = S_2 - S_1 = 10 - 3 = 7$$
$$a_3 = S_3 - S_2 = 21 - 10 = 11$$
$$a_4 = S_4 - S_3 = 36 - 21 = 15$$
and so on.

Hence, the required AP is
$$3, 7, 11, 15, ...$$

Example 30. *If the ratio of the sums of n terms of two AP's is $n + 7 ; 3n + 1$, then find the ratio the 7th terms of the series.*

Sol. Let a_1, d_1 and a_2, d_2 be the first terms and the common difference of the first and second AP's respectively.

Then, according to the question,

$$\frac{\text{Sum of } n \text{ terms of the first AP}}{\text{Sum of } n \text{ terms of the second AP}}$$
$$= \frac{n+7}{3n+1}$$

$$\Rightarrow \frac{\frac{n}{2}[2a_1 + (n-1)d_1]}{\frac{n}{2}[2a_2 + (n-1)d_2]} = \frac{n+7}{3n+1}$$

$$\Rightarrow \frac{2a_1 + (n-1)d_1}{2a_2 + (n-1)d_2} = \frac{n+7}{3n+1}$$

$$\Rightarrow \frac{a_1 + (n-1)\frac{d_1}{2}}{a_2 + (n-1)\frac{d_2}{2}} = \frac{n+7}{3n+1}$$

Put $n = 13$ in both sides, we get

$$\frac{a_1 + (13-1)\frac{d_1}{2}}{a_2 + (13-1)\frac{d_2}{2}} = \frac{13+7}{3(13)+1}$$

$$\Rightarrow \frac{a_1 + 6d_1}{a_2 + 6d_2} = \frac{20}{40}$$

$$\Rightarrow \frac{a_1 + 6d_1}{a_2 + 6d_2} = \frac{1}{2}$$

$$\Rightarrow \frac{a_1 + (7-1)d_1}{a_2 + (7-1)d_2} = \frac{1}{2}$$

$$\Rightarrow \frac{\text{7th term of the first AP}}{\text{7th term of the second AP}} = \frac{1}{2}$$

Hence, the required ratio is 1 : 2.

TEST YOUR KNOWLEDGE

1. Find the sum of the first 22 terms of the AP : 8, 3, – 2, ...

2. If the sum of the first 14 terms of an AP is 1050 and its first term is 10, find the 20th term.

3. How many terms of the AP : 24, 21, 18, ... must be taken so that their sum is 78 ?
4. Find the sum of
   (i) the first 1000 positive integers
   (ii) the first n positive integers.
5. Find the sum of first 24 terms of the list of numbers whose nth term is given by
   $a_n = 3 + 2n$.
6. A manufacturer of TV sets produced 600 sets in the third year and 700 sets in the seventh year. Assuming that the production increases uniformly by a fixed number every year, find
   (i) the production in the first year.
   (ii) the production in the 10th year.
   (iii) the total production in first 7 years.
7. Find the number of terms of the AP : 64, 60, 56, ... so that their sum is 544. *(CBSE 2005)*
8. Find the number of terms of the AP : 63, 60, 57, ... so that their sum is 693. *(CBSE 2005)*
9. Find the sum of all multiples of a lying between 300 and 700. *(AI CBSE 2005)*
10. Find the sum of all multiples of 7, lying between 500 and 800. *(AI CBSE 2005)*
11. Find the sum of the first 25 terms of an AP whose nth term is given by $t_n = 2 - 3n$. *(CBSE 2004)*
12. Find the sum of the first 25 terms of an AP whose nth term is given by $t_n = 7 - 3n$. *(AI CBSE 2004)*
13. Find the sum of the first 100 natural numbers.
14. Find the sum of the following arithmetic progression :
    $1 + 3 + 7 + ... + 199$.
15. How many terms of the reries $24 + 20 + 16 + ...$ give the sum 72 ? Give reason for the two answers.

Answers

1. -979
2. 200
3. 4 or 13
4. (i) 500 500 (ii) $\dfrac{n(n+1)}{2}$
5. 672
6. (i) 550 (ii) 775 (iii) 4375
7. 16, 17
8. 21, 22
9. 21978
10. 27993
11. -925
12. -800
13. 5050
14. 10000
15. 4, 9; sum of 5th for term to 9th term is zero.

ILLUSTRATIVE EXAMPLES

[Exercise 5.4 (Optional)]*
(Page No. 115)

Example 1. *Which term of the AP : 121, 117, 113, ..., is the first negative term ?*

[Hint. Find n for $a_n < 0$]

Sol. The given AP is
121, 117, 113, ...
Here, $a = 121$
$d = 117 - 121 = -4$

Let the nth term of the AP be the first negative term. Then,
$a_n < 0$
$\Rightarrow a + (n-1)d < 0$
$\Rightarrow 121 + (n-1)(-4) < 0$
$\Rightarrow 121 < (n-1)4$
$\Rightarrow (n-1)4 > 121$
$\Rightarrow (n-1) > \dfrac{121}{4}$
$\Rightarrow n > \dfrac{121}{4} + 1$
$\Rightarrow n > \dfrac{125}{4}$
$\Rightarrow n > 31\dfrac{1}{4}$.

Least integral value of $n = 32$. Hence, 32nd term of the given AP is the first negative term.9

---

*These exercises are not from the examination point of view.

Example 2. *The sum of the third and the seventh terms of an AP is 6 and their product is 8. Find the sum of first sixteen terms of the AP.*

Sol. Let the first term and the common difference of the AP be a and d respectively.

According to the question,

Third term + seventh term = 6

$\Rightarrow \quad [a + (3-1)d] + [a + (7-1)d] = 6$

$\qquad |\because a_n = a + (n-1)d$

$\Rightarrow \quad (a + 2d) + (a + 6d) = 6$

$\Rightarrow \quad 2a + 8d = 6$

$\Rightarrow \quad a + 4d = 3 \qquad ...(1)$

$\qquad |\because$ Dividing throughout by 2

and (third term)(seventh term) = 8

$\Rightarrow \quad (a + 2d)(a + 6d) = 8$

$\Rightarrow \quad (a + 4d - 2d)(a + 4d + 2d) = 8$

$\Rightarrow \quad (3 - 2d)(3 + 2d) = 8$

$\qquad |$ Using (1)

$\Rightarrow \quad a - 4d^2 = 8$

$\Rightarrow \quad 4d^2 = 9 - 8$

$\Rightarrow \quad 4d^2 = 1$

$\Rightarrow \quad d^2 = \frac{1}{4}$

$\Rightarrow \quad d = \pm \frac{1}{2}$

Case I. When $d = \frac{1}{2}$

Then from (1),

$a + 4\left(\frac{1}{2}\right) = 3$

$\Rightarrow \quad a + 2 = 3$

$\Rightarrow \quad a = 3 - 2$

$\Rightarrow \quad a = 1$

$\therefore$ Sum of first sixteen terms of the AP

$= S_{16}$

$= \frac{16}{2}[2a + (16-1)d]$

$\qquad |\because S_n = \frac{n}{2}[2a + (n-1)d]$

$= 8[2a + 15d]$

$= 8\left[2(1) + (15)\left(\frac{1}{2}\right)\right]$

$= 8\left[2 + \frac{15}{2}\right]$

$= 8\left[\frac{9}{2}\right]$

$= 76$

Case II. When $d = -\frac{1}{2}$

Then from (1),

$a + 4\left(-\frac{1}{2}\right) = 3$

$\Rightarrow \quad a - 2 = 3$

$\Rightarrow \quad a = 3 + 2$

$\Rightarrow \quad a = 5$

$\therefore$ Sum of first sixteen terms of the AP

$= S_{16}$

$= \frac{16}{2}[2a + (16-1)d]$

$\qquad |\because S_n = \frac{n}{2}[2a + (n-1)d]$

$= 8[2a + 15d]$

$= 8\left[2(5) + 15\left(-\frac{1}{2}\right)\right]$

$= 8\left[10 - \frac{15}{2}\right]$

$= 8\left[\frac{5}{2}\right] = 20.$

Example 3. *A ladder has rungs 25 cm apart (see figure). The rungs decrease uniformly in length from 45 cm at the bottom to 25 cm at the top. If the top and bottom rungs* are $2\frac{1}{2}$ m apart, what is the length of the wood required for the rungs?

[**Hint.** Number of rungs = $\frac{250}{25}$]

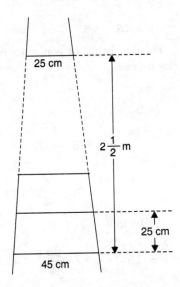

Sol. Number of rungs $(n) = \frac{250}{25}$

$= 10$

$\left| 2\frac{1}{2} \text{ m} = 250 \text{ cm} \right.$

Hence, there are 10 rungs.
The length of the wood required for the rungs (S_{10}).

$S_{11} = \frac{10}{2}(45 + 25) = 350$ cm.

$\left| \because S_n = \frac{n}{2}(a + l) \right.$

Example 4. *The houses of a row are numbered consecutively from 1 to 49. Show that there is a value of x such that the sum of the numbers of the houses preceding the house numbered x is equal to the sum of the numbers of the following it. Find this value of x.*

[**Hint.** $S_{x-1} = S_{49} - S_x$]

Sol. The consecutive numbers on the houses of a row are 1, 2, 3, ..., 49
Clearly this list of number forming an AP.

Here, $a = 1$
$d = 2 - 1 = 1$
According to the question,
$$S_{x-1} = S_{49} - S_x$$

$\Rightarrow \frac{x-1}{2}[2a + (x-1-1)d]$

$= \frac{49}{2}[2a + (49-1)d]$

$- \frac{x}{2}[2a + (x-1)d]$

$\left| \because S_n = \frac{n}{2}[2a + (n-1)d] \right.$

$\Rightarrow \frac{x-1}{2}[2(1) + (x-2)(1)]$

$= \frac{49}{2}[2(1) + (48)(1)]$

$- \frac{x}{2}[2(1) + (x-1)(1)]$

$\Rightarrow \frac{x-1}{2}[x] = 1225 - \frac{x(x+1)}{2}$

$\Rightarrow \frac{(x-1)(x)}{2} + \frac{x(x+1)}{2} = 1225$

$\Rightarrow \frac{x}{2}(x - 1 + x + 1) = 1225$

$\Rightarrow x^2 = 1225$

$\Rightarrow x = \sqrt{1225}$

$\Rightarrow x = 35$

Hence, the required values of x is 35.

Example 5. *A small terrace at a football ground comprises of 15 steps each of which is 50m long and built of solid concrete. Each step has a rise of $\frac{1}{4}$ m and a tread of $\frac{1}{2}$ m (see figure). Calculate total volume of concrete required to build the terrace.*

[**Hint.** Volume of concrete required to build the first step = $\frac{1}{4} \times \frac{1}{2} \times 50 \, m^3$]

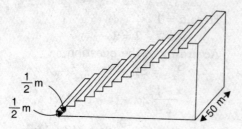

Sol. Volume of concrete required to build the first step = $\frac{1}{4} \times \frac{1}{2} \times 50 \, m^3 = \frac{25}{4} \, m^3$.

Volume of concrete required to build the second step = $\left(\frac{1}{4} + \frac{1}{4}\right) \times \frac{1}{2} \times 50 \, m^3 = \frac{25}{2} \, m^3$

Volume of concrete required to build the third step = $\left(\frac{1}{4} + \frac{1}{4} + \frac{1}{4}\right) \times \frac{1}{2} \times 50 \, m^3 = \frac{75}{4} \, m^3$
and so on.

Thus, the volumes (in m^3) of concrete required to build the various steps are

$$\frac{25}{4}, \frac{25}{2}, \frac{75}{4}, \ldots$$

Clearly this list of numbers forms an AP.

Here, $a = \frac{25}{4}$

$d = \frac{25}{2} - \frac{25}{4} = \frac{25}{4}$

$n = 15$

∴ Total volume of concrete required to build the terrace

$= S_n$
$= S_{15}$
$= \frac{15}{2}[2a + (15-1)d]$

$\left| \because S_n = \frac{n}{2}[2a + (n-1)d] \right.$

$= 15(a + 7d)$

$= (15)\left(\frac{25}{4} + 7 \times \frac{25}{4}\right)$

$= (15)(50)$

$= 750 \, m^3$.

MISCELLANEOUS EXERCISE

1. Find the sum of all two digit and positive numbers. *(AI CBSE 2005)*
2. Which term of the arithmetic progression 3, 10, 17, ... will be 84 more than its 13th term ? *(AI CBSE 2004)*
3. Which term of the AP. 3, 15, 27, 39, ... is 132 more than its 54th term?
4. Find the sum of the following arithmetic progression:
   $1 + 3 + 5 + 7 + \ldots + 199$.
5. Find the Arithmetic progression whose third term is 16 and the seventh term exceeds its fifth term by 12.
6. The pth term of an AP is $\frac{1}{q}$ and qth term is $\frac{1}{p}$. Prove that the sum of its pq term is $\frac{1}{2}(pq+1)$.
7. The sum of p terms of an AP is q and sum of q terms is p. Then prove that the sum of $(p+q)$ terms is $-(p+q)$.
8. If the sum of n terms of an AP is S_n and if $S_{2n} = 3S_n$, then prove that
   $$S_{3n} = 6S_n.$$
9. Find the number of terms common to two AP' is
   3, 7, 11, ..., 407
   and   2, 9, 16, ..., 709.
10. If the sum of n terms of an AP is $3n^2 + 3n$, find the rth term of the series.
11. If three consecutive terms of an AP are $3x$, $x+2$ and 8, then find the value of x. Also, find its 4th term.
12. How many terms of the series 17 + 15 + 13 + ... give the sum 727 ? Give reason for two answers.

13. In an arithmetic progression, $S_n = n^2 p$ and $S_n = m^2 p$ and $m \neq n$. Prove that $S_p = p^3$.
14. Find the number of numbers of two digits which are divisible by 5.
15. Show that the sequence $\log a$, $\log (ab)$, $\log (ab^2)$, $\log (ab^3)$, ... is an AP. Find its nth term.
16. Is 310 a term of the AP
    3, 8, 13, 18, ... ?
17. The last term of the sequence 27, 24, 21, 18, ... is 0. Find the number of terms.
18. Write the sequence with nth term:
    (i) $3 + 4n$ (ii) $5 + 2n$
    (iii) $6 - n$ (iv) $9 - 5n$.
    Show that all of the above sequences form an AP.
    Find the sum of the first 15 terms of each.
19. If 7 times the 7th term of an AP is equal to 11 times is 11th term, show that the 18th term of the AP is zero.
20. An arithmetic sequence is 1, 3, 5, 7, 9, ... to $(n + 1)$ terms. Show that the sum of $1 + 5 + 9 + ...$ bears the ratio $(n + 1) : n$ to the sum of $3 + 7 + 11 + ...$
21. If S_n denotes the sum of n terms of an AP and if $S_1 = 6$, $S_7 = 105$, then show that $S_n : S_{(n-3)} = (n + 3) : (n - 3)$.
22. The ratio of the sum of n terms of two AP is $(7n + 1) : (4n + 27)$. Find the ratio of their nth terms.
23. The sums of n terms of three arithmetic progressions are S_1, S_2 and S_3. The first term of each is unity and the common differences are 1, 2 and 3 respectively. Prove that $S_1 + S_3 = 2S_2$.
24. Two AP's have the same common difference. The first term of one of there is 3, and that of the other is 8. What is the difference between their (i) 2nd terms? (ii) 4th terms? (iii) 10th term? (iv) 30th terms?
25. A man saves Rs 32 during the first year, Rs 36 in the next year and Rs 40 in the third year. If he continues his saving in this sequence, in how many years will he save Rs 2000?

Answers

1. 2475   2. 25th term
3. 65th term   4. 10,000
5. 4, 10, 16, 22, ... 9. 14
10. 62   11. 10, 18
12. 6, 12 ; The sum of terms from 7th to 12th is zero.
14. 18   15. $\log (ab^{n-1})$
16. No   17. 10
18. (i) 525 (ii) 315 (iii) -30 (iv) -465
22. $\dfrac{14m - 6}{8m + 23}$
24. (i) 5 (ii) 5 (iii) 5 (iv) 5
25. 25 years.

SUMMARY

1. An **arithmetic progression** (AP) is a list of numbers in which each term is obtained by adding a fixed number d to the preceding term, except the first term. The fixed number d is called the **common difference**.
   The general form of an AP is a, $a + d$, $a + 2d$, $a + 3d$, ...
2. A given list of numbers $a_1, a_2, a_3, ...$ is an AP, if the differences $a_2 - a_1, a_3 - a_2, a_4 - a_3$, ..., give the same value, i.e., if $a_{k+1} - a_k$ is the same for different values of k.
3. In an AP with first term a and common difference d, the nth term (or the general term) is given by $a_n = a + (n - 1)d$
4. The sum of the first n terms of AP is given by
   $$S = \frac{n}{2}[2a + (n - 1)d]$$
5. If l is the last term of the finite AP, say the nth term, then the sum of all terms of the AP is given by
   $$S = \frac{n}{2}(n + l).$$

6

Triangles

IMPORTANT POINTS

1. Congruent Figures. Two figures (in particular, two triangles) are said to be congruent, if they have the same shape and the same size.

2. Similar Figures. Two figures (in particular, two triangles) having the same shape (and not necessarily the same size) are called similar figures.

3. Need of Similarity. The heights of mountains (say Mount Everest) or distances of some long distant objects (say moon) can not be measured directly with the help of a measuring tape. In fact, all these heights and distances are found out using the idea of indirect measurements, which is based on the principle of similarity of figures.

4. Similar Figures. We know that all circles with the same radii are congruent, all squares with the same side lengths are congruent and all equilateral triangles with the same side lengths are congruent.

Now consider any two (or more) circles as shown below:

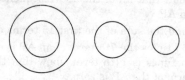

Since all of them do not have the same radius. They are not congruent to each other. Note that some are congruent and some are not, but all of them have the same shape. So they all are, what we call, *similar*.

Two similar figures have the same shape but not necessarily the same size. Therefore, all circles are similar.

As observed in the case of circles, all squares are similar and all equilateral triangles are similar as shown below:

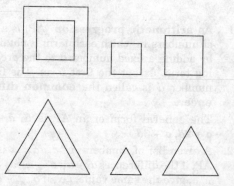

From the above, we can say *that all congruent figures are similar but the similar figures need not be congruent.*

TRIANGLES

Not similar. A circle and a square cannot be similar. Similarly, a triangle and a square cannot be similar. This is evident just by looking at the figures.

5. Need of definition of similarity of figures. Let us consider the two quadrilaterals ABCD and PQRS shown below:

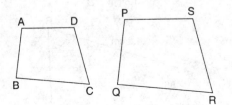

These figures appear to be similar but we cannot be certain about it. Therefore, we must have some definition of similarity of figures and based on this definition some rules to decide whether the two given figures are similar or not.

6. Essence of the similarity of two figures.

(*i*) Let us look at the photographs given below:

We will at once say that they are the photographs of the same monument (Taj Mahal) but are in different sizes. We would say, therefore, that the three photographs are similar.

(*ii*) Again, look at the two photographs of the same size of the same person one at the age of 10 years and the other at the age of 40 years. These photographs are of the same size but certainly they are not of the same shape. So, they are not similar.

(*iii*) Let us see what a photographer does when she prints photographs of different sizes from the same negative. We have heard about the stamp size, passport size and postcard size photographs.

She generally takes a photograph on a small size film, say of 35 mm size and then enlarges it into a bigger size, say 45 mm (or 55 mm). Thus, if we consider any line segment in the smaller photograph (figure), its corresponding line segment in the bigger photograph (figure) will be $\frac{45}{35}\left(\text{or } \frac{55}{35}\right)$ of that of the line segment. This really means that every line segment of the smaller photograph is enlarged (increased) *in the ratio* 35 : 45 (or 35 : 55). It can also be said that every line segment of the bigger photograph is reduced (decreased) in the ratio 45 : 35 (or 55 : 35). Further, if we consider inclinations (or angles) between any pair of corresponding line segments in the two photographs of different sizes, we shall see that these inclinations (or angles) *are always equal*.

This is the essence of the similarity of two figures and in particular of two polygons. We say that:

Two polygons of the same number of sides are similar, if (i) their corresponding angles are equal and (ii) their corresponding sides are in the same ratio (or proportion).

Note that, the same ratio of the corresponding sides is referred to as *the scale factor* (or the *Representative Fraction*) for the polygons. We have heard that world maps (*i.e.*, global maps) and blue prints for the construction of a building are prepared using a suitable scale factor and observing certain conventions.

7. An Activity. In order to understand similarity of figures more clearly, let us perform the following activity:

Activity 1. Place a lighted bulb at a point O on the ceiling and directly below it a table in your classroom. Let us cut a polygon, say a quadrilateral ABCD, from a plane cardboard and place this cardboard parallel to the ground between the lighted bulb and the table. Then a shadow of ABCD is cast on the table. Mark the outline of this shadow as A'B'C'D' shown below:

Note that the quadrilateral A'B'C'D' is an enlargement (or magnification) of the quadrilateral ABCD. This is because of the property of light that light propagates in a straight line. We may also note that A' lies on ray OA, B' lies on ray OB, C' lies on OC and D' lies on OD. Thus, quadrilaterals A'B'C'D' and ABCD are of the same shape but of different sizes.

So, quadrilateral A'B'C'D' is similar to quadrilateral ABCD. We can also say that quadrilateral ABCD is similar to the quadrilateral A'B'C'D'.

Note. Here, we can also note that vertex A' corresponds to vertex A, vertex B' corresponds to vertex B, vertex C' corresponds to vertex C and vertex D' corresponds to vertex D. Symbolically, these correspondences are represented as A' ↔ A, B' ↔ B, C' ↔ C and D' ↔ D. By actually measuring the angles and the sides of the two quadrilaterals, we may verify that

(i) $\angle A = \angle A'$, $\angle B = \angle B'$,
$\angle C = \angle C'$, $\angle D = \angle D'$ and

(ii) $\dfrac{AB}{A'B'} = \dfrac{BC}{B'C'} = \dfrac{CD}{C'D'} = \dfrac{DA}{D'A'}$.

Inference. This again emphasizes that *two polygons of the same number of sides are similar, if (i) all the corresponding angles are equal and (ii) all the corresponding sides are in the same ratio (or proportion).*

8. Certain Illustrations

1. From the above inference, we can easily say that quadrilaterals ABCD and PQRS shown below are similar.

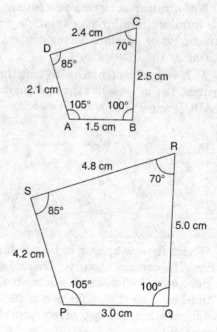

Remark. We can verify that if one polygon is similar to another polygon and this second polygon is similar to a third polygon, then the first polygon is similar to the third polygon.

2. We may note that in the two quadrilaterals (a square and a rectangle) shown below, corresponding angles are equal, but their corresponding sides are not in the same ratio.

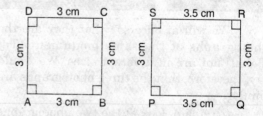

So, the two quadrilaterals are not similar.

3. Similarly, we may note that in the two quadrilaterals (a square and a rhombus) shown below, corresponding sides are in the same ratio, but their corresponding angles are not equal. So, again, the two polygons (quadrilaterals) are not similar.

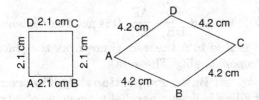

An important inference. Based on the above illustration, we may infer that either of the above two conditions (i) and (ii) of similar of two polygons is not sufficient for them to be similar.

ILLUSTRATIVE EXAMPLES

[NCERT Exercise 6.1]
(Page No. 122)

Example 1. *Fill in the blanks using the correct word given in brackets:*

(i) *All circles are ......* (congruent, similar.)

(ii) *All squares are ......* (similar, congruent).

(iii) *All ...... triangles are similar.* (isosceles, equilateral)

(iv) *Two polygons of the same number of sides are similar, if (a) their corresponding angles are ...... and (b) their corresponding sides are ......* (equal, proportional).

Sol. (i) All circles are **similar**.

(ii) All squares are **similar**.

(iii) All **equilateral** triangles are similar.

(iv) Two polygon of the same number of sides are similar, if (a) their corresponding angles are **equal** and (b) their corresponding sides are **proportional**.

Example 2. *Give two different examples of pair of*

(i) *similar figures*

(ii) *non-similar figures.*

Sol.

(i) (a) Any two circles

(b) Any two squares

(ii) (a) An equilateral triangle and a scalene triangle

(b) An equilateral triangle and a right-angled triangle.

Example 3. *State whether the following quadrilaterals are similar or not:*

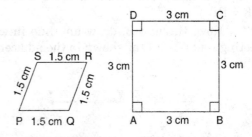

Sol. No, because the corresponding angles are not equal.

SIMILARITY OF TRIANGLES

IMPORTANT POINTS

1. Similarity of two triangles. We know that triangle is also a polygon. So, we can state the same conditions for the similarity of two triangles. That is:

Two triangles are similar, if

(i) their corresponding angles are equal and

(ii) their corresponding sides are in the same ratio (or proportion).

Note. If corresponding angles of two triangles are equal, then they are known as equiangular triangles.

2. Truth relating to two equiangular triangles. A famous Greek mathematician Thales gave an important truth relating to two equiangular triangles which is as follows:

The ratio of any two corresponding sides in two equiangular triangles is always the same.

Thales
(640–546 B.C.)

It is believed that he had used a result called the *Basic Proportionality Theorem* (now known as the *Thales Theorem*) for the same.

3. An Activity. To understand the Basic Proportionality Theorem, we perform the following activity:

Activity. Draw any angle XAY and on its one arm AX, mark points (say five points) P, Q, D, R and B such that AP = PQ = QD = DR = RB.

Now, through B, draw any line intersecting arm AY at C as shown in the adjacent figure.

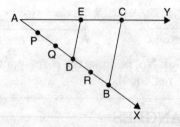

Also, through the point D, draw a line parallel to BC to intersect AC at E. We observe from our constructions that $\frac{AD}{DB} = \frac{3}{2}$.

Measure AE and EC. We observe that $\frac{AE}{EC}$ is also equal to $\frac{3}{2}$. Thus, we can see that in $\triangle ABC$, DE ∥ BC and $\frac{AD}{DB} = \frac{AE}{EC}$. It is not a coincidence. It is due to a theorem (known as the Basic Proportionality Theorem).

4. Basic Proportionality Theorem. If a line is drawn parallel to one side of a triangle to intersect the other two sides in distinct points, the other two sides are divided in the same ratio.

Given: A triangle ABC in which a line parallel to side BC intersects the other two sides AB and AC at D and E respectively.

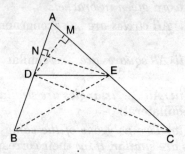

To Prove: $\frac{AD}{DB} = \frac{AE}{EC}$.

Construction: Join BE and CD and then draw DM ⊥ AC and EN ⊥ AB.

Proof: Area of $\triangle ADE = \frac{1}{2}$ base × height

$$= \frac{1}{2} AD \times EN$$

$$\Rightarrow \text{ar (ADE)} = \frac{1}{2} AD \times EN$$

Similarly, ar (BDE) = $\frac{1}{2}$ DB × EN,

ar (ADE) = $\frac{1}{2}$ AE × DM

and ar (DEC) = $\frac{1}{2}$ EC × DM.

Therefore, $\frac{\text{ar(ADE)}}{\text{ar(BDE)}}$

$$= \frac{\frac{1}{2} AD \times EN}{\frac{1}{2} DB \times EN} = \frac{AD}{DB} \qquad ...(1)$$

TRIANGLES

and $\dfrac{\text{ar(ADE)}}{\text{ar(DEC)}} = \dfrac{\frac{1}{2} AE \times DM}{\frac{1}{2} EC \times DM} = \dfrac{AE}{EC}$ ...(2)

Note that $\triangle$BDE and DEC are on the same base DE and between the same parallels BC and DE.

So, ar (BDE) = ar (DEC)   ...(3)

Therefore, from (1), (2) and (3), we have:

$$\dfrac{AD}{DB} = \dfrac{AE}{EC}.$$

5. An Activity. To examine whether the converse of Basic Proportionality Theorem is also true, we perform the following activity:

Activity. Draw an angle XAY on the notebook and on ray AX, mark points B_1, B_2, B_3, B_4 and B such that $AB_1 = B_1B_2 = B_2B_3 = B_3B_4 = B_4B$.

Similarly, on ray AY, mark points C_1, C_2, C_3, C_4 and C such that $AC_1 = C_1C_2 = C_2C_3 = C_3C_4 = C_4C$. Then join B_1C_1 and BC as shown in the given figure.

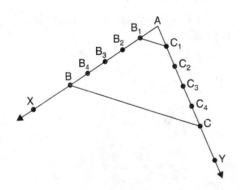

We note that $\dfrac{AB_1}{B_1B} = \dfrac{AC_1}{C_1C}$

(Each equal to $\dfrac{1}{4}$).

We can also see that lines B_1C_1 and BC are parallel to each other, i.e.,

$B_1C_1 \parallel BC$ ...(1)

Similarly, by joining B_2C_2, B_3C_3 and B_4C_4, we can see that:

$\dfrac{AB_2}{B_2B} = \dfrac{AC_2}{C_2C} \left(= \dfrac{2}{3}\right)$ and $B_2C_2 \parallel BC$ ...(2)

$\dfrac{AB_3}{B_3B} = \dfrac{AC_3}{C_3C} \left(= \dfrac{3}{2}\right)$ and $B_3C_3 \parallel BC$ ...(3)

$\dfrac{AB_4}{B_4B} = \dfrac{AC_4}{C_4C} \left(= \dfrac{4}{1}\right)$ and $B_4C_4 \parallel BC$ ...(4)

From (1), (2), (3) and (4), we can observe that if a line divides two sides of a triangle in the same ratio, then the line is parallel to the third side.

We can repeat this activity by drawing any angle XAY of different measure and taking any number of equal parts on arms AX and AY. Each time, we will arrive at the same result. Thus, we obtain the following theorem, which is the converse of Basic Proportionality Theorem.

6. Converse of Basic Proportionality Theorem. If a line divides any two sides of a triangle in the same ratio, then the line is parallel to the third side.

Proof: Take a line DE such that

$$\dfrac{AD}{DB} = \dfrac{AE}{EC} \qquad ....(1)$$

Let us assume that DE is not parallel to BC.

If DE is not parallel to BC, draw a line DE' parallel to BC.

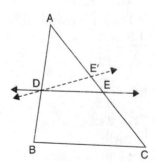

∵ DE' ∥ BC

∴ By Basic Proportionality Theorem,

$$\dfrac{AD}{DB} = \dfrac{AE'}{E'C} \qquad ...(2)$$

From (1) and (2), we get

$$\frac{AE}{EC} = \frac{AE'}{E'C}$$

Adding 1 to both sides of above, we get

$$\frac{AE}{EC} + 1 = \frac{AE'}{E'C} + 1$$

$$\Rightarrow \frac{AE + EC}{EC} = \frac{AE' + E'C}{E'C}$$

$$\Rightarrow \frac{AC}{EC} = \frac{AC}{E'C}$$

$$\Rightarrow EC = E'C$$

$$\Rightarrow \text{E and E' coincide.}$$

$$\therefore DE \parallel BC.$$

ILLUSTRATIVE EXAMPLES

[NCERT Exercise 6.2]
(Page No. 128)

Example 1. *In figure (i) and (ii), DE ∥ BC. Find EC in (i) and AD in (ii).*

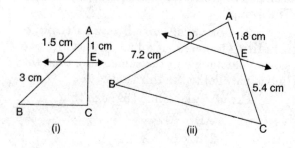

(i) (ii)

Sol. In $\triangle ABC$,

(i) $\because$ DE ∥ BC

$\therefore \quad \dfrac{AD}{DB} = \dfrac{AE}{EC}$

| By Basic Proportionality Theorem

$\Rightarrow \quad \dfrac{1.5}{3} = \dfrac{1}{EC}$

$\Rightarrow \quad EC = \dfrac{3}{1.5}$

$\Rightarrow \quad EC = 2$ cm.

(ii) In $\triangle ABC$,

$\because$ DE ∥ BC

$\therefore \quad \dfrac{AD}{DB} = \dfrac{AE}{EC}$

| By Basic Proportionality Theorem

$\Rightarrow \quad \dfrac{AD}{7.2} = \dfrac{1.8}{5.4}$

$\Rightarrow \quad AD = \dfrac{7.2 \times 1.8}{5.4}$

$\Rightarrow \quad AD = 2.4$ cm.

Example 2. *E and F are points on the sides PQ and PR respectively of a $\triangle PQR$. For each of the following cases, state whether EF ∥ QR :*

(i) PE = 3.9 cm, EQ = 3 cm, PF = 3.6 cm and FR = 2.4 cm.

(ii) PE = 4 cm, QE = 4.5 cm, PF = 8 cm and RF = 9 cm.

(iii) PQ = 1.28 cm, PR = 2.56 cm, PE = 0.18 cm and PF = 0.36 cm.

Sol. (*i*) We have,

$$\dfrac{PE}{EQ} = \dfrac{3.9}{3} = \dfrac{1.3}{1} \qquad \ldots(1)$$

$$\dfrac{PF}{FR} = \dfrac{3.6}{2.4} = \dfrac{3}{2} = \dfrac{1.5}{1} \qquad \ldots(2)$$

From (1) and (2),

$$\dfrac{PE}{EQ} \neq \dfrac{PF}{FR}$$

Therefore, EF is not parallel to QR.

| By converse of Basic Proportionality Theorem

(*ii*) We have,

$$\dfrac{PE}{QE} = \dfrac{4}{4.5} = \dfrac{40}{45} = \dfrac{8}{9} \qquad \ldots(1)$$

TRIANGLES

$$\frac{PF}{RF} = \frac{8}{9} \quad ...(2)$$

From (1) and (2),

$$\frac{PE}{QE} = \frac{PF}{RF}$$

Therefore, EF ∥ QR

| By Converse of Basic Proportionality Theorem

(*iii*) We have,
 PQ = 1.28 cm
 PR = 2.56 cm
 PE = 0.18 cm
 PF = 0.36 cm

∴ EQ = PQ − PE
 = 1.28 − 0.18 = 1.10 cm

and FR = PR − PF
 = 2.56 − 0.36 = 2.20 cm

Now, $\dfrac{PE}{EQ} = \dfrac{0.18}{1.10} = \dfrac{9}{55}$

and $\dfrac{PF}{PR} = \dfrac{0.36}{2.20} = \dfrac{9}{55}$

∴ $\dfrac{PE}{EQ} = \dfrac{PF}{FR}$

∴ EF ∥ QR

| By Converse of Basic Proportionality Theorem

Example 3. *In figure, if LM ∥ CB and LN ∥ CD, prove that*

$$\frac{AM}{AB} = \frac{AN}{AD}.$$

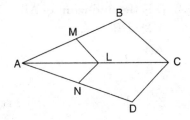

Sol. In △ACB,
∵ LM ∥ CB

∴ $\dfrac{AM}{MB} = \dfrac{AL}{LC} \quad ...(1)$

| By Basic Proportionality Theorem

In △ACD,
∵ LN ∥ CD

∴ $\dfrac{AL}{LC} = \dfrac{AN}{ND} \quad ...(2)$

| By Basic Proportionality Theorem

From (1) and (2), we get

$$\frac{AM}{MB} = \frac{AN}{ND}$$

Example 4. *In figure, DE ∥ AC and DF ∥ AE. Prove that*

$$\frac{BF}{FE} = \frac{BE}{EC}.$$

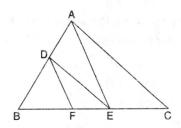

Sol. In △ABE,
∵ DF ∥ AE

∴ $\dfrac{AD}{DB} = \dfrac{FE}{BF} \quad ...(1)$

| By Basic Proportionality Theorem

In △ABC,
∵ DE ∥ AC

∴ $\dfrac{AD}{DB} = \dfrac{EC}{BE} \quad ...(2)$

| By Basic Proportionality Theorem

From (1) and (2),

$$\frac{FE}{BF} = \frac{EC}{BE}$$

⇒ $\dfrac{BF}{FE} = \dfrac{BF}{EC}$ | Taking reciprocals

Example 5. *In figure, DE ∥ OQ and DF ∥ OR. Show that EF ∥ QR.*

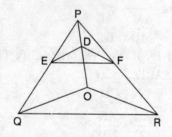

Sol. In $\triangle PQO$,

$\because \quad DE \parallel OQ$

$\therefore \quad \dfrac{PD}{DO} = \dfrac{PE}{EQ}$ ...(1)

| By Basic Proportionality Theorem

In $\triangle PRO$,

$\because \quad DF \parallel OR$

$\therefore \quad \dfrac{PD}{DO} = \dfrac{PF}{FR}$ ...(2)

| By Basic Proportionality Theorem

From (1) and (2),

$\dfrac{PE}{EQ} = \dfrac{PF}{FR}$

$\therefore \quad EF \parallel QR$

| By Converse of Basic Proportionality Theorem

Example 6. *In figure, A, B and C are points on OP, OQ and OR respectively such that AB $\parallel$ PQ and AC $\parallel$ PR. Show that BC $\parallel$ QR.*

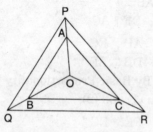

Sol. In $\triangle OPQ$,

$\because \quad AB \parallel PQ$

$\therefore \quad \dfrac{OA}{AP} = \dfrac{OB}{BQ}$ ...(1)

| By Basic Proportionality Theorem

In $\triangle OPR$,

$\because \quad AC \parallel PR$

$\therefore \quad \dfrac{OA}{AP} = \dfrac{OC}{CR}$ ...(2)

| By Basic Proportionality Theorem

From (1) and (2),

$\dfrac{OB}{BQ} = \dfrac{OC}{OR}$

$\therefore \quad BC \parallel QR$

| By Converse of Basic Proportionality Theorem

Example 7. *Using Theorem 6.1 (Textbook), prove that a line draw through the mid-point of one side of a triangle parallel to another side bisects the third side. (Recall that you have proved it in Class IX).*

Sol. Given: A $\triangle ABC$ in which D is the mid-point of AB and DE $\parallel$ BC.

To Prove: E is the mid-point of AC.

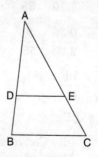

Proof:

$\because \quad DE \parallel BC$

$\therefore \quad \dfrac{AD}{DB} = \dfrac{AE}{EC}$ ...(1)

| By Basic Proportionality Theorem

$\because \quad$ D is the mid-point of AB

$\therefore \quad AD = DB$

$\therefore \quad \dfrac{AD}{DB} = 1$

$\therefore \quad \dfrac{AE}{EC} = 1 \quad$ | From (1)

$\therefore \quad AE = EC$

$\therefore \quad$ E is the mid-point of AC.

Example 8. *Using Theorem 6.2 (Textbook), prove that the line joining the mid-points of any two sides of a triangle is parallel to the third side. (Recall that you have done it in Class IX).*

Sol. Given: A △ABC in which D and E are the mid-points of sides AB and AC respectively. DE is the line joining D and E.

To Prove: DE ∥ BC

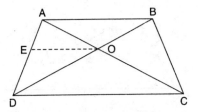

To Prove: $\dfrac{AD}{BD} = \dfrac{CO}{DO}$

Construction: Through O, draw a line OE parallel to AB or DC intersecting AD at E.

Proof: In △ADC,

∵ OE ∥ DC

∴ $\dfrac{AO}{CO} = \dfrac{AE}{DE}$ ...(1)

| By Basic Proportionality Theorem

In △DBA,

∵ OE ∥ AB

∴ $\dfrac{DE}{AE} = \dfrac{DO}{BO}$

| By Basic Proportionality Theorem

⇒ $\dfrac{AE}{DE} = \dfrac{BO}{DO}$ ...(2)

| By Invertendo

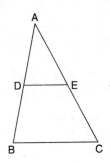

Proof:

∵ D is the mid-point of AB

∴ AD = DB

∴ $\dfrac{AD}{DB} = 1$ ...(1)

∵ E is the mid-point of AC

∴ AE = EC

∴ $\dfrac{AE}{EC} = 1$ ...(2)

From (1) and (2),

$\dfrac{AD}{DB} = \dfrac{AE}{EC}$

∴ DE ∥ BC

| By Converse of Basic Proportionality Theorem

Example 9. *ABCD is a trapezium in which AB ∥ DC and its diagonals intersect each other at the point O. Show that $\dfrac{AO}{BO} = \dfrac{CO}{DO}$.*

Sol. Given: ABCD is a trapezium in which AB ∥ DC. Its diagonals intersect each other at the point O.

From (1) and (2),

$\dfrac{AO}{CO} = \dfrac{BO}{DO}$

⇒ $\dfrac{AO}{BO} = \dfrac{CO}{DO}$.

Example 10. *The diagonals of a quadrilateral ABCD intersect each other at the point O such that $\dfrac{AO}{BO} = \dfrac{CO}{DO}$. Show that ABCD is a trapezium.*

Sol. Given: The diagonals of a quadrilateral ABCD intersect each other at the point O such that $\dfrac{AO}{BO} = \dfrac{CO}{DO}$.

To Prove: ABCD is a trapezium.

Construction: Through O, draw a line OE ∥ BA intersecting AD at E.

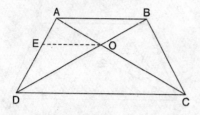

Proof: In $\triangle DBA$,

$\because \quad OE \parallel BA$

$\therefore \quad \dfrac{DO}{BO} = \dfrac{DE}{AE}$

$\Rightarrow \quad \dfrac{CO}{AO} = \dfrac{DE}{AE}$

$\left| \begin{array}{l} \because \dfrac{AO}{BO} = \dfrac{CO}{DO} \text{ (given)} \\ \Rightarrow \dfrac{DO}{BO} = \dfrac{CO}{AO} \end{array} \right.$

$\Rightarrow \quad \dfrac{AO}{CO} = \dfrac{AE}{DE}$ | Taking reciprocals

$\therefore$ In $\triangle ADC$,

$OE \parallel CD$

| By Converse of Basic Proportionality Theorem

But $\quad OE \parallel BA$ | by construction

$\therefore \quad BA \parallel CD$

$\therefore$ The quadrilateral ABCD is a trapezium.

ADDITIONAL EXAMPLES

Example 11. *If three or more parallel lines are intersected by two transversals prove that the intercepts made by them on the transversals are proportional.*

[This result is generally referred to as the Proportional Intercept Property.]

Sol. Given: AB, CD and EF are three parallel lines. They are intersected by two transversals LM and NQ at the points G, H, K and R, S, T respectively.

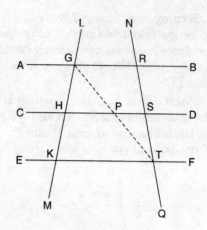

To Prove: $\dfrac{GH}{HK} = \dfrac{RS}{ST}$.

Construction: Join G and T. Let GT intersect CD at P.

Proof: In $\triangle GKT$,

$\because \quad HP \parallel KT$

$\therefore \quad \dfrac{GH}{HK} = \dfrac{GP}{PT}$ ...(1)

| By Basic Proportionality Theorem

Also, in $\triangle TGR$,

$\because \quad PS \parallel GR$

$\therefore \quad \dfrac{TP}{GP} = \dfrac{ST}{RS}$

| By Basic Proportionality Theorem

$\therefore \quad \dfrac{GP}{PT} = \dfrac{RS}{ST}$ ...(2)

| Taking reciprocals

From (1) and (2),

$\dfrac{GH}{HK} = \dfrac{RS}{ST}$.

Example 12. *In the given figure, $\angle A = \angle B$ and $AD = BE$. Show that $DE \parallel AB$.*

Sol. Given: In $\triangle CAB$, $\angle A = \angle B$ and $AD = BE$.

TRIANGLES

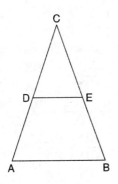

To Prove: DE ∥ AB.
Proof: ∠A = ∠B | Given
∴   CB = CA

| Sides opposite to equal angles
| of a triangle are equal

⇒   CA = CB
⇒   CD + AD = CE + BE
⇒   CD = CE | ∵ AD = BE (Given)
Now, CD = CE | Proved above
     AD = BE | Given

∴ $\dfrac{CD}{AD} = \dfrac{CE}{BE}$

∴   DE ∥ AB

| By Converse of Basic
| Proportionality Theorem

Example 13. *In the following figure, △ABC and △DBC lie on the same side of the base BC. From a point P on BC, PQ ∥ AB and PR ∥ BD are drawn. They meet AC in Q and DC in R respectively. Prove that QR ∥ AD.*

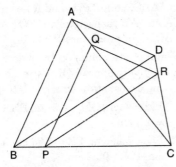

Sol. Given: △ABC and △DBC lie on the same side of the base BC. From a point P on BC, PQ ∥ AB and PR ∥ BD are drawn. They meet AC in Q and DC in R respectively.

To Prove: QR ∥ AD.
Proof: In △ABC,
∵   PQ ∥ AB
∴ $\dfrac{CQ}{QA} = \dfrac{CP}{PB}$ ...(1)

| By Basic Proportionality Theorem
In △CBD,
∵   PR ∥ BD
∴ $\dfrac{CP}{PB} = \dfrac{CR}{RD}$ ...(2)

| By Basic Proportionality Theorem
From (1) and (2),

$\dfrac{CQ}{QA} = \dfrac{CR}{RD}$

∴   QR ∥ AD

| By Converse of Basic
| Proportionality Theorem

Example 14. *The bisector of the exterior angle A of a triangle ABC intersects the side BC produced in D. Prove that*

$\dfrac{AB}{AC} = \dfrac{BD}{CD}.$

Sol. Given: In △ABC, AD is the external bisector of ∠A. It meets BC produced in D.

To Prove: $\dfrac{AB}{AC} = \dfrac{BD}{CD}.$

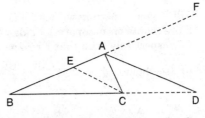

Construction: Through C, draw CE ∥ DA. Let CE meet AB in E.

Proof: CE ∥ DA | by construction
and AC intersects them
∴   ∠DAC = ∠ACE   ...(1)
| Alt. ∠s

Again, $\because$ CE $\parallel$ DA | by construction
and   AE intersects them
$\therefore \quad \angle DAF = \angle CEA$ ...(2)
| Corres. $\angle$s
But $\angle DAC = \angle DAF$ ...(3) | Given
$\therefore$ From (1), (2) and (3),
$\angle ACE = \angle CEA$
$\therefore \quad AE = AC$ (4)
| Sides opposite to equal angles of a triangle are equal

Now, in $\triangle$ ABD,
$\because$ CE $\parallel$ AD | by construction

$\therefore \dfrac{BC}{CD} = \dfrac{BE}{EA}$

| By Basic Proportionality Theorem

$\Rightarrow \dfrac{BC}{CD} + 1 = \dfrac{BE}{EA} + 1$

$\Rightarrow \dfrac{BC + CD}{CD} = \dfrac{BE + EA}{EA}$

$\Rightarrow \dfrac{BD}{CD} = \dfrac{BA}{EA}$

$\Rightarrow \dfrac{BD}{CD} = \dfrac{AB}{AC}$ | Using (4)

$\Rightarrow \dfrac{AB}{AC} = \dfrac{BD}{CD}$.

TEST YOUR KNOWLEDGE

1. If a line intersects sides AB and AC of a $\triangle$ABC at D and E respectively and is parallel to BC, prove that $\dfrac{AD}{AB} = \dfrac{AE}{AC}$ (see figure)
| Given

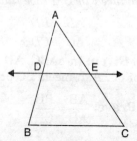

2. ABCD is a trapezium with AB $\parallel$ DC. E and F are points on non-parallel sides AD and BC respectively such that EF is parallel to AB (see figue). Show that $\dfrac{AE}{ED} = \dfrac{BF}{FC}$.

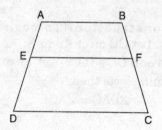

3. In figure, $\dfrac{PS}{SQ} = \dfrac{PT}{TR}$ and $\angle$ PST = $\angle$ PRQ. Prove that PQR is an isosceles triangle.

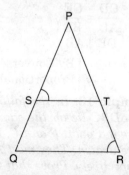

4. In $\triangle$ABC, points P, Q and R lie on sides BC, CA and AB respectively. If PQ $\parallel$ AB and QR $\parallel$ BC, then prove that RP $\parallel$ CA.

5. M and N are points on the side PQ and PR respectively of a $\triangle$PQR. For each of the following cases, state whether MN $\parallel$ QR:
   (i) PM = 4 cm, QM = 4.5 cm, PN = 4 cm, NR = 4.5 cm
   (ii) PQ = 1.28 cm, PR = 2.56 cm, BM = 0.16 cm, PN = 0.32 cm.

6. If in figures (i) and (ii), PQ $\parallel$ BC, find QC in (i) and AQ in (ii).

TRIANGLES

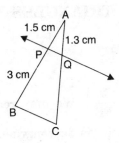

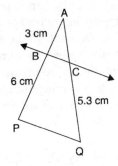

7. In the following figure, DE ∥ BC and $\dfrac{AD}{DB} = \dfrac{3}{5}$. If AC = 4.8 m, then find AE.

(CBSE 2003)

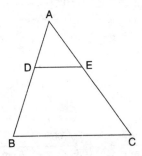

8. The shadow of a vertical pole 4m high measures 3m. At the same time the shadow of a vertical flag-staff measures 15m. Find the length of the flag-staff.

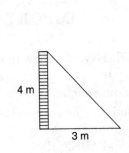

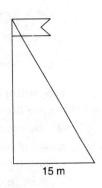

9. P and Q are points on sides AB and AC respectively of △ABC. If AP = 3 cm, PB = 6 cm, AQ = 5 cm and QC = 10 cm, show that BC = 3PQ.

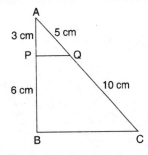

10. △ABC, D is the mid-point of BC and E is the mid-point of AD. BE when produced meets AC in Q. Prove that BE : EQ = 3 : 1.

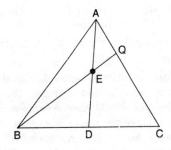

Answers

5. (*i*) Yes (*ii*) Yes
6. (*i*) 2.6 cm (*ii*) 2.65 cm
7. 1.8 cm
8. 20 cm.

CRITERIA FOR SIMILARITY OF TRIANGLES

IMPORTANT POINTS

1. Symbol of similarity. We know that two triangles are similar, if

(i) their corresponding angles are equal

and (ii) their corresponding sides are in the same ratio (or proportion).

That is, in $\triangle ABC$ and $\triangle DEF$, if

(i) $\angle A = \angle D$, $\angle B = \angle E$, $\angle C = \angle F$ and

and (ii) $\dfrac{AB}{DE} = \dfrac{BC}{EF} = \dfrac{CA}{FE}$, then the two triangles are similar (see figure).

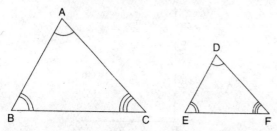

Here, we can see that A corresponds to D, B corresponds to E and C corresponds to F. Symbolically, we write the similarity of these two triangles as '$\triangle ABC \sim \triangle DEF$' and read it as 'triangle ABC is similar to triangle DEF'. So, the symbol '$\sim$' stands for 'is similar to'. Here, we may recall that we use the symbol '$\cong$' for 'is congruent to'.

Note. It must be noted that as done in the case of congruency of two triangles, the similarity of two triangles should also be expressed symbolically, using correct correspondence of their vertices. For example, for the triangles ABC and DEF of figure, we cannot write $\triangle ABC \sim \triangle EDF$ or $\triangle ABC \sim \triangle FED$. However, we can write $\triangle BAC \sim \triangle EDF$.

2. Minimum essential requirements for similarity of two triangles. Now, we shall examine that for checking the similarity of two triangles, say ABC and DEF, whether we should always look for all the equality relations of their corresponding angles ($\angle A = \angle D$, $\angle B = \angle E$, $\angle C = \angle F$) and all the equality relations of the ratios of their corresponding sides $\left(\dfrac{AB}{DE} = \dfrac{BC}{EF} = \dfrac{CA}{FE}\right)$. We may recall that, we have some criteria for congruency of two triangles involving only three pairs of corresponding parts (or elements) of the two triangles. Here also, we shall make an attempt to arrive at certain criteria for similarity of two triangles involving relationship between less number of pairs of corresponding parts of the two triangles, instead of all the six pairs of corresponding parts. For this, let us perform the following activity:

3. An Activity. Draw the line segments BC and EF of two different lengths, say 3 cm and 5 cm respectively. Then, at the points B and C respectively, construct angles PBC and QCB of some measures, say, 60° and 40°. Also, at the points E and F, construct angles REF and SFE of 60° and 40° respectively (see figure).

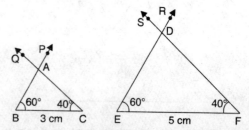

Let rays BP and CQ intersect each other at A and rays ER and FS intersect each other at D. In the two triangles ABC and DEF, we can see that $\angle B = \angle E$, $\angle C = \angle F$ and $\angle A = \angle D$. That is, corresponding angles of these two triangles are equal. Regarding corresponding sides, we note that $\dfrac{BE}{EF} = \dfrac{3}{5} = 0.6$. On measuring AB, DE, CA and FD, we find that $\dfrac{AB}{DE}$ and $\dfrac{CA}{FD}$ are also equal to 0.6 (or nearly equal to 0.6, if there is some error in the measurement).

TRIANGLES

Thus, $\dfrac{AB}{DE} = \dfrac{BC}{EF} = \dfrac{CA}{FD}$. We can repeat this activity by constructing several pairs of triangles having their corresponding angles equal. Every time, we find that their corresponding sides are in the same ratio (or proportion). This activity leads us to the following criterion for similarity of two triangles.

4. Theorem. *If in two triangles, corresponding angles are equal, then their corresponding sides are in the same ratio (or proportion) and hence the two triangles are similar.*

This criterion is referred to as the AAA (Angle–Angle–Angle) criterion of similarity of two triangles.

Proof: (*i*) Take two triangles ABC and DEF such that $\angle A = \angle D$, $\angle B = \angle E$ and $\angle C = \angle F$ (see figure)

Cut DP = AB and DQ = AC and join PQ.

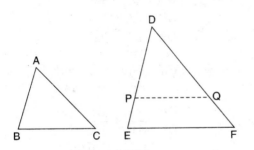

In △'s ABC and DPQ,
$$AB = DP$$
$$AC = DQ$$
$$\angle BAC = \angle PDQ$$
$$\therefore \quad \triangle ABC \cong \triangle DPQ$$
| SAS congruence criterion

This gives
$$\angle B = \angle P \quad \text{| CPCT}$$
But $\angle B = \angle E$ | Given
$$\therefore \quad \angle P = \angle E$$

But these form a pair of corresponding angles.
$$\therefore \quad PQ \parallel EF$$
$$\therefore \quad \dfrac{DP}{PE} = \dfrac{DQ}{QF}$$
| By Basic Proportionality Theorem

$\Rightarrow \dfrac{PE}{DP} = \dfrac{QF}{DQ}$ | Taking reciprocals

$\Rightarrow \dfrac{PE}{DP} + 1 = \dfrac{QF}{DQ} + 1$
| Adding 1 to both sides

$\Rightarrow \dfrac{PE + DP}{DP} = \dfrac{QF + DQ}{DQ}$

$\Rightarrow \dfrac{DE}{DP} = \dfrac{DF}{DQ}$

$\Rightarrow \dfrac{DE}{AB} = \dfrac{DF}{AC}$ | By construction

$\Rightarrow \dfrac{AB}{DE} = \dfrac{AC}{DF}$ | Taking reciprocals

Similarly,
$$\dfrac{AB}{DE} = \dfrac{BC}{EF} \text{ and so } \dfrac{AB}{DE} = \dfrac{BC}{EF} = \dfrac{AC}{DF}.$$
$$\therefore \quad \triangle ABC \sim \triangle DEF.$$

Remark. If two angles of a triangle are respectively equal to two angles of another triangle, then by the angle sum property of a triangle their third angles will also be equal. Therefore, AAA similarity criterion can also be stated as follows:

If two angles of one triangle are respectively equal to two angles of another triangle, then the two triangles are similar.

This is referred to as the AA *similarity criterion* for two triangles.

5. Converse of above theorem. *If the sides of a triangle are respectively proportional to the sides of another triangle, then their corresponding angles are equal.*

We shall examine it through an activity.

6. An Activity. Draw two triangles ABC and DEF such that AB = 3 cm, BC = 6 cm, CA = 8 cm, DE = 4.5 cm, EF = 9 cm and FD = 12 cm (see figure).

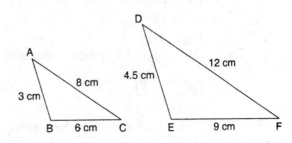

So, we have

$$\frac{AB}{DE} = \frac{BC}{EF} = \frac{CA}{FD} \text{ (each equal to } \frac{2}{3}\text{)}$$

Now measure $\angle A, \angle B, \angle C, \angle D, \angle E$ and $\angle F$. We observe that $\angle A = \angle D$, $\angle B = \angle E$ and $\angle C = \angle F$, *i.e.*, the corresponding angles of the two triangles are equal.

We can repeat this activity by drawing several such triangles (having their sides in the same ratio). Every time we see that their corresponding angles are equal. It is due to the following criterion of similarity of two triangles:

7. Theorem. *If in two triangles, sides of one triangle are proportional to (i.e., in the same ratio of) the sides of the other triangle, then their corresponding angles are equal and hence the two triangles are similar.*

This criterion is referred to as the SSS (Side–Side–Side) *similar criterion for two triangles.*

Proof: Take two triangles ABC and DEF such that $\frac{AB}{DE} = \frac{BC}{EF} = \frac{CA}{FD}$ (< 1) (see figure):

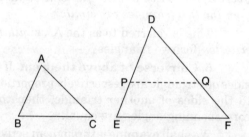

Cut DP = AB and DQ = AC and join PQ.

Now, $\frac{AB}{DE} = \frac{CA}{FD}$

$\Rightarrow \frac{DP}{DE} = \frac{DQ}{FD}$

$\Rightarrow \frac{DE}{DP} = \frac{FD}{DQ}$ | Taking reciprocals

$\Rightarrow \frac{DE}{DP} - 1 = \frac{FD}{DQ} - 1$

| Subtracting 1 from both sides

$\Rightarrow \frac{DE - DP}{DP} = \frac{FD - DQ}{DQ}$

$\Rightarrow \frac{PE}{DP} = \frac{QF}{DQ}$

$\Rightarrow \frac{DP}{PE} = \frac{DQ}{QF}$ | Taking reciprocals

$\therefore$ PQ ∥ EF

| By converse of Basic Proportionality Theorem

So, $\angle P = \angle E$

and $\angle Q = \angle F$ | Corresponding Angles

Therefore,

$$\frac{DP}{DE} = \frac{DQ}{DF} = \frac{PQ}{EF}$$

| By preceding Theorem

$\Rightarrow \frac{AB}{DE} = \frac{AC}{DF} = \frac{PQ}{EF}$

But $\frac{AB}{DE} = \frac{AC}{DF} = \frac{BC}{EF}$

$\therefore \frac{PQ}{EF} = \frac{BC}{EF}$

$\therefore$ BC = PQ

and $\frac{DP}{DE} = \frac{DQ}{DF} = \frac{BC}{EF}$

Now, in $\triangle ABC$ and $\triangle DPQ$,

AB = DP | By construction
AC = DQ | By construction
BC = PQ | Proved above

$\therefore \triangle ABC \cong \triangle DPQ$

| SSS congruence criteria

So, $\angle A = \angle D$
$\angle B = \angle P = \angle E$
$\angle C = \angle Q = \angle F$

$\therefore \triangle ABC \sim \triangle DEF$.

Remark. We may recall the either of the two conditions namely, (*i*) corresponding angles are equal and (*ii*) corresponding sides are in the same ratio is not sufficient for two polygons to be similar. However, on the basis of Theorems 4 and 7, we can now say that in case of similarity of the two triangles, it is not necessary to check both the conditions as one condition implies the other.

TRIANGLES

8. A similarity criterion comparable to SAS congruency criterion of triangles. We observe that SSS similarity criterion can be compared with the SSS congruency criterion. This suggests us to look for a similarity criterion comparable to SAS congruency criterion of triangles. For this, let us perform an activity.

9. An Activity. Draw two triangles ABC and DEF such that AB = 2 cm, ∠A = 50°, AC = 4 cm, DE = 3 cm, ∠D = 50° and DF = 6 cm (see figure)

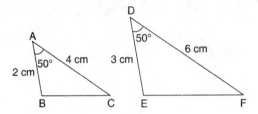

Here, we observe that $\dfrac{AB}{DE} = \dfrac{AC}{DF}$ (each equal to $\dfrac{2}{3}$) and ∠A (included between the sides AB and AC) = ∠D (included between the sides DE and DF). That is, one angle of a triangle is equal to one angle of another triangle and sides including these angles are in the same ratio (*i.e.*, proportion). Now let us measure ∠B, ∠C, ∠E and ∠F.

We find that ∠B = ∠E and ∠C = ∠F. That is, ∠A = ∠D, ∠B = ∠E and ∠C = ∠F. So, by AAA similarity criterion, △ABC ~ △DEF. We may repeat this activity by drawing several pairs of such triangles with one angle of a triangle equal to one angle of another triangle and the sides including these angles are proportional. Everytime, you will find that the triangles are similar. It is due to the following criterion of similarity of triangles:

10. Theorem. *If one angle of a triangle is equal to one angle of the other triangle and the sides including these angles are proportional, then the two triangles are similar.*

This criterion is referred to as the SAS (Side–Angle–Side) similarity criterion for two triangles.

Proof: Take two triangles ABC and DEF such that $\dfrac{AB}{DE} = \dfrac{AC}{DF}$ (< 1) and ∠A = ∠D (see figure). Cut DP = AB, DQ = AC and join PQ.

Now, $\dfrac{AB}{DE} = \dfrac{AC}{DF}$

⇒ $\dfrac{DP}{DE} = \dfrac{DQ}{DF}$

⇒ $\dfrac{DE}{DP} = \dfrac{DF}{DQ}$ | Taking reciprocals

⇒ $\dfrac{DP + PE}{DP} = \dfrac{DQ + QF}{DQ}$

⇒ $1 + \dfrac{PE}{DP} = 1 + \dfrac{QF}{DQ}$

⇒ $\dfrac{PE}{DP} = \dfrac{QF}{DQ}$

⇒ $\dfrac{DP}{PE} = \dfrac{DQ}{QF}$ | Taking reciprocals

∴ PQ ∥ EF

| By converse of Basic Proportionality Theorem

So, ∠P = ∠E
     ∠Q = ∠F | Corresponding angles

and

In △ABC and △DPQ,
     AB = DP | By construction
     AC = DQ | by construction
     ∠A = ∠D | Given
∴ △ABC ≅ △DPQ
     | SAS congruence criteria
So, ∠A = ∠D
     ∠B = ∠P = ∠E
     ∠C = ∠Q = ∠F
Therefore, △ABC ~ △DEF.
     | By theorem 4

ILLUSTRATIVE EXAMPLES

[NCERT Exercise 6.3]
(Page No. 138)

Example 1. *State which pairs of triangles in figure are similar. Write the similarity criterion used by you for answering the question and also write the pairs of similar triangles in the symbolic form.*

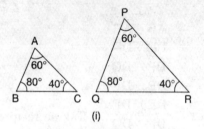
(i)

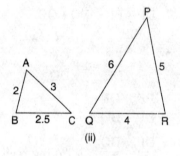
(ii)

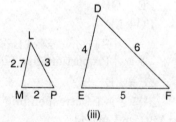
(iii)

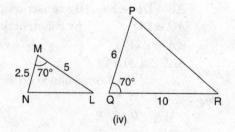
(iv)

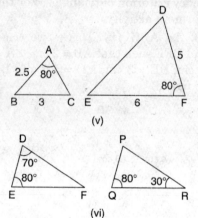
(v)

(vi)

Sol. (*i*) In $\triangle ABC$ and $\triangle PQR$,
$\angle A = \angle P$
$\angle B = \angle Q$
$\angle C = \angle R$
$\therefore \quad \triangle ABC \sim \triangle PQR$
             | AAA similarity criterion

(*ii*) In $\triangle ABC$ and $\triangle QRP$,
$$\frac{AB}{QR} = \frac{BC}{RP} = \frac{CA}{PQ}$$
$\therefore \quad \triangle ABC \sim \triangle PQR$
             | SSS similarity criterion

(*iii*) No

(*iv*) In $\triangle MNL$ and $\triangle QPR$,
$$\frac{ML}{QR} = \frac{MN}{QP} \left(= \frac{1}{2}\right)$$
and $\quad \angle NML = \angle PQR$
$\therefore \quad \triangle MNL \sim \triangle QPR$
             | SAS similarity criterion

(*v*) No

(*vi*) In $\triangle DEF$ and $\triangle PQR$,
$\angle D = \angle P \; (= 70°)$
$\angle E = \angle Q \; (= 80°)$
$\angle F = \angle R \; (= 30°)$
$\therefore \quad \triangle DEF \sim \triangle PQR$
             | AAA similarity criterion

Example 2. *In figure, △ODC ~ △ OBA, ∠BOC = 125° and ∠CDO = 70°. Find ∠DOC, ∠DCO and ∠OAB.*

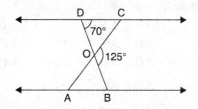

Sol. ∠DOC + ∠BOC = 180°
　　　　　| Linear Pair Axum
⇒　∠DOC + 125° = 180°
⇒　∠DOC = 180° − 125°
⇒　∠DOC = 55°　　　　　...(1)
In △DOC,
　∠DOC + ∠ODC + ∠DCO = 180°

　　| ∵ The sum of the three angles of a triangle in 180°

⇒　55° + 70° + ∠DCO = 180°
⇒　125° + ∠DCO = 180°
⇒　∠DCO = 180° − 125°
⇒　∠DCO = 55°　　　　　...(2)
∵　△ODC ~ △OBA　　| Given
∴　∠OCD = ∠OAB
　　| Corresponding angles of two similar triangles are equal
⇒　∠DCO = ∠OAB
⇒　∠OAB = ∠DCO
⇒　∠OAB = 55°　　　　　...(3)

Example 3. *Diagonals AC and BD of a trapezium ABCD with AB ∥ DC intersect each other at the point O. Using a similarity criterion for two triangles, show that $\dfrac{OA}{OC} = \dfrac{OB}{OD}$.*

Sol. Given: Diagonals AC and BD of a trapezium ABCD with AB ∥ DC intersect each other at the point O.

To Prove: $\dfrac{OA}{OC} = \dfrac{OB}{OD}$

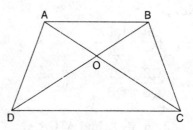

Proof: In △OAB and △OCD,
　　　　　∠OAB = ∠OCD

　| Alternate Angles
　| (∵ AB ∥ DC and AC intersects them)

　　　　　∠OBA = ∠ODC

　| Alternate Angles
　| (∵ AB ∥ DC and BD intersects them)

∴　△OAB ~ △OCD
　　　　　| AA similarity criterion

Example 4. *In Figure, $\dfrac{QR}{QS} = \dfrac{QT}{PR}$ and ∠1 = ∠2. Show that △PQS ~ △TQR.*

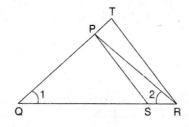

Sol. Given: In figure, $\dfrac{QR}{QS} = \dfrac{QT}{PR}$ and
　∠1 = ∠2.
To Prove: △PQS ~ △TQR
Proof: In △PQR
∵　∠1 = ∠2
∴　PR = QP　　　　　...(1)

　| ∵ Sides opposite to equal angles of a triangle are equal

Now,

$$\frac{QR}{QS} = \frac{QT}{PR} \qquad \text{| Given}$$

$$\Rightarrow \frac{QR}{QS} = \frac{QT}{QP} \qquad \text{...(2)}$$

| Using (1)

Again, in $\triangle PQS$ and $\triangle TQR$,

∵ $\dfrac{QR}{QS} = \dfrac{QT}{QP}$ | From (2)

∴ $\dfrac{QS}{QR} = \dfrac{QP}{QT}$

and $\angle SQP = \angle RQT$

∴ $\triangle PQS \sim \triangle TQR$

| SAS similarity criterion

Example 5. *S and T are points on sides PR and QR of $\triangle PQR$ such that $\angle P = \angle RTS$. Show that $\triangle RPQ \sim \triangle RTS$.*

Sol. Given: S and T are points on sides PR and QR of $\triangle PQR$ such that $\angle P = \angle RTS$.

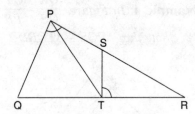

To Prove: $\triangle RPQ \sim \triangle RTS$

Proof: In $\triangle RPQ$ and $\triangle RTS$,

$\angle RPQ = \angle RTS$ | Given

$\angle QRP = \angle SRT$ | Common angle

∴ $\triangle RPQ \sim \triangle RTS$

| AA criterion of similarity.

Example 6. *In figure, if $\triangle ABE \cong \triangle ACD$, show that $\triangle ADE \sim \triangle ABC$.*

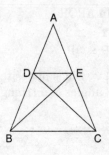

Sol. Given: In figure,
$\triangle ABE \cong \triangle ACD$.

To Prove: $\triangle ADE \sim \triangle ABC$

Proof:

∵ $\triangle ABE \cong \triangle ACD$ | Given

∴ AB = AC | CPCT

and AE = AD | CPCT

∴ $\dfrac{AB}{AD} = \dfrac{AC}{AE}$ ...(1)

Also, $\angle DAE = \angle BAC$ | common $\angle$
...(2)

In view of (1) and (2),

$\triangle ADE \sim \triangle ABC$

| SAS similarity criterion

Example 7. *In figure, altitudes AD and CE of $\triangle ABC$ intersect each other at the point P. Show that:*

(i) $\triangle AEP \sim \triangle CDP$

(ii) $\triangle ABD \sim \triangle CBE$

(iii) $\triangle AEP \sim \triangle ADB$

(iv) $\triangle PDC \sim \triangle BEC$.

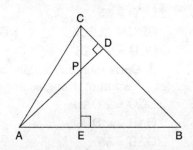

Sol. Given: In figure, altitudes AD and CE of $\triangle ABC$ intersect each other at the point P.

To Prove:

(i) $\triangle AEP \sim \triangle CDP$

(ii) $\triangle ABD \sim \triangle CBE$

(iii) $\triangle AEP \sim \triangle ADB$

(iv) $\triangle PDC \sim \triangle BEC$.

Proof: *(i)* In $\triangle AEP$ and $\triangle CDP$,

$\angle AEP = \angle CDF$ ...(1)

| Each equal to 90°

$\angle EPA = \angle DPC$ ...(2)

| Vert. opp. $\angle$s

In view of (1) and (2),

$\triangle AEP \sim \triangle CDP$

| AA similarity criterion

(ii) In $\triangle ABD$ and $\triangle CBE$,

$\angle ADB = \angle CEB$ ...(1)

| Each equal to 90°

$\angle ABD = \angle CBE$ ...(2)

| Common angle

In view of (1) and (2),

$\triangle ABD \sim \triangle CBE$

| AA similarity criterion

(iii) In $\triangle AEP$ and $\triangle ADB$,

$\angle AEP = \angle ADB$ ...(1)

| Each equal to 90°

$\angle EAP = \angle DAB$ ...(2)

| Common angle

In view of (1) and (2),

$\triangle AEP \sim \triangle ADB$

| AA similarity criterion

(iv) In $\triangle PDC$ and $\triangle BEC$,

$\angle PDC = \angle BEC$ ...(1)

| Each equal to 90°

$\angle DCP = \angle ECB$ ...(2)

| Common angle

In view of (1) and (2),

$\triangle PDC \sim \triangle BEC$

| AA similarity criterion

Example 8. *E is a point on the side AD produced of a parallelogram ABCD and BE intersects CD at F. Show that $\triangle ABE \sim \triangle CFB$.*

Sol. Given: E is a point on the side AD produced of a parallelogram ABCD and BE intersects CD at F.

To Prove: $\triangle ABE \sim \triangle CFB$

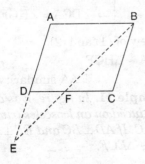

Proof: In $\triangle ABE$ and $\triangle CFB$,

$\angle BAE = \angle FCB$ ...(1)

| Opp. $\angle$ s of a || gram

$\angle AEB = \angle CBF$ ...(2)

| At Int. $\angle$ s
($\because$ AE || BC and BE intersects them)

In view of (1) and (2),

$\triangle ABE \sim \triangle CFB$.

Example 9. *In figure, ABC and AMP are two right triangles, right angled at B and M respectively. Prove that :*

(i) $\triangle ABC \sim \triangle AMP$

(ii) $\dfrac{CA}{PA} = \dfrac{BC}{MP}$

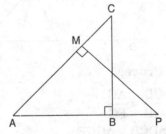

Sol. Given: In figure, ABC and AMP are two right triangles, right angled at B and M respectively.

To Prove: (i) $\triangle ABC \sim \triangle AMP$

(ii) $\dfrac{CA}{PA} = \dfrac{BC}{MP}$.

Proof: (i) In $\triangle ABC$ and $\triangle AMP$,

$\angle ABC = \angle AMP$ ...(1)

| Each equal to 90°

$\angle BAC = \angle MAP$ ...(2)

| Common angle

In view of (1) and (2),

$\triangle ABC \sim \triangle AMP$

| AA similarity criterion

(ii) $\because$ $\triangle ABC \sim \triangle AMP$

| Proved above in (i)

$\therefore$ $\dfrac{CA}{PA} = \dfrac{BC}{MP}$

| Corresponding sides of two similar triangles are proportional

Example 10. *CD and GH are respectively the bisectors of ∠ACB and ∠EGF such that D and H lie on sides AB and FE of △ABC and △EFG respectively. If △ABC ~ △FEG, show that:*

(i) $\dfrac{CD}{GH} = \dfrac{AC}{FG}$

(ii) △DCB ~ △HGE

(iii) △DCA ~ △HGF.

Sol. Given: CD and GH are respectively the bisectors of ∠ACB and ∠EGF such that D and H lie on sides AB and FE of △ABC and △EFG respectively. Also, △ABC ~ △FEG.

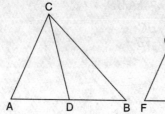

To Prove:

(i) $\dfrac{CD}{GH} = \dfrac{AC}{FG}$

(ii) △DCB ~ △HGE

(iii) △DCA ~ △HGF.

Proof: (i) In △ACD and △FGH,

∠CAD = ∠GFH   ...(1)

| ∵ △ABC ~ △FEG
| ∴ ∠CAB = ∠GFE
| ⇒ ∠CAD = ∠GFH

∠ACD = ∠FGH   ...(2)

| ∵ △ABC ~ △FEG
| ∴ ∠ACB = ∠FGE
| ⇒ $\dfrac{1}{2}$∠ACB = $\dfrac{1}{2}$∠FGE
| (Halves of equals are equal)
| ⇒ ∠ACD = ∠FGH

In view of (1) and (2),
△ACD ~ △FGH

| AA similarity criterion

∴ $\dfrac{CD}{GH} = \dfrac{AC}{FG}$

| ∵ Corresponding sides of two similar trangles are proportional

(ii) In △DCB and △HGE,

∠DBC = ∠HEG   ...(1)

| ∵ △ABC ~ △FEG
| ∴ ∠ABC = ∠FEG
| ⇒ ∠DBC = ∠HEG

∠DCB = ∠HGE   ...(3)

| ∵ △ABC ~ △FEG
| ∴ ∠ACB = ∠FGE
| ⇒ $\dfrac{1}{2}$∠ACB = $\dfrac{1}{2}$∠FGE
| (Halves of equals are equal)
| ⇒ ∠DCB = ∠HGE

In view of (1) and (2),
△DCB ~ △HGE

| AA similarity criterion

(iii) In △DCA and △HGF,

∠DAC = ∠HFG   ...(1)

| ∵ △ABC ~ △FEG
| ∴ ∠CAB = ∠GFE
| ⇒ ∠CAD = ∠GFH
| ⇒ ∠DAC = ∠HFG

∠DCA = ∠HGF   ...(2)

| ∵ △ABC ~ △FEG
| ∴ ∠ACB = ∠FGE
| ⇒ $\dfrac{1}{2}$∠ACB = $\dfrac{1}{2}$∠FGE
| (Halves of equals are equal)
| ⇒ ∠DCA = ∠HGF

In view of (1) and (2),
△DCA ~ △HGF

| AA similarity criterion

Example 11. *In figure, E is a point on side CB produced on an isosceles triangle ABC with AB = AC. If AD ⊥ BC and EF ⊥ AC, prove that △ABD ~ △ECF.*

TRIANGLES

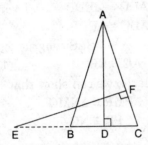

Sol. Given: E is a point on side CB produced of an isosceles triangle ABC with AB = AC. Also, AD ⊥ BC and EF ⊥ AC.

To Prove: △ABD ~ △ECF

Proof: In △ABD and △ECF,

∵ AB = AC | Given

∴ ∠ACB = ∠ABC

| Angles opposite to equal sides of a triangle are equal

⇒ ∠ABC = ∠ACB

⇒ ∠ABD = ∠ECF   ...(1)

∠ADB = ∠EFC   ...(2)

| Each equal to 90°

In view of (1) and (2),

△ABD ~ △ECF

| AA similarity criterion.

Example 12. *Sides AB and BC and median AD of a triangle ABC are respectively proportional to sides PQ and QR and median PM of △PQR (see figure). Show that △ABC ~ △PQR.*

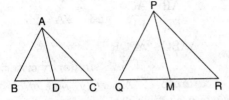

Sol. Given: Sides AB and BC and median AD of a triangle ABC are respectively proportional to sides PQ and QR and median PM of △PQR i.e.,

$$\frac{AB}{PQ} = \frac{BC}{QR} = \frac{AD}{PM}.$$

To Prove: △ABC ~ △PQR.

Proof: $\frac{AB}{PQ} = \frac{BC}{QR} = \frac{AD}{PM}$ | Given

⇒ $\frac{AB}{PQ} = \frac{\frac{1}{2}BC}{\frac{1}{2}QR} = \frac{AD}{PM}$

⇒ $\frac{AB}{PQ} = \frac{BD}{QM} = \frac{AD}{PM}$

⇒ △ABD ~ △PQM

| SSS similarity criterion

∴ ∠ABD = ∠PQM

| ∵ Corresponding angles of two similar triangles are equal.

⇒ ∠ABC = ∠PQR

Now, in △ABC and △PQR,

$\frac{AB}{PQ} = \frac{BC}{QR}$ ...(1)

| Given

∠ABC = ∠PQR   ...(2)

| Proved above

In view of (1) and (2),

△ABC ~ △PQR.

| SAS similarity criterion

Example 13. *D is a point on the side BC of a triangle ABC such that ∠ADC = ∠BAC. Show that $CA^2 = CB \cdot CD$.* (CBSE 2004)

Sol. Given: *D is a point on the side BC of a triangle ABC such that ∠ADC = ∠BAC.*

To Prove: $CA^2 = CB \cdot CD$.

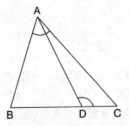

Proof: In △BAC and △ADC,

∠BAC = ∠ADC | Given

∠BCA = ∠DCA | Common angle
∴ ΔBAC ~ ΔADC
          | AA similarity criterion
∴ $\dfrac{CA}{CD} = \dfrac{CB}{CA}$

| ∵ Corresponding sides of two similar triangles are proportional

⇒ $CA^2 = CB \cdot CD$.

Example 14. *Sides AB and AC and median AD of a triangle ABC are respectively proportional to sides PQ and PR and median PM of another triangle PQR. Show that ΔABC ~ ΔPQR.*

Sol. Given: Sides AB and AC and median AB of a triangle ABC are respective proportional to sides PQ and PR and median PM of another triangle PQR, i.e., in ΔABC are ∠PQR,

⇒ $\dfrac{AB}{PQ} = \dfrac{AC}{PR} = \dfrac{AD}{PM}$.

To Prove: ΔABC ~ ΔPQR.

Construction: Produce AD to a point E such that AD = DE and produce PM to a point N such that PM = MN. Join BE and QN.

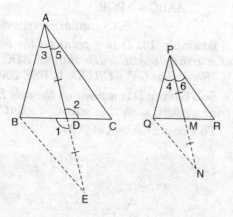

Proof: In ΔABC and ΔEDB,
  DC = DB | ∵ AD is a median
  AD = ED | By construction
  ∠ADC = ∠EDB | vert. opp. ∠s
∴ ΔADC ≅ ΔEDB
        | SAS congruence criterion
∴ AC = EB   ...(1) | CPCT
Similarly, we can show that
  ΔPMR = ΔNMQ
∴ PR = NQ   ...(2)
               | CPCT
Now,
  $\dfrac{AB}{PQ} = \dfrac{AC}{PR} = \dfrac{AD}{PM}$ | Given

⇒ $\dfrac{AB}{PQ} = \dfrac{EB}{NQ} = \dfrac{AD}{PM}$
               | From (1) and (2)

⇒ $\dfrac{AB}{PQ} = \dfrac{EB}{NQ} = \dfrac{AD}{PM} = \dfrac{2AD}{2PM} = \dfrac{AE}{PN}$

∴ ΔABE ~ PQN
         | SSS similarity criterion
∴ ∠ABE = ∠PQN
  | ∵ Corresponding angles of two similar triangles are equal

⇒ ∠3 = ∠4   ...(3)
Similarly, we can prove that
  ∠5 = ∠6   ...(4)
Adding (3) and (4), we get
  ∠3 + ∠5 = ∠4 + ∠6
⇒ ∠A = ∠P
Now, in ΔABC and ΔPQR,
  $\dfrac{AB}{PQ} = \dfrac{AC}{PR}$ and ∠A = ∠P
∴ ΔABC ~ ΔPQR
         | SAS similarity criterion

Example 15. *A vertical pole of length 6 m casts a shadow 4 m long on the ground and at the same time a tower casts a shadow 28 m long. Find the height of the tower.*

Sol.

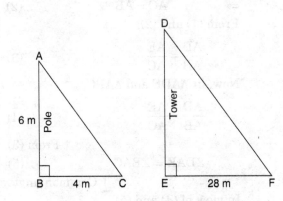

Let AB denote the vertical pole of length 6 m. BC is the shadow of the pole on the ground BC = 4 m.

Let DE denote the Tower. EF is the shadow of the tower on the ground.

EF = 28 m.

Let the height of the tower be h m.

In $\triangle ABC$ and $\triangle DEF$,

$\angle B = \angle E$

| Each euqal to 90° because pole and tower are standing vertical to the ground

$\angle A = \angle D$ | $\because$ Shadow are cast at the same time

$\therefore$ $\triangle ABC \sim \triangle DEF$

| AA similarly criterion

$\therefore \quad \dfrac{AB}{DE} = \dfrac{BC}{EF}$

| $\because$ Corresponding sides of two similar triangles are proportional

$\Rightarrow \quad \dfrac{6}{h} = \dfrac{4}{28}$

$\Rightarrow \quad h = \dfrac{6 \times 28}{4}$

$\Rightarrow \quad h = 42$

Hence, the height of the tower is 42 m.

Example 16. *If AD and PM are medians of triangles ABC and PQR, respectively where $\triangle ABC \sim \triangle PQR$, prove that $\dfrac{AB}{PQ} = \dfrac{AD}{PM}$.*

Sol. Given: AD and PM are median of triangles ABC and PQR respectively where $\triangle ABC \sim \triangle PQR$

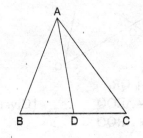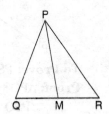

To Prove: $\dfrac{AB}{PQ} = \dfrac{AD}{PM}$.

Proof: $\triangle ABC \sim \triangle PQR$ | Given

$\therefore \quad \dfrac{AB}{PQ} = \dfrac{BC}{QR} = \dfrac{CA}{RP}$ ...(1)

| $\because$ Corresponding sides of two similar triangles are proportional

and $\quad \angle A = \angle P$ | $\because$ Corresponding
$\angle B = \angle Q$ | angles of two similar
$\angle C = \angle R$ | triangles are equal
...(2)

But BC = 2BD and QR = 2QM

| $\because$ AD and PM are medians

So, from (1),

$\dfrac{AB}{PQ} = \dfrac{2BD}{2QM}$

$\Rightarrow \quad \dfrac{AB}{PQ} = \dfrac{BD}{QM}$ ...(3)

Also, $\angle ABD = \angle PQM$ ...(4)
| From (2)

$\therefore \quad \triangle ABD \sim \triangle PQM$

| SAS similarity criterion

$\therefore \quad \dfrac{AB}{PQ} = \dfrac{AD}{PM}$

| $\because$ Corresponding sides of two similar triangles are proportional

ADDITIONAL EXAMPLES

Example 17. *In figure, if $\triangle POS \sim \triangle ROQ$, prove that PS $\parallel$ QR.*

Sol. Given: In figure, $\triangle PQS \sim \triangle ROQ$

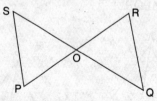

To Prove: PS $\parallel$ QR.
Proof: $\triangle POS \sim \triangle ROQ$ | Given
$\therefore \quad \angle PSO = \angle RQO$

   | $\because$ Corresponding angles of two similar triangles are equal

But these form a pair of alternate angles
$\therefore \quad$ PS $\parallel$ QR.

Example 18. *In figure, $\triangle FEC \cong \triangle GBD$ and $\angle 1 = \angle 2$. Prove that $\triangle ADE \sim \triangle ABC$.*

Sol. Given: In figure, $\triangle FEC \cong \triangle GBD$ and $\angle 1 = \angle 2$.

To Prove: $\triangle ADE \sim \triangle ABC$.

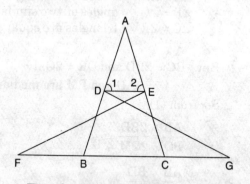

Proof: $\triangle FEC \cong \triangle GBD$ | Given
$\therefore$ CPT
        $\angle F = \angle G$
        $\angle E = \angle B$
        $\angle C = \angle D$
        CE = DB
        $\angle 1 = \angle 2$ | Given
$\therefore$ AE = AD   ...(1)

Sides opposite to equal angles of a triangle are equal

$\Rightarrow$ CE + AE = DB + AD
$\Rightarrow$ AC = AB   ...(2)
From (1) and (2),
$$\frac{AD}{AB} = \frac{AE}{AC} \quad ...(3)$$
Now, in $\triangle ADE$ and $\triangle ABC$,
$$\frac{AD}{AB} = \frac{AE}{AC} \quad ...(4)$$
                               | From (3)
$\angle DAE = \angle BAC$ ...(5)
                             | Common angle
In view of (4) and (5),
    $\triangle ADE \sim \triangle ABC$
                  | SAS similarity criterion

Aliter. AB = AC | Proved above
$\therefore \quad \angle C = \angle B$

   | Angles opposite to equal sides of a triangle are equal

But $\angle B = \angle E$ | Given
$\therefore \quad \angle C = \angle E$
Now, in $\triangle ADE$ and $\triangle ABC$,
       $\angle DAE = \angle ABC$ | Common angles
       $\angle AED = \angle ACB$ | Proved above
$\therefore \quad \triangle ADE \sim \triangle ABC$
                  | AA similarly criterion

Example 19. *ABC is an isosceles triangle with AB = AC and D is a pointer AC such that $BC^2 = AC \times CD$. Prove that BD = BC.*

Sol. Given: ABC is an isosceles triangle with AB = AC and D is a point on AC such that $BC^2 = AC \times CD$.

To Prove: BD = BC.

Proof: $BC^2 = AC \times CD$ | Given

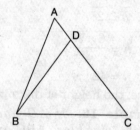

TRIANGLES

$$\Rightarrow \frac{AC}{BC} = \frac{BC}{CD} \qquad ...(1)$$

Also, $\angle ACB = \angle BCD$ ...(2)

| Common angle

In view of (1) and (2),

$\triangle ABC \sim \triangle BDC$

| SAS similarity criterion

$$\therefore \frac{AC}{BC} = \frac{AB}{BD}$$

| ∵ Corresponding sides of two similar triangles are proportional

But AB = AC | Given

$\therefore$ BD = BC.

Example 20. *Through the vertex D of a parallelogram ABCD, a line is drawn to intersect the sides AB and CB produced at E and F respectively. Prove that*

$$\frac{DA}{AE} = \frac{FB}{BE} = \frac{FC}{CD}.$$

Sol. Given: Through the vertex D of a parallelogram ABCD, a line is drawn to intersect the sides AB and CB produced at E and F respectively.

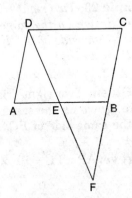

To Prove: $\frac{DA}{AE} = \frac{FB}{BE} = \frac{FC}{CD}$.

Proof: In $\triangle FBE$ and $\triangle FCD$,

$\angle FBE = \angle FCD$

| Corresponding angles

$\angle FEB = \angle FDC$

| Corresponding angles

$\therefore \triangle FBE \sim \triangle FCD$

| AA similarity criterion

$$\therefore \frac{FB}{FC} = \frac{BE}{CD}$$

| ∵ Corresponding sides of two similar triangles are proportional

$$\Rightarrow \frac{FB}{BE} = \frac{FC}{CD} \qquad ...(1)$$

In $\triangle FBE$ and $\triangle DAE$,

$\angle EFB = \angle EDA$ | Alt. Int. $\angle$s

$\angle BEF = \angle AED$ | Vert. opp. $\angle$s

$\therefore \triangle FBE \sim \triangle DAE$

| AA similarity criterion

$$\therefore \frac{FB}{DA} = \frac{BE}{AE}$$

| ∵ Corresponding sides of two similar triangles are proportional

$$\Rightarrow \frac{FB}{BE} = \frac{DA}{AE} \qquad ...(2)$$

Form (1) and (2),

$$\frac{DA}{AE} = \frac{FB}{BE} = \frac{FC}{CD}.$$

Example 21. *The diagonal BD of a parallelogram ABCD intersects the line-segment AE at the point F, where E is any point on the side BC. Prove that*

$$DF \times EF = FB \times FA.$$

Sol. Given: The diagonal BD of a parallelogram ABCD intersects the line segment AE at the point F. where E is any point on the side BC.

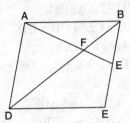

To Prove: $DF \times EF = FB \times FA$.

Proof: In $\triangle FBE$ and $\triangle FDA$,

$\angle FBE = \angle FDA$ ...(1)

∠FEB = ∠AFD | Alt. Int. ∠s
                    ...(2)
                    | Vert. opp. ∠s

In view of (1) and (2),

△FBE ~ △FDA

| AA similarity criterion

∴ $\dfrac{EF}{AF} = \dfrac{FB}{FD}$

| ∵ Corresponding sides of two similar triangles are proportional

⇒ $\dfrac{EF}{FA} = \dfrac{FB}{DF}$

⇒ DF × EF = FB × FA.

Example 22. *In figure, AD and BE are respectively perpendiculars to BC and AC. Show that*

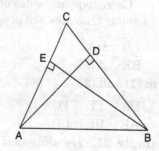

(i) △ADC ~ △BEC
(ii) CA × CE = CB × CD
(iii) △ABC ~ △DEC
(iv) CD × AB = CA × DE.

Sol. Given: In figure, AD and BE are respectively perpendiculars to BC and AC.

To Prove:
(i) △ADC ~ △BEC
(ii) CA × CE = CB × CD
(iii) △ABC ~ △DEC
(iv) CD × AB = CA × DE.

Proof:
(i) In △ADC and △BEC,
∠ADC = ∠BEC
| Each equal to 90°
∠ACD = ∠BCE | Common angle

∴ △ADC ~ △BEC
| AA similarity criterion

(ii) △ADC ~ △BEC | Prove in (i)

∴ $\dfrac{CA}{CB} = \dfrac{CD}{CE}$

| ∵ Corresponding sides of two similar triangles are proportional

⇒ CA × CE = CB × CD

(iii) CA × CE = CB × CD | Proved in (ii)

⇒ $\dfrac{CA}{CB} = \dfrac{CD}{CE}$

∠ACB = ∠DCE | Common angle

∴ △ABC ~ △DEC
| SAS similarity criterion

(iv) △ABC ~ △DEC | Proved in (iii)

∴ $\dfrac{CA}{CD} = \dfrac{AB}{DE}$

| ∵ Corresponding sides of two similar triangles are proportional

⇒ CD × AB = CA × DE.

Example 23. *Two right triangles ABC and DBC are drawn on the same hypotenuse BC and on the same side of BC. If AC and DB intersect at P, prove that*

AP × PC = BP × PD.

Sol. Given: Two right triangles ABC and DBC are drawn on the same hypotenuse BC and on the same side of BC. AC and BD intersect at P.

To Prove: AP × PC = BP × PD.

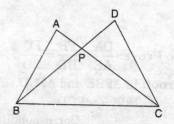

Proof: In △APB and △DPC,
∠BAP = ∠CDP | Each equal to 90°
∠APB = ∠DPC | vert. opp. ∠s

∴ △APB ~ △DPC
　　　　| AA similarity criterion
∴ $\dfrac{AP}{DP} = \dfrac{BP}{CP}$

| ∵ Corresponding sides of two similar triangles are proportional

⇒ AP × CP = BP × DP
⇒ AP × PC = BP × PD.

Example 24. *In a triangle ABC, P, Q are points on AB, AC respectively and PQ ∥ BC. Prove that the median AD bisects PQ.*

Sol. Given: In a triangle ABC, P, Q are points on AB, AC respectively and PQ ∥ BC.

To Prove: The median AD bisects PQ.

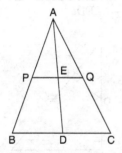

Proof: In △APE and △ABD,
　∠APE = ∠ABD　　| Corres. ∠s
　∠AEP = ∠ADB　　| Corres. ∠s
∴ △APE ~ △ABD
　　　　| AA similarity criterion
∴ $\dfrac{AP}{AB} = \dfrac{AE}{AD} = \dfrac{PE}{BD}$ 　...(1)

| ∵ Corresponding sides of two similar triangles are proportional

Similarly, we can show that
　△AEQ ~ △ADC
∴ $\dfrac{AE}{AD} = \dfrac{AQ}{AC} = \dfrac{EQ}{DC}$ 　...(2)

| ∵ Corresponding sides of two similar triangles are similar

From (1) and (2), we get
$\dfrac{PE}{BD} = \dfrac{EQ}{DC}$
But　BD = DC　　| ∵ AD is a median
∴　PE = EQ
⇒ The median AD bisects PQ.

Example 25. *Prove that the line segments joining the mid-points of the sides of a triangle form four triangles each of which is similar to the original one.*

Sol. Given: In △ABC, D, E and F are the mid-points of the sides AB, AC and BC respectively.

To Prove: △I ~ △II ~ △III ~ △IV ~ ABC.

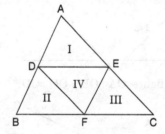

Proof: In △ABC,
　∵ D is the mid-point of AB
∴　$\dfrac{AD}{AB} = \dfrac{1}{2}$ 　　　　...(1)
and E is the mid-point of AC
∴　$\dfrac{AE}{AC} = \dfrac{1}{2}$ 　　　　...(2)
From (1) and (2),
　$\dfrac{AD}{AB} = \dfrac{AE}{AC}$
Now, in △I and △ABC,
　$\dfrac{AD}{AB} = \dfrac{AE}{AC}$　　| Proved above
and　∠A is common
∴　△I ~ △ABC
　　　　| SAS similarity criterion
Similarly, we can prove the other results.

TEST YOUR KNOWLEDGE

1. In figure, if PQ ∥ PS, prove that ΔPOQ ~ ΔSOR.

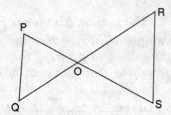

2. In figure, find ∠P.

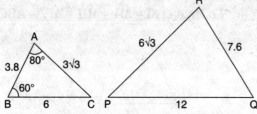

3. In figure, OA . OB = OC . OD

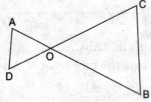

   Show that ∠A = ∠C and ∠B = ∠D

4. A girl of height 90 cm is walking away from the base of a lamp-post at a speed of 1.2 m/second. If the lamp is 3.6 m above ground, find the length of her shadow after 4 seconds.

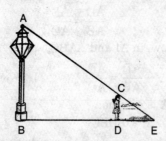

5. In figure, CM and RN are respectively the medians of ΔABC and ΔPQR. If ΔABC ~ ΔPQR, prove that:

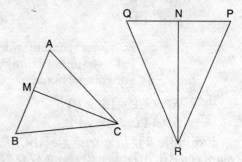

   (i) ΔAMC ~ ΔPNR

   (ii) $\dfrac{CM}{RN} = \dfrac{AB}{PQ}$

   (iii) ΔCMB ~ ΔRNQ.

6. In figure, AB ⊥ BC and DE ⊥ AC. Prove that ΔABC ~ ΔAED.

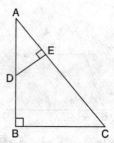

7. In the given figure, ΔABC is right-angled at B. BD is perpendicular to AC. Prove that ΔADB is similar to ΔABC.

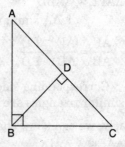

8. In the given figure, if $\dfrac{EA}{EC} = \dfrac{EB}{ED}$, prove that

   (i) ΔEAB ~ ΔECD   and   (ii) AB ∥ CD.

TRIANGLES

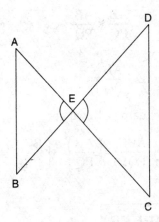

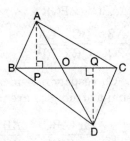

10. Through the mid-point M of the side CD of a parallelogram ABCD, the line BM is drawn intersecting AC in L and AD produced in E. Prove that EL = 2BL.

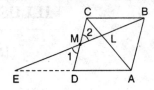

9. In the figure, ABC and DBC are two triangles on the same base BC. Prove that

$$\frac{ar\,(\triangle ABC)}{ar\,(\triangle DBC)} = \frac{AO}{DO}.$$

Answers

2. 40°   4. 1.6 m.

AREAS OF SIMILAR TRIANGLES

IMPORTANT POINTS

1. Theorem. *The ratio of the areas of two similar triangles is equal to the square of the ratio of their corresponding sides.*

Given: Two triangles ABC and PQR such that $\triangle ABC \sim \triangle PQR$.

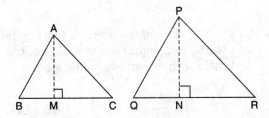

To Prove:

$$\frac{ar\,(ABC)}{ar\,(PQR)} = \left(\frac{AB}{PQ}\right)^2 = \left(\frac{BC}{QR}\right)^2 = \left(\frac{CA}{RP}\right)^2$$

Construction: Draw altitudes AM and PN of the triangles ABC and PQR respectively.

Proof:

$$ar\,(ABC) = \frac{1}{2} BC \times AM$$

and

$$ar\,(PQR) = \frac{1}{2} QR \times PN$$

So,

$$\frac{ar\,(ABC)}{ar\,(PQR)} = \frac{\frac{1}{2} \times BC \times AM}{\frac{1}{2} \times QR \times PN}$$

$$= \frac{BC \times AM}{QR \times PN} \qquad \ldots(1)$$

Now, in $\triangle ABM$ and $\triangle PQN$,

$\angle B = \angle Q$ (As $\triangle ABC \sim \triangle PQR$)

and $\angle M = \angle N$ (Each is of 90°)

So, $\triangle ABM \sim \triangle PQN$

| AA similarity criterion

Therefore, $\dfrac{AM}{PN} = \dfrac{AB}{PQ}$ ...(2)

∵ Corresponding sides of two similar triangles are proportional

Also, $\triangle ABC \sim \triangle PQR$ (Given)

So, $\dfrac{AB}{PQ} = \dfrac{BC}{QR} = \dfrac{CA}{RP}$ ...(3)

| $\because$ Corresponding sides of two similar triangles are proportional

Therefore, $\dfrac{ar(ABC)}{ar(PQR)} = \dfrac{AB}{PQ} \times \dfrac{AM}{PN}$

| [From (1) and (3)]

$= \dfrac{AB}{PQ} \times \dfrac{AB}{PQ}$ [From (2)]

$= \left(\dfrac{AB}{PQ}\right)^2$

Now using (3), we get

$\dfrac{ar(ABC)}{ar(PQR)} = \left(\dfrac{AB}{PQ}\right)^2 \times \left(\dfrac{BC}{QR}\right)^2 = \left(\dfrac{CA}{RP}\right)^2$.

ILLUSTRATIVE EXAMPLES

[NCERT Exercise 6.4]
(Page No. 143)

Example 1. *Let $\triangle ABC \sim \triangle DEF$ and their areas be, respectively, $64\ cm^2$ and $121\ cm^2$. If $EF = 15.4\ cm$, find BC.*

Sol.

$\triangle ABC \sim \triangle PQR$ | Given

$\therefore \dfrac{ar(\triangle ABC)}{ar(\triangle PQR)} = \left(\dfrac{BC}{QR}\right)^2$

| $\because$ The ratio of the areas of two similar triangles is equal to the square of the ratio of their corresponding sides

$\Rightarrow \dfrac{64}{121} = \left(\dfrac{BC}{15.4}\right)^2$

$\Rightarrow \left(\dfrac{8}{11}\right)^2 = \left(\dfrac{BC}{15.4}\right)^2$

$\Rightarrow \dfrac{8}{11} = \dfrac{BC}{15.4}$ | Taking square root on both sides

$\Rightarrow BC = \dfrac{8 \times 15.4}{11}$

$\Rightarrow BC = 11.2\ cm.$

Example 2. *Diagonals of a trapezium ABCD with AB || DC intersect each other at the point O. If AB = 2CD, find the ratio of the areas of triangles AOB and COD.*

Sol. In $\triangle AOB$ and $\triangle COD$,

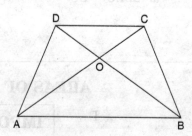

$\angle AOB = \angle COD$ | Vert. opp. $\angle s$
$\angle OAB = \angle OCD$ | Alt. Int. $\angle s$
$\therefore \triangle AOB \sim \triangle COD$
| AA similarity criterion

$\therefore \dfrac{ar(\triangle AOB)}{ar(\triangle COD)} = \left(\dfrac{AB}{CD}\right)^2$

| $\because$ The ratio of the areas of two similar triangles is equal to the square of the ratio of their corresponding sides

$= \left(\dfrac{2CD}{CD}\right)^2$ | $\because$ AB = 2CD

$= \dfrac{4}{1}$

$= 4 : 1.$

Example 3. *In figure, ABC and DBC are two triangles on the same base BC. If AD intersects BC at O, show that $\dfrac{ar(ABC)}{ar(DBC)} = \dfrac{AO}{DO}$.*

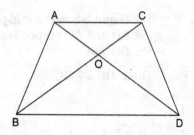

Sol. Given: In figure, ABC and DBC are two triangles on the same base BC. AD intersects BC at O.

To Prove: $\dfrac{\text{ar}(\triangle ABC)}{\text{ar}(\triangle DBC)} = \dfrac{AO}{DO}$

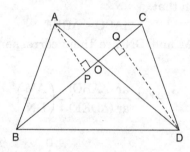

Construction: Draw AP ⊥ BC and DQ ⊥ BC.

Proof: In △AOP and △DOQ

∠APO = ∠DQO
| Each equal to 90°

∠AOP = ∠DOQ | Vert. opp. ∠s

∴ △AOP ~ △DOQ
| AA similarity criterion

∴ $\dfrac{AP}{DQ} = \dfrac{AO}{DO}$ ...(1)

| ∵ Corresponding sides of two similar triangles are proportional

Now, $\dfrac{\text{ar}(\triangle ABC)}{\text{ar}(\triangle DBC)} = \dfrac{\frac{1}{2}(BC)(AP)}{\frac{1}{2}(BC)(DQ)}$

$= \dfrac{AP}{DQ}$

$= \dfrac{AO}{DO}.$ | From (1)

Example 4. *If the areas of two similar triangles are equal, prove that they are congruent.*

Sol. Given: ABC and DEF are two similar triangles such that

ar (△ABC) = ar (△DEF)

To Prove: △ABC ≅ △DEF

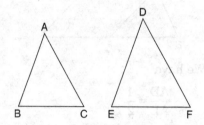

Proof: △ABC ~ △DEF | Given

∴ $\dfrac{\text{ar}(\triangle ABC)}{\text{ar}(\triangle DEF)} = \left(\dfrac{BC}{EF}\right)^2$

| ∵ The ratio of the areas of two similar triangles is equal to the square of the ratio of their corresponding sides

$\Rightarrow$ $1 = \left(\dfrac{BC}{EF}\right)^2$

| ∵ ar (△ABC) = ar (△DEF)

$\Rightarrow$ $1 = \dfrac{BC^2}{EF^2}$

$\Rightarrow$ $BC^2 = EF^2$

$\Rightarrow$ BC = EF   ...(1)

Also, ∠B = ∠E   ...(2)

And ∠C = ∠F   ...(3)

| ∵ △ABC ≅ △DEF

In view of (1), (2) and (3),

△ABC ≅ △DEF

| ASA congruence criterion

Example 5. *D, E and F are respectively the mid-points of sides AB, BC and CA of △ ABC. Find the ratio of the areas of △DEF and △ ABC.*

Sol. Given: D, E and F are respectively the mid-points of sides AB, BC and CA of △ABC.

To determine. Ratio of the areas of ΔDEF and ΔABC.

Determination.

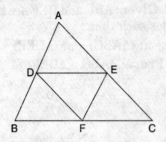

We have

$$\frac{AD}{AB} = \frac{1}{2} \qquad ...(1)$$

| ∵ D is the mid-point of AB

and $\quad \dfrac{AE}{AC} = \dfrac{1}{2} \qquad ...(2)$

| ∵ E is the mid-point of AC

From (1) and (2),

$$\frac{AD}{AB} = \frac{AE}{AC}$$

∴ DE ∥ BC

| By Converse of Basic Proportionality theorem

∴ ∠ADE = ∠ABC ...(3)

| Corresponding angles

and ∠AED = ∠ACB ...(4)

| Corresponding angles

In view of (3) and (4),

ΔABC ~ ΔDEF

| AA similarity criterion

∴ $\dfrac{ar(\Delta DEF)}{ar(\Delta ABC)} = \left(\dfrac{DE}{BC}\right)^2$

| ∵ The ratio of the areas of two similar triangles is equal to the square of the ratio of their corresponding sides

$$= \left(\frac{\frac{1}{2}BC}{BC}\right)^2$$

| ∵ D and E are the mid-points of AB and AC respectively
∴ DE ∥ BC
and DE = $\frac{1}{2}$BC

$$= \frac{1}{4}$$

∴ ar(ΔDEF) : ar(ΔABC) = 1 : 4.

Example 6. *Prove that the ratio of the areas of two similar triangles is equal to the square of the ratio of their corresponding medians.*

Sol. Given: Two triangles ABC and DEF such that

ΔABC ~ ΔDEF

AM and DN are their corresponding medians.

To Prove: $\dfrac{ar(\Delta ABC)}{ar(\Delta DEF)} = \left(\dfrac{AM}{DN}\right)^2$

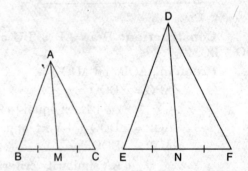

Proof: Δ ABC ~ ΔDEF | Given

∴ $\dfrac{ar(\Delta ABC)}{ar(\Delta DEF)} = \left(\dfrac{AB}{DE}\right)^2 \qquad ...(1)$

| ∵ The ratio of the areas of two similar triangles is equal to the square of the ratio of their corresponding sides

and $\quad \dfrac{AB}{DE} = \dfrac{BC}{EF} = \dfrac{CA}{FD} \qquad ...(2)$

| ∵ The corresponding sides of two similar triangles are proportional

From first two of (2),

$$\frac{AB}{DE} = \frac{BC}{EF}$$

$$= \frac{2BM}{2EN}$$ | ∵ AM and DN are the medians

$$= \frac{BM}{EN} \qquad \ldots(3)$$

Also, ∠ABM = ∠DEN   ...(4)

| ∵ △ABC ~ △DEF
| ∴ ∠ABC = ∠DEF
| (corresponding angles of two similar triangles are equal)

In view of (3) and (4),
△ABM ~ △DEN
| SAS similarity criterion

∴ $\frac{AB}{DE} = \frac{BM}{EN} = \frac{AM}{DN}$ ...(5)

| ∵ Corresponding sides of two similar triangles are proportional

From (1) and (5),

$$\frac{ar(\triangle ABC)}{ar(\triangle DEF)} = \left(\frac{AM}{DN}\right)^2.$$

Example 7. *Prove that the area of an equilateral triangle described on one side of a square is equal to half the area of the equilateral triangle described on one of its diagonals.*

Sol. Given: ABCD is a square whose one diagonal is AC. △APC and △BQC are two equilateral triangles described on the diagonal AC and side BC of the square ABCD.

To Prove: ar(△BQC) = $\frac{1}{2}$ ar(△APC)

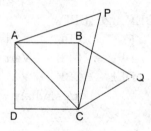

Proof: ∵ △APC and △BQC are both equilateral triangles

∴ △APC ~ △BQC
| AAA similarity criterion

∴ $\frac{ar(\triangle APC)}{ar(\triangle BQC)} = \left(\frac{AC}{BC}\right)^2$

| ∵ The ratio of the areas of two similar triangles is equal to the square of the ratio of their corresponding sides

$$= \frac{AC^2}{BC^2}$$

$$= \left(\frac{\sqrt{2}\ BC}{BC}\right)^2$$

| ∵ Diagonal = $\sqrt{2}$ side

= 2

⇒ ar(△BQC) = $\frac{1}{2}$ ar(△APC).

Tick the correct answer and justify:

Example 8. *ABC and BDE are two equilateral triangles such that D is the mid-point of BC. Ratio of the areas of triangles ABC and BDE is*

(A) 2 : 1 (B) 1 : 2 (C) 4 : 1 (D) 1 : 4.

Sol. ∵ △ABC and △BDE are both equilateral triangles

∴ △ABC ~ △BDE
| AAA similarity criterion

∴ $\frac{ar(\triangle ABC)}{ar(\triangle BDE)} = \left(\frac{AB}{BD}\right)^2$

| ∵ The ratio of the areas of two similar triangles is equal to the square of the ratio of their corresponding sides

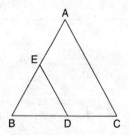

$$= \left(\frac{BC}{BD}\right)^2 \quad | \because AB = BC = CA$$

$$= \left(\frac{2BD}{BD}\right)^2$$

| $\because$ D is the mid-point of BC

$$= \frac{4}{1}$$

$\Rightarrow$ ar($\triangle$ABC) : ar($\triangle$BDE) = 4 : 1

Hence, (C) 4 : 1 is the correct answer.

Example 9. *Sides of two similar triangles are in the ratio 4 : 9. Areas of these triangles are in the ratio*

(A) *2 : 3* (B) *4 : 9* (C) *81 : 16* (D) *16 : 81*

Sol. Ratio of the areas of these triangles = $4^2 : 9^2$

| $\because$ The ratio of the areas of two similar triangles is equal to the square of the ratio of their corresponding sides

Hence, (D) 16 : 81 is the correct answer.

ADDITIONAL EXAMPLES

Example 10. *ABCD is a trapezium with AB $\parallel$ DC. If $\triangle$ AED is similar to $\triangle$BEC, prove that AD = BC.*

Sol. Given: ABCD is a trapezium with AB $\parallel$ DC. Also $\triangle$AED ~ $\triangle$BEC.

To Prove: AD = BC

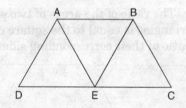

Proof: $\dfrac{\text{ar}(\triangle AED)}{\text{ar}(\triangle BEC)} = 1$

| Similar triangles between the same parallels

$\Rightarrow \left(\dfrac{AD}{BC}\right)^2 = 1$

| $\because$ The ratio of the areas of two similar triangles is equal to the square of the ratio of their corresponding sides

$\Rightarrow$ AD^2 = BC^2

$\Rightarrow$ AD = BC.

Example 11. *D and E are points on the sides AB and AC respectively of $\triangle$ ABC such that DE is parallel to BC and AD : DB = 4 : 5. CD and BE intersect each other at F. Find the ratio of the areas of $\triangle$ DEF and $\triangle$ BCF.*

(AI CBSE 2003)

Sol. Given. D and E are points on the sides AB and AC respectively of $\triangle$ABC such that DE is parallel to BC and DE : DB = 4 : 5. CD and BE intersect each other at F.

To Determine. $\dfrac{\text{ar}(\triangle DEF)}{\text{ar}(\triangle BCF)}$.

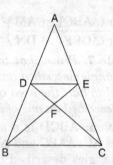

Determination. $\because$ DE $\parallel$ BC and transversal BE intersects them

$\therefore \angle$FED = $\angle$FBC ...(1)

| Alt. Int. $\angle$s

In $\triangle$DEF and $\triangle$CBF,

$\angle$FED = $\angle$FBC | From (1)

$\angle$DFE = $\angle$CFB | Vert. Opp. $\angle$s

$\therefore \quad \triangle$DEF ~ $\triangle$CBF

| AA similarity criterion

TRIANGLES

∴ $\dfrac{ar(\triangle DEF)}{ar(\triangle CBF)} = \left(\dfrac{DE}{CB}\right)^2$ ...(2)

| ∵ The ratio of the areas of two similar triangles is equal to the square of the ratio of their corresponding sides

In △ADE and △ABC,

| ∠ADE = ∠ABC | Corres. ∠s
| ∠AED = ∠ACB
| AA similarity criterion

∴ △ADE ~ △ABC
| AA similarity criterion

∴ $\dfrac{AD}{AB} = \dfrac{DE}{BC} = \dfrac{AE}{AC}$ ...(3)

| ∵ Corresponding sides of two similar triangles are proportional

$\dfrac{AD}{DB} = \dfrac{4}{5}$ | Given

⇒ $\dfrac{DB}{AD} = \dfrac{5}{4}$ | By invertendo

⇒ $\dfrac{DB}{AD} + 1 = \dfrac{5}{4} + 1$
| Adding 1 to both sides

⇒ $\dfrac{DB + AD}{AD} = \dfrac{5+4}{4}$

⇒ $\dfrac{AB}{AD} = \dfrac{9}{4}$

⇒ $\dfrac{AD}{AB} = \dfrac{4}{9}$ ...(4)

| By invertendo

(3) and (4) give

$\dfrac{DE}{BC} = \dfrac{4}{9}$ ...(5)

(2) and (3) give

$\dfrac{ar(\triangle DEF)}{ar(\triangle CBF)} = \left(\dfrac{4}{9}\right)^2 = \dfrac{16}{81}$

⇒ $\dfrac{ar(\triangle DEF)}{ar(\triangle BCF)} = \dfrac{16}{81}$

⇒ $ar(\triangle DEF) : ar(\triangle BCF) = 16 : 81$.

Example 12. *Equilateral triangles are drawn on the sides of a right angled triangle. Show that the area of the triangle on the hypotenuse is equal to the sum of the areas of the triangle on the other two sides.*

Sol. Given: Equilateral triangles ACM, ABN and BCL are drawn on the sides AC, AB and BC respectively of the right-angled triangle ABC.

To Prove: $ar(\triangle ABN) + ar(\triangle BCL)$
$= ar(\triangle ACM)$

Proof: ∵ △ABN and △ACM are both equilateral triangles

∴ △ABN ~ △ACM
| AAA similarity criterion

∴ $\dfrac{ar(\triangle ABN)}{ar(\triangle ACM)} = \left(\dfrac{AB}{AC}\right)^2$ ...(1)

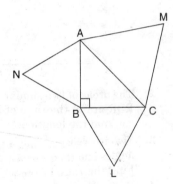

| ∵ The ratio of the areas of two similar triangles is equal to the square of the ratio of their corresponding sides

Similarly,

$\dfrac{ar(\triangle BCL)}{ar(\triangle ACM)} = \left(\dfrac{BC}{AC}\right)^2$ ...(2)

Adding (1) and (2), we get

$\dfrac{ar(\triangle ABN)}{ar(\triangle ACM)} + \dfrac{ar(\triangle BCL)}{ar(\triangle ACM)}$

$= \left(\dfrac{AB}{AC}\right)^2 + \left(\dfrac{BC}{AC}\right)^2$

$$= \frac{AB^2}{AC^2} + \frac{BC^2}{AC^2}$$

$$= \frac{AB^2 + BC^2}{AC^2}$$

$$= \frac{AC^2}{AC^2} = 1$$

[$\because AB^2 + BC^2 = A$
Pythagoras th]

$$\Rightarrow \frac{ar(\triangle ABN) + ar(\triangle BCL)}{ar(\triangle ACM)} = 1$$

$$\Rightarrow ar(\triangle ABN) + ar(\triangle BCL) = ar(\angle$$

TEST YOUR KNOWLEDGE

1. In figure, the line segment XY is parallel to side AC of $\triangle ABC$ and it divides the triangle into two parts of equal areas. Find the ratio $\frac{AX}{AB}$.

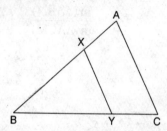

2. The areas of two similar triangles ABC and PQR are in the ratio of 9 : 16. If BC = 4.5 cm, find the length of QR. (CBSE 2004)

3. Two isosceles triangles have equal vertical angles and their areas are in the ratio 9 : 16. Find the ratio of their altitudes.

4. Vertical angles of two isosceles triangles one equal their corresponding altitudes are in the ratio 4 : 9. Find the ratio of their areas.

5. Prove that the areas of two similar triangles are in the ratio of the squares of the corresponding (i) altitudes (ii) angle tor segments.

6. In the trapezium ABCD, AB ∥ CD a 2CD. If area of $\triangle AOB = 84$ cm², find t of $\triangle COD$.

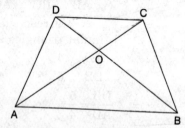

7. In the given figure, PB and QA one p ...culars to segment AB. If PO = 5 cm 7 cm and area of $\triangle POB = 150$ cm², fi area of $\triangle QOA$.

Answers

1. $\frac{2-\sqrt{2}}{2}$ 2. 6 cm
3. 3 : 4   4. 16 : 81
6. 21 cm²   7. 210 cm².

PYTHAGORAS THEOREM

IMPORTANT POINTS

1. **An important result related to similarity of two triangles formed by the perpendicular to the hypotenuse from the opposite vertex of the right triangle.**

Theorem. *If a perpendicular is drawn from the vertex of the right angle of a right tri-* angle to the hypotenuse then triangles on sides of the perpendicular are similar t whole triangle and to each other.

Given: A right triangle ABC, righ gled at B. BD is the perpendicular fron vertex B to the hypotenuse AC.

TRIANGLES

To Prove:
(i) $\triangle ADB \sim \triangle ABC$
(ii) $\triangle BDC \sim \triangle ABC$
(iii) $\triangle ADB \sim \triangle BDC$.

Given: A right triangle ABC right angled at B
To Prove: $AC^2 = AB^2 + BC^2$.
Construction: Draw $BD \perp AC$.

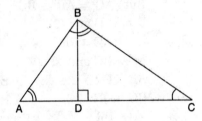

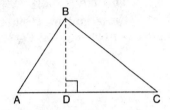

Proof: (i) In $\triangle ADB$ and $\triangle ABC$,
$\angle ADB = \angle ABC$
| Each equal to 90°
$\angle DAB = \angle BAC$
| Common angle
$\therefore \triangle ADB \sim \triangle ABC$
| AA similarity criterion

(ii) In $\triangle BDC$ and $\triangle ABC$,
$\angle BDC = \angle ABC$
| Each equal to 90°
$\angle DCB = \angle BCA$
| Common angle
$\therefore \triangle BDC \sim \triangle ABC$

(iii) $\because \triangle ADB \sim \triangle ABC$
and $\triangle BDC \sim \triangle ABC$
and $\triangle ADB \sim \triangle BDC$

| Two triangles similar to the same triangle are similar to each other.

Note. We shall apply this theorem in proving the Pythagoras Theorem.

2. Pythagoras Theorem : *In a right triangle, the square of the hypotenuse is equal to the sum of the squares of the other two sides.*

Pythagoras

Proof: $\triangle ADB \sim \triangle ABC$
| By previous theorem
So, $\dfrac{AD}{AB} = \dfrac{AB}{AC}$

| $\because$ Corresponding sides of two similar triangles are proportional

$\Rightarrow AD \cdot AC = AB^2$ ...(1)

Also, $\triangle BDC \sim \triangle ABC$
| By previous theorem
So, $\dfrac{DC}{BC} = \dfrac{BC}{AC}$

| $\because$ Corresponding sides of two similar triangles are proportional

$\Rightarrow DC \cdot AC = BC^2$ ...(2)

Adding (1) and (2),
$AD \cdot AC + DC \cdot AC = AB^2 + BC^2$
$\Rightarrow AC (AD + DG) = AB^2 + BC^2$
$\Rightarrow AC \cdot AC = AB^2 + BC^2$
$\Rightarrow AC^2 = AB^2 + BC^2$

The above theorem was earlier given by an ancient Indian mathematician Baudhayan (about 800 B.C.) in the following form:

The diagonal of a rectangle produces by itself the same area as produced by its both sides (i.e., length and breadth).

For this reason, this theorem is sometimes also referred to as the *Baudhayan Theorem.*

3. Converse of the Pythagoras Theorem. *In a triangle, if square of one side is equal to the sum of the squares of the other two sides* then the angle opposite the first side is a right angle.

Given: A triangle ABC in which
$AC^2 = AB^2 + BC^2$.

To Prove: $\angle B = 90°$

Construction: Construct a $\triangle PQR$ right angled at Q such that PQ = AB and QR = BC

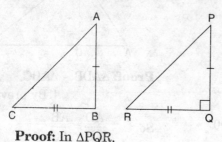

Proof: In $\triangle PQR$,
$\therefore \quad \angle Q = 90°$

$\therefore$ By Pythagoras theorem,
$PR^2 = PQ^2 + QR^2$
$\Rightarrow PR^2 = AB^2 + BC^2$ ...(1)
| By construction
But $AC^2 = AB^2 + BC^2$ ...(2) | Given
From (1) and (2),
$AC = PR$ ...(3)
In $\triangle ABC$ and $\triangle PQR$,
AB = PQ | By construction
BC = QR | By construction
AC = PR | Proved in (3) above
$\therefore \triangle ABC \cong \triangle PQR$
                | SSS congruence criterion
Therefore $\angle B = \angle Q$ | CPCT
But $\angle Q = 90°$ | By construction
So, $\angle B = 90°$.

ILLUSTRATIVE EXAMPLES

[NCERT Exercise 6.5]
(Page No. 150)

Example 1. *Sides of triangles are given below. Determine which of them are right triangles. In case of a right triangle, write the length of its hypotenuse.*

(i) 7 cm, 24 cm, 25 cm
(ii) 3 cm, 8 cm, 6 cm
(iii) 50 cm, 80 cm, 100 cm
(iv) 13 cm, 12 cm, 5 cm

Sol. (i) **7 cm, 24 cm, 25 cm**
$7^2 = 49$
$24^2 = 576$
$25^2 = 625$
We see that $7^2 + 24^2 = 25^2$
$\therefore$ The given triangle is right angled.
Hypotenuse = 25 cm

(ii) **3 cm, 8 cm, 6 cm**
$3^2 = 9$
$8^2 = 64$
$6^2 = 36$

$\therefore 3^2 + 6^2 \neq 8^2$
$\therefore$ The given triangle is not right angled.

(iii) **50 cm, 80 cm, 100 cm**
$50^2 = 2500$
$80^2 = 6400$
$100^2 = 10000$
$\therefore 50^2 + 80^2 \neq 100^2$
$\therefore$ The given triangle is not right angled.

(iv) **13 cm, 12 cm, 5 cm**
$13^2 = 169$
$12^2 = 144$
$5^2 = 25$
$\therefore 5^2 + 12^2 = 13^2$
$\therefore$ The given triangle is right angled.
Hypotenuse = 13 cm.

TRIANGLES

Example 2. *PQR is a triangle right angled at P and M is a point on QR such that $PM \perp QR$. Show that $PM^2 = QM \cdot MR$.*

Sol. Given: PQR is a triangle right angled at P and M is a point on QR such that $PM \perp QR$.

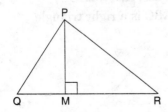

To Prove: $PM^2 = QM \cdot MR$.

Proof: In right triangle PQR,

$QR^2 = PQ^2 + PR^2$ ...(1)

| By Pythagoras theorem

In right triangle PMQ,

$PQ^2 = PM^2 + MQ^2$ ...(2)

| By Pythagoras theorem

On right triangle PMR,

$PR^2 = PM^2 + MR^2$ ...(3)

| By Pythagoras theorem

Using (2) and (3), (1) gives

$QR^2 = (PM^2 + MQ^2) + (PM^2 + MR^2)$

$\Rightarrow QR^2 = 2PM^2 + MQ^2 + MR^2$

$= (MQ + MR)^2$

$= 2PM^2 + MQ^2 + MR^2$

$\Rightarrow MQ^2 + MR^2 + 2MQ \cdot MR$

$= 2PM^2 + MQ^2 + MR^2$

$\Rightarrow PM^2 = QM \cdot MR$

Aliter.

In $\triangle QMP$ and $\triangle PMR$,

$\angle QMP = \angle PMR$

| Each equal to 90°

$\angle MQP = 90° - \angle MPQ = \angle MPR$

$\therefore \triangle QMP \sim \triangle PMR$

| AA similarity criterion

$\therefore \dfrac{QM}{PM} = \dfrac{PM}{RM}$

| ∵ Corresponding sides of two similar triangles are proportional

$\Rightarrow PM^2 = QM \cdot RM$

$\Rightarrow PM^2 = QM \cdot MR$

Example 3. *In figure, ABD is a triangle right angled at A and $AC \perp BD$. Show that*

(i) $AB^2 = BC \cdot BD$
(ii) $AC^2 = BC \cdot DC$
(iii) $AD^2 = BD \cdot CD$

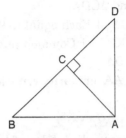

Sol. Given: In figure, ABD is a triangle right angled at A and $AC \perp BD$.

To Prove:

(i) $AB^2 = BC \cdot BD$
(ii) $AC^2 = BC \cdot DC$
(iii) $AD^2 = BD \cdot CD$.

Proof: (i) In $\triangle BAC$ and $\triangle BDA$,

$\angle BAC = \angle BDA$ ...(1)

In $\triangle ABC$,

$\angle BAC + \angle CBA = 90°$ ...(2)

In $\triangle ABD$,

$\angle BDA + \angle CBA = 90°$ ...(3)

In view of (2) and (3),

$\angle BAC = \angle BDA$

$\angle ACB = \angle DAB$ | Each equal to 90°

$\therefore \triangle BAC \sim \triangle BDA$

| AA similarity criterion

$\therefore \dfrac{BA}{BD} = \dfrac{BC}{BA}$

| ∵ Corresponding sides of two similar triangles and propotional

$\Rightarrow \quad BA^2 = BC \cdot BD$

$\Rightarrow \quad AB^2 = BC \cdot BD.$

(ii) In $\triangle ACB$ and $\triangle DCA$,

$\angle ACB = \angle DCA$ | Each equal to 90°

$\angle BAC = \angle ADC$ | Proved above

$\therefore \quad \angle ACB \sim \angle DCA$

  | AA similarity criterion

$\therefore \quad \dfrac{AC}{DC} = \dfrac{BC}{AC}$

| $\because$ Corresponding sides of two similar triangles are proportional

$\Rightarrow \quad AC^2 = BC \times DC.$

(iii) In $\triangle ADB$ and $\triangle CDA$,

$\angle DAB = \angle DCA$ | Each equal to 90°

$\angle BDA = \angle ADC$ | Common angle

$\therefore \quad \triangle ADB \sim \triangle CDA$

  | AA similarity criterion

$\therefore \quad \dfrac{AD}{CD} = \dfrac{BD}{AD}$

| $\because$ Corresoponding sides of two similar triangls are proportional

$\Rightarrow \quad AD^2 = BD \times CD.$

Example 4. *ABC is an isosceles triangle right angled at C. Prove that $AB^2 = 2\,AC^2$.*

Sol. Given: ABC is an isosceles triangle right angled at C.

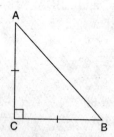

To Prove: $AB^2 = 2AC^2$.

Proof: $\because$ ABC is an isosceles triangle right angled at C.

$\therefore \quad AC = BC$ ...(1)

$\angle C = 90°$ ...(2)

In view of (2) in $\triangle ACB$,

$AB^2 = AC^2 + BC^2$

| By Pythagoras theorem

$= AC^2 + AC^2$ | From (1)

$= 2AC^2.$

Example 5. *ABC is an isosceles triangle with AC = BC. If $AB^2 = 2AC^2$, prove that ABC is a right triangle.*

Sol. Given: ABC is an isosceles triangle with AC = BC. Also, $AB^2 = 2AC^2$.

To Prove: ABC is a right triangle.

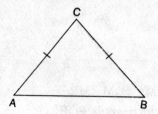

Proof:

$AC^2 + BC^2 = AC^2 + AC^2$ | $\because \quad AC = BC$

$= 2AC^2$

$= AB^2$ | $\because \; AB^2 = 2AC^2$

$\therefore$ By converse of Pythagoras theorem

$\angle ACB = 90°$

$\Rightarrow$ ABC is a right triangle.

Example 6. *ABC is an equilateral triangle of side 2a. Find each of its altitudes.*

Sol. Given: ABC is an equilateral traingle in which AB = BC = CA = 2a. Also, AD $\perp$ BC, BE $\perp$ CA and CF $\perp$ AB.

To determine. AD, BE and CF.

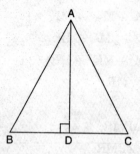

Determination. In right triangles ADB and ADC.

Hypotenuse AB = Hypotenuse AC

  | Given

AD = AD | common side

$\triangle ADB \cong \triangle ADC$
| RHS congruence criterion
$BD = CD$ | CPCT
$= \frac{1}{2} BC$
| $\because$ D is the mid-point of BC
$= \frac{1}{2}(2a) = a$

In right triangle ADB,
$AD^2 + BD^2 = AB^2$
$\Rightarrow AD^2 = AB^2 - BD^2$
$= (2a)^2 - (a)^2$
$= 4a^2 - a^2 = 3a^2$
$\Rightarrow AD = \sqrt{3}a$

Similarly,
$BE = \sqrt{3}a$
$CF = \sqrt{3}a$.

Example 7. *Prove that the sum of the squares of the sides of a rhombus is equal to the sum of the squares of its diagonals.*

Sol. Given: ABCD is a rhombus whose diagonals AC and BD intersect at O.

To Prove: $AB^2 + BC^2 + CD^2 + DA^2 = AC^2 + BD^2$

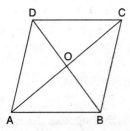

Proof: We know that the diagonals of a rhombus bisect each over at right angles
$\therefore OA = OC$
$OB = OD$
$\angle AOB = \angle BOC$
$= \angle COD = \angle DOA = 90°$

In right triangle AOB,
$AB^2 = OA^2 + OB^2$ ...(1)
| By Pythagoras theorem

In right triangle BOC,
$BC^2 = OB^2 + OC^2$ ...(2)
| By Pythagoras theorem

In right triangle COD,
$CD^2 = OC^2 + OD^2$ ...(3)
| By Pythagoras theorem

In right traingle DOA,
$DA^2 = OD^2 + OA^2$ ...(4)
| By Pythagoras theorem

Adding (1), (2), (3) and (4), we get
$AB^2 + BC^2 + CD^2 + DA^2$
$= 2(OA^2 + OB^2 + OC^2 + OD^2)$
$= 2(OA^2 + OB^2 + OA^2 + OB^2)$
| $\because$ OC = OA and OD = OB
$= 4(OA^2 + OB^2)$
$= 4OA^2 + 4OB^2$
$= (2OA)^2 + (2OB)^2$
$= AC^2 + BD^2$.

Example 8. *In figure, O is a point in the interior of a triangle ABC, $OD \perp BC$, $OE \perp AC$ and $OF \perp AB$. Show that*
(i) $OA^2 + OB^2 + OC^2 - OD^2 - OE^2 - OF^2$
$= AF^2 + BD^2 + CE^2$,
(ii) $AF^2 + BD^2 + CE^2 = AE^2 + CD^2 + BF^2$.

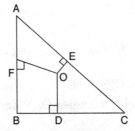

Sol. Given: In figure, O is a point in the interior of a triangle ABC, $OD \perp BC$, $OE \perp AC$ and $OF \perp AB$.

To Prove:
(i) $OA^2 + OB^2 + OC^2 - OD^2 - OE^2 - OF^2$
(ii) $AF^2 + BD^2 + CE^2 = AE^2 + CD^2 + BF^2$

Construction: Join OA, OB and OC

Proof: (i) In right triangle OAF,
$OA^2 = OF^2 + AF^2$
| By Pythagoras theorem
$\Rightarrow AF^2 = OA^2 - OF^2$ ...(1)

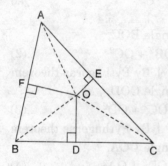

In right triangle OBD,
$$OB^2 = OD^2 + BD^2$$
| By Pythagoras theorem
$$\Rightarrow BD^2 = OB^2 - OD^2 \quad ...(2)$$
In right trangle OCE,
$$OC^2 = OE^2 + CE^2$$
| By Pythagoras theorem
$$\Rightarrow CE^2 = OC^2 - OE^2 \quad ...(3)$$
Adding (1), (2) and (3), we get
$$AF^2 + BD^2 + CE^2 = OA^2 + OB^2$$
$$+ OC^2 - OA^2 - OE^2 - OF^2$$

(ii) In right triangle OBD,
$$OB^2 = OD^2 + BD^2$$
| By Pythagores theorem
In right trangle OCD,
$$OC^2 = OD^2 + CD^2$$
| By Pythagores theorem
$$\therefore OB^2 - OC^2 = BD^2 - CD^2 \quad ...(4)$$
Similarly, by considering right triangles OCE and OAE, we get
$$OC^2 - OA^2 = CE^2 - AE^2 \quad ...(5)$$
and, by considering right triangles OAF and OBF, we get
$$OA^2 - OB^2 = AF^2 - BF^2 \quad ...(6)$$
Adding (4), (5) and (6), we get
$$0 = BD^2 + CE^2 + AF^2 - CD^2 - AE^2 - BF^2$$
$$\Rightarrow AF^2 + BD^2 + CE^2 = AE^2 + BF^2 + CD^2.$$

Example 9. *A ladder 10 m long reaches a window 8 m above the ground. Find the distance of the foot of the ladder from base of the wall.*

Sol. RP is the ladder. R is the foot of the ladder
$$RP = 10 \text{ m}$$
P is the window
PQ is the wall
Q is the base of the wall

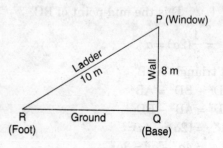

In right triangle PQR,
$$RP^2 = RQ^2 - PQ^2$$
| By Pythagoras theorem
$$\Rightarrow (10)^2 = RQ^2 + (8)^2$$
$$\Rightarrow RQ^2 = (10)^2 - (8)^2$$
$$\Rightarrow RQ^2 = 100 - 64$$
$$\Rightarrow RQ^2 = 36$$
$$\Rightarrow RQ = \sqrt{36}$$
$$\Rightarrow RQ = 6 \text{ m}$$

Hence, the distance of the foot of the ladder from the base to the wall is 6 m.

Example 10. *A guy wire attached to a vertical pole of height 18 m, is 24 m long and has a stake attached to the other end. How far from the base of the pole should the stake be driven so that the wire will be taut?*

Sol. In right triangle ABC
$$AC^2 = AB^2 + BC^2$$
| By Pythagoras theorem
$$\Rightarrow (24)^2 = (18)^2 + BC^2$$
$$\Rightarrow 576 = 324 + BC^2$$
$$\Rightarrow BC^2 = 576 - 324$$
$$\Rightarrow BC^2 = 252$$
$$\Rightarrow BC = \sqrt{252}$$
$$\Rightarrow BC = 6\sqrt{7} \text{ m}$$

TRIANGLES

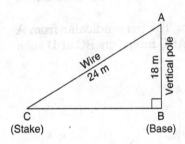

Hence, the stake should be driven $6\sqrt{7}$ m for from the base of the pole so that the wire will be taut.

Example 11. *An aeroplane leaves an airport and flies due north at a speed of 1000 km per hour. At the same time, another aeroplane leaves the same airport and flies due west at a speed of 1200 km per hour. How far apart will be the two planes after $1\frac{1}{2}$ hours ?*

Sol. Distance of the aeroplane leaving the airport and flying due north at a speed of 1000 km per hour after $1\frac{1}{2}$ hours = OA.

$$= 1000 \times 1\frac{1}{2} \text{ km}$$
$$= 1500 \text{ km}$$

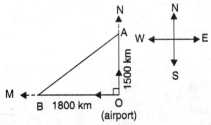

Distance of the aeroplane leaving the airport and flying due west at a speed of 1200 km per hour after $1\frac{1}{2}$ hours.

$$= OB = 1200 \times 1\frac{1}{2} \text{ km} = 1800 \text{ km}$$

In right triangle AOB,
$$AB^2 = OA^2 + OB^2$$
$$= (1500)^2 + (1800)^2$$
$$= 2250000 + 3240000$$

$$= 5490000$$
$$\Rightarrow AB = \sqrt{5490000}$$
$$\Rightarrow AB = \sqrt{61 \times 90000}$$
$$\Rightarrow AB = 300\sqrt{61} \text{ km}$$

Hence, the two planes will be $300\sqrt{61}$ km apart after $1\frac{1}{2}$ hours.

Example 12. *Two poles of heights 6 m and 11 m stand on a plane ground. If the distance between the feet of the poles is 12 m, find the distance between their tops.*

Sol. Given: AD and BE are two poles of height 6 cm and 11 m respectively. D and E are their tops respectively. A and B are their feet respectively such that AB = 12 m.

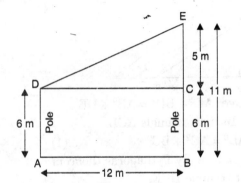

Required: To find out DE.

Construction: Draw DC ⊥ BE. Join DE.

Determination: ABCD is a rectangle and BC and AD are its opposite sides

∴ BC = AD

| ∵ Opposite sides of a rectangle one equal

= 6 m

Similarly, DC = AB = 12 cm

Now, CE = BE – BC
= 11 m – 6 m
= 5 m

Again, in right triangle DCE,
$$DE^2 = DC^2 + CE^2$$
| By Pythagoras theorem

$$= (12)^2 + (5)^2$$
$$= 144 + 25$$
$$= 169$$
$$\Rightarrow \quad DE = \sqrt{169} = 13 \text{ cm}$$

Hence, the distance between their tops is 13 m.

Example 13. *D and E are points on the sides CA and CB respectively of a triangle ABC right angled at C. Prove that*
$$AE^2 + BD^2 = AB^2 + DE^2.$$

Sol. Given: D and E are points on the sides CA and CB respectively of a traingle ABC right angled at C.

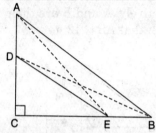

To Prove: $AE^2 + BD^2 = AB^2 + DE^2$.
Proof: In right triangle ACB,
$$AB^2 = AC^2 + BC^2 \qquad ...(1)$$
| By Pythagoras theorem
In right triangle DCE,
$$DE^2 = CD^2 + CE^2 \qquad ...(2)$$
| By Pythagoras theorem
Adding (1) and (2), we get
$$AB^2 + DE^2 = (AC^2 + BC^2) + (CD^2 + CE^2)$$
$$= (AC^2 + CE^2) + (BC^2 + CD^2)$$
$$= AE^2 + BD^2$$
∵ In right triangle ACE,
$$AC^2 + CE^2 = AE^2$$
(By Pythagoras theorem)
and in right triangle BCD,
$$BC^2 + CD^2 = BD^2$$
(By Pythagoras theorem).

Example 14. *The perpendicular from A on side BC of a △ABC intersects BC at D such that DB = 3 CD (see figure). Prove that $AB^2 = 2 AC^2 + BC^2$.*

Sol. Given: The perpendicular from A on side BC of a △ABC intersects BC at D such that DB = 3CD.

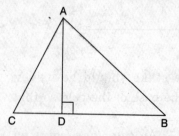

To Prove: $2 AB^2 = 2 AC^2 + BC^2$.
Proof: In right triangle ADB,
$$AB^2 = AD^2 + BD^2 \qquad ...(1)$$
| By Pythagoras theorem
$$AC^2 = AD^2 + CD^2 \qquad ...(2)$$
| By Pythagoras theorem
In right triangle ADC,
Subtracting (2) from (1), we get
$$AB^2 - AC^2 = BD^2 - CD^2$$
$$= (BD + CD)(BD - CD)$$
$$= (BC)(3\ CD - CD)$$
| ∵ BD = 3 CD (given)
$$= (BC)(2\ CD)$$
$$= 2(BC)(CD)$$
$$= 2\ (BC)\left(\frac{1}{4}BC\right)$$

DB = 3 CD
$$\Rightarrow \quad \frac{DB}{CD} = 3$$
$$\Rightarrow \quad \frac{DB}{CD} + 1 = 3 + 1$$
$$\Rightarrow \quad \frac{DB + CD}{CD} = 4$$
$$\Rightarrow \quad \frac{BC}{CD} = 4$$
$$\Rightarrow \quad CD = \frac{1}{4} BC$$
$$= \frac{1}{2} BC^2$$

$\Rightarrow \quad 2(AB^2 - AC^2) = BC^2$

$\Rightarrow \quad 2AB^2 - 2AC^2 = BC^2$

$\Rightarrow \quad 2AB^2 = 2AC^2 + BC^2.$

Example 15. *In an equilateral triangle D is a point on side BC such that $BD = \dfrac{1}{3} BC$. Prove that $9 AD^2 = 7 AB^2$.*

Sol. Given: In an equilateral triangle D is a point on side BC such that $BD = \dfrac{1}{3} BC$.

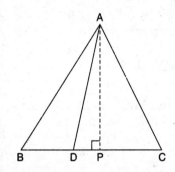

To Prove: $9 AD^2 = 7 AB^2$.

Construction: Draw $AD \perp BC$.

Proof: In right triangle APB,

$AB^2 = AP^2 + BP^2$ ...(1)

| By Pythagoras theorem

In right triangle APD,

$AD^2 = AP^2 + DP^2$ ...(2)

| By Pythagoras theorem

From (2),

$AP^2 = AD^2 - DP^2$ ...(3)

From (3), putting the value of AP^2 in get

$AB^2 = AD^2 - DP^2 + BP^2$

$= AD^2 - DP^2 + \left(\dfrac{BC}{2}\right)^2.$

In right triangle APB and APC,

Hyp. AB = Hyp. AC

AP = AP | Common side

$\therefore \quad \triangle APB \cong \triangle ABC$

| RHS congruence criterion

$\therefore \quad BP = CP$ | CPCT

$\therefore \quad BP = CP = \dfrac{BC}{2}$

$= AD^2 - DP^2 + \dfrac{BC^2}{4}$

$= AD^2 - (BP - BD)^2 + \dfrac{BC^2}{4}$

$= AD^2 - (BP^2 + BD^2 - 2 BP \cdot BD) + \dfrac{BC^2}{4}$

$= AD^2 - BP^2 - BD^2 + 2 BP \cdot BD + \dfrac{BC^2}{4}$

$= AD^2 - \left(\dfrac{BC}{2}\right)^2 - \left(\dfrac{BC}{3}\right)^2$

$+ 2\left(\dfrac{BC}{2}\right)\left(\dfrac{BC}{3}\right) + \dfrac{BC^2}{4}$ $\quad \left| \begin{array}{l} \because \ BP = \dfrac{BC}{2} \\ \text{and } BD = \dfrac{BC}{3} \end{array} \right.$

$= AD^2 - \dfrac{BC^2}{4} - \dfrac{BC^2}{9} + \dfrac{BC^2}{3} + \dfrac{BC^2}{4}$

$= AD^2 + \dfrac{2}{9} BC^2$

$= AD^2 + \dfrac{2}{9} AB^2$ | $\because$ AB = BC

$\Rightarrow \quad AB^2 \left(1 - \dfrac{2}{9}\right) = AD^2$

$\Rightarrow \quad \dfrac{7}{9} AB^2 = AD^2$

$\Rightarrow \quad 7 AB^2 = 9 AD^2.$

Example 16. *In an equilateral triangle, prove that three times the square of one side is equal to four times the square of one of its altitudes.*

Sol. Given: In $\triangle ABC$,

$AB = BC = CA$

and $\quad AD \perp BC.$

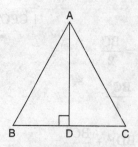

To Prove: $3 AB^2 = 4 AD^2$.

Proof: In right triangle ADB and ADC,

Hyp. AB = Hyp. AC | Given

AD = AD | Common side

∴ △ADB ≅ △ADC

        | RHS congruence criterion

∴ BD = CD | CPCT

$= \frac{1}{2} BC$

       | ∵ D is the mid-point of BC

$= \frac{1}{2} AB$

        | ∵ BC = AB (given)

Now, in right triangle ADB,

∵ ∠ADB = 90°

∴ $AB^2 = AD^2 + BD^2$

        | By Pythagoras theorem

$= AD^2 + \left(\frac{1}{2} AB\right)^2$

$= AD^2 + \frac{1}{4} AB^2$

$\Rightarrow \left(1 - \frac{1}{4}\right) AB^2 = AD^2$

$\Rightarrow \frac{3}{4} AB^2 = AD^2$

$\Rightarrow 3 AB^2 = 4 AD^2$.

Example 17. *Tick the correct answer and justify:* In △ABC, $AB = 6\sqrt{3}$ cm, $AC = 12$ cm and $BC = 6$ cm. The angles is B:

(A) 120°   (B) 60°
(C) 90°   (D) 45°.

Sol. $AB^2 = (6\sqrt{3})^2 = 108$

$AC^2 = (12)^2 = 144$

$BC^2 = (6)^2 = 36$

We see that

$AB^2 + BC^2 = AC^2$

∴ △ABC is right angled with

∠B = 90°

Hence, the correct answer is (C) 90°.

ADDITIONAL EXAMPLES

Example 18. In △BAC, ∠BCA, is a right angle and Q is the mid-point of BC. Prove that

(i) $BC^2 = 4(AQ^2 - AC^2)$

(ii) $AB^2 = 4 AQ^2 - 3 AC^2$.

Sol. Given: In △ABC, ∠BCA = 90° and Q is the mid-point of BC.

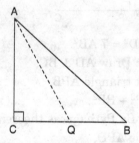

To Prove: (i) $BC^2 = 4(AQ^2 - AC^2)$

(ii) $AB^2 = 4 AQ^2 - 3 AC^2$

Construction: Join AQ.

Proof: (i) In right triangle ACQ,

∵ ∠C = 90°

∴ $AQ^2 = AC^2 + CQ^2$

        | By Pythagoras theorem

$\Rightarrow AQ^2 - AC^2 = CQ^2$

$\Rightarrow 4(AQ^2 - AC^2) = 4 CQ^2$

$= (2 CQ)^2$

$= BC^2$

        | ∵ Q is the mid-point of BC

(ii) In right triangle ABC,
∵ ∠C = 90°
∴ AB² = AC² + BC²   ...(1)
| By Pythagoras theorem
In right trangle ACQ,
∵ ∠C = 90°
∴ AQ² = AC² + CQ²
| By Pythagoras theorem
⇒ 4 AQ² = 4 AC² + 4 CQ²   ...(2)
Subtracting (2) from (1), we get
AB² − 4 AQ² = (AC² + BC²)
                    − (4 AC² + 4 CQ²)
= − 3 AC² + BC² − 4 CQ²
= − 3 AC² + BC² − (2 CQ)²
= − 3 AC² + BC² − (BC)²
| ∵ Q is the mid-point of BC
= − 3 AC²
⇒ AB² = 4 AQ² − 3 AC².

Example 19. *In a quadrilateral ABCD, ∠B = 90°, AD² = AB² + BC² + CD². Prove that ∠ACD = 90°.*

Sol. Given: In a quadrilateral ABCD, ∠B = 90° and AD² = AB² + BC² + CD².

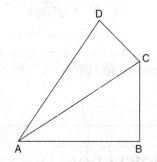

To Prove: ∠ACD = 90°.
Proof: In right triangle ABC,
∵ ∠B = 90°
∴ AC² = AB² + BC²   ...(1)
| By Pythagoras theorem
But AD² = AB² + BC² + CD² | Given
        = AC² + CD² | from (1)
∴ ∠ACD = 90°
| By converse of Phythagoras theorem

Example 20. *A right triangle has hypotenuse of length p cm and one side of length q cm. If p − q = 1, express the length of the third side of the right triangle in terms of p.*

Sol. Let ABC be a right triangle in which ∠B = 90°. Let AC = p cm and AB = q cm.
Then, By Pythagoras theorem in △ABC,
AC² = AB² + BC²
⇒ $p^2 = q^2 + BC^2$
⇒ $BC^2 = p^2 - q^2$

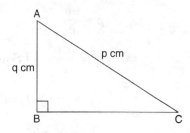

⇒ $BC^2 = p^2 - (p-1)^2$ | ∵ p − q = 1
                          | ∴ q = p − 1
⇒ $BC^2 = p^2 - (p^2 - 2p + 1)$
⇒ $BC^2 = 2p - 1$
⇒ $BC = \sqrt{2p-1}$ cm.

Example 21. *ABC is a right triangle, right angled at C. It p is the length of the prependicular from C to AB and AB = c, BC = a and CA = b, than prove that*

(i) $pc = ab$ (ii) $\dfrac{1}{p^2} = \dfrac{1}{a^2} + \dfrac{1}{b^2}$.

Sol. Given: ABC is a right triangle, right angled at c. p is the length of the prependicular from c to AB. Also, AB = c, BC = a and CA = b.

To Prove: (i) pc = ab

(ii) $\dfrac{1}{p^2} = \dfrac{1}{a^2} + \dfrac{1}{b^2}$

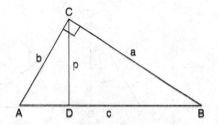

Proof: (i) The base of ABC is c and altitude is p.

$$\therefore \quad ar(\triangle ABC) = \frac{1}{2}cp \quad ...(1)$$

Also, $ar(\triangle ABC) = \frac{1}{2}ab \quad ...(2)$

From (1) and (2),

$$\frac{1}{2}cp = \frac{1}{2}ab$$

$$\Rightarrow \quad cp = ab \quad ...(3)$$

(ii) From (3),

$$p = \frac{ab}{c}$$

$$\Rightarrow \quad p^2 = \frac{a^2b^2}{c^2} \quad | \text{ Squaring both sides}$$

$$\Rightarrow \quad \frac{1}{p^2} = \frac{c^2}{a^2b^2} \quad | \text{ Taking reciprocals}$$

$$\Rightarrow \quad \frac{1}{p^2} = \frac{a^2+b^2}{a^2b^2}$$

$\because$ In right triangle angle ACB.

| By Pythagoras theorem

$$AB^2 = BC^2 + AC^2$$

$$\Rightarrow \quad c^2 = a^2 + b^2$$

$$\Rightarrow \quad \frac{1}{p^2} = \frac{a^2}{a^2b^2} + \frac{b^2}{a^2b^2}$$

$$\Rightarrow \quad \frac{1}{p^2} = \frac{1}{b^2} + \frac{1}{a^2}$$

$$\Rightarrow \quad \frac{1}{p^2} = \frac{1}{a^2} + \frac{1}{b^2}.$$

Example 22. *A person goes 10 East and then 30 m due North. Find tance from the starting point.*

Sol. Let A be the starting point.

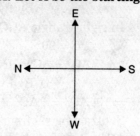

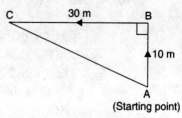

In right triangle ABC,

$\because \quad \angle B = 90°$

$\therefore \quad AC^2 = AB^2 + BC^2$

| By Pythagoras th

$= (10)^2 + (30)^2$

$= 100 + 900 = 1000$

$\Rightarrow \quad AC = \sqrt{1000}$

$\Rightarrow \quad AC = 10\sqrt{10}$ m

Hence, his distance from the st point is $10\sqrt{10}$ m.

TEST YOUR KNOWLEDGE

1. In figure, $\angle ACB = 90°$ and $CD \perp AB$. Prove that $\dfrac{BC^2}{AC^2} = \dfrac{BD}{AD}$.

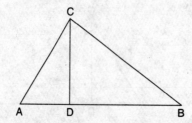

2. A ladder is placed against a wall su its foot is at a distance of 2.5 m fr wall and its top reaches a window 6 n the ground. Find the length of the la

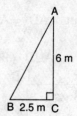

3. In figure, if AD ⊥ BC, prove that
   $AB^2 + CD^2 = BD^2 + AC^2$.

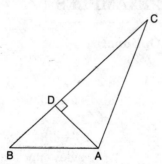

4. BL and CM are medians of a triangle ABC right angled at A. Prove that
   $4(BL^2 + CM^2) = 5BC^2$.

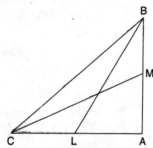

5. O is any point divides a rectangle ABCD (see figure). Prove that
   $OB^2 + OD^2 = OA^2 + OC^2$.

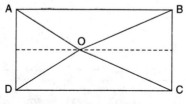

6. If △ABC, AB = BC and AD ⊥ BC. Prove that
   $AD^2 = 3BD^2$

7. A ladder reaches a window which is 12 m above the ground on one side of the street keeping its foot at the same point, the ladder is turned to the other side of the street to reach a window 9 m high. Find the width of the street if the length of the ladder is 15 cm.

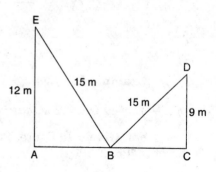

8. In △ABC, ∠C is a right angle and $AC = \sqrt{3}$ BC prove that ∠ABC = 60°.

9. ABC is a triangle in which AB = AC and D is any point on BC. Prove that
   $AB^2 - AD^2 = BD \cdot CD$.

10. In isosceles triangle ABC with AB = AC, BD is perpendicular from B to the side AC. Prove that
    $BD^2 - CD^2 = 2CD \cdot AD$.

11. In an equilateral triangle of side a, prove that

    (i) the altitude is of length $\dfrac{\sqrt{3}a}{2}$,

    (ii) the area of the triangle is $\dfrac{\sqrt{3}a^2}{4}$.

12. In a right angled triangle of a perpendicular is drawn from the right angle to the hypotenuse, the square on the perpendicular is equal to the rectangle contained by the two segments of the hypotenuse prove.

Answers

2. 6.5 cm   7. 21 m.

ILLUSTRATIVE EXAMPLES

[Exercise 6.6 (Optional)*]
(Page No. 152)

Example 1. *In figure, PS is the bisector of $\angle QPR$ of $\triangle PQR$. Prove that $\dfrac{QS}{SR} = \dfrac{PQ}{PR}$.*

Sol. Given: In figure, PS is the bisector of $\angle QPR$ of $\triangle PQR$.

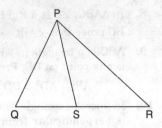

To Prove: $\dfrac{QS}{SR} = \dfrac{PQ}{PR}$.

Construction: Draw RT ∥ SP to meet QP produced in T.

Proof: ∵ RT ∥ SP and transversal PR intersects them

∴ $\angle 1 = \angle 2$ ...(1) | Alt.Int. $\angle s$

∵ RT ∥ SP and transversal QT intersects them

∴ $\angle 3 = \angle 4$ ...(2)
| Corres $\angle s$

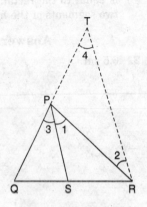

But $\angle 1 = \angle 3$ | Given

∴ $\angle 2 = \angle 4$ | From (1) and (2)

∴ PT = PR ...(3)

| ∵ sides opposite to equal angles of a triangle are equal

Now, in $\triangle QRT$,

PS = RT | By construction

∴ $\dfrac{QS}{SR} = \dfrac{PQ}{PT}$

| By basic proportionally theorem

⇒ $\dfrac{QS}{SR} = \dfrac{PQ}{PR}$ | From (3)

Example 2. *In figure, D is a point on hypotenuse AC of $\triangle ABC$, DM ⊥ BC and DN ⊥ AB. Prove that :*

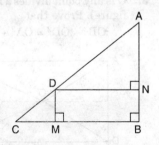

(i) $DM^2 = DN \cdot MC$

(ii) $DN^2 = DM \cdot AN$

Sol. Given: D is a point on hypotenuse AC of $\triangle ABC$, DM ⊥ BC and DN ⊥ AB.

To Prove: (i) $DM^2 = DN \cdot MC$

(ii) $DN^2 = DM \cdot AN$.

Construction: Join NM. Let BD and NM intersect at O.

Proof: (i) In $\triangle DMC$ and $\triangle NDM$,

$\angle DMC = \angle NDM$

| Each equal to 90°

$\angle MCD = \angle DMN$

---

*These exercises are not from examination point of view.

TRIANGLES

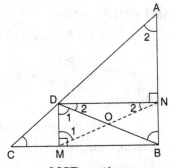

Let MCD = ∠1
Then, ∠MDC = 90° − ∠1
∴ ∠ODM = 90° − (90° − ∠1)
 = ∠1
∴ ∠DMO = ∠ODM = ∠1
⇒ ∠DMN = ∠1
∴ ΔDMC ~ ΔNDM
 | AA similarity criterion

∴ $\dfrac{DM}{ND} = \dfrac{MC}{DM}$

 | Corresponding sides of the similar triangles are proportional

⇒ $DM^2 = DN \times MC$

(ii) In ΔDNM and ΔNAD,
 ∠NDM = ∠AND
 | Each equal to 90°
 ∠DNM = ∠NAD
Let ∠NAD = ∠2
Then, ∠NDA = 90° − ∠2
∴ ∠ODN = 90° − (90° − ∠2) = ∠2
∴ ∠DNO = ∠2
⇒ ∠DNM = ∠2
∴ ΔDNM ~ ΔNAD
 | AA similarity criterion

∴ $\dfrac{DN}{NA} = \dfrac{DM}{ND}$

⇒ $\dfrac{DN}{AN} = \dfrac{DM}{DN}$

⇒ $DN^2 = DM \times AN$.

Example 3. *In figure, ABC is a triangle in which* ∠ABC > 90° *and AD* ⊥ *CB produced. Prove that* $AC^2 = AB^2 + BC^2 + 2BC \cdot BD$.

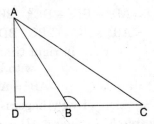

Sol. Given: In figure, ABC is a triangle in which ∠ABC > 90° and AD ⊥ CB produced.

To Prove: $AC^2 = AB^2 + BC^2 + 2BC \cdot BD$

Proof: In right triangle ABC,
∵ ∠D = 90°
∴ $AC^2 = AD^2 + DC^2$
 | By Pythagoras theorem
 $= AD^2 + (DB + BC)^2$
 $= AD^2 + DB^2 + BC^2 = 2DB \cdot BC$
 $= AB^2 + BC^2 + 2BC^2 \cdot BD$
 ∵ In right triangle ADB
 with ∠D = 90°,
 $AB^2 = AD^2 + DB^2$.
 By Pythagoras theorem

Example 4. *In figure, ABC is a triangle in which* ∠ABC < 90° *and AD* ⊥ *BC. Prove that :*

$AC^2 = AB^2 + BC^2 − 2BC \cdot BD$.

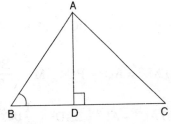

Sol. Given: In figure, ABC is a triangle in which ∠ABC < 90° and AD ⊥ BC.

To Prove:
$AC^2 = AB^2 + BC^2 − 2BC \cdot BD$.

Proof: In right triangle ADC,
∴ ∠D = 90°
∴ $AC^2 = AD^2 + DC^2$
 | By Pythagoras theorem
 $= AD^2 + (BC − BD)^2$
 $= AD^2 + BC^2 + BD^2 − 2BC \cdot BD$

$= AD^2 + BD^2 + BC^2 - 2BC \cdot BD$

$= AB^2 + BC^2 - 2BC \cdot BD$

∴ In right triangle ADB with $\angle D = 90°$, $AB^2 = AD^2 + BD^2$

| By Pythagoras theorem

Example 5. *In figure, AD is a median of a triangle ABC and $AM \perp BC$. Prove that :*

(i) $AC^2 = AD^2 + BC \cdot DM + \left(\dfrac{BC}{2}\right)^2$

(ii) $AB^2 = AD^2 - BC \cdot DM + \left(\dfrac{BC}{2}\right)^2$

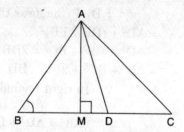

(iii) $AC^2 + AB^2 = 2AD^2 + \dfrac{1}{2} BC^2$

Sol. Given: In figure, AD is a median of a triangle ABC and $AM \perp BC$.

To Prove:

(i) $AC^2 = AD^2 + BC \cdot DM + \left(\dfrac{BC}{2}\right)^2$

(ii) $AB^2 = AD^2 - BC \cdot DM + \left(\dfrac{BC}{2}\right)^2$

(iii) $AC^2 + AB^2 = 2AD^2 + \dfrac{1}{2} BC^2$.

Proof: (i) In right triangle AMC,

∵ $\angle M = 90°$

∴ $AC^2 = AM^2 + MC^2$

| By Pythagoras theorem

$= AM^2 + (MD^2 + DC^2)$

$= AM^2 + MD^2 + DC^2 + 2MD \cdot DC$

$= AD^2 + \left(\dfrac{BC}{2}\right)^2 + 2DC \cdot DM$.

In right triangle AMD with $\angle M = 90°$,

$AM^2 + MD^2 = AD^2$

| By Pythagoras theorem

$= AD^2 + \left(\dfrac{BC}{2}\right)^2 + BC \cdot DM$

∵ $2DC = BC$

(∵ AD is a median of $\triangle ABC$)

(ii) In right triangle AMB,

∵ $\angle M = 90°$

∴ $AB^2 = AM^2 + MB^2$

$= AM^2 + (BD - MD)^2$

$= AM^2 + BD^2 + MD^2 - 2BD \cdot MD$

$= AM^2 + MD^2 + BD^2 - (2BD) MD$

$= AD^2 + \left(\dfrac{BC}{2}\right)^2 - BC \cdot DM$

In right triangle AMD with $\angle M = 90°$

$AM^2 + MD^2 = AD^2$

| By Pythagoras theorem

$2BD = BC$

(∵ AD is a median of $\triangle ABC$)

$= AD^2 - BC \cdot DM + \left(\dfrac{BC}{2}\right)^2$

(iii) From (i) and (ii), on adding, we get

$AC^2 + AB^2 = 2AD^2 + \dfrac{1}{2} BC^2$.

Example 6. *Prove that the sum of the squares of the diagonals of parallelogram is equal to the sum of the squares of its sides.*

Sol. Given: ABCD is a parallelogram whose dragonals are AC and BD.

To Prove: $AB^2 + BC^2 + CD^2 + DA^2 = AC^2 + BD^2$

Construction: Draw $AM \perp DC$ and $BN \perp D$ (Produced).

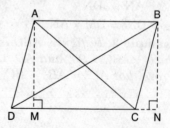

TRIANGLES

Proof: In right triangles AMD and BNC,

$$AD = BC$$
| Opp. sides of a ||gm
$$AM = BN$$
| Both are altitudes of the same Parallelogram to the same base

∴ $\triangle AMD \cong \triangle BNC$
| RHS congruence criterion

∴ $MD = NC$ ...(1) | CPCT

In right triangle BND,

∵ $\angle N = 90°$

∴ $BD^2 = BN^2 + DN^2$
| By Pyathagoras theorem
$= BN^2 + (DC + CN)^2$
$= BN^2 + DC^2 + CN^2 + 2DC \cdot CN$
$= (BN^2 + CN^2) + DC^2 + 2DC \cdot CN$
$= BC^2 + DC^2 + 2DC \cdot CN$ ...(2)

∵ In right triangle BNC with $\angle N = 90°$,

$$BN^2 + CN^2 = BC^2$$
| By Pythagoras theorem

In right triangle AMC

∵ $\angle M = 90°$

∴ $AC^2 = AM^2 + MC^2$
$= AM^2 + (DC - DM)^2$
$= AM^2 + DC^2 + DM^2 - 2DC \cdot DM$
$= (AM^2 + DM^2) + DC^2 - 2DC \cdot DM$
$= AD^2 + DC^2 - 2DC \cdot DM$

| ∵ In right triangle AMD with $\angle M = 90°$,
$$AD^2 = AM^2 + DM^2$$
(By Pythagoras theorem)

$= AD^2 + AB^2 - 2DC \cdot CN$ ...(3)

∵ $DC = AB$ (opp. sides of || gm)

and $BM = CN$ | From (1)

Adding (3) and (2), we get

$AC^2 + BD^2 = (AD^2 + AB^2) + (BC^2 + DC^2)$
$= AB^2 + BC^2 + CD^2 + DA^2$.

Example 7. *In figure, two chords AB and CD intersect each other at the point P. Prove that :*

(i) $\triangle APC \sim \triangle DPB$

(ii) $AP \cdot PB = CP \cdot DP$.

Sol. Given: In figure, two chords AB and CD intersect each other at the point P.

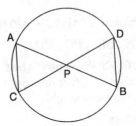

To Prove: (i) $\triangle APC \sim \triangle DPB$

(ii) $AP \cdot PB = CP \cdot DP$.

Proof: (i) $\triangle APC$ and $\triangle DPB$

$\angle APC = \angle DPB$
| Vert. opp. $\angle s$

$\angle CDP = \angle BDP$
| Angles in the same segment

∴ $\triangle APC \sim \triangle DPB$
| AA similarity criterion

(ii) $\triangle APC \sim \triangle DPB$ | Proved above in (1)

∴ $\dfrac{AP}{DP} = \dfrac{CP}{BP}$

∵ Corresponding sides of two similar triangles are proportional.

⇒ $AP \cdot BP = CP \cdot DP$

⇒ $AP \cdot PB = CP \cdot DP$.

Example 8. *In figure, two chords AB and CD of a circle intersect each other at the point P (when produced) outside the circle. Prove that*

(i) $\triangle PAC \sim \triangle PDB$

(ii) $PA \cdot PB = PC \cdot PD$.

Sol. Given: In figure, two chords AB and CD of a circle intersect each other at the point P (when produced) out the circle.

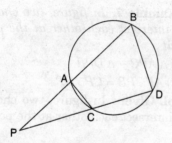

To Prove: (i) ΔPAC ~ ΔPDB

(ii) PA . PB = PC . PD.

Proof: (i) We know that in a cyclic quadrilaterals the exterior angle is equal to the interior opposite angle.

Therefore,

∠PAC = ∠PDB   ...(1)

and ∠PCA = ∠PBD   ...(2)

In view of (1) and (2),

ΔPAC ~ ΔPDB

| AA similarity criterion

(ii) ΔPAC ~ ΔPDB | Proved above in (1)

∴ $\dfrac{PA}{PD} = \dfrac{PC}{PB}$

| ∵ Corresponding sides of the similar triangles are proportional

⇒ PA . PB ~ PC . PD.

Example 9. *In figure, D is a point on side BC of ΔABC such that $\dfrac{BD}{CD} = \dfrac{AB}{AC}$. Prove that AD is the bisector of ∠BAC.*

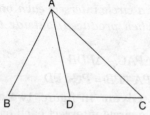

Sol. Given: In figure, D is a point on side BC of ΔABC such that $\dfrac{BD}{CD} = \dfrac{AB}{AC}$.

To Prove: AD is the bisector of ∠BAC i.e. ∠BAD = ∠CAD.

Construction: From BA produce cut off AE = A. Join CE.

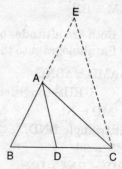

Proof: $\dfrac{BD}{CD} = \dfrac{AB}{AC}$ | Given

⇒ $\dfrac{BD}{CD} = \dfrac{AB}{AE}$

| ∵ AC = AE (by construction

∴ In ΔBCE,

AD ∥ CE

| By converse of basic proportionality theorem

∴ ∠BAD = ∠AEC   ...(1)

| Corres. ∠s

and ∠CAD = ∠ACE   ...(2)

| Alt. Int. ∠s

∵ AC = AE | By construction

∴ ∠AEC = ∠ACE   ...(3)

| Angles opposite equal sides of a triangle are equal

Using (3), (1) and (2) give

∠BAD = ∠CAD.

Example 10. *Nazima is fly fishing in a stream. The tip of her fishing rod is 1.8 m above the surface of the water and the fly at the end of the string rests on the water 3.6 m away and 2.4 m from a point directly under the tip of the rod. Assuming that her string (from the tip of her rod to the fly) is taut, how much string does she have out (see figure) ? If she pulls in the string at the rate of 5 cm per second, what will the horizontal distance of the fly from her after 12 seconds ?*

TRIANGLES

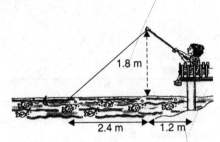

Sol. Length of the string that she has out

$= \sqrt{(1.8)^2 + (2.4)^2}$

| Using Pythagoras theorem

$= \sqrt{3.24 + 5.76}$

$= 3$ m

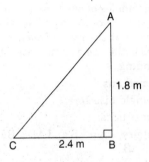

Hence, she has 3m string out.

Length of the string pulled in 12 seconds $= 5 \times 12 = 60$ cm $= 0.6$ cm.

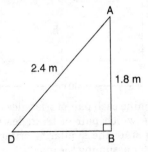

∴ Length of remaining string left out
$= 3.0 - 0.6 = 2.4$ m
$BD^2 = AD^2 - AB^2$
| By Pythagoras theorem
$= (2.4)^2 - (1.8)^2$
$= 5.76 - 3.24 = 2.52$

$\Rightarrow \quad BD = \sqrt{2.52} = 1.59$ m (approx.)

Hence, the horizontal distance of the fly from Nazima after 12 seconds
$= 1.2 + 1.59 = 2.79$ m
(approx.)

MISCELLANEOUS EXERCISE

1. In the following figure, DE ∥ BC. Prove that

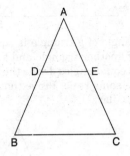

(i) $\dfrac{AB}{AD} = \dfrac{AC}{AE}$ (ii) $\dfrac{AB}{DB} = \dfrac{AC}{EC}$

2. P and Q are the points on the sides AB and AC respectively of a triangle ABC. Determine whether PQ is parallel to BC if AP = 4 cm, PB = 6 cm, AQ = 6 cm and QC = 8 cm.

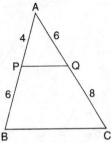

3. A vertical stick 12 cm long casts a shadow 8 cm long on the ground. At the same time a tower casts the shadow 40 cm long on the ground. Determine the height of the tower.

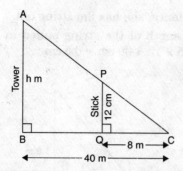

4. Examine each pair of triangles in figure and state which pair of triangles are similar. Also state the similarity criterion used by you for assuming the question and write the similarity relation in symbolic form.

(i)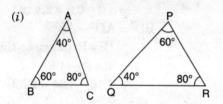

(ii)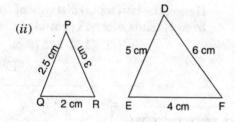

(iii)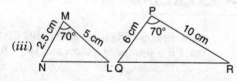

(iv)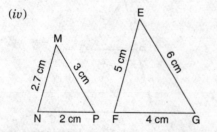

(v)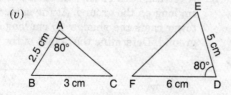

(vi)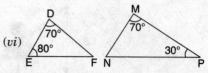

5. In figure, $\triangle EDC \sim \triangle EBA$, $\angle BEC = 115°$ and $\angle EDC = 70°$. Find $\angle DEC$, $\angle DCE$, $\angle EAB$, $\angle AEB$ and $\angle EBA$.

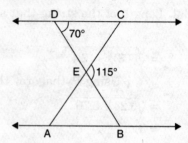

6. If the angles of one triangle on respectively equal to the angles of another triangle. Prove that the ratio of their corresponding sides is the same as the ratio of their corresponding
   (i) medians
   (ii) angle bisectors
   (iii) altitudes.

7. In figure, $\angle A = \angle B$ and $AD = BE$. Prove that $DE \parallel AB$.

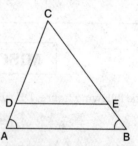

8. One angle of a triangle is equal to one angle of another triangle and the bisectors of these equal angles divide the opposite sides in the same ratio. Prove that the triangles are similar.

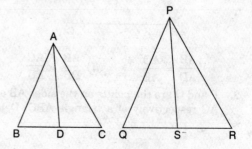

TRIANGLES

9. In △ABC, ∠A is acute, BD and CE are perpendiculars on AC and AB respectively. Prove that
$$AB \times AE = AC \times AD$$

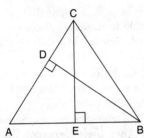

10. In △ABC, AD ⊥ BC and $AD^2 = BD \times DC$ or $\dfrac{AD}{DC} = \dfrac{BD}{AD}$. Prove that ∠BAC = 90°.

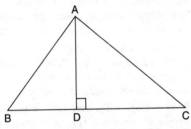

11. In the given figure, $\dfrac{AO}{OC} = \dfrac{BO}{OD} = \dfrac{1}{2}$ and AB = 4 cm. Find the value of DC.

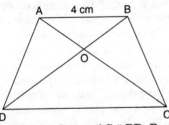

12. In the given figure, AC ∥ BD. Prove that
    (i) △ACE ~ △BDE
    (ii) $\dfrac{AE}{CE} = \dfrac{BE}{DE}$

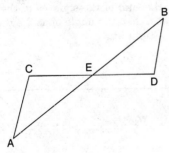

13. In the given figure, AD ∥ BC. Find the value of x.

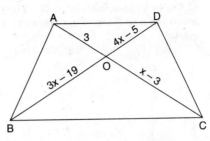

14. In a trapezium ABCD, AB ∥ DC and DC = 2AB. EF is drawn parallel to AB and cuts in F and BC in E such that $\dfrac{BE}{EC} = \dfrac{3}{4}$. Diagonal DB intersects FE at G. Prove that 7 FE = 10 AB.

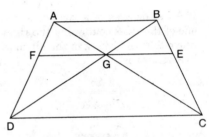

15. In △ABC, a line PQ is drawn parallel to the base BC, which meets the sides AB and AC in the points P and Q respectively. If AP = $\dfrac{1}{4}$ AB. Then find ar (△APQ) : ar (△ABC).

16. In the figure, ∠ABD = ∠CDB = ∠PQB = 90°. If AB = x units, CD = y units and PQ = z units, then prove that
$$\dfrac{1}{x} + \dfrac{1}{y} = \dfrac{1}{z}.$$

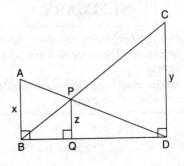

17. ABC is a right-angled triangle, right-angled at A. A circle is inscribed in it. The lengths of the two sides containing the right angle are 6 cm and 8 cm. Find the radius of the circle.

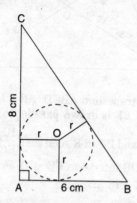

18. If A be the area of a right triangle and b one of the sides containing the right angle, prove that the length of the altitude on the hypotenuse is

$$\frac{2Ab}{\sqrt{b^2 + 4A^2}}.$$

Answers

3. 60 m
4. (i) $\triangle ABC \sim \triangle PQR$ |AA
   (ii) $\triangle PQR \sim \triangle DEF$ |SSS
   (iii) No (iv) No (v) No
   (vi) $\triangle DEF \sim \triangle MNP$ |AA
5. 65°, 45°, 45°, 65°, 70°
11. 8 cm   12. 8 or 9
15. 1 : 16   17. 2 cm.

SUMMARY

1. Two figures having the same shape but not necessarily the same size are called similar figures.
2. All the congruent figures are similar but the converse is not true.
3. Two polygons of the same number of sides are similar, if (i) their corresponding angles are equal and (ii) their corresponding sides are in the same ratio (i.e. proportional).
4. If a line is drawn parallel to one side of a triangle to intersect the other two sides in distinct points, then the other two sides are divided in the same ratio.
5. If a line divides any two sides of a triangle in the same ratio, then the line is parallel to the third side.
6. If in two triangles, corresponding angles are equal, then their corresponding sides are in the same ratio and hence the two triangles are similar (AAA similarity criterion).
7. If in two triangles, two angles of one triangle are respectively equal to the two angles of the other triangle, then the two triangles are similar (AAA similarity criterion).
8. If in two triangles, corresponding sides are in the same ratio, then their corresponding angles are equal and hence the triangles are similar (SSS similarity criterion).
9. If one angle of a triangle is equal to one angle of another triangle and the sides including these angles are in the same ratio (proportional), then the triangles are similar (SAS similarity criterion).
10. The ratio of the areas of two similar triangles is equal to the square of the ratio of their corresponding sides.
11. If a perpendicular is drawn from the vertex of the right angle of a right triangle to the hypotenuse, then the triangles on both sides of the perpendicular are similar to the whole triangle and also to each other.
12. In a right triangle, the square of the hypotenuse is equal to the sum of the squares of the other two sides (Pythagoras Theorem).
13. If in a triangle, square of one side is equal to the sum of the squares of the other two sides, then the angle opposite the first side is a right angle.

7

Coordinate Geometry

IMPORTANT POINTS

1. Abscissa and ordinate. To locate the position of a point on a plane, we require a pair of coordinate axes. The distance of a point from the y-axis is called its **x-coordinate**, or **abscissa**. The distance of a point from the x-axis is called its **y-coordinate**, or **ordinate**. The coordinates of a point on the x-axis are of the form $(x, 0)$, and of a point on the y-axis are of the form $(0, y)$.

2. A play. Draw a set of perpendicular axes on graph paper. Now plot the following points and join them as directed : Join the point A(4, 8) to B(3, 9) to C(3, 8) to D(1, 6) to E(1, 5)

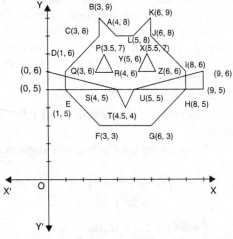

to F(3, 3) to G(6, 3) to H(8, 5) to I(8, 6) to J(6, 8) to K(6, 9) to L(5, 8) to A. Then join the points P(3.5, 7), Q(3, 6) and R(4, 6) to form a triangle. Also join the points X(5.5, 7), Y(5, 6) and Z(6, 6) to form a triangle. Now join S(4, 5), T(4.5, 4) and U(5, 5) to form a triangle. Lastly join S to the points (0, 5) and (0, 6) and join U to the points (9, 5) and (9, 6). The picture we get is that of a cat's face.

3. Scope. We know that a linear equation in two variables of the form $ax + by + c = 0$, (a, b are not simultaneously zero), when represented graphically, gives a straight line. Further, we know that the graph of $y = ax^2 + bx + c (a \neq 0)$, is a parabola. In fact, coordinate geometry has been developed as an algebraic tool for studying geometry of figures. It helps us to study geometry using algebra, and understand algebra with the help of geometry. Because of this, coordinate geometry is widely applied in various fields such as physics, engineering, navigation, seismology and art !

4. Certain situation. (*i*) Let us consider the following situation:

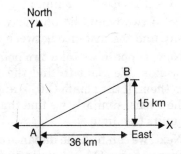

A town B is located 36 km east and 15 km north of the town A, which is densely populated. We want to find the distance from town A to town B without actually measuring it. This situation can be represented graphically as shown in figure. We may use the Pythagoras Theorem to calculate this distance.

287

$$AB = \sqrt{(15)^2 + (36)^2}$$
$$= \sqrt{225 + 1296}$$
$$= \sqrt{1521} = 39 \text{ km}$$

(ii) Let us consider the following situation:

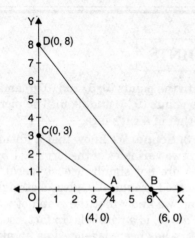

Suppose two points lie on the x-axis. We want to find the distance between them. For instance, consider two points A(4, 0) and B(6, 0) in figure. The points A and B lie on the x-axis.

From the figure we can see that OA = 4 units and OB = 6 units.

Therefore, the distance of B from A, *i.e.*, AB = OB – OA = 6 – 4 = 2 units.

So, if two points lie on the x-axis, we can easily find the distance between them.

Now, suppose we take two points lying on the y-axis. We want to find the distance between them. If the points C(0, 3) and D(0, 8) lie on the y-axis, similarly we find that CD = 8 – 3 = 5 units (see figure).

Next, we can find the distance of A from C (in figure). Since OA = 4 units and OC = 3 units, the distance of A from C, *i.e.*, AC = $\sqrt{3^2 + 4^2}$ = 5 units. Similarly, we can find the distance of B from D = BD = $\sqrt{6^2 + 8^2} = \sqrt{100}$ = 10 units.

(iii) Let us consider the following situation:

Let us consider two points not lying on coordinate axis, we want to find the distance between them. We shall use Pythagoras theorem to do so. Let us see an example.

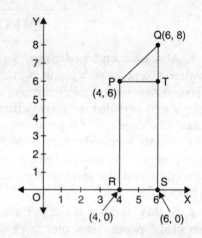

In figure, the points P(4, 6) and Q(6, 8) lie in the first quadrant. To use Pythagoras theorem to find the distance between them, we draw PR and QS perpendicular to the x-axis from P and Q respectively. Also, draw a perpendicular from P on QS to meet QS at T. Then the coordinates of R and S are (4, 0) and (6, 0), respectively. So, RS = 2 units. Also, QS = 8 units and TS = PR = 6 units.

Therefore, QT = 2 units and PT = RS = 2 units.

Now, using the Pythagoras theorem, we have
$$PQ^2 = PT^2 + QT^2$$
$$= 2^2 + 2^2 = 8$$

So, PQ = $2\sqrt{2}$ units

(iv) Let us consider the following situation:

COORDINATE GEOMETRY

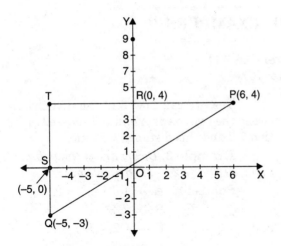

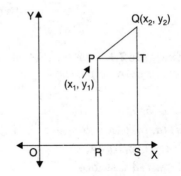

Also, $SQ = y_2$, $ST = PR = y_1$.
So, $QT = y_2 - y_1$.

We want to find the distance between two points in different quadrants.

Consider the points P(6, 4) and Q(– 5, – 3) (see figure). Draw QS perpendicular to the x-axis. Also draw a perpendicular PT from the point P on QS (extended) to meet y-axis at the point R.

Then PT = 11 units and QT = 7 units (Why ?).

Using the Pythagoras Theorem to the right triangle PTQ, we get

$$PQ = \sqrt{11^2 + 7^2}$$
$$= \sqrt{170} \text{ units.}$$

Distance Formula. Now we want to find the distance between any two points $P(x_1, y_1)$ and $Q(x_2, y_2)$. Draw PR and QS perpendicular to the x-axis. A perpendicular from the point P on QS is drawn to meet it at the point T (see figure).

Then, $OR = x_1$, $OS = x_2$.
So, $RS = x_2 - x_1 = PT$.

Now, applying the Pythagoras theorem in $\triangle$ PTQ, we get

$$PQ^2 = PT^2 + QT^2$$
$$= (x_2 - x_1)^2 + (y_2 - y_1)^2$$
$$PQ = \sqrt{(x_2 - x_1)^2 + (y_2 - y_1)^2}$$

Note that since distance is always non-negative, we take only the positive square root, So, the distance between the points $P(x_1, y_1)$ and $Q(x_2, y_2)$ is

$$PQ = \sqrt{(x_2 - x_1)^2 + (y_2 - y_1)^2},$$

which is called the **distance formula**.

Remarks : 1. In particular, the distance of a point $P(x, y)$ from the origin $O(0, 0)$ is given by

$$OP = \sqrt{x^2 + y^2}.$$

2. We can also write,

$$PQ = \sqrt{(x_1 - x_2)^2 + (y_1 - y_2)^2}$$
$$\because \quad (x_1 - x_2)^2 = (x_2 - x_1)^2$$
$$\text{and} \quad (y_1 - y_2)^2 = (y_2 - y_1)^2$$

ILLUSTRATIVE EXAMPLES

[NCERT Exercise 7.1]
(Page No. 161)

Example 7.1. *Find the distance between the following pairs of points*
(i) (2, 3), (4, 1)
(ii) (– 5, 7), (– 1, 3)
(iii) (a, b), (– a, – b)

Sol. (i) **(2, 3), (4, 1)**
Required distance
$$= \sqrt{(4-2)^2 + (1-3)^2}$$
$$= \sqrt{4+4} = \sqrt{8} = 2\sqrt{2}$$

(ii) **(– 5, 7), (– 1, 3)**
Required distance
$$= \sqrt{\{-1-(-5)\}^2 + (3-(-1))^2}$$
$$= \sqrt{16+16} = \sqrt{32}$$
$$= 4\sqrt{2}$$

(iii) **(a, b), (– a, – b)**
Required distance
$$= \sqrt{(-a-a)^2 + (-b-b)^2}$$
$$= \sqrt{(-2a)^2 + (-2b)^2}$$
$$= \sqrt{4a^2 + 4b^2}$$
$$= \sqrt{4(a^2+b^2)}$$
$$= 2\sqrt{a^2+b^2}.$$

Example 2. *Find the distance between the points (0, 0) and (36, 15). Can you now find the distance between the two towns A and B discussed in Section 7.2.*

Sol. Required distance
$$= \sqrt{(36-0)^2 + (15-0)^2}$$
$$= \sqrt{(36)^2 + (15)^2}$$
$$= \sqrt{1296 + 225} = \sqrt{1521}$$
$$= 39.$$

Yes, we can now find the distance between the two towns A and B discussed in section 7.2 and this distance = 39 km.

Example 3. *Determine if the points (1, 5), (2, 3) and (– 2, – 11) are collinear.*

Sol. Let $A \to (1, 5)$
$B \to (2, 3)$
and, $C \to (-2, -11)$

Then $AB = \sqrt{(2-1)^2 + (3-5)^2}$
$$= \sqrt{1+4} = \sqrt{5}$$

$BC = \sqrt{(-2-2)^2 + (-11-3)^2}$
$$= \sqrt{(-4)^2 + (-18)^2}$$
$$= \sqrt{16 + 324}$$
$$= \sqrt{340}$$

$CA = \sqrt{\{1-(-2)\}^2 + \{5-(-11)\}^2}$
$$= \sqrt{(3)^2 + (16)^2}$$
$$= \sqrt{9 + 256}$$
$$= \sqrt{265}$$

We see that
$AB + BC \ne CA$
$BC + CA \ne AB$
and $CA + AB \ne BC$

Hence, the given points are not collinear.

Example 4. *Check whether (5, – 2), (6, 4) and (7, – 2) are the vertices of an isosceles triangle.*

Sol. Let $A \to (5, -2)$, $B \to (6, 4)$ and $C \to (7, -2)$.

Then,

$AB = \sqrt{(6-5)^2 + \{4-(-2)\}^2}$

$$= \sqrt{(1)^2 + (6)^2} = \sqrt{1+36}$$
$$= \sqrt{37}$$
$$BC = \sqrt{(7-6)^2 + (-2-4)^2}$$
$$= \sqrt{(1)^2 + (-6)^2}$$
$$= \sqrt{1+36} = \sqrt{37}$$

We see that $AB = BC$

Therefore, ABC is an isosceles triangle.

Example 5. *In a classroom, 4 friends are seated at the points A, B, C and D as shown in figure. Champa and Chameli walk into the class and after observing for a few minutes Champa asks Chameli, "Don't you think ABCD is a square ?" Chameli disagrees. Using distance formula, find which of them is correct.*

Sol. We see that
$A \to (3, 4)$
$B \to (6, 7)$
$C \to (9, 4)$
$D \to (6, 1)$

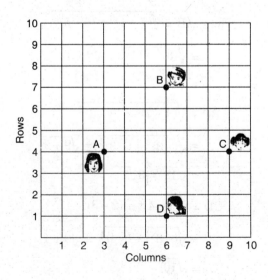

Now, $AB = \sqrt{(6-3)^2 + (7-4)^2}$
$$= \sqrt{(3)^2 + (3)^3}$$
$$= \sqrt{9+9} = \sqrt{18} = 3\sqrt{2}$$

$$BC = \sqrt{(9-6)^2 + (4-7)^2}$$
$$= \sqrt{(3)^2 + (-3)^2}$$
$$= \sqrt{9+9} = \sqrt{18} = 3\sqrt{2}$$
$$CD = \sqrt{(6-9)^2 + (1-4)^2}$$
$$= \sqrt{(-3)^2 + (-3)^2}$$
$$= \sqrt{9+9} = \sqrt{18}$$
$$= 3\sqrt{2}$$
$$DA = \sqrt{(3-6)^2 + (4-1)^2}$$
$$= \sqrt{(-3)^2 + (3)^2}$$
$$= \sqrt{9+9} = \sqrt{18} = 3\sqrt{2}$$
$$AC = \sqrt{(9-3)^2 + (4-4)^2} = 6$$
$$BD = \sqrt{(6-6)^2 + (1-7)^2} = 6$$

We see that
$AB = BC = CD = DA$
[*i.e.*, all the four sides are equal]
and $AC = BD$
[*i.e.*, the diagonals are equal]
Therefore, ABCD is a square.
Hence, Champa is correct.

Aliter. We have, $AB = 3\sqrt{2}$,
$BC = 3\sqrt{2}$, $CD = 3\sqrt{2}$, $DA = 3\sqrt{2}$
and $AC = 6$
$\therefore$ $AD^2 + DC^2 = 18 + 18 = 36 = AC^2$
Therefore, by the converse of Pythagoras
Therefore, $\angle D = 90°$
A quadrilateral with all four sides equal and one angle 90° is a square.
So, ABCD is a square.
Hence, Champa is correct.

Example 6. *Name the type of quadrilateral formed, if any, by the following points, and give reasons for your answer:*
(i) (– 1, – 2), (1, 0), (– 1, 2), (– 3, 0)

(ii) (– 3, 5), (3, 1), (0, 3), (– 1, – 4)
(iii) (4, 5), (7, 6), (4, 3), (1, 2)

Sol. (i) (– 1, – 2), (1, 0), (– 1, 2), (– 3, 0)
Let $A \to (-1, -2), B \to (1, 0),$
$C \to (-1, 2)$ and $D \to (-3, 0)$
Then,

$$AB = \sqrt{\{1-(-1)\}^2 + (0-(-2))^2}$$
$$= \sqrt{(2)^2 + (2)^2} = \sqrt{4+4}$$
$$= \sqrt{8} = 2\sqrt{2}$$

$$BC = \sqrt{(-1-1)^2 + (2-0)^2}$$
$$= \sqrt{(-2)^2 + (2)^2} = \sqrt{4+4}$$
$$= \sqrt{8} = 2\sqrt{2}$$

$$CD = \sqrt{\{-3-(-1)\}^2 + (0-2)^2}$$
$$= \sqrt{(-2)^2 + (-2)^2}$$
$$= \sqrt{4+4} = \sqrt{8} = 2\sqrt{2}$$

$$DA = \sqrt{\{-1-(-3)\}^2 + (-2-0)^2}$$
$$= \sqrt{(2)^2 + (-2)^2} = \sqrt{4+4}$$
$$= \sqrt{8} = 2\sqrt{2}$$

$$AC = \sqrt{\{-1-(-1)\}^2 + \{-2-(-2)\}^2}$$
$$= 4$$

$$BD = \sqrt{(-3-1)^2 + (0-0)^2} = 4$$

Since AB = BC = CD = DA (*i.e.*, all the four sides of the quadrilateral ABCD are equal) and AC = BD (*i.e.*, diagonals of the quadrilateral ABCD are equal)

Therefore, ABCD is a square.

(ii) (– 3, 5), (3, 1), (0, 3), (– 1, – 4)
Let $A \to (-3, 5), B \to (3, 1), C \to (0, 3)$ and $D(-1, -4)$.
Then, $AB = \sqrt{\{3-(-3)\}^2 + (1-5)^2}$
$$= \sqrt{(6)^2 + (-4)^2}$$
$$= \sqrt{36+16} = \sqrt{52} = 2\sqrt{13}$$

$$BC = \sqrt{(0-3)^2 + (3-1)^2}$$
$$= \sqrt{9+4} = \sqrt{13}$$

$$CD = \sqrt{(-1-0)^2 + (-4-3)^2}$$
$$= \sqrt{1+49} = \sqrt{50}$$

$$DA = \sqrt{\{-3-(-1)\}^2 + \{5-(-4)\}^2}$$
$$= \sqrt{4+81} = \sqrt{85}$$

$$AC = \sqrt{\{0-(-3)\}^2 + (3-5)^2} = \sqrt{13}$$

$$BD = \sqrt{(-1-3)^2 + (-4-1)^2} = \sqrt{41}$$

We see that
$$BC + AC = AB$$

Hence, the points A, B and C are collinear.

So, ABCD is not a quadrilateral.

(iii) **(4, 5), (7, 6), (4, 3) , (1, 2)**

Let $A \to (4, 5), B \to (7, 6), C \to (4, 3)$ and $D \to (1, 2)$.

Then, $AB = \sqrt{(7-4)^2 + (6-5)^2}$
$$= \sqrt{(3)^2 + (1)^2} = \sqrt{9+1}$$
$$= \sqrt{10}$$

$$BC = \sqrt{(4-7)^2 + (3-6)^2}$$
$$= \sqrt{(-3)^2 + (-3)^2} = \sqrt{9+9}$$
$$= \sqrt{18} = 3\sqrt{2}$$

$$CD = \sqrt{(1-4)^2 + (2-3)^2}$$
$$= \sqrt{(-3)^2 + (-1)^2} = \sqrt{9+1}$$
$$= \sqrt{10}$$

$$DA = \sqrt{(4-1)^2 + (5-2)^2}$$
$$= \sqrt{9+9} = \sqrt{18} = 3\sqrt{2}$$

$AC = \sqrt{(4-4)^2 + (3-5)^2}$

$ = 2$

$BD = \sqrt{(1-7)^2 + (2-6)^2}$

$ = \sqrt{36 + 16} = \sqrt{52}$

We see that
$\quad AB = CD$, Opposite sides are equal
$\quad BC = DA$
and $\quad AC \neq BD$
$\qquad$ | Diagonals are unequal

Hence, the quadrilateral ABCD is a parallelogram.

Example 7. *Find the point on the x-axis which is equidistant from (2, – 5) and (– 2, 9).*

Sol. We know that a point on the x-axis is of the form $(x, 0)$. So, let the point $P(x, 0)$ be equidistant from $A(2, -5)$ and $B(-2, 9)$. Then
$\qquad PA = PB$
$\Rightarrow PA^2 = PB^2$
$\Rightarrow (2 - x)^2 + (-5 - 0)^2$
$\qquad = (-2 - x)^2 + (9 - 0)^2$
$\Rightarrow 4 + x^2 - 4x + 25 = 4 + x^2 + 4x + 81$
$\Rightarrow 8x = -56$
$\Rightarrow x = \dfrac{-56}{8} = -7$

Hence, the required point is $(-7, 0)$

Check

$PA = \sqrt{\{2 - (-7)\}^2 + (-5 - 0)^2}$

$ = \sqrt{81 + 25} = \sqrt{106}$

$PB = \sqrt{\{-2 - (-7)\}^2 + (9 - 0)^2}$

$ = \sqrt{25 + 81} = \sqrt{106}$

$\therefore \quad PA = PB$
$\therefore \quad$ Our solution is checked.

Example 8. *Find the values of y for which the distance between the points P (2, – 3) and Q(10, y) is 10 units.*

Sol. $PQ = 10$ $\qquad$ | Given
$\qquad PQ^2 = 10^2 = 100$

$\Rightarrow (10 - 2)^2 + \{y - (-3)\}^2 = 100$
$\Rightarrow (8)^2 + (y + 3)^2 = 100$
$\Rightarrow 64 + y^2 + 6y + 9 = 100$
$\Rightarrow y^2 + 6y - 27 = 0$
$\Rightarrow y^2 + 9y - 3y - 27 = 0$
$\Rightarrow y(y + 9) - 3(y + 9) = 0$
$\Rightarrow (y + 9)(y - 3) = 0$
$\Rightarrow y + 9 = 0 \text{ or } y - 3 = 0$
$\Rightarrow y = -9 \text{ or } y = 3$
$\Rightarrow y = -9, 3$

Hence, the required value of y is -9 or 3.

Example 9. *If Q(0, 1) is equidistant from P(5, – 3) and R(x, 6), find the values of x. Also find the distances QR and PR.*

Sol. $PQ = RQ$. $\qquad$ | Given
$\Rightarrow PQ^2 = RQ^2$
$\Rightarrow (0 - 5)^2 + [1 - (-3)]^2$
$\qquad = (0 - x)^2 + (1 - 6)^2$
$\Rightarrow 25 + 16 = x^2 + 25$
$\Rightarrow x^2 = 16$
$\Rightarrow x = \pm 4$
$\therefore R \rightarrow (\pm 4, 6)$

$QR = \sqrt{(0 \pm 4)^2 + (1 - 6)^2} = \sqrt{41}$

$PR = \sqrt{(\pm 4 - 5)^2 + \{6 - (-3)\}^2}$

$ = \sqrt{(4-5)^2 + 81} \text{ or } \sqrt{(-4-5)^2 + 81}$

$ = \sqrt{82} \text{ or } 9\sqrt{2}$.

Example 10. *Find a relation between x and y such that the point (x, y) is equidistant from the point (3, 6) and (– 3, 4).*

Sol. Let $P \rightarrow (x, y)$, $A \rightarrow (3, 6)$ and $B \rightarrow (-3, 0)$

Then, $PA = PB$ $\qquad$ | Given
$\Rightarrow PA^2 = PB^2$
$\Rightarrow (3 - x)^2 + (6 - y)^2$
$\qquad = (-3 - x)^2 + (4 - y)^2$
$\Rightarrow 9 + x^2 - 6x + 36 + y^2 - 12y$
$\qquad = 9 + x^2 + 6x + 16 + y^2 - 8y$
$\Rightarrow 12x + 4y - 20 = 0$

$\Rightarrow \quad 3x + y - 5 = 0$

| Dividing throughout by 4

This is the required relation.

ADDITIONAL EXAMPLES

Example 11. *Show that the points A(2, –2), B(14, 10), C(11, 13) and D(–1, 1) are the vertices of a rectangle.* (AI CBSE 2004)

Sol. We have

$AB = \sqrt{(14-2)^2 + \{10-(-2)\}^2}$

$= \sqrt{144 + 144} = \sqrt{288} = 12\sqrt{2}$

$BC = \sqrt{(11-14)^2 + (13-10)^2}$

$= \sqrt{9+9} = 3\sqrt{2}$

$CD = \sqrt{(-1-11)^2 + (1-13)^2}$

$= \sqrt{144 + 144} = \sqrt{288}$

$DA = \sqrt{\{2-(-1)\}^2 + (-2-1)^2}$

$= \sqrt{9+9} = 3\sqrt{2}$

$AC = \sqrt{(11-2)^2 + (13-(-2))^2}$

$= \sqrt{81 + 225} = \sqrt{306}$

$BD = \sqrt{(-1-14)^2 + (1-10)^2}$

$= \sqrt{225 + 81}$

$= \sqrt{306}$.

We see that

AB = CD and BC = DA

$\Rightarrow$ Opposite sides are equal

Also, AC = BD

$\Rightarrow$ Diagonals are equal.

Hence, the given points A, B, C and D are the vertices of a rectangle.

Example 12. *Find the coordinates of the point equidistant from three given points A(5, 1), B(–3, –7) and C(7, –1).* (CBSE 2006)

Sol. We have

$A \to (5, 1)$

$B \to (-3, -7)$

$C \to (7, -1)$

Let the point $P(x, y)$ be equidistant from the three given points A, B and C.

Then PA = PB = PC

$\Rightarrow \quad PA^2 = PB^2 = PC^2$

First two give

$PA^2 = PB^2$

$\Rightarrow (x-5)^2 + (y-1)^2 = (x+3)^2 + (y+7)^2$

$\Rightarrow x^2 - 10x + 25 + y^2 - 2y + 1$
$\quad = x^2 + 6x + 9 + y^2 + 14y + 49$

$\Rightarrow 16x + 16y + 32 = 0$

$\Rightarrow \quad x + y + 2 = 0$

| Dividing throughout by 16 ...(1)

Last two give

$PB^2 = PC^2$

$\Rightarrow (x+3)^2 + (y+7)^2 = (x-7)^2 + (y+1)^2$

$\Rightarrow x^2 + 6x + 9 + y^2 + 14y + 49$
$\quad = x^2 - 14x + 49 + y^2 + 2y + 1$

$\Rightarrow 20x + 12y + 8 = 0$

$\Rightarrow \quad 5x + 3y + 2 = 0$

| Dividing throughout by 4 ...(2)

Multiplying equation (1) by 3, we get

$3x + 3y + 6 = 0$ ...(3)

Subtracting equation (3) from equation (2), we get

$2x - 4 = 0$

$\Rightarrow \quad 2x = 4$

$\Rightarrow \quad x = \dfrac{4}{2} = 2$

Substituting $x = 2$ in equation (1), we get

$2 + y + 2 = 0$

$\Rightarrow \quad y + 4 = 0$

$\Rightarrow \quad y = -4$

Hence, the required points is (2, –4).

Example 13. *Prove that the points (0, 0), (5, 5) and (–5, 5) are the vertices of a right isosceles triangle.* (AI CBSE 2005)

Sol. Let $O \to (0, 0)$, $A \to (5, 5)$

and $\quad B(-5, 5)$

Then, $OA = \sqrt{(5-0)^2 + (5-0)^2} = 5\sqrt{2}$

$AB = \sqrt{(-5-5)^2 + (5-5)^2} = 10$

$OB = \sqrt{(-5-0)^2 + (5-0)^2} = 5\sqrt{2}$

We see that

$OA = OB$

$\Rightarrow \triangle OAB$ is isosceles.

Also, $OA^2 + OB^2 = 50 + 50$
$= 100 = AB^2$

$\Rightarrow \triangle OAB$ is right angled with $\angle A = 90°$

Hence, $\triangle OAB$ is a right isosceles triangle.

Example 14. *Prove that the points (a, b + c), (b, c + a) and (c, a + b) are collinear.*

Sol. Let $A \to (a, b+c)$, $B \to (b, c+a)$ and $C \to (c, a+b)$

Then,

$AB = \sqrt{(b-a)^2 + \{(c+a)-(b+c)\}^2}$

$= \sqrt{(b-a)^2 + (a-b)^2}$

$= \sqrt{(a-b)^2 + (a-b)^2}$

$= \sqrt{2(a-b)^2} = \sqrt{2}\,(a-b)$

$BC = \sqrt{(c-b)^2 + \{(a+b)-(c+a)\}^2}$

$= \sqrt{(c-b)^2 + (b-c)^2}$

$= \sqrt{(b-c)^2 + (b-c)^2}$

$= \sqrt{2(b-c)^2} = \sqrt{2}\,(b-c)$

$CA = \sqrt{(a-c)^2 + \{(b+c)-(a+b)\}^2}$

$= \sqrt{(a-c)^2 + (c-a)^2}$

$= \sqrt{(a-c)^2 + (a-c)^2}$

$= \sqrt{2(a-c)^2} = \sqrt{2}\,(a-c)$

We see that

$AB + BC = \sqrt{2}\,(a-b) + \sqrt{2}\,(b-c)$

$= \sqrt{2}\,(a-b+b-c)$

$= \sqrt{2}\,(a-c) = CA$

Hence, the given points A, B and C are collinear.

TEST YOUR KNOWLEDGE

1. Do the points (3, 2), (– 2, – 3) and (2, 3) form a triangle? If so, name the type of triangle formed.
2. Show that the points (1, 7), (4, 2), (– 1, – 1) and (– 4, 4) are the vertices of a square.
3. Figure shows the arrangement of desks in a class-room. Ashima, Bharti and Camella are seated at A(3, 1), B(6, 4) and C(8, 6). Do you think they are seated in a line ? Give reasons for your answer.

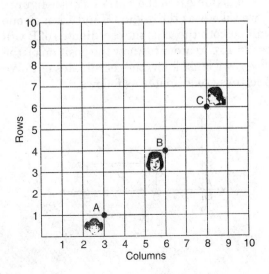

4. Find a relation between x and y such that the point (x, y) is equidistant from the points $(7, 1)$ and $(3, 5)$.

5. Find a point on the y-axis which is equidistant from the points A(6, 5) and B(– 4, 3).

6. Show that the points A(1, 2), B(5, 4), C(3, 8) and D(–1, 6) are the vertices of a square. *(CBSE 2006)*

7. Show that the points A(6, 2), B(2, 1), C(1, 5) and (5, 6) are the vertices of a square. *(AI CBSE 2006)*

8. Find the coordinates of the point equidistant from three given points A(5, 3), B(5, – 5) and C(1, – 5). *(AI CBSE 2006)*

9. Find the value of x such that PQ = QR where the coordinates of P, Q and R are (6, – 1), (1, 3) and $(x, 8)$ respectively. *(CBSE 2005)*

10. Find a point on x-axis which is equidistant from the points (7, 6) and (– 3, 4). *(CBSE 2005)*

11. If the point P(x, y) is equidistant from the points A(5, 1) and B(– 1, 5), prove that
$$3x = 2y.$$ *(AI CBSE 2005)*

12. Find a point on the x-axis which is equidistant from the points (– 2, 5) and (2, – 3). *(AI CBSE 2004)*

13. Show that the points A(5, 6), B(1, 5), C(2, 1) and D(6, 2) are the vertices of a square. *(CBSE 2004)*

14. Prove that the points (0, 9), $\left(\dfrac{b}{2}, \dfrac{a}{2}\right)$ and $(b, 0)$ are collinear.

Answers
1. yes, right triangle.
3. yes because A, B and C are collinear.
4. $x - y = 2$ 5. (0, 9)
8. (3, 1)   9. 5, – 3
10. (3, 0)   12. (– 2, 0)

SECTION FORMULA

IMPORTANT POINTS

1. A situation. Let A and B be two towns. Let the town B be located 36 km east and 15 km north of the town A.

Suppose a telephone company wants to position a relay tower at P between A and B is such a way that the distance of the tower from B is twice its distance from A. If P lies on AB, it will divide AB in the ratio 1 : 2 (see figure). If we take A as the origin O, and 1 km as one unit on both the axis, the coordinates of B will be (36,15). In order to know the position of the tower, we must know the coordinates of P. We find these coordinates as follows :

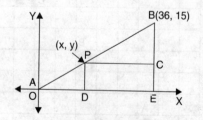

Let the coordinates of P be (x, y). Draw perpendiculars from P and B to the x-axis, meeting it in D and E, respectively. Draw PC perpendicular to BE. Then, by the AA similarity criterion, ΔPOD and ΔBPC are similar.

Therefore, $\dfrac{OD}{PC} = \dfrac{OP}{PB} = \dfrac{1}{2}$,

and $\dfrac{PD}{BC} = \dfrac{OP}{PB} = \dfrac{1}{2}$

∵ Corresponding sides of two similar triangles are proportional

So, $\dfrac{x}{36 - x} = \dfrac{1}{2}$ and $\dfrac{y}{15 - y} = \dfrac{1}{2}$.

These equations give $x = 12$ and $y = 5$.

We can check that P(12, 5) meets the condition that OP : PB = 1 : 2.

2. Section Formula (General). Consider any two points A (x_1, y_1) and B(x_2, y_2) and assume that P(x, y) divides AB

COORDINATE GEOMETRY

internally in the ratio $m_1 : m_2$, i.e., $\dfrac{PA}{PB} = \dfrac{m_1}{m_2}$ (see figure).

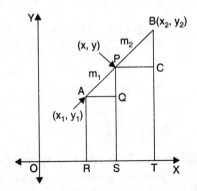

Draw AR, PS and BT perpendicular to the x-axis. Draw AQ and PC parallel to the x-axis. Then, by the AA similarity criterion,

$$\triangle PAQ \sim \triangle BPC$$

Therefore, $\dfrac{PA}{BP} = \dfrac{AQ}{PC} = \dfrac{PQ}{BC}$ ...(1)

Now, $AQ = RS = OS - OR = x - x_1$
$PC = ST = OT - OS = x_2 - x$
$PQ = PS - QS = PS - AR = y - y_1$
$BC = BT - CT = BT - PS = y_2 - y$

Substituting these values in (1), we get

$$\dfrac{m_1}{m_2} = \dfrac{x - x_1}{x_2 - x} = \dfrac{y - y_1}{y_2 - y}$$

Taking $\dfrac{m_1}{m_2} = \dfrac{x - x_1}{x_2 - x}$, we get

$$x = \dfrac{m_1 x_2 + m_2 x_1}{m_1 + m_2}$$

Similarly, taking $\dfrac{m_1}{m_2} = \dfrac{y - y_1}{y_2 - y}$, we get

$$y = \dfrac{m_1 y_2 + m_2 y_1}{m_1 + m_2}$$

So, the coordinates of the point $P(x, y)$ which divides the line segment joining the points $A(x_1, y_1)$ and $B(x_2, y_2)$, internally, in the ratio $m_1 : m_2$ are

$$\left(\dfrac{m_1 m_2 + m_2 x_1}{m_1 + m_2}, \dfrac{m_1 y_2 + m_2 y_1}{m_1 + m_2} \right) \quad ...(2)$$

This is known as the **section formula.**

Note 1. This can also be derived by drawing perpendiculars from A, P and B on the y-axis and proceeding as above.

Note 2. If the ratio in which P divides AB is $k : 1$, then the coordinates of the point P will be

$$\left(\dfrac{k x_2 + x_1}{k + 1}, \dfrac{k y_2 + y_1}{k + 1} \right).$$

Special Case. The mid-point of a line segment divides the line segment in the ratio $1 : 1$. Therefore, the coordinates of the mid-point P of the join of the points $A(x_1, y_1)$ and $B(x_2, y_2)$ is

$$\left(\dfrac{1 \cdot x_1 + 1 \cdot x_2}{1 + 1}, \dfrac{1 \cdot y_1 + 1 \cdot y_2}{1 + 1} \right)$$

$$= \left(\dfrac{x_1 + x_2}{2}, \dfrac{y_1 + y_2}{2} \right)$$

ILLUSTRATIVE EXAMPLES

[NCERT Exercies 7.2]
(Page No. 167)

Example 1. *Find the coordinates of the point which divides the join of (–1, 7) and (4, – 3) in the ratio 2 : 3.*

Sol. Let the coordinates of the required point be (x, y). Then,

$$x = \dfrac{m_1 x_2 + m_2 x_1}{m_1 + m_2}$$

$$= \dfrac{(2)(4) + (3)(-1)}{2 + 3}$$

$$= \frac{8-3}{5} = \frac{5}{5} = 1$$

$$y = \frac{m_1 y_2 + m_2 y_1}{m_1 + m_2}$$

$$= \frac{(2)(-3) + (3)(7)}{2+3}$$

$$= \frac{-6+21}{5} = \frac{15}{5} = 3$$

Hence, the required point is (1, 3).

Example 2. *Find the coordinates of the points of trisection of the line segment joining (4, –1) and (–2, –3).*

Sol. Let $A \to (4, -1)$ and $B \to (-2, -3)$.

```
    A         P         Q         B
  (4,–1)    (x,y)     (x,y)    (–2,–3)
```

Let the points of trisection of the line segment AB be $P(x, y)$ and $Q(x, y)$ respectively.

Then, AP = PQ = QB

Clearly P divides AB in the ratio 1 : 2 internally and Q divides AB in the ratio 2 : 1 internally.

Therefore,

$$x = \frac{m_1 x_2 + m_2 x_1}{m_1 + m_2}$$

$$= \frac{(1)(-2) + (2)(4)}{1+2}$$

$$= \frac{-2+8}{3} = \frac{6}{3} = 2$$

$$y = \frac{m_1 y_2 + m_2 y_1}{m_1 + m_2}$$

$$= \frac{(1)(-3) + (2)(-1)}{1+2}$$

$$= \frac{-3-2}{3} = \frac{-5}{3}$$

$$\therefore P \to \left(2, \frac{-5}{3}\right)$$

$$X = \frac{m_1 x_2 + m_2 x_1}{m_1 + m_2}$$

$$= \frac{(2)(-2) + (1)(4)}{2+1}$$

$$= \frac{-4+4}{3} = 0$$

$$Y = \frac{m_1 y_2 + m_2 y_1}{m_1 + m_2}$$

$$= \frac{(2)(-3) + (1)(-1)}{2+1}$$

$$= \frac{-6-1}{3} = -\frac{7}{3}$$

$$\therefore Q \to \left(0, -\frac{7}{3}\right)$$

Example 3. *To conduct Sports Day activities, in your rectangular shaped school ground ABCD, lines have been drawn with chalk powder at a distance of 1m each. 100 flower pots have been placed at a distance of 1m from each other along AD, as shown in figure. Niharika runs $\frac{1}{4}$ th the distance AD on the 2nd line and posts a green flag. Preet runs $\frac{1}{5}$ th the distance AD on the eight line and posts a red flag. What is the distance between both the flags ? If Rashmi has to post a blue flag exactly halfway between the line segment joining the two flags, where should she post her flag ?*

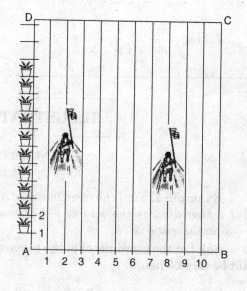

COORDINATE GEOMETRY

Sol. Take A as origin, AB as x-axis and AD as y-axis. Position of the green flag post $\to \left(2, \dfrac{100}{4}\right)$ or (2, 25).

Position of the red flag post $\to \left(8, \dfrac{100}{5}\right)$ or (8, 20).

∴ Distance between both the flags

$$= \sqrt{(8-2)^2 + (20-25)^2}$$
$$= \sqrt{(6)^2 + (-5)^2}$$
$$= \sqrt{61} \text{ m}$$

Position of the blue flag post $\to \left(\dfrac{2+8}{2}, \dfrac{25+20}{2}\right)$ or $\left(\dfrac{10}{2}, \dfrac{45}{2}\right)$ or (5, 22.5)

So, she should post her blue flag on the 5th line at a distance of 22.5 km from AB.

Example 4. *Find the ratio in which the segment joining the points (– 3,10) and (6, – 8) is divided by (– 1, 6).*

Sol. Let A $\to$ (– 3, 10), B $\to$ (6, – 8) and P $\to$ (– 1, 6).

Let P divide AB in the ratio K : 1.

```
A         K      P      1      B
(-3, 10)         (-1, 6)       (6, -8)
```

Then,

$$P \to \left\{\dfrac{(K)(6) + (1)(-3)}{K+1}, \dfrac{(K)(-8) + (1)(10)}{K+1}\right\}$$

or $\quad P \to \left(\dfrac{6K-3}{K+1}, \dfrac{-8K+10}{K+1}\right)$

But $\quad P \to$ (– 1, 6)

∴ $\quad \dfrac{6K-3}{K+1} = -1$

$\Rightarrow \quad 6K - 3 = -K - 1$
$\Rightarrow \quad\quad 7K = 2$
$\Rightarrow \quad\quad K = \dfrac{2}{7}$

and, $\quad \dfrac{-8K+10}{K+1} = 6$

$\Rightarrow \quad -8K + 10 = 6K + 6$
$\Rightarrow \quad\quad 14K = 4$
$\Rightarrow \quad\quad K = \dfrac{4}{14} = \dfrac{2}{7}$

Hence, the required ratio is 2 : 7.

Example 5. *Find the ratio in which the segment joining A(1, – 5) and B(–4, 5) is divided by the x-axis. Also find the coordinates of the point of division.*

Sol. Let the point of division be P. Let the ratio be K : 1.

```
A        K     P      1      B
(1, -5)                      (-4, 5)
```

Then,
P $\to$

$$\left\{\dfrac{(K)(-4) + (1)(1)}{K+1}, \dfrac{(K)(5) + (1)(-5)}{K+1}\right\}$$

$\Rightarrow \quad P \to \left\{\dfrac{-4K+1}{K+1}, \dfrac{5K-5}{K+1}\right\}$

∵ P lies on the x-axis and we know that on the x-axis the ordinate is 0.

∴ $\quad \dfrac{5K-5}{K+1} = 0$

$\Rightarrow \quad 5K - 5 = 0$
$\Rightarrow \quad\quad 5K = 5$
$\Rightarrow \quad\quad K = \dfrac{5}{5} = 1$

Hence, the required ratio is 1 : 1.
Putting K = 1, we get

$$P \to \left\{-\dfrac{3}{2}, 0\right\}$$

Example 6. *If (1, 2), (4, y), (x, 6) and (3, 5) are the vertices of a parallelogram taken in order, find x and y.*

Sol. Let A $\to$ (1, 2), B(4, y), C $\to$ (4, 6) and D $\to$ (3, 5).

We know that the diagonals of parallelogram bisect each other. So,

Coordinates of the mid-point of diagonal AC

= Coordinates of the mid-point of diagonal BD

$$\Rightarrow \left(\frac{1+x}{2}, \frac{2+6}{2}\right) = \left(\frac{4+3}{2}, \frac{y+5}{2}\right)$$

$$\Rightarrow \left(\frac{1+x}{2}, 4\right) = \left(\frac{7}{2}, \frac{y+5}{2}\right)$$

$$\Rightarrow \frac{1+x}{2} = \frac{7}{2}$$

$$\Rightarrow 1+x = 7$$

$$\Rightarrow x = 6$$

and $\quad 4 = \frac{y+5}{2}$

$$\Rightarrow y+5 = 8 \Rightarrow y = 3$$

Example 7. *Find the coordinates of a point A, where AB is the diameter of a circle whose centre is (2, – 3) and B is (1, 4).*

Sol. Let C be the centre of the circle.
Then, $C \to (2, -3)$.

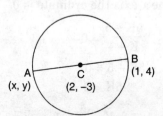

Let the coordinates of the point A be (x, y).

$\because$ AB is a diameter of a circle whose centre is C.

$\therefore$ C is the mid-point of AB.

$\therefore \left(\frac{x+1}{2}, \frac{y+4}{2}\right) = (2, -3)$

$$\Rightarrow \frac{x+1}{2} = 2$$

$$\Rightarrow x+1 = 4 \Rightarrow x = 3$$

and $\quad \frac{y+4}{2} = -3$

$$\Rightarrow y+4 = -6 \Rightarrow y = -10$$

Hence, the coordinates of the points A are (3, – 10).

Example 8. *If A and B are (– 2, – 2) and (2, – 4), respectively, find the coordinates of P such that $AP = \frac{3}{7} AB$ and P lies on the line segment AB.*

Sol. $AP = \frac{3}{7} AB$

```
●————3————●————4————●
A          P          B
(-2,-2)   (x, y)    (2,-4)
```

$\Rightarrow$ 7 AP = 3 AB

$\Rightarrow$ 7 AP = 3(AP + PB)

$\because$ P lies in the line segment AB

$\Rightarrow$ 7 AP = 3 AP + 3 PB

$\Rightarrow$ 7 AP – 3 AP = 3 PB

$\Rightarrow$ 4 AP = 3 PB

$\Rightarrow \quad \frac{AP}{PB} = \frac{3}{4}$

Let the coordinates of P be (x, y). Then,

$$x = \frac{m_1 x_2 + m_2 x_1}{m_1 + m_2}$$

$$= \frac{(3)(2) + (4)(-2)}{3+4}$$

$$= \frac{6-8}{7} = -\frac{2}{7}$$

$$y = \frac{m_1 y_2 + m_2 y_1}{m_1 + m_2}$$

$$= \frac{(3)(-4) + (4)(-2)}{3+4}$$

$$= \frac{-12-8}{7} = -\frac{20}{7}$$

Hence, the coordinates of the point P are $\left(-\dfrac{2}{7}, -\dfrac{20}{7}\right)$.

Example 9. *Find the coordinates of the points which divide the line segment joining A(– 2, 2) and B(2, 8) into four equal parts.*

Sol. Let P (x_1, y_1) Q(x_2, y_2) and R(x_3, y_3) be the points which divide the line segment AB into four equal parts.

```
      P       Q       R
A •———•———•———•———• B
(-2, 2)                (2, 8)
```

Then, P divides AB in the ratio 1 : 3 internally.

$\therefore \quad x_1 = \dfrac{(1)(2) + (3)(-2)}{1+3}$

$= \dfrac{2 - 6}{4}$

$= -\dfrac{4}{4} = -1$

$y_1 = \dfrac{(1)(8) + (3)(2)}{1+3}$

$= \dfrac{8+6}{4} = \dfrac{14}{4}$

$= \dfrac{7}{2}$

So, P $\to \left(-1, \dfrac{7}{2}\right)$

Also, Q divides AB in the ratio 1 : 1 *i.e.*, Q is the mid-point of AB

$x_2 = \dfrac{-2+2}{2} = 0$

$y_2 = \dfrac{2+8}{2} = \dfrac{10}{2} = 5$

So, Q $\to (0, 5)$

and, R divides AB in the ratio 3 : 1

$\therefore \quad x_3 = \dfrac{(3)(2) + (1)(-2)}{3+1}$

$= \dfrac{6-2}{4} = \dfrac{4}{4} = 1$

$y_3 = \dfrac{(3)(8) + (1)(2)}{3+1}$

$= \dfrac{24+2}{4} = \dfrac{26}{4} = \dfrac{13}{2}$

So, R $\to \left(1, \dfrac{13}{2}\right)$.

Example 10. *Find the area of a rhombus if its vertices are (3, 0), (4, 5), (– 1, 4) and (– 2, – 1) taken in order.* [**Hint:** *Area of a rhombus* $= \dfrac{1}{2}$ *(product of its diagonals)*]

Sol. Let A $\to$ (3, 0), B $\to$ (4, 5), C $\to$ (– 1, 4) and D $\to$ (– 2, – 1).

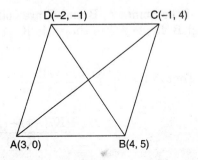

We know that the area of a rhombus = $\dfrac{1}{2}$ (product of its diagonals)

$\therefore$ Area of the rhombus ABCD.

$= \dfrac{1}{2}$ (AC × BD)

$$= \frac{1}{2}(\sqrt{(-1-3)^2+(4-0)^2})$$
$$(\sqrt{(-2-4)^2+(-1-5)^2})$$
$$= \frac{1}{2}\sqrt{16+16}\sqrt{36+36}$$
$$= \frac{1}{2}\sqrt{32}\sqrt{72}$$
$$= \frac{1}{2}(4\sqrt{2})(6\sqrt{2})$$
$$= 24 \text{ square units}$$

ADDITIONAL EXAMPLES

Example 11. *Find the value of p for which the points (– 1, 3), (2, p) and (5, – 1) are collinerar.* (CBSE 2006)

Sol. Let $A \to (-1, 3)$
$B \to (2, p)$ and $C \to (5, -1)$.

```
    k     B     1
A ●─────●─────● C
(-1,3)         (5,-1)
```

If the points A, B and C are collinear, then let B divide AC in the ratio K : 1 internally.

Then, $B \to \left\{ \dfrac{(K)(5)+(1)(-1)}{K+1}, \right.$

$$\left. \dfrac{(K)(-1)+(1)(3)}{K+1} \right\}$$

$\Rightarrow B \to \left(\dfrac{5K-1}{K+1}, \dfrac{-K+3}{K+1} \right)$

But B is given to be (2, p)

$\therefore \quad \dfrac{5K-1}{K+1} = 2$

$\Rightarrow \quad 5K - 1 = 2(K+1)$
$\Rightarrow \quad 5K - 1 = 2K + 2$
$\Rightarrow \quad 5K - 2K = 2 + 1$
$\Rightarrow \quad 3K = 3$

$\Rightarrow \quad K = \dfrac{3}{3} = 1$ and, $\dfrac{-K+3}{K+1} = p$

$\Rightarrow \quad \dfrac{-1+3}{1+1} = p$

$\Rightarrow \quad p = 1.$

Hence, the required value of p is 1.

Example 12. *Find the value of p for which the points (– 5, 1), (1, p) and (4, – 2) are collinear.* (AI CBSE 2006)

Sol. Let $A \to (-5, 1)$, $B \to (1, p)$ and $C \to (4, -2)$.

```
       k        B       1
A ●───────────●─────────● C
(-5,1)       (1,p)       (4,-2)
```

Let A, B, C be collinear. Let B divide AC in the ratio K : 1, Then,

$B \to \left\{ \dfrac{(K)(4)+(1)(-5)}{K+1}, \right.$

$$\left. \dfrac{(K)(-2)+(1)(1)}{K+1} \right\}$$

$\Rightarrow B \to \left(\dfrac{4K-5}{K+1}, \dfrac{-2K+1}{K+1} \right)$

But B is given to be (1, p)

$\therefore \quad \dfrac{4K-5}{K+1} = 1$

$\Rightarrow \quad 4K - 5 = K + 1$
$\Rightarrow \quad 4K - K = 1 + 5$
$\Rightarrow \quad 3K = 6$

$\Rightarrow \quad K = \dfrac{6}{3} = 2$

and, $\dfrac{-2K+1}{K+1} = p$

$\Rightarrow \quad \dfrac{-2(2)+1}{2+1} = p$

$\Rightarrow \quad p = -1.$

Hence, the required value of p is – 1.

COORDINATE GEOMETRY

Example 13. *The line-segment joining the points (3, – 4) and (1, 2) in trisected at the points P and Q. If the coordinates of P and Q are (p, – 2) and $\left(\frac{5}{3}, q\right)$ respectively, find the values of p and q.* (CBSE 2005)

Sol. Clearly P divides the line segment AB internally in the ratio 1 : 2

```
A           P           Q           B
(3,–4)     (p,–2)      (5/3, q)    (1, 2)
```

∴ By section formula,

$$\frac{(1)(1) + (2)(3)}{1+2} = p \Rightarrow p = \frac{7}{3}$$

and $\frac{(1)(2) + (2)(-4)}{1+2} = -2$ which is true.

Hence, the required value of p is $\frac{7}{3}$.

Again, Q divides the line segment AB internally in the ratio 2 : 1

∴ By section formula,

$$\frac{(2)(1) + (1)(3)}{2+1} = \frac{5}{3} \text{ which is true}$$

and $\frac{(2)(2) + (1)(-4)}{2+1} = q \Rightarrow q = 0.$

Hence, the required value of q is 0.

Example 14. *The line joining the points (2, 1) and (5, – 8) is trisected at the points P and Q. If point P lies on the line $2x - y + K = 0$, find the value of K.* (CBSE 2005)

Sol. Let A → (2, 1) and B → (5, – 8)

```
        P       Q
A                       B
(2, 1)                  (5,–8)
```

Since P and Q are the points of trisection, therefore d divides AB internally in the ratio 2 : 2.
Therefore,

$$P \to \left\{\frac{(1)(5) + (2)(2)}{1+2}, \frac{(1)(-8) + (2)(1)}{1+2}\right\}$$

$\Rightarrow \quad P \to (3, -2)$

∵ P lies on the line $2x - y + K = 0$ | Given

∴ $2(3) - (-2) + K = 0$
$\Rightarrow \quad 6 + 2 + K = 0$
$\Rightarrow \quad 8 + K = 0$
$\Rightarrow \quad K = -8$

Hence, the required value of K is – 8.

Example 15. *Determine the ratio in which the point P(m, 6) divides the join of A(– 4, 3) and B(2, 8). Also find the value of m.* (CBSE 2004)

Sol. Let the point P (m, 6) divide the join of A(– 4, 3) and B(2, 8) in the ratio K : 1. Then,

```
          k           1
A                 P                 B
(-4, 3)         (m, 6)             (2, 8)
```

$$P \to \left\{\frac{(K)(2) + (1)(-4)}{K+1}, \frac{(K)(8) + (1)(3)}{K+1}\right\}$$

$\Rightarrow \quad P \to \left(\frac{2K-4}{K+1}, \frac{8K+3}{K+1}\right)$

But $P \to (m, 6)$ | Given

∴ $\frac{2K-4}{K+1} = m$ ...(1)

and $\frac{8K+3}{K+1} = 6$ ...(2)

From (2), we get
$8K + 3 = 6(K+1)$
$\Rightarrow \quad 8K + 3 = 6K + 6$
$\Rightarrow \quad 8K - 6K = 6 - 3$
$\Rightarrow \quad 2K = 3$
$\Rightarrow \quad K = \frac{3}{2}$

Hence, the required ratio is 3 : 2.
Again, from (1)

$$m = \frac{2\left(\frac{3}{2}\right) - 4}{\frac{3}{2} + 1}$$

$$= \frac{-\frac{1}{5}}{\frac{5}{2}} = -\frac{2}{5}$$

Hence, the value of m is $-\frac{2}{5}$.

Example 16. *Prove that the coordinates of the centroid of a $\triangle ABC$, with vertices $A(x_1, y_1)$, $B(x_2, y_2)$ and $C(x_3, y_3)$ are given by*
$$\left(\frac{x_1 + x_2 + x_3}{3}, \frac{y_1 + y_2 + y_3}{3}\right)$$

(AI CBSE 2004)

Sol. We know that the centroid of a triangle divides each median of the triangle from the vertex in the ratio 2 : 1. Let D be the mid-point of BC. Then

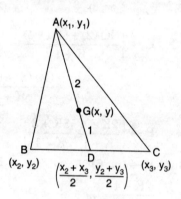

$$D \to \left(\frac{x_2 + x_3}{2}, \frac{y_2 + y_3}{2}\right)$$

Let $G(x, y)$ be the centroid of $\triangle ABC$. Then, G will divide AB internally in the ratio 2 : 1.

$$\therefore \quad X = \frac{2\left(\frac{x_2 + x_3}{2}\right) + 1(x_1)}{2 + 1}$$

$$= \frac{x_1 + x_2 + x_3}{3}$$

and $\quad y = \dfrac{2\left(\dfrac{y_2 + y_3}{2}\right) + 1(y_1)}{2 + 1}$

$$= \frac{y_1 + y_2 + y_3}{3}$$

Hence, the coordinates of the centroid are given by $\left(\dfrac{x_1 + x_2 + x_3}{3}, \dfrac{y_1 + y_2 \; y_3}{3}\right)$.

Example 17. *$A(b, 2)$ and $B(-2, 1)$ are two vertices of a triangle ABC, whose centroid G has a coordinates $\left(\dfrac{5}{3}, -\dfrac{1}{3}\right)$. Find the coordinates of the third vertex C of the triangle.*

(CBSE 2004)

Sol. We have,

$$A \to (3, 2) \text{ and } B(-2, 1).$$

Let the coordinates of the third vertex C of the triangle be (x, y).

Then,

$$G \to \left\{\frac{(3) + (-2) + (x)}{3}, \frac{(2)(1) + (y)}{3}\right\}$$

$$\Rightarrow \quad G \to \left(\frac{x + 1}{3}, \frac{y + 3}{3}\right)$$

But $\quad G \to \left(\dfrac{5}{3}, -\dfrac{1}{3}\right)$ | Given

$$\therefore \quad \frac{x + 1}{3} = \frac{5}{3} \Rightarrow x + 1 = 5$$

$$\Rightarrow \quad x = 5 - 1 = 4$$

and, $\quad \dfrac{y + 3}{3} = -\dfrac{2}{3}$

$$\Rightarrow \quad y + 3 = -1$$

$$\Rightarrow \quad y = -1 - 3 = -4$$

Hence, $\quad C \to (4, -4)$.

Example 18. *Determine the ratio in which the point $(-6, a)$ divides the join of $A(-3, -1)$ and $B(-8, a)$. Also find the value of a.*

(AI CBSE 2004)

Sol. Let the point $P(-6, a)$ divide the join of $A(-3, -1)$ and $B(-8, a)$ in the ratio $\lambda : 1$ internally then,

```
         λ              1
  •──────────────•──────────────•
  A              P              B
(-3,-1)        (-6, a)        (-8, a)
```

$$P \to \left\{\frac{(\lambda)(-8)(1)(-3)}{\lambda + 1}, \frac{(\lambda)(9) + (1)(-1)}{\lambda + 1}\right\}$$

COORDINATE GEOMETRY

$$= P \to \left(\frac{-8\lambda - 3}{\lambda + 1}, \frac{9\lambda - 1}{\lambda + 1}\right)$$

But $P \to (-6, a)$ | Given

$$\therefore \quad \frac{-8\lambda - 3}{\lambda + 1} = -6$$

$$\Rightarrow -8\lambda - 3 = -6(\lambda + 1)$$

$$\Rightarrow -8\lambda - 3 = -6\lambda - 6$$

$$\Rightarrow 8\lambda - 6\lambda = -3 + 6$$

$$\Rightarrow 2\lambda = 3$$

$$\Rightarrow \lambda = \frac{3}{2}$$

Hence, the required ratio is 3 : 2.

Again, $a = \dfrac{a\lambda - 1}{\lambda + 1}$

$$= \frac{a\left(\frac{3}{2}\right) - 1}{\frac{3}{2} + 1} = \frac{\frac{27}{2} - 1}{\frac{3}{2} + 1}$$

$$= \frac{\frac{25}{2}}{\frac{5}{2}} = \frac{25}{5} = 5$$

Hence, the required value of a is 5.

TEST YOUR KNOWLEDGE

1. Find the coordinates of the point which divides the line segment joining the points (4, – 3) and (8, 5) in the ratio 3 : 1 internally.
2. In what ratio does the point (– 4, 6) divide the line segment joining the points A (– 6, 10) and B (3, – 8).
3. Find the coordinates of the points of trisection (*i.e.*, points dividing in three equal parts) of the line segment joining the points A (2, – 2) and B (– 3, 4).
4. Find the ratio in which the *y*-axis divides the line segment joining the points (5, – 6) and (– 1, – 4). Also find the point of intersection.
5. If the points A (6, 1), B (8, 2), C (9, 4) and D (*p*, 3) and the vertices of a parallelogram, find the value of *p*.
6. Find the ratio in which the point (– 3, *p*) divides the line segment joining the points (– 3, – 4) and (– 2, 3). Hence find the value of *p*.
7. Three coordinates of the vertices of a parallelogram ABCD are A (1, 2), B (1, 0) and C (4, 0). Find the fourth vertex D.
8. If A (4, – 8), B (– 9, 7) and C (18, 13) are the vertices of a triangle ABC, find the length of the median through A and coordinates of centroid of the triangle.
9. The middle points of the sides of a triangle are (2, 1), (– 1, – 3) and (4, 5) respectively. Find the coordinates of the vertices of the triangle.
10. The line joining the points (3, – 2) and (– 3, – 4) is trisected equally. Find the coordinates of the points of trisection.
11. The coordinates of one end of the diameter of a centre are (3, 5) and the coordinates of its centre are (6, 6); then find out the coordinates of the other end of the diameter.
12. Prove that the diagonals of a rectangle bisect each other and one equal.
13. Find the third vertex of a triangle, if two of its vertices are at (– 3, 1) and (0, – 2) and the centriod is at the origin.
14. Find the coordinates of the points which divide the line segment joining the points (– 4, 0) and (0, 6) in four equal parts.
15. Find the centroid of the triangle whose vertices are given below:
    (*i*) (4, – 8) (– 9, 7), (8, 13)
    (*ii*) (3, – 5), (– 7, 4), (10, – 2)
    (*iii*) (2, 1), (5, 2), (3, 4).

Answers

1. (7, 3)   2. 2 : 7
3. (– 1, 0), (– 4, 2)   4. 5 : 1, $\left(0, \dfrac{-13}{3}\right)$

5. 7
6. $2:1, \frac{2}{3}$
7. (4, 2)
8. $\frac{\sqrt{1297}}{7}, \left(\frac{13}{3}, 4\right)$
9. (7, 9), (−3, −7), (1, 1)
10. $\left(1, -\frac{8}{3}\right), \left(-1, -\frac{10}{3}\right)$
11. (9, 7)
13. (3, 1)
14. (−3, 1.5), (−2, 3), (−1, 4.5)
15. (i) (1, 4) (ii) (2, −1)
(iv) $\left(\frac{10}{3}, \frac{7}{3}\right)$

AREA OF A TRIANGLE

IMPORTANT POINTS

1. Area of a triangle = $\frac{1}{2}$ × base × corresponding height (altitude).

2. If the coordinates of the vertices of a triangle and given, then we can find its area by first finding the lengths of the three sides using the distance formula and then using Heron's formula. But this could be tedious, particularly if the lengths of the sides any irrational numbers.

3. An easier way. Let ABC be any triangle whose vertices are A (x_1, y_1), B (x_2, y_2) and C (x_3, y_3). Draw AP, BQ and CR perpendiculars from A, B and C, respectively, to the x-axis. Clearly ABQP, APRC and BQRC are all trapezia (see figure).

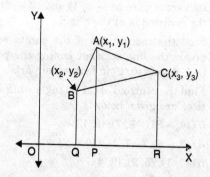

Now, from figure, it is clear that
area of △ ABC = area of trapezium ABQP + area of trapezium APRC − area of trapezium BQRC.

We also know that the

area of a trapezium = $\frac{1}{2}$ (sum of parallel sides) (distance between them)

Therefore,

Area of △ ABC = $\frac{1}{2}$ (BQ + AP) QP

$+ \frac{1}{2}$ (AP + CR) PR $- \frac{1}{2}$ (BQ + CR) QR

$= \frac{1}{2} (y_2 + y_1)(x_1 - x_2) + \frac{1}{2}(y_1 + y_3)$

$(x_3 - x_1) - \frac{1}{2}(y_2 + y_3)(x_3 - x_2)$

$= \frac{1}{2} [x_1(y_2 - y_3) + x_2(y_3 - y_1)$

$+ x_3(y_1 - y_2)]$

Thus, the area of △ ABC is the numerical value of the expression

$\frac{1}{2} [x_1(y_2 - y_3) + x_2(y_3 - y_1) + x_3(y_1 - y_2)]$

Note 1. If the area comes out to be negative, then, we take the positive value as area is a measure, which can not be negative.

Note 2. If the area of a triangle comes out to be 0 square units, then its vertices will be collinear.

Note 3. To find the area of a polygon, we divide it into triangular regions, which have no common area, and add the areas of base regions.

ILLUSTRATIVE EXAMPLES

[NCERT Exercise 7.3]
(*Page No. 170*)

Example 1. *Find the area of the triangle whose vertices are:*
(i) (2, 3), (– 1, 0), (2, – 4)
(ii) (– 5, – 1), (3, – 5), (5, 2)

Sol. (i) (2, 3), (– 1, 0), (2, – 4)
Area of the triangle
$= \frac{1}{2} [2 [0 - (- 4)] + (- 1) [- 4 - 3]$
$\qquad + 2[3 - 0]]$
$= \frac{1}{2} (8 + 7 + 6)$
$= \frac{21}{2}$ square units.

(ii) (– 5, – 1), (3, – 5), (5, 2)
Area of the triangle
$= \frac{1}{2} [(- 5) \{- 5 - 2\} + (3) \{2 - (- 1)\}$
$\qquad + (5) \{(- 1) - (- 5)\}]$
$= \frac{1}{2} [35 + 9 + 20] = 32$ square units.

Example 2. *In each of the following find the value of 'k', for which the points are collinear.*
(i) (7, – 2), (5, 1), (3, k)
(ii) (8, 1), (k, – 4), (2 – 5)

Sol. (i) (7, – 2), (5, 1) , (3, k)
Area of the triangle
$= \frac{1}{2} [7 (1 - k) + 5 \{k - (- 2)\} + 3 (- 2 - 1)]$
$= \frac{1}{2} [7 - 7k + 5k + 10 - 9]$
$= \frac{1}{2} [8 - 2k] = 4 - k$

If the points are collinear, then area of the triangle = 0
$\Rightarrow \quad 4 - k = 0$
$\Rightarrow \quad k = 4$

(ii) (8, 1), (k, – 4), (2, – 5)
Area of the triangle
$= \frac{1}{2} [8 \{- 4 - (- 5)\} + k (- 5 - 1)$
$\qquad + 2 \{1 - (- 4)\}]$
$= \frac{1}{2} [8 - 6k + 10]$
$= \frac{1}{2} [18 - 6k] = 9 - 3k$

If the points are collinear, then area of the triangle = 0
$\Rightarrow \quad 9 - 3k = 0$
$\Rightarrow \quad 3k = 9$
$\Rightarrow \quad k = \frac{9}{3} = 3.$

Example 3. *Find the area of the triangle formed by joining the mid-points of the sides of the triangle whose vertices are (0, – 1), (2, 1) and (0, 3). Find the ratio of this area to the area of the given triangle.*

Sol. Let A → (0, – 1), B → (2, 1) and C → (0, 3) be the vertices of the triangle ABC. Let D, E and F be the mid-points of sides BC, CA and AB respectively. Then,

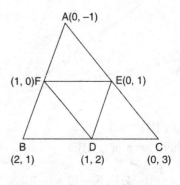

$\Rightarrow \quad D \to \left(\frac{2+0}{2}, \frac{1+3}{2}\right)$
$\Rightarrow \quad D \to (1, 2)$

$$E \to \left\{ \frac{0+0}{2}, \frac{3+(-1)}{2} \right\}$$

$\Rightarrow \quad E \to (0, 1)$

$$F \to \left\{ \frac{2+0}{2}, \frac{1+(-1)}{2} \right\}$$

$\Rightarrow \quad F \to (1, 0)$

∴ Area of the triangle DEF

$$= \frac{1}{2} [1(1-0) + 0(0-2) + 1(2-1)]$$

$$= \frac{1}{2} [1 + 0 + 1]$$

= 1 squre unit.

Again, area of the triangle ABC

$$= \frac{1}{2} [0(1-3) + 2\{3-(-1)\} + 0(-1-1)]$$

= 4 square units

∴ Ratio of the area of the triangle formed to the area of the given triangle

= 1 : 4.

Example 4. *Find the area of the quadrilateral whose vertices, taken in order, are (– 4, – 2), (– 3, – 5), (3, – 2) and (2, 3).*

Sol. Let $A \to (-4, -2)$, $B \to (-3, -5)$, $C \to (3, -2)$ and $D \to (2, 3)$ be the vertices of the quadrilateral ABCD.

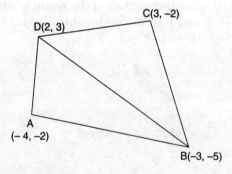

Join BD

Then, area of △ ABD

$$= \frac{1}{2} [(-4)\{-5, -3\} + (-3)\{3-(-2)\}$$

$$+ 2\{(-2)-(-5)\}]$$

$$= \frac{1}{2} [32 - 15 + 6]$$

$$= \frac{23}{2} \text{ square units}$$

and, Area of △ CBD

$$= \frac{1}{2} [3(-5-3) + (-3)$$

$$\{3-(-2)\} + (2)\{(-2)-(-5)\}]$$

$$= \frac{1}{2} [-24 - 15 + 6]$$

$$= -\frac{33}{2}$$

$$= \frac{33}{2} \text{ square units | numerically}$$

So, Area of the quadrilateral ABCD

= Area of the triangle ABD
 + Area of the triangle CBD

$$= \frac{23}{2} \text{ square units}$$

$$+ \frac{33}{2} \text{ square units}$$

= 28 square units.

Example 5. *You have studied in Class IX, (Chapter 9, Example 3) that, a median of a triangle divides it into two triangles of equal areas. Verify this result for △ ABC whose vertices are A (4 – 6), B (3, – 2) and C (5, 2).*

Sol. Let D be the mid-point of the side BC of the triangle ABC. Then

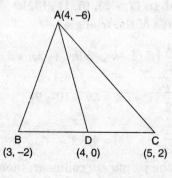

$$D \to \left\{ \frac{3+5}{2}, \frac{(-2)+2}{2} \right\}$$

or, D → (4, 0)
AD is a median.
Area of △ ABD

$$= \frac{1}{2} [4\{(-2) - 0\} + 3 \{-(-6)\} + 4\{(-6) - (-2)\}]$$

$$= \frac{1}{2} [-8 + 18 - 16]$$

$= -3$ square units

$= 3$ square units | numerically

and, Area of △ ACD

$$= \frac{1}{2} [4(2-0) + 5\{0 - (-6)\} + 4(-6-2)]$$

$$= \frac{1}{2} [8 + 30 - 32]$$

$= 3$ square units

Clearly, Area of △ ABD = Area of △ ACD

⇒ A median of a triangle divides it into two triangles of equal areas.

ADDITIONAL EXAMPLES

Example 6. *For what value(s) of x, the area of the triangle formed by the points (5, –1), (x, 4) and (6, 3) is 5.5 square units?*

and

Sol. Let A → (5, –1), B → (x, 4)
            C → (6, 3).
Then, Area of △ ABC

$$= \left| \frac{1}{2} [5(4-3) + x\{3-(-1)\} + 6(-1-4)] \right|$$

$$= \left| \frac{1}{2} [5 + 4x - 30] \right|$$

$$= \left| \frac{1}{2} (4x - 25) \right|$$

If this area is 5.5 square units, then

$$\left| \frac{1}{2}(4x - 25) \right| = 5.5$$

⇒ $\left| \frac{1}{2}(4x - 25) \right| = \frac{11}{2}$

⇒ $|4x - 25| = 11$

⇒ $4x - 25 = 11$,   i.e., $x = 9$

or

$4x - 25 = -11$,   i.e., $x = \frac{7}{2}$.

TEST YOUR KNOWLEDGE

1. Find the area of a triangle whose vertices are (1, –1), (–4, 6) and (–3, –5).
2. Find the area of a triangle formed by the points A (5, 2), B (4, 7) and C (7, –4).
3. Find the area of the triangle formed by the points P (–1.5, 3), Q (6, –2) and R (–3, 4).
4. Find the value of k if the points A (2, 3), B (4, k) and C (6, –3) are collinear.
5. If A (–5, 7), B (–4, –5), C (–1, –6) and D (4, 5) are the vertices of a quadrilateral, find the area of the quadrilateral ABCD.
6. Find the condition that the point (x, y) may lie on the line joining points (3, 4) and (–5, –6).
7. Find the area of the triangle formed by the mid-points of sides of the triangle whose vertices are (2, 1), (–2, 3), (4, –3).
8. Show that the points (–1, –1), (2, 3) and (8, 11) lie on a line.
9. For what value of x will the following points lie on a line?

   $(x, –1), (2, 1), (4, 5)$.
10. Find the area of the quadrilateral whose vertices are (2, 1), (6, 0), (5, –2) and (–3, –1).

Answers

1. 3 square units
2. 2 square units
3. 0 square units
4. 0
5. 72 square units
6. $5x - 4y + 1 = 0$
7. 1.5 square units
9. 1
10. 15 square units

ILLUSTRATIVE EXAMPLES

[NCERT Exercise 7.4 (Optional)*]
(Page No. 171)

Example 1. *Determine the ratio in which the line $2x + y - 4 = 0$ divides the line segment joining the points A (2, – 2) and B (3, 7).*

Sol. Let the line $2x + y - 4 = 0$ divide the line segment joining the points A (2, – 2) and B (3, 7) in the ratio $\lambda : 1$. Let the point of intersection be P. Then,

```
      λ      P       1
  •──────────•──────────•
  A                     B
(2, –2)               (3, 7)
```

$$P \to \left\{\frac{(\lambda)(3) + (1)(2)}{\lambda + 1}, \frac{(\lambda)(7) + (1)(-2)}{\lambda + 1}\right\}$$

$$\Rightarrow P \to \left(\frac{3\lambda + 2}{\lambda + 1}, \frac{7\lambda - 2}{\lambda + 1}\right)$$

∵ P lies on the line $2x + y - 4 = 0$

∴ $2\left(\frac{3\lambda + 2}{\lambda + 1}\right) + \left(\frac{7\lambda - 2}{\lambda + 1}\right) - 4 = 0$

$\Rightarrow 2(3\lambda + 2) + (7\lambda - 2) - 4(\lambda + 1) = 0$

$\Rightarrow 6x + y + 7\lambda - 2 - 4\lambda - 4 = 0$

$\Rightarrow 9\lambda - 2 = 0$

$\Rightarrow \lambda = \frac{2}{9}$

Hence, the required ratios is 2 : 9.

Example 2. *Find a relation between x and y if the points (x, y), (1, 2) and (7, 0) are collinear.*

Sol. If the given points are collinear, then the area of the triangle with these points as vertices will be zero.

∴ $\frac{1}{2}[x(2 - 0) + 1(0 - y) + 7(y - 2)] = 0$

$\Rightarrow \frac{1}{2}[2x - y + 7y - 14] = 0$

$\Rightarrow \frac{1}{2}[2x + 6y - 14] = 0$

$\Rightarrow 2x + 6y - 14 = 0$

$\Rightarrow x + 3y - 7 = 0$

| Dividing throughout by 2

This is the required relation between x and y.

Example 3. *Find the centre of a circle passing through the points (6, – 6), (3, – 7) and (3, 3).*

Sol. Let A $\to$ (6, – 6), B $\to$ (3, – 7) and C $\to$ (3, 3).

Let the centre of the circle be I(x, y)

Then, IA = IB = IC

| By definition of a circle

$\Rightarrow$ IA² = IB² = IC²

$\Rightarrow (x - 6)^2 + (y + 6)^2 = (x - 3)^2 + (y + 7)^2$
$\qquad = (x - 3)^2 + (y - 3)^2$

Taking first two, we get

$(x - 6)^2 + (y + 6)^2 = (x - 3)^2 + (y + 7)^2$

$\Rightarrow x^2 - 12x + 36 + y^2 + 12y + 36$
$\qquad = x^2 - 6x + 9 + y^2 + 14y + 49$

$\Rightarrow 6x + 2y = 14$

$\Rightarrow 3x + y = 7$ ...(1)

| Dividing throughout by 2

Taking last two, we get

$(x - 3)^2 + (y + 7)^2 = (x - 3)^2 + (y - 3)^2$

$\Rightarrow (y + 7)^2 = (y - 3)^2$

$\Rightarrow (y + 7) = \pm (y - 3)$

---

*These exercises are not from examination point of view.

Taking + ve sign, we get
$$y + 7 = y - 3$$
$$\Rightarrow \quad 7 = -3$$
which is impossible

Taking – ve sign, we get
$$y + 7 = -(y - 3)$$
$$\Rightarrow \quad y + 7 = 3y$$
$$\Rightarrow \quad 2y = -4$$
$$\Rightarrow \quad y = \frac{-4}{2} = -2$$

Putting $y = -2$ in equation (1), we get
$$3x - 2 = 7$$
$$\Rightarrow \quad 3x = 9$$
$$\Rightarrow \quad x = 3$$

Thus, I → (3, – 2).

Hence, the centre of the circle is (3, – 2).

Example 4. *The two opposite vertices of a square are (– 1, 2) and (3, 2). Find the coordinates of the other two vertices.*

Sol. Let A (– 1, 2) and C (3, 2) be the two opposite vertices of a square ABCD. Let B(x, y) be the unknown vertex.

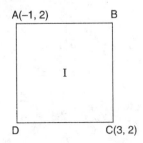

Then, AB = BC
$$\Rightarrow \quad AB^2 = BC^2$$
$$\Rightarrow \quad (x + 1)^2 + (y - 2)^2 = (x - 3)^2 + (y - 2)^2$$
$$\Rightarrow \quad x^2 + 2x + 1 + y^2 - 4y + 4$$
$$= x^2 - 6x + 9 + y^2 - 4y + 4$$
$$\Rightarrow \quad 8x = 8$$
$$\Rightarrow \quad x = 1$$

Also, $AB^2 + BC^2 = AC^2$

| ∵ ∠B = 90° and therefore using Pythagoras theorem

$$\Rightarrow (x + 1)^2 + (y - 2)^2 + (x - 3)^2 + (y - 2)^2$$
$$= (3 + 1)^2 + (2 - 2)^2$$
$$\Rightarrow x^2 + 2x + 1 + y^2 - 4y + 4 + x^2 - 6x$$
$$+ 9 + y^2 - 4y + 4 = 16$$
$$\Rightarrow \quad 2x^2 + 2y^2 - 4x - 8y + 2 = 0$$
$$\Rightarrow \quad x^2 + y^2 - 2x - 4y + 1 = 0$$

| Dividing throughout by 2

Putting $x = 1$, we get
$$1 + y^2 - 2 - 4y + 1 = 0$$
$$\Rightarrow \quad y(y - 4) = 0$$
$$\Rightarrow \quad y = 0, 4$$

Hence, the other vertices are (1, 0) and (1, 4).

Example 5. *The Class X students of a secondary school in Krishinagar have been allotted a rectangular plot of land for their gardening activity. Sapling of Gulmohar are planted on the boundary at a distance of 1m from each other. There is a triangular grassy lawn in the plot as shown in figure. The students are to sow seeds of flowering plants on the remaining area of the plot.*

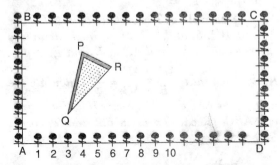

(i) Taking A as origin, find the coordinates of the vertices of the triangle.

(ii) What will be the coordinates of the vertices of ∆PQR if C is the origin ? Also calculate the area of the triangles in these cases. What do you observe ?

Sol. (i) Taking A as origin, AD and AB as coordinate axes, the coordinates of the vertices of the triangle PQR are
P → (4, 6)
Q → (3, 2)
R → (6, 5)

(ii) Taking C as origin, CB and CD as coordinate axes, the coordinates of the vertices of the triangle PQR are

$$P \to (12, 2)$$
$$Q \to (13, 6)$$
$$R \to (10, 3)$$

Area of the triangle PQR in the first case

$$= \frac{1}{2}[4(2-5) + 3(5-6) + 6(6-2)]$$
$$= \frac{1}{2}[-12 - 3 + 24]$$
$$= \frac{9}{2} \text{ sq. units}$$

Area of the triangle PQR in the second case

$$= \frac{1}{2}[12(6-3) + 13(3-2) + 10(2-6)]$$
$$= \frac{1}{2}[36 + 13 - 40]$$
$$= \frac{9}{2} \text{ sq. units}$$

We observe that areas are the same in both the cases.

Example 6. *The vertices of a △ABC are A(4, 6), B(1, 5) and C(7, 2). A line is drawn to intersect sides AB and AC at D and E respectively, such that $\frac{AD}{AB} = \frac{AE}{AC} = \frac{1}{4}$. Calculate the area of the △ADE and compare it with the area of △ABC. (Recall Theorem 6.2 and Theorem 6.6).*

Sol. $\frac{AD}{AB} = \frac{1}{4}$

$\Rightarrow \frac{AB}{AD} = \frac{4}{1}$

$\Rightarrow \frac{AD + DE}{AD} = \frac{4}{1}$

$\Rightarrow 1 + \frac{DE}{AD} = \frac{4}{1}$

$\Rightarrow \frac{DE}{AD} = \frac{3}{1}$

$\Rightarrow \frac{AD}{DE} = \frac{1}{3}$

$\Rightarrow AD : DE = 1 : 3$

$\therefore D \to \left\{ \frac{(1)(1) + (3)(4)}{1+3}, \frac{(1)(5) + (3)(6)}{1+3} \right\}$

$\Rightarrow D \to \left(\frac{13}{4}, \frac{23}{4} \right)$

Similarly,

$AE : EC = 1 : 3$

$\therefore E \to \left\{ \frac{(1)(7) + (3)(4)}{1+3}, \frac{(1)(2) + (3)(6)}{1+3} \right\}$

$\Rightarrow E \to \left\{ \frac{19}{4}, 5 \right\}$

Now,

Area of △ ADE $= \frac{1}{2}\left[4\left(\frac{23}{4} - 5\right) + \frac{13}{4}(5-6) + \frac{19}{4}\left(6 - \frac{23}{4}\right) \right]$

$= \frac{1}{2}\left[3 - \frac{13}{4} + \frac{19}{16} \right]$

$= \frac{1}{2}\left[\frac{48 - 52 + 19}{16} \right]$

$= \frac{15}{32}$ sq. units.

and, Area of △ ABC

$= \frac{1}{2}[4(5-2) + 1(2-6) + 7(6-5)]$

COORDINATE GEOMETRY

$$= \frac{1}{2}[12 - 4 + 7]$$

$$= \frac{15}{2} \text{ sq.units}$$

$$\therefore \quad \frac{\text{ar}(\triangle ADE)}{\text{ar}(\triangle ABC)} = \frac{\frac{15}{32}}{\frac{15}{2}} = \frac{1}{16}$$

Hence, the required ratio is 1 : 16.

Example 7. *Let A(4, 2), B(6, 5) and C(1, 4) be the vertices of $\triangle$ ABC.*

(i) The median from A meets BC at D. Find the coordinates of the point D.

(ii) Find the coordinates of the point P on AD such that AP : PD = 2 : 1.

(iii) Find the coordinates of points Q and R on medians BE and CF respectively such that BQ : QE = 2 : 1 and CR : RF = 2 : 1.

(iv) What do you observe ?

[Note : The point which is common to all the three medians is called the centroid and this point divides each median in the ratio 2 : 1].

(v) If $A(x_1, y_1)$, $B(x_2, y_2)$ and $C(x_3, y_3)$ are the vertices of $\triangle$ ABC, find the coordinates of the centroid of the triangle.

Sol. (i) $\because$ The median from A meets BC at D

$\therefore$ D is the mid-point of BC

$$\therefore \quad D \to \left(\frac{6+1}{2}, \frac{5+4}{2}\right)$$

| Using mid-point formula

$$\Rightarrow \quad D \to \left(\frac{7}{2}, \frac{9}{2}\right)$$

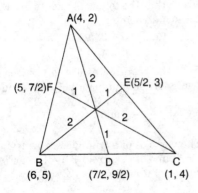

(ii) $P \to \left(\dfrac{(2)\left(\dfrac{7}{2}\right) + (1)(4)}{2+1}, \dfrac{(2)\left(\dfrac{9}{2}\right) + (1)(2)}{2+1}\right)$

| Using section formula

$$\Rightarrow \quad P \to \left(\frac{11}{3}, \frac{11}{3}\right)$$

$$E \to \left(\frac{4+1}{2}, \frac{2+4}{2}\right)$$

$$\Rightarrow \quad E \to \left(\frac{5}{2}, 3\right)$$

(iii) $Q \to$

$$\left\{\dfrac{(2)\left(\dfrac{5}{2}\right) + (1)(6)}{2+1}, \dfrac{(2)(3) + (1)(5)}{2+1}\right\}$$

$$\Rightarrow \quad Q \to \left(\frac{11}{3}, \frac{11}{3}\right)$$

$$F \to \left(\frac{4+6}{2}, \frac{2+5}{2}\right)$$

$$\Rightarrow \quad F \to \left(5, \frac{7}{2}\right)$$

$$R \to \left\{\dfrac{(2)(5) + (1)(1)}{2-1}, \dfrac{(2)\left(\dfrac{7}{2}\right) + (1)(4)}{2+1}\right\}$$

$$\Rightarrow \quad R \to \left(\frac{11}{3}, \frac{11}{3}\right)$$

(iv) We observe that P, Q, R are the same point.

(v) See Example 16 on page 291.

Example 8. *ABCD is a rectangle formed by the points A(– 1, – 1), B(– 1, 4), C(5, 4) and D(5, – 1). P, Q, R and S are the mid-points of AB, BC, CD and DA respectively. Is the quadrilateral PQRS a square ? a rectangle ? or a rhombus ? Justify your answer.*

Sol. We have,

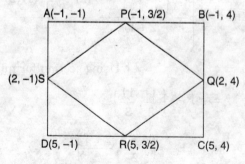

A → (− 1, − 1)
B → (− 1, 4)
C → (5, 4)
D → (5, − 1)

Therefore,

$$P \to \left(\frac{-1-1}{2}, \frac{-1+4}{2}\right)$$

$$P \to \left(-1, \frac{3}{2}\right)$$

$$Q \to \left(\frac{-1+5}{2}, \frac{4+4}{2}\right)$$

$$Q \to (2, 4)$$

$$R \to \left(\frac{5+5}{2}, \frac{-1+4}{2}\right)$$

$$R \to \left(5, \frac{3}{2}\right)$$

$$S \to \left(\frac{-1+5}{2}, \frac{-1-1}{2}\right) \Rightarrow S \to (2, -1)$$

$$\therefore PQ = \sqrt{(2+1)^2 + \left(4 - \frac{3}{2}\right)^2}$$

$$= \sqrt{9 + \frac{25}{4}} = \frac{\sqrt{61}}{2}$$

$$QR = \sqrt{(5-2)^2 + \left(\frac{3}{2} - 4\right)^2}$$

$$= \sqrt{9 + \frac{25}{4}} = \frac{\sqrt{61}}{2}$$

$$RS = \sqrt{(2-5)^2 + \left(-1 - \frac{3}{2}\right)^2}$$

$$= \sqrt{9 + \frac{25}{4}} = \frac{\sqrt{61}}{2}$$

$$SP = \sqrt{(2+1)^2 + \left(-1 - \frac{3}{2}\right)^2}$$

$$= \sqrt{9 + \frac{25}{4}} = \frac{\sqrt{61}}{2}$$

$$PR = \sqrt{(5+1)^2 + \left(\frac{3}{2} - \frac{3}{2}\right)^2} = 6$$

$$QS = \sqrt{(2-2)^2 + (4+1)^2} = 5$$

We see that
$$PQ = QR = RS = SP$$
(all the sides are equal)

and $PR \ne QS$
(diagonals are not equal)

Therefore, PQRS is a rhombus.

MISCELLANEOUS EXERCISE

1. The distance of the point $P(x, y)$ from the points $(a + b, b - a)$ and $(a - b, a + b)$ are equal. Then show that $ay = bx$.

2. Test whether the points $(-3, 4), (2, -5)$ and $(11, 18)$ are collinear or not.

3. Find out the distance between the following pairs of points :
   (i) $(a \cos \alpha, a \sin \alpha)$ and $(a \cos \beta, a \sin \beta)$
   (ii) $(a m_1^2, 2 a m_1)$ and $(a m_2^2, 2am_2)$
   (iii) $\left(6a, \frac{a}{2}\right)$ and $\left(-4a, \frac{5a}{2}\right)$.

4. Prove that the following points are the vertices of a rhombus :
   $(a, b), (a + 3, b + 4), (a - 1, b + 7)$
   and $(a - 4, b + 3)$.

5. Find the coordinates of the vertices of the triangle, the mid-points of whose sides are
   $\left(0, \frac{1}{2}\right), \left(\frac{1}{2}, \frac{1}{2}\right)$ and $\left(\frac{1}{2}, 0\right)$.

6. The vertices of a triangle are $(-4, 2), (9, 7)$ $(2, b)$ and its centroid is $(-1, 3)$. Find the values of a and b.

7. If G is the centroid of a triangle ABC and O is any other point, then prove that
$$OA^2 + OB^2 + OC^2 = GA^2 + GB^2 + GC^2 + 3\,GO^2$$

8. Show that the points A(1, 0), B(5, 3), C(2, 7) and D(– 2, 4) are the vertices of a parallelogram.

   [**Hint**: Diagonals of a parallelogram bisect each other.]

9. In figure, a right triangle BOA is given. C is the mid-point of the hypotenuse AB. Show that it is equidistant from the vertices O, A and B.

10. Find the lengths of the medians of the triangle whose vertices are (1, – 1), (0, 4) and (5, 3).

11. Find the distance of the point (1, 2) from the mid-point of the line-segment joining the points (6, 8) and (2, 4).

12. Show that the mid-point of the line-segment joining the points (5, 7) and (3, 9) is also the mid-point of the line-segment joining the points (8, 6) and (– 0, 10).

13. Prove that the triangle whose vertices are given by the points $(2a, 4a)$, $(2a, 6a)$ and $(2a + \sqrt{3}\,a, 5a)$ is equilateral.

14. The point (9, 6) divides the line joining (a, b) and (5, 7) in the ratio 2 : 1. Find the values of a and b.

15. Prove that no value of x exists for which the points A(x, 4), B(1, – 2) and C(– 3, 2) form an isosceles triangle at the vertex B.

16. If (x, y) is a point on the line joining (a, 0) and (0, b), then show that $\dfrac{x}{a} + \dfrac{y}{b} = 1$.

17. Find the area of the triangle formed by the points $(p + 1, 1)$, $(2p + 1, 3)$ and $(2p + 1, 2p)$. Show that these points are collinear if $p = 2$ or $-\dfrac{1}{2}$.

18. If $a \ne b \ne c$, prove that the points (a, a^2), (b, b^2), (c, c^2) can never be collinear.

19. If three points (x_1, y_1), (x_2, y_2), (x_3, y_3) lie on the same line, prove that
$$\dfrac{y_2 - y_3}{x_2 x_3} + \dfrac{y_3 - y_1}{x_3 x_1} + \dfrac{y_1 - y_2}{x_1 x_2} = 0.$$

20. Two vertices of an equilateral triangle are (0, 0) and (0, $2\sqrt{3}$). Find the third vertex.

21. If the points (a, b) and (b, a) are equidistant from the point (x, y), then prove that $x = y$.

22. Find the coordinates of the point of trisection A the line-segment joining the points (3, – 3) and (6, 9).

Answers

2. No.

3. (i) $2a \sin \dfrac{\alpha - \beta}{2}$

   (ii) $a(m_1 - m_2)\sqrt{(m_1 + m_2)^2 + 4}$

   (iii) $\sqrt{104}\ a$

5. (0, 0), (1, 0), (0, 1)   6. $a = -1, b = 0$

10. $\dfrac{\sqrt{130}}{2}, \sqrt{13}, \dfrac{\sqrt{130}}{2}$ 11. 5

14. $a = 2, b = 4$ 17. $\dfrac{1}{2}(2p^2 - 3p - 2)$

20. $(3, \sqrt{3})(-3, \sqrt{3})$.   22. (4, 1), (5, 5)

SUMMARY

1. The distance between $P(x_1, y_1)$ and $Q(x_2, y_2)$ is $\sqrt{(x_2 - x_1)^2 + (y_2 - y_1)^2}$.

2. The distance of a point $P(x, y)$ from the origin is $\sqrt{x^2 + y^2}$.

3. The coordinates of the point $P(x, y)$ which divides the line segment joining the points $A(x_1, y_1)$ and $B(x_2, y_2)$ internally in the ratio $m_1 : m_2$ are $\left(\dfrac{m_1 x_2 + m_2 x_1}{m_1 + m_2}, \dfrac{m_1 y_2 + m_2 y_1}{m_1 + m_2}\right)$.

4. The midpoint of the line segment joining the points $P(x_1, y_1)$ and $Q(x_2, y)$ is $\left(\dfrac{x_1 + x_2}{2}, \dfrac{y_1 + y_2}{2}\right)$.

5. The area of the triangle formed by the points (x_1, y_1), (x_2, y_2) and (x_3, y_3) is the numerical value of the expression.
$$\dfrac{1}{2}[x_1(y_2 - y_3) + x_2(y_3 - y_1) + x_3(y_1 - y_2)].$$

8

Introduction to Trigonometry

IMPORTANT POINTS

1. Right triangles in our surroundings. Let us take some examples from own surroundings where right triangles can be imagined to be formed. For instance :

1. Suppose the students of a school are visiting Qutub Minar. Now, if a student is looking at the top of the Minar, a right triangle can be imagined to be made, as shown in figure. Can the student find out the height of the Minar, without actually measuring it ?

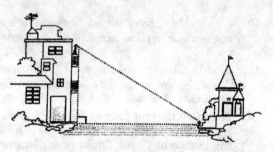

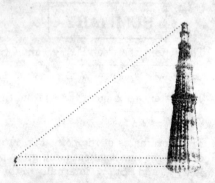

2. Suppose a girl is sitting on the balcony of her house located on the bank of a river. She is looking down at a flower pot placed on a stair of a temple situated nearby on the other bank of the river. A right triangle is imagined to be made in this situation as shown in figure. If we know the height at which the person is sitting, can we find the width of the river ?

3. Suppose a hot air balloon is flying in the air. A girl happens to spot the balloon in the sky and runs to her mother to tell her about it. Her mother rushes out of the house to look at the balloon. Now when the girl had spotted the balloon initially it was at point A. When both the mother an daughter came out to see it, it had already travelled to another point B. Can we find the altitude of B from the ground where they are ?

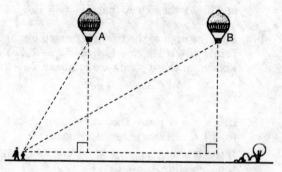

In all the situations given above, the distances or heights can be found by using some mathematical techniques, which come

INTRODUCTION TO TRIGONOMETRY

under a branch of mathematics called 'trigonometry'.

2. Historical Facts. The word 'trigonometry' is derived from the Greek words 'tri' (meaning three), 'gon' (meaning sides) and 'metron' (meaning measure). In fact, **trigonometry** is the study of relationships between the sides and angles of a triangle. The earliest known work on trigonometry was recorded in Egypt and Babylon. Early astronomers used it to find out the distances of the stars and planets from the Earth. Even today, most of the technologically advanced methods used in Engineering and Physical Sciences are still based on trigonometrical concepts.

3. Trigonometric Ratios. There are some ratios of the side of a right triangle with respect to its acute angles, called trigonometric ratios of the angle. We have seen above some right triangles imagined to be formed in different situations.

Let us take a right triangle ABC as shown in figure.

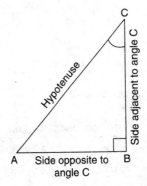

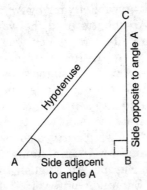

Here, $\angle CAB$ (or, in brief, angle A) is an acute angle. Note the position of the side BC with respect to angle A. It faces $\angle A$. We call it the *side opposite* to angle A. AC is the *hypotenuse* of the right triangle and the side AB is *adjacent* to $\angle A$. So, we call it the *side adjacent* to angle A.

Note. The position of sides changes when we consider angle C in place of A as shown in figure.

We are familiar with the concept of "ratio".

We now define certain ratios involving the sides of a right triangle, and call them trigonometric ratios.

The trigonometric ratios of the angle A in right triangle ABC (see figure) are defined as:

$$\text{sine of } \angle A = \frac{\text{side opposite to angle A}}{\text{hypotenuse}}$$

$$= \frac{BC}{AC}$$

$$\text{cosine of } \angle A = \frac{\text{side adjacent to angle A}}{\text{hypotenuse}}$$

$$= \frac{AB}{AC}$$

tangent of $\angle A$

$$= \frac{\text{side opposite to angle A}}{\text{side adjacent to angle A}}$$

$$= \frac{BC}{AB}$$

cosecant of $\angle A$

$$= \frac{1}{\text{sine of } \angle A}$$

$$= \frac{\text{hypotenuse}}{\text{side opposite to angle A}}$$

$$= \frac{AC}{BC}$$

$$\text{secant of } \angle A = \frac{1}{\text{cosine of } \angle A}$$

$$= \frac{\text{hypotenuse}}{\text{side adjacent to angle A}}$$

$$= \frac{AC}{BC}$$

$$\text{cotangent of } \angle A = \frac{1}{\text{tangent of } \angle A}$$

$$= \frac{\text{side adjacent to angle A}}{\text{side opposite to angle A}}$$

$$= \frac{AB}{BC}$$

The ratios defined above are abbreviated as sin A, cos A, tan A, cosec A, sec A and cot A respectively. Note that the ratios **cosec A, sec A and cot A** are respectively, the reciprocals of the ratios sin A, cos A and tan A.

Also, observe that

$$\tan A = \frac{BC}{AB} = \frac{\frac{BC}{AC}}{\frac{AB}{AC}} = \frac{\sin A}{\cos A}.$$

So, the **trigonometric ratios** of an acute angle in a right triangle express the relationship between the angle and the length of its sides.

Trigonometric ratio for angle C

$$\text{sine of } \angle C = \frac{\text{side opposite to angle C}}{\text{hypotenuse}}$$

$$= \frac{AB}{AC}$$

$$\text{cosine of } \angle C = \frac{\text{side adjacent to angle C}}{\text{hypotenuse}}$$

$$= \frac{BC}{AC}$$

$$\text{tangent of } \angle C$$

$$= \frac{\text{side opposite to angle C}}{\text{side adjacent to angle C}}$$

$$= \frac{AB}{BC}$$

$$\text{cosecant of } \angle C = \frac{1}{\text{sine of } \angle C}$$

$$= \frac{\text{hypotenuse}}{\text{side opposite to angle C}}$$

$$= \frac{AC}{AB}$$

$$\text{secant of } \angle C = \frac{1}{\text{cosine of } \angle C}$$

$$= \frac{\text{hypotenuse}}{\text{side adjacent to angle C}}$$

$$= \frac{AC}{BC}$$

$$\text{cotangent of } \angle C = \frac{1}{\text{tangent of } \angle C}$$

$$= \frac{\text{side adjacent to angle C}}{\text{side opposite to angle C}}$$

$$= \frac{BC}{AB}.$$

4. Historical Facts. The first use of the idea of "sine" in the way we use it today was in the work 'Aryabhatiyam' by Aryabhatta, in A.D. 500. Aryabhatta used the word *ardha-jya* for the half-chord, which was shortened to *jya* or *jiva* in due course. When the Aryabhatiyam was translated into Arabic the word *jiva* was retained as it is. The word *jiva* was translated into *sinus*, which means curve, when the Arabic version was translated into Latin. Soon the word *sinus* also used as *sine*, became common in mathematical texts throughout Europe. An English Professor of astronomy Edmund Gunter (1581-1626), first used the abbreviated notation '*sin*'.

The origin of the terms '**cosine**' and '**tangent**' is much later. The cosine function arose from the need to compute the sine of the complementary angle. Aryabhatta called it **kotijya**. The name *cosinus* originated with Edmund Gunter. In 1674, the English Mathematician Sir Jonas Moore first used the abbreviated notation '*cos*'.

INTRODUCTION TO TRIGONOMETRY

Aryabhatta
A.D. 476—550

5. Remarks. (1) Note that the symbol sin A is used as an abbreviation for 'the sine of the angle A'. Sin A *is not* the product of 'sin' and A. 'Sin' separated from A has no meaning. Similarly, cos A is *not* the product of 'cos' and A. Similar interpretations follow for other trigonometric ratios also.

(2) Let us take a point P on the hypotenuse AC or a point Q on AC extended of the right triangle ABC and draw PM perpendicular to AB and QN perpendicular to AB extended (see figure)

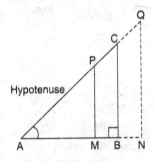

Then using AA similarity criterion, we see that

$$\triangle \text{PAM} \sim \triangle \text{CAB}$$

$$\therefore \quad \frac{AM}{AB} = \frac{AP}{AC} = \frac{MP}{BC}$$

∵ Corresponding sides of two similar triangles are proportional

From this, we find $\frac{MP}{AP} = \frac{BC}{AC} = \sin A$.

Similarly, $\frac{AM}{AP} = \frac{AB}{AC} = \cos A$,

$$\frac{MP}{AM} = \frac{BC}{AB} = \tan A \text{ and so on.}$$

This shows that the trigonometric ratios of angle A in $\triangle$ PAM do not differ from those of angle A in $\triangle$ CAB.

In the same way, we may check that the value of sin A, cos A, tan A and so on remains the same in $\triangle$ QAN also.

From our observations, it is now clear that **the values of the trigonometric ratios of an angle do not vary with the lengths of the sides of the triangle, if the angle remains the same.**

(3) For the sake of convenience, we may write $\sin^2 A$, $\cos^2 A$, etc., in place of $(\sin A)^2$, $(\cos A)^2$, etc., respectively. But cosec $A = (\sin A)^{-1} \neq \sin^{-1} A$ (it is called sine inverse A). $\sin^{-1} A$ has different meaning. Similar conventions hold for the other trigonometric ratios as well.

6. To obtain other five ratios when any one of the six trigonometric ratios is known. Suppose that in a right triangle ABC, $\sin A = \frac{1}{3}$, then this means that $\frac{BC}{AC} = \frac{1}{3}$, i.e., the lengths of the sides BC and AC of the triangle ABC are in the ratio 1 : 3 see figure. So if BC is equal to k, then AC will be $3k$, where k is any positive number. To determine other trigonometric ratios for the angle A, we need to find the length of the third side AB. We shall use the Pythagoras theorem to determine the required length AB.

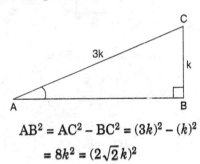

$$AB^2 = AC^2 - BC^2 = (3k)^2 - (k)^2$$
$$= 8k^2 = (2\sqrt{2}\,k)^2$$

Therefore, $AB = 2\sqrt{2}\,k$

Now, $\cos A = \dfrac{AB}{AC} = \dfrac{2\sqrt{2}k}{3k} = \dfrac{2\sqrt{2}}{3}$

Similarly, we can obtain the other trigonometric ratios of the angle A.

Remark. Since the hypotenuse is the longest side in a right triangle, the value of sin A or cos A is always less than 1 (or, in particular, equal to 1).

ILLUSTRATIVE EXAMPLES

[NCERT Exercise 8.1]
(Page No. 181)

Example 1. *In $\triangle ABC$, right angled at B, AB = 24 cm, BC = 7 cm. Determine:*

(i) sin A, cos A

(ii) sin C, cos C

Sol. (i) **sin A, cos A**

In $\triangle ABC$,

$\because \quad \angle B = 90°$ | Given

$\therefore \quad AC^2 = AB^2 + BC^2$

  | By Pythagoras theorem

$= (24)^2 + (7)^2$

$= 576 + 49$

$= 625$

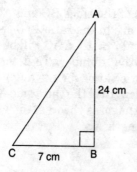

$\Rightarrow \quad AC = \sqrt{625} = 25$ cm

$\therefore \quad \sin A = \dfrac{BC}{AC} = \dfrac{7}{25}$

and $\quad \cos A = \dfrac{AB}{AC} = \dfrac{24}{25}$.

(ii) **sin C, cos C**

$\sin C = \dfrac{AB}{AC} = \dfrac{24}{25}$

$\cos C = \dfrac{BC}{AC} = \dfrac{7}{25}$.

Example 2. *In figure, find $\tan P - \cot R$.*

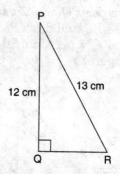

Sol. In $\triangle PQR$,

$\therefore \quad \angle Q = 90°$

$\therefore \quad PR^2 = PQ^2 + QR^2$

  | By Pythagoras theorem

$\Rightarrow \quad (13)^2 = (12)^2 + QR^2$

$\Rightarrow \quad 169 = 144 + QR^2$

$\Rightarrow \quad QR^2 = 169 - 144$

$\Rightarrow \quad QR^2 = 25$

$\Rightarrow \quad QR = \sqrt{25} = 5$ cm

$\therefore \quad \tan P - \cot R$

$= \dfrac{QR}{PQ} - \dfrac{QR}{PQ} = 0$.

INTRODUCTION TO TRIGONOMETRY

Example 3. *If $\sin A = \dfrac{3}{4}$, calculate $\cos A$ and $\tan A$.*

Sol. Let us draw a right triangle ABC.

$$\sin A = \frac{3}{4} \quad \text{| Given}$$

$$\Rightarrow \quad \frac{BC}{AC} = \frac{3}{4}$$

$$\Rightarrow \quad \frac{BC}{3} = \frac{AC}{4} = k \text{ (say)}$$

where k is a positive number.

$$\Rightarrow \quad BC = 3k$$
$$\quad AC = 4k$$

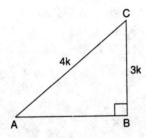

By using the Pythagoras theorem, we have

$$AC^2 = AB^2 + BC^2$$
$$\Rightarrow (4k)^2 = AB^2 + (3k)^2$$
$$\Rightarrow 16k^2 = AB^2 + 9k^2$$
$$\Rightarrow AB^2 = 16k^2 - 9k^2$$
$$\Rightarrow AB^2 = 7k^2$$
$$\Rightarrow AB = \sqrt{7}\,k$$

Now, $\cos A = \dfrac{AB}{AC} = \dfrac{\sqrt{7}\,k}{4k} = \dfrac{\sqrt{7}}{4}$

and $\tan A = \dfrac{BC}{AB} = \dfrac{3k}{\sqrt{7}\,k} = \dfrac{3}{\sqrt{7}}$.

Example 4. *Given $15 \cot A = 8$, find $\sin A$ and $\sec A$.*

Sol. Let us draw a right triangle ABC.

$$15 \cot A = 8 \quad \text{| Given}$$

$$\Rightarrow \quad \cot A = \frac{8}{15}$$

$$\Rightarrow \quad \frac{AB}{BC} = \frac{8}{15}$$

$$\Rightarrow \quad \frac{AB}{8} = \frac{BC}{15} = k \text{ (say)}$$

where k is a positive number

$$\Rightarrow \quad AB = 8k$$
$$\quad BC = 15k$$

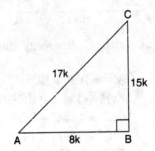

By using the Pythagoras theorem, we have

$$AC^2 = AB^2 + BC^2$$
$$\Rightarrow AC^2 = (8k)^2 + (15k)^2$$
$$\Rightarrow AC^2 = 64k^2 + 225k^2$$
$$\Rightarrow AC^2 = 289k^2$$
$$\Rightarrow AC = \sqrt{289k^2}$$
$$\Rightarrow AC = 17k$$

Now, $\sin A = \dfrac{BC}{AC} = \dfrac{15k}{17k} = \dfrac{15}{17}$

and, $\sec A = \dfrac{AC}{AB} = \dfrac{17k}{8k} = \dfrac{17}{8}$.

Example 5. *Given $\sec \theta = \dfrac{13}{12}$, calculate all other trigonometric ratios.*

Sol. Let us draw a right triangle ABC in which $\angle BAC = \theta$.

$$\sec \theta = \frac{13}{12} \quad \text{| Given}$$

$$\Rightarrow \quad \frac{AC}{AB} = \frac{13}{12}$$

$$\Rightarrow \quad \frac{AC}{13} = \frac{AB}{12} = k \text{ (say)},$$

where k is a positive number.

$\Rightarrow$ AC = 13k
   AB = 12k

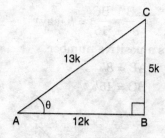

By using the Pythagoras theorem, we have

$$AC^2 = AB^2 + BC^2$$
$\Rightarrow$ $(13k)^2 = (12k)^2 + BC^2$
$\Rightarrow$ $169k^2 = 144k^2 + BC^2$
$\Rightarrow$ $BC^2 = 169k^2 - 144k^2$
$\Rightarrow$ $BC^2 = 25k^2$
$\Rightarrow$ $BC = \sqrt{25}\,k^2$
$\Rightarrow$ $BC = 5k$

Therefore,

$$\sin\theta = \frac{BC}{AC} = \frac{5k}{13k} = \frac{5}{13}$$

$$\cos\theta = \frac{AB}{AC} = \frac{12k}{13k} = \frac{12}{13}$$

or $\cos\theta = \dfrac{1}{\sec\theta} = \dfrac{12}{13}$

$$\tan\theta = \frac{BC}{AB} = \frac{5k}{12k} = \frac{5}{12}$$

or $\tan\theta = \dfrac{\sin\theta}{\cos\theta} = \dfrac{\tfrac{5}{13}}{\tfrac{12}{13}} = \dfrac{5}{12}$

$$\csc\theta = \frac{AC}{BC} = \frac{13k}{5k} = \frac{13}{5}$$

or $\csc\theta = \dfrac{1}{\sin\theta} = \dfrac{13}{5}$

$$\cot\theta = \frac{AB}{BC} = \frac{12k}{5k} = \frac{12}{5}$$

or $\cot\theta = \dfrac{1}{\tan\theta} = \dfrac{1}{\tfrac{5}{12}} = \dfrac{12}{5}$.

Example 6. *If $\angle A$ and $\angle B$ are acute angles such that $\cos A = \cos B$, then show that $\angle A = \angle B$.*

Sol. Let us consider two right triangles LMN and PQR such that $\angle LNM = \angle A$ and $\angle PRQ = \angle B$.

$\cos A = \cos B$ | Given

$\Rightarrow$ $\dfrac{NM}{NL} = \dfrac{RQ}{RP}$

$\Rightarrow$ $\dfrac{NM}{RQ} = \dfrac{NL}{RP} = k$ (say),   ...(1)

where k is a positive number.

$\Rightarrow$ NM = k RQ
   NL = k RP

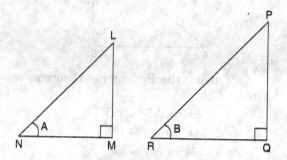

Now, using Pythagoras theorem,

$$ML = \sqrt{NL^2 - NM^2}$$
$$= \sqrt{(k\,RP)^2 - (k\,RQ)^2}$$
$$= \sqrt{k^2\,RP^2 - k^2\,RQ^2}$$
$$= k\,\sqrt{RP^2 - RQ^2}$$

and $QP = \sqrt{RP^2 - RQ^2}$

So, $\dfrac{ML}{QP} = \dfrac{k\,\sqrt{RP^2 - RQ^2}}{\sqrt{RP^2 - RQ^2}} = k$

...(2)

INTRODUCTION TO TRIGONOMETRY

From (1) and (2), we have

$$\frac{NM}{RQ} = \frac{NL}{RP} = \frac{ML}{QP}$$

∴ △LMN ~ △PQR
 | SSS similarity criterion

∴ ∠LNM = ∠PRQ
 | ∵ Corresponding angles of two similar triangles are equal

⇒ ∠A = ∠B.

Example 7. If $\cot \theta = \frac{7}{8}$, evaluate :

(i) $\dfrac{(1+\sin\theta)(1-\sin\theta)}{(1+\cos\theta)(1-\cos\theta)}$, (ii) $\cot^2 \theta$

Sol. Let us draw a right triangle ABC in which ∠BAC = θ.

$\cot \theta = \dfrac{7}{8}$ | Given

⇒ $\dfrac{AB}{BC} = \dfrac{7}{8}$

⇒ $\dfrac{AB}{7} = \dfrac{BC}{8} = k$ (say),

where k is a positive number.

⇒ AB = 7k
 BC = 8k

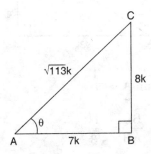

By using the Pythagoras theorem, we have

$AC^2 = AB^2 + BC^2$
$= (7k)^2 + (8k)^2$
$= 49k^2 + 64k^2$
$= 113k^2$

⇒ $AC = \sqrt{113k^2}$

⇒ $AC = \sqrt{113}\,k$

Therefore,

$\sin \theta = \dfrac{BC}{AC} = \dfrac{8k}{\sqrt{113}\,k} = \dfrac{8}{\sqrt{113}}$

$\cos \theta = \dfrac{AB}{AC} = \dfrac{7k}{\sqrt{113}\,k} = \dfrac{7}{\sqrt{113}}$

(i) $\dfrac{(1+\sin\theta)(1-\sin\theta)}{(1+\cos\theta)(1-\cos\theta)}$

$= \dfrac{\left(1+\dfrac{8}{\sqrt{113}}\right)\left(1-\dfrac{8}{\sqrt{113}}\right)}{\left(1+\dfrac{7}{\sqrt{113}}\right)\left(1-\dfrac{7}{\sqrt{113}}\right)}$

$= \dfrac{(1)^2 - \left(\dfrac{8}{\sqrt{113}}\right)^2}{(1)^2 - \left(\dfrac{7}{\sqrt{113}}\right)^2}$

$= \dfrac{1 - \dfrac{64}{113}}{1 - \dfrac{49}{113}}$

$= \dfrac{113-64}{113-49} = \dfrac{49}{64}.$

(ii) $\cot^2 \theta = (\cot \theta)^2$

$= \left(\dfrac{7}{8}\right)^2$

$= \dfrac{49}{64}.$

Example 8. If $3 \cot A = 4$, check whether $\dfrac{1-\tan^2 A}{1+\tan^2 A} = \cos^2 A - \sin^2 A$ or not.

Sol. Let us draw a right triangle ABC.

3 cot A = 4 | Given

⇒ $\cot A = \dfrac{4}{3}$

$\Rightarrow \quad \dfrac{AB}{BC} = \dfrac{4}{3}$

$\Rightarrow \quad \dfrac{AB}{4} = \dfrac{BC}{3} = k$ (say),

where k is a positive number.

$\Rightarrow \quad AB = 4k$

$BC = 3k$

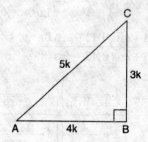

By using the Pythagoras theorem, we have

$AC^2 = AB^2 + BC^2$

$\Rightarrow \quad AC^2 = (4k)^2 + (3k)^2$

$\Rightarrow \quad AC^2 = 16k^2 + 9k^2$

$\Rightarrow \quad AC^2 = 25k^2$

$\Rightarrow \quad AC = \sqrt{25k^2}$

$\Rightarrow \quad AC = 5k$

So, $\tan A = \dfrac{BC}{AB} = \dfrac{3k}{4k} = \dfrac{3}{4}$

$\cos A = \dfrac{AB}{BC} = \dfrac{4k}{5k} = \dfrac{4}{5}$

$\sin A = \dfrac{BC}{AC} = \dfrac{3k}{5k} = \dfrac{3}{5}$

Now, L.H.S. $= \dfrac{1 - \tan^2 A}{1 + \tan^2 A}$

$= \dfrac{1 - \left(\dfrac{3}{4}\right)^2}{1 + \left(\dfrac{3}{4}\right)^2}$

$= \dfrac{1 - \dfrac{9}{16}}{1 + \dfrac{9}{16}} = \dfrac{\dfrac{16-9}{16}}{\dfrac{16+9}{16}}$

$= \dfrac{\dfrac{7}{16}}{\dfrac{25}{16}} = \dfrac{7}{25}$ ...(1)

R.H.S. $= \cos^2 A - \sin^2 A$

$= \left(\dfrac{4}{5}\right)^2 - \left(\dfrac{3}{5}\right)^2 = \dfrac{16}{25} - \dfrac{9}{25}$

$= \dfrac{16-9}{25} = \dfrac{7}{25}$ ...(2)

From (1) and (2), we get

L.H.S = R.H.S

$\Rightarrow \quad \dfrac{1 - \tan^2 A}{1 + \tan^2 A} = \cos^2 A - \sin^2 A.$

Example 9. *In triangle ABC, right-angled at B, if* $\tan A = \dfrac{1}{\sqrt{3}}$, *find the value of:*

(i) $\sin A \cos C + \cos A \sin C$

(ii) $\cos A \cos C - \sin A \sin C$

Sol. Let us draw a right-angled triangle ABC.

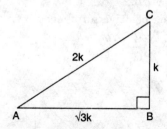

$\tan A = \dfrac{1}{\sqrt{3}}$ | Given

$\Rightarrow \quad \dfrac{BC}{AB} = \dfrac{1}{\sqrt{3}}$

$\Rightarrow \quad \dfrac{BC}{1} = \dfrac{AB}{\sqrt{3}} = k$ (say),

where k is a positive number.

$\Rightarrow \quad BC = k$

$AB = \sqrt{3}\,k$

In $\triangle ABC$,

$\because \quad \angle B = 90°$

∴ $AC^2 = AB^2 + BC^2$
    | By Pythagoras theorem
⇒ $AC^2 = (\sqrt{3}k)^2 + (k)^2$
⇒ $AC^2 = 3k^2 + k^2$
⇒ $AC^2 = 4k^2$
⇒ $AC = \sqrt{4k^2}$
⇒ $AC = 2k$

Therefore,

$$\sin A = \frac{BC}{AC} = \frac{k}{2k} = \frac{1}{2}$$

$$\cos A = \frac{AB}{AC} = \frac{\sqrt{3}k}{2k} = \frac{\sqrt{3}}{2}$$

$$\sin C = \frac{AB}{AC} = \frac{\sqrt{3}k}{2k} = \frac{\sqrt{3}}{2}$$

$$\cos C = \frac{BC}{AC} = \frac{k}{2k} = \frac{1}{2}$$

Now,

(i) $\sin A \cos C + \cos A \sin C$

$$= \frac{1}{2} \cdot \frac{1}{2} + \frac{\sqrt{3}}{2} \cdot \frac{\sqrt{3}}{2}$$

$$= \frac{1}{4} + \frac{3}{4} = 1$$

(ii) $\cos A \cos C - \sin A \sin C$

$$= \frac{\sqrt{3}}{2} \cdot \frac{1}{2} - \frac{1}{2} \cdot \frac{\sqrt{3}}{2} = 0.$$

Example 10. *In △ PQR, right angled at Q, PR + QR = 25 cm and PQ = 5 cm. Determine the values of sin P, cos P and tan P.*

Sol. In △ PQR,

∵ ∠Q = 90° | Given
∴ $PR^2 = PQ^2 + QR^2$
    | By Pythagoras theorem
⇒ $(25 - QR)^2 = (5)^2 + QR^2$
    | ∵ PR + QR = 25 (given)
⇒ $625 + QR^2 - 50\,QR = 25 + QR^2$
⇒ $50\,QR = 600$

⇒ $QR = \dfrac{600}{50} = 12$ cm

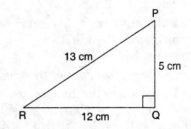

Now, PR + QR = 25
⇒ PR + 12 = 25
⇒ PR = 25 - 12
⇒ PR = 13 cm

So, $\sin P = \dfrac{QR}{PR} = \dfrac{12}{13}$,

$\cos P = \dfrac{PQ}{PR} = \dfrac{5}{13}$,

and, $\tan P = \dfrac{QR}{PQ} = \dfrac{12}{5}$.

Example 11. *State whether the following are true or false. Justify your answer.*

(i) *The value of tan A is always less than 1.*

(ii) $\sec A = \dfrac{12}{5}$ *for some value of angle A.*

(iii) *cos A is the abbreviation used for the cosecant of angle A.*

(iv) *cot A is the product of cot and A.*

(v) $\sin \theta = \dfrac{4}{3}$ *for some angle* θ.

Sol. (i) **False** since

$$\tan A = \frac{\text{Perpendicular}}{\text{Base}}$$ and perpendicular may be longer than base.

(ii) **True** since $\sec A = \dfrac{\text{Hypotenuse}}{\text{Base}}$ and hypotenuse being the longest side may be $\frac{12}{5}$ times the base.

(iii) **False** since cos A is the abbreviation used for the cosine of angle A.

(iv) **False** since cot A is used as an abbreviation for 'the cotangent of the angle A'.

(v) **False** since the hypotenuse is the longest side in a right triangle. As such the value of sin A is always less than 1 (or, in particular equal to 1).

ADDITIONAL EXAMPLES

Example 12. *If* $\sin \theta = \dfrac{a^2 - b^2}{a^2 + b^2}$, *find the value of other trigonometric ratios.*

Sol. Let us draw a right triangle ABC in which $\angle BAC = \theta$.

$$\sin \theta = \frac{a^2 - b^2}{a^2 + b^2} \quad | \text{ Given}$$

$$\Rightarrow \quad \frac{BC}{AC} = \frac{a^2 - b^2}{a^2 + b^2}$$

$$\Rightarrow \quad \frac{BC}{a^2 - b^2} = \frac{AC}{a^2 + b^2} = k(\text{say}),$$

where k is a positive number.

$$\Rightarrow \quad BC = k(a^2 - b^2)$$
$$AC = k(a^2 + b^2)$$

In $\triangle ABC$,
$\because \quad \angle B = 90°$
$\therefore \quad AC^2 = AB^2 + BC^2$
  | By Pythagoras theorem
$\Rightarrow \quad k^2(a^2 + b^2)^2 = AB^2 + k^2(a^2 - b^2)^2$
$\Rightarrow \quad AB^2 = k^2\{(a^2 + b^2)^2 - (a^2 - b^2)^2\}$
$\Rightarrow \quad AB^2 = k^2(4a^2b^2)$
$\Rightarrow \quad AB = 2abk$

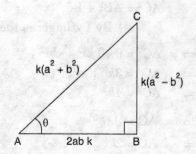

Therefore,

$$\cos \theta = \frac{AB}{AC} = \frac{2abk}{k(a^2 + b^2)} = \frac{2ab}{a^2 + b^2}$$

$$\tan \theta = \frac{BC}{AB} = \frac{k(a^2 - b^2)}{2abk} = \frac{a^2 - b^2}{2ab}$$

$$\operatorname{cosec} \theta = \frac{AC}{BC} = \frac{k(a^2 + b^2)}{k(a^2 - b^2)} = \frac{a^2 + b^2}{a^2 - b^2}$$

$$\sec \theta = \frac{AC}{AB} = \frac{k(a^2 + b^2)}{2abk} = \frac{a^2 + b^2}{2ab}$$

$$\cot \theta = \frac{AB}{BC} = \frac{2abk}{k(a^2 - b^2)} = \frac{2ab}{a^2 - b^2}.$$

Example 13. *In a triangle ABC right angled at C, if* $\tan A = \dfrac{1}{\sqrt{3}}$, *show that sin A cos B + cos A sin B = 1.*

Sol. Let ABC be a triangle right angled at C such that $\tan A = \dfrac{1}{\sqrt{3}}$ and $\tan B = \sqrt{3}$.

$$\tan A = \frac{1}{\sqrt{3}}$$

$$\Rightarrow \quad \frac{BC}{AC} = \frac{1}{\sqrt{3}}$$

$$\Rightarrow \quad \frac{BC}{1} = \frac{AC}{\sqrt{3}} = k \text{ (say)},$$

where k is a positive number

$\Rightarrow$ $\left.\begin{array}{r}BC = k \\ AC = \sqrt{3}\,k\end{array}\right\}$ ...(1)

Again, $\tan B = \sqrt{3}$

$\Rightarrow \dfrac{AC}{BC} = \sqrt{3}$

$\Rightarrow \dfrac{AC}{\sqrt{3}} = \dfrac{BC}{1} = k$ (say),

where k is a positive number.

$\left.\begin{array}{r}AC = \sqrt{3}\,k \\ BC = k\end{array}\right\}$ ...(2)

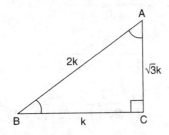

From (1) and (2),

$AC = \sqrt{3}\,k$

$BC = k$

In $\triangle ABC$,

$\because$ $\angle C = 90°$

$\therefore$ $AB^2 = AC^2 + BC^2$

| By Pythagoras theorem

$\Rightarrow AB^2 = (\sqrt{3}\,k)^2 + (k)^2$

$\Rightarrow AB^2 = 3k + k^2$

$\Rightarrow AB^2 = 4k^2$

$\Rightarrow AB = \sqrt{4k^2}$

$\Rightarrow AB = 2k$

Therefore,

$\sin A = \dfrac{BC}{AB} = \dfrac{k}{2k} = \dfrac{1}{2}$

$\cos B = \dfrac{BC}{AB} = \dfrac{k}{2k} = \dfrac{1}{2}$

$\cos A = \dfrac{AC}{AB} = \dfrac{\sqrt{3}\,k}{2k} = \dfrac{\sqrt{3}}{2}$

$\sin B = \dfrac{AC}{AB} = \dfrac{\sqrt{3}\,k}{2k} = \dfrac{\sqrt{3}}{2}$

Now, $\sin A \cos B + \cos A \sin B$

$= \dfrac{1}{2} \cdot \dfrac{1}{2} + \dfrac{\sqrt{3}}{2} \cdot \dfrac{\sqrt{3}}{2}$

$= \dfrac{1}{4} + \dfrac{3}{4} = 1.$

Example 14. *If $\operatorname{cosec} A = 2$, find the value of $\cot A + \dfrac{\sin A}{1 + \cos A}$.*

Sol. Let us draw a right triangle ABC in which $\angle ACB = 90°$

$\operatorname{cosec} A = 2$ | Given

$\Rightarrow \dfrac{AB}{BC} = 2$

$\Rightarrow \dfrac{AB}{2} = \dfrac{BC}{1} = k$ (say),

where k is a positive number.

$\Rightarrow AB = 2k$

$BC = k$

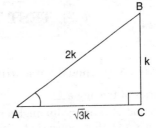

In $\triangle ABC$,

$\because$ $\angle C = 90°$

$\therefore$ $AB^2 = AC^2 + BC^2$

| By Pythagoras theorem

$\Rightarrow (2k)^2 = AC^2 + (k)^2$

$\Rightarrow 4k^2 = AC^2 + k^2$

$\Rightarrow AC^2 = 4k^2 - k^2$

$\Rightarrow AC^2 = 3k^2$

$\Rightarrow AC = \sqrt{3}\,k$

$$\therefore \quad \cot A = \frac{AC}{BC} = \frac{\sqrt{3}k}{k} = \sqrt{3}$$

$$\sin A = \frac{BC}{AB} = \frac{k}{2k} = \frac{1}{2}$$

$$\cos A = \frac{AC}{AB} = \frac{\sqrt{3}k}{2k} = \frac{\sqrt{3}}{2}$$

Now, $\cot A + \dfrac{\sin A}{1 + \cos A}$

$$= \sqrt{3} + \frac{\dfrac{1}{2}}{1 + \dfrac{\sqrt{3}}{2}}$$

$$= \sqrt{3} + \frac{1}{2 + \sqrt{3}}$$

$$= \frac{\sqrt{3}(2 + \sqrt{3}) + 1}{2 + \sqrt{3}}$$

$$= \frac{2\sqrt{3} + 3 + 1}{2 + \sqrt{3}}$$

$$= \frac{2\sqrt{3} + 4}{2 + \sqrt{3}}$$

$$= \frac{2(\sqrt{3} + 2)}{2 + \sqrt{3}} = 2.$$

Example 15. *If* $\tan A = \sqrt{2} - 1$, *show that* $\dfrac{\tan A}{1 + \tan^2 A} = \dfrac{\sqrt{2}}{4}$.

Sol. $\dfrac{\tan A}{1 + \tan^2 A}$

$$= \frac{\sqrt{2} - 1}{1 + (\sqrt{2} - 1)^2}$$

$$= \frac{\sqrt{2} - 1}{1 + 2 + 1 - 2\sqrt{2}}$$

$$= \frac{\sqrt{2} - 1}{4 - 2\sqrt{2}} = \frac{\sqrt{2} - 1}{2\sqrt{2}(\sqrt{2} - 1)} = \frac{1}{2\sqrt{2}}$$

$$= \frac{1}{2\sqrt{2}} \cdot \frac{\sqrt{2}}{\sqrt{2}} = \frac{\sqrt{2}}{4}.$$

TEST YOUR KNOWLEDGE

1. Given $\tan A = \dfrac{4}{3}$, find the other trigonometric ratios of the angle A.
2. If $\angle B$ and $\angle Q$ are acute angles such that $\sin B = \sin Q$, then prove that $\angle B = \angle Q$.
3. Consider $\triangle ACB$, right-angled at C, in which AB = 29 units, BC = 21 units and $\angle ABC = \theta$. Determine the values of

   (i) $\cos^2 \theta + \sin^2 \theta$
   (ii) $\cos^2 \theta - \sin^2 \theta$.

4. In a right-triangle ABC, right-angled at B, if $\tan A = 1$, then verify that

   $2 \sin A \cos A = 1$.

5. In $\triangle OPQ$, right-angled as P, OP = 7 cm and OQ – PQ = 1 cm. Determine the values of $\sin Q$ and $\cos Q$.
6. If $\cos \theta = \dfrac{3}{5}$, calculate the value of $\sin \theta$ and $\tan \theta$.
7. Given $\cot \theta = \dfrac{20}{21}$, determine $\cos \theta$ and $\csc \theta$.
8. If $\sec \theta = \dfrac{25}{7}$, find the values of $\tan \theta$ and $\csc \theta$.

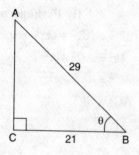

INTRODUCTION TO TRIGONOMETRY

9. If $\csc \theta = \dfrac{41}{40}$, find the values of $\sin \theta$ and $\sec \theta$.

10. If $\sin \theta = \dfrac{1}{\sqrt{2}}$, find the values of other trigonometric ratios.

11. $\triangle ABC$ has a right angle at A. In each of the following, find $\sin B$, $\cos C$ and $\tan B$.
    (i) $AB = AC = 1$ cm
    (ii) $AB = 5$ cm, $BC = 13$ cm
    (iii) $AB = 20$ cm, $AC = 21$ cm.

12. If $\sec \theta = \dfrac{5}{4}$, verify that
$$\dfrac{\tan \theta}{1 + \tan^2 \theta} = \dfrac{\sin \theta}{\sec \theta}$$

13. If $\cot B = \dfrac{12}{5}$, show that
$$\tan^2 B - \sin^2 B = \sin^2 B \tan^2 B.$$

14. If $\cot \theta = \dfrac{1}{\sqrt{3}}$, show that $\dfrac{1 - \cos^2 \theta}{2 - \sin^2 \theta} = \dfrac{3}{5}$.

15. If $\sec \theta = \dfrac{13}{5}$, show that $\dfrac{2 \sin \theta - 3 \cos \theta}{4 \sin \theta - 9 \cos \theta} = 3$.

16. If $\sin B = \dfrac{1}{2}$, show that $3 \cos B - 4 \cos^3 B = 0$.

Answers

1. $\sin A = \dfrac{4}{5}$
   $\cos A = \dfrac{3}{5}$
   $\cot A = \dfrac{3}{4}$

   $\csc A = \dfrac{5}{4}$
   $\sec A = \dfrac{5}{3}$

3. (i) 1 (ii) $\dfrac{41}{841}$

5. $\sin Q = \dfrac{7}{25}$
   $\cos Q = \dfrac{24}{25}$

6. $\sin \theta = \dfrac{4}{5}$
   $\tan \theta = \dfrac{4}{3}$

7. $\cos \theta = \dfrac{20}{29}$, $\csc \theta = \dfrac{29}{21}$

8. $\tan \theta = \dfrac{24}{7}$, $\csc \theta = \dfrac{25}{24}$

9. $\sin \theta = \dfrac{40}{41}$, $\sec \theta = \dfrac{41}{9}$

10. $\cos \theta = \dfrac{1}{\sqrt{2}}$, $\tan \theta = 1$, $\csc \theta = \sqrt{2}$, $\sec \theta = \sqrt{2}$, $\cot \theta = 1$.

11. (i) $\sin B = \dfrac{1}{\sqrt{2}}$, $\cos C = \dfrac{1}{\sqrt{2}}$, $\tan B = 1$.
    (ii) $\sin B = \dfrac{12}{13}$, $\cos C = \dfrac{12}{13}$, $\tan B = \dfrac{12}{5}$.
    (iii) $\sin B = \dfrac{21}{29}$, $\cos C = \dfrac{21}{29}$, $\tan B = \dfrac{21}{20}$.

TRIGONOMETRIC RATIOS OF SOME SPECIFIC ANGLES

IMPORTANT POINTS

1. Trigonometric Ratios of 45°. In $\triangle ABC$, right-angled at B, if one angle is 45°, then the other angle is also 45°
i.e., $\angle A = \angle C = 45°$ (see figure)
So, BC = AB
| Sides opposite to equal angles of a triangle are equal

Now, Suppose BC = AB = a.

Then by Pythagoras Theorem

$AC^2 = AB^2 + BC^2 = a^2 + a^2 = 2a^2$,

and therefore, AC = $a\sqrt{2}$.

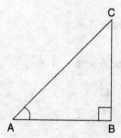
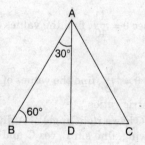

Using the definitions of the trigonometric ratios, we have:

$$\sin 45° = \frac{\text{side opposite to angle } 45°}{\text{hypotenuse}}$$

$$= \frac{BC}{AC} = \frac{a}{a\sqrt{2}} = \frac{1}{\sqrt{2}}$$

$$\cos 45° = \frac{\text{side adjacent to angle } 45°}{\text{hypotenuse}}$$

$$= \frac{AB}{AC} = \frac{a}{a\sqrt{2}} = \frac{1}{\sqrt{2}}$$

$$\tan 45° = \frac{\text{side opposite to angle } 45°}{\text{side adjacent to angle } 45°}$$

$$= \frac{BC}{AB} = \frac{a}{a} = 1$$

Also, $\operatorname{cosec} 45° = \dfrac{1}{\sin 45°} = \sqrt{2}$,

$\sec 45° = \dfrac{1}{\cos 45°} = \sqrt{2}$,

$\cot 45° = \dfrac{1}{\tan 45°} = 1$.

2. Trigonometric Ratios of 30° and 60°. Consider an equilateral triangle ABC. Since each angle in an equilateral triangle is 60°, therefore, $\angle A = \angle B = \angle C = 60°$.

Draw the perpendicular AD from A to the side BC (see figure).

In $\triangle ABD$ and $\triangle ACD$,

$AB = AC$ | Given
$AD = AD$ | Common side
$\angle BAD = \angle CAD$

| $\because$ $\angle ABD = \angle ACD = 60°$,
  $\angle ADB = \angle ADC = 90°$,
  and the sum of the three
  angles of a triangle is 180°

$\therefore$ $\triangle ABD \cong \triangle ACD$

| SAS congruence criterion

Therefore,
$BD = CD$ | CPCT
and $\angle BAD = \angle CAD$ | CPCT

Now we observe that :

$\triangle ABD$ is a right triangle, right angled at D with $\angle BAD = 30°$ and $\angle ABD = 60°$ (see figure).

Let us suppose that $AB = 2a$.

Then $BD = \dfrac{1}{2} BC = a$

and $AD^2 = AB^2 - BD^2$
$= (2a)^2 - (a)^2 = 3a^2$,

Therefore, $AD = a\sqrt{3}$

Now, we have

$$\sin 30° = \frac{BD}{AB} = \frac{a}{2a} = \frac{1}{2},$$

$$\cos 30° = \frac{AD}{AB} = \frac{a\sqrt{3}}{2a} = \frac{\sqrt{3}}{2}$$

$$\tan 30° = \frac{BD}{AD} = \frac{a}{a\sqrt{3}} = \frac{1}{\sqrt{3}}.$$

INTRODUCTION TO TRIGONOMETRY

Also, $\csc 30° = \dfrac{1}{\sin 30°} = 2$,

$\sec 30° = \dfrac{1}{\sin 30°} = \dfrac{2}{\sqrt{3}}$

$\cos 30° = \dfrac{1}{\sin 30°} = \sqrt{3}$.

Similarly,

$\sin 60° = \dfrac{AD}{AB} = \dfrac{a\sqrt{3}}{2a} = \dfrac{\sqrt{3}}{2}$,

$\cos 60° = \dfrac{1}{2}$, $\tan 60° = \sqrt{3}$,

$\csc 60° = \dfrac{2}{\sqrt{3}}$, $\sec 60° = 2$

and $\cot 60° = \dfrac{1}{\sqrt{3}}$.

3. Trigonometric Ratios of 0° and 90°. We shall see what happens to the trigonometric ratios of angle A, if it is made smaller and smaller in the right triangle ABC (see figure), till it becomes zero. As ∠A gets smaller and smaller, the length of the side BC decreases. The point C gets closer to point B, and finally when ∠A becomes very close to 0°, AC becomes almost the same as AB (see figure).

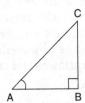

When ∠A is very close to 0, BC gets very close to 0 and so the value of $\sin A = \dfrac{BC}{AC}$ is very close to 0. Also, when ∠A is very close to 0, AC is nearly the same as AB and so the value of $\cos A = \dfrac{AB}{AC}$ is very close to 1.

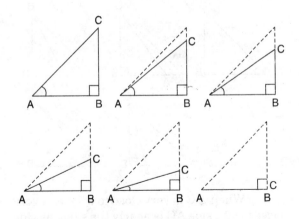

This helps us to see how we can define the values of sin A and cos A when A = 0°. We define : **sin 0° = 0 and cos 0° = 1**.

Using these, we have :

$\tan 0° = \dfrac{\sin 0°}{\cos 0°} = 0$,

$\cot 0° = \dfrac{1}{\tan 0°}$,

which is not defined as division by 0 is not defined

$\sec 0° = \dfrac{1}{\cos 0°} = 1$

and $\csc 0° = \dfrac{1}{\sin 0°}$,

which is again not defined as division by 0 is not defined.

Now, we shall see what happens to the trigonometric ratios of ∠A when it is made larger and larger in △ABC till it becomes 90°. As ∠A gets larger and larger, ∠C gets smaller and smaller. Therefore, as in the case above, the length of the side AB goes on decreasing. The point A gets closer to point B. Finally when ∠A is very close to 90°, ∠C becomes very close to 0° and the side AC almost coincides with side BC (see figure).

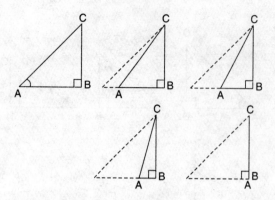

When ∠C is very close to 0°, ∠A is very close to 90°, side AC is nearly the same as side BC, and so sin A is very close to 1. Also when ∠A is very close to 90°, ∠C is very close to 0°, and the side AB is nearly zero, so cos A is very close to 0.

So, we define: **sin 90° = 1**
and **cos 90° = 0**.

Using these, we have

$$\tan 90° = \frac{\sin 90°}{\cos 90°} = \frac{1}{0}$$

which is not defined as division by 0 is not defined.

$$\cot 90° = \frac{\cos 90°}{\sin 90°} = \frac{0}{1} = 0$$

$$\operatorname{cosec} 90° = \frac{1}{\sin 90°} = \frac{1}{1} = 1$$

and $$\sec 90° = \frac{1}{\cos 90°} = \frac{1}{0},$$

which is not defined as division by 0 is not defined.

4. A table for ready reference. The following table gives the values of all the trigonometric ratios of 0°, 30°, 45°, 60° and 90° for ready reference :

Table

∠A	0°	30°	45°	60°	90°
sin A	0	$\frac{1}{2}$	$\frac{1}{\sqrt{2}}$	$\frac{\sqrt{3}}{2}$	1
cos A	1	$\frac{\sqrt{3}}{2}$	$\frac{1}{\sqrt{2}}$	$\frac{1}{2}$	0
tan A	0	$\frac{1}{\sqrt{3}}$	1	$\sqrt{3}$	Not defined
cosec A	Not defined	2	$\sqrt{2}$	$\frac{2}{\sqrt{3}}$	1
sec A	1	$\frac{2}{\sqrt{3}}$	$\sqrt{2}$	2	Not defined
cot A	Not defined	$\sqrt{3}$	1	$\frac{1}{\sqrt{3}}$	0

Remark. From the table we can observe that as ∠A increases from 0° to 90°, sin A increases from 0 to 1 and cos A decreases from 1 to 0.

5. If one of the sides and any other part (either an acute angle or any side) of a right triangle is known, the remaining sides and angles of the triangle can be determined by using Pythagoras theorem or a suitable trigono-metric ratios as the case may be.

ILLUSTRATIVE EXAMPLES

[NCERT Exercise 8.2]
(Page No. 187)

Example 1. *Evaluate the following*
(i) $\sin 60° \cos 30° + \sin 30° \cos 60°$
(ii) $2 \tan^2 45° + \cos^2 30° - \sin^2 60°$
(iii) $\dfrac{\cos 45°}{\sec 30° + \operatorname{cosec} 30°}$
(iv) $\dfrac{\sin 30° + \tan 45° - \operatorname{cosec} 60°}{\sec 30° + \cos 60° + \cot 45°}$
(v) $\dfrac{5 \cos^2 60° + 4 \sec^2 30° - \tan^2 45°}{\sin^2 30° + \cos^2 30°}$

INTRODUCTION TO TRIGONOMETRY

Sol. (i) $\sin 60° \cos 30° + \sin 30° \cos 60°$

$$= \frac{\sqrt{3}}{2} \cdot \frac{\sqrt{3}}{2} + \frac{1}{2} \cdot \frac{1}{2}$$

$$= \frac{3}{4} + \frac{1}{4} = 1$$

(ii) $2 \tan^2 45° + \cos^2 30° - \sin^2 60°$

$$= 2(1)^2 + \left(\frac{\sqrt{3}}{2}\right)^2 - \left(\frac{\sqrt{3}}{2}\right)^2$$

$$= 2 + \frac{3}{4} - \frac{3}{4} = 2$$

(iii) $\dfrac{\cos 45°}{\sec 30° + \csc 30°}$

$$= \frac{\frac{1}{\sqrt{2}}}{\frac{2}{\sqrt{3}} + 2} = \frac{\frac{1}{\sqrt{2}}}{\frac{2 + 2\sqrt{3}}{\sqrt{3}}}$$

$$= \frac{\sqrt{3}}{\sqrt{2}(2 + 2\sqrt{3})}$$

$$= \frac{\sqrt{3}}{2\sqrt{2}(1 + \sqrt{3})}$$

$$= \frac{\sqrt{3}}{2\sqrt{2}(1+\sqrt{3})} \cdot \frac{\sqrt{2}}{\sqrt{2}} \cdot \frac{1-\sqrt{3}}{1-\sqrt{3}}$$

$$= \frac{\sqrt{2}(\sqrt{3}-3)}{4(1-3)} = \frac{\sqrt{6} - 3\sqrt{2}}{-8}$$

$$= \frac{3\sqrt{2} - \sqrt{6}}{8}$$

(iv) $\dfrac{\sin 30° + \tan 45° - \csc 60°}{\sec 30° + \cos 60° + \cot 45°}$

$$= \frac{\frac{1}{2} + 1 - \frac{2}{\sqrt{3}}}{\frac{2}{\sqrt{3}} + \frac{1}{2} + 1}$$

$$= \frac{\frac{\sqrt{3} + 2\sqrt{3} - 4}{2\sqrt{3}}}{\frac{4 + \sqrt{3} + 2\sqrt{3}}{2\sqrt{3}}}$$

$$= \frac{\frac{3\sqrt{3} - 4}{2\sqrt{3}}}{\frac{4 + 3\sqrt{3}}{2\sqrt{3}}} = \frac{3\sqrt{3} - 4}{4 + 3\sqrt{3}}$$

$$= \frac{(3\sqrt{3} - 4)(4 - 3\sqrt{3})}{(4 + 3\sqrt{3})(4 - 3\sqrt{3})}$$

$$= \frac{12\sqrt{3} - 27 - 16 + 12\sqrt{3}}{16 - 27}$$

$$= \frac{24\sqrt{3} - 43}{-11} = \frac{43 - 24\sqrt{3}}{11}$$

(v) $\dfrac{5\cos^2 60° + 4\sec^2 30° - \tan^2 45°}{\sin^2 30° + \cos^2 30°}$

$$= \frac{5\left(\frac{1}{2}\right)^2 + 4\left(\frac{2}{\sqrt{3}}\right)^2 - (1)^2}{\left(\frac{1}{2}\right)^2 + \left(\frac{\sqrt{3}}{2}\right)^2}$$

$$= \frac{\frac{5}{4} + \frac{16}{3} - 1}{\frac{1}{4} + \frac{3}{4}} = \frac{15 + 64 - 12}{12}$$

$$= \frac{67}{12}.$$

Example 2. *Choose the correct option and justify your choice:*

(i) $\dfrac{2\tan 30°}{1 + \tan^2 30°} =$

(A) $\sin 60°$ (B) $\cos 60°$
(C) $\tan 60°$ (D) $\sin 30°$

(ii) $\dfrac{1 - \tan^2 45°}{1 + \tan^2 45°} =$

(A) $\tan 90°$ (B) 1
(C) $\sin 45°$ (D) 0

(iii) $\sin 2A = 2 \sin A$ is true when $A =$
(A) $0°$ (B) $30°$
(C) $45°$ (D) $60°$

(iv) $\dfrac{2 \tan 30°}{1 - \tan^2 30°} =$

(A) $\cos 60°$ (B) $\sin 60°$
(C) $\tan 60°$ (D) $\sin 30°$

Sol. (i) $\dfrac{2 \tan 30°}{1 + \tan^2 30°}$

$= \dfrac{2\left(\dfrac{1}{\sqrt{3}}\right)}{1 + \left(\dfrac{1}{\sqrt{3}}\right)^2}$

$= \dfrac{\dfrac{2}{\sqrt{3}}}{1 + \dfrac{1}{3}} = \dfrac{\dfrac{2}{\sqrt{3}}}{\dfrac{4}{3}} = \dfrac{2}{\sqrt{3}} \cdot \dfrac{3}{4} = \dfrac{\sqrt{3}}{2}$

$= \sin 60°$

Hence, the correct option is **(A) sin 60°**.

(ii) $\dfrac{1 - \tan^2 45°}{1 + \tan^2 45°} = \dfrac{1 - (1)^2}{1 + (1)^2} = 0$

Hence, the correct option is **(D) 0**

(iii) When $A = 0°$,
$\sin 2A = \sin 2(0°) = \sin 0° = 0$
$2 \sin A = 2 \sin (0°) = 2(0) = 0$
$\therefore \quad \sin 2A = 2 \sin 2A$

Hence, the correct option is **(A) 0°**

(iv) $\dfrac{2 \tan 30°}{1 - \tan^2 30°}$

$= \dfrac{2\left(\dfrac{1}{\sqrt{3}}\right)}{1 - \left(\dfrac{1}{\sqrt{3}}\right)^2} = \dfrac{\dfrac{2}{\sqrt{3}}}{1 - \dfrac{1}{3}} = \dfrac{\dfrac{2}{\sqrt{3}}}{\dfrac{2}{3}} = \dfrac{2}{\sqrt{3}} \cdot \dfrac{3}{2}$

$= \sqrt{3} = \tan 60°$

Hence, the correct option is **(C) tan 60°**.

Example 3. If $\tan (A + B) = \sqrt{3}$ and $\tan (A - B) = \dfrac{1}{\sqrt{3}}$; $0° < A + B \leq 90°; A > B$, find A and B.

Sol. $\tan (A + B) = \sqrt{3} = \tan 60°$
$\Rightarrow \quad A + B = 60°$ ...(1)

$\tan (A - B) = \dfrac{1}{\sqrt{3}} = \tan 30°$

$\Rightarrow \quad A - B = 30°$ ...(2)

Following (1) and (2), we get
$\angle A = 45°$ and $\angle B = 15°$.

Example 4. State whether the following are true or false. Justify your answer.

(i) $\sin (A + B) = \sin A + \sin B$.
(ii) The value of $\sin \theta$ increases as θ increases.
(iii) The value of $\cos \theta$ increases as θ increases.
(iv) $\sin \theta = \cos \theta$ for all values of θ.
(v) $\cot A$ is not defined for $A = 0°$.

Sol. (i) Take $A = 30°$ and $B = 30°$. Then
$\sin (A + B) = \sin (30° + 30°)$

$= \sin 60° = \dfrac{\sqrt{3}}{2}$

and, $\sin A + \sin B = \sin 30° + \sin 30°$

$= \dfrac{1}{2} + \dfrac{1}{2} = 1$

So, $\sin (A + B) \neq \sin A + \sin B$

Hence, this statement is **false**.

(ii) **True** as is clear form the ready reference table.

(iii) **False** as we observe from the ready reference table that the value of $\cos \theta$ decreases as θ increases

(iv) Take $\theta = 60°$

Then $\sin \theta = \sin 60° = \dfrac{\sqrt{3}}{2}$

and $\cos \theta = \cos 60° = \dfrac{1}{2}$

So, $\sin \theta \neq \cos \theta$ when $\theta = 60°$

Hence, this statement is **false**.

(v) **True** as is clear from the ready reference table.

ADDITIONAL EXAMPLES

Examples 5. *ABC is a triangle right angled at C. If $\angle A = 30°$, $AB = 12$ cm, determine BC and AC.*

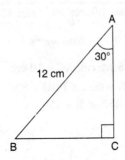

Sol. In right triangle ABC,

$$\sin 30° = \frac{BC}{AC}$$

$$\Rightarrow \quad \frac{1}{2} = \frac{BC}{12}$$

$$\Rightarrow \quad BC = \frac{12}{2} = 6 \text{ cm}$$

Again, $\cos 30° = \dfrac{AC}{AB}$

$$\Rightarrow \quad \frac{\sqrt{3}}{2} = \frac{AC}{12}$$

$\Rightarrow \quad AC = 6\sqrt{3}$ cm

or $\quad AB^2 = AC^2 + BC^2$

| By Pythagoras theorem

$\Rightarrow \quad (12)^2 = AC^2 + (6)^2$

$\Rightarrow \quad 144 = AC^2 + 36$

$\Rightarrow \quad AC^2 = 144 - 36$

$\Rightarrow \quad AC^2 = 108$

$\Rightarrow \quad AC = \sqrt{108}$

$\Rightarrow \quad AC = 6\sqrt{3}$ cm.

Example 6. *If $\cos(40° + x) = \sin 30°$, find the value of x.*

Sol. $\cos(40° + x) = \sin 30°$

$\Rightarrow \quad \cos(40° + x) = \dfrac{1}{2}$

$\Rightarrow \quad \cos(40° + x) = \cos 60°$

$\Rightarrow \quad 40° + x = 60°$

$\Rightarrow \quad x = 60° - 40° = 20.$

Example 7. *If $\sin(A + B) = 1$ and $\cos(A - B) = 1$, find A and B.*

Sol. $\sin(A + B) = 1 = \sin 90°$

$\Rightarrow \quad A + B = 90°$ ...(1)

$\cos(A - B) = 1 = \cos 0°$

$\Rightarrow \quad A - B = 0°$ ...(2)

Solving (1) and (2), we get

$A = 45°$

$B = 45°.$

TEST YOUR KNOWLEDGE

1. In $\triangle$ ABC, right-angled at B, AB = 5 cm and $\angle ACB = 30°$ (see figure). Determine the length of the sides BC and AC.

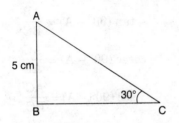

2. In $\triangle$ PQR, right angled at Q (see figure), PQ = 3 cm and PR = 6 cm. Determine $\angle$ QPR and $\angle$ PRQ.

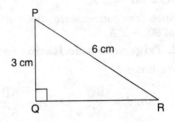

3. If $\sin(A - B) = \frac{1}{2}$, $\cos(A + B) = \frac{1}{2}$, $0° < A + B \leq 90°$, $A > B$, find A and B.

4. Find the value of $\frac{\tan 60° - \tan 30°}{1 + \tan 60° \tan 30°}$.

5. If A = 30° and B = 60°, prove that $\sin(A + B) = \sin A \cos B + \cos A \sin B$.

6. Prove that
$4(\sin^2 30° + \cos^2 60°) - 3(\cos^2 45° - \tan^2 45°) = \frac{7}{2}$.

7. If A = 30°, verify that
$\cos 2A = \frac{1 - \tan^2 A}{1 + \tan^2 A} = 1 - 2\sin^2 A$.

8. Find the value of
$\sin^2 30° - 3\cos^2 45° + 4\tan^2 60°$.

9. Prove that
$4\cos^2 60° + 4\tan^2 45° - \sin^2 30° = 4\frac{3}{4}$.

10. Find the value of
$\frac{5\sin^2 30° + \cos^2 45° - 4\tan^2 30°}{2\sin 30° \cos 30° + \tan 45°}$.

Answers

1. BC = $5\sqrt{3}$ cm, AC = 10 cm
2. ∠QPR = 60°, ∠PRQ = 30°
3. ∠A = 45°, ∠B = 15°
4. $\frac{1}{\sqrt{3}}$ 8. $10\frac{3}{4}$
10. $\frac{5}{6}(2 - \sqrt{3})$.

TRIGONOMETRIC RATIOS OF COMPLEMENTARY ANGLES

IMPORTANT POINTS

1. Complementary Angles. We know that two angles are said to be complementary if their sum equals 90°. For example, in △ABC, right-angled at B, ∠A and ∠C form such a pair.

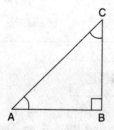

because ∠A + ∠C = 90°
i.e., ∠C = 90° − ∠A.

Note. For convenience, the write 90° − A instead of 90° − ∠A.

2. Trigonometric Ratios for ∠A
We have
$$\sin A = \frac{BC}{AC} \quad \cos A = \frac{AB}{AC}$$

$$\tan A = \frac{BC}{AB}$$
$$\text{cosec } A = \frac{AC}{BC} \quad \sec A = \frac{AC}{AB} \quad \ldots(1)$$
$$\cot A = \frac{AB}{BC}$$

3. Trigonometric Ratios for ∠C = 90° − ∠A. We see that AB is the side opposite and BC is the side adjacent to the angle 90° − A.
Therefore,

$$\sin(90° - A) = \frac{AB}{AC},$$
$$\cos(90° - A) = \frac{BC}{AC},$$
$$\tan(90° - A) = \frac{AB}{BC},$$
$$\text{cosec}(90° - A) = \frac{AC}{AB}, \quad \ldots(2)$$
$$\sec(90° - A) = \frac{AC}{BC},$$
$$\cot(90° - A) = \frac{BC}{AB}$$

INTRODUCTION TO TRIGONOMETRY

4. Comparison of the two trigonometric ratios. Comparing the ratios in (1) and (2). Observe that:

$$\sin(90° - A) = \frac{AB}{AC} = \cos A$$

and $\cos(90° - A) = \dfrac{BC}{AC} = \sin A$

Also, $\tan(90° - A) = \dfrac{AB}{BC} = \cot A$,

$\cot(90° - A) = \dfrac{BC}{AB} = \tan A$

$\sec(90° - A) = \dfrac{AC}{BC} = \operatorname{cosec} A$,

$\operatorname{cosec}(90° - A) = \dfrac{AC}{AB} = \sec A$

So, $\sin(90° - A) = \cos A$, $\cos(90° - A) = \sin A$.

$\tan(90° - A) = \cot A$, $\cot(90° - A) = \tan A$

$\sec(90° - A) = \operatorname{cosec} A$,

$\operatorname{cosec}(90° - A) = \sec A$

for all values of angle A lying between 0° and 90°.

5. For A = 0° or A = 90°. Let us check whether this holds for A = 0° or A = 90°.

(i) We have

$\sin(90° - A) = \cos A$ for $0° < A < 90°$

Also, $\cos 0° = 1 = \sin 90°$

and $\cos 90° = 0 = \sin 0°$

Thus, we have

$\sin(90° - A) = \cos A$ for $0° \le A \le 90°$

(ii) We have

$\cos(90° - A) = \sin A$ for $0° < A < 90°$

Also, $\sin 0° = 0 = \cos 90°$

$\sin 90° = 1 = \cos 0°$

Thus, we have

$\cos(90° - A) = \sin A$ for $0° \le A \le 90°$

(iii) For $0° < A \le 90°$, we have

$$\tan(90° - A) = \frac{\sin(90° - A)}{\cos(90° - A)} = \frac{\cos A}{\sin A}$$

$= \cot A$

(iv) For $0° \le A < 90°$, we have

$$\cot(90° - A) = \frac{\cos(90° - A)}{\sin(90° - A)} = \frac{\sin A}{\cos A}$$

$= \tan A$

(v) For $0° < A \le 90°$, we have

$$\sec(90° - A) = \frac{1}{\cos(90° - A)} = \frac{1}{\sin A}$$

$= \operatorname{cosec} A$

(vi) For $0° \le A < 90°$, we have

$$\operatorname{cosec}(90° - A) = \frac{1}{\sin(90° - A)} = \frac{1}{\cos A}$$

$= \sec A$.

Note. $\tan 0° = 0 = \cot 90°$, $\sec 0° = 1 = \operatorname{cosec} 90°$ and $\sec 90°$, $\operatorname{cosec} 0°$, $\tan 90°$ and $\cot 0°$ are not defined.

ILLUSTRATIVE EXAMPLES

[NCERT Exercise 8.3]
(Page No. 189)

Example 1. *Evaluate:*

(i) $\dfrac{\sin 18°}{\cos 72°}$

(ii) $\dfrac{\tan 26°}{\cot 64°}$

(iii) $\cos 48° - \sin 42°$

(iv) $\operatorname{cosec} 31° - \sec 59°$.

Sol. (i) $\dfrac{\sin 18°}{\cos 72°} = \dfrac{\sin 18°}{\cos(90° - 18°)}$

$= \dfrac{\sin 18°}{\sin 18°}$

$= 1$ [$\because \cos(90° - \theta) = \sin \theta$]

(ii) $\dfrac{\tan 26°}{\cot 64°} = \dfrac{\tan 26°}{\cot(90° - 26°)}$

$$= \frac{\tan 26°}{\tan 26°}$$

$$= 1 \quad | \because \cot(90° - \theta) = \tan \theta$$

(iii) $\cos 48° - \sin 42°$

$= \cos(90° - 42°) - \sin 42°$

$= \sin 42° - \sin 42°$

$\quad | \because \cos(90° - \theta) = \sin \theta$

$= 0$

(iv) $\csc 31° - \sec 59°$

$= \csc 31° - \sec(90° - 31°)$

$= \csc 31° - \csc 31°$

$\quad | \because \sec(90° - \theta) = \csc \theta$

$= 0$.

Example 2. *Show that*:

(i) $\tan 48° \tan 23° \tan 42° \tan 67° = 1$

(ii) $\cos 38° \cos 52° - \sin 38° \sin 52° = 0$

Sol. (i) $\tan 48° \tan 23° \tan 42° \tan 67°$

$= \tan(90° - 42°) \tan 23°$

$\qquad \tan 42° \tan(90° - 23°)$

$= \cot 42° \tan 23° \tan 42° \cot 23°$

$\quad | \because \tan(90° - \theta) = \cot \theta$

$= \dfrac{1}{\tan 42°} \tan 23° \tan 42° \dfrac{1}{\tan 23°}$

$\quad \left| \because \cos \theta = \dfrac{1}{\tan \theta} \right.$

$= 1$

(ii) $\cos 38° \cos 52° - \sin 38° \sin 52°$

$= \cos 38° \cos(90° - 38°) - \sin 38° \sin(90° - 38°)$

$= \cos 38° \sin 38° - \sin 38° \cos 38°$

$\quad \left| \begin{array}{l} \because \cos(90° - \theta) = \sin \theta \\ \text{and } \sin(90° - \theta) = \cos \theta \end{array} \right.$

$= 0$

Example 3. *If* $\tan 2A = \cot(A - 18°)$, *where 2A is an acute angle, find the value of A.*

Sol. $\tan 2A = \cot(A - 18°)$

$\Rightarrow \cot(90° - 2A) = \cot(A - 18°)$

$\quad | \because \cot(90° - \theta) = \tan \theta$

$\Rightarrow 90° - 2A = A - 18°$

$\quad | \because 90° - 2A \text{ and } A - 18°$ are both acute angles

$\Rightarrow 3A = 108°$

$\Rightarrow A = \dfrac{108°}{3} = 36°$.

Example 4. *If* $\tan A = \cot B$, *prove that* $A + B = 90°$.

Sol. $\tan A = \cot B$

$\Rightarrow \tan A = \tan(90° - B)$

$\quad | \because \tan(90° - \theta) = \cot \theta$

$\Rightarrow A = 90° - B$

$\quad | \because A \text{ and } 90° - B \text{ are both acute angles}$

$\Rightarrow A + B = 90°$.

Example 5. *If* $\sec 4A = \csc(A - 20°)$, *where 4A is an acute angle, find the value of A.*

Sol. $\sec 4A = \csc(A - 20°)$

$\Rightarrow \csc(90° - 4A) = \csc(A - 20°)$

$\quad | \because \csc(90° - \theta) = \sec \theta$

$\Rightarrow 90° - 4A = A - 20°$

$\quad | \because 90° - 4A \text{ and } A - 20°$ are both acute angles

$\Rightarrow 5A = 110°$

$\Rightarrow A = \dfrac{110°}{5} = 22°$.

Example 6. *If A, B and C are interior angles of a triangle ABC, then show that*

$$\sin\left(\dfrac{B + C}{2}\right) = \cos \dfrac{A}{2}.$$

Sol. L.H.S $= \sin\left(\dfrac{B + C}{2}\right)$

$= \sin\left(\dfrac{180° - A}{2}\right)$

$\quad | \because A + B + C = 180°$ (the sum of the interior angles of a triangle is 180°)

$= \sin\left(90° - \dfrac{A}{2}\right)$

$= \cos \dfrac{A}{2}$

$\quad | \because \sin(90° - \theta) = \cos \theta$

$= $ R.H.S.

Example 7. *Express* $\sin 67° + \cos 75°$ *in terms of trigonometric ratios of angles between 0° and 45°.*

Sol. $\sin 67° + \cos 75°$
$= \sin (90° - 23°) + \cos (90° - 15°)$
$= \cos 23° + \sin 15°$

$$\because \quad \sin(90° - \theta) = \cos \theta$$
$$\text{and } \cos(90° - \theta) = \sin \theta$$

ADDITIONAL EXAMPLES

Example 8. *Without using trigonometric angles, evaluate the following:*

$$\frac{2 \sin 68°}{\cos 22°} - \frac{2 \cot 15°}{5 \tan 75°}$$

$$- \frac{3 \tan 45° \tan 20° \tan 40° \tan 50° \tan 70°}{5}$$

(CBSE 2004)

Sol. $\dfrac{2 \sin 68°}{\cos 22°} - \dfrac{2 \cot 15°}{5 \tan 75°}$

$$- \frac{3 \tan 45° \tan 20° \tan 40° \tan 50° \tan 70°}{5}$$

$= \dfrac{2 \sin (90° - 22°)}{\cos 22°} - \dfrac{2 \cot 15°}{5 \tan (90° - 15°)}$

$$- \frac{3.1.\tan 20° \tan 40° \tan(90° - 40°) \tan(90° - 20°)}{5}$$

$= \dfrac{2 \cos 22°}{\cos 22°} - \dfrac{2 \cot 15°}{5 \cot 15°}$

$$- \frac{3 \tan 20° \tan 40° \cot 40° \cot 20°}{5}$$

$$\because \quad \sin(90° - \theta) = \cos \theta$$
$$\tan(90° - \theta) = \cot \theta$$

$= 2 - \dfrac{2}{5}$

$$- \frac{3 \tan 20° \tan 40° \dfrac{1}{\tan 40°} \dfrac{1}{\tan 20°}}{5}$$

$$\because \quad \cot \theta = \frac{1}{\tan \theta}$$

$= 2 - \dfrac{2}{5} - \dfrac{3}{5}$
$= 2 - 1 = 1.$

Example 9. *Without using trigonometric tables, evaluate:*

$$\frac{\cos 35°}{\sin 55°} + \frac{\tan 27° \tan 63°}{\sin 30°} - 3 \tan^2 60°.$$

(CBSE SP1)

Sol. $\dfrac{\cos 35°}{\sin 55°} + \dfrac{\tan 27° \tan 63°}{\sin 30°}$

$$- 3 \tan^2 60°$$

$= \dfrac{\cos 35°}{\sin(90° - 35°)}$

$$+ \frac{\tan 27° \tan(90° - 27°)}{\left(\dfrac{1}{2}\right)} - 3(\sqrt{3})^2$$

$= \dfrac{\cos 35°}{\cos 35°} + 2 \tan 27° \cot 23° - 9$

$$\because \quad \sin(90° - \theta) = \cos \theta$$
$$\tan(90° - \theta) = \cos \theta$$

$= 1 + 2 \tan 27° \dfrac{1}{\tan 27°} - 9$

$$\because \quad \cot \theta = \frac{1}{\tan \theta}$$

$= 1 + 2 - 9$
$= -6.$

Example 10. *Without using trigonometric tables, evaluate the following:*

$$\frac{2 \cos 65°}{\sin 25°} - \frac{\tan 20°}{\cot 30°} - \sin 90° + \tan 5° +$$
$$\tan 35° \tan 60° \tan 55° \tan 85°.$$

(CBSE 2003)

Sol. $\dfrac{2 \cos 65°}{\sin 25°} - \dfrac{\tan 20°}{\cot 30°} - \sin 90°$
$+ \tan 5° \tan 35° \tan 60° \tan 55° \tan 85°$

$= \dfrac{2 \cos(90° - 25°)}{\sin 25°} - \dfrac{\tan 20°}{\cot(90° - 20°)}$

$$- 1 + \tan 5° + \tan 35° \sqrt{3}$$
$$\tan(90° - 35°) \tan(90° - 5°)$$

$$= \frac{2 \sin 25°}{\sin 25°} - \frac{\tan 20°}{\tan 20°} - 1$$

$$+ \tan 5° \tan 35° \sqrt{3} \cot 35° \cot 5°$$

$$\because \quad \cos(90° - \theta) = \sin \theta$$
$$\cot(90° - \theta) = \tan \theta$$
$$\tan(90° - \theta) = \cot \theta$$

$$= 2 - 1 - 1 + \tan 5° \tan 35° \cdot \sqrt{3} \cdot$$

$$\frac{1}{\tan 35°} \cdot \frac{1}{\tan 5°}$$

$$\because \quad \cot \theta = \frac{1}{\tan \theta}$$

$$= \sqrt{3}.$$

TEST YOUR KNOWLEDGE

1. Evaluate $\dfrac{\tan 65°}{\cot 25°}$.

2. If $\sin 3A = \cos(A - 26°)$, where $3A$ is an acute angle, find the value of A.

3. Express $\cot 85° + \cos 75°$ in terms of trigonometric ratios of angles between $0°$ and $45°$.

4. If A and B are acute angles and $\sin A = \cos B$, prove that $A + B = 90°$.

5. If $\sin 3\theta = \cos(\theta - 60)$ where 3θ and $\theta - 60$ are acute angles, find the value of θ.

6. If A, B, C are the interior angles of a triangle ABC, prove that

$$\frac{\cos \dfrac{B+C}{2} \cos \dfrac{C+A}{2} \cos \dfrac{A+B}{2}}{\sin \dfrac{A}{2} \sin \dfrac{B}{2} \sin \dfrac{C}{2}} = 1.$$

7. Without using trigonometric tables, evaluate the following:

$$\frac{2 \cos 67°}{\sin 23°} - \frac{\tan 40°}{\cot 50°} - \cos 0°$$

$$+ \tan 15° \tan 25° \tan 60° \tan 65° \tan 75°.$$
(AI CBSE 2003)

8. Without using trigonometric tables, evaluate the following:

$$\frac{\cos 58°}{\sin 32°} + \frac{\sin 22°}{\cos 68°}$$

$$- \frac{\cos 38° \operatorname{cosec} 52°}{\tan 18° \tan 35° \tan 60° \tan 72° \tan 55°}.$$
(CBSE 2003)

Answers

1. 1 2. $29°$
3. $\tan 5° + \sin 15°$ 5. $21°$
7. $\sqrt{3}$ 8. $2 - \dfrac{1}{\sqrt{3}}$.

TRIGONOMETRIC IDENTITIES

IMPORTANT POINTS

1. Trigonometric Identity. We know that an equation is called an identity when it is true for all values of the variables involved. Similarly, an equation involving trigonometric ratios of an angle is called a **trigonometric identity**, if it is true for all values of the angle(s) involved.

2. Three Fundamental Trigonometric Identities

(i) $\cos^2 A + \sin^2 A = 1; \; 0° \leq A \leq 90°$

(ii) $1 + \tan^2 A = \sec^2 A; \; 0° \leq A < 90°$

(iii) $\cot^2 A + 1 = \operatorname{cosec}^2 A; \; 0° < A \leq 90°$.

Proof. (i) $\cos^2 A + \sin^2 A = 1;$
$0° \leq A \leq 90°$

Consider a $\triangle ABC$, right angled at B. Then we have:

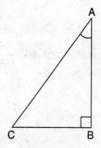

INTRODUCTION TO TRIGONOMETRY

$$AB^2 + BC^2 = AC^2 \quad \ldots(1)$$

| By Pythagoras theorem

Dividing each term of (1) by AC^2, we get

$$\frac{AB^2}{AC^2} + \frac{BC^2}{AC^2} = \frac{AC^2}{AC^2}$$

i.e., $\left(\frac{AB}{AC}\right)^2 + \left(\frac{BC}{AC}\right)^2 = \left(\frac{AC}{AC}\right)^2$

i.e., $(\cos A)^2 + (\sin A)^2 = 1$

i.e., $\mathbf{\cos^2 A + \sin^2 A = 1} \quad \ldots(2)$

This is true for all A such that $0° \leq A \leq 90°$. So, this is a trigonometric identity.

(ii) $\mathbf{1 + \tan^2 A = \sec^2 A; \; 0° \leq A < 90°}$

Let us now divide (1) by AB^2. We get

$$\frac{AB^2}{AB^2} + \frac{BC^2}{AB^2} = \frac{AC^2}{AB^2}$$

or, $\left(\frac{AB}{AB}\right)^2 + \left(\frac{BC}{AB}\right)^2 = \left(\frac{AC}{AB}\right)^2$

i.e., $\mathbf{1 + \tan^2 A = \sec^2 A} \quad \ldots(3)$

This equation is true for $A = 0°$. Since $\tan A$ and $\sec A$ are not defined for $A = 90°$, so (3) is true for all A such that $0° \leq A < 90°$.

(iii) $\mathbf{\cot^2 A + 1 = \csc^2 A; \; 0° < A \leq 90°}$

Again, let us divide (1) by BC^2, we get

$$\frac{AB^2}{BC^2} + \frac{BC^2}{BC^2} = \frac{AC^2}{BC^2}$$

$\Rightarrow \left(\frac{AB}{BC}\right)^2 + \left(\frac{BC}{BC}\right)^2 = \left(\frac{AC}{BC}\right)^2$

$\Rightarrow \cot^2 A + 1 = \csc^2 A \quad \ldots(4)$

Since $\csc A$ and $\cot A$ are not defined for $A = 0°$, therefore (4) is true for all A such that $0° < A \leq 90°$.

3. A use of the above trigonometric identities. Using the above trigonometric identities, we can express each trigonometric ratio in term of the other trigonometric ratios, *i.e.*, if any one of the ratios is known, we can also determine the values of other trigonometric ratios.

An example

Suppose we know that

$$\tan A = \frac{1}{\sqrt{3}}$$

then, $\cot A = \sqrt{3}$

Since $\sec^2 A = 1 + \tan^2 A$

$\therefore \sec^2 A = 1 + \left(\frac{1}{\sqrt{3}}\right)^2 = 1 + \frac{1}{3} = \frac{4}{3}$

$\Rightarrow \sec A = \frac{2}{\sqrt{3}}$

$\therefore \cos A = \frac{\sqrt{3}}{2}$

Again,

$\sin A = \sqrt{1 - \cos^2 A} = \sqrt{1 - \frac{3}{4}} = \sqrt{\frac{1}{4}} = \frac{1}{2}$

$\therefore \csc A = 2.$

ILLUSTRATIVE EXAMPLES

[NCERT Exercise 8.4]
(Page No. 193)

Example 1. *Express the trigonometric ratios $\sin A$, $\sec A$ and $\tan A$ in terms of $\cot A$.*

Sol. $\sin A = \dfrac{1}{\csc A} = \dfrac{1}{\sqrt{\csc^2 A}}$

$= \dfrac{1}{\sqrt{1 + \cot^2 A}}$

$\sec A = \sqrt{\sec^2 A} = \sqrt{1 + \tan^2 A}$

$= \sqrt{1 + \dfrac{1}{\cot^2 A}}$

$= \dfrac{\sqrt{1 + \cot^2 A}}{\cot A}$

$$\tan A = \frac{1}{\cot A}.$$

Example 2. *Write all the other trigonometric ratios of $\angle A$ in terms of sec A.*

Sol. $\sin A = \dfrac{1}{\operatorname{cosec} A}$

$= \dfrac{1}{\sqrt{\operatorname{cosec}^2 A}}$

$= \dfrac{1}{\sqrt{1 + \cot^2 A}}$

$= \dfrac{1}{\sqrt{1 + \dfrac{1}{\tan^2 A}}}$

$= \dfrac{1}{\sqrt{\dfrac{\tan^2 A + 1}{\tan^2 A}}}$

$= \dfrac{1}{\dfrac{\sqrt{\tan^2 A + 1}}{\tan A}}$

$= \dfrac{\tan A}{\sqrt{\tan^2 A + 1}} = \dfrac{\tan A}{\sec A}$

$= \dfrac{\sqrt{\tan^2 A}}{\sec A} = \dfrac{\sqrt{\sec^2 A - 1}}{\sec A}$

Aliter. $\sin A = \dfrac{\sin A}{1}$

$= \dfrac{\dfrac{\sin A}{\cos A}}{\dfrac{1}{\cos A}} = \dfrac{\tan A}{\sec A}$

$= \dfrac{\sqrt{\tan^2 A}}{\sec A} = \dfrac{\sqrt{\sec^2 A - 1}}{\sec A}$

$\cos A = \dfrac{1}{\sec A}$

$\tan A = \sqrt{\tan^2 A} \quad \sqrt{\sec^2 A - 1}$

$\operatorname{cosec} A = \dfrac{1}{\sin A} = \dfrac{\sec A}{\sqrt{\sec^2 A - 1}}$

$\cot A = \dfrac{1}{\tan A} = \dfrac{1}{\sqrt{\sec^2 A - 1}}.$

Example 3. *Evaluate:*

(i) $\dfrac{\sin^2 63° + \sin^2 27°}{\cos^2 17° + \cos^2 73°}$

(ii) $\sin 25° \cos 65° + \cos 25° \sin 65°$

Sol. (i) $\dfrac{\sin^2 63° + \sin^2 27°}{\cos^2 17° + \cos^2 73°}$

$= \dfrac{\sin^2 (90° - 27°) + \sin^2 27°}{\cos^2 17° + \cos^2 (90° - 17°)}$

$= \dfrac{\cos^2 27° + \sin^2 27°}{\cos^2 17° + \sin^2 17°}$

$\quad \left| \begin{array}{l} \because \sin (90° - \theta) = \cos \theta \\ \cos (90° - \theta) = \sin \theta \end{array} \right.$

$= \dfrac{1}{1} \quad | \because \sin^2 \theta + \cos^2 \theta = 1$

$= 1.$

(ii) $\sin 25° \cos 65° + \cos 25° \sin 65°$

$= \sin 25° \cos (90° - 25°)$

$\qquad + \cos 25° \sin (90° - 25°)$

$= \sin 25° \sin 25° + \cos 25° \cos 25°$

$\quad \left| \begin{array}{l} \because \cos (90° - \theta) = \sin \theta \\ \sin (90° - \theta) = \cos \theta \end{array} \right.$

$= \sin^2 25° + \cos^2 25°$

$= 1. \quad | \because \sin^2 \theta + \cos^2 \theta = 1$

Example. 4. *Choose the correct option. Justify your choice.*

(i) $9 \sec^2 A - 9 \tan^2 A =$

(A) 1   (B) 9

(C) 8   (D) 0

(ii) $(1 + \tan \theta + \sec \theta)$

$\qquad (1 + \cot \theta - \operatorname{cosec} \theta) =$

(A) 0   (B) 1

(C) 2   (D) –1

INTRODUCTION TO TRIGONOMETRY

(iii) $(\sec A + \tan A)(1 - \sin A) =$
  (A) $\sec A$ (B) $\sin A$
  (C) $\operatorname{cosec} A$ (D) $\cos A$

(iv) $\dfrac{1 + \tan^2 A}{1 + \cot^2 A} =$

  (A) $\sec^2 A$ (B) -1
  (C) $\cot^2 A$ (D) $\tan^2 A$

Sol. (i) $9 \sec^2 A - 9 \tan^2 A$
$= 9 (\sec^2 A - \tan^2 A)$
$= 9 (1) = 9$

Hence, the correct choice is **(B) 9**.

(ii) $(1 + \tan\theta + \sec\theta)(1 + \cot\theta - \operatorname{cosec}\theta)$

$= (1 + \tan\theta + \sec\theta)\left(1 + \dfrac{1}{\tan\theta} - \operatorname{cosec}\theta\right)$

$= (1 + \tan\theta + \sec\theta) \cdot \dfrac{(\tan\theta + 1 - \tan\theta \operatorname{cosec}\theta)}{\tan\theta}$

$= \dfrac{(1 + \tan\theta + \sec\theta)(\tan\theta + 1 - \sec\theta)}{\tan\theta}$

$\because \tan\theta \operatorname{cosec}\theta = \dfrac{\sin\theta}{\cos\theta} \cdot \dfrac{1}{\sin\theta} = \dfrac{1}{\cos\theta} = \sec\theta$

$= \dfrac{(1 + \tan\theta)^2 - \sec^2\theta}{\tan\theta}$

$= \dfrac{1 + \tan^2\theta + 2\tan\theta - \sec^2\theta}{\tan\theta}$

$= \dfrac{1 + 2\tan\theta - (\sec^2\theta - \tan^2\theta)}{\tan\theta}$

$= \dfrac{1 + 2\tan\theta - 1}{\tan\theta} = 2$

Hence, the correct choice is **(C) 2**.

(iii) $(\sec A + \tan A)(1 - \sin A)$

$= \left(\dfrac{1}{\cos A} + \dfrac{\sin A}{\cos A}\right)(1 - \sin A)$

$= \left(\dfrac{1 + \sin A}{\cos A}\right)(1 - \sin A)$

$= \dfrac{1 - \sin^2 A}{\cos A}$

$= \dfrac{\cos^2 A}{\cos A} = \cos A$

Hence, the correct choice is **(D) cos A**.

(iv) $\dfrac{1 + \tan^2 A}{1 + \cot^2 A} = \dfrac{1 + \tan^2 A}{1 + \dfrac{1}{\tan^2 A}}$

$= \dfrac{1 + \tan^2 A}{\dfrac{\tan^2 A + 1}{\tan^2 A}}$

$= (1 + \tan^2 A)\left(\dfrac{\tan^2 A}{\tan^2 A + 1}\right)$

$= \tan^2 A$

Hence, the correct choice is **(D) $\tan^2 A$**.

Example 5. *Prove the following identities, where the angles involved are acute angles for which the expressions are defined.*

(i) $(\operatorname{cosec}\theta - \cot\theta)^2 = \dfrac{1 - \cos\theta}{1 + \cos\theta}$

(ii) $\dfrac{\cos A}{1 + \sin A} + \dfrac{1 + \sin A}{\cos A} = 2\sec A$

(iii) $\dfrac{\tan\theta}{1 - \cot\theta} + \dfrac{\cot\theta}{1 - \tan\theta} = 1 + \sec\theta \operatorname{cosec}\theta$

[**Hint:** Write the expression in terms of $\sin\theta$ and $\cos\theta$]

(iv) $\dfrac{1 + \sec A}{\sec A} = \dfrac{\sin^2 A}{1 - \cos A}$

[**Hint:** Simplify LHS and RHS separately]

(v) $\dfrac{\cos A - \sin A + 1}{\cos A + \sin A - 1} = \operatorname{cosec} A + \cot A$,
using the identity
$\operatorname{cosec}^2 A = 1 + \cot^2 A.$

(vi) $\sqrt{\dfrac{1+\sin A}{1-\sin A}} = \sec A + \tan A$

(vii) $\dfrac{\sin\theta - 2\sin^3\theta}{2\cos^3\theta - \cos\theta} = \tan\theta$

(viii) $(\sin A + \operatorname{cosec} A)^2 + (\cos A + \sec A)^2$
$= 7 + \tan^2 A + \cot^2 A$

(ix) $(\operatorname{cosec} A - \sin A)(\sec A - \cos A)$
$= \dfrac{1}{\tan A + \cot A}$

[**Hint:** Simplify LHS and RHS separately]

(x) $\left(\dfrac{1+\tan^2 A}{1+\cot^2 A}\right) = \left(\dfrac{1-\tan A}{1-\cot A}\right)^2$
$= \tan^2 A$

Sol. (i) L.H.S.
$= (\operatorname{cosec}\theta - \cot\theta)^2$
$= \left(\dfrac{1}{\sin\theta} - \dfrac{\cos\theta}{\sin\theta}\right)^2$
$= \left(\dfrac{1-\cos\theta}{\sin\theta}\right)^2 = \dfrac{(1-\cos\theta)^2}{\sin^2\theta}$
$= \dfrac{(1-\cos\theta)^2}{1-\cos^2\theta}$
$= \dfrac{(1-\cos\theta)^2}{(1-\cos\theta)(1+\cos\theta)}$
$= \dfrac{1-\cos\theta}{1+\cos\theta} = $ R.H.S.

(ii) L.H.S.
$= \dfrac{\cos A}{1+\sin A} + \dfrac{1+\sin A}{\cos A}$
$= \dfrac{\cos^2 A + (1+\sin A)^2}{(1+\sin A)\cos A}$
$= \dfrac{\cos^2 A + 1 + \sin^2 A + 2\sin A}{(1+\sin A)\cos A}$
$= \dfrac{1 + 1 + 2\sin A}{(1+\sin A)\cos A}$
$\qquad |\because \sin^2 A + \cos^2 A = 1$
$= \dfrac{2+2\sin A}{(1+\sin A)\cos A}$
$= \dfrac{2(1+\sin A)}{(1+\sin A)\cos A}$
$= \dfrac{2}{\cos A} = 2\cdot\dfrac{1}{\cos A}$
$= 2\sec A$
$= $ R.H.S.

(iii) L.H.S.
$= \dfrac{\tan\theta}{1-\cot\theta} + \dfrac{\cot\theta}{1-\tan\theta}$
$= \dfrac{\dfrac{\sin\theta}{\cos\theta}}{1-\dfrac{\cos\theta}{\sin\theta}} + \dfrac{\dfrac{\cos\theta}{\sin\theta}}{1-\dfrac{\sin\theta}{\cos\theta}}$
$= \dfrac{\sin^2\theta}{\cos\theta(\sin\theta - \cos\theta)}$
$\qquad + \dfrac{\cos^2\theta}{\sin\theta(\cos\theta - \sin\theta)}$
$= \dfrac{\sin^3\theta - \cos^3\theta}{\sin\theta\cos\theta(\sin\theta - \cos\theta)}$
$= \dfrac{(\sin\theta - \cos\theta)(\sin^2\theta + \cos^2\theta + \sin\theta\cos\theta)}{\sin\theta\cos\theta(\sin\theta - \cos\theta)}$
$= \dfrac{1+\sin\theta\cos\theta}{\sin\theta\cos\theta}$
$\qquad |\because \sin^2\theta + \cos^2\theta = 1$
$= \dfrac{1}{\sin\theta\cos\theta} + \dfrac{\sin\theta\cos\theta}{\sin\theta\cos\theta}$
$= \dfrac{1}{\sin\theta}\dfrac{1}{\cos\theta} + 1 = \operatorname{cosec}\theta\ \sec\theta + 1$
$= 1 + \sec\theta\operatorname{cosec}\theta$
$= $ R.H.S.

INTRODUCTION TO TRIGONOMETRY

(*iv*) L.H.S.

$$= \frac{1+\sec A}{\sec A}$$

$$= \frac{1+\dfrac{1}{\cos A}}{\dfrac{1}{\cos A}}$$

$$= \frac{\dfrac{\cos A + 1}{\cos A}}{\dfrac{1}{\cos A}}$$

$$= \cos A + 1$$
$$= 1 + \cos A$$

$$= \frac{(1+\cos A)(1-\cos A)}{1-\cos A}$$

$$= \frac{1-\cos^2 A}{1-\cos A}$$

$$= \frac{\sin^2 A}{1-\cos A}$$

$$\mid \because \ \sin^2 A + \cos^2 A = 1$$

= R.H.S.

(*v*) L.H.S. $= \dfrac{\cos A - \sin A + 1}{\cos A + \sin A - 1}$

$$= \frac{\dfrac{\cos A}{\sin A} - \dfrac{\sin A}{\sin A} + \dfrac{1}{\sin A}}{\dfrac{\cos A}{\sin A} + \dfrac{\sin A}{\sin A} - \dfrac{1}{\sin A}}$$

| Dividing the numerator and denominator by sin A

$$= \frac{\cot A - 1 + \csc A}{\cot A + 1 - \csc A}$$

$$= \frac{\cot A + \csc A - 1}{\cot A - \csc A + 1}$$

$$= \frac{\{(\cot A + \csc A) - 1\}(\cot A + \csc A)}{\{(\cot A - \csc A) + 1\}(\cot A + \csc A)}$$

$$= \frac{\{(\cot A + \csc A) - 1\}(\cot A + \csc A)}{\cot^2 A - \csc^2 A + (\cot A + \csc A)}$$

$$= \frac{\{(\cot A + \csc A) - 1\}(\cot A + \csc A)}{-1 + (\cot A + \csc A)}$$

$$= \cot A + \csc A$$
= R.H.S.

(*vi*) L.H.S.

$$= \sqrt{\frac{1+\sin A}{1-\sin A}}$$

$$= \sqrt{\frac{1+\sin A}{1-\sin A} \cdot \frac{1+\sin A}{1+\sin A}}$$

$$= \sqrt{\frac{(1+\sin A)^2}{1-\sin^2 A}}$$

$$= \sqrt{\frac{(1+\sin A)^2}{\cos^2 A}}$$

$$= \frac{1+\sin A}{\cos A}$$

$$= \frac{1}{\cos A} + \frac{\sin A}{\cos A}$$

$$= \sec A + \tan A$$
= R.H.S.

(*vii*) L.H.S.

$$= \frac{\sin\theta - 2\sin^3\theta}{2\cos^3\theta - \cos\theta}$$

$$= \frac{\sin\theta(1 - 2\sin^2\theta)}{\cos\theta(2\cos^2\theta - 1)}$$

$$= \frac{\sin\theta(\cos^2\theta + \sin^2\theta - 2\sin^2\theta)}{\cos\theta(2\cos^2\theta - \cos^2\theta - \sin^2\theta)}$$

$$\mid \because \ \cos^2\theta + \sin^2\theta = 1$$

$$= \frac{\sin\theta(\cos^2\theta - \sin^2\theta)}{\cos\theta(\cos^2\theta - \sin^2\theta)}$$

$$= \tan\theta = \text{R.H.S.}$$

(viii) L.H.S.
$= (\sin A + \text{cosec } A)^2 + (\cos A + \sec A)^2$
$= (\sin^2 A + \text{cosec}^2 A + 2 \sin A \text{ cosec } A)$
$\quad + (\cos^2 A + \sec^2 A + 2 \cos A \sec A)$
$= (\sin^2 A + \cos^2 A) + \text{cosec}^2 A + \sec^2 A$
$\quad + 2 \sin A \text{ cosec } A + 2 \cos A \sec A$
$= 1 + (1 + \cot^2 A) + (1 + \tan^2 A) + 2 \sin A \cdot \dfrac{1}{\sin A} + 2 \cos A \cdot \dfrac{1}{\cos A}$
$= 1 + (1 + \cot^2 A) + (1 + \tan^2 A) + 2 + 2$
$= 7 + \tan^2 A + \cot^2 A = $ R.H.S.

(ix) L.H.S.
$= (\text{cosec } A - \sin A)(\sec A - \cos A)$
$= \left(\dfrac{1}{\sin A} - \sin A\right)\left(\dfrac{1}{\cos A} - \cos A\right)$
$= \dfrac{1 - \sin^2 A}{\sin A} \cdot \dfrac{1 - \cos^2 A}{\cos A}$
$= \dfrac{\cos^2 A}{\sin A} \cdot \dfrac{\sin^2 A}{\cos A}$
$\qquad |\because \sin^2 A + \cos^2 A = 1$
$= \sin A \cos A$
$= \dfrac{\sin A \cos A}{\sin^2 A + \cos^2 A}$
$\qquad |\because \sin^2 A + \cos^2 A = 1$
$= \dfrac{\dfrac{\sin A \cos A}{\sin A \cos A}}{\dfrac{\sin^2 A}{\sin A \cos A} + \dfrac{\cos^2 A}{\sin A \cos A}}$

| Dividing the numerator and denominator by $\sin A \cos A$

$= \dfrac{1}{\tan A + \cot A} = $ L.H.S.

(x) $\dfrac{1 + \tan^2 A}{1 + \cot^2 A}$

$= \dfrac{1 + \tan^2 A}{1 + \dfrac{1}{\tan^2 A}} \quad \left|\because \cot A = \dfrac{1}{\tan A}\right.$

$= \dfrac{1 + \tan^2 A}{\dfrac{\tan^2 A + 1}{\tan^2 A}}$

$= \tan^2 A \qquad \ldots(1)$

$\left(\dfrac{1 - \tan A}{1 - \cot A}\right)^2 = \left(\dfrac{1 - \tan A}{1 - \dfrac{1}{\tan A}}\right)^2$

$= \left\{\dfrac{1 - \tan A}{\left(\dfrac{\tan A - 1}{\tan A}\right)}\right\}^2 = (-\tan A)^2$

$= \tan^2 A \qquad \ldots(2)$

(1) and (2) taken together give the result.

ADDITIONAL EXAMPLES

Example 6. *Prove that:*

$\dfrac{\sin \theta + \cos \theta}{\sin \theta - \cos \theta} + \dfrac{\sin \theta - \cos \theta}{\sin \theta + \cos \theta} = \dfrac{2 \sec^2 \theta}{\tan^2 \theta - 1}.$

(CBSE 2006)

Sol. L.H.S.

$= \dfrac{\sin \theta + \cos \theta}{\sin \theta - \cos \theta} + \dfrac{\sin \theta - \cos \theta}{\sin \theta + \cos \theta}$

$= \dfrac{(\sin \theta + \cos \theta)^2 (\sin \theta - \cos \theta)^2}{(\sin \theta - \cos \theta)(\sin \theta + \cos \theta)}$

$= \dfrac{(\sin^2 \theta + \cos^2 \theta + 2 \sin \theta \cos \theta) + (\sin^2 \theta + \cos^2 \theta - 2 \sin \theta \cos \theta)}{\sin^2 \theta - \cos^2 \theta}$

$= \dfrac{2(\sin^2 \theta + \cos^2 \theta)}{\sin^2 \theta - \cos^2 \theta}$

$= \dfrac{2}{\sin^2 \theta - \cos^2 \theta}$

$\qquad |\because \sin^2 \theta + \cos^2 \theta = 1$

INTRODUCTION TO TRIGONOMETRY

$$= \frac{\frac{2}{\cos^2\theta}}{\frac{\sin^2\theta}{\cos^2\theta} - \frac{\cos^2\theta}{\cos^2\theta}}$$

| Dividing the numerator and denominator by $\cos^2\theta$

$$= \frac{2\sec^2\theta}{\tan^2\theta - 1} = \text{R.H.S.}$$

Example 7. *Evaluate without using trigonometric tables :*

$$\frac{\sec^2(90° - \theta) - \cot^2\theta}{2(\sin^2 25° + \sin^2 65°)}$$

$$+ \frac{2\cos^2 60° \tan^2 28° \tan^2 62°}{3(\sec^2 43° - \cot^2 47°)}.$$

(CBSE 2006)

Sol. $\dfrac{\sec^2(90° - \theta) - \cot^2\theta}{2(\sin^2 25° + \sin^2 65°)}$

$$+ \frac{2\cos^2 60° \tan^2 28° \tan^2 62°}{3(\sec^2 43° - \cot^2 47°)}$$

$$= \frac{\csc^2\theta - \cot^2\theta}{2\{\sin^2 25° + \sin^2 (90° - 25°)\}}$$

$$+ \frac{2\left(\dfrac{1}{2}\right)^2 \tan^2 28° \tan^2 (90° - 28°)}{3\{\sec^2 43° - \cot^2 (90° - 43°)\}}$$

| $\because$ $\sec(90° - \theta) = \csc\theta$

$$= \frac{1}{2(\sin^2 25° + \cos^2 25°)}$$

$$+ \frac{2 \cdot \dfrac{1}{4} \cdot \tan^2 28° \cdot \cot^2 28°}{3(\sec^2 43° - \tan^2 43°)}$$

| $\because$ $\csc^2\theta - \cot^2\theta = 1$
$\sin(90° - \theta) = \cos\theta,$
$\tan(90° - \theta) = \cot\theta,$
$\cot(90° - \theta) = \tan\theta$

$$= \frac{1}{2(1)} + \frac{2 \cdot \dfrac{1}{4} \cdot \tan^2 28° \cdot \dfrac{1}{\tan^2 28°}}{3(1)}$$

| $\because$ $\sin^2\theta + \cos^2\theta = 1,$
$\cot\theta = \dfrac{1}{\tan\theta},$
$\sec^2\theta - \tan^2\theta = 1$

$$= \frac{1}{2} + \frac{1}{6} = \frac{3+1}{6}$$

$$= \frac{4}{6} = \frac{2}{3}.$$

Example 8. *Prove that*

$$\frac{1}{\csc\theta - \cot\theta} - \frac{1}{\sin\theta} = \frac{1}{\sin\theta}$$

$$- \frac{1}{\csc\theta + \cot\theta}.$$

(AI CBSE 2006)

Sol. $\dfrac{1}{\csc\theta - \cot\theta} - \dfrac{1}{\sin\theta} = \dfrac{1}{\sin\theta}$

$$- \frac{1}{\csc\theta + \cot\theta}$$

$$\Rightarrow \frac{1}{\csc\theta - \cot\theta} + \frac{1}{\csc\theta + \cot\theta}$$

$$= \frac{2}{\sin\theta} \qquad \ldots(1)$$

L.H.S. of (1)

$$= \frac{1}{\csc\theta - \cot\theta} + \frac{1}{\csc\theta + \cot\theta}$$

$$= \frac{(\csc\theta + \cot\theta) + (\csc\theta - \cot\theta)}{(\csc\theta - \cot\theta)(\csc\theta + \cot\theta)}$$

$$= \frac{2\csc\theta}{\csc^2\theta - \cot^2\theta}$$

$$= \frac{2\csc\theta}{1} \quad | \because \csc^2\theta - \cot^2\theta = 1$$

$$= 2\csc\theta$$

$$= \frac{2}{\sin\theta} = \text{R.H.S. of (1)}.$$

Example 9. *Evaluate without using trigonometric tables :*

$$\frac{\csc^2(90°-\theta)-\tan^2\theta}{4(\cos^2 48°+\cos^2 42°)}$$

$$-\frac{2\tan^2 30° \sec^2 52° \sin^2 38°}{\csc^2 70°-\tan^2 20°}.$$

(AI CBSE 2006)

Sol. $\dfrac{\csc^2(90°-\theta)-\tan^2\theta}{4(\cos^2 48°+\cos^2 42°)}$

$-\dfrac{2\tan^2 30° \sec^2 52° \sin^2 38°}{\csc^2 70°-\tan^2 20°}$

$=\dfrac{\sec^2\theta-\tan^2\theta}{4\{\cos^2(90°-42°)+\cos^2 42°\}}$

$-\dfrac{2\left(\dfrac{1}{\sqrt{3}}\right)^2 \sec^2(90°-38°)\sin^2 38°}{\csc^2(90°-20°)-\tan^2 20°}$

| $\because$ $\csc(90°-\theta)=\sec\theta$

$=\dfrac{1}{4(\sin^2 42°+\cos^2 42°)}$

$-\dfrac{\dfrac{2}{3}\csc^2 38° \sin^2 38°}{\sec^2 20°-\tan^2 20°}$

| $\because$ $\sec^2\theta-\tan^2\theta=1,$
$\cos(90°-\theta)=\sin\theta$
$\sec(90°-\theta)=\csc\theta$
$\csc(90°-\theta)=\sec\theta$

$=\dfrac{1}{4(1)}-\dfrac{\dfrac{2}{3}\cdot\dfrac{1}{\sin^2 38°}\cdot\sin^2 38°}{1}$

| $\because$ $\sin^2\theta+\cos^2\theta=1,$
$\sec^2\theta-\tan^2\theta=1$
$\csc\theta=\dfrac{1}{\sin\theta}$

$=\dfrac{1}{4}-\dfrac{2}{3}=\dfrac{3-8}{12}=-\dfrac{5}{12}.$

Example 10. *Prove that*

$$\tan^2 A-\tan^2 B=\frac{\sin^2 A-\sin^2 B}{\cos^2 A \cos^2 B}.$$

(CBSE 2003)

Sol. L.H.S. $=\tan^2 A-\tan^2 B$

$=\dfrac{\sin^2 A}{\cos^2 A}-\dfrac{\sin^2 B}{\cos^2 B}$

$=\dfrac{\sin^2 A \cos^2 B-\sin^2 B \cos^2 A}{\cos^2 A \cos^2 B}$

$=\dfrac{\sin^2 A(1-\sin^2 B)-\sin^2 B(1-\sin^2 A)}{\cos^2 A \cos^2 B}$

$=\dfrac{\sin^2 A-\sin^2 A \sin^2 B-\sin^2 B+\sin^2 B \sin^2 A}{\cos^2 A \cos^2 B}$

$=\dfrac{\sin^2 A-\sin^2 B}{\cos^2 A \cos^2 B}=$ R.H.S.

TEST YOUR KNOWLEDGE

1. Express the ratios $\cos A$, $\tan A$ and $\sec A$ in terms of $\sin A$.
2. Prove that
   $$\sec(1-\sin A)(\sec A+\tan A)=1.$$
3. Prove that
   $$\frac{\cot A-\cos A}{\cot A+\cos A}=\frac{\csc A-1}{\csc A+1}.$$
4. Prove that
   $$\frac{\sin\theta-\cos\theta+1}{\sin\theta+\cos\theta-1}=\frac{1}{\sec\theta-\tan\theta},$$
   using the identity $\sec^2\theta=1+\tan^2\theta$.

INTRODUCTION TO TRIGONOMETRY

5. Prove that

$$\frac{1}{\sec x - \tan x} - \frac{1}{\cos x} = \frac{1}{\cos x} - \frac{1}{\sec x + \tan x}.$$

(AI CBSE 2005)

6. Evaluate :

$$\frac{\sec^2 54° - \cot^2 36°}{\csc^2 57° - \tan^2 33°} + 2 \sin^2 38° \sec^2 52° - \sin^2 45°.$$

(AI CBSE 2005)

7. Find the value of

$$\frac{-\tan \theta \cot (90° - \theta) + \sec \theta \csc (90° - \theta) + \sin^2 35° + \sin^2 55°}{\tan 10° \tan 20° \tan 30° \tan 70° \tan 80°}.$$

(CBSE 2005)

8. If $\sec \theta + \tan \theta = p$, prove that

$$\sin \theta = \frac{p^2 - 1}{p^2 + 1}.$$

(CBSE 2004)

9. Prove that

$$\frac{\tan \theta}{1 - \cot \theta} + \frac{\cot \theta}{1 - \tan \theta} = 1 + \tan \theta + \cot \theta.$$

(AI CBSE 2004)

10. Without using trigonometric tables, evaluate the following :

$$\frac{3 \tan 25° \tan 40° \tan 50° \tan 65° - \frac{1}{2} \tan^2 60°}{4 (\cos^2 29° + \cos^2 61°)}.$$

(AI CBSE 2004)

11. Prove the identity :

$(1 + \cot \theta - \csc \theta)(1 + \tan \theta + \sec \theta) = 2$

(CBSE SP Set 1)

12. Prove that

$$\frac{\tan \theta}{1 - \cot \theta} + \frac{\cot \theta}{1 - \tan \theta} = 1 + \sec \theta \csc \theta$$

(CBSE SP Set 2)

13. Evaluate :

$$\frac{\sin 39°}{\cos 51°} + 2 \tan 11° \tan 31° \tan 45° \tan 59° \tan 79° - 3 (\sin^2 21° + \sin^2 69°).$$

(CBSE SP Set 2)

Answers

1. $\cos A = \sqrt{1 - \sin^2 A}$,

$\tan A = \dfrac{\sin A}{\sqrt{1 - \sin^2 A}}$,

$\sec A = \dfrac{1}{\sqrt{1 - \sin^2 A}}$

6. $\dfrac{5}{2}$ 7. $2\sqrt{3}$

10. $\dfrac{3}{8}$ 13. 0.

MISCELLANEOUS EXERCISE

1. In $\triangle$ ABC, right angled at B, if AB = 12 cm and BC = 5 cm, find

   (i) sin A and tan A

   (ii) sin C and cot C.

2. ABC is a triangle, right angled at B. If AB = 4 cm, BC = 3 cm, find all the trigonometric ratios of angle A.

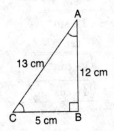

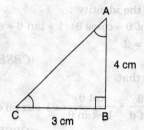

3. If $\cos\theta = \dfrac{3}{5}$, evaluate $\dfrac{\sin\theta - \cot\theta}{2\tan\theta}$.

4. Given $\cos\theta = \dfrac{21}{29}$, determine the value of

$\dfrac{\sec\theta}{\tan\theta - \sin\theta}$.

5. Evaluate $\dfrac{\csc\theta - \cot\theta}{2\cot\theta}$ when $\sin\theta = \dfrac{3}{5}$.

6. Given that $\cos\theta = \dfrac{p}{q}$, find the value of $\tan\theta$.

7. If $\sin A = \dfrac{1}{3}$, evaluate

$\cos A \csc A + \tan A \sec A$.

8. If $\tan\theta = \dfrac{4}{3}$, find the value of

$\dfrac{3\sin\theta + 2\cos\theta}{3\sin\theta - 2\cos\theta}$.

9. If $\tan\theta = \dfrac{a}{b}$, find the value of the expression $\dfrac{\cos\theta + \sin\theta}{\cos\theta - \sin\theta}$.

10. If $\tan\theta = 2$, evaluate $\sin\theta\sec\theta + \tan^2\theta - \csc\theta$.

11. If $\sec\theta = \dfrac{5}{4}$, evaluate $\dfrac{\sin\theta - 2\cos\theta}{\tan\theta - \cot\theta}$.

12. Evaluate :
   (i) $\sin 30° \cos 45° + \cos 30° \sin 45°$
   (ii) $\cos 30° \cos 45° - \sin 30° \sin 45°$
   (iii) $\tan^2 60° + 4\cos^2 45° + 3\sec^2 30° + 5\cos^2 90°$
   (iv) $4\cot^2 45° - \sec^2 60° + \sin^2 60° + \cos^2 90°$
   (v) $\csc^2 30° \sin 45° - \sec^2 60°$
   (vi) $\dfrac{\sin 60°}{\cos^2 45°} + 5\cos 90° - \cot 30°$
   (vii) $\dfrac{\tan 45°}{\sin 30° + \cos 30°}$
   (viii) $\dfrac{\tan 60°}{\sec 60° + \csc 60°}$
   (ix) $\dfrac{5\sin^2 30° + \cos^2 45° + 4\tan^2 60°}{2\sin 30° \cos 60° + \tan 45°}$.

13. Verify each of the following :
   (i) $\cos 60° = 1 - 2\sin^2 30° = 2\cos^2 30° - 1$
   (ii) $\cos 90° = 1 - 2\sin^2 45° = 2\cos^2 45° - 1$
   (iii) $\sin 30° \cos 60° + \cos 30° \sin 60° = \sin 90°$
   (iv) $\dfrac{\tan 60° - \tan 30°}{1 + \tan 60° \tan 30°} = \tan 30°$
   (v) $\cos^2 30° - \sin^2 30° = \cos 60°$
   (vi) $1 + \cot^2 30° = \csc^2 30°$
   (vii) $\dfrac{\cos 30° + \sin 60°}{1 + \sin 30° + \cos 60°} = \cos 30°$
   (viii) $\dfrac{\sin 60°}{1 + \cos 60°} = \tan 30°$.

14. Show that :
   (i) $\sin^2 45° + \sin^2 30° + \sin^2 60° = \dfrac{3}{2}$
   (ii) $\sin^2 45° + \cos^2 45° = 1$
   (iii) $2\sin^2 60° \cos 60° = \dfrac{3}{4}$
   (iv) $\cos^2 30° + \sin^2 30° + \tan 45° = 2$.

15. Given $A = 30°$, verify
   (i) $\sin 2A = 2\sin A \cos A$
   (ii) $\sin A = \sqrt{\dfrac{1 - \cos 2A}{2}}$
   (iii) $\sin 3A = 3\sin A - 4\sin^3 A$
   (iv) $\cos 3A = 4\cos^3 A - 3\cos A$
   (v) $\cos A = \dfrac{1}{\sqrt{1 + \tan^2 A}}$
   (vi) $\tan A = \dfrac{\sin A}{\sqrt{1 - \sin^2 A}}$
   (vii) $\sin A = \dfrac{\tan A}{\sqrt{1 + \tan^2 A}}$
   (viii) $\tan A = \dfrac{\sqrt{1 - \cos^2 A}}{\cos A}$.

16. Find the remaining parts of the triangle ABC, right angled at B, in which
    (i) $\angle C = 45°$, AB = 5 cm
    (ii) $\angle A = 30°$, AC = 8 cm
    (iii) $\angle C = 60°$, BC = 3 cm
    (iv) $\angle C = 60°$, AC = 5 cm
    (v) $\angle A = 45°$, BC = 7.5 cm
    (vi) $\angle A = 60°$, AB = 11 cm.

17. In the triangle PMO, right angled at M, determine the remaining parts in the following cases :
    (i) PM = 3 cm, OP = 6 cm
    (ii) PM = 5 cm, OP = $5\sqrt{2}$ cm
    (iii) PM = 8 cm, OP = $\dfrac{16\sqrt{3}}{3}$ cm
    (iv) OM = 4 cm, OP = 8 cm
    (v) PM = 5 cm, OM = 5 cm.

18. Without using trigonometric tables, evaluate the following :
    $$2 \times \left(\dfrac{\cos^2 20° + \cos^2 70°}{\sin^2 25° + \sin^2 65°}\right) - \tan 45° + \tan 13° \tan 23° \tan 30° \tan 63° \tan 77°.$$
    (AI CBSE 2003)

19. Without using the trigonometric tables, evaluate the following :
    $$\left(\dfrac{\tan 20°}{\text{cosec } 70°}\right)^2 + \left(\dfrac{\cot 20°}{\sec 70°}\right)^2 + 2 \tan 15° \tan 37° \tan 53° \tan 60° \tan 75°.$$
    (CBSE 2003)

20. Without using tables, evaluate :
    $\sin(50° + \theta) - \cos(40° - \theta) + \tan 1° \tan 10° \tan 20° \tan 70° \tan 80° \tan 89°$.
    (AI CBSE 2002)

21. If $\cos \theta + \sin \theta = \sqrt{2} \cos \theta$, prove that $\cos \theta - \sin \theta = \sqrt{2} \sin \theta$. *(AI CBSE 2003)*

22. Prove that
    $$\dfrac{1 + \cos \theta - \sin^2 \theta}{\sin \theta (1 + \cos \theta)} = \cos \theta.$$
    (CBSE 2003)

23. Prove that
    $\sec A (1 - \sin A)(\sec A + \tan A) = 1$. *(CBSE 2003)*

24. Prove that
    $$\dfrac{\cos A}{1 - \tan A} - \dfrac{\sin^2 A}{\cos A - \sin A} = \sin A + \cos A.$$

25. If $a = x \sin \theta + y \cos \theta$ and $b = x \cos \theta - y \sin \theta$, then show that $a^2 + b^2 = x^2 + y^2$.

26. Prove that
    $\sec^2 \theta + \text{cosec}^2 \theta = \sec^2 \theta \, \text{cosec}^2 \theta$.

27. If $x = r \sin \theta \cos \phi$, $y = r \sin \theta \sin \theta$ and $z = r \cos \theta$, then prove that
    $x^2 + y^2 + z^2 = r^2$.

28. If $\tan \theta + \sin \theta = m$ and $\tan \theta - \sin \theta = n$, show that $m^2 - n^2 = 4\sqrt{mn}$.

29. If $7 \sin^2 \theta + 3 \cos^2 \theta = 4$, find the value of $\sec \theta + \text{cosec } \theta$.

30. Find the value of when
    $$\dfrac{\cos \theta}{\text{cosec} + 1} + \dfrac{\cos \theta}{\text{cosec} - 1} = 2.$$

Answers

1. (i) $\dfrac{5}{13}, \dfrac{5}{12}$ (ii) $\dfrac{12}{13}, \dfrac{5}{12}$

2. $\sin A = \dfrac{3}{5}$, $\cos A = \dfrac{4}{5}$, $\tan A = \dfrac{3}{4}$,
   $\text{cosec } A = \dfrac{5}{3}$, $\sec A = \dfrac{5}{4}$, $\cot A = \dfrac{4}{3}$.

3. $\dfrac{3}{160}$ 4. $\dfrac{841}{160}$

5. $\dfrac{1}{8}$ 6. $\sqrt{\dfrac{q^2 - p^2}{p}}$

7. $2\sqrt{2} + \dfrac{3}{8}$ 8. 3

9. $\dfrac{b + a}{b - a}$ 10. $6 - \dfrac{\sqrt{5}}{2}$

11. $\dfrac{12}{7}$

12. (i) $\dfrac{\sqrt{2} + \sqrt{6}}{4}$ (ii) $\dfrac{\sqrt{3} - 1}{2\sqrt{2}}$ (iii) 9
    (iv) $\dfrac{3}{4}$ (v) $2\sqrt{2} - 4$ (vi) 0
    (vii) $\sqrt{3} - 1$ (viii) $\dfrac{3}{4}(\sqrt{3} - 1)$ (ix) $\dfrac{55}{6}$

16. (i) $\angle A = 45°$, $AC = 5\sqrt{2}$ cm, $BC = 5$ cm

(ii) $\angle C = 60°$, $BC = 4$ cm, $AB = 4\sqrt{3}$ cm

(iii) $\angle A = 30°$, $AC = 6$ cm, $AB = 3\sqrt{3}$ cm

(iv) $\angle A = 30°$, $BC = 2.5$ cm, $AB = \dfrac{5\sqrt{3}}{2}$ cm

(v) $\angle C = 45°$, $AB = 7.5$ cm, $AC = \dfrac{15\sqrt{2}}{2}$ cm

(vi) $\angle C = 30°$, $BC = 11\sqrt{3}$ cm, $AC = 22$ cm.

17. (i) $\angle POM = 30°$, $\angle OPM = 60°$,

$OM = 3\sqrt{3}$ cm

(ii) $\angle POM = 45°$, $\angle OPM = 45°$, $OM = 5$ cm

(iii) $\angle POM = 60°$, $\angle OPM = 30°$,

$OM = \dfrac{8\sqrt{3}}{3}$ cm

(iv) $\angle POM = 60°$, $\angle OPM = 30°$,

$PM = 4\sqrt{3}$ cm

(v) $POM = \angle 45°$, $\angle OPM = 45°$,

$OP = 5\sqrt{2}$ cm.

18. $1 + \dfrac{1}{\sqrt{3}}$ **19.** $1 + 2\sqrt{3}$

20. 1 **21.** $\dfrac{2}{\sqrt{3}} + 2$

30. $\dfrac{\pi}{4}$.

SUMMARY

1. In a right angle triangle ABC, right angled at B,

$\sin A = \dfrac{\text{side opposite to angle A}}{\text{hypotenuse}}$,

$\cos A = \dfrac{\text{side adjacent to angle A}}{\text{hypotenuse}}$,

$\tan A = \dfrac{\text{side opposite to angle A}}{\text{side adjacent to angle A}}$.

2. $\operatorname{cosec} A = \dfrac{1}{\sin A}$;

$\sec A = \dfrac{1}{\cos A}$; $\tan A = \dfrac{1}{\cot A}$,

$\tan A = \dfrac{\sin A}{\cos A}$.

3. The values of trigonometric ratios for angles $0°$, $30°$, $45°$, $60°$ and $90°$.

4. The value of sin A or cos A never exceeds 1, whereas the value of sec A or cosec A is always greater or equal to 1.

5. $\sin(90° - A) = \cos A$, $\cos(90° - A) = \sin A$;

$\tan(90° - A) = \cot A$, $\cot(90° - A) = \tan A$;

$\sec(90° - A) = \operatorname{cosec} A$, $\operatorname{cosec}(90° - A) = \sec A$.

6. If one of the trigonometric ratios of an acute angle is known, the remaining trigonometric ratios of the angle can be easily determined.

7. $\sin^2 A + \cos^2 A = 1$,

$\sec^2 A - \tan^2 A = 1$ for $0° \leq A \leq 90°$

$\operatorname{cosec}^2 A = 1 + \cot^2 A$ for $0° < A \leq 90°$.

9
Some Applications of Trigonometry

IMPORTANT POINTS

1. Use of Trigonometry. Trigonometry is used in the life around us. Trigonometry is one of the most ancient subjects studied by scholars all over the world. Trigonometry was invented because its need arose in astronomy. Since then the astronomers have used it, for instance, to calculate distances from the earth to the planets and stars. Trigonometry is also used in geography and in navigation. The knowledge of trigono-metry is used to construct maps, determine the position of an island in relation to the longitudes and latitudes. Trigonometry is also used for finding the heights and distances of various objects, without actually measuring them.

2. Historical Facts. Surveyors have used trigonometry for centuries. One such large surveying project of the 1800's was the **'Great Trigonometric Survey'** of British India for which the two largest-ever theodolites were built. During the survey in 1852, the highest mountain in the world was discovered. From a distance of over 160 km, the peak was observed from six different stations. In 1856, this peak was named after Sir George Everest, who had commissioned and first used the giant theodolites (see the figure alongside). The theodolites are now on display in the Museum of the Survey of India in Dehradun.

A Theodolite
(Surveying instrument which is based on the Principles of trigonometry, is used for measuring angles with a rotating telescope)

3. Heights and Distances. (*i*) Consider the following figures :

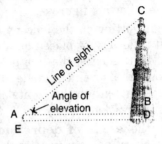

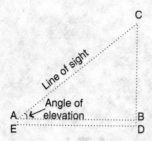

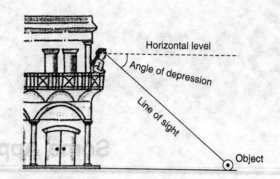

In this figure, the line AC drawn from the eye of the student to the top of the minar is called the *line of sight*. The student is looking at the top of the Minar. The angle BAC, so formed by the line of sight with the horizontal, is called the *angle of elevation* of the top of the Minar from the eye of the student.

Thus, the **line of sight** is the line drawn from the eye of an observer to the point in the object viewed by the observer. The **angle of elevation** of the point viewed is the angle formed by the line of sight with the horizontal when the point being viewed is above the horizontal level, *i.e.*, the case when we raise our head to look at the object (see Fig.).

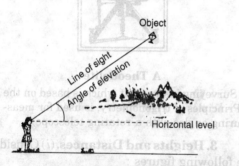

(*ii*) Next, consider the situation given in the following figure :

The girl sitting on the balcony is *looking down* at a flower pot placed on a stair of the temple. In this case, the line of sight is *below* the horizontal level. The angle so formed by the line of sight with the horizontal is called the *angle of depres*sion.

Thus, the **angle of depression** of a point on the object being viewed is the angle formed by the line of sight with the horizontal when the point is below the horizontal level, *i.e.*, the case when we lower our head to look at the point being viewed (see Fig.)

4. Determination of Height. If we want to find the height CD of the Minar without actually measuring it, then we need to know the following :

(*i*) The distance DE at which the student is standing from the foot of the minar.

(*ii*) the angle of elevation, ∠ BAC, of the top of the Minar.

(*iii*) the height AE of the student.

Suppose that the above three conditions are known to us. Then, in the Fig.

$$CD = CB + BD$$

Here, BD = AE, which is the height of the student.

To find BC, we will use trigonometric ratios of ∠ BAC (or ∠ A).

In ∠ ABC, the side BC is the opposite side in relation to the known ∠ A. Now, we shall see the trigonometric ratios which we can use. For this we will have to search which one of them has the two values that we have and the one we need to determine. Our search narrows down to using either tan A or cot A, as these ratios involve AB and BC.

Therefore, $\tan A = \dfrac{BC}{AB}$ or $\cot A = \dfrac{AB}{BC}$, which on solving would give us BC. By adding AE to BC, we will get the height of the Minar.

ILLUSTRATIVE EXAMPLES

[NCERT Exercise 9.1]
(Page No. 203)

Example 1. *A circus artist is climbing a 20 m long rope, which is tightly stretched and tied from the top of a vertical pole to the ground. Find the height of the pole, if the angle made by the rope with the ground level is 30° (see figure).*

Sol. In right triangle ABC,

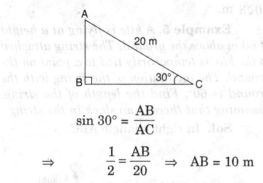

$$\sin 30° = \frac{AB}{AC}$$

$$\Rightarrow \quad \frac{1}{2} = \frac{AB}{20} \quad \Rightarrow \quad AB = 10 \text{ m}$$

Hence, the height of the pole is 10 m.

Example 2. *A tree breaks due to storm and the broken part bends so that the top of the tree touches the ground making an angle 30° with it. The distance between the foot of the tree to the point where the top touches the ground is 8 m. Find the height of the tree.*

Sol. In right triangle ABC,

$$\cos 30° = \frac{BC}{AC}$$

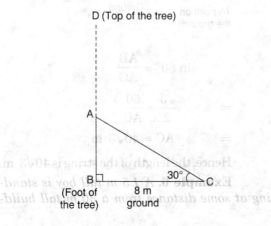

$$\Rightarrow \quad \frac{\sqrt{3}}{2} = \frac{8}{AC}$$

$$\Rightarrow \quad AC = \frac{16}{\sqrt{3}} \text{ m}$$

$$\tan 30° = \frac{AB}{BC}$$

$$\Rightarrow \quad \frac{1}{\sqrt{3}} = \frac{AB}{8} \Rightarrow AB = \frac{8}{\sqrt{3}} \text{ m}$$

∴ Height of the tree = AB + AC

$$= \left(\frac{8}{\sqrt{3}} + \frac{16}{\sqrt{3}}\right) \text{ m}$$

$$= \frac{24}{\sqrt{3}} \text{ m} = 8\sqrt{3} \text{ m}$$

Hence, the height of the tree is $8\sqrt{3}$ m.

Example 3. *A contractor plans to install two slides for the children to play in a park. For the children below the age of 5 years, she prefers to have a slide whose top is at a height of 1.5 m, and is inclined at an angle of 30° to the ground, whereas for elder children, she wants to have a steep slide at a height of 3 m, and inclined at an angle of 60° to the ground. What should be the length of the slide in each case ?*

Sol. In right triangle ABC,

$$\sin 30° = \frac{AB}{AC}$$

$$\Rightarrow \quad \frac{1}{2} = \frac{1.5}{AC}$$

$$\Rightarrow \quad AC = 3 \text{ m}$$

In right triangle PQR,

$$\sin 60° = \frac{PQ}{PR}$$

$$\Rightarrow \quad \frac{\sqrt{3}}{2} = \frac{3}{PR}$$

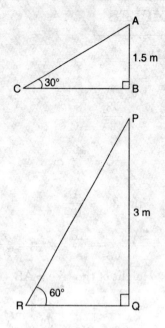

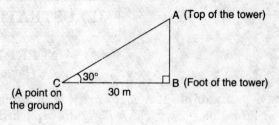

$$\Rightarrow \quad AB = \frac{30}{\sqrt{3}}$$

$$\Rightarrow \quad AB = 10\sqrt{3} \text{ m}$$

Hence, the height of the tower is $10\sqrt{3}$ m.

Example 5. *A kite is flying at a height of 60 m above the ground. The string attached to the kite is temporarily tied to a point on the ground. The inclination of the string with the ground is 60°. Find the length of the string, assuming that there is no slack in the string.*

Sol. In right triangle ABC,

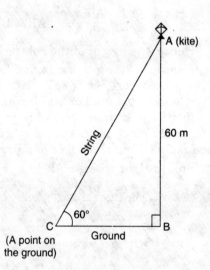

$$\sin 60° = \frac{AB}{AC}$$

$$\Rightarrow \quad \frac{\sqrt{3}}{2} = \frac{60}{AC}$$

$$\Rightarrow \quad AC = 40\sqrt{3} \text{ m}$$

Hence, the length of the string is $40\sqrt{3}$ m.

Example 6. *A 1.5 m tall boy is standing at some distance from a 30 m tall build-*

$$\Rightarrow \quad PR = 2\sqrt{3} \text{ m}$$

$$\sin 30° = \frac{AB}{AC}$$

$$\Rightarrow \quad \frac{1}{2} = \frac{1.5}{AC}$$

$$\Rightarrow \quad AC = 3 \text{ m}$$

In right triangle PQR,

$$\sin 60° = \frac{PQ}{PR}$$

$$\Rightarrow \quad \frac{\sqrt{3}}{2} = \frac{3}{PR}$$

$$\Rightarrow \quad PR = 2\sqrt{3} \text{ m}$$

Hence, the lengths of the slides are 3 m and $2\sqrt{3}$ m respectively.

Example 4. *The angle of elevation of the top of a tower from a point on the ground, which is 30 m away from the foot of the tower, is 30°. Find the height of the tower.*

Sol. In right triangle ABC,

$$\tan 30° = \frac{AB}{BC}$$

$$\Rightarrow \quad \frac{1}{\sqrt{3}} = \frac{AB}{30}$$

ing. *The angle of elevation from his eyes to the top of the building increases from 30° to 60° as he walks towards the building. Find the distance he walked towards the building.*

Sol. AB = 30 m
PR = 1.5 m

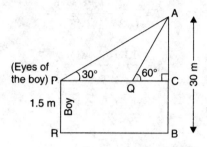

∴ AC = AB − BC = AB − PR
= 30 m − 1.5 m = 28.5 m

In right triangle ACQ,

$$\tan 60° = \frac{AC}{QC}$$

⇒ $\sqrt{3} = \dfrac{28.5}{QC}$

⇒ $QC = \dfrac{28.5}{\sqrt{3}}$ m

In right triangle ACP,

$$\tan 30° = \frac{AC}{PC}$$

⇒ $\tan 30° = \dfrac{AC}{PQ + QC}$

⇒ $\dfrac{1}{\sqrt{3}} = \dfrac{28.5}{PQ + \dfrac{28.5}{\sqrt{3}}}$

⇒ $\dfrac{1}{\sqrt{3}} = \dfrac{(28.5)\sqrt{3}}{PQ\sqrt{3} + 28.5}$

⇒ $PQ\sqrt{3} + 28.5 = 85.5$

⇒ $PQ\sqrt{3} = 85.5 − 28.5$

⇒ $PQ\sqrt{3} = 57$

⇒ $PQ = \dfrac{57}{\sqrt{3}}$

⇒ $PQ = 19\sqrt{3}$

Hence, the distance walked towards the building is $19\sqrt{3}$ m.

Example 7. *From a point on the ground, the angles of elevation of the bottom and top of a transmission tower fixed at the top of a 20 m high building are 45° and 60° respectively. Find the height of the tower.*

Sol. Let the height of the tower be h m
Then, in right triangle CBP,

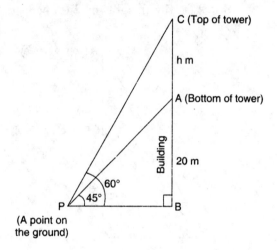

$$\tan 60° = \frac{BC}{BP}$$

⇒ $\sqrt{3} = \dfrac{AB + AC}{BP}$

⇒ $\sqrt{3} = \dfrac{20 + h}{BP}$ ...(1)

In right triangle ABP,

$$\tan 45° = \frac{AB}{BP}$$

⇒ $1 = \dfrac{20}{BP}$ ...(2)

Dividing (1) by (2), we get

$$\sqrt{3} = \frac{20 + h}{20}$$

⇒ $20\sqrt{3} = 20 + h$

$\Rightarrow \qquad h = 20\sqrt{3} - 20$

$\Rightarrow \qquad h = 20(\sqrt{3} - 1)$

Hence, the height of the tower is $20(\sqrt{3} - 1)$ m.

Example 8. *A statue 1.6 m tall stands on the top of a pedestal. From a point on the ground, the angle of elevation of the top of the statue is 60° and from the same point the angle of elevation of the top of the pedestal is 45°. Find the height of the pedestal.*

Sol. Let the height of the pedestal be h m.

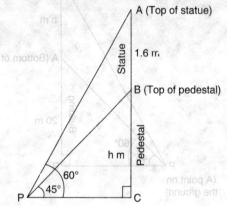

Then $BC = h$ m

In right triangle ACP,

$$\tan 60° = \frac{AC}{PC}$$

$\Rightarrow \qquad \sqrt{3} = \frac{AB + BC}{PC}$

$\Rightarrow \qquad \sqrt{3} = \frac{1.6 + h}{PC}$ ...(1)

In right triangle BCP,

$$\tan 45° = \frac{BC}{PC}$$

$\Rightarrow \qquad 1 = \frac{h}{PC}$ ...(2)

Dividing (1) by (2), we get

$$\frac{\sqrt{3}}{1} = \frac{1.6 + h}{h}$$

$\Rightarrow \qquad h\sqrt{3} = 1.6 + h$

$\Rightarrow \qquad h(\sqrt{3} - 1) = 1.6$

$\Rightarrow \qquad h = \frac{1.6}{\sqrt{3} - 1}$

$\Rightarrow \qquad h = \frac{1.6(\sqrt{3} + 1)}{(\sqrt{3} - 1)(\sqrt{3} + 1)}$

$\Rightarrow \qquad h = \frac{1.6(\sqrt{3} + 1)}{3 - 1}$

$\Rightarrow \qquad h = \frac{1.6(\sqrt{3} + 1)}{2}$

$\Rightarrow \qquad h = 0.8(\sqrt{3} + 1)$

Hence, the height of the pedestal is $0.8(\sqrt{3} + 1)$ m.

Example 9. *The angle of elevation of the top of a building from the foot of the tower is 30° and the angle of elevation of the top of the tower from the foot of the building is 60°. If the tower is 50 m high, find the height of the building.*

Sol. Let the height of the building be h m. Then, $AB = h$ m.

In right triangle PQB,

$$\tan 60° = \frac{PQ}{BQ}$$

$\Rightarrow \qquad \sqrt{3} = \frac{50}{BQ}$

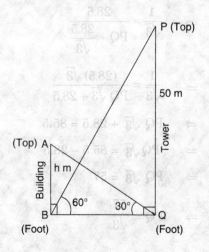

$\Rightarrow \quad BQ = \dfrac{50}{\sqrt{3}}$ m ...(1)

In right triangle ABQ,

$\tan 30° = \dfrac{AB}{BQ}$

$\Rightarrow \quad \dfrac{1}{\sqrt{3}} = \dfrac{h}{BQ}$

$\Rightarrow \quad h = \dfrac{BQ}{\sqrt{3}}$

$\Rightarrow \quad h = \dfrac{\frac{50}{\sqrt{3}}}{\sqrt{3}}$ | From (1)

$\Rightarrow \quad h = \dfrac{50}{3}$ m $= 16\dfrac{2}{3}$ m

Hence, the height of the building is $16\dfrac{2}{3}$ m.

Example 10. *Two poles of equal heights are standing opposite each other on either side of the road, which is 80 m wide. From a point between them on the road, the angles of elevation of the top of the poles are 60° and 30°, respectively. Find the height of the poles and the distances of the point from the poles.*

Sol. In right triangle PRQ,

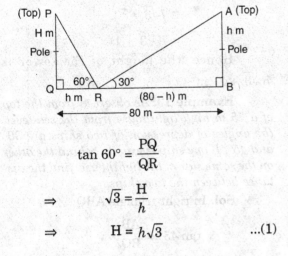

$\tan 60° = \dfrac{PQ}{QR}$

$\Rightarrow \quad \sqrt{3} = \dfrac{H}{h}$

$\Rightarrow \quad H = h\sqrt{3}$ ...(1)

In right triangle ABR,

$\tan 30° = \dfrac{AB}{BR}$

$\Rightarrow \quad \dfrac{1}{\sqrt{3}} = \dfrac{H}{80-h}$

$\Rightarrow \quad \dfrac{1}{\sqrt{3}} = \dfrac{h\sqrt{3}}{80-h}$ | From (1)

$\Rightarrow \quad 80 - h = 3h$

$\Rightarrow \quad 4h = 80$

$\Rightarrow \quad h = \dfrac{80}{4} = 20$ ...(2)

Again from (1)

$H = h\sqrt{3}$

$\Rightarrow \quad H = 20\sqrt{3}$ | From (2)

Also, $80 - h = 80 - 20 = 60$ | From (2)

Hence, the heights of the poles are $20\sqrt{3}$ m each and the distances of the point from the poles are 20 m and 60 m respectively.

Example 11. *A TV tower stands vertically on a bank of a canal. From a point on the other bank directly opposite the tower, the angle of elevation of the top of the tower is 60°. From another point 20 m away from this point on the line joining this point to the foot of the tower, the angle of elevation of the top of the tower is 30° (see figure). Find the height of the tower and the width of the canal.*

Sol. In right triangle ABC,

$\tan 60° = \dfrac{AB}{BC}$

$\Rightarrow \quad \sqrt{3} = \dfrac{AB}{BC}$ ...(1)

In right triangle ABD,

$$\tan 30° = \frac{AB}{BD}$$

$\Rightarrow \quad \dfrac{1}{\sqrt{3}} = \dfrac{AB}{BC + CD}$

$\Rightarrow \quad \dfrac{1}{\sqrt{3}} = \dfrac{AB}{BC + 20}$ ...(2)

Dividing (1) by (2), we get

$$\frac{\sqrt{3}}{\frac{1}{\sqrt{3}}} = \frac{BC + 20}{BC}$$

$\Rightarrow \quad 3 = \dfrac{BC + 20}{BC}$

$\Rightarrow \quad 3BC = BC + 20$

$\Rightarrow \quad 3BC - BC = 20$

$\Rightarrow \quad 2BC = 20$

$\Rightarrow \quad BC = \dfrac{20}{2} = 10$ ...(3)

From (1),

$$\sqrt{3} = \frac{AB}{BC}$$

$\Rightarrow \quad \sqrt{3} = \dfrac{AB}{10}$ | From (3)

$\Rightarrow \quad AB = 10\sqrt{3}$

Hence, the height of the tower is $10\sqrt{3}$ m and the width of the canal is 10 m.

Example 12. *From the top of a 7 m high building, the angle of elevation of the top of a cable tower is 60° and the angle of depression of its foot is 45°. Determine the height of the tower.*

Sol. In right triangle ABD,

$$\tan 45° = \frac{AB}{BD}$$

$\Rightarrow \quad 1 = \dfrac{7}{BD}$

$\Rightarrow \quad BD = 7$

$\Rightarrow \quad AE = 7$

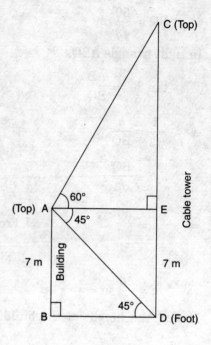

In right triangle AEC,

$$\tan 60° = \frac{CE}{AE}$$

$\Rightarrow \quad \sqrt{3} = \dfrac{CE}{7}$

$\Rightarrow \quad CE = 7\sqrt{3}$

$\therefore \quad CD = CE + ED$
$ = CE + AB$
$ = 7\sqrt{3} + 7$
$ = 7(\sqrt{3} + 1)$

Hence, the height of the tower is $7(\sqrt{3} + 1)$ m.

Example 13. *As observed from the top of a 75 m high lighthouse from the sea-level, the angles of depression of two ships are 30° and 45°. If one ship is exactly behind the other on the same side of the lighthouse, find the distance between the two ships.*

Sol. In right triangle ABQ,

$$\tan 45° = \frac{AB}{BQ}$$

SOME APPLICATIONS OF TRIGONOMETRY

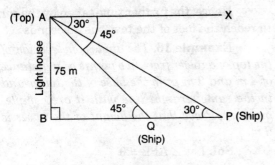

$\Rightarrow \quad 1 = \dfrac{75}{BQ}$

$\Rightarrow \quad BQ = 75$ m ...(1)

In right triangle ABP,

$\Rightarrow \quad \tan 30° = \dfrac{AB}{BP}$

$\Rightarrow \quad \dfrac{1}{\sqrt{3}} = \dfrac{AB}{BQ + QP}$

$\Rightarrow \quad \dfrac{1}{\sqrt{3}} = \dfrac{75}{75 + QP}$ | From (1)

$\Rightarrow \quad 75 + QP = 75\sqrt{3}$

$\Rightarrow \quad QP = 75(\sqrt{3} - 1)$

Hence, the distance between the two ships is $75(\sqrt{3} - 1)$ m.

Example 14. *A 1.2 m tall girl spots a balloon moving with the wind in a horizontal line at a height of 88.2 m from the ground. The angle of elevation of the balloon from the eyes of the girl at any instant is 60°. After some time, the angle of elevation reduces to 30° (see figure). Find the distance travelled by the balloon during the interval.*

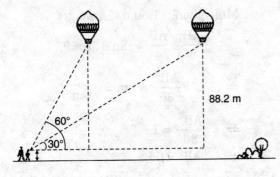

Sol. In right triangle ABC,

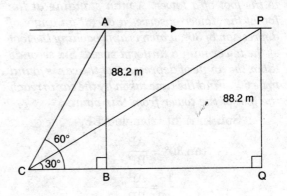

$\tan 60° = \dfrac{AB}{BC}$

$\Rightarrow \quad \sqrt{3} = \dfrac{88.2}{BC}$

$\Rightarrow \quad BC = \dfrac{88.2}{\sqrt{3}}$ ...(1)

In right triangle PQC,

$\tan 30° = \dfrac{PQ}{CQ}$

$\Rightarrow \quad \tan 30° = \dfrac{PQ}{CB + BQ}$

$\Rightarrow \quad \dfrac{1}{\sqrt{3}} = \dfrac{88.2}{\dfrac{88.2}{\sqrt{3}} + BQ}$ | From (1)

$\Rightarrow \quad \dfrac{1}{\sqrt{3}} = \dfrac{88.2\sqrt{3}}{88.2 + BQ\sqrt{3}}$

$\Rightarrow \quad 88.2 + BQ\sqrt{3} = 264.6$

$\Rightarrow \quad BQ\sqrt{3} = 264.6 - 88.2 = 176.4$

$\Rightarrow \quad BQ = \dfrac{176.4}{\sqrt{3}} = \dfrac{(176.4)\sqrt{3}}{\sqrt{3}\sqrt{3}}$

$\quad = \dfrac{(176.4)\sqrt{3}}{3} = (58.8)\sqrt{3}$

$\quad = \dfrac{58.8}{10}\sqrt{3} = \dfrac{294}{5}\sqrt{3}$

Hence, the distance travelled by the balloon during the interval is $\dfrac{294}{5}\sqrt{3}$ m.

Example 15. *A straight highway leads to the foot of a tower. A man standing at the top of the tower observes a car at an angle of depression of 30°, which is approaching the foot of the tower with a uniform speed. Six seconds later, the angle of depression of the car is found to be 60°. Find the time taken by the car to reach the foot of the tower from this point.*

Sol. In right triangle ABP,

$$\tan 30° = \frac{AB}{BP}$$

$$\Rightarrow \quad \frac{1}{\sqrt{3}} = \frac{AB}{BP}$$

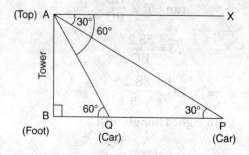

$$\Rightarrow \quad BP = AB\sqrt{3} \qquad ...(1)$$

In right triangle ABQ,

$$\tan 60° = \frac{AB}{BQ}$$

$$\Rightarrow \quad \sqrt{3} = \frac{AB}{BQ} \Rightarrow BQ = \frac{AB}{\sqrt{3}} \qquad ...(2)$$

$$\therefore \quad PQ = BP - BQ$$

$$= AB\sqrt{3} - \frac{AB}{\sqrt{3}}$$

$$= \frac{3AB - AB}{\sqrt{3}}$$

$$= \frac{2AB}{\sqrt{3}} = (2BQ) \qquad | \text{ From (2)}$$

$$\Rightarrow \quad BQ = \frac{1}{2} PQ$$

∵ Time taken by the car to travel a distance PQ = 6 seconds.

∴ Time taken by the car to travel a distance BQ, *i.e.*, $\frac{1}{2}$ PQ = $\frac{1}{2} \times 6$ seconds = 3 seconds.

Hence, the further time taken by the car to reach the foot of the tower is 3 seconds.

Example 16. *The angles of elevation of the top of a tower from two points at a distance of 4 m and 9 m from the base of the tower and in the same straight line with it are complementary. Prove that the height of the tower is 6 m.*

Sol. Let ∠ APB = θ

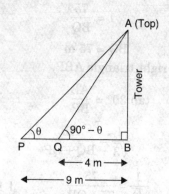

Then ∠ AQB = 90° − θ

| ∵ ∠ APB and ∠ AQB are complementary

In right triangle ABP,

$$\tan \theta = \frac{AB}{PB}$$

$$\Rightarrow \quad \tan \theta = \frac{AB}{9} \qquad ...(1)$$

In right triangle ABQ,

$$\tan (90° - \theta) = \frac{AB}{QB}$$

$$\Rightarrow \quad \cot \theta = \frac{AB}{4} \qquad ...(2)$$

Multiplying (1) and (2), we get

$$\frac{AB}{9} \cdot \frac{AB}{4} = \tan \theta \cdot \cot \theta$$

$$\Rightarrow \quad \frac{AB^2}{36} = \tan \theta \cdot \frac{1}{\tan \theta}$$

$$\Rightarrow \quad \frac{AB^2}{36} = 1$$

$$\Rightarrow \quad AB^2 = 36$$

$\Rightarrow \quad AB = \sqrt{36} = 6$

Hence the height of the tower is 6 m.

ADDITIONAL EXAMPLES

Example 17. *The angles of depression of the top and the bottom of a building 50 metres high as observed from the top of a tower are 30° and 60° respectively. Find the height of the tower and also the horizontal distance between the building and the tower.*

(CBSE 2006)

Sol. AB is the tower.

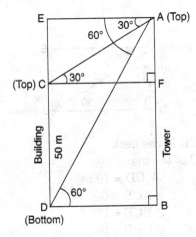

A is the top of the tower. CD is the building. C and D are respectively the top and bottom of the building.

$CD = 50$ m

$\angle CAE = 30°$

$\therefore \quad \angle ACF = 30°$

$\angle DAE = 60°$

$\therefore \quad \angle ADB = 60°$

In right triangle AFC,

$\tan 30° = \dfrac{AF}{CF}$

$\Rightarrow \quad \dfrac{1}{\sqrt{3}} = \dfrac{AF}{CF}$ ...(1)

In right triangle ABD,

$\tan 60° = \dfrac{AB}{BD}$

$\Rightarrow \quad \sqrt{3} = \dfrac{AF + FB}{BD}$

$\Rightarrow \quad \sqrt{3} = \dfrac{AF + CD}{CF}$

$\Rightarrow \quad \sqrt{3} = \dfrac{AF + 50}{CF}$ ...(2)

Dividing (2) by (1), we get

$3 = \dfrac{AF + 50}{AF}$

$\Rightarrow \quad 3AF = AF + 50$

$\Rightarrow \quad 3AF = AF + 50$

$\Rightarrow \quad 2AF = 50$

$\Rightarrow \quad AF = \dfrac{50}{2}$

$\Rightarrow \quad AF = 25$ m ...(3)

$\therefore$ Height of the tower = AB

$= AF + FB$

$= AF + CD$

$= 25 + 50$

$= 75$ m

and, the horizontal distance between the building and the tower

$= CF$

$= \sqrt{3}\ AF$ | From (1)

$= 25\sqrt{3}$ m. | From (3)

Example 18. *The angle of elevation of the top of a tower as observed from a point on the ground is 'α' and on moving a metres towards the tower, the angle of elevation is β. Prove that the height of the tower is*

$$\dfrac{a \tan \alpha \tan \beta}{\tan \beta - \tan \alpha}$$

(CBSE 2006)

Sol. AB is the tower.

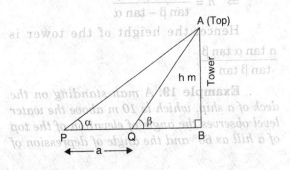

A is the top of the tower.

$\angle APB = \alpha$

$\angle AQB = \beta$

$PQ = a$ m.

Let the height of the tower be h m.
In right triangle ABP,

$$\tan \alpha = \frac{AB}{PB}$$

$\Rightarrow \quad \tan \alpha = \frac{h}{PQ + QB}$

$\Rightarrow \quad \tan \alpha = \frac{h}{a + QB}$

$\Rightarrow \quad a + QB = \frac{h}{\tan \alpha}$

$\Rightarrow \quad QB = \frac{h}{\tan \alpha} - a$ ...(1)

In right triangle ABQ,

$$\tan \beta = \frac{AB}{QB}$$

$\Rightarrow \quad \tan \beta = \frac{h}{QB}$

$\Rightarrow \quad QB = \frac{h}{\tan \beta}$ ...(2)

From (1) and (2), we get

$$\frac{h}{\tan \alpha} - a = \frac{h}{\tan \beta}$$

$\Rightarrow h \left(\frac{1}{\tan \alpha} - \frac{1}{\tan \beta} \right) = a$

$= h \left(\frac{\tan \beta - \tan \alpha}{\tan \alpha \tan \beta} \right) = a$

$\Rightarrow h = \frac{a \tan \alpha \tan \beta}{\tan \beta - \tan \alpha}$

Hence, the height of the tower is $\frac{a \tan \alpha \tan \beta}{\tan \beta \tan \alpha}$.

Example 19. *A man standing on the deck of a ship, which is 10 m above the water level observes the angle of elevation of the top of a hill as 60° and the angle of depression of the base of the hill as 30°. Calculate the distance of the hill from the ship and the height of the hill.* (AI CBSE 2006, 2005, 2004)

Sol. AB is the hill whose top is A and the base is B.

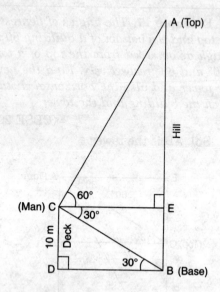

CD is the deck.
C is the man.

$CD = 10$ m

$\angle ACE = 60°$

$\angle BCE = 30°$

$\Rightarrow \quad \angle CBD = 30°$

In right triangle CDB,

$$\tan 30° = \frac{CD}{BD}$$

$\Rightarrow \quad \frac{1}{\sqrt{3}} = \frac{10}{BD}$

$\Rightarrow \quad BD = 10\sqrt{3}$ m ...(1)

In right triangle AEC,

$$\tan 60° = \frac{AE}{CE}$$

$\Rightarrow \quad \sqrt{3} = \frac{AE}{BD}$ | $\because$ CE = BD

$\Rightarrow \quad \sqrt{3} = \frac{AE}{10\sqrt{3}}$ | From (1)

$\Rightarrow \quad AE = 10\sqrt{3} \times \sqrt{3} = 30$ m

∴ AB = AE + EB = AE + CD
= 30 + 10 = 40 m

Hence, the distance of the hill from the ship is $10\sqrt{3}$ m and the height of the hill is 40 m.

Example 20. *From a window x metres high above the ground in a street, the angles of elevation and depression of the top and the foot of the other house on the opposite side of the street are α and β respectively. Show that the height of the opposite house is $x(1 + \tan\alpha \cot\beta)$ metres.* (AI CBSE 2006)

Sol. A is the window x metres high above the ground. CD is the other house whose top and foot are C and D respectively.

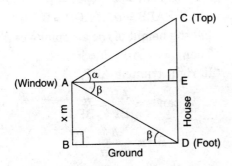

∠ CAE = α
∠ EAD = β
∠ ADB = β

In right triangle ABD,

$$\tan\beta = \frac{AB}{BD} = \frac{x}{BD}$$

⇒ BD = $x \cot\beta$ ...(1)

In right triangle AEC,

$$\tan\alpha = \frac{CE}{AE} = \frac{CE}{BD} = \frac{CE}{x\cot\beta}$$

| From (1)

⇒ CE = $x \tan\alpha \cot\beta$

∴ CD = CE + ED = CE + AB
= AB + CE = $x + x \tan\alpha \cot\beta$
= $x(1 + \tan\alpha \cot\beta)$ m

Hence, the height of the opposite house is $x(1 + \tan\alpha \cot\beta)$ m.

Examples 21. *If the angle of elevation of a cloud from a point h metres above a lake is α and the angle of depression of its reflection in the lake is β, prove that the distance of the cloud from the point of observation is*

$$\frac{2h\sec\alpha}{\tan\beta - \tan\alpha}$$ (CBSE 2004)

Sol. A is the cloud.

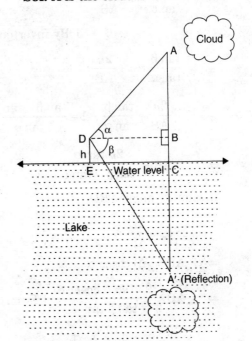

D is a point h metres above the lake

∠ ADB = α

A′ is the reflection of A in the lake

∠ A′DB = β

EC is the water level.

In right triangle ABD,

$$\tan\alpha = \frac{AB}{DB}$$ ...(1)

In right triangle A′BD,

$$\tan\beta = \frac{A'B}{DB} = \frac{A'C + CB}{DB}$$

$$= \frac{AC + CB}{DB}$$

| ∵ AC = A′C (by law of reflection)

$$= \frac{AB + BC + CB}{DB} = \frac{AB + h + h}{DB}$$

$$= \frac{AB + 2h}{DB}$$ ...(2)

Dividing (1) by (2), we get

$$\frac{\tan \alpha}{\tan \beta} = \frac{AB}{AB + 2h}$$

$$\Rightarrow \frac{\tan \beta}{\tan \alpha} = \frac{AB + 2h}{AB}$$

| By invertendo

$$\Rightarrow \frac{\tan \beta}{\tan \alpha} = 1 + \frac{2h}{AB}$$

$$\Rightarrow \frac{2h}{AB} = \frac{\tan \beta}{\tan \alpha} - 1 = \frac{\tan \beta - \tan \alpha}{\tan \alpha}$$

$$\Rightarrow AB = \frac{2h \tan \alpha}{\tan \beta - \tan \alpha} \quad ...(3)$$

In right triangle ABD,

$$\text{cosec } \alpha = \frac{AD}{AB}$$

$$\Rightarrow AD = AB \text{ cosec } \alpha$$

$$\Rightarrow AD = \frac{2h \tan \alpha \text{ cosec } \alpha}{\tan \beta \tan \alpha} \quad | \text{ From (2)}$$

$$\Rightarrow AD = \frac{2h \dfrac{\sin \alpha}{\cos \alpha} \cdot \dfrac{1}{\sin \alpha}}{\tan \beta \tan \alpha}$$

$$\Rightarrow AD = \frac{2h \sec \alpha}{\tan \beta - \tan \alpha}$$

Hence, the distance of the cloud from the point of observation is

$$\frac{2h \sec \alpha}{\tan \beta - \tan \alpha}.$$

Example 22. *From an aeroplane vertically above a straight horizontal plane, the angles of depression of two consecutive kilometre stones on the opposite sides of the aeroplane are found to be α and β. Show that the height of the aeroplane is $\dfrac{\tan \alpha \tan \beta}{\tan \alpha + \tan \beta}$.*

(CBSE 2004)

Sol. QP is a straight horizontal plane. A is an aeroplane vertically above QP. P and Q are two consecutive kilometre stones on the opposite sides of the aeroplane.

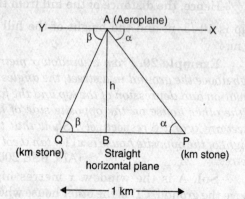

$\angle PAX = \alpha \angle QAY = \beta$ | Given

$\Rightarrow \angle APB = \alpha, \angle AQB = \beta$

Let the height of the aeroplane be h km
Then, $AB = h =$ km
In right triangle ABP,

$$\tan \alpha = \frac{AB}{BP} = \frac{h}{BP}$$

$$\Rightarrow BP = \frac{h}{\tan \alpha} \quad ...(1)$$

In right triangle ABQ,

$$\tan \beta = \frac{AB}{BQ} = \frac{h}{BQ}$$

$$\Rightarrow BQ = \frac{h}{\tan \beta} \quad ...(2)$$

Now, PQ = 1 km

$$\Rightarrow BP + BQ = 1$$

$$\Rightarrow \frac{h}{\tan \alpha} + \frac{h}{\tan \beta} = 1$$

$$\Rightarrow h \left(\frac{1}{\tan \alpha} + \frac{1}{\tan \beta} \right) = 1$$

$$\Rightarrow h \left(\frac{\tan \beta + \tan \alpha}{\tan \alpha \tan \beta} \right) = 1$$

$$\Rightarrow h = \frac{\tan \alpha \tan \beta}{\tan \alpha + \tan \beta}$$

Hence, the height of the aeroplane is

$$\frac{\tan \alpha \tan \beta}{\tan \alpha + \tan \beta}.$$

TEST YOUR KNOWLEDGE

1. A tower stands vertically on the ground. From a point on the ground, which is 15 m away from the foot of the tower, the angle of elevation of the top of the tower is found to be 60°. Find the height of the tower.

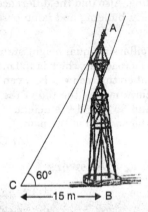

2. An electrician has to repair an electric fault on a pole of height 4 m. She needs to reach a point 1.3 m below the top of the pole to undertake the repair work (see figure).

   What should be the length of the ladder that she should use which, when inclined at an angle of 60° to the horizontal, would enable her to reach the required position ? Also, how far from the foot of the pole should she place the foot of the ladder ? (You may take $\sqrt{3} = 1.73$)

3. An observer 1.5 m tall is 28.5 m away from a chimney. The angle of elevation of the top of the chimney from her eyes is 45°. What is the height of the chimney ?

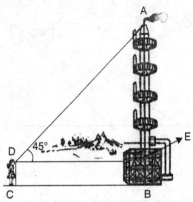

4. From a point P on the ground the angle of elevation of the top of a 10 m tall building is 30°. A flag is hoisted at the top of the building and the angle of elevation of the top of the flagstaff from P is 45°. Find the length of the flagstaff and the distance of the building from the point P. (You may take $\sqrt{3} = 1.732$)

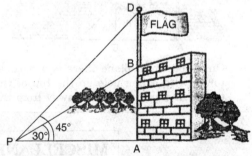

5. The shadow of a tower standing on a level ground is found to be 40 m longer when the Sun's altitude is 30° than when it is 60°. Find the height of the tower.

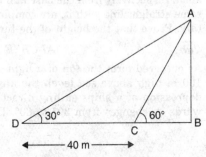

6. The angles of depression of the top and bottom of an 8 m tall building from the top of a multi-storeyed building are 30° and 45°, respectively. Find the height of the multi-storeyed building and the distance between the two buildings.

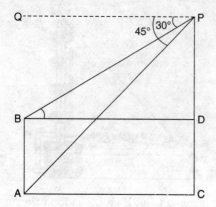

7. From a point on a bridge across a river, the angles of depression of the banks on opposite sides of the river are 30° and 45°, respectively. If the bridge is at a height of 3 m from the banks, find the width of the river.

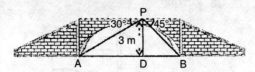

8. On a horizontal plane there is a vertical tower with a flag pole on the top of the tower. At a point a metres away from the foot of the tower, the angles of elevation of the top and the bottom of the flag pole are 60° and 30° respectively. Find the heights of the tower and flag pole mounted on it.

(CBSE 2005)

9. From a building 60 metres high, the angles of depression of the top and the bottom of lamp post are 30° and 60° respectively. Find the distance between lamp post and building. Also find the difference of heights between building and lamp post.

(CBSE 2005)

10. Two pillars of equal height stand on either side of a roadway which is 150 m wide. From a point on the roadway between the pillars, the elevations of the top of the pillars are 60° and 30°. Find the height of the pillars and the position of the point.

(AI CBSE 2003)

Answers

1. $15\sqrt{3}$ m
2. 3.11 m, 1.56 m
3. 30 m
4. 7.32 m, 17.32 m
5. $20\sqrt{3}$ m
6. $4(3+\sqrt{3})$ m, $4(3+\sqrt{3})$ m
7. $3(\sqrt{3}+1)$ m
8. $3\sqrt{3}$ m, $6\sqrt{3}$ m
9. $20\sqrt{3}$ m, 20 m
10. $\dfrac{75\sqrt{3}}{2}$ m ; 37.5 m, 112.5 m.

MISCELLANEOUS EXERCISE

1. The angles of elevation of the top of a tower from two points P and Q at distances of a and b respectively from the base and in the same straight line with it, are complementary. Prove that the height of the tower is $\sqrt{ab}$. (AI CBSE 2004)

2. As observed form the top of a light-house 100 m high above sea level, the angle of depression of a ships sailing directly towards it, changes from 30° to 60°. Determine the distance travelled by the ship during the period of observation.

(Use $\sqrt{3}$ = 1.732) (AI CBSE 2004)

3. From the top of a tower 60 m high, the angles of depression of the top and bottom of a building whose base is in the same straight line with the base of the tower are observed to be 30° and 60° respectively. Find the height of the building.

(CBSE Sample Paper Set 1)

SOME APPLICATIONS OF TRIGONOMETRY

4. An aeroplane flying horizontally at a height of 1.5 km above the ground is observed at a certain point on earth to subtend an angle of 60°. After 15 seconds, its angle of elevation at the same point is observed to be 30°. Calculate the speed of the aeroplane in km/h.   *(CBSE Sample Paper Set 1)*

5. A vertical tower is surmounted by a flag staff of height h metres. At a point on the ground, the angles of elevation of the bottom and top of the flag staff are α and β respectively. Prove that the height of the tower is $\dfrac{h \tan \alpha}{\tan \beta - \tan \alpha}$.   *(CBSE 2002)*

6. The angles of elevation and depression of the top and bottom of a light house from the top of a building, 60 m high, are 30° and 60° respectively. Find
   (i) the difference between the heights of the light-house and the building.
   (ii) the distance between the light-house and the building.   *(AI CBSE 2003)*

7. From a window (60 metres high above the ground) of a house in a street, the angles of elevation and depression of the top and the foot of another house on opposite side of the street are 60° and 45° respectively. Show that the height of the opposite house is $60(1+\sqrt{3})$ metres.   *(CBSE 2003)*

8. The height of a tower is half the height of the flag-staff at its top. The angle of elevation of the top of the tower as seen from a distance of 10 metres from its foot is 30°. Find the angle of elevation of the top of the flag-staff from the same point.

9. A round balloon of radius 'a' subtends an angle θ at the eye of the observer while the angle of elevation of its centre is ϕ. Prove that the height of the centre of the balloon is $a \sin \phi \, \mathrm{cosec} \dfrac{\theta}{2}$.

10. From the top of a light-house, the angles of depression of two boats on the opposite sides of it are observed to be θ and ϕ. If the height of the light-house be h metres and the line joining the boats passes through the foot of light-house, then show that the distance between the two boats is
$$\dfrac{h(\tan \theta + \tan \phi)}{\tan \theta \tan \phi}.$$

Answers

2. 115.46 m   3. 40 m
4. $240\sqrt{3}$ km/h   6. 20 m, 34.6 m
8. 60°.

SUMMARY

1. (i) The **line of sight** is the line drawn from the eye of an observer to the point in the object viewed by the observer.

   (ii) The **angle of elevation** of an object viewed, is the angle formed by the line of sight with the horizontal when it is above the horizontal level, *i.e.,* the case when we raise our head to look at the object.

   (iii) The **angle of depression** of an object viewed is the angle formed by the line of sight with the horizontal when it is below the horizontal level, *i.e.,* the case when we lower our head to look at the object.

2. The height or length of an object or the distance between two distant objects can be determined with the help of trigonometric ratios.

10
Circles

IMPORTANT POINTS

1. Circle. A circle is a collection of all points in a plane which are at a constant distance (radius) from a fixed point (centre).

2. A circle and a line in a plane. Let us consider a circle and a line PQ. There can be three possibilities given in figure below :

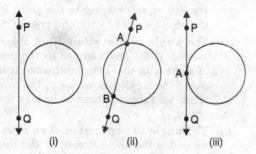

In figure (i), the line PQ and the circle have no common point. In this case, PQ is called a **non-intersecting** line with respect to the circle. In figure (ii), there are two common points A and B that the line PQ and the circle have. In this case, we call the line PQ a **secant** of the circle. In figure (iii), there is only one point A which is common to the line PQ and the circle. In this case, the line is called a **tangent** to the circle.

There cannot be any other type of position of the line with respect to the circle.

3. Practical example of tangent to the circle. We have seen a pulley fitted over a well which is used in taking out water from the well. Look at figure. Here the rope on both sides of the pulley, if considered as a ray, is like a tangent to the circle representing the pulley.

4. A tangent to a circle. A tangent* to a circle is a line that intersects the circle in only one point.

5. Certain activities. To understand the existence of the tangent to a circle at a point, let us perform the following activities :

Activity 1. Take a circular wire and attach a straight wire AB at a point P of the circular wire so that it can rotate about the point P in a plane. Put the system on a table and gently rotate the wire AB about the point P to get different positions of the straight wire [see figure (i)]

In various positions, the wire intersects the circular wire at P and at another point Q_1 or Q_2 or Q_3, etc. In one position, we will see that it will intersect the circle at the point P only (see position A'B' of AB). This shows that a tangent exists at the point P of the circle. On rotating further, we can observe that in all other positions of AB, it will intersect the circle at P and at another point, say R_1 or R_2 or R_3 etc. So, we can observe that **there is only one tangent at a point of the circle.**

---

*The word 'tangent' comes from the Latin word 'tangere', which means to touch and was introduced by the Danish mathematican Thomas Fineke in 1583.

CIRCLES

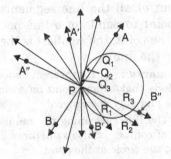

While doing activity above, we observe that as the position AB moves towards the position A'B', the common point, say Q_1, of the line AB and the circle gradually comes nearer and nearer to the common point P. Ultimately, it coincides with the point P in the position A'B' of A"B". Again we note, that if 'AB' is rotated rightwards about P, the common point R_3 gradually comes nearer and nearer to P and ultimately coincides with P. So, what we see is :

The tangent to a circle is a special case of the secant, when the two end points of its corresponding chord coincide.

Activity 2. On a paper, draw a circle and a secant PQ of the circle. Draw various lines parallel to the secant on both sides of it. We will find that after some steps, the length of the chord cut by the lines will gradually decrease, *i.e.*, the two points of intersection of the line and the circle are coming closer and closer [see figure (*ii*)]. In one case, it becomes zero on one side of the secant and in another case, it becomes zero on the other side of the secant. See the positions P'Q' and P"Q" of the secant in figure (*ii*). These are the tangents to the circle parallel to the given secant PQ. This also helps us to see that there cannot be more than two tangents parallel to a given secant.

This activity also establishes that a tangent is the secant when both of the end points of the corresponding chord coincide.

The common point of the tangent and the circle is called the **point of contact** and the tangent is said to **touch** the circle at the common point.

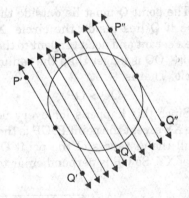

6. Moving bicycle. Let us look around us. We have seen a moving bicycle or a cart. If we look at its wheels we find that all the spokes of a wheel are along its radii. Now let us note the position of the wheel with respect to its movement on the ground. (see figure). In fact, the wheel moves along a line which is a tangent to the circle representing the wheel. Also, notice that in all positions, the radius through the point of contact with the ground appears to be at right angles to the tangent (see figure).

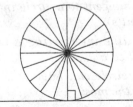

7. Theorem. The tangent at any point of a circle is perpendicular to the radius through the point of contact.

Given. A circle with centre O and a tangent XY to the circle at a point P.

To prove. OP is perpendicular to XY.

Construction. Take a point Q to XY other than P and join OQ.

Proof.

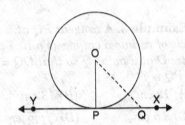

The point Q must lie outside the circle because if Q lies inside the circle, XY will become a secant and not a tangent to the circle. Therefore, OQ is longer than the radius OP cf the circle. That is,

OQ > OP.

Since this happens for every point on the line XY except the point P, OP is the shortest of all the distances of the point O to the points of XY. So OP is perpendicular to XY.

[Out of all the line segments, drawn from a point to points of a line not passing through the point, the smallest is the perpendicular to the line].

Remarks : 1. By theorem above, we can also conclude that at any point on a circle there can be one and only one tangent.

2. The line containing the radius through the point of contact is also sometimes called the 'normal' to the circle at the point.

ILLUSTRATIVE EXAMPLES

[NCERT Exercise 10.1]
(Page No. 209)

Example 1. *How many tangents can a circle have ?*

Sol. A circle can have infinitely many tangents since there are infinitely many points on a circle and at each point of it, it has a unique tangent.

Example 2. *Fill in the blanks:*

(i) A tangent to a circle intersects it in .... point(s).

(ii) A line intersecting a circle in two points is called a ....

(iii) A circle can have .... parallel tangents at the most.

(iv) The common point of a tangent to a circle and the circle is called ....

Sol. (*i*) A tangent to a circle intersects it in **one** point.

(*ii*) A line intersecting a circle in two points is called a **secant.**

(*iii*) A circle can have **two** parallel tangents at the most.

(*iv*) The common point of a tangent to a circle and the circle is called **point of contact.**

Example 3. *A tangent PQ at a point P of a circle of radius 5 cm meets a line through the centre O at a point Q so that OQ = 12 cm. Length PQ is :*

(A) *12 cm*   (B) *13 cm*

(C) *8.5 cm*   (D) $\sqrt{119}$ *cm.*

Sol. ∵ PQ is the tangent and OP is the radius through the point of contact.

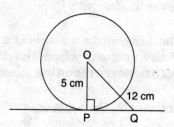

∴ ∠OPQ = 90°

| The tangent at any point of a circle is perpendicular to the radius through the point of contact

∴ By Pythagoras theorem in right △OPQ,

$OQ^2 = OP^2 + PQ^2$

⇒ $(12)^2 = (5)^2 + PQ^2$

⇒ $144 = 25 + PQ^2$

⇒ $PQ^2 = 144 - 25$

⇒ $PQ^2 = 119$

⇒ $PQ = \sqrt{119}$ cm

Hence, the length PQ is $\sqrt{119}$ cm.

So, the correct choice is (D) $\sqrt{119}$ cm.

Example 4. *Draw a circle and two lines parallel to a given line such that one is a tangent and other a secant to the circle.*

Sol.

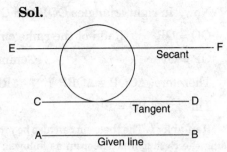

Number of tangents from a point on a circle

1. An activity. To get an idea of the number of tangents from a point to a circle, we perform the following activity :

Activity. Draw a circle on a paper. Take a point P inside it. We find that all the lines through this point intersect the circle in two points. So, it is not possible to draw any tangent to a circle to a point inside it (see figure (*i*)].

Next take a point P on the circle and draw tangents through this point. We have already observed that there is only one tangent to the circle at such a point [see figure (*ii*)].

Finally, take a point P outside the circle and try to draw tangents to the circle. We find that we can draw exactly two tangents to the circle through this point [see figure (*iii*)].

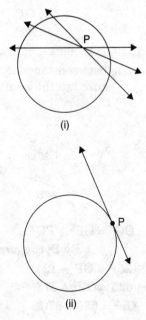

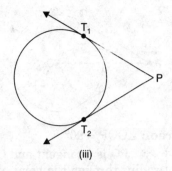

2. Summary. We can summarise the above facts as follows :

Case 1. There is no tangent to a circle passing through a point lying inside the circle.

Case 2. There is one and only one tangent to a circle passing through a point lying on the circle.

Case 3. There are exactly two tangents to a circle through a point lying outside the circle.

3. Length of the tangent. In figure (*iii*), T_1 and T_2 are the points of contact of the tangents PT_1 and PT_2 respectively.

The length of the segment of the tangent from the external point P to the point of contact with the circle is called the **length of the tangent** from the point P to the circle.

We note that in figure (*iii*), PT_1 and PT_2 are the lengths of the tangents from P to the circle. The lengths PT_1 and PT_2 have a common property. If we measure PT_1 and PT_2 we find that these are equal. In fact, this is always so.

4. Theorem. The length of tangents drawn from an external point to a circle are equal.

Given: A circle with centre O, a point P lying out side the circle and the two tangents PQ, PR on the circle from P.

To prove: PQ = PR

Construction: Join OP, OQ and OR.

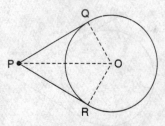

Proof: ∠OQP = 90°

∵ PQ is a tangent and OP is the radius through the point of contact (the tangent at any point of a circle is perpendicular to the radius through the point of contact)

Similarly,
∠ORP = 90°

∵ PR is a tangent and OR is the radius through the point of contact

Now, In right triangles OQP and ORP,

OQ = OR   (Radii of the same circle)

OP = OP   (Common)

Therefore, ΔOQP ≅ ΔORP   (RHS)

This gives PQ = PR   (CPCT)

Remark : 1. The theorem can also be proved by using the Pythagoras Theorem as follows :

$PQ^2 = OP^2 - OQ^2$
$= OP^2 - OR^2$
$= PR^2$ (As OQ = OR)

which gives   PQ = PR.

2. Note also that ∠OPQ = ∠OPR (CPCT). Therefore, OP is the angle bisector of ∠QPR, i.e., the centre lies on the bisector of the angle between the two tangents.

ILLUSTRATIVE EXAMPLES

[NCERT Exercise 10.2]
(Page No. 213)

Example. *In Q. 1 to 3, choose the correct option and give justification.*

1. From a point Q, the length of the tangent to a circle is 24 cm and the distance of Q from the centre is 25 cm. The radius of the circle is
(A) 7 cm   (B) 12 cm
(C) 15 cm   (D) 24.5 cm

2. In figure, if TP and TQ are the two tangents to a circle with centre O so that ∠POQ = 110°, then ∠PTQ is equal to
(A) 60°   (B) 70°
(C) 80°   (D) 90°

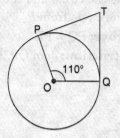

3. If tangents PA and PB from a point P to a circle with centre O are inclined to each other at angle of 80°, then ∠POA is equal to
(A) 50°   (B) 60°
(C) 70°   (D) 80°

Sol. 1. ∠OPQ = 90°

Angle between tangent and radius through the point of contact

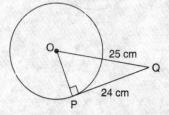

∴ $OQ^2 = OP^2 + PQ^2$

| By Pythagoras theorem

⇒ $(25)^2 = OP^2 + (24)^2$

⇒ $625 = OP^2 + 576$

⇒ $OP^2 = 625 - 576$

$\Rightarrow$ $OP^2 = 49$

$\Rightarrow$ $OP = \sqrt{49} = 7$ cm

$\Rightarrow$ Radius of the circle is 7 cm

Hence, the correct option is **(A) 7 cm.**

2. $\angle POQ = 110°$

$\angle OPT = 90°$

| Angle between tangent and radius through the point of contact

$\angle OQT = 90°$

| Angle between tangent and radius through the point of contact

In quadrilateral OPTQ,

$\angle POQ + \angle OPT + \angle OQT + \angle PTQ$
$= 360°$

| ∵ The sum of all the angles of a quadrilateral is 360°

$\Rightarrow$ $110° + 90° + 90° + \angle PTQ = 360°$

$\Rightarrow$ $290° + \angle PTQ = 360°$

$\Rightarrow$ $\angle PTQ = 360° - 290° = 70°$

Hence, the correct choice is **(B) 70°.**

3. $\angle OAP = 90°$

| Angle between tangent and radius through the point of contact

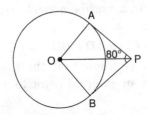

$\angle OPA = \dfrac{1}{2} \angle BPQ$

| The centre lies on the bisector of the angle between the two tangents

$= \dfrac{1}{2}(80°) = 40°$

In $\triangle$ OPQ,

$\angle OAP + \angle OPA + \angle POA = 180°$

| ∵ The sum of the three angle of a triangle is 180°

$\Rightarrow$ $90° + 40° + \angle POA = 180°$

$\Rightarrow$ $130° + \angle POA = 180°$

$\Rightarrow$ $\angle POA = 180° - 130°$

$\Rightarrow$ $\angle POA = 50°$

Hence, the correct choice is **(A) 50°.**

Example 4. *Prove that the tangents drawn at the end of a diameter of a circle are parallel.*

Sol. Given: PQ is a diameter of a circle with centre O. The lines AB and CD are the tangents at P and Q respectively.

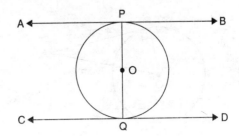

To prove: AB ∥ CD

Proof: ∵ AB is a tangent to the circle at P and OP is the radius through the point of contact

∴ $\angle OPA = 90°$ ...(1)

| The tangent at any point of a circle is perpendicular to the radius through the point of contact

∵ CD is a tangent to the circle at Q and OQ is the radius through the point of contact

∴ $\angle OQD = 90°$ ...(2)

| The tangent at any point of a circle is perpendicular to the radius through the point of contact

From (1) and (2),

$\angle OPA = \angle OQD$

But these form a pair of equal alternate angles

∴ AB ∥ CD.

Example 5. *Prove that the perpendicular at the point of contact to the tangent to a circle passes through the centre.*

Sol. We know that the tangent at any point of a circle is perpendicular to the radius through the point of contact and the radius essentially passes through the centre of the circle, therefore the perpendicular at the point of contact to the tangent to a circle passes through the centre.

Example 6. *The length of a tangent from a point A at distance 5 cm from the centre of the circle is 4 cm. Find the radius of the circle.*

Sol. We know that the tangent at any point of a circle is perpendicular to the radius through the point of contact.

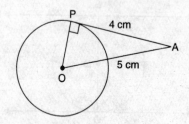

Therefore, $\angle OPA = 90°$

$\therefore \quad OA^2 = OP^2 + AP^2$

| By Pythagoras theorem

$\Rightarrow \quad (5)^2 = OP^2 + (4)^2$

$\Rightarrow \quad 25 = OP^2 + 16$

$\Rightarrow \quad OP^2 = 25 - 16 = 9$

$\Rightarrow \quad OP = \sqrt{9} = 3$ cm

Hence, the radius of the circle is 3 cm.

Example 7. *Two concentric circles are of radii 5 cm and 3 cm. Find the length of the chord of the larger circle which touches the smaller circle.*

Sol. Let O be the common centre of the two concentric circles.

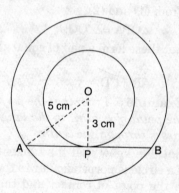

Let AB be a chord of the larger circle which touches the smaller circle at P.

Join OP and OA

Then, $\angle OPA = 90°$

∵ The tangent at any point of a circle is perpendicular to the radius through the point of contact

$\therefore \quad OA^2 = OP^2 + AP^2$

| By Pythagoras theorem

$\Rightarrow \quad (5)^2 = (3)^2 + AP^2$

$\Rightarrow \quad 25 = 9 + AP^2$

$\Rightarrow \quad AP^2 = 25 - 9$

$\Rightarrow \quad AP^2 = 16$

$\Rightarrow \quad AP = \sqrt{16} = 4$ cm

Since the perpendicular from the centre of a circle to a chord bisects the chord, therefore,

$\quad AP = BP = (4$ cm$)$

$\therefore \quad AB = AP + BP$

$\quad = AP + AP = 2\ AP$

$\quad = 2(4) = 8$ cm

Hence, the required length is 8 cm.

Example 8. *A quadrilateral ABCD is drawn to circumscribe a circle (see figure). Prove that*

$AB + CD = AD + BC$

Sol. We know that the tangent segments from an external point to a circle are equal

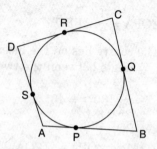

$\therefore \quad AP = AS \qquad \qquad ...(1)$

$\quad BP = BQ \qquad \qquad ...(2)$

$\quad CR = CQ \qquad \qquad ...(3)$

$\quad DR = DS \qquad \qquad ...(4)$

Adding (1), (2), (3) and (4), we get
(AP + BP) + (CR + DR)
= (AS + BQ + CQ + DS)
⇒ AB + CD = (AS + DS) + (BQ + CQ)
⇒ AB + CD = AD + BC.

Example 9. *In figure, XY and X'Y' are two parallel tangents to a circle with centre O and another tangent AB with point of contact C intersecting XY at A and X'Y' at B. Prove that* $\angle AOB = 90°$.

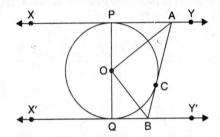

Sol. Given: In figure. XY and X'Y' are two parallel tangents to a circle with centre O and another tangent AB with point of contact C intersecting XY at A and X'Y' at B.

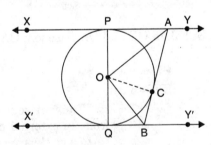

To prove: $\angle AOB = 90°$
Construction: Join OC
Proof: $\angle OPA = 90°$ ...(1)
| The tangent at any point of a circle is perpendicular to the radius through the point of contact
$\angle OCA = 90°$ ...(2)
| The tangent at any point of a circle is perpendicular to the radius through the point of contact

∴ In right triangles OPA and OCA,
OA = OA | Common
AP = AC
| Tangent segments from an external point to a circle are equal
∴ ΔOPA ≅ ΔOCA
| RHS congruence criterion
∴ $\angle OAP = \angle OAC$ | CPCT
⇒ $\angle OAC = \frac{1}{2} \angle PAB$ ...(3)
Similarly, $\angle OBQ = \angle OBC$
⇒ $\angle OBC = \frac{1}{2} \angle QBA$ ...(4)
∵ XY ∥ X'Y'
and a transversal AB intersects them
∴ $\angle PAB + \angle QBA = 180°$
| ∵ Sum of the consecutive interior angles on the same side of the transversal is 180°
⇒ $\frac{1}{2} \angle PAB + \frac{1}{2} \angle QBA = \frac{1}{2}(180°)$
| Halves of equals are equal
⇒ $\angle OAC + \angle OBC = 90°$ ...(5)
| From (3) and (4)
In Δ AOB,
$\angle OAC + \angle OBC + \angle AOB = 180°$
| Angle from property of a triangle
⇒ $90° + \angle AOB = 180°$
| From (5)
⇒ $\angle AOB = 90°$.

Example 10. *Prove that the angle between the two tangents drawn from an external point to a circle is supplementary to the angle subtended by the line-segment joining the points of contact at the centre.*

Sol. $\angle OAP = 90°$ ...(1)
| Angle between tangent and radius through the point of contact is 90°
$\angle OBP = 90°$ ...(2)
| Angle between tangent and radius through the point of contact is 90°

∴ OAPB is quadrilateral

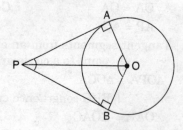

∴ ∠APB + ∠AOB + ∠OAP
   + ∠OBP = 360°
   | Angle from property
     of a quadrilateral
⇒ ∠APB + ∠AOB + 90° + 90° = 360°
   | From (1) and (2)
⇒ ∠APB + ∠AOB = 180°
⇒ ∠APB and ∠AOB and supplementary.

Example 11. *Prove that the parallelogram circumscribing a circle is a rhombus.*

Sol. Given: ABCD is a parallelogram circumscribing a circle

To prove: ABCD is a rhombus

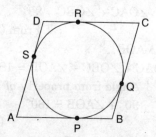

Proof: ∵ The tangent segments from an external point to a circle are equal.
∴   AP = AS
    BP = BQ
    CR = CQ
and DR = DS
⇒ (AP + BP) + (CR + DR)
      = (AS + DS) + (BQ + CQ)
⇒ AB + CD = AD + BC
⇒ AB + AB = AD + AD
   | ∵ Opposite sides of a
     parallelogram are equal

⇒ 2 AB = 2 AD
⇒ AB = AD
But AB = CD and AD = BC
   | Opposite sides of parallelogram
∴ AB = BC = CD = AD
∴ Parallelogram ABCD is a rhombus.

Example 12. *A triangle ABC is drawn to circumscribe a circle of radius 4 cm such that the segments BD and DC into which BC is divided by the point of contact D are of lengths 8 cm and 6 cm respectively (see figure). Find the sides AB and AC.*

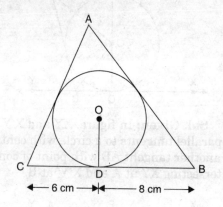

Sol. Join OE and OF. Also join OA, OB and OC
∵ BD = 8 cm
∴ BE = 8 cm
   | ∵ Tangent segments
     from an external point
     to a circle are equal

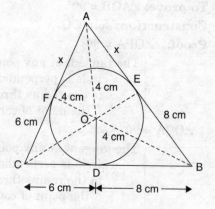

CIRCLES

∵ CD = 6 cm
∴ CF = 6 cm
| ∵ Tangent segments from an external point to a circle are equal
Let AE = AF = x
| ∵ Tangent segments from an external point to a circle are equal
OD = OE = OF = 4 cm
| ∵ Radii of a circle are equal

Semi-perimeter of $\triangle ABC$

$$= \frac{(x+6)+(x+8)+(6+8)}{2}$$

$= (x + 14)$ cm

∴ Area of $\triangle ABC$

$= \sqrt{s(s-a)(s-b)(s-c)}$

$= \sqrt{(x+14)(x+14-14)(x+14-\overline{x+8})(x+14-\overline{x+6})}$

$= \sqrt{(x+14)(x)(6)(8)}$

Now, Area of $\triangle ABC$
 = Area of $\triangle OBC$ + Area of $\triangle OCA$ + Area of $\triangle OAB$

$\Rightarrow \sqrt{(x+14)(x)(6)(8)} = \frac{(6+8)4}{2} + \frac{(x+6)4}{2} + \frac{(x+8)4}{2}$

$\Rightarrow \sqrt{(x+14)(x)(6)(8)}$
$= 28 + 2x + 12 + 2x + 16$

$= \sqrt{(x+14)(x)(6)(8)} = 4x + 56$

$= \sqrt{(x+14)(x)(6)(8)} = 4(x + 14)$

Squaring both sides, we get
$(x+14)(x)(6)(8) = 16(x+14)^2$

$\Rightarrow \quad 3x = x + 14$
$\Rightarrow \quad 3x - x = 14$
$\Rightarrow \quad 2x = 14$
$\Rightarrow \quad x = \frac{14}{2} = 7$

∴ AB = x + 8 = 7 + 8 = 15 cm
and AC = x + 6 = 7 + 6 = 13 cm.

Example 13. *Prove that opposite sides of a quadrilateral circumscribing a circle subtend supplementary angles at the centre of the circle.*

Sol. Given: ABCD is a quadrilateral circumscribing a circle whose centre is O

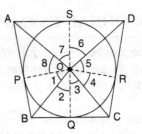

To prove: (i) $\angle AOB + \angle COD = 180°$
(ii) $\angle BOC + \angle AOD = 180°$

Construction: Join OP, OQ, OR and OS

Proof: Since tangent segments from area external point to a circle are equal, therefore

AP = AS
BP = BQ
CQ = CR   ...(1)
DR = DS

In $\triangle OBP$ and $\triangle OBQ$,
OP = OQ | Radii of the same circle
OB = OB | Common circle
BP = BQ | From (1)
∴ $\triangle OBP \cong \triangle OBQ$
       | SSS congruence criterion
∴ $\angle 1 = \angle 2$ | CPCT

Similarly,
$\angle 3 = \angle 4$
$\angle 5 = \angle 6$ ...(2)
and $\angle 7 = \angle 8$

Since the sum of all the angles round a point is equal to 360°, therefore,
∴ $\angle 1 + \angle 2 + \angle 3 + \angle 4 + \angle 5 + \angle 6 + \angle 7 + \angle 8 = 360°$

$\Rightarrow \angle 1 + \angle 1 + \angle 4 + \angle 4 + \angle 5 + \angle 5 + \angle 8 + \angle 8 = 360°$ | From (2)

$\Rightarrow 2(\angle 1 + \angle 4 + \angle 5 + \angle 8) = 360°$

⇒ ∠1 + ∠4 + ∠5 + ∠8 = 180°
⇒ (∠1 + ∠8) + (∠4 + ∠5) = 180°
⇒ ∠AOB + ∠COD = 180°
Similarly, we can prove that
∠BOC + ∠AOD = 180°.

ADDITIONAL EXAMPLES

Example 14. *Prove that the line-segment joining the points of contact of two parallel tangents to a circle is a diameter of the circle.*

Sol. Let PQ and RS be two parallel tangents of a circle where centre is O. Let their points of contact be A and B respectively. Join OA and OB.

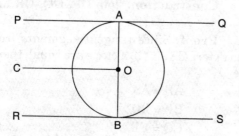

Draw OC ∥ PQ or RS
Now, PQ ∥ OC
and transversal AO intersects them
∴ ∠PAO + ∠COA = 180°
| ∵ The sum of consecutive interior angles on the same side of the transversal is 180°
⇒ 90° + ∠COA = 180°
| ∵ PA is a tangent and OA is the radius through the point of contact
∴ ∠PAO = 90° (the tangent of any point of a circle is perpendicular to the radius through the property contact)
⇒ ∠COA = 90°
Similarly, ∠COB = 90°
∴ ∠COA + ∠COB = 180°
⇒ AB is a straight line through O.

Example 15. *Prove that the tangents drawn at the end of a chord of a circle make equal angles with the chord.*

Sol. Let AB be a chord of a circle whose centre is O. Let PA and PB be tangents to the circle at A and B respectively. Then,

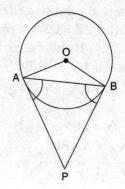

PA = PB
| ∵ Tangent segments from an external point to a circle are equal
In ΔPAB,
∠PAB = ∠PBA
| Angles opposite to equal sides of a triangle are equal

Example 16. *Two circles touch internally at a point P and from a point T on the common tangent at P, tangent segments TQ, TR are drawn to the two circles. Prove that TQ = TR.*

Sol. Given: Two circles touch internally at a point P.

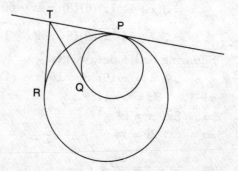

From a point T on the common tangent at P, tangent segments TQ, TR are drawn in the two circles.

CIRCLES

To prove: TQ = TR

Proof: Since the tangent segments from an external point to a circle are equal

∴   TQ = TP   ...(1)

and,   TR = TP   ...(2)

From (1) and (2), we get

TQ = TR.

Example 17. *A circle touches the side BC of a △ABC at P and AB and AC produced at Q and R respectively. Prove that AQ is half the perimeter of △ABC.*

Sol. Given: A circle touches the side BC of a △ABC at P and AB and AC produced at Q and R respectively.

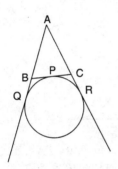

To prove: $AQ = \frac{1}{2}$ perimeter of △ABC.

Proof: ∵ Length of the two tangents from an external point to a circle are equal

∴   AQ = AR   ...(1)

     BQ = BP   ...(2)

     CP = CR   ...(3)

∴ Perimeter of △ABC

= AB + BC + AC

= AB + (BP + CP) + AC

= (AB + BQ) + (CR + AC)

             | Using (2) and (3)

= AQ + AR

= AQ – AR

= AQ + AQ | From (1)

= 2AQ

⇒ $AQ = \frac{1}{2}$ perimeter of △ABC.

TEST YOUR KNOWLEDGE

1. Prove that in two concentric circles, the chord of the larger circle, which touches the smaller circle, is bisected at the point of contact.

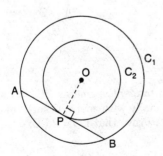

2. Two tangents TP and TQ are drawn to a circle with centre O from an external point T. Prove that ∠PTQ = 2∠OPQ.

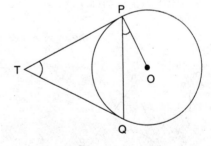

3. PQ is a chord of length 8 cm of a circle of radius 5 cm. The tangents at P and Q intersect at a point T (see figure). Find the length TP.

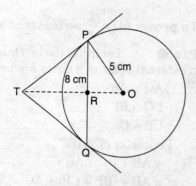

4. The radii of two concentric circles are 13 cm and 8 cm. AB is a diameter of the bigger circle. BD is a tangent to the smaller circle touching it at D. Find the length AD.

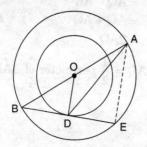

[**Hint.** Let line BD intersect the bigger circle at E. Join AE. AE = 2 × 8 = 16 cm].

5. The in circle of a $\triangle ABC$ touches the sides AB, BC and CA at the points F, D and E respectively. Show that

   AF + BD + CE = FB + DC + EA

   $= \frac{1}{2}$ perimeter of $\triangle ABC$.

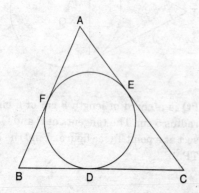

6. From an external point A, two tangents are drawn to the circle with centre O. Prove that OA is the perpendicular bisector of BC.

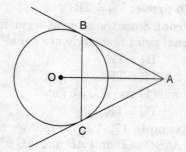

7. In the given figure, prove that BD = DC if AB = AC.

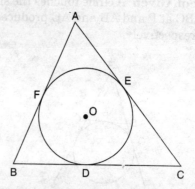

8. In the given figure, PA and PB are tangents from P to the circle with centre O. At M, a tangent is drawn cutting PA at K and PB at N. Prove that

   KN = AK + BN

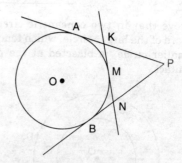

9. Two tangent segments BC, BD are drawn to a circle with centre O such that $\angle DBC = 120°$. Prove that BO = 2 BC.

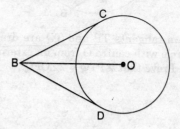

10. From a point P, two tangents PA and PB are drawn to a circle with centre O. If the distance between P and O is equal to the diameter of the circle, show that △APB is equilateral.

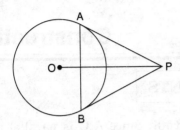

Answers

3. $\dfrac{20}{3}$ cm   4. 19 cm.

SUMMARY

1. The meaning of a tangent to a circle.
2. The tangent to a circle is perpendicular to the radius through the point of contact.
3. The lengths of the two tangents from an external point to a circle are equal.

11

Constructions

IMPORTANT POINTS

1. Division of a line segment. Suppose we are given a line segment and we have to divide it in a given ratio, say 3 : 2. We may do it by measuring the length and then marking a point on it that divides it in the given ratio. But suppose we do not have any way of measuring it precisely. Then, we have two ways for finding such a point. These are given below.

2. Construction 1. *To divide a line segment in a given ratio*

Given a line segment AB. We want to divide it in the ratio $m : n$, where both m and n are positive integers. In particular, we take $m = 3$ and $n = 2$.

Steps of Construction:

1. Draw any ray AX, making an acute angle with AB.

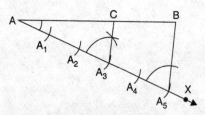

2. Locate 5 ($= m + n$) points A_1, A_2, A_3, A_4 and A_5 on AX so that $AA_1 = A_1A_2 = A_2A_3 = A_3A_4 = A_4A_5$.
3. Join BA_5.
4. Through the point A_3 ($m = 3$), draw a line parallel to A_5B (by making an angle equal to $\angle AA_5B$ at A_3) intersecting AB at the point C (see figure). Then,
 $AC : CB = 3 : 2$.

Proof: Since A_3C is parallel to A_5B, therefore,

$$\frac{AA_3}{A_3A_5} = \frac{AC}{CB}$$

(By the basic Proportionality Theorem)

By construction, $\dfrac{AA_3}{A_3A_5} = \dfrac{3}{2}$. Therefore,

$$\frac{AC}{CB} = \frac{3}{2}.$$

This shows that C divides AB in the ratio 3 : 2.

Alternative Method

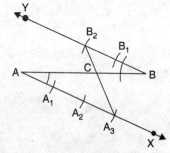

Steps of Construction:

1. Draw any ray AX making an acute angle with AB.
2. Draw a ray BY parallel to AX by making $\angle ABY$ equal to $\angle BAX$.
3. Locate the points A_1, A_2, A_3 ($m = 3$) on AX and B_1, B_2 ($n = 2$) on BY such that $AA_1 = A_1A_2 = A_2A_3 = BB_1 = B_1B_2$.
4. Join A_3B_2. Let it intersect AB at a point C (see figure).

Then AC : CB = 3 : 2.

Proof: ∵ △AA₃C is similar to △BB₂C

| AA similarity criterion

∴ $\dfrac{AA_3}{BB_2} = \dfrac{AC}{BC}$

| By the Basic Proportionality theorem

Since by construction,

$\dfrac{AA_3}{BB_2} = \dfrac{3}{2}$,

Therefore, $\dfrac{AC}{BC} = \dfrac{3}{2}$.

Note. In fact, the methods given above work for dividing the line segment in any ratio.

Construction 2. *To construct a triangle similar to a given triangle as per given scale factor.*

This construction involves two different situations :

(i) The triangle to be constructed is smaller than the given triangle.

(ii) The triangle to be constructed is larger than the given triangle.

Scale Factor. Scale factor means the ratio of the sides of the triangle to be constructed with the corresponding sides of the given triangle.

ILLUSTRATIVE EXAMPLES

Example 1. *Construct a triangle similar to a given triangle ABC with its sides equal to $\dfrac{3}{4}$ of the corresponding sides of the triangle ABC $\left(i.e., \text{ of scale factor } \dfrac{3}{4}\right)$.*

Sol. Given a triangle ABC, we are required to construct another triangle whose sides are $\dfrac{3}{4}$ of the corresponding sides of the triangle ABC.

Steps of Construction:

1. Draw any ray BX making an acute angle with BC on the side opposite to the vertex A.

2. Locate 4 $\left(\text{the greater of 3 and 4 in } \dfrac{3}{4}\right)$ points B_1, B_2, B_3 and B_4 on BX so that $BB_1 = B_1B_2 = B_2B_3 = B_3B_4$.

3. Join B_4C and draw a line through B_3 $\left(\text{the 3rd point, 3 being smaller of 3 and 4 in } \dfrac{3}{4}\right)$ parallel to B_4C to intersect BC at C'.

4. Draw a line through C' parallel to the line CA to intersect BA at A' (see figure).

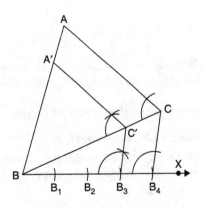

Then, $\triangle A'BC'$ is the required triangle.
Proof: By construction 1,

$$\frac{BC'}{C'C} = \frac{3}{1} \qquad ...(1)$$

$$\therefore \frac{BC}{BC'} = \frac{BC' + C'C}{BC'} = \frac{BC'}{BC'} + \frac{C'C}{BC'}$$

$$= 1 + \frac{C'C}{BC'}$$

$$= 1 + \frac{1}{3} \qquad | \text{ From (1)}$$

$$= \frac{4}{3}$$

$$\Rightarrow \frac{BC'}{BC} = \frac{3}{4} \qquad ...(2)$$

$\because \quad C'A' \parallel CA$

$\therefore \quad \triangle A'BC' \sim \triangle ABC$

| AA similarity criterion

$$\therefore \frac{A'B}{AB} = \frac{A'C'}{AC} = \frac{BC'}{BC} \left(= \frac{3}{4}\right)$$

| From (2)
| By the Basic Proportionality Theorem

Example 2. *Construct a triangle similar to a given triangle ABC with its sides equal to $\frac{5}{3}$ of the corresponding sides of the triangle ABC $\left(i.e., \text{ of scale factor } \frac{5}{3}\right)$.*

Sol. Given a triangle ABC, we are required to construct a triangle whose sides are $\frac{5}{3}$ of the corresponding sides of $\triangle ABC$.

Steps of Construction:
1. Draw any ray BX making an acute angle with BC on the side opposite to the vertex A.
2. Locate 5 points $\Big($ the greater of 5 and 3 in $\frac{5}{3}\Big)$ B_1, B_2, B_3, B_4 and B_5 on BX so that $BB_1 = B_1B_2 = B_2B_3 = B_3B_4 = B_4B_5$.
3. Join $B_3 \Big($ the 3rd point, 3 being smaller of 3 and 5 in $\frac{5}{3}\Big)$ to C and draw a line through B_5 parallel to B_3C, intersecting the extended line segment BC at C'.
4. Draw a line through C' parallel to CA intersecting the extended line segment BA at A' (see figure).

Then A'BC' is the required triangle.

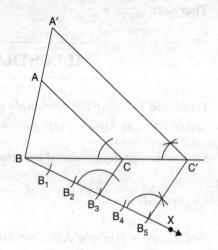

Justification of the construction

$\because \quad C'A' \parallel CA$

$\therefore \quad \triangle ABC \sim \triangle A'BC'$

| AA similarity criterion

$$\therefore \frac{AB}{A'B} = \frac{AC}{A'C'} = \frac{BC}{BC'}$$

| $\because$ Corresponding sides of two similar triangles are proportional

$\because \quad B_5C' \parallel B_3C$

$\therefore \quad \triangle BB_5C' \sim \triangle BB_3C$

| AA similarity criterion

$$\therefore \frac{BC}{BC'} = \frac{BB_3}{BB_5}$$

| By Basic Proportionality Theorem

CONSTRUCTIONS

But $\dfrac{BB_3}{BB_5} = \dfrac{3}{5}$ (by construction)

$\therefore \quad \dfrac{BC}{BC'} = \dfrac{3}{5}$

$\therefore \quad \dfrac{BC'}{BC} = \dfrac{5}{3}$

and, therefore,

$\dfrac{A'B}{AB} = \dfrac{A'C'}{AC} = \dfrac{BC'}{BC} = \dfrac{5}{3}$

Remark 1. In Examples 1 and 2, we could take a ray making an acute angle with AB or AC and proceed similarly.

Remark 2. The same methods would apply for the general case also.

ILLUSTRATIVE EXAMPLES

[NCERT Exercise 11.1]
(Page No. 219)

In each of the following, give the justification of the construction also :

Example 1. *Draw a line segment of length 7.6 cm and divide it in the ratio 5 : 8. Measure the two parts.*

Sol. Given: A line segment of length 7.6 cm.

Required: To divide it in the ratio 5 : 8 and to measure the two parts.

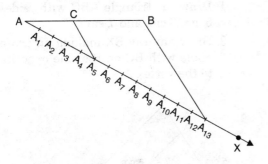

Steps of Construction:

1. From any ray AX, making an acute angle with AB.

2. Locate 13(= 5 + 8) points $A_1, A_2, A_3, A_4, A_5, A_6, A_7, A_8, A_9, A_{10}, A_{11}, A_{12}$ and A_{13}, on AX such that
$AA_1 = A_1A_2 = A_2A_3 = A_3A_4$
$= A_4A_5 = A_5A_6 = A_6A_7$
$= A_7A_8 = A_8A_9 = A_9A_{10}$
$= A_{10}A_{11} = A_{11}A_{12} = A_{12}A_{13}$

3. Join BA_{13}.

4. Through the point A_5, draw a line parallel to $A_{13}B$ intersecting AB at the point C.

Then, AC : CB = 5 : 8

On measurement, AC = 3.1 cm, CB = 4.5 cm.

Justification:

$\because \quad A_5C \parallel A_{13}B$ | By Construction

$\therefore \quad \dfrac{AA_5}{A_5A_{13}} = \dfrac{AC}{CB}$

| By the Basic Proportionality Theorem

But $\dfrac{AA_5}{A_5A_{13}} = \dfrac{5}{8}$ | By construction

Therefore, $\dfrac{AC}{CB} = \dfrac{5}{8}$

This shows that C divides AB in the ratio 5 : 8.

Example 2. *Construct a triangle of sides 4 cm, 5 cm and 6 cm and then a triangle similar to it whose sides are $\dfrac{2}{3}$ of the corresponding sides of the first triangle.*

Sol. Required: To construct a triangle of sides 4 cm, 5 cm and 6 cm and then a triangle similar to it whose sides are $\dfrac{2}{3}$ of the corresponding sides of the first triangle.

Steps of Construction:
1. Draw a triangle ABC of sides 4 cm, 5 cm and 6 cm.
2. Draw any ray BX making an acute angle with BC on the side opposite to the vertex A.
3. Locate 3 points B_1, B_2 and B_3 on BX such that $BB_1 = B_1B_2 = B_2B_3$.

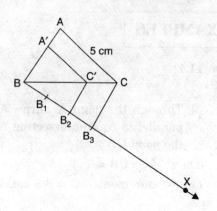

4. Join B_3C and draw a line through B_2 parallel to B_3C intersecting BC at C'.
5. Draw a line through C' parallel to the line CA to intersect BA at A'.

Then, $\triangle A'BC'$ is the required triangle.

Justification:

$\because$ $B_3C \parallel B_2C'$ | By construction

$\therefore$ $\dfrac{BB_2}{B_2B_3} = \dfrac{BC'}{C'C}$

| By the Basic Proportionality Theorem

But $\dfrac{BB_2}{B_2B_3} = \dfrac{2}{1}$ | By construction

$\therefore$ $\dfrac{BC'}{C'C} = \dfrac{2}{1}$

$\therefore$ $\dfrac{C'C}{BC'} = \dfrac{1}{2}$

$\Rightarrow$ $\dfrac{C'C}{BC'} + 1 = \dfrac{1}{2} + 1$

$\Rightarrow$ $\dfrac{C'C + BC'}{BC'} = \dfrac{1+2}{2}$

$\Rightarrow$ $\dfrac{BC}{BC'} = \dfrac{3}{2}$

$\Rightarrow$ $\dfrac{BC'}{BC} = \dfrac{2}{3}$ ...(1)

$\because$ $CA \parallel C'A'$ | By construction

$\therefore$ $\triangle BC'A' \sim \triangle BCA$

| AA similarity criterion

$\therefore$ $\dfrac{A'B}{AB} = \dfrac{A'C'}{AC} = \dfrac{BC'}{BC} \left(= \dfrac{2}{3}\right)$

| From (1)
| By the Basic Proportionality Theorem

Example 3. *Construct a triangle with sides 5 cm, 6 cm and 7 cm and then another triangle whose sides are $\dfrac{7}{5}$ of the corresponding sides of the first triangle.*

Sol. Required: To construct a triangle with sides 5 cm, 6 cm and 7 cm and then another triangle whose sides are $\dfrac{7}{5}$ of the corresponding sides of the first triangle.

Steps of Construction:
1. Draw a triangle ABC with sides 5 cm, 6 cm, and 7 cm.
2. Draw any ray BX making an acute angle with BC on the side opposite to the vertex A.

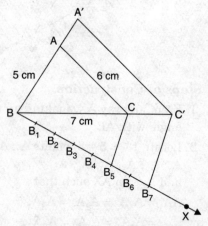

3. Locate 7 points B_1, B_2, B_3, B_4, B_5, B_6 and B_7 on BX such that $BB_1 = B_1B_2 = B_2B_3 = B_3B_4 = B_4B_5 = B_5B_6 = B_6B_7$.

4. Join B_5 to C and draw a line through B_7 parallel to B_5 C, intersecting the extended line segment BC at C'.

5. Draw a line through C' parallel to CA intersecting the extended line segment BA at A'.

Then, $\triangle$ A'BC' is the required triangle.

Justification:

$\because$ C'A' $\parallel$ CA | By construction

$\therefore$ $\triangle$ ABC ~ $\triangle$ A'BC'

          | AA similarity criterion

$\therefore \quad \dfrac{A'B}{AB} = \dfrac{A'C'}{AC} = \dfrac{BC'}{BC}$

    $\because$ Corresponding sides of two similar triangles are proportional

$\because \quad B_7 C' \parallel B_5 C$

          | By construction

$\therefore \quad \triangle BB_7 C' \sim \triangle BB_5 C$

          | AA similarity criterion

$\therefore \quad \dfrac{BC}{BC'} = \dfrac{BB_5}{BB_7}$

    | By Basic Proportionality Theorem

But $\quad \dfrac{BB_5}{BB_7} = \dfrac{5}{7}$ | By construction

$\therefore \quad \dfrac{BC}{BC'} = \dfrac{5}{7}$

$\therefore \quad \dfrac{BC'}{BC} = \dfrac{7}{5}$

$\therefore \quad \dfrac{A'B}{AB} = \dfrac{A'C'}{AC} = \dfrac{BC'}{BC} = \dfrac{7}{5}.$

Example 4. *Construct an isosceles triangle whose base is 8 cm and altitude 4 cm and then another triangle whose sides are* $1\dfrac{1}{2}$ *times the corresponding sides of the isosceles triangle.*

Sol. Required: To construct an isosceles triangle whose base is 8 cm and altitude 4 cm and then another triangle whose sides are $1\dfrac{1}{2}$ $\left(\text{or } \dfrac{3}{2}\right)$ times the corresponding sides of the isosceles triangle.

Steps of Construction:

1. Draw BC = 8 cm.
2. Draw perpendicular bisector of BC. Let it meet BC at D.
3. Mark a point A on the perpendicular bisector such that AD = 4 cm.
4. Join AB and AC.

Then $\triangle$ ABC is the required isosceles triangle.

5. Draw any ray BX making an acute angle with BC on the side opposite to the vertex A.
6. Locate 3 points B_1, B_2 and B_3 on BX such that $BB_1 = B_1 B_2 = B_2 B_3$.
7. Join B_2 to C and draw a line through B_3 parallel to $B_2 C$, intersecting the extended line segment BC at C'.

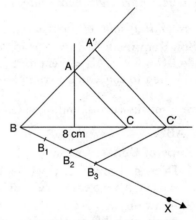

8. Draw a line through C' parallel to CA intersecting the extended line segment BA at A'.

Then, A'BC' is the required triangle.

Justification:

$\because$ C'A' $\parallel$ CA | By construction

$\therefore$ $\triangle$ ABC ~ $\triangle$ A'BC'

          | AA similarity criterion

$\therefore \quad \dfrac{A'B}{AB} = \dfrac{A'C'}{AC} = \dfrac{BC'}{BC}$

    $\because$ Corresponding sides of two similar triangles are proportional

$\because \quad B_3 C' \parallel B_2 C$

          | By construction

∴ Δ BB₃C' ~ Δ BB₂C

　　　　| AA similarly criterion

∴ $\dfrac{BC'}{BC} = \dfrac{BB_3}{BB_2}$

　　　　| By the Basic Proportionality Theorem

But $\dfrac{BB_3}{BB_2} = \dfrac{3}{2}$　　　| By construction

∴ $\dfrac{BC'}{BC} = \dfrac{3}{2}$

∴ $\dfrac{A'B}{AB} = \dfrac{A'C'}{AC} = \dfrac{BC'}{BC} = \dfrac{3}{2}$.

Example 5. *Draw a triangle ABC with side BC = 6 cm, AB = 5 cm and ∠ ABC = 60°. Then construct a triangle whose sides are $\dfrac{3}{4}$ of the corresponding sides of the triangle ABC.*

Sol. Required: To draw a triangle ABC with side BC = 6 cm, AB = 5 cm and ∠ ABC = 60° and then to construct a triangle whose sides are $\dfrac{3}{4}$ of the corresponding sides of the triangle ABC.

Steps of Construction:

1. Draw a triangle ABC with side BC = 6 cm, AB = 5 cm and ∠ ABC = 60°.
2. Draw any ray BX making an acute angle with BC on the side opposite to the vertex A.

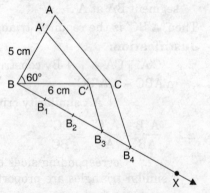

3. Locate 4 points B₁, B₂, B₃ and B₄ on BX such that BB₁ = B₁B₂ = B₂B₃ = B₃B₄.

4. Join B₄C and draw a line through B₃ parallel to B₄C intersecting BC to C'.
5. Draw a line through C' parallel to the line CA to intersect BA at A'.

Then, Δ A'BC' is the required triangle.

Justification:

∵　　B₄C ∥ B₃C'　　　| By construction

∴ $\dfrac{BB_3}{BB_4} = \dfrac{BC'}{BC}$

　　　　| By the Basic Proportionality Theorem

But $\dfrac{BB_3}{BB_4} = \dfrac{3}{4}$　　　| By construction

∴ $\dfrac{BC'}{BC} = \dfrac{3}{4}$　　　...(1)

∵　　CA ∥ C'A'　　　| By construction

∴ Δ BC'A' ~ Δ BCA

　　　　| AA similarity criterion

∴ $\dfrac{A'B}{AB} = \dfrac{A'C}{AC} = \dfrac{BC'}{BC} = \dfrac{3}{4}$.

　　　　| From (1)
　　　　| By the Basic Proportionality Theorem

Example 6. *Draw a triangle ABC with side BC = 7 cm, ∠B = 45°, ∠A = 105°. Then, construct a triangle whose sides are $\dfrac{4}{3}$ times the corresponding sides of Δ ABC.*

Required: To draw a triangle ABC with side BC = 7 cm CB = 45°, CA = 105° and then to construct a triangle whose sides are $\dfrac{4}{3}$ times the corresponding sides of ΔABC.

Steps of Construction:

1. Draw a triangle ABC with side BC = 7 cm, ∠B = 45°, ∠A = 105°.
2. Draw any ray BX making an acute angle with BC on the side opposite to the vertex A.
3. Locate 4 points B₁, B₂, B₃ and B₄ on BX such that BB₁ = B₁B₂ = B₂B₃ = B₃B₄.

4. Join B_3 to C and draw a line through B_4 parallel to B_3C, intersecting the extended line segment BC at C'.
5. Draw a line through C' parallel to CA intersecting the extended line segment BA at A'.

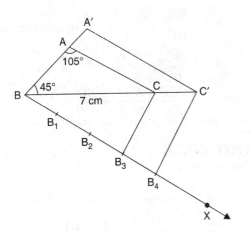

Then, $\triangle$ A'BC' is the required triangle.
Justification:
$\because$ C'A' ∥ CA | By construction
$\therefore$ $\triangle$ ABC ~ $\triangle$ A'BC'
               | AA similarity criterion

$\therefore$ $\dfrac{A'B}{AB} = \dfrac{A'C'}{AC} = \dfrac{BC'}{BC}$

     | $\because$ Corresponding sides of two similar triangles are proportional

$\therefore$ $B_4C' \parallel B_3C$
               | By construction
$\therefore$ $\triangle BB_4C' \sim \triangle BB_3C$
               | AA similarity criterion

$\therefore$ $\dfrac{BC'}{BC} = \dfrac{BB_4}{BB_3}$

     | By the Base Proportionality Theorem

But $\dfrac{BB_4}{BB_3} = \dfrac{4}{3}$ | By construction

$\therefore$ $\dfrac{BC'}{BC} = \dfrac{4}{3}$

$\therefore$ $\dfrac{A'B}{AB} = \dfrac{A'C'}{AC} = \dfrac{BC'}{BC} = \dfrac{4}{3}$

Example 7. *Draw a right triangle in which the sides (other than hypotenuse) are of lengths 4 cm and 3 cm. Then construct another triangle whose sides are $\dfrac{5}{3}$ times the corresponding sides of the given triangle.*

Sol. Required: To draw a right triangle in which the sides (other than hypotenuse) are of lengths 4 cm and 3 cm and then construct another triangle whose sides are $\dfrac{5}{3}$ times the corresponding sides of the given triangle.

Steps of Construction:
1. Draw a right triangle ABC in which the sides (other than hypotenuse) are of lengths 4 cm and 3 cm.

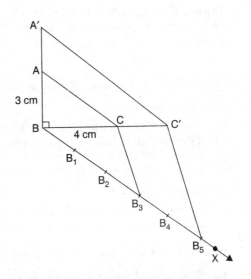

2. Draw any ray BX making an acute angle with BC on the side opposite to the vertex A.
3. Locate 5 points B_1, B_2, B_3, B_4 and B_5 on BX such that $BB_1 = B_1B_2 = B_2B_3 = B_3B_4 = B_4B_5$.
4. Join B_3 to C and draw a line through B_5 parallel to B_3C, intersecting the extended line segment BC at C'.
5. Draw a line through C' parallel to CA intersecting the extended line segment BA at A'.

Then, $\triangle$ A'BC' is the required triangle.

Justification:

$\because$ C'A' ∥ CA

| By construction

$\therefore$ $\triangle$ ABC ~ $\triangle$ A'BC'

| AA similarity criterion

$\Rightarrow \dfrac{A'B}{AB} = \dfrac{A'C'}{AC} = \dfrac{BC'}{BC}$

| $\because$ Corresponding sides of two similar triangles are proportional

$\therefore$ $B_5 C' \parallel B_3 C$

| By construction

$\therefore$ $\triangle BB_5C' \sim \triangle BB_3C$

| AA similarity criterion

$\therefore \dfrac{BC'}{BC} = \dfrac{BB_5}{BB_3}$

| By the Basic Proportionality Theorem

But $\dfrac{BB_5}{BB_3} = \dfrac{5}{3}$

| By construction

$\therefore \dfrac{BC'}{BC} = \dfrac{5}{3}$

$\therefore \dfrac{A'B}{AB} = \dfrac{A'C'}{AC} = \dfrac{BC'}{BC} = \dfrac{5}{3}.$

TEST YOUR KNOWLEDGE

1. Draw a line segment AB of length 10 cm and divide it internally in the ratio 3 : 4.
2. Construct a triangle of sides 3 cm, 4 cm and 5 cm and then a triangle similar to it whose sides are $\dfrac{3}{4}$ of the corresponding sides of the first triangle.
3. Construct a triangle with sides 5 cm, 12 cm and 13 cm and then another triangle whose sides are $\dfrac{8}{5}$ of the corresponding sides of the first triangle.
4. Construct an isosceles triangle whose base is 10 cm and altitude 6 cm and then another triangle whose sides are $\dfrac{4}{3}$ times the corresponding sides of the isosceles triangle.
5. Draw a triangle ABC with side BC = 8 cm, AB = 6 cm and $\angle ABC = 75°$. Then construct a triangle whose sides are $\dfrac{2}{3}$ of the corresponding sides of the triangle ABC.
6. Draw a triangle ABC with side BC = 6 cm, $\angle B = 55°$ and $\angle C = 70°$. Then construct a triangle whose sides are $\dfrac{5}{4}$ times the corresponding sides of $\triangle$ ABC.
7. Draw a right triangle in which the sides (other than hypotenuse) are of length 12 cm and 5 cm. Then construct another triangle whose sides are $\dfrac{3}{2}$ times the corresponding side of the given triangle.
8. Draw an equilateral triangle whose one side is 6 cm and then construct another triangle whose sides are $\dfrac{4}{3}$ times the corresponding sides of the given triangle.

CONSTRUCTION OF TANGENTS TO A CIRCLE

IMPORTANT POINTS

1. Tangent at a point of a circle. We know that if a point lies inside a circle, there cannot be a tangent to the circle through this point. However, if a point lies on the circle, then there is only one tangent to the circle at this point and it is perpendicular to the radius through this point. Therefore, if we want to draw a tangent at a point of a circle, simply draw the radius through this point and draw a line perpendicular to this radius through this point and this will be the required tangent at the point.

Note. If the point lies outside the circle, then there will be two tangents to the circle from this point.

2. Construction 3. *To construct the tangents to a circle from a point outside it.*

We are given a circle with centre O and a point P outside it. We have to construct the two tangents from P to the circle.

Steps of Construction:
1. Join PO and bisect it. Let M be the mid-point of PO.

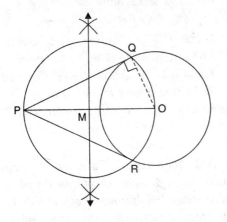

2. Taking M as centre and MO as radius, draw a circle. Let it intersect the given circle at the points Q and R.
3. Join PQ and PR.

Then PQ and PR are the required two tangents (see figure).

Proof: Join OQ. Then ∠PQO is an angle in the semicircle and, therefore,

$$\angle PQO = 90°$$
$$\Rightarrow \quad PQ \perp OQ$$

Since, OQ is a radius of the given circle, PQ has to be a tangent to the circle. Similarly, PR is also a tangent to the circle.

Note. If centre of the circle is not given, we may locate its centre first by taking any two non-parallel chords and then finding the point of intersection of their perpendicular bisectors. Then we can proceed as above.

ILLUSTRATIVE EXAMPLES

[NCERT Exercise 11.2]
(Page No. 221)

In each of the following, give also the justification of the construction :

Example 1. *Draw a circle of radius 6 cm. From a point 10 cm away from its centre, construct the pair of tangents to the circle and measure their lengths.*

Sol. Given: A circle whose centre is O and radius is 6 cm and a point P is 10 cm away from the its centre.

Required: To construct the pair of tangents to the circle and measure their lengths.

Steps of Construction:
1. Join PO and bisect it. Let M be the mid-point of PO.
2. Taking M as centre and MO as radius, draw a circle. Let it intersect the given circle at the points Q and R.
3. Join PQ and PR.

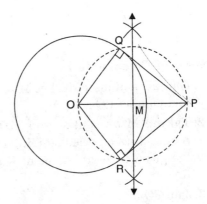

Then PQ and PR are the required two tangents.

By measurement, PQ = PR = 8 cm

Justification: Join OQ and OR.

Since ∠OQP and ∠ORP are the angles in semi-circles.

∴ ∠OQP = 90° = ∠ORP

Also, since OQ, OR are radii of the circle, PQ and PR will be the tangents to the circle at Q and R respectively.

We may see that the circle with OP as diameter intersects the given circle in two points. Therefore, only two tangents can be drawn.

Example 2. *Construct a tangent to a circle of radius 4 cm from a point on the concentric circle of radius 6 cm and measure its length. Also verify the measurement by actual calculation.*

Sol. Required: To construct a tangent to a circle of radius 4 cm from a point on the concentric circle of radius 6 cm and measure its length. Also to verify the measurement by actual calculation.

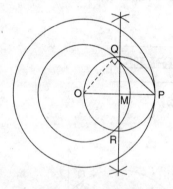

Steps of Construction:

1. Join PO and bisect it. Let M be the mid-point of PO.
2. Taking M as centre and MO as radius, draw a circle. Let it intersect the given circle at the point Q and R.
3. Join PQ

Then PQ is the required tangent.

By measurement, PQ = 4.5 cm

By actual calculation,

$$PQ = \sqrt{OP^2 - OQ^2}$$

| By Pythagoras Theorem

$$= \sqrt{(6)^2 - (4)^2}$$
$$= \sqrt{36 - 16}$$
$$= \sqrt{20}$$
$$= 4.47 \text{ cm.}$$

Justification: Join OQ. Then ∠PQO is an angle in the semicircle and, therefore,

∠PQO = 90°

⇒ PQ ⊥ OQ

Since, OQ is a radius of the given circle, PQ has to be a tangent to the circle.

Example 3. *Draw a circle of radius 3 cm. Take two points P and Q on one of its extended diameter each at a distance of 7 cm from its centre. Draw tangents to the circle from these two points P and Q.*

Sol. Required: To draw a circle of radius 3 cm and take two points P and Q on one of its extended diameter each at a distance of 7 cm from its centre and then draw tangents to the circle from these two points P and Q.

Steps of Construction:

1. Bisect PO. Let M be the mid-point of PO.
2. Taking M as centre and MO as radius, draw a circle. Let it intersect the given circle at the points A and B.
3. Join PA and PB.

Then, PA and PB are the required two tangents.

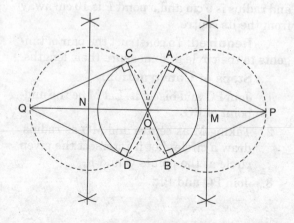

4. Bisect QO. Let N be the mid-point of QO.
5. Taking N as centre and NO as radius, draw a circle. Let it intersect the given circle at the points C and D.
6. Join QC and QD.

Then QC and QD are the required two tangents.

Justification: Join OA and OB.

Then ∠PAO is an angle in the semicircle and, therefore,

∠PAO = 90°

⇒   PA ⊥ OA

Since, OA is a radius of the given circle, PA has to be a tengent to the circle. Similarly, PB is also a tangent to the circle.

Again, Join OC and OD.

Then ∠QCO is an angle on the semicircle and, therefore,

∠QCO = 90°

Since, OC is a radius of the given circle, QC has to be a tangent to the circle.

Similarly, QD is also a tangent to the circle.

Example 4. *Draw a pair of tangents to a circle of radius 5 cm which are inclined to each other at an angle of 60°.*

Sol. Required: To draw a pair of tangents to a circle of radius 5 cm which are inclined to each other at an angle of 60°.

Steps of Construction:
1. Draw a circle of radius 5 cm with centre O.
2. Draw an angle AOB of 120°.
3. At A and B, draw 90° angles which meet at C.

Then, AC and BC are the required tangents which are inclined to each other at an angle of 60°.

Justification:

∵ ∠OAC = 90° | By construction

and OA is a radius

∴ AC is a tangent to the circle.

∵ ∠OBC = 90°

| By construction

and OB is a radius.

∴ BC is a tangent to the circle.

Now, In quadrilateral OACB

∠AOB + ∠OAC + ∠OBC + ∠ACB = 360°

| Angle sum Property of a quadrilateral

⇒ 120° + 90° + 90° + ∠ACB = 360°

⇒   300° + ∠ACB = 360°

⇒   ∠ACB = 360° − 300° = 60°.

Example 5. *Draw a line segment AB of length 8 cm. Taking A as centre, draw a circle of radius 4 cm and taking B as centre, draw another circle of radius 3 cm. Construct tangents to each circle from the centre of the other circle.*

Sol. Required: To draw a line segment of length 8 cm and taking A as centre, to draw a circle of radius 4 cm and taking B as centre, draw another circle of radius 3 cm. Also, to construct tangents to each circle from the centre of the other circle.

Steps of Construction:
1. Bisect BA. Let M be the mid-point of BA.

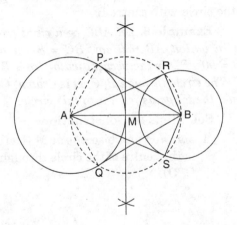

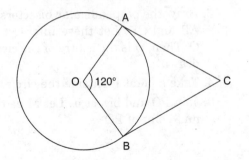

2. Taking M as centre and MA as radius, draw a circle. Let it intersect the circle with centre A at the points P and Q.
3. Join BP and BQ.

Then, BP and BQ are the required two tangents from B to the circle with centre A.

4. Again, let M be the mid-point of AB.
5. Taking M as centre and MB as radius, draw a circle. Let it intersect the circle with center B at the points R and S.
6. Join AR and AS.

Then, AR and AS are the required two tangents from A to the circle with centre B.

Justification: Join BP and BQ.

Then ∠APB, being an angle in the semi-circle, is 90°.

So, BP ⊥ AP.

Since AP is a radius of the circle with centre A, BP has to be a tangent to the circle with centre A. Similarly, BQ is also a tangent to the circle with centre A.

Again, Join AR and AS.

Then ∠ARB = 90°, it being an angle in the semi circle.

So, AR ⊥ BR

Since BR is a radius of the circle with centre B, AR has to be a tangent to the circle with centre B. Similarly, AS is also a tangent to the circle with centre B.

Example 6. *Let ABC be a right triangle in which AB = 6 cm, BC = 8 cm and ∠B = 90°. BD is the perpendicular from B on AC. The circle through B, C, D is drawn. Construct the tangents from A to this circle.*

Sol. Steps of Construction:
1. Join AO and bisect it at M (Here O is the centre of the circle through B, C, D).
2. Taking M as centre and MA as radius, draw a circle. Let it intersect the given circle (circle through B, C, D) at the points B and E.
3. Join AB and AE.

Then, AB and AE are the required two tangents.

Justification: Join OE. Then ∠AEO is an angle in the semicircle and, therefore,

∠AEO = 90°
⇒   AE ⊥ OE

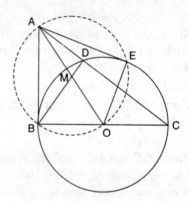

Since, OE is a radius of the given circle, AE has to be a tangent to the circle. Similarly, AB is also a tangent to the circle.

Example 7. *Draw a circle with the help of a bangle. Take a point outside the circle. Construct the pair of tangents from this point to the circle.*

Sol. Steps of construction:
1. Draw a circle with the help of a bangle.
2. Take two non-parallel chords AB and CD of this circle.
3. Draw the perpendicular bisectors of AB and CD. Let these intersect at O. Then, O is the centre of the circle drawn.
4. Take a point P outside the circle.
5. Join PO and bisect it. Let M be the mid-point of PO.

CONSTRUCTIONS

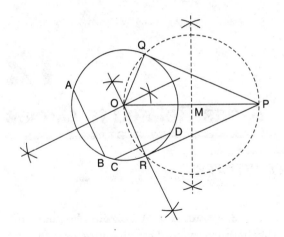

6. Taking M as centre and MO as radius, draw a circle. Let it intersect the given circle at the points Q and R.

7. Join PQ and PR.

Then, PQ and PR are the required two tangents.

Justification: Join OQ and OR.

Then ∠PQO is an angle in the semicircle and, therefore,

$$\angle PQO = 90°$$
$$\Rightarrow \quad PQ \perp OQ$$

Since OQ is a radius of the given circle, PQ has to be a tangent to the circle. Similarly, PR is also a tangent to the circle.

TEST YOUR KNOWLEDGE

In each of the following, give also the justification of the construction :

1. Construct the pair of tangents from a point 5 cm away from the centre of a circle of radius 2 cm.

2. Draw a pair of tangents to a circle of radius 6 cm which are inclined to each other at 60°.

   [**Hint :** Draw tangents at the ends of two radii which are inclined to each other at 120°].

3. Construct two tangents to a circle of radius 3 cm from a point on the concentric circle of radius 6 cm.

4. Draw a circle of radius 4 cm. Take two points P and Q on one of its extended diameter at a distance of 6 cm from its centre. Draw tangents to the circle from these two points P and Q.

5. Draw a line segment AB of length 10 cm. Taking A as centre, draw of a circle of radius 5 cm and taking B as centre, draw another while of radius 4 cm. Construct tangents to each circle from the centre of the other circle.

6. Let ABC be a right triangle in which AB = 3 cm, BC = 4 cm and ∠B = 90°. BD is the perpendicular from B and AC. Construct the tangents from A to the circle through B, C, D.

7. Draw a circle of radius OA = 6 cm. Produce OA and take two points P and Q on the extended part at distances 8 cm and 10 cm respectively from O. Draw tangents to the circle from these two point P and Q.

SUMMARY

1. To divide a line segment in a given ratio.
2. To construct a triangle similar to a given triangle as per a given scale factor which may be less than 1 or greater than 1.
3. To construct the pair of tangents from an external point to a circle.

A Note to The Reader

Construction of a quadrilateral (or a polygon) similar to a given quadrilateral (or a polygon) with a given scale factor can also be done following the similar steps as used in Examples 1 and 2 of Construction 11.2.

(Page 222 NCERT Textbook)

12

Areas Related to Circles

IMPORTANT POINTS

1. Importance of finding perimeters and areas related to circular figures. We are already familiar with some methods of finding perimeters and areas of simple plane figures such as rectangles, squares, parallelograms, triangles and circles. Many objects that we come across in our daily life are related to the circular shape in some form or the other. Cycle wheels, wheel barrow (thela), dartboard, round cake, *papad*, drain cover, various designs, bangles, broaches, circular paths, washers, flower beds etc. are some examples of such objects (see figure). So, the problem of finding perimeters and areas related to circular figures is of great practical importance.

Wheel   Design   Washer   Cake

Wheel barrow

2. Perimeter of a circle. We know that the distance covered by travelling once around a circle is its *perimeter*, usually called its *circumference*. We also know that circumference of a circle bears a constant ratio with its diameter. This constant ratio is denoted by the Greek latter π (read as 'pi'). In other words,

$$\frac{\text{circumference}}{\text{diameter}} = \pi$$

or, circumference = π × diameter
$\qquad\qquad\quad = \pi \times 2r$
(where r is the radius of the circle)
$\qquad\qquad\quad = 2\pi r$

3. Historical Facts. The great Indian mathematician Aryabhatta (A.D. 476–550) gave an approximate value of π. He stated that $\pi = \dfrac{62832}{20000}$, which is nearly equal to 3.1416. It is also interesting to note that using an identity of the great mathematical genius Srinivas Ramanujan (1887–1920) of India, mathematicians have been able to calculate the value of π correct to million places of decimals. Also we know that π is an irrational number and its decimal expansion is non-terminating and non-recurring (non-repeating). However, for practical purposes, we generally take the value of π as $\dfrac{22}{7}$ or 3.14, approximately.

AREAS RELATED TO CIRCLES

4. Area of a circle. We know that area of a circle is πr^2, where r is the radius of the circle. We may verify it by cutting a circle into a number of sectors and rearranging them as shown in figure.

We can see that the shape in figure (*ii*) is nearly a rectangle of length $\frac{1}{2} \times 2\pi r$ and breadth r. This suggests that the area of the circle = $\frac{1}{2} \times 2\pi r \times r = \pi r^2$.

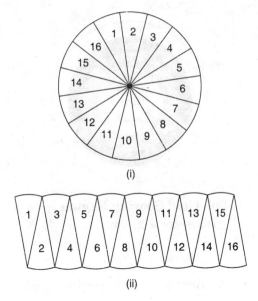

ILLUSTRATIVE EXAMPLES

[NCERT Exercise 12.1]
(*Page No. 225*)

$\left(\text{Unless stated otherwise, use } \pi = \frac{22}{7}\right)$

Example 1. *The radii of two circles are 19 cm and 9 cm respectively. Find the radius of the circle which has circumference equal to the sum of the circumferences of the two circles.*

Sol. Circumference of the circle whose radius is 19 m = $2\pi(19)$ cm.

Circumference of the circle whose radius is 9 cm = $2\pi(9)$ cm

Let R cm be the radius of the circle which has circumference equal to the sum of the circumferences of the two circles. Then,

$2\pi R = 2\pi(19) + 2\pi(9)$

$\Rightarrow$ R = 19 + 9

| Cancelling 2π throughout

$\Rightarrow$ R = 28

Hence, the required radius is 28 cm.

Example 2. *The radii of two circles are 8 cm and 6 cm respectively. Find the radius of the circle having area equal to the sum of the areas of the two circles.*

Sol. Area of a circle where radius is 8 cm = $\pi(8)^2$ cm^2. Area of a circle where radius is 6 cm = $\pi(6)^2$ cm^2. Let R cm be the radius of the circle whose area is equal to the sum of the area of the two circles. Then,

$\pi R^2 = \pi(8)^2 + \pi(6)^2$

$\Rightarrow$ $R^2 = (8)^2 + (6)^2$

| cancelling π throughout

$\Rightarrow$ $R^2 = 64 + 36$

$\Rightarrow$ $R^2 = 100$

$\Rightarrow$ R = $\sqrt{100}$

$\Rightarrow$ R = 10

Hence, the required radius is 10 cm.

Example 3. *Figure depicts an archery target marked with its five scoring areas from the centre outwards as Gold, Red, Blue, Black and White. The diameter of the region representing Gold score is 21 cm and each of the other bands is 10.5 cm wide. Find the area of each of the five scoring regions.*

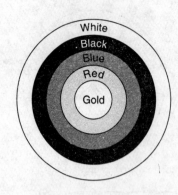

Sol. Gold Scoring Region

Diameter = 21 cm

$\Rightarrow$ Radius = $\frac{21}{2}$ cm

$\therefore$ Area of the gold scoring region

$= \pi \left(\frac{21}{2}\right)^2$

$= \frac{22}{7} \left(\frac{21}{2}\right)^2$

$= \frac{22}{7} \times \frac{21}{2} \times \frac{21}{2}$

$= \frac{793}{2} = 346.5$ cm^2.

Red Scoring Region. Area of the red scoring region

$= \pi \left(\frac{21}{2} + 10.5\right)^2 - \pi \left(\frac{21}{2}\right)^2$

$= \pi(21)^2 - 346.5$

$= \frac{22}{7}(21)^2 - 346.5$

$= \frac{22}{7} \times 21 \times 21 - 346.5$

$= 1386 - 346.5$

$= 1039.5$ cm^2

Blue Scoring Region. Area of the blue scoring region

$= \pi(21 + 10.5)^2 - (1039.5 + 346.5)$

$= \pi(31.5)^2 - 1386$

$= \frac{22}{7}(31.5)^2 - 1386$

$= \frac{22}{7} \times 31.5 \times 31.5 - 1386$

$= 3118.5 - 1386 = 1732.5$ cm^2.

Black Scoring Region. Area of the black scoring region

$= \pi(31.5 + 10.5)^2$
$\qquad - \{1732.5 + 1039.5 + 346.5\}$

$= \pi(42)^2 - 3118.5$

$= \frac{22}{7}(42)^2 - 3118.5$

$= \frac{22}{7} \times 42 \times 42 - 3118.5$

$= 5544 - 3118.5$

$= 2425.5$ cm^2.

White Scoring Region. Area of the white scoring region

$= \pi(42 + 10.5)^2 - (2425.5 + 1732.5$
$\qquad + 1039.5 + 346.5)$

$= \pi(52.5)^2 - 5544$

$= \frac{22}{7}(52.5)^2 - 5544$

$= \frac{22}{7} \times 52.5 \times 52.5 - 5544$

$= 8662.5 - 5544$

$= 3118.5$ cm^2.

Example 4. *The wheels of a car are of diameter 80 cm each. How many complete revolutions does each wheel make in 10 minutes when the car is travelling at a speed of 66 km per hour?*

Sol. $\because$ Diameter of the wheel = 80 cm

$\therefore$ Radius of the wheel = $\frac{80}{2}$ = 40 cm

$\therefore$ Circumference of the wheel

$= 2\pi(40)$ cm

$= 80\pi$ cm

$\therefore$ Distance covered by the wheel in one complete revolution

AREAS RELATED TO CIRCLES

$= 80\pi$ cm $= 80 \times \dfrac{22}{7}$ cm

$= \dfrac{1760}{7}$ cm

∵ Distance covered by the wheel in 1 hour (= 60 minutes) = 66 km

$= 6600000$ cm

∴ Distance covered by the wheel in 10 minutes

$= \dfrac{6600000}{60} \times 10$

$= 1100000$ cm

∴ Required number of complete revolution

$= \dfrac{1100000}{\frac{1760}{7}} = \dfrac{1100000 \times 7}{1760} = 4375.$

Example 5. *Tick the correct answer in the following and justify your choice : If the perimeter and area of a circle are numerically equal, then the radius of the circle is*

(A) *2 units* (B) *π units* (C) *4 units*

(D) *7 units.*

Sol. Let the radius of the circle be r units.

Then,

Its circumference = $2\pi r$ units and its area = πr^2 sq. units.

According to the question,

$2\pi r = \pi r^2$

$\Rightarrow \quad r = 2$ units

Hence the correct answer is **(A) 2 units**

ADDITIONAL EXAMPLES

Example 6. *A wheel of a cart is making 4 revolutions per second. If the diameter of the wheel be 84 cm, find the speed of the wheel in cm/sec.*

Sol. ∵ Diameter of the wheel = 84 cm

∴ Radius of the wheel = $\dfrac{84}{2} = 42$ cm

∴ Circumference of the wheel = $2\pi(42)$

$= 2 \times \dfrac{22}{7} \times 42$

$= 264$ cm

∴ Distance covered by the wheel in 1 revolution = 264 cm.

∴ Distance covered by the wheel in 4 revolutions $264 \times 4 = 1056$ cm.

∴ Distance covered by the wheel in 1 second

$= 1056$ cm

∴ Speed of the wheel = 1056 cm/sec

Hence the speed of the wheel is 1056 cm/sec.

Example 7. *Two circles touch internally. The sum of their area is 116π cm² and the distance between their centre is 6 cm. Find the radii of the circles.*

Sol. Let the radius of the circle having centre O be R and the radius of the circle having centre O' be r.

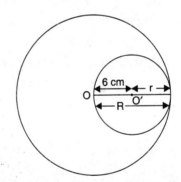

Then, sum of their areas

$= \pi R^2 + \pi r^2$

According to the question,

$= \pi R^2 + \pi r^2 = 116\pi$

$\Rightarrow \quad R^2 + r^2 = 116$ ...(1)

and, Distance between their centres = 6 cm

$\Rightarrow \quad R - r = 6$ cm ...(2)

Now, $(R - r)^2 + (R + r)^2 = 2(R^2 + r^2)$

$\Rightarrow \quad (6)^2 + (R + r)^2 = 2(116)$

| From (1) and (2)

$\Rightarrow \quad 36 + (R + r)^2 = 232$

$\Rightarrow$ $(R + r)^2 = 232 - 36$
$\Rightarrow$ $(R + r)^2 = 196$
$\Rightarrow$ $R + r = \sqrt{196}$
$\Rightarrow$ $R + r = 14$ ...(3)

Solving (1) and (2), we get
$R = 10$
and $r = 4$

Hence, the radii of the given circles are 10 cm and 4 cm respectively.

Example 8. *The sum of radii of two circles is 140 cm and the difference of their circumferences is 88 cm. Find the diameters of the circles.*

Sol. Let the radius of the first circle be r cm.

Then the radius of the second circle
$= (140 - r)$ cm

Circumference of the first circle
$= 2\pi r$ cm

Circumference of the second circle
$= 2\pi(140 - r)$ cm

$\therefore$ Difference of their circumferences
$= 2\pi r - 2\pi(140 - r)$
$= 2\pi(r - 140 + r)$
$= 2\pi(2r - 140)$

According to the question
$2\pi(2r - 140) = 88$

$\Rightarrow$ $2 \times \dfrac{22}{7} \times (2r - 140) = 88$

$\Rightarrow$ $\dfrac{2r - 140}{7} = 2$

$\Rightarrow$ $2r - 140 = 14$

$\Rightarrow$ $2r = 140 + 14$

$\Rightarrow$ $2r = 154$

$\Rightarrow$ $r = \dfrac{154}{2} = 77$

$\therefore$ Radius of the first circle = 77 cm
$\therefore$ Diameter of the first circle
$= 2 \times 77 = 154$ cm

Radius of the second circle $= 140 - r$
$= 140 - 77 = 63$ cm

$\therefore$ Diameter of the second circle
$= 2 \times 63 = 126$ cm

Hence, the diameter of the two circles are 154 cm and 126 cm respectively.

TEST YOUR KNOWLEDGE

1. The cost of fencing a circular field at the rate of Rs. 24 per metre is Rs. 5280. The field is to be ploughed at the rate of Rs. 0.50 per m². Find the cost of ploughing the field $\left(\text{Take } \pi = \dfrac{22}{7}\right)$.

2. Find the cost of fencing a circular field of 560 meters radius at Rs. 332 per 10 metres.

3. The driving wheel of a locomotive engine 2.1 m in radius makes 75 revolutions in one minute. Find the speed of the train in km per hour.

4. Find the area of a ring whose outer and inner radii are respectively 20 cm and 15 cm.

5. A road which is 7 m wide surrounds a circular park whose circumference is 352 m. Find the area of the road.

6. The circumference of a circular park is 314 m. A 20 m wide concrete track runs round it. Calculate the cost of laying turf in the park at Rs. 1.25 per sq. m and the cost of concrete track at Rs. 50 per sq. m.

Answers

1. Rs. 1925 2. Rs. 116864
3. 59.4 km per hour 4. 550 cm²
5. 2618 m²
6. Rs 9812.50, Rs 376800

AREAS OF SECTOR AND SEGMENT OF A CIRCLE

IMPORTANT POINTS

1. Sector and segment of a circle. We know that the portion (or part) of the circular region enclosed by two radii and the corresponding arc is called a *sector* of the circle and the portion (or part) of the circular region enclosed between a chord and the corresponding arc is called a *segment* of the circle. Thus, in figure shaded region OAPB is a *sector* of the circle with centre O. ∠AOB is called the *angle* of the sector. Note that in this figure, unshaded region OAQB is also a sector of the circle. For obvious reasons, OAPB is called the *minor sector* and OAQB is called the *major sector*. We can also see that angle of the major sector is 360° − ∠AOB.

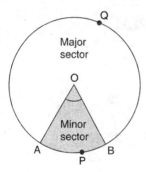

Now, let us look at figure in which AB is a chord of the circle with centre O. So, shaded region APB is a segment of the circle. We can also note that unshaded region AQB is another segment of the circle formed by the chord AB. For obvious reasons, APB is called the *minor segment* and AQB is called the *major segment*.

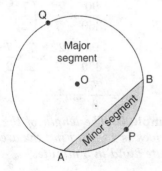

Remark. When we write 'segment' and 'sector' we will mean 'the minor segment' and minor sector respectively, unless stated otherwise.

2. Area of the sector of angle θ. Let OAPB be a sector of a circle with centre O and radius r (see figure (*i*)). Let the degree measure of ∠AOB be θ.

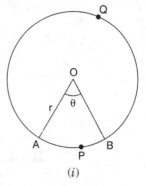

(*i*)

We know that area of a circle (in fact of a circular region or disc) is πr^2.

In a way, we can consider this circular region to be a sector forming an angle of 360° (*i.e.*, of degree measure 360) at the centre O. Now by applying the Unitary Method, we can arrive at the area of the sector OAPB as follows:

When degree measure of the angle at the centre is 360, area of the sector = πr^2

So, when the degree measure of the angle at the centre is 1, area of the sector

$$= \frac{\pi r^2}{360°}$$

Therefore, when the degree measure of the angle at the centre is θ, area of the sector

$$= \frac{\pi r^2}{360°} \times \theta = \frac{\theta}{360°} \times \pi r^2$$

Thus, we obtain the following relation (or formula) for area of a sector of a circle:

Area of the sector of angle θ

$$= \frac{\theta}{360°} \times \pi r^2$$

where r is the radius of the circle and θ the angle of the sector in degrees.

3. Length of an arc of sector of angle θ. We can also find the length of the arc APB corresponding to this sector by applying the Unitary Method and taking the whole length of the circle (of angle 360°) as $2\pi r$. Thus, we can obtain the required length of the arc APB as $\dfrac{\theta}{360°} \times 2\pi r$.

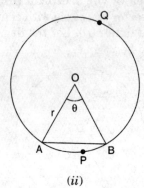

(ii)

So, **length of an arc of sector of angle** $\theta = \dfrac{\theta}{360°} \times 2\pi r$.

4. Area of the segment. Area of the segment APB of a circle with centre O and radius r (see figure (ii)).

= Area of the sector OAPB
    − Area of ΔOAB

$= \dfrac{\theta}{360°} \times \pi r^2 -$ area of ΔOAB

Note. From figure (i) and figure (ii) respectively, we can observe that Area of the major sector OAQB = πr^2 − Area of the minor sector OAPB and Area of major segment AQB = πr^2 − Area of the minor sector APB.

ILLUSTRATIVE EXAMPLE

[NCERT Exercise 12.2]
(Page No. 230)

$\left(\text{Unless stated otherwise, use } \pi = \dfrac{22}{7}\right)$

Example 1. *Find the area of a sector of a circle with radius 6 cm, if angle of the sector is 60°.*

Sol. $r = 6$ cm
$\theta = 60°$

∴ Area of the sector

$= \dfrac{\theta}{360} \times \pi r^2$

$= \dfrac{60}{360} \times \dfrac{22}{7} \times (6)^2$

$= \dfrac{132}{7}$ cm².

Example 2. *Find the area of a quadrant of a circle whose circumference is 22 cm.*

Sol. Let the radius of the circle be r cm.
Then, circumference of the circle = $2\pi r$ cm
According to the question,
$2\pi r = 22$

$\Rightarrow 2 \times \dfrac{22}{7} \times r = 22$

$\Rightarrow r = \dfrac{22 \times 7}{2 \times 22}$

$\Rightarrow r = \dfrac{7}{2}$ cm

For a quadrant of a circle,
$\theta = 90°$

∴ Area of the quadrant

$= \dfrac{\theta}{360} \times \pi r^2$

$= \dfrac{90}{360} \times \dfrac{22}{7} \times \left(\dfrac{7}{2}\right)^2$

$= \dfrac{90}{360} \times \dfrac{22}{7} \times \dfrac{7}{2} \times \dfrac{7}{2}$

$= \dfrac{77}{8}$ cm².

Example 3. *The length of the minute hand of a clock is 14 cm. Find the area swept by the minute hand in 5 minutes.*

Sol. $r = 14$ cm

$$\theta = \frac{90°}{3} = 30°$$

∴ Area swept

$$= \frac{\theta}{360} \times \pi r^2$$

$$= \frac{30}{360} \times \frac{22}{7} \times 14 \times 14$$

$$= \frac{154}{3} \text{ cm}^2.$$

Example 4. *A chord of a circle of radius 10 cm subtends a right angle at the centre. Find the area of the corresponding : (i) minor segment (ii) major sector. (Use $\pi = 3.14$).*

Sol. (i) $r = 10$ cm

$$\theta = 90°$$

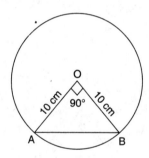

Area of minor sector

$$= \frac{\theta}{360} \times \pi r^2$$

$$= \frac{90}{360} \times 3.14 \times 10 \times 10$$

$$= 78.5 \text{ cm}^2$$

Area of $\triangle$OAB

$$= \frac{OA \times OB}{2}$$

$$= \frac{10 \times 10}{2}$$

$$= 50 \text{ cm}^2$$

∴ Area of the minor segment
= Area of minor sector
 − Area of $\triangle$OAB
= 78.5 cm^2 − 50 cm^2

$$= 28.5 \text{ cm}^2$$

(ii) Area of major sector

$$= \pi r^2 - 78.5$$

$$= 3.14 \times 10 \times 10 - 78.5$$

$$= 314 - 78.5$$

$$= 235.5 \text{ cm}^2.$$

Alternative Method

Area of major sector

$$= \frac{360 - \theta}{360} \times \pi r^2$$

$$= \frac{360 - 90}{360} \times 3.14 \times (10)^2$$

$$= 235.5 \text{ cm}^2.$$

Example 5. *In a circle of radius 21 cm, an arc subtends an angle of 60° at the centre. Find :*

(i) *The length of the arc*

(ii) *Area of the sector formed by the arc*

(iii) *Area of the segment formed by the corresponding chord*

Sol. $r = 21$ cm

$$\theta = 60°$$

(i) The length of the arc

$$= \frac{\theta}{360} \times 2\pi r$$

$$= \frac{60}{360} \times 2 \times \frac{22}{7} \times 21$$

$$= 22 \text{ cm.}$$

(ii) Area of the sector formed by the arc

$$= \frac{\theta}{360} \times \pi r^2$$

$$= \frac{60}{360} \times \frac{22}{7} \times 21 \times 21$$

$$= 231 \text{ cm}^2.$$

(iii) Area of the segment formed by the corresponding chord

$$= \frac{\theta}{360} \times \pi r^2 - \text{Area of } \triangle\text{OAB}$$

$$= 231 - \text{Area of } \triangle\text{OAB} \quad ...(1)$$

Draw OM ⊥ AB

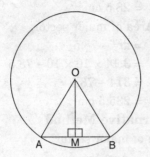

In right triangle OMA and OMB,

OA = OB

| Radii of the same circle

OM = OM | Common side

∴ ΔOMA ≅ ΔOMB

| RHS congruence criterion.

∴ AM = BM | CPCT

⇒ M is the mid-point of AB

and ∠AOM = ∠BOM | CPCT

⇒ ∠AOM = ∠BOM = $\frac{1}{2}$∠AOB

$= \frac{1}{2} \times 60° = 30°$

So, in right triangle OMA,

$\cos 30° = \frac{OM}{OA}$

⇒ $\frac{\sqrt{3}}{2} = \frac{OM}{21}$

⇒ $OM = \frac{21\sqrt{3}}{2}$ cm

Also, $\sin 30° = \frac{AM}{OA}$

⇒ $\frac{1}{2} = \frac{AM}{21}$

⇒ $AM = \frac{21}{2}$ cm

Therefore, AB = 2 AM = $2 \times \frac{21}{2} = 21$ cm

So, area of ΔOAB = $\frac{1}{2}$ AB × OM

$= \frac{1}{2} \times 21 \times \frac{21\sqrt{3}}{2}$

$= \frac{441\sqrt{3}}{2}$ cm^2

∴ From (1),

Area of the segment formed by the corresponding chord

$= \left(231 - \frac{441\sqrt{3}}{2}\right)$ cm^2.

Example 6. *A chord of a circle of radius 15 cm subtends an angle of 60° at the centre. Find the areas of the corresponding minor and major segment of the circle.*

(Use π = 3.14 and $\sqrt{3}$ = 1.73).

Sol. r = 15 cm

θ = 60°

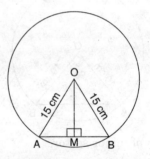

Area of the minor sector

$= \frac{\theta}{360} \times \pi r^2$

$= \frac{60}{360} \times 3.14 \times (15)^2$

= 117.75 cm^2.

Area of ΔAOB

Draw OM ⊥ AB

In right triangles OMA and OMB,

OA = OB

| Radii of the same circle

OM = OM | Common side

∴ ΔOMA ≅ ΔOMB

| RHS congruence criterion

∴ AM = BM | CPCT

$\Rightarrow$ AM = BM = $\frac{1}{2}$ AB

and $\angle$AOM = $\angle$BOM | CPCT

$\Rightarrow \angle AOM = \angle BOM$

$= \frac{1}{2} \angle AOB = \frac{1}{2} \times (60°) = 30°$

∴ In right triangle OMA,

$\cos 30° = \frac{OM}{OA}$

$\Rightarrow \frac{\sqrt{3}}{2} = \frac{OM}{15}$

$\Rightarrow OM = \frac{15\sqrt{3}}{2}$ cm

$\sin 30° = \frac{AM}{OA}$

$\Rightarrow \frac{1}{2} = \frac{AM}{15}$

$\Rightarrow AM = \frac{15}{2}$

$\Rightarrow 2AM = 2\left(\frac{15}{2}\right) = 15$

$\Rightarrow$ AB = 15 cm

∴ Area of $\triangle$AOB

$= \frac{1}{2} \times AB \times OM$

$= \frac{1}{2} \times 15 \times \frac{15\sqrt{3}}{2}$

$= \frac{225\sqrt{3}}{4}$

$= \frac{225 \times 1.73}{4}$

$= 97.3125$ cm^2

∴ Area of the corresponding minor segment of the circle

= Area of the minor sector
 − Area of $\triangle$AOB

= 117.75 − 97.3125

= 20.4375 cm^2

and, area of the corresponding major segment of the circle

= πr^2 − area of the corresponding minor segment of the circle

= 3.14 × 15 × 15 − 20.4375

= 706.5 − 20.4375

= 686.0625 cm^2.

Example 7. *A chord of a circle of radius 12 cm subtends an angle of 120° at the centre. Find the area of the corresponding segment of the circle.*

(Using $\pi = 3.14$ and $\sqrt{3} = 1.73$)

Sol. $r = 12$ cm

$\theta = 120°$

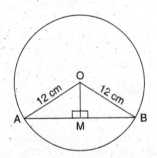

∴ Area of the corresponding sector of the circle

$= \frac{\theta}{360} \times \pi r^2$

$= \frac{120}{360} \times 3.14 \times 12 \times 12$

$= 150.72$ cm^2.

Area of $\triangle$AOB

Draw OM ⊥ AB

In right triangles OMA and OMB,

OA = OB

| Radii of the same circle

OM = OM | Common side

∴ $\triangle$OMA ≅ $\triangle$OMB

| RHS congruence criterion

∴ AM = BM | CPCT

$\Rightarrow$ AM = BM = $\frac{1}{2}$ AB

and $\angle$AOM = $\angle$BOM | CPCT

$\Rightarrow \angle AOM = \angle BOM = \dfrac{1}{2} \angle AOB$

$= \dfrac{1}{2} \times (120)° = 60°$

∴ In right triangle OMA,

$\cos 60° = \dfrac{OM}{OA}$

$\Rightarrow \dfrac{1}{2} = \dfrac{OM}{12}$

$\Rightarrow OM = 6$ cm

$\sin 60° = \dfrac{AM}{OA}$

$\Rightarrow \dfrac{\sqrt{3}}{2} = \dfrac{AM}{12}$

$\Rightarrow AM = 6\sqrt{3}$ cm

$\Rightarrow 2AM = 12\sqrt{3}$ cm

$\Rightarrow AB = 12\sqrt{3}$ cm

∴ Area of $\triangle AOB = \dfrac{1}{2} \times AB \times OM$

$= \dfrac{1}{2} \times 12\sqrt{3} \times 6$

$= 36\sqrt{3}$ cm²

$= 36 \times 1.73$ cm²

$= 62.28$ cm².

So, area of the corresponding segment of the circle

= Area of the corresponding sector of the circle – Area of $\triangle AOB$

$= 150.72 - 62.28$

$= 88.44$ cm².

Example 8. *A horse is tied to a peg at one corner of a square shaped grass field of side 15 m by means of a 5 m long rope (see figure). Find*

(i) the area of that part of the field in which the horse can graze.

(ii) the increase in the grazing area if the rope were 10 m long instead of 5 m (Use π = 3.14).

Sol. (i) The area of that part of the field in which the horse can graze if the length of the rope is 5 m

$= \dfrac{1}{4}\pi r^2 = \dfrac{1}{4} \times 3.14 \times (5)^2$

$= \dfrac{1}{4} \times 78.5 = 19.625$ m².

(ii) The area of that part of the field in which the horse can graze if the length of the rope is 10 m

$= \dfrac{1}{4}\pi r^2 = \dfrac{1}{4} \times 3.14 \times (10)^2$

$= 78.5$ m²

∴ The increase in the grazing area

$= 78.5 - 19.625 = 58.875$ cm².

Example 9. *A brooch is made with silver wire in the form of a circle with diameter 35 mm. The wire is also used in making 5 diameters which divide the circle into 10 equal sectors as shown in figure. Find*

(i) the total length of the silver wire required.

(ii) the area of each sector of the brooch.

Sol. (i) ∵ Diameter = 35 mm

∴ Radius = $\frac{35}{2}$ mm

∴ Circumference = $2\pi r$

$= 2 \times \frac{22}{7} \times \frac{35}{2}$

$= 110$ mm ...(1)

Length of 5 diameters

$= 35 \times 5 = 175$ mm ...(2)

∴ The total length of the silver wire required

$= 110 + 175$

$= 285$ mm

(ii) $r = \frac{35}{2}$ mm

$\theta = \frac{360°}{10} = 36°$

∴ The area of each sector of the brooch

$= \frac{\theta}{360} \times \pi r^2$

$= \frac{36}{360} \times \frac{22}{7} \times \frac{35}{2} \times \frac{35}{2}$

$= \frac{385}{4}$ mm^2.

Example 10. *An umbrella has 8 ribs which are equally spaced (see figure). Assuming umbrella to be a flat circle of radius 45 cm, find the area between the two consecutive ribs of the umbrella.*

Sol. $r = 45$ cm

$\theta = \frac{360°}{8} = 45°$

∴ Area between the two consecutive ribs of the umbrella

$= \frac{\theta}{360} \times \pi r^2$

$= \frac{45}{360} \times \frac{22}{7} \times 45 \times 45$

$= \frac{22275}{28}$ cm^2.

Example 11. *A car has two wipers which do not overlap. Each wiper has a blade of length 25 cm sweeping through an angle of 115°. Find the total area cleaned at each sweep of the blades.*

Sol. $r = 25$ cm

$\theta = 115°$

∴ The total area cleaned at each sweep of the blades

$= 2 \times \left[\frac{\theta}{360} \times \pi r^2 \right]$

$= 2 \times \left[\frac{115}{360} \times \frac{22}{7} \times (25)^2 \right]$

$= \frac{158125}{126}$ cm^2.

Example 12. *To warn ships for underwater rocks, a lighthouse spreads a red coloured light over a sector of angle 80° to a distance of 16.5 km. Find the area of the sea over which the ships are warned. (Use $\pi = 3.14$)*

Sol. $\theta = 80°$

$r = 16.5$ km

∴ The area of the sea over which the ships are warned

$= \frac{\theta}{360} \times \pi r^2$

$= \frac{80}{360} \times 31.4 \times (16.5)^2$

$= 189.97$ km^2.

Example 13. *A round table cover has six equal designs as shown in figure. If the radius of the cover is 28 cm, find the cost of making the designs at the rate of Rs. 0.35 per cm^2. (Use $\sqrt{3} = 1.7$)*

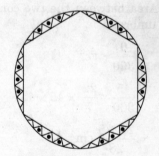

Sol. $r = 28$ cm

$$\theta = \frac{360°}{6} = 60°$$

Area of minor sector

$$= \frac{\theta}{360} \times \pi r^2$$

$$= \frac{60}{360} \times \frac{22}{7} \times (28)^2$$

$$= \frac{1232}{3} \text{ cm}^2$$

$$= 410.67 \text{ cm}^2$$

Area of $\triangle AOB$

Draw $OM \perp AB$

In right triangles OMA and OMB

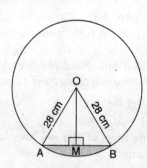

$OA = OB$

         | Radii of the same circle

$OM = OM$

         | Common side

$\therefore \quad \triangle OMA \cong \triangle OMB$

         | RHS congruence criterion

$\therefore \quad AM = BM$ | CPCT

$\Rightarrow \quad AM = BM = \frac{1}{2} AB$

and, $\quad \angle AOM = \angle BOM$

$\Rightarrow \quad \angle AOM = \angle BOM = \frac{1}{2} \angle AOB$

$$= \frac{1}{2}(60°) = 30°$$

In right triangle OMA,

$$\cos 30° = \frac{OM}{OA}$$

$\Rightarrow \quad \frac{\sqrt{3}}{2} = \frac{OM}{28}$

$\Rightarrow \quad OM = 14\sqrt{3}$ cm

$$\sin 30° = \frac{AM}{OA}$$

$\Rightarrow \quad \frac{1}{2} = \frac{AM}{28}$

$\Rightarrow \quad AM = 14$ cm

$\Rightarrow \quad 2AM = 28$ cm

$\Rightarrow \quad AB = 28$ cm

$\therefore$ Area of $\triangle AOB$

$$= \frac{1}{2} \times AB \times OM$$

$$= \frac{1}{2} \times 28 \times 14\sqrt{3}$$

$$= 196\sqrt{3} \text{ cm}^2$$

$$= 196 \times 1.7 \text{ cm}^2$$

$$= 333.2 \text{ cm}^2$$

$\therefore$ Area of minor segment

    = Area of minor sector

                      – Area of $\triangle AOB$

    = 410.67 – 333.2

    = 77.47 cm^2

$\therefore$ Area of one design = 77.47 cm^2

$\therefore$ Area of six design = 77.47 × 6

    = 464.82 cm^2

$\therefore$ Cost of making the designs at the rate of Rs 0.35 per cm^2

    = 464.82 × 0.35

    = Rs 162.68

AREAS RELATED TO CIRCLES

Example 14. *Tick the correct answer in the following :*

Area of a sector of angle p (in degrees) of a circle with radius R is

(A) $\dfrac{p}{180} \times 2\pi R$ (B) $\dfrac{p}{180} \times \pi R^2$

(C) $\dfrac{p}{360} \times 2\pi R$ (D) $\dfrac{p}{720} \times 2\pi R^2$

Sol. Area of a sector of angle $p°$ of a circle with radius R

$$= \dfrac{\theta}{360} \times \pi r^2$$

$$= \dfrac{p}{360} \times \pi R^2$$

$$= \dfrac{p}{2(360)} \times 2\pi R^2$$

$$= \dfrac{p}{720} \times 2\pi R^2$$

Hence, the correct answer is (D) $\dfrac{p}{720} \times 2\pi R^2$.

ADDITIONAL EXAMPLES

Example 15. *The minute hand of a clock is 10 cm long. Find the area on the face of the clock described by the minute hand between 9 A.M. and 9.35 A.M.*

Sol. Time between 9 A.M. and 9.35 A.M.
= 35 minutes

∵ Angle described by the minute hand in one hour (60 minutes) = 360°

∴ Angle described by the minute hand in 35 minutes = $\dfrac{360°}{60} \times 35 = 210°$

∴ $\theta = 210°$

Length of the minute hand = 10 cm

∴ Radius of the circle described by the minute hand (r) = 10 cm

∴ Area on the face of the clock described by the minute hand

$$= \dfrac{\theta}{360} \times \pi r^2$$

$$= \dfrac{210}{360} \times \dfrac{22}{7} \times (10)^2$$

$$= \dfrac{550}{3} \text{ cm}^2.$$

Example 16. *The radius of a circle is 14 cm and the area of the sector is 102.7 cm². Find the central angle of the sector.*

Sol. $r = 14$ cm

Let the central angle of the sector be θ.

We know that,

Area of the sector = $\dfrac{\theta}{360} \times \pi r^2$

⇒ $102.7 = \dfrac{\theta}{360} \times \dfrac{22}{7} \times (14)^2$

⇒ $102.7 = \dfrac{\theta}{360} \times 616$

⇒ $102.7 = \dfrac{77\theta}{45}$

⇒ $\theta = \dfrac{102.7 \times 45}{77} = 60°$ (nearly)

Hence, the central angle of the sector is nearly 60°.

Example 17. *The perimeter of a sector of a circle of radius 5.7 cm is 27.2 cm. Find the area of the sector.*

Sol. Let O be the centre of the circle whose radius is 5.7 cm.

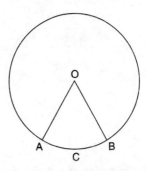

Let OAB be the sector with perimeter 27.2 cm. Then,

$$OA + OB + \text{arc } ACB = 27.2$$
$$\Rightarrow \quad 5.7 + 5.7 + \text{arc } ACB = 27.2$$
$$\Rightarrow \quad 11.4 + \text{arc } ACB = 27.2$$
$$\Rightarrow \quad \text{arc } ACB = 27.2 - 11.4 = 15.8 \text{ cm}$$
$$\Rightarrow \quad \frac{\theta}{360} \times 2\pi r = 15.8$$
$$\Rightarrow \quad \frac{\theta}{360} \times 2 \times \frac{22}{7} \times 5.7 = 15.8$$

$$\Rightarrow \quad \theta = \frac{15.8 \times 360 \times 7}{2 \times 22 \times 5.7}$$

∴ Area of the sector
$$= \frac{\theta}{360} \times \pi r^2$$
$$= \frac{15.8 \times 360 \times 7}{2 \times 22 \times 5.7 \times 360} \times \frac{22}{7} \times (5.7)^2$$
$$= 45.03 \text{ cm}^2.$$

TEST YOUR KNOWLEDGE

1. Find the area of the sector of a circle with radius 4 cm and of angle 30°. Also, find the area of the corresponding major sector (Use $\pi = 3.14$).

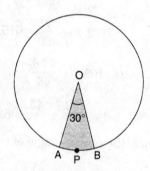

2. Find the area of the segment AYB shown in figure, if radius of the circle is 21 cm and $\angle AOB = 120°$ $\left(\text{Use } \pi = \frac{22}{7}\right)$.

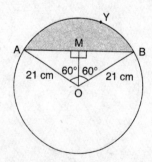

3. Find the radius of a circle if an arc of central angle 40° has length 4π cm. Hence, find the area of the sector formed by this arc.

4. Find the area of a sector of a circle where central angle is 30° and the radius of the circle is 42 cm.

5. A sector is cut from a circle of radius 21 cm. The angle of the sector is 150°. Find the length of its arc and area.

6. A horse is tethered to a corner of a field which is of the shape of an equilateral triangle. The length of the rope by which it is tied be 7 m. Find the area of the field over which it can graze.

7. In figure are shown sectors of two consecutive circles of radii 7 cm and 3.5 cm. Find the area of the shaded region. $\left(\text{Use } \pi = \frac{22}{7}\right)$

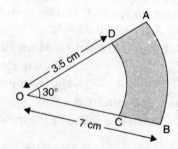

8. A chord AB of a circle of radius 15 cm makes an angle of 60° at the centre of the circle. Find the area of the major and minor segments.

9. A chord 10 cm long is drawn in a circle whose radius is $\sqrt{50}$ cm. Find the areas of both the segments.

AREAS RELATED TO CIRCLES   413

10. A chord AB of a circle of radius 10 cm makes a right angle at the centre of the circle. Find the area of the major and minor segments.

Answers

1. 4.19 cm² (approx.), 46.1 cm² (approx.)
2. $\frac{21}{4}(88 - 21\sqrt{3})$ cm²
3. 18 cm, 36π cm²
4. 462 cm²
5. 55 cm, $\frac{1155}{2}$ cm²
6. $\frac{77}{3}$ cm²
7. $\frac{77}{8}$ cm²
8. 686.205 cm², 20.295 cm²
9. 14.285 cm², 142.85 cm²
10. 285.5 cm², 28.5 cm²

AREAS OF COMBINATIONS OF PLANE FIGURES

IMPORTANT POINTS

1. Occurrence in daily life. We come across those types of figures in the daily life which are some combinations of plane figures. Also, we come across such figures in the form of various interesting designs. Flower beds, drain covers, window designs, designs on table covers, are some of such examples.

ILLUSTRATIVE EXAMPLES

[NCERT Exercise 12.3]
(Page No. 234)

Unless stated otherwise, use $\pi = \frac{22}{7}$.

Example 1. *Find the area of the shaded region in figure, if PQ = 24 cm, PR = 7 cm and O is the centre of the circle.*

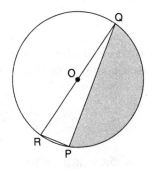

Sol. ∠RPQ = 90°

| ∵ ∠ in a semicircle is 90°

∴ RQ² = PR² + PQ²

| By Pythagoras theorem

= $(7)^2 + (24)^2$

= 49 + 576 = 625

⇒ RQ = $\sqrt{625}$ = 25 cm

⇒ Diameter of the circle = 25 cm

∴ Radius of the circle $(r) = \frac{25}{2}$ cm

Area of the semicircle = $\frac{1}{2} \times \pi r^2$

$= \frac{1}{2} \cdot \pi \left(\frac{25}{2}\right)^2$

$= \frac{625}{8}\pi = \frac{625}{8} \times \frac{22}{7}$

$= \frac{6875}{28}$ cm²

Area of right triangle RPQ

$= \frac{1}{2} \times PQ \times PR$

$= \frac{1}{2} \times 24 \times 7$

= 84 cm²

∴ Area of the shaded region

= Area of the semicircle
  − Area of right triangle RPQ

$$= \frac{6875}{28} - 84 = \frac{6875 - 2352}{28}$$

$$= \frac{4523}{28} \text{ cm}^2$$

Hence, the area of the shaded region is $\frac{4523}{28}$ cm².

Example 2. *Find the area of the shaded region in figure, if radii of the two concentric circles with centre O are 7 cm and 14 cm respectively and* $\angle AOC = 40°$.

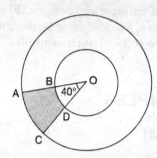

Sol. Area of the shaded region
= Area of the sector OAC
  − Area of the sector OBD

$$= \frac{40°}{360°} \pi(14)^2 - \frac{40°}{360°} \pi(7)^2$$

$$= \frac{40°}{360°} \cdot \pi \cdot \{(14)^2 - (7)^2\}$$

$$= \frac{1}{9} \cdot \frac{22}{7} (14 - 7)(14 + 7)$$

$$= \frac{154}{3} \text{ cm}^2.$$

Example 3. *Find the area of the shaded region in figure, if ABCD is a square of side 14 cm and APD and BPC are semicircles.*

Sol. Area of the shaded region
   = Area of the square ABCD
      − (Area of semicircle APD
         + Area of semicircle BPC)

$$= 14 \times 14 - \left[\frac{1}{2}\pi\left(\frac{14}{2}\right)^2 + \frac{1}{2}\pi\left(\frac{14}{2}\right)^2\right]$$

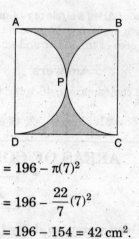

$$= 196 - \pi(7)^2$$

$$= 196 - \frac{22}{7}(7)^2$$

$$= 196 - 154 = 42 \text{ cm}^2.$$

Example 4. *Find the area of the shaded region in figure, where a circular arc of radius 6 cm has been drawn with vertex O of an equilateral triangle OAB of side 12 cm as centre.*

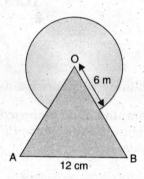

Sol. Area of the shaded region = Area of circle + Area of equilateral triangle OAB − Sectorial area common to the circle and the triangle

$$= \pi(6)^2 + \frac{\sqrt{3}}{4}(12)^2 - \frac{60°}{360°}\pi(6)^2$$

$$= 36\pi + 36\sqrt{3} - 6\pi$$

$$= 30\pi + 36\sqrt{3}$$

$$= 30 \times \frac{22}{7} + 36\sqrt{3}$$

$$= \left(\frac{660}{7} + 36\sqrt{3}\right) \text{ cm}^2.$$

AREAS RELATED TO CIRCLES

Example 5. *From each corner of a square of side 4 cm a quadrant of a circle of radius 1 cm is cut and also a circle of diameter 2 cm is cut as shown in figure. Find the area of the remaining portion of the square.*

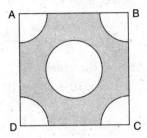

Sol. Area of the remaining portion of the square

= Area of the square
— [4 Area of a quadrant
+ Area of a circle]

$$= 4 \times 4 - \left[4 \times \frac{90°}{360°} \pi(1)^2 + \pi\left(\frac{2}{2}\right)^2\right]$$

$$= 16 - 2\pi$$

$$= 16 - 2 \times \frac{22}{7} = 16 - \frac{44}{7}$$

$$= \frac{112 - 44}{7} = \frac{68}{7} \text{ cm}^2.$$

Example 6. *In a circular table cover of radius 32 cm, a design is formed leaving an equilateral triangle ABC in the middle as shown in figure. Find the area of the design (shaded region).*

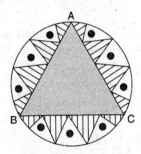

Sol. Area of the design (shaded region)
= Area of the circular table cover – Area of the equilateral triangle ABC

$$= \pi(32)^2 - \frac{\sqrt{3}}{4} a^2 \qquad ...(1)$$

where a cm is the side of the equilateral triangle ABC.

Let h cm be the height of $\triangle ABC$. Since the centre of the circle coincides with the combined of the equilateral triangle,

Therefore, Radius of the circumscribed circle $= \frac{2}{3} h$ cm

According to the question,

$$\frac{2}{3} h = 32$$

$$\Rightarrow \qquad h = 48$$

Again, $a^2 = h^2 + \left(\frac{a}{2}\right)^2$

| By Pythagoras theorem

$$\Rightarrow \qquad a^2 = h^2 + \frac{a^2}{4}$$

$$\Rightarrow \qquad a^2 - \frac{a^2}{4} = h^2 = \frac{3a^2}{4} = h^2$$

$$\Rightarrow \qquad a^2 = \frac{4h^2}{3}$$

$$\Rightarrow \qquad a^2 = \frac{4(48)^2}{3}$$

$$\Rightarrow \qquad a^2 = 3072$$

$$\Rightarrow \qquad a = \sqrt{3072}$$

∴ From (1)
Required area

$$= \pi(32)^2 - \frac{\sqrt{3}}{4}(3072)$$

$$= \frac{22}{7}(1024) - 768\sqrt{3}$$

$$= \left(\frac{22528}{7} - 768\sqrt{3}\right) \text{cm}^2.$$

Example 7. *In figure, ABCD is a square of side 14 cm. With centres A, B, C and D, four circles are drawn such that each circle touch externally two of the remaining three circles. Find the area of the shaded region.*

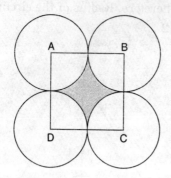

Sol. Area of the shaded region = Area of the square of side 14 cm − 4[Area of a sector of central angle 90°]

$$= 14 \times 14 - 4 \times \frac{90}{360} \pi \left(\frac{14}{2}\right)^2$$

$$= 196 - \frac{22}{7}(7)^2$$

$$= 196 - 154$$

$$= 42 \text{ cm}^2.$$

Example 8. *Figure depicts a racing track whose left and right ends are semi-circular.*

The distance between the two inner parallel line segments is 60 m and they are each 106 m long. If the track is 10 m wide, find

(i) the distance around the track along its inner edge

(ii) the area of the track.

Sol. (*i*) The distance around the track along its inner edge.

$$= 106 + 106 + 2\left[\frac{180}{360} 2\pi \left(\frac{60}{2}\right)\right]$$

$$= 212 + 60\pi$$

$$= 212 + 60 \times \frac{22}{7}$$

$$= 212 + \frac{1320}{7} = \frac{2804}{7} \text{ m}$$

(*ii*) Area of the track

$$= 106 \times 10 + 106 \times 10 + 2$$

$$\times \left[\frac{1}{2}\pi(30+10)^2 - \frac{1}{2}\pi(30)^2\right]$$

$$= 1060 + 1060 + \pi[(40)^2 - (30)^2]$$

$$= 2120 + \frac{22}{7} \times 700$$

$$= 2120 + 2200$$

$$= 4320 \text{ m}^2.$$

Example 9. *In figure, AB and CD are two diameters of a circle (with centre O) perpendicular to each other and OD is the diameter of the smaller circle. If OA = 7 cm, find the area of the shaded region.*

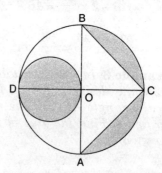

Sol. Area of the shaded region = Area of the circle where diameter is OD + Area of semicircle ACB − Area of △ACB

$$= \pi \left(\frac{7}{2}\right)^2 + \frac{1}{2}\pi(7)^2 - \left(\frac{7 \times 7}{2} + \frac{7 \times 7}{2}\right)$$

$$= \frac{49}{4}\pi + \frac{49}{2}\pi - 49$$

$$= \frac{147}{4}\pi - 49$$

AREAS RELATED TO CIRCLES

$$= \frac{147}{4} \cdot \frac{22}{7} - 49 = \frac{231}{2} - 49$$

$$= \frac{133}{2} = 66.5 \text{ cm}^2.$$

Example 10. *The area of an equilateral triangle ABC is 17320.5 cm². With each vertex of the triangle as centre, a circle is drawn with radius equal to half the length of the side of the triangle (see figure). Find the area of the shaded region. (Use $\pi = 3.14$ and $\sqrt{3} = 1.73205$).*

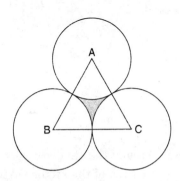

Sol. Let the length of the side of the equilateral triangle ABC be a cm. Then,

its area $= \frac{\sqrt{3}}{4} a^2$ cm²

According to the question,

$$\frac{\sqrt{3}}{4} a^2 = 17320.5$$

$\Rightarrow \qquad a^2 = \dfrac{17320.5 \times 4}{\sqrt{3}}$

$\Rightarrow \qquad a^2 = \dfrac{17320.5 \times 4}{1.73205}$

$\Rightarrow \qquad a^2 = 10000 \times 4$

$\Rightarrow \qquad a^2 = 40000$

$\Rightarrow \qquad a = \sqrt{40000}$

$\Rightarrow \qquad a = 200$ cm

Area of the shaded region = Area of the equilateral triangle ABC

$$- 3 \left[\frac{60}{360} \pi \left(\frac{200}{2} \right)^2 \right]$$

$$= 17320.5 - \frac{\pi}{2}(10000)$$

$$= 17320.5 - 3.14 \times 5000$$

$$= 17320.5 - 15700$$

$$= 1620.5 \text{ cm}^2.$$

Example 11. *On a square handkerchief, nine circular designs each of radius 7 cm are made (see figure). Find the area of the remaining portion of the handkerchief.*

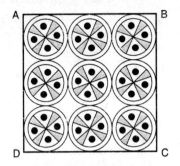

Sol. Area of the remaining portion of the handkerchief

= Area of the square ABCD
  − Area of nine circular design

$= 42 \times 42 - 9\pi(7)^2$

$= 1764 - 9 \times \dfrac{22}{7} \times (7)^2$

$= 1764 - 1386$

$= 378$ cm².

Example 12. *In figure, OACB is a quadrant of a circle with centre O and radius 3.5 cm. If OD = 2 cm, find the area of the*

(i) *quadrant OACB,*

(ii) *shaded region.*

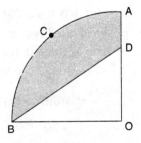

Sol. (*i*) Area of the quadrant OACB

$= \dfrac{90}{360} \pi (3.5)^2$

$= \dfrac{1}{4} \cdot \dfrac{22}{7} \cdot \dfrac{35}{10} \cdot \dfrac{35}{10}$

$= \dfrac{77}{8}$ cm^2.

(*ii*) Area of the shaded region

$= $ Area of the quadrant OACB

$\quad - $ Area of the $\triangle$OBD

$= \dfrac{77}{8} - \dfrac{OB \times OD}{2}$

$= \dfrac{77}{8} - \dfrac{3.5 \times 2}{2}$

$= \dfrac{77}{8} - \dfrac{35}{10}$

$= \dfrac{77}{8} - \dfrac{7}{2} = \dfrac{49}{8}$ cm^2.

Example 13. *In figure, a square OABC is inscribed in a quadrant OPBQ. If OA = 20 cm, find the area of the shaded region. (Use $\pi = 3.14$)*

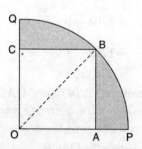

Sol. OB $= \sqrt{OA^2 + AB^2} = \sqrt{OA^2 + OA^2}$

$= \sqrt{2}$ OA $= \sqrt{2}(20) = 20\sqrt{2}$ cm

Area of the shaded region

$= $ Area of the quadrant OPBQ

$\quad - $ Area of the square OABC

$= \dfrac{90}{360} \pi (20\sqrt{2})^2 - 20 \times 20$

$= 200\pi - 400$

$= 200 \times 3.14 - 400$

$= 628 - 400$

$= 228$ cm^2.

Example 14. *AB and CD are respectively arcs of two concentric circles of radii 21 cm and 7 cm and centre O (see figure). If $\angle AOB = 30°$, find the area of the shaded region.*

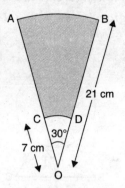

Sol. Area of the shaded region

$= $ Area of the sector OAB

$\quad - $ Area of the sector OCD

$= \dfrac{30}{360} \pi (21)^2 - \dfrac{30}{360} \pi (7)^2$

$= \dfrac{1}{12} \cdot \dfrac{22}{7} \cdot 21 \cdot 21 - \dfrac{1}{12} \cdot \dfrac{22}{7} \cdot 7 \cdot 7$

$= \dfrac{231}{2} - \dfrac{77}{6} = \dfrac{693 - 77}{6}$

$= \dfrac{616}{6} = \dfrac{308}{3}$ cm^2.

Example 15. *In figure, ABC is a quadrant of a circle of radius 14 cm and a semicircle is drawn with BC as diameter. Find the area of the shaded region.*

Sol. In right triangle BAC,

$BC^2 = AB^2 + AC^2$

| By Pythagoras theorem

$= (14)^2 + (14)^2$

$= 2(14)^2$

AREAS RELATED TO CIRCLES

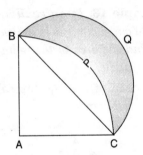

$\Rightarrow \quad BC = 14\sqrt{2}$ cm

$\Rightarrow$ Radius of the semicircle

$$= \frac{14\sqrt{2}}{2} \text{ cm}$$

$$= 7\sqrt{2}$$

Required area

= Area BPCQB
= Area BCQB − Area BCPB
= Area BCQB − (Area BACPB
  − Area of △BAC)

$$= \frac{180}{360}\pi(7\sqrt{2})^2 - \left[\frac{90}{360}\pi(14)^2 - \frac{14 \times 14}{2}\right]$$

$$= \frac{1}{2} \times \frac{22}{7} \times 98 - \left(\frac{1}{4} \times \frac{22}{7} \times 14 \times 14 - 98\right)$$

$$= 154 - (154 - 98)$$

$$= 98 \text{ cm}^2.$$

Example 16. *Calculate the area of the designed region in figure common between the two quadrants of circles of radius 8 cm each.*

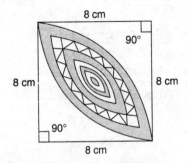

Sol. In right triangle ADC,
$AC^2 = AD^2 + CD^2$

| By Pythagoras theorem

$$= 8^2 + 8^2$$
$$= 64 + 64$$
$$= 128$$
$\Rightarrow \quad AC = \sqrt{128}$
$\Rightarrow \quad AC = 8\sqrt{2}$

Draw BM ⊥ AC

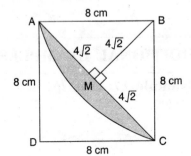

Then, $AM = MC = \frac{1}{2} AC$

$$= \frac{1}{2}(8\sqrt{2}) = 4\sqrt{2} \text{ cm}$$

In right triangle AMB
$AB^2 = AM^2 + BM^2$
| By Pythagoras theorem
$\Rightarrow \quad (8)^2 = (4\sqrt{2})^2 + BM^2$
$\Rightarrow \quad 64 = 32 + BM^2$
$\Rightarrow \quad BM^2 = 64 - 32$
$\Rightarrow \quad BM^2 = 32$
$\Rightarrow \quad BM = \sqrt{32} = 4\sqrt{2}$ cm

∴ Area of △ABC = $\dfrac{AC \times BM}{2}$

$$= \frac{8\sqrt{2} \times 4\sqrt{2}}{2}$$

$$= 32 \text{ cm}^2$$

∴ Shaded Area = $\dfrac{90}{360}\pi(8)^2 - 32$

$$= 16\pi - 32$$

$$= 16 \times \frac{22}{7} - 32$$

$$= \frac{352}{7} - 32$$

$$= \frac{352 - 224}{7}$$

$$= \frac{128}{7} \text{ cm}^2$$

∴ Area of the designed region

$$= 2 \times \frac{128}{7}$$

$$= \frac{256}{7} \text{ cm}^2.$$

ADDITIONAL EXAMPLES

Example 17. *Find the area of the shaded region*

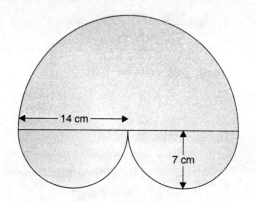

Sol. Radius of the larger semicircle = 14 cm

∴ Area of the larger semicircle

$$= \frac{1}{2} \pi (14)^2 = \frac{1}{2} \times \frac{22}{7} \times 196$$

$$= 308 \text{ cm}^2$$

Radius of each smaller semicircle
= 7 cm

∴ Area of two smaller semicircles

$$= 2 \left[\frac{1}{2} \times \pi (7)^2 \right]$$

$$= \frac{22}{7} \times (7)^2$$

$$= 154 \text{ cm}^2$$

∴ The area of the shaded region
= 308 cm² + 154 cm²
= 462 cm².

Example 18. *In an equilateral triangle of side 24 cm, a circle is inscribed touching its sides. Find the area of the remaining portion of the triangle.*

Sol. Area of the equilateral triangle ABC with side 24 cm

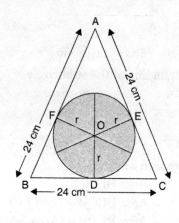

$$= \frac{\sqrt{3}}{4} (24)^2$$

$$= 144 \sqrt{3} \text{ cm}^2$$

Let r cm be the radius of the inscribed circle.

Area of △ ABC = Area of △ ABC + Area of △ OCA + Area of △ OAB

$$\Rightarrow 144 \sqrt{3} = \frac{24 \times r}{2} + \frac{24 \times r}{2} \times \frac{24 \times r}{2}$$

$$\Rightarrow 144\sqrt{3} = 36r$$

$$\Rightarrow r = \frac{144 \sqrt{3}}{36}$$

$$\Rightarrow r = 4\sqrt{3}$$

∴ Area of the inscribed circle

$$= \pi (4\sqrt{3})^2$$

$$= 48 \pi \text{ cm}^2$$

∴ Area of the remaining portion of the triangle

$$= (144 \sqrt{3} - 48 \pi) \text{ cm}^2$$

Example 19. *In the square ABCD with side 2a, with the vertices of the square as centre and $\frac{1}{2}$ the side of the square as radius, four sectors have been drawn. Find the area of the shaded portion.*

Sol. Area of square ABCD = $(2a)^2 = 4a^2$

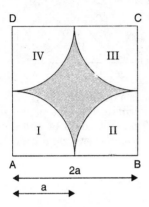

Area of sector I = Area of sector II = Area of sector III = Area of sector IV

$$= \frac{90}{360} \pi (a)^2 = \frac{\pi a^2}{4}$$

∴ Area of the shaded portion

$$= 4a^2 - 4 \times \frac{\pi a^2}{4}$$

$$= 4a^2 - \pi a^2 = (4 - \pi) a^2$$

Example 20. *It is proposed to add to a square lawn with each side 58 cm, two circular ends, the centre of each circle being the point of intersection of the diagonals of the square. Find the area of the whole lawn.*

Sol. Diagonal of the square = $58\sqrt{2}$ cm

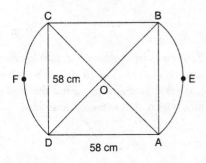

∴ Radius of the circular end

$$= \frac{58\sqrt{2}}{2}$$

$$= 29\sqrt{2} \text{ cm.}$$

Area of one circular end

= Area of sector OAEB

− Area of △ OAB

$$= \frac{90}{360} \times \pi (29\sqrt{2})^2$$

$$- \frac{(29\sqrt{2})(29\sqrt{2})}{2}$$

$$= \frac{1}{4} \times \frac{22}{7} \times 29\sqrt{2} \times 29\sqrt{2} - 29.29$$

$$= \frac{9251}{7} - 841$$

$$= \frac{9251 - 5887}{7}$$

$$= \frac{3364}{7} \text{ cm}^2$$

∴ Area of the whole lawn

= Area of the square

+ 2 (Area of one circular end)

$$= 58 \times 58 + 2 \times \frac{3364}{7}$$

$$= 3364 + \frac{2 \times 3364}{7}$$

$$= 3364 \left(1 + \frac{2}{7}\right)$$

$$= \frac{3364 \times 9}{7}$$

$$= \frac{30276}{7} \text{ cm}^2.$$

TEST YOUR KNOWLEDGE

1. In figure, two circular flower beds have been shown on two sides of a square lawn ABCD of side 56 m. If the centre of each circular flower bed is the point of intersection O of the diagonals of the square lawn, find the sum of the areas of the lawn and the flower beds.

   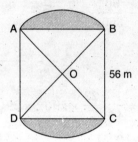

2. Find the area of the shaded region in figure, where ABCD is a square of side 14 cm.

   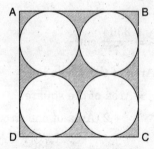

3. Find the area of the shaded design in figure, where ABCD is a square of side 10 cm and semicircles are drawn with each side of the square as diameter. (User π = 3.14)

   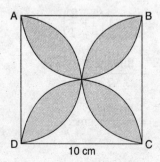

   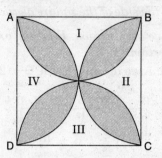

4. AOBC is a quadrant of a circle of radius 10 cm. Calculate the area of the shaded portion. Take π to be 3.14 and give your answer correct to two significant figures.

   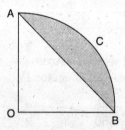

5. In figure, ABC is an equilateral triangle inscribed in a circle of radius 4 cm. Find the area of the shaded portion.

   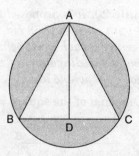

6. Four cows are tethered at four corners of a square plot of side 50 m so that they can just reach one another. Find the area of the plot which still remains ungrazed.

AREAS RELATED TO CIRCLES

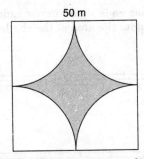

7. A circular disc of 6 cm in radius is divided into three sectors with central angles 120°, 150° and 90°. What part of the whole area is the sector with central angle 120°? Also give the ratio of the area of the sectors.

8. PQRS represents a flower bed. If OP = 21 m and OR = 14 m, find the area of the bed.

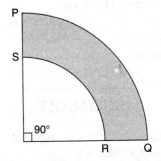

9. Three equal circles, each of radius 2 cm, touch one another. Find the area enclosed in between them.

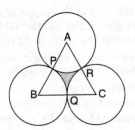

10. A square water tank has its sides equal to 40 m each. There are four semicircular grassy plots all around it. Find the cost of turfing the plots at Rs 1.25 per sq.m.

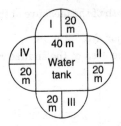

Answers

1. 4032 m^2 2. 42 cm^2
3. 57 cm^2 4. 28.5 cm^2
5. $(16\pi - 12\sqrt{3})$ cm^2 6. 535.714 m^2
7. $\frac{1}{3}$; 4 : 5 : 3   8. 192.5 m^2
9. 0.643 cm^2 10. Rs. 3140.

MISCELLANEOUS EXERCISE

1. Two circles touch externally. The sum of their area is 130 π sq. cm and the distance between their centres is 14 cm. Find the radii of the circles.

2. The circumference of a circle exceeds its diameter by 16.8 cm. Find the radius of the circles.

3. The diameters of two circles are in the ratio 3 : 4 and the sum of the areas of the circles is equal to the area of a circle where diameter measures 30 cm. Find the diameters of the given circles.

4. A paper is in the form of a rectangle ABCD, where AB = 18 cm and BC = 14 cm. A semicircular portion with BC as diameter is cut off. Find the area of the remaining paper.

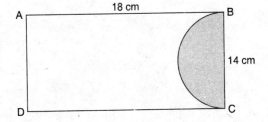

5. Find the area of the sector of a circle when the angle of the sector is 63° and the diameter of the circle is 20 cm.

6. The perimeter of a certain sector of a circle of radius 5.6 cm is 27.2 cm. Find the area of the sector.

7. The radius of a circle with centre O is 5 cm. Two radii OA and OB are drawn at right angles to each other. Find the area of the two segments made by the chord AB.

8. In a circle of diameter 21 cm, find the area of the sector corresponding to the major arc of the measure of the mirror arc of the circle is $\frac{1}{5}$ th of the measure of the corresponding major arc.

9. Find the area of the shaded region :

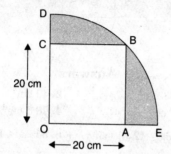

10. The short and long hands of a clock are 4 cm and 6 cm long respectively. Find the sum of the distances travelled their by tips in 2 days.

11. The diameters of the front and rear wheels of a tractor are 80 cm and 2 m respectively. Find the number of revolutions that a rear wheel makes to cover the distance which the front wheel covers in 800 revolutions.

12. An athletic tracks 14 m wide consists of two straight sections 120 m long joining semi-circular ends where minor radius is 35 m. Calculate the area of the shaded region.

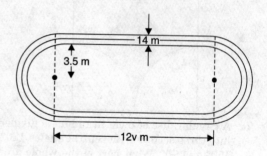

Answers

1. 11 cm, 3 cm   2. 3.92 cm.
3. 18 cm, 24 cm   4. 175 cm²
5. 55 cm²   6. 44.8 cm²
7. 7.125 cm², 71.375 cm²
8. 227.20 cm²   9. 228 cm²
10. 1910.57 cm.   11. 320
12. 7056 m²

SUMMARY

1. Circumference of a circle = $2\pi r$.
2. Area of a circle = πr^2.
3. Length of an arc of a sector of a circle with radius r and angle with degree measure θ is $\frac{\theta}{360} \times 2\pi r$.
4. Area of a sector of a circle with radius r and angle with degrees measure θ is $\frac{\theta}{360} \times \pi r^2$.
5. Area of segment of a circle
   = Area of the corresponding sector
   − Area of the corresponding triangle.

13

Surface Areas And Volumes

IMPORTANT POINTS

1. Pre-knowledge. We are familiar with some of the solids like cuboid, cone, cylinder, and sphere (see figure). We also know how to find their surface areas and volumes.

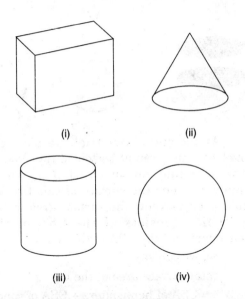

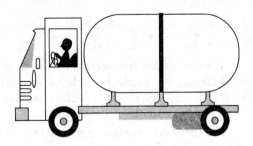

(2) Again, we have seen an object like the one in figure. Rightly, we can name it as a test tube. We have used one in your science laboratory. This tube is also a combination of a cylinder and a hemisphere

2. Combinations of solids. In our day-to-day life, we come across a number of solids made up of combinations of two or more of basic solids as shown above.

Examples :

(1) We have seen a truck with a container fitted on its back (see figure), carrying oil or water from one place to another. It is not in the shape of any of the four basic solids mentioned above. We may guess that it is made of a cylinder with two hemispheres as its ends.

Similarly, while travelling, we see some big and beautiful buildings or monuments made up of a combination of solids mentioned above.

If for some reason we want to find the surface areas, or volumes, or capacities of such objects, then it is not easy for us because we cannot classify these under any of the solids we have already studied.

3. Surfaces Area of a Combination of Solids.

(1) **Situation.** Let us consider the container see in figure. We want to find the surface area of such a solid. Now, whenever we come across a new problem, we first try to see, if we can break it down into smaller problems, we have earlier solved. We can see that this solid is made up of a cylinder with two hemispheres stuck at either and. It would look like what we have in figure after we put the pieces all together.

If we consider the surface of the newly formed object, we would be able to see only the curved surfaces of the two hemispheres and the curved surface of the cylinder.

So, the *total* surface area of the new solid is the sum of the *curved* surface areas of each of the individual parts. This gives,

TSA of new solid

= CSA of one hemisphere

+ CSA of cylinder

+ CSA of other hemisphere

where TSA, CSA stand for 'Total Surface Area' and 'Curved Surface Area' respectively.

(2) **Situation 2.** Let us now consider another situation. Suppose we are making a toy by putting together a hemisphere and a cone. Let us see the steps that we would be going through.

First we would take a cone and a hemisphere and bring their flat faces together. Here, of course, we would take the base radius of the cone equal to the radius of the hemisphere, for the toy is to have a smooth surface. So, the steps would be as shown in figure.

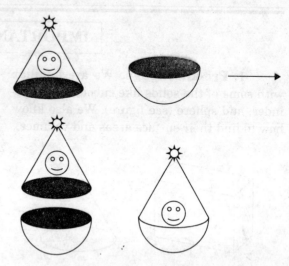

At the end of our trial, we have got ourselves a nice round bottomed toy. Now if we want to find how much paint we would require to colour the surface of this toy, we would need to know the surface area of the toy, which consists of the CSA of the hemisphere and the CSA of the cone.

So, we can say:

Total surface area of the toy

= CSA of hemisphere + CSA of cone.

ILLUSTRATIVE EXAMPLES

[NCERT Exercise 13.1]
(Page No. 244)

Unless stated otherwise, take $\pi = \dfrac{22}{7}$.

Example 1. *2 cubes each of volume 64 cm^3 are joined end to end. Find the surface area of the resulting cuboid.*

Sol. Let the length of the edge of a cube be a cm.

Then, its volume = $a^3 \text{ cm}^3$

According to the question,

$a^3 = 64$

$\Rightarrow \quad a^3 = 4^3$

$\Rightarrow \quad a = 4$

SURFACE AREAS AND VOLUMES

Hence, the length of the edge of each cube is 4 cm.

For the resulting cuboid

length $(l) = 4 + 4 = 8$ cm

breadth $(b) = 4$ cm

height $(h) = 4$ cm

∴ Surface area of the resulting cuboid

$= 2(lb + bh + hl)$

$= 2(8 \times 4 + 4 \times 4 + 4 \times 8)$

$= 2(32 + 16 + 32)$

$= 2(80)$

$= 160 \text{ cm}^2$.

Example 2. *A vessel is in the form of a hollow hemisphere mounted by a hollow cylinder. The diameter of the hemisphere is 14 cm and the total height of the vessel is 13 cm. Find the inner surface area of the vessel.*

Sol. ∵ Diameter of the hollow hemisphere = 14 cm

∴ Radius of the hollow hemisphere

$= \dfrac{14}{2} = 7$ cm

∴ Radius of the base of the hollow cylinder = 7 cm.

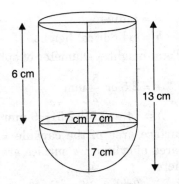

Total height of the vessel = 13 cm

∴ Height of the hollow cylinder

$= 13 - 7 = 6$ cm

Therefore,

Inner surface area of the vessel

= Inner surface area of the hollow hemisphere
+ Inner surface area of the hollow cylinder

$= 2\pi(7)^2 + 2\pi(7)(6)$

$= 98\pi + 84\pi$

$= 182\pi$

$= 182 \times \dfrac{22}{7}$

$= 26 \times 22$

$= 572 \text{ cm}^2$.

Example 3. *A toy is in the form of a cone of radius 3.5 cm mounted on a hemisphere of same radius. The total height of the toy is 15.5 cm. Find the total surface area of the toy.*

Sol. Radius of the cone = 3.5 cm

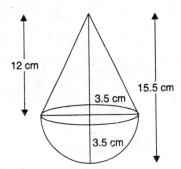

∴ Radius of the hemisphere = 3.5 cm

Total height of the toy = 15.5 cm

∴ Height of the cone

$= 15.5 - 3.5 = 12$ cm

Slant height of the cone

$= \sqrt{(3.5)^2 + (12)^2} = \sqrt{12.25 + 144}$

$= \sqrt{156.25} = 12.5$ cm

Therefore,

Total surface area of the toy

= curved surface area of the hemisphere + curved surface area of the cone.

$$= 2\pi(3.5)^2 + \pi(3.5)(12.5)$$
$$= 24.5\pi + 43.75\pi$$
$$= 68.25\pi$$
$$= \frac{68.25 \times 22}{7}$$
$$= 214.5 \text{ cm}^2.$$

Example 4. *A cubical block of side 7 cm is surmounted by a hemisphere. What is the greatest diameter the hemisphere can have ? Find the surface area of the solid.*

Sol. Greatest diameter of the hemisphere
$$= \text{Side of the cubical block}$$
$$= 7 \text{ cm}$$
Therefore,
Surface area of the solid
= External surface area of the cubical block + Surface area of the hemisphere

$$= \left\{6(7)^2 = \pi\left(\frac{7}{2}\right)^2\right\} + 2\pi\left(\frac{7}{2}\right)^2$$
$$= \left\{294 - \frac{49}{4}\pi\right\} + \frac{49}{2}\pi$$
$$= 294 + \frac{49}{4}\pi$$
$$= 294 + \frac{49}{4} \times \frac{22}{7}$$
$$= 294 + \frac{77}{2}$$
$$= 294 + 38.5$$
$$= 332.5 \text{ cm}^2.$$

Example 5. *A hemispherical depression is cut out from one face of a cubical wooden block such that the diameter l of the hemisphere is equal to the edge of the cube. Determine the surface area of the remaining solid.*

Sol. ∵ Diameter of the hemisphere
$$= l$$
∴ Radius of the hemisphere $= \frac{l}{2}$

Also, length of the edge of the cube $= l$.
Therefore, surface area of the remaining solid

$$= 2\pi\left(\frac{l}{2}\right)^2 + 6l^2 - \pi\left(\frac{l}{2}\right)^2$$
$$= \pi\left(\frac{l}{2}\right)^2 + 6l^2$$
$$= \frac{\pi l^2}{4} + 6l^2$$
$$= \frac{l^2}{4}(\pi + 24).$$

Example 6. *A medicine capsule is in the shape of a cylinder with two hemispheres stuck to each of its ends (see figure below). The length of the entire capsule is 14 mm and the diameter of the capsule is 5 mm. Find its surface area.*

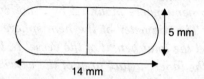

Sol. Radius of the hemisphere $= \frac{5}{2}$ mm

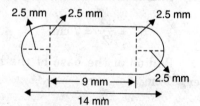

r = radius
h = cylindrical height
(Total height − Diameter of sphere)
$r = 2.5$ or $\frac{5}{2}$ mm
$h = 14 - (2.5 + 2.5) = 9$ mm

Surface area of the capsule = curved surface area of cylinder + surface area of the hemisphere.
$$= 2\pi r h + 2(2\pi r^2).$$
Surface area of the capsule
$$= 2\pi\left(\frac{5}{2}\right)(9) + 2\left[2.\pi.\left(\frac{5}{2}\right)^2\right]$$

= 45π + 25π

= 70π = 70 × $\frac{22}{7}$ = 220 mm².

Example 7. *A tent is in the shape of a cylinder surmounted by a conical top. If the height and diameter of the cylindrical part are 2.1 m and 4m respectively, and the slant height of the top is 2.8 m, find the area of the canvas used for making the tent. Also, find the cost of the canvas of the tent at the rate of Rs. 500 per m² (Note that the base of the tent will not be covered with canvas.)*

Sol. ∵ Diameter of the cylindrical part

= 4 cm

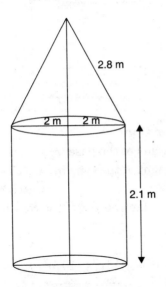

∴ Radius of the cylindrical part

= $\frac{4}{2}$ = 2 m

∴ Radius of the base of the conical top

= 2 cm

Therefore,

Total surface area of the tent

= curved surface area of the cylindrical part

+ curved surface area of the conical top

= 2π(2)(2.1) + π(2)(2.8)

= 8.4π + 5.6π

= 14π

= 14 × $\frac{22}{7}$

= 44 m²

∴ Cost of the canvas of the tent at the rate of Rs. 500 per m².

= 44 × 500

= Rs. 22000.

Example 8. *From a solid cylinder whose height is 2.4 cm and diameter 1.4 cm, a conical cavity of the same height and same diameter is hollowed out. Find the total surface area of the remaining solid to the nearest cm².*

Sol. ∵ Diameter of the solid cylinder

= 1.4 cm

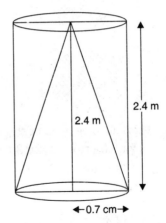

∴ Radius of the solid cylinder

= $\frac{1.4}{2}$ cm = 0.7 cm

∴ Radius of the base of the conical cavity

= 0.7 cm

∵ Height of the solid cylinder = 2.4 cm

∴ Height of the conical cavity = 2.4 cm

∴ Slant height of the conical cavity

= $\sqrt{(0.7)^2 + (2.4)^2}$

= $\sqrt{0.49 + 5.76}$

= $\sqrt{6.25}$

= 2.5 cm

Therefore,
Total surface area of the remaining solid
$= 2\pi(0.7)(2.4) + \pi(0.7)^2 + \pi(0.7)(2.5)$
$= 3.36\pi + 0.49\pi + 1.75\pi$
$= 5.6\pi$
$= (5.6) \times \dfrac{22}{7} = 17.6$ cm^2
$= 18$ cm^2 (to the nearest cm^2).

Example 9. *A wooden article was made by scooping out a hemisphere from each end of a solid cylinder, as shown in figure. If the height of the cylinder is 10 cm, and its base is of radius 3.5 cm, find the total surface area of the article.*

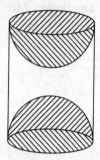

Sol. Total surface area of the article
$= 2\pi(3.5)(10) + 2[2\pi(3.5)^2]$
$= 70\pi + 49\pi$
$= 119\pi$
$= 119 \times \dfrac{22}{7}$
$= 17 \times 22$
$= 374$ cm^2.

ADDITIONAL EXAMPLES

Example 10. *A tent is in the shape of a right circular cylinder upto a height of 3m and conical above it. The total height of the tent is 13.5 m and radius of base is 14m. Find the cost of cloth required to make the tent at the rate of Rs. 80 per sq. m.* (CBSE 2005)

Sol. For right circular cylinder
Radius of the base (r) = 14 m
Height of the cylinder (h) = 3 m

∴ Curved surface area of the cylinder
$= 2\pi rh$
$= 2 \times \dfrac{22}{7} \times 14 \times 3$
$= 264$ m^2

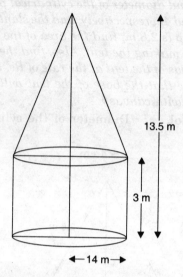

For Cone
Radius of the base (r) = 14 m
Height of the cone (h) = 13.5 − 3
$= 10.5$ m
∴ Slant height of the cone (l)
$= \sqrt{r^2 + h^2}$
$= \sqrt{(14)^2 + (10.5)^2}$
$= \sqrt{196 + 110.25}$
$= \sqrt{306.25}$
$= 17.5$ m
∴ Curved surface area of the cone
$= \pi rl$
$= \dfrac{22}{7} \times 14 \times 17.5$
$= 770$ m^2
∴ Total curved surface area of the tent
= curved surface area of the cylinder
  + curved surface area of the cone
$= 264$ m^2 + 770 m^2

= 1034 m²

∴ Cloth required to make the tent
= 1034 m²

∴ Cost of cloth required to make the tent
= 1034 × 80
= Rs. 82720.

Example 11. *A circus tent is cylindrical to a height of 3 and conical above it. If its base radius is 52.5 m and slant height of the conical portion is 53 m, find the area of the canvas needed to make the tent.* $\left(Use\ \pi = \dfrac{22}{7}\right)$

(AI CBSE 2004)

Sol. For cylindrical portion

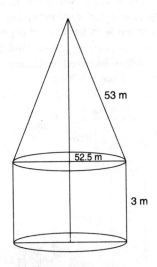

Base radius (r) = 52.5 m
Height (h) = 3 m
∴ Curved surface area
= $2\pi rh$
= $2 \times \dfrac{22}{7} \times 52.5 \times 3$
= 990 m²

For conical portion
Base radius (r) = 52.5 m
Slant height (l) = 53 m
∴ Curved surface area
= πrl

= $\dfrac{22}{7} \times 52.5 \times 53$
= 8745 m²

∴ Total surface area = Curved surface area of the cylindrical portion + Curved surface area of the conical portion
= 990 m² + 8745 m²
= 9735 m²

Hence, the area of the canvas needed to make the tent is 9735 m².

Example 12. *A circus tent is in the shape of a cylinder surrounded by a cone. The diameter of the cylindrical part is 24 m and its height is 11 m. If the vertex of the tent is 16 m above the ground, find the area of the canvas required to make the tent.*

(CBSE SP Set 2)

Sol. For cylindrical part

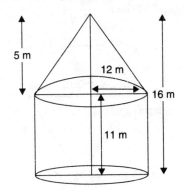

Radius of the base (r)
= 12 m
Height (h) = 11 m
∴ Curved surface area
= $2\pi rh$
= $2\pi (12)(11)$
= 264 π m²

For conical part
Radius of the base (r) = 12 m
Height (h) = 16 – 11 = 5 m

∴ Slant height (l) = $\sqrt{r^2 + h^2}$

= $\sqrt{(12)^2 + 5^2}$

= $\sqrt{144 + 25} = \sqrt{119}$ = 13 m

∴ Curved surface area
$= \pi r l = \pi (12)(13) = 156 \pi \text{ m}^2$

∴ Total surface area of the tent
= Curved surface area of the cylindrical part
+ Curved surface area of the conical part
$= 264\pi + 156\pi$
$= 420\pi$
$= 420 \times \dfrac{22}{7} = 1320 \text{ m}^2$

Hence, the area of the canvas required to make the tent is 1320 m².

TEST YOUR KNOWLEDGE

1. Rasheed got a playing top (*lattu*) as his birthday present, which surprisingly had no colour on it. He wanted to colour it with his crayons. The top is shaped like a cone surmounted by a hemisphere (figure). The entire top is 5 cm in height and the diameter of the top is 3.5 cm. Find the area he has to colour. $\left(\text{Take } \pi = \dfrac{22}{7}\right)$.

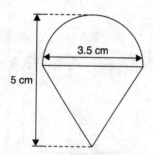

2. The decorative block shown in figure is made of two solids—a cube and a hemisphere. The base of the block is a cube with edge 5 cm, and the hemisphere fixed on the top has a diameter of 4.2 cm. Find the total surface area of the block. $\left(\text{Take } \pi = \dfrac{22}{7}\right)$

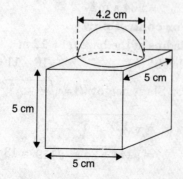

3. A wooden toy rocket is in the shape of a cone mounted on a cylinder, as shown in figure. The height of the entire rocket is 26 cm, while the height of the conical part is 6 cm. The base of the conical portion has a diameter of 5 cm, while the base diameter of the cylindrical portion is 3 cm. If the conical portion is to be painted orange and the cylindrical portion yellow, find the area of the rocket painted with each of these colours. (Take π = 3.14)

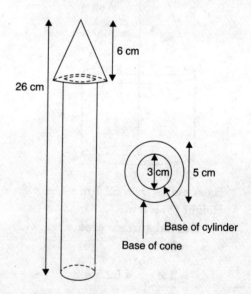

4. Mayank made a bird-bath for his garden in the shape of a cylinder with a hemispherical depression at one end (see figure). The height of the cylinder is 1.45 m and its radius is 30 cm. Find the total surface area of the bird-bath. $\left(\text{Take } \pi = \dfrac{22}{7}\right)$

SURFACE AREAS AND VOLUMES

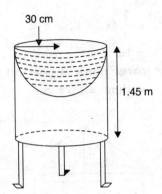

5. A tent of height 8.25 m is in the form of a right circular cylinder with diameter of base 30 m and height 5.5 m, surmounted by a right circular cone of the same base. Find the cost of the canvas of the tent at the rate of Rs. 45 per m².

6. A solid is composed of a cylinder with hemispherical ends. If the whole length of the solid is 108 cm and the diameter of each of the hemispherical ends is 36 cm, find the cost of polishing the surface at the rate of 7 paise per sq. m. (Take $\pi = 3.1416$).

7. A toy is in the form of a cone mounted on a hemisphere. The diameter of the base of the cone is 6 cm and its height is 4 cm. Calculate the surface area of the toy. (Use $\pi = 3.14$).

8. A tent is in the shape of a right circular cylinder upto a height of 3 metres and then becomes a right circular cone with a maximum height of 13.5 metres above the ground. Calculate the cost of painting the inner side of the tent @ Rs. 2 per sq. m, if the radius of the base is 14 metres.

Answers

1. 39.6 cm² (approx.)
2. 163.86 cm²
3. Orange → 63.585 cm², yellow → 195.465 cm².
4. 3.3 cm²
5. Rs. 55687.50
6. Rs. 855.02
7. 103.62 cm²
8. Rs. 2068.

VOLUME OF A COMBINATION OF SOLIDS

IMPORTANT POINTS

1. Volume of the combination of solids as the sum of the volumes of the constituents. We have discussed how to find the surface area of solids made up of combination to two basic solids. Now, we shall learn to calculate their volumes. It is noteworthy that in calculating the surface area, we do not add the surface areas of the two constituents, because some part of the surface area disappeared in the process of joining them. However, this will not be the case when we calculate the volume. The volume of the solid formed by joining two basic solids will actually be the sum of the volumes of the constituents.

ILLUSTRATIVE EXAMPLES

[NCERT Exercise 13.2]
(Page No. 247)

Unless stated otherwise, take $\pi = \dfrac{22}{7}$.

Example 1. *A solid is in the shape of a cone standing on a hemisphere with both their radii being equal to 1 cm and the height of the cone is equal to its radius. Find the volume of the solid in terms of π.*

Sol. For Hemisphere

Radius $(r) = 1$ cm

$\therefore$ Volume $= \dfrac{2}{3}\pi r^3$

$= \dfrac{2}{3}\pi(1)^3$

$= \frac{2}{3} \pi \text{ cm}^3$

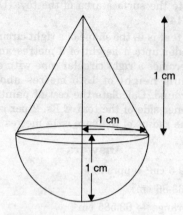

For cone

Radius of the base (r) = 1 cm

Height (h) = 1 cm

$\therefore$ Volume $= \frac{1}{3}\pi r^2 h$

$= \frac{1}{3}\pi (1)^2 (1) = \frac{1}{3}\pi \text{ cm}^3$

Therefore, volume of the solid
= volume of the hemisphere
  + volume of cone

$= \frac{2}{3}\pi + \frac{1}{3}\pi$

$= \pi \text{ cm}^3$.

Example 2. *Rachel, an engineering student, was asked to make a model shaped like a cylinder with two cones attached at its two ends by using a thin aluminium sheet. The diameter of the model is 3 cm and its length is 12 cm. If each cone has a height of 2 cm, find the volume of air contained in the model that Rachel made. (Assume the outer and inner dimensions of the model to be nearly the same.)*

Sol. For upper conical portion

Radius of the base (r) = 1.5 cm

Height (h_1) = 2 cm

$\therefore$ Volume $= \frac{1}{3}\pi r^2 h_1$

$= \frac{1}{3}(1.5)^2 (2)$

$= 1.5 \pi \text{ cm}^3$

For lower conical portion

volume = $1.5 \pi \text{ cm}^3$

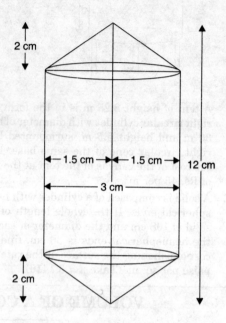

For central cylindrical portion

Radius of the base (r) = 1.5 cm

Height (h_2) = 12 − (2 + 2)

= 8 cm

$\therefore$ Volume

$= \pi r^2 h_2$

$= \pi (1.5)^2 (8)$

$= 18 \pi \text{ cm}^3$

Therefore, volume of the model

$= 1.5 \pi + 1.5 \pi + 18 \pi$

$= 21 \pi$

$= 21 \times \frac{22}{7}$

$= 66 \text{ cm}^3$

Hence, the volume of the air contained in the model that Rachel made is 66 cm³.

Example 3. *A gulab jamun, contains sugar syrup up to about 30% of its volume. Find* approximately how much syrup would be found in 45 gulab jamuns, each shaped like a cylinder with two hemispherical ends with length 5 cm and diameter 2.8 cm (see figure).

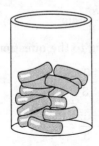

Sol. Volume of a gulab jamun

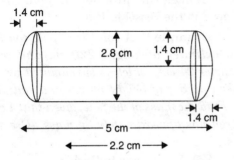

$$= \frac{2}{3}\pi(1.4)^2 + \pi(1.4)^2(2.2) + \frac{2}{3}\pi(1.4)^3$$

$$= \frac{4}{3}\pi(1.4)^3 + \pi(1.4)^2(2.2)$$

$$= \pi(1.4)^2\left[\frac{4 \times 1.4}{3} + 2.2\right]$$

$$= \pi(1.96)\left[\frac{5.6 + 6.6}{3}\right]$$

$$= \frac{\pi(1.96)(12.2)}{3}\text{ cm}^3$$

∴ Volume of 45 gulab jamuns

$$= 45 \times \frac{\pi(1.96)(12.2)}{3}$$

$$= 15\pi(1.96)(12.2)$$

$$= 15 \times \frac{22}{7} \times 1.96 \times 12.2$$

$$= 15 \times 22 \times 2.8 \times 12.2$$

$$= 1127.28 \text{ cm}^3$$

∴ Volume of syrup

$$= 1127.28 \times \frac{30}{100}$$

$$= 338.184 \text{ cm}^3$$

$$= 338 \text{ cm}^3 \text{ (approximately)}.$$

Example 4. *A pen stand made of wood is in the shape of a cuboid with four conical depressions to hold pens. The dimensions of the cuboid are 15 cm by 10 cm by 3.5 cm. The radius of each of the depressions is 0.5 cm and the depth is 1.4 cm. Find the volume of wood in the entire stand (see figure).*

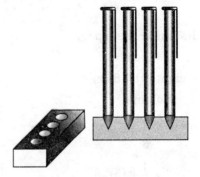

Sol. Volume of the cuboid

$$= 15 \times 10 \times 3.5 = 525 \text{ cm}^3$$

Volume of a conical depression

$$= \frac{1}{3}\pi(0.5)^2(1.4)$$

$$= \frac{1}{3} \times \frac{22}{7} \times 0.25 \times \frac{14}{10}$$

$$= \frac{11}{30} \text{ cm}^3$$

∴ Volume of four conical depressions

$$= 4 \times \frac{11}{30} \text{ cm}^3 = \frac{22}{15} \text{ cm}^3$$

$$= 1.47 \text{ cm}^3$$

∴ Volume of the wood in the entire stand

$$= 525 - 1.47 = 523.53 \text{ cm}^3.$$

Example 5. *A vessel is in the form of an inverted cone. Its height is 8 cm and the radius of its top, which is open, is 5 cm. It is filled with water up to the brim. When lead shots, each of which is a sphere of radius 0.5 cm* are dropped into the vessel, one-fourth of the water flows out. Find the number of lead shots dropped in the vessel.

Sol. For cone

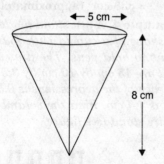

Radius of the top (r)
$= 5$ cm

Height $(h) = 8$ cm

∴ Volume

$= \dfrac{1}{3} \pi r^2 h$

$= \dfrac{1}{3} \pi (5)^2 \, 8$

$= \dfrac{200}{3} \pi \text{ cm}^3$

For spherical lead shot

Radius (R) = 0.5 cm

∴ Volume of a spherical lead shot

$= \dfrac{4}{3} \pi R^3$

$= \dfrac{4}{3} \pi (0.5)^3$

$= \dfrac{\pi}{6} \text{ cm}^3$

Volume of water that flows out

$= \dfrac{1}{4}$ volume of the cone

$= \dfrac{1}{4} \left(\dfrac{200 \, \pi}{3} \right) \text{ cm}^3$

$= \dfrac{50 \, \pi}{3} \text{ cm}^3$

Let the number of lead shots dropped in the vessel be n.

Then,

Volume of n lead shots

$= \dfrac{n\pi}{6} \text{ cm}^3$

According to the question,

$\dfrac{n\pi}{6} = \dfrac{50 \, \pi}{3}$

$\Rightarrow \quad n = \dfrac{50\pi}{3} \cdot \dfrac{6}{\pi}$

$\Rightarrow \quad n = 100$

Hence, the number of lead shots dropped in the vessel is 100.

Example 6. *A solid iron pole consists of a cylinder of height 220 cm and base diameter 24 cm, which is surmounted by another cylinder of height 60 cm and radius 8 cm. Find the mass of the pole, given that 1 cm³ of iron has approximately 8g mass. (Use π = 3.14).*

Sol. For lower cylinder

Base radius $(r) = \dfrac{24}{2} = 12$ cm

Height $(h) = 220$ cm

∴ Volume $= \pi r^2 h$

$= \pi (12)^2 \, (220)$

$= 31680 \, \pi \text{ cm}^3$

For upper cylinder

Base radius (R) = 8 cm

Height (H) = 60 cm

∴ Volume

$= \pi R^2 H$

$= \pi (8)^2 \, (60)$

$= 3840 \, \pi \text{ cm}^3$

∴ Volume of the solid iron pole
 = volume of the lower cylinder
   + volume of the upper cylinder
 = 31680 π + 3840 π

= 35520 π
= 35520 × 3.14
= 111532.8 cm³

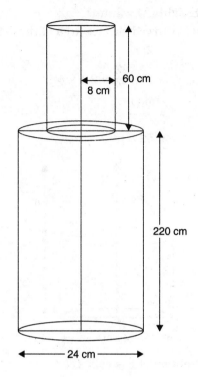

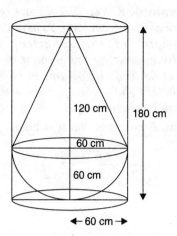

∴ Volume = $\frac{1}{3}\pi r^2 h_1$

= $\frac{1}{3}\pi (60)^2 (120)$

= 144000 π cm³

For hemisphere

Radius of the base (r) = 60 cm

∴ Volume = $\frac{2}{3}\pi r^3$

= $\frac{2}{3}\pi (60)^3$

= 144000 π cm³

For right Circular Cylinder

Radius of the base (r) = 60 cm

Height (h_2) = 180 cm

∴ Volume = $\pi r^2 h_2$

= π (60)² (180)

= 648000π cm³

∴ Volume of water left in the cylinder = Volume of the right circular cylinder − [volume of the right circular cone + volume of the hemisphere]

= 648000π − [144000π + 144000π]
= 648000π − 288000π
= 360000π cm³
= $\frac{360000}{100 \times 100 \times 100}\pi$ cm³
= 0.36π cm³
= 0.36 × $\frac{22}{7}$
= 1.131 m³ (approx.)

∴ Mass of the pole
= 111532.8 × 8 g
= 892262.4 g
= 892.2 6 kg

Hence, the mass of the pole is 892.26 kg (approximately).

Example 7. *A solid consisting of a right circular cone of height 120 cm and radius 60 cm standing on a hemisphere of radius 60 cm is placed upright in a right circular cylinder full of water such that it touches the bottom. Find the volume of water left in the cylinder, if the radius of the cylinder is 60 cm and its height is 180 cm.*

Sol. For right circular cone

Radius of the base (r) = 60 cm

Height (h_1) = 120 cm

Example 8. *A spherical glass vessel has a cylindrical neck 8 cm long, 2 cm in diameter; the diameter of the spherical part is 8.5 cm. By measuring the amount of water it holds, a child finds its volume to be 345 cm³. Check whether she is correct, taking the above as the inside measurements, and π = 3.14.*

Sol. Amount of water it holds

$$= \frac{4}{3}\pi\left(\frac{8.5}{2}\right)^3 + \pi\left(\frac{2}{2}\right)^2 (8)$$

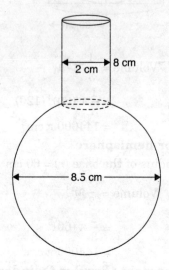

$$= \frac{4}{3} \times 3.14 \times 4.25 \times 4.25 \times 4.25$$

$$+ 8 \times 3.14$$

$$= 321.39 + 25.12$$

$$= 346.51 \text{ cm}^3$$

Hence, she is not correct. The correct volume is 346.51 cm³.

ADDITIONAL EXAMPLES

Example 9. *A tent is in the form of a cylinder of diameter 4.2 m and height 4m, surmounted by a cone of equal base and height 2.8 m. Find the capacity of the tent and the cost of canvas for making the tent at Rs. 100 per sq. m.* (AI CBSE 2005)

Sol. For cylinder
Diameter = 4.2 m
∴ Radius $(r) = \frac{4.2}{2}$ m = 2.1 m
Height (h) = 4 m
∴ Curved surface area of the cylinder
$$= 2\pi rh$$
$$= 2 \times \frac{22}{7} \times 2.1 \times 4$$
$$= \frac{264}{5}$$
$$= 52.8 \text{ m}^2$$

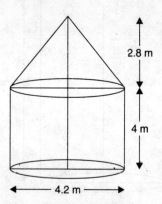

Volume of the cylinder
$$= \pi r^2 h$$
$$= \frac{22}{7} \times (2.1)^2 \times 4$$
$$= 55.44 \text{ m}^3$$

For cone

Radius of the base $(r) = \frac{4.2}{2} = 2.1$ m
Height (H) = 2.8 m
∴ Slant height (L) $= \sqrt{r^2 + H^2}$
$$= \sqrt{(2.1)^2 + (2.8)^2}$$
$$= \sqrt{4.41 + 7.84}$$
$$= \sqrt{12.25}$$
$$= 3.5 \text{ m}$$
∴ Curved surface area of the cone
$$= \pi r L$$
$$= \frac{22}{7} \times (2.1) \times (3.5)$$
$$= 23.1 \text{ m}^2$$

SURFACE AREAS AND VOLUMES

Volume of the cone = $\frac{1}{3}\pi r^2 H$

$= \frac{1}{3} \times \frac{22}{7} \times (2.1)^2 \times (2.8)$

$= 12.936 \text{ m}^3$

∴ Capacity of the tent
= Volume of the cylinder
    + Volume of the cone
= 55.44 + 12.936
= 68.376 m³

Curved surface area of the tent
= Curved surface area of the cylinder + curved surface area of the cone
= 52.8 + 23.1
= 75.9 m²

∴ Cost of canvas for making the tent at Rs. 100 per sq. m = 75.9 × 100 = Rs. 7590.

Example 10. *A solid toy is in the form of a hemisphere surmounted by a right circular cone of the height of the cone is 4 cm and diameter of the base is 6 cm, calculate :*

(i) *The volume of the toy*

(ii) *Surface area of the toy. (Use* $\pi = 3.14$)

(CBSE SP SET 1)

Sol. For hemisphere

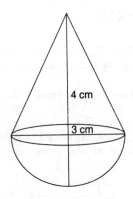

Radius (r) = 3 cm

∴ Volume = $\frac{2}{3}\pi r^3$

$= \frac{2}{3}\pi(3)^3$

$= 18\pi \text{ cm}^3$

∴ Curved surface area

$= 2\pi r^2$
$= 2\pi(3)^2$
$= 18\pi \text{ cm}^2$

For right circular cone

Radius of the base (r) = 3 cm
Height (h) = 4 cm

∴ Slant height (l) = $\sqrt{r^2 + h^2}$

$= \sqrt{(3)^2 + (4)^2}$

$= \sqrt{9 + 16} = \sqrt{25} = 5 \text{ cm}$

∴ Volume = $\frac{1}{3}\pi r^2 h$

$= \frac{1}{3}\pi(3)^2(4) = 12\pi \text{ cm}^3$

Curved surface area = $\pi r l$

$= \pi(3)(5)$
$= 15\pi \text{ cm}^2$

(i) The volume of the toy = Volume of the hemisphere + Volume of the right circular cone

$= 18\pi + 12\pi$
$= 30\pi$
$= 30(3.14)$
$= 94.2 \text{ cm}^3$

(ii) Surface area of the toy = Curved surface area of the hemisphere + Curved surface area of the right circular cone

$= 18\pi + 15\pi$
$= 33\pi$
$= 33 \times 3.14 \text{ cm}^2$
$= 103.62 \text{ cm}^2.$

Example 11. *An 'ice-cream cone' is the union of a right circular cone and a hemisphere that has the same (circular) base as the cone. Find the volume of the ice-cream if the height of the cone is 9 cm and the radius of its base is 2.5 cm.*

Sol. For cone

Radius of the base (r)

$= 2.5 \text{ cm} = \frac{5}{2} \text{ cm}$

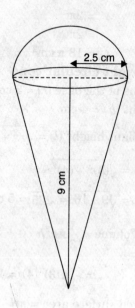

Height $(h) = 9$ cm

∴ Volume

$= \frac{1}{3} \pi r^2 h$

$= \frac{1}{3} \times \frac{22}{7} \times \frac{5}{2} \times \frac{5}{2} \times 9$

$= \frac{825}{14}$ cm^3

For hemisphere

Radius $(r) = 2.5$ cm $= \frac{5}{2}$ cm

∴ Volume $= \frac{2}{3} \pi r^3$

$= \frac{2}{3} \times \frac{22}{7} \times \frac{5}{2} \times \frac{5}{2} \times \frac{5}{2}$

$= \frac{1375}{42}$ cm^3

∴ Volume of the ice-cream = Volume of the cone + Volume of the hemisphere

$= \frac{825}{14} + \frac{1375}{42}$

$= \frac{2475 + 1375}{42}$

$= \frac{3850}{42} = \frac{275}{3} = 91\frac{2}{9}$ cm^3.

TEST YOUR KNOWLEDGE

1. Shanta runs an industry in a shed which is in the shape of a cuboid surmounted by a half cylinder (see figure). If the base of the shed is of dimension 7 m × 15 m, and the height of the cuboidal portion is 8 m, find the volume of air that the shed can hold.

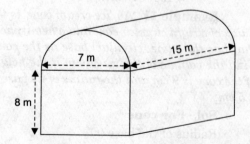

Further, suppose the machinery in the shed occupies a total space of 300 m^3, and there are 20 workers, each of whom occupy about 0.08 m^3 space on an average. Then, how much air is in the shed ? $\left(\text{Take } \pi = \frac{22}{7}\right)$.

2. A juice seller was serving his customers using glasses as shown in figure. The inner diameter of the cylindrical glass was

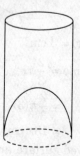

SURFACE AREAS AND VOLUMES

5 cm, but the bottom of the glass had a hemispherical raised portion which reduced the capacity of the glass. If the height of a glass was 10 cm, find the apparent capacity of the glass and its actual capacity. (Use $\pi = 3.14$).

3. A solid toy is in the form of a hemisphere surmounted by a right circular cone. The height of the cone is 2 cm and the diameter of the base is 4 cm. Determine the volume of the toy. If a right circular cylinder circumscribes the toy, find the difference of the volumes of the cylinder and the toy. (Take $\pi = 3.14$).

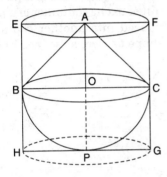

4. A circus tent has cylindrical shape surmounted by conical roof. The radius of the cylindrical base is 20 m. The heights of the cylindrical and conical portions are 4.2 m and 2.1 m, respectively. Find the volume of the tent.

5. Find the volume of a solid in the form of a right circular cylinder with hemispherical ends whose total length is 2.7 m and the diameter of each hemispherical end is 0.7 m.

6. An iron pole consisting of a cylindrical portion 110 cm high and of base diameter 12 cm is surmounted by a cone 9 cm high. Find the mass of the pole, given that 1 cm³ of iron has 8 g mass (approx.).

$$\left(\text{Use } \pi = \frac{355}{113}\right).$$

7. A petrol tank is a cylinder of base diameter 21 cm and length 18 cm fitted with conical ends each of axis length 9 cm. Determine the capacity of the tank.

8. A tent is in the form of a cylinder of diameter 20 m and height 2.5 m, surmounted by a cone of equal base and height 7.5 m. Find the capacity of the tent and the cost of the canvas at Rs. 100 per square metre.

9. A boiler is in the form of a cylinder 2 m long with hemispherical ends each of 2 m diameter. Find the volume of the boiler.

10. A vessel is a hollow cylinder fitted with a hemispherical bottom of the same base. The depth of the cylinder is $4\frac{2}{3}$ m and the diameter of the hemisphere is 3.5 m. Calculate the volume and the internal surface area of the solid.

11. A cylindrical vessel of diameter 14 m and height 42 cm is fixed symmetrically in side a similar vessel of diameter 16 cm and height 42 cm. The total space between the two vessels is filled with cork dust for heat insulation proposes. How many cubic centimetres of cork dust will be required ?

Answers

1. 827.15 m³ 2. 163.54 cm³
3. 25.12 cm³, 25.12 cm³ 4. 6160 m³
5. 0.95 m³ 6. 102.24 kg
7. 8316 cm³
8. 500 π cm³, Rs. 55,000
9. $10\frac{10}{21}$ m³
10. 56.15 m³, $7\frac{7}{12}$ m²
11. 1980 cm³.

CONVERSION OF SOLID FROM ONE SHAPE TO ANOTHER

IMPORTANT POINTS

1. Volume remains unaltered in conversion. We have seen candles. Generally, they are in the shape of a cylinder. We have also seen some candles shaped like an animal (see figure).

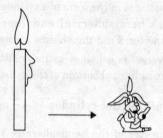

Let us see how they are made. If we want a candle of any special shape, we will have to heat the wax in a metal container till it becomes completely liquid. Then we will have to pour it into another container which has the special shape that we want. For example, take a candle in the shape of a solid cylinder, melt it and pour whole of the melten wax into another container shaped like a rabbit. On cooling, we will obtain a candle in the shape of the rabbit. **The volume of the new candle will be the same as the volume of the earlier candle. This is what we have to remember when we come across objects which are converted from one shape to an other,** or when a liquid which is originally filled in one container of a particular shape is poured into another container of a different shape or size, as we see in figure.

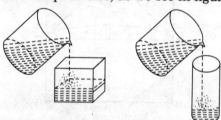

ILLUSTRATIVE EXAMPLES

[NCERT Exercise 13.3]
(Page No. 251)

Take $\pi = \dfrac{22}{7}$, unless stated otherwise.

Example 1. *A metallic sphere of radius 4.2 cm is melted and recast into the shape of a cylinder of radius 6 cm. Find the height of the cylinder.*

Sol. For sphere

Radius $(r) = 4.2$ cm

$\therefore$ Volume $= \dfrac{4}{3} \pi r^3$

$= \dfrac{4}{3} \pi (4.2)^3$ cm^3

For cylinder

Radius $(R) = 6$ cm

Let the height of the cylinder be H cm. Then,

Volume $= \pi R^2 H$

$= \pi (6)^2 H$ cm^3

According to the question,

Volume of the sphere

= Volume of the cylinder

$\Rightarrow \dfrac{4}{3} \pi (4.2)^3 = \pi (6)^2 H$

$\Rightarrow H = \dfrac{4(4.2)^3}{3(6)^2}$

$\Rightarrow H = 2.74$

Hence, the height of the cylinder is 2.74 cm.

Example 2. *Metallic spheres of radii 6 cm, 8 cm and 10 cm, respectively, are melted to form a single solid sphere. Find the radius of the resulting sphere.*

SURFACE AREAS AND VOLUMES

Sol. Volume of sphere of radius 6 cm

$$= \frac{4}{3} \pi (6)^3 \text{ cm}^3$$

Volume of sphere of radius 8 cm

$$= \frac{4}{3} \pi (8)^3 \text{ cm}^3$$

Volume of sphere of radius 10 cm

$$= \frac{4}{3} \pi (10)^3 \text{ cm}^3$$

Let the radius of the resulting sphere be R cm.

Then, volume of the resulting sphere

$$= \frac{4}{3} \pi R^3 \text{ cm}^3$$

According to the question,

$$\frac{4}{3} \pi R^3 = \frac{4}{3} \pi (6)^3 + \frac{4}{3} \pi (8)^3 + \frac{4}{3} \pi (10)^3$$

$\Rightarrow \quad R^3 = (6)^3 + (8)^3 + (10)^3$

$\Rightarrow \quad R^3 = 216 + 512 + 1000$

$\Rightarrow \quad R^3 = 1728$

$\Rightarrow \quad R = (1728)^{1/3}$

$\Rightarrow \quad R = 12$

Hence, the radius of the resulting sphere is 12 cm.

Example 3. *A 20 m deep well with diameter 7m is dug and the earth from digging is evenly spread out to form a platform 22 m by 14 m. Find the height of the platform.*

Sol. For well Diameter = 7m

$\therefore \quad$ Radius $(r) = \dfrac{7}{2}$ m

Depth $(h) = 20$ m

$\therefore \quad$ Volume $= \pi r^2 h$

$$= \pi \left(\frac{7}{2}\right)^2 (20)$$

$= 245 \pi \text{ cm}^3$

For Platform

Length (L) = 22 m

Breadth (B) = 14 m

Let the height of the platform be H m.

Then, Volume of the platform

= LBH

= 22 × 14 × H

= 308 H m³

According to the question,

308H = 245 π

$\Rightarrow \quad H = \dfrac{245 \pi}{308}$

$\Rightarrow \quad H = \dfrac{245 \times 22}{308 \times 7}$

$\Rightarrow \quad H = 2.5$

Hence, the height of the platform is 2.5 m.

Example 4. *A well of diameter 3 m is dug 14 m deep. The earth taken out of it has been spread evenly all around it in the shape of a circular ring of width 4 m to form an embankment. Find the height of the embankment.*

Sol. For well

Diameter = 3 m

Radius $(r) = \dfrac{3}{2}$ m

Depth $(h) = 14$ m

$\therefore \quad$ Volume $= \pi r^2 h$

$$= \pi \left(\frac{3}{2}\right)^2 (14)$$

$$= \frac{63}{2} \pi \text{ m}^3$$

Width of the embankment = 4m

Let the height of the embankment be H m.

$\therefore \quad$ Radius of the well with embankment (R)

$= \dfrac{3}{2} + 4 = \dfrac{11}{2}$ m

$\therefore \quad$ Volume of the embankment

$= \pi R^2 H - \pi r^2 H$

$= \pi (R^2 - r^2) H$

$$= \pi\left\{\left(\frac{11}{2}\right)^2 - \left(\frac{3}{2}\right)^2\right\} H$$

$$= \pi\left(\frac{121}{4} - \frac{9}{4}\right) H$$

$$= 28\pi H \text{ m}^3$$

According to the question,

$$28\pi H = \frac{63}{2}\pi$$

$$\Rightarrow \quad H = \frac{63}{2 \times 28}$$

$$\Rightarrow \quad H = \frac{9}{8}$$

$$\Rightarrow \quad H = 1.125$$

Hence, the height of the embankment is 1.125 m.

Example 5. *A container shaped like a right circular cylinder having diameter 12 cm and height 15 m is full of ice-cream. The ice cream is to be filled into cones of height 12 cm and diameter 6 cm, having a hemispherical shape on the top. Find the number of such cones which can be filled with ice-cream.*

Sol. For right circular cylinder

Diameter = 12 cm

$$\therefore \quad \text{Radius } (r) = \frac{12}{2} = 6 \text{ cm}$$

Height (h) = 15 cm

$$\therefore \quad \text{Volume} = \pi r^2 h$$

$$= \pi (6)^2 (15)$$

$$= 540\pi \text{ cm}^3$$

For a cone

Diameter = 6 cm

$$\therefore \quad \text{Radius } (R) = \frac{6}{2} = 3 \text{ cm}$$

Height (H) = 12 cm

$$\therefore \quad \text{Volume} = \frac{1}{3}\pi r^2 h$$

$$= \frac{1}{3}\pi (3)^2 (12)$$

$$= 36\pi \text{ cm}^3$$

Let n cones be filled with ice-cream.
Then,
Volume of n cones = $n(36\pi)$ cm³
According to the question,

$$n(36\pi) = 540\pi$$

$$\Rightarrow \quad n = \frac{540\pi}{36\pi}$$

$$\Rightarrow \quad n = 15$$

Hence, 15 cones can be filled with ice-cream.

Example 6. *How many silver coins, 1.75 cm in diameter and of thickness 2 mm, must be melted to form a cuboid of dimensions 5.5 cm × 10 cm × 3.5 cm ?*

Sol. For a silver coin

Diameter = 1.75 cm

$$\therefore \quad \text{Radius } (r) = \frac{1.75}{2} \text{ cm} = \frac{7}{8} \text{ cm}$$

Thickness (h) = 2 mm = $\frac{2}{10}$ cm = $\frac{1}{5}$ cm

$$\therefore \quad \text{Volume of a silver coin}$$

$$= \pi r^2 h$$

$$= \pi \left(\frac{7}{8}\right)^2 \left(\frac{1}{5}\right)$$

$$= \frac{49}{320}\pi \text{ cm}^3$$

Let n coins be melted.

Then, volume of n coins = $n\dfrac{49\pi}{320}$ cm³

For cuboid

length (L) = 5.5 cm
breath (B) = 10 cm
height (H) = 3.5 cm

$\therefore$ Volume of the cuboid

$$= l\, b\, h$$

$$= 5.5 \times 10 \times 3.5$$

$$= 192.5$$

$$= \frac{1925}{10}$$

SURFACE AREAS AND VOLUMES

$= \dfrac{385}{2}$ cm^3

According to the question,

$n \dfrac{49\pi}{320} = \dfrac{385}{2}$

$\Rightarrow \quad n = \dfrac{385}{2} \cdot \dfrac{320}{49\pi}$

$\Rightarrow \quad n = \dfrac{385}{2} \cdot \dfrac{320}{49} \cdot \dfrac{7}{22}$

$\Rightarrow \quad n = 400$

Hence, 400 coins must be melted.

Example 7. *A cylindrical bucket, 32 cm high and with radius of base 18 cm, is filled with sand. This bucket is emptied on the ground and a conical heap of sand is formed. If the height of the conical heap is 24 cm, find the radius and slant height of the heap.*

Sol. For cylindrical bucket

Radius of base (r) =18 cm

Height (h) = 32 cm

$\therefore$ Volume $= \pi r^2 h$

$\qquad = \pi (18)^2 (32)$

$\qquad = 10368\, \pi$ cm^3

For conical heap

Height (H) = 24 cm

Let the radius be R cm

Then, volume $= \dfrac{1}{3} \pi R^2 H$

$\qquad = \dfrac{1}{3} \pi R^2 (24)$

$\qquad = 8\, \pi R^2$ cm^3

According to the question,

$8\pi R^2 = 10368\, \pi$

$\Rightarrow \quad 8R^2 = 10368$

$\Rightarrow \quad R^2 = \dfrac{10368}{8}$

$\Rightarrow \quad R^2 = 1296$

$\Rightarrow \quad R = \sqrt{1296}$

$\Rightarrow \quad R = 36$

Hence, the radius of the heap is 36 cm.

Again, slant height (L) $= \sqrt{R^2 + H^2}$

$= \sqrt{(36)^2 + (24)^2}$

$= \sqrt{1296 + 576}$

$= \sqrt{1872}$

$= \sqrt{12 \times 12 \times 13}$

$= 12\sqrt{13}$

Hence, the slant height of the heap is $12\sqrt{13}$ cm.

Example 8. *Water in a canal, 6 m wide and 1.5 m deep is flowing with a speed of 10 km/h. How much area will it irrigate in 30 minutes, if 8 cm of standing water is needed ?*

Sol. For canal

Width = 6 m

Depth = 1.5 m $= \dfrac{15}{10}$ m $= \dfrac{3}{2}$ m

Speed of flow of water = 10 km/h

$= 10 \times 1000$ m/h

$= 10000$ m/h

$= \dfrac{10000}{60}$ m/min

$= \dfrac{500}{3}$ m/min

$= \dfrac{500 \times 30}{3}$ m/30 minutes

$= 5000$ m/30 minutes

$\therefore$ Volume of water that flows in 30 minutes

$= 6 \times \dfrac{3}{2} \times 5000$ m^3

$= 45000$ m^3

$\therefore$ The area it will irrigate

$= \dfrac{45000}{\left(\dfrac{8}{100}\right)}$

$= \dfrac{4500000}{8}$

$= 562500 \text{ m}^2$

$= \dfrac{562500}{10000}$ hectares

$= 56.25$ hectares.

Example 9. *A farmer connects a pipe of internal diameter 20 cm from a canal into a cylindrical tank in her field, which is 10 m in diameter and 2 m deep. If water flows through the pipe at the rate of 3 km/h, in how much time will the tank be filled?*

Sol. For cylindrical tank

Diameter = 10 m

$\therefore$ Radius $(r) = \dfrac{10}{2} = 5$ m

Depth $(h) = 2$ m

$\therefore$ Volume $= \pi r^2 h$

$= \pi (5)^2 (2)$

$= 50 \pi \text{ m}^3$

Rate of flow of water = 3 km/h

$= 3 \times 1000$ m/h

$= 3000$ m/h

$= \dfrac{3000}{60}$ m/min

$= 50$ m/min

For pipe

Internal diameter = 20 cm

$\therefore$ Internal radius (R) $= \dfrac{20}{2} = 10$ cm

$= \dfrac{10}{100}$ m $= 0.1$ m

$\therefore$ Volume of water that flow per minute

$= \pi R^2 (50)$

$= \pi (0.1)^2 (50) \text{ m}^3$

$= 0.5 \pi \text{ m}^3$

$= \dfrac{5\pi}{10} \text{ m}^3$

$= \dfrac{\pi}{2} \text{ m}^3$

$\therefore$ Required time $= \dfrac{50\pi}{\left(\dfrac{\pi}{2}\right)} = 100$ minutes

Hence, the tank will be filled in 100 minutes.

ADDITIONAL EXAMPLES

Example 10. *The rain water from a roof 22 m × 20 m drains into a cylindrical vessel having diameter of base 2 m and height 3.5 m. If the vessel is just full, find the rain fall in cm.* **(CBSE 2006)**

Sol. For a cylindrical vessel

Diameter of the base = 2 m

$\therefore$ Radius of the base $(r) = \dfrac{2}{2}$ m $= 1$ m

Height $(h) = 3.5$ m

$\therefore$ Volume $= \pi r^2 h$

$= \pi (1)^2 (3.5)$

$= \pi (3.5)$

$= \dfrac{22}{7} \times 3.5$

$= 11 \text{ m}^3$

Let the rainfall be x m.

Then, according to the question,

$22 \times 20 \times x = 11$

$\Rightarrow \qquad x = \dfrac{11}{22 \times 20} = \dfrac{1}{40}$

$= 0.025$ m $= 2.5$ cm

Hence, the rainfall is 2.5 cm.

Example 11. *Water is flowing at the rate of 15 km per hour through a pipe of diameter 14 cm into a rectangular tank which is 50 m long and 44 m wide. Find the time in which the level of water in the tank will rise by 21 cm.* **(AI CBSE 2006)**

Sol. For pipe

Diameter = 14 cm

$\therefore$ Radius $(r) = \dfrac{14}{2} = 7$ cm $= \dfrac{7}{100}$ m

SURFACE AREAS AND VOLUMES

Rate of flow of water = 15 km/h
= 15000 m/h

∴ Volume of the water that flows through the pipe in 1 hour

$$= \frac{22}{7}\left(\frac{7}{100}\right)^2 (15000) \text{ m}^3$$

$$= 231 \text{ m}^3$$

For water in the tank
Length (L) = 50 m
Breath (B) = 44 m
Height (H) = 21 cm = $\frac{21}{100}$ m

∴ Volume of water = $l\,b\,h$

$$= 50 \times 44 \times \frac{21}{100}$$

$$= 462 \text{ m}^3$$

∴ Required time = $\frac{462}{231}$ = 2 hours.

Hence, the level of water in the tank will rise by 21 cm in 2 hours.

Example 12. *A hemispherical bowl of internal radius 9 cm is full of liquid. The liquid is to be filled into cylindrical shaped small bottles each of diameter 3 cm and height 4 cm. How many bottles are needed to empty the bowl?* (CBSE 2005)

Sol. For hemispherical bowl
Internal radius (r) = 9 cm

∴ Volume = $\frac{2}{3}\pi r^3 = \frac{2}{3}\pi (9)^3$ cm^3

For a cylindrical shaped small bottle
Diameter = 3 cm

∴ Radius (R) = $\frac{3}{2}$ cm

Height (H) = 4 cm

∴ Volume = $\pi R^2 H$

$$= \pi \left(\frac{3}{2}\right)^2 (4)$$

$$= 9\pi \text{ cm}^3$$

Let n bottles needed. Then, volume of n bottles = $9n\pi$ cm^3

According to the question,

$$9n\pi = \frac{2}{3}\pi(9)^3$$

$$\Rightarrow 9n = \frac{2}{3}(9)^3$$

$$\Rightarrow n = \frac{2}{3}(9)^2$$

$$\Rightarrow n = 54$$

Hence, 54 bottles are needed to empty the bowl.

Example 13. *The base radius and height of a right circular solid cone are 2 cm and 8 cm respectively. It is melted and recast into spheres of diameter 2 cm each. Find the number of sphere formed.* (AI CBSE 2005)

Sol. For right circular cone
Base radius (r) = 2 cm
Height (h) = 8 cm

∴ Volume = $\frac{1}{3}\pi r^2 h = \frac{1}{3}\pi(2)^2(8)$

$$= \frac{32\pi}{3} \text{ cm}^3$$

For a sphere
Diameter = 2 cm

Radius (R) = $\frac{2}{2}$ = 1 cm

∴ Volume = $\frac{4}{3}\pi R^3$

$$= \frac{4}{3}\pi(1)^3 = \frac{4}{3}\pi \text{ cm}^3$$

Let the number of spheres formed be n.

Then, volume of n spheres = $n\left(\frac{4}{3}\pi\right)$ cm^3

According to the question,

$$n\left(\frac{4}{3}\pi\right) = \frac{32\pi}{3}$$

$$\Rightarrow n = 8$$

Hence, the number of spheres formed is 8.

TEST YOUR KNOWLEDGE

1. A cone of height 24 cm and radius of base 6 cm is made up of modelling clay. A child reshapes it in the form of a sphere. Find the radius of the sphere.

2. Selvi's house has an overhead tank in the shape of a cylinder. This is filled by pumping water from a sump (an underground tank) which is in the shape of a cuboid. The sump has dimensions 1.57 m × 1.44 m × 95 cm. The overhead tank has its radius 60 cm and height 95 cm. Find the height of the water left in the sump after the overhead tank has been completely filled with water from the sump which had been full. Compare the capacity of the tank with that of the sump. (Use π = 3.14).

3. A copper rod of diameter 1 cm and length 8 cm is drawn into a wire of length 18 m of uniform thickness. Find the thickness of the wire.

4. A hemispherical tank full of water is emptied by a pipe at the rate of $3\frac{4}{7}$ litres per second. How much time will it take to empty half the tank, if it is 3 m in diameter? $\left(\text{Take } \pi = \frac{22}{7}\right)$.

5. A solid metallic sphere of diameter 21 cm is melted and recast into a number of smaller cones, each of diameter 7 cm and height 3 cm. Find the number of cones so formed. (CBSE 2004)

6. A solid metallic sphere of diameter 28 cm in melted and recast into a number of smaller cones, each of diameter $4\frac{2}{3}$ cm and height 3 cm. Find the number of cones so formed. (CBSE 2004)

7. A solid metallic sphere of diameter 21 cm is melted and recast into a number of smaller cones, each of diameter 3.5 cm and height 3 cm. Find the number of cones so formed. (CBSE 2004)

8. A hemispherical bowl of internal diameter 30 cm contains some liquid. This liquid is to be filled into cylindrical shaped bottles each of diameter 5 cm and height 6 cm. Find the number of bottles necessary to empty the bowl. (AI CBSE 2004)

9. How many balls, each of radius 1 cm can be made from a solid sphere of lead of radius 8 cm?

10. The diameter of a metallic sphere is 6 cm. It is melted and drawn into a wire having diameter of cross-section as 0.2 cm. Find the length of the wire.

11. A conical flask is full of water. The flask has base-radius r and height h. The water is poured into a cylindrical flask of base-radius mr. Find the height of water in the cylindrical flask.

12. The largest sphere is carved out of a cube of side 7 cm. Find the volume of the sphere.

Answers

1. 6 cm
2. 47.5 cm, 1 : 2
3. 0.67 mm (approx.)
4. 16.5 minutes
5. 126
6. 672
7. 504
8. 60
9. 512
10. 36 m
11. $\dfrac{h}{3 m^2}$ units
12. $\dfrac{539}{3}$ cm^3.

FRUSTUM OF A CONE

IMPORTANT POINTS

1. Removal of a portion of a right circular cone. We have observed the objects that are formed when two basic solids were joined together. Now we shall do something different. We will take a right circular cone and *remove* a portion of it. There are so many ways in which we can do this. But one particular case that we are interested in is the removal of a smaller right circular cone by cutting the given cone by a plane parallel to its base. We may observe that the glasses (tumblers), in general, used for drinking water, are of this shape. (see figure)

2. An Activity. Take some clay, or any other such material (like plasticine, etc.) and form a cone. Cut it with a knife parallel to its base. Remove that smaller cone. Then, we are left with a solid called a frustum of the cone. We can see that this has two circular ends with different radii.

So, given a cone, when we slice (or cut) through it with a plane parallel to its base (see figure) and remove the cone that is formed on one side of that plane, the part that is now left over on the other side of the plane is called a **frustum* of the cone.**

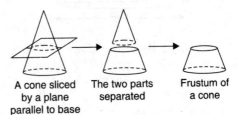

A cone sliced by a plane parallel to base   The two parts separated   Frustum of a cone

3. Formulae. Let h be the height, l the slant height and r_1 and r_2 the radii of the ends ($r_1 > r_2$) of the frustum of a cone. Then we can directly find the volume, the curved surface area and the total surface area of frustum by using the formulae given below :

(i) Volume of the frustum of the cone
$$= \frac{1}{3}\pi h(r_1^2 + r_2^2 + r_1 r_2).$$

(ii) the curved surface area of the frustum of the cone = $\pi(r_1 + r_2)l$

where $l = \sqrt{h^2 + (r_1 - r_2)^2}$.

(iii) Total surface area of the frustum of the cone = $\pi(r_1 + r_2) + \pi r_1^2 + \pi r_2^2$,

where $l = \sqrt{h^2 + (r_1 - r_2)}$.

---

*'Frustum' is a latin word meaning 'piece cut off', and its plural is 'frusta'.

ILLUSTRATIVE EXAMPLES

[NCERT Exercise 13.4]
(Page No. 257)

Use $\pi = \dfrac{22}{7}$, unless stated otherwise.

Example 1. *A drinking glass is in the shape of a frustum of a cone of height 14 cm. The diameters of its two circular ends are 4 cm and 2 cm. Find the capacity of the glass.*

Sol. Here,

$r_1 = \dfrac{4}{2} = 2$ m

$r_2 = \dfrac{2}{2} = 1$ cm

$h = 14$ cm

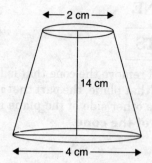

∴ Capacity of the glass

$= \frac{1}{3}\pi h(r_1^2 + r_2^2 + r_1 r_2)$

$= \frac{1}{3} \times \frac{22}{7} \times 14[(2)^2 + (1)^2 + (2)(1)]$

$= \frac{1}{3} \times \frac{22}{7} \times 14 \times 7$

$= \frac{308}{3} = 102\frac{2}{3}$ cm^2.

Example 2. *The slant height of a frustum of a cone is 4 cm and the perimeters (circumference) of its circular ends are 18 cm and 6 cm. Find the curved surface area of the frustum.*

Sol. Let r_1 cm and r_2 cm be the radii of the ends $(r_1 > r_2)$ of the frustum of a cone.

Then, $l = 4$ cm

$2\pi r_1 = 18$ cm $\Rightarrow$ $\pi r_1 = 9$ cm

$2\pi r_2 = 6$ cm $\Rightarrow$ $\pi r_2 = 3$ cm

Now, curved surface area of the frustum

$= \pi(r_1 + r_2)l$

$= (\pi r_1 + \pi r_2)l$

$= (9 + 3)4$

$= 48$ cm^2.

Example 3. *A fez, the cap used by the Turks, is shaped like the frustum of a cone (see figure). If its radius on the open side is 10 cm, radius at the upper base is 4 cm and its slant height is 15 cm, find the area of material used for making it.*

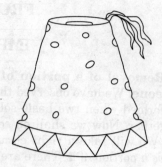

Sol. Here,

$r_1 = 10$ cm

$r_2 = 4$ cm

$l = 15$ cm

∴ Surface area

$= \pi(r_1 + r_2)l + \pi r_2^2$

$= \frac{22}{7}(10 + 4)(15) + \pi(4)^2$

$= 660 + 16\pi$

$= 660 + 16 \times \frac{22}{7}$

$= 660 + \frac{352}{7}$

$= \frac{4620 + 352}{7}$

$= \frac{4972}{7} = 710\frac{2}{7}$ cm^2.

Hence, the area of material used for making it is $710\frac{2}{7}$ cm^2.

Example 4. *A container, opened from the top is made up of a metal sheet, is in the form of a frustum of a cone of height 16 cm with radii of its lower and upper ends as 8 cm and 20 cm, respectively. Find the cost of the milk which can completely fill the container, at the rate of Rs. 20 per litre. Also find the cost of metal sheet used to make the container, if it costs Rs. 8 per 100 cm^2. (Take $\pi = 3.14$)*

Sol. Here,
$h = 16$ cm
$r_1 = 20$ m
$r_2 = 8$ cm

∴ Volume of the container

$= \frac{1}{3}\pi h(r_1^2 + r_2^2 + r_1 r_2)$

$= \frac{1}{3}(3.14)(16)\{(20)^2 + (8)^2 + (20)(8)\}$

$= \frac{1}{3}(3.14)(16)(400 + 64 + 160)$

$= \frac{1}{3}(3.14)(16)(624)$

$= (3.14)(16)(208)$

$= 10449.92$ cm³ $= 10.44992$ litres

∴ Cost of the milk $= 10.44992 \times 20$
$= 208.9984$
$=$ Rs. 209

Surface area

$= \pi(r_1 + r_2)\sqrt{h^2 + (r_1 - r_2)^2} + \pi r_2^2$

$= (3.14)(20 + 8)\sqrt{(16)^2 + (20 - 8)^2}$
$\qquad + (3.14)(8)^2$

$= (3.14)(28)\sqrt{256 + 144} + (3.14)(64)$

$= (3.14)(28)(20) + 200.96$

$= 1158.4 + 200.96$

$= 1959.36$ cm²

∴ Area of the metal sheet used
$= 1959.36$ cm²

∴ Cost of metal sheet

$= 1959.36 \times \frac{8}{100}$

$= 156.7488$

$=$ Rs. 156.75.

Example 5. *A metallic right circular cone 20 cm high and whose vertical angle is 60° is cut into two parts at the middle of its height by a plane parallel to its base. If the frustum so obtained be drawn into a wire of diameter $\frac{1}{16}$ cm, find the length of the wire.*

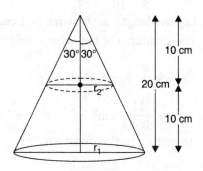

Sol. $\tan 30° = \frac{r_2}{10}$

$\Rightarrow \quad \frac{1}{\sqrt{3}} = \frac{r_2}{10}$

$\Rightarrow \quad r_2 = \frac{10}{\sqrt{3}}$ cm

$\tan 30° = \frac{r_1}{20}$

$\Rightarrow \quad \frac{1}{\sqrt{3}} = \frac{r_1}{20}$

$\Rightarrow \quad r_1 = \frac{20}{\sqrt{3}}$ cm

$h = 10$ cm

∴ Volume $= \frac{1}{3}\pi h(r_1^2 + r_2^2 + r_1 r_2)$

$= \frac{1}{3} \times \frac{22}{7} \times 10$

$\quad \times \left\{\left(\frac{20}{\sqrt{3}}\right)^2 + \left(\frac{10}{\sqrt{3}}\right)^2 + \left(\frac{20}{\sqrt{3}}\right)\left(\frac{10}{\sqrt{3}}\right)\right\}$

$= \frac{1}{3} \times \frac{22}{7} \times 10 \times \left(\frac{400}{3} + \frac{100}{3} + \frac{200}{3}\right)$

$= \frac{1}{3} \times \frac{22}{7} \times 10 \times \frac{700}{3}$

$= \frac{2200}{9}$ cm³

Diameter of the wire $= \frac{1}{16}$ cm

∴ Radius of the wire

$r = \dfrac{1}{2} \times \dfrac{1}{16} = \dfrac{1}{32}$ cm

Let the length is the wire be l cm.

Then, volume of the wire = $\pi r^2 l$

$= \dfrac{22}{7} \left(\dfrac{1}{32}\right)^2 l$

$= \dfrac{11l}{3584}$ cm^3

According to the question,

$\dfrac{11l}{3584} = \dfrac{22000}{9}$

$\Rightarrow l = \dfrac{22000 \times 3584}{11 \times 9}$

$\Rightarrow l = \dfrac{2000 \times 3584}{9}$

$\Rightarrow l = \dfrac{7168000}{9}$ cm

$\Rightarrow l = 796444.44$ cm

$\Rightarrow l = 7964.4$ m

Hence, the length of the wire is 7964.4 m.

ADDITIONAL EXAMPLES

Example 6. *A bucket made up of a metal sheet is in the form of a frustum of a cone of height 16 cm with radii of its lower and upper ends as 8 cm and 20 cm respectively. Find the cost of the bucket if the cost of metal sheet used is Rs. 15 per 100 m^2. (Use $\pi = 3.14$)*

(CBSE 2006)

Sol. Here, $h = 16$ cm

$r_1 = 20$ cm

$r_2 = 6$ cm

∴ Slant height (l)

$= \sqrt{h^2 + (r_1 - r_2)^2}$

$= \sqrt{(16)^2 + (20 - 8)^2}$

$= \sqrt{(16)^2 + (12)^2}$

$= \sqrt{256 + 144}$

$= \sqrt{400}$

$= 20$ cm

Total surface area of the bucket (excluding the upper end)

$= \pi l (r_1 + r_2) + \pi r_2^2$

$= 3.14 + 20 \times (20 + 8) + 3.14 \times (8)^2$

$= 1758.40 + 200.96$

$= 1959.36$ cm^2

∴ Area of the metal sheet used

$= 1959.36$ cm^2

∴ Cost of the metal sheet used at Rs. 15 per 100 m^2

$= \dfrac{1959.36 \times 15}{100}$

$= $ Rs. 293.90

Hence, the cost of the bucket is Rs. 293.90.

Example 7. *A bucket made up of a metal sheet is in the form of a frustum of a cone. Its depth is 24 cm and the diameters of the top and bottom are 30 cm and 10 cm respectively. Find the cost of milk which can completely fill the bucket at the rate of Rs. 20 per litre and the cost of the metal sheet used, if it costs Rs. 10 per 100 cm^2. (Use $\pi = 3.14$)*

(AI CBSE 2006)

Sol. Here, $h = 24$ cm

$r_1 = \dfrac{30}{2} = 15$ cm

$r_2 = \dfrac{10}{2} = 5$ cm

∴ Slant height (l)

$= \sqrt{h^2 + (r_1 - r_2)^2}$

$= \sqrt{(24)^2 + (15 - 5)^2}$

$= \sqrt{(24)^2 + (10)^2}$

SURFACE AREAS AND VOLUMES

$= \sqrt{576 + 100}$

$= \sqrt{676} = 26$ cm

∴ Volume of the bucket

$= \frac{1}{3} \pi h (r_1^2 + r_2^2 + r_1 r_2)$

$= \frac{1}{3} \times 3.14 \times 24 \{(15)^2 + (5)^2 + (15)(5)\}$

$= \frac{1}{3} \times 3.14 \times 24 \times (225 + 25 + 75)$

$= \frac{1}{3} \times 3.14 \times 24 \times (325)$

$= 8164$ cm^3

$= 8.164$ litres

∴ Quantity of milk $= 8.164$ litres

∴ Cost of milk $= 8.164 \times 20$

$=$ Rs. 163.28

Total surface area of the bucket (excluding the upper end)

$= \pi (r_1 + r_2) l + \pi r_2^2$

$= (3.14)(15 + 5)(26) + (3.14)(5)^2$

$= 1632.8 + 78.5$

$= 1711.3$ cm^2

∴ Surface area of the metal sheet used $= 1711.3$ cm^2

∴ Cost of the metal sheet used

$= \frac{1711.3 \times 10}{100} =$ Rs. 171.13.

Example 8. *The radii of circular ends of a solid frustum of a cone are 33 cm and 27 cm and its slant height is 10 cm. Find its total surface area.* (CBSE 2005)

Sol. Here, $r_1 = 33$ cm

$r_2 = 27$ cm

$l = 10$ cm

∴ Total surface area of the solid frustum of cone

$= \pi (r_1 + r_2) l + \pi r_1^2 + \pi r_2^2$

$= \pi (33 + 27)(10) + \pi (33)^2 + \pi (27)^2$

$= 600 \pi + 1089 \pi + 729 \pi$

$= 2418 \pi$

$= 2418 \times \frac{22}{7}$

$= \frac{53196}{7}$ cm^2.

Example 9. *A hollow cone is cut by a plane parallel to the base and the upper portion is removed. If the curved surface of the remainder is $\frac{8}{9}$ of the curved surface of the whole cone, find the ratio of the line segments in which the altitude of the cone is divided by the plane.*

(AI CBSE 2004)

Sol. Curved surface area of the remainder $= \frac{8}{9}$ curved surface area of the whole cone

⇒ curved surface area of the cut-cut cone $= \frac{1}{9}$ curved surface area of the whole cone

...(1)

∵ Δ VFB ~ Δ VEB

| AA similarity criterion

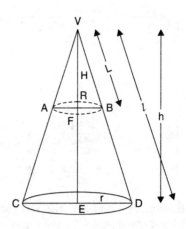

∴ $\frac{R}{r} = \frac{H}{h} = \frac{L}{l}$

| ∵ Corresponding sides of two similar triangles are proportional

...(2)

From (1), $\pi RL = \frac{1}{9} \pi r l$

$$\Rightarrow \quad \frac{RL}{rl} = \frac{1}{9}$$

$$\Rightarrow \quad \left(\frac{R}{r}\right)\left(\frac{L}{l}\right) = \frac{1}{9}$$

$$\Rightarrow \quad \left(\frac{H}{h}\right)\left(\frac{H}{h}\right) = \frac{1}{9} \qquad | \text{ Using (2)}$$

$$\Rightarrow \quad \left(\frac{H}{h}\right)^2 = \left(\frac{1}{3}\right)^2$$

$$\Rightarrow \quad \frac{H}{h} = \frac{1}{3}$$

$$\Rightarrow \quad H = \frac{h}{3}$$

$$\therefore \quad EF = h - H = h - \frac{h}{3} = \frac{2h}{3}$$

$$\therefore \quad \frac{H}{EF} = \frac{h/3}{2h/3} = \frac{1}{2}$$

Hence, the required ratio is 1 : 2.

TEST YOUR KNOWLEDGE

1. The radii of the ends of a frustum of a cone 45 cm high are 28 cm and 7 cm (see figure). Find its volume, the curved surface area and the total surface area $\left(\text{Take } \pi = \frac{22}{7}\right)$.
   (AI CBSE 2005, CBSE 2004)

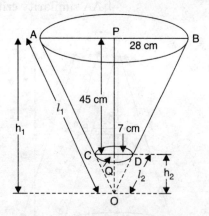

2. Hanumappa and his wife Gangamma are busy making jaggery out of sugarcane juice. They have processed the sugarcane juice to make the molasses, which is poured into moulds in the shape of a frustum of a cone having the diameters of its two circular faces as 30 cm and 35 cm and the vertical height of the mould is 14 cm (see figure). If each cm³ of molasses has mass about 1.2 g, find the mass of the molasses that can be poured into each mould. $\left(\text{Take } \pi = \frac{22}{7}\right)$.

3. An open metal bucket is in the shape of a frustum of a cone, mounted on a hollow cylindrical bases made of the same metallic sheet (see figure). The diameters of the two circular ends of the bucket are 45 cm and 25 cm, the total vertical height of the bucket is 40 cm and that of the cylindrical base is 6 cm. Find the area of the metallic sheet used to make the bucket, where we do not take into account the handle of the bucket. Also, find the volume of water the bucket can hold. $\left(\text{Take } \pi = \frac{22}{7}\right)$.

4. A bucket of height 8 cm and made up of copper sheet is in the form of a frustum of a right circular cone with radii of its lower and upper ends as 3 cm and 9 cm respectively. Calculate :
   (i) The height of the cone of which the bucket is a part.
   (ii) The volume of water which can be filled in the bucket.
   (iii) The area of copper sheet required to make the bucket.
   (Leave the answer in term of π).
   (CBSE SP SET 1)

5. The height of a cone is 30 cm. A small cone is cut off at the top by a plane parallel to the base. If its volume be $\frac{1}{27}$ of the volume of the given cone, at what height above the base is the reaction made ?

6. A container made up of a metal sheet is in the form of a frustum of a cone of height 16 cm with radii of its lower and upper ends as 8 cm and 20 cm respectively. Find the cost of milk which can completely fill the container at the rate of Rs. 15 per litre and the cost of the metal sheet used, if its costs Rs. 5 per 100 m² (Use $\pi = 3.14$)

Answers

1. 48510 cm³, 5461.5 cm², 8079.5 cm²
2. 15 kg (approx.)
3. 4860.9 cm², 33.62 litres (approx.)
4. (i) 12 cm
   (ii) 312 π cm³
   (iii) 129 π cm³
5. 20 cm
6. Rs. 156.75, Rs. 97.97 (approx.).

ILLUSTRATIVE EXAMPLES

[NCERT Exercise 13.5 (Optional)*]
(Page No. 258)

Example 1. *A copper wire, 3 mm in diameter, is wound about a cylinder whose length is 12 cm, and diameter 10 cm, so as to cover the curved surface of the cylinder. Find the length and mass of the wire, assuming the density of copper to be 8.88 g per cm³.*

Sol. It is clear that one round of wire covers 3 mm of thickness of the cylinder.

Length of the cylinder
$$= 12 \text{ cm} = 120 \text{ mm}$$

∴ Number of rounds to cover 120 mm
$$= \frac{120}{3} = 40$$

For cylinder
Diameter = 10 cm
Radius $(r) = \frac{10}{2} = 5$ cm

∴ Length of the wire in completing one round = $2\pi r$
$$= 2\pi(5) = 10\pi \text{ cm}$$

∴ Length of the wire in covering the whole surface = Length of the wire in completing 40 rounds
$$= 10\pi \times 40 = 400\pi \text{ cm}$$

Radius of the copper wire
$$= \frac{3}{2} \text{ mm} = \frac{3}{20} \text{ cm}$$

∴ Volume of wire
$$= \pi \left(\frac{3}{20}\right)^2 (400\pi)$$
$$= 9\pi^2 \text{ cm}^3$$

∴ Mass of wire = $9\pi^2 \times 8.88$
$$= 9 \times (3.14)^2 \times 8.88$$
$$= 787.98 \text{ g}.$$

Example 2. *A right triangle, whose sides are 3 cm and 4 cm (other than hypotenuse) is made to revolve about its hypotenuse. Find the volume and surface area of the double cone so formed. (Choose value of π as found appropriate.)*

---

*These exercise are not from the examination point of veiw

Sol.

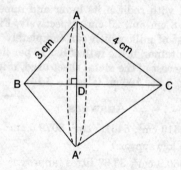

Hypotenuse = $\sqrt{3^2 + 4^2}$

| By Pythagoras theorem

= 5 cm

Here, AD or A'D is the radius of the common base of the double cone formed by revolving the right triangle BAC about the hypotenuse BC.

Now, $\triangle ADB \sim \triangle CAB$

| AA similarly criterion

$\therefore \quad \dfrac{AD}{CA} = \dfrac{AB}{CB}$

$\Rightarrow \quad \dfrac{AD}{4} = \dfrac{3}{5}$

$\Rightarrow \quad AD = \dfrac{12}{5}$ cm

Also, $\dfrac{DB}{AB} = \dfrac{AB}{CB}$

$\Rightarrow \quad \dfrac{DB}{3} = \dfrac{3}{5}$

$\Rightarrow \quad DB = \dfrac{3 \times 3}{5} = \dfrac{9}{5}$ cm

$\therefore \quad CD = BC - DB$

$= 5 - \dfrac{9}{5}$

$= \dfrac{16}{5}$ cm

$\therefore$ Volume of the double cone

$= \left[\dfrac{1}{3}\pi\left(\dfrac{12}{5}\right)^2\left(\dfrac{9}{5}\right) + \dfrac{1}{3}\pi\left(\dfrac{12}{5}\right)^2\left(\dfrac{16}{5}\right)\right]$

$= \dfrac{1}{3}\pi\left(\dfrac{12}{5}\right)^2 (5)$

$= \dfrac{1}{3} \times 3.14 \times \dfrac{144}{25} \times 5 = 30.14$ cm^3

and, surface area of the double cone

$= \pi \times \dfrac{12}{5} \times 3 + \pi \times \dfrac{12}{5} \times 4$

$= \pi \times \dfrac{12}{5} (3 + 4)$

$= 3.14 \times \dfrac{12}{5} \times 7$

$= 52.75$ cm^2.

Example 3. *A cistern, internally measuring 150 cm × 120 cm × 110 cm, 129600 cm³ of water in it. Porous bricks are placed in the water until the cistern is full to the brim. Each brick absorbs one-seventeenth of its own volume of water. How many bricks can be put in without overflowing the water, each brick being 22.5 cm × 7.5 cm × 6.5 cm ?*

Sol. Volume of cistern

$= 150 \times 120 \times 110 = 1980000$ cm^3

Volume of water = 129600 cm^3

$\therefore$ Volume of cistern to be filled

$= 1980000 - 129600$

$= 1850400$ cm^3

Volume of a brick

$= 22.5 \times 7.5 \times 6.5 = 1096.875$ cm^3

Let n bricks be needed. Then, Water absorbed by n bricks

$= n \left(\dfrac{1096.875}{17}\right)$ cm^3

$\therefore \quad 1850400 + n \dfrac{1096.875}{17}$

$= n\, (1096.875)$

$\Rightarrow \quad \dfrac{16n}{17} \times 1096.875 = 1850400$

$\Rightarrow \quad n = \dfrac{1850400 \times 17}{16 \times 1096.875}$

$\Rightarrow \quad n = 1792.4102 = 1792$ (approx.)

Hence, 1792 bricks can be put in.

SURFACE AREAS AND VOLUMES

Example 4. *In one fortnight of a given month, there was a rainfall of 10 cm in a river valley. If the area of the valley is 97280 km², show that the total rainfall was approximately equivalent to the addition to the normal water of three rivers each 1072 km long, 75 m wide and 3 m deep.*

Sol. Volume of rainfall

$$= 97280 \times \frac{10}{100 \times 1000}$$

$$= 9.728 \text{ km}^3$$

Volume of three rivers

$$= 3 \times 1072 \times \frac{75}{1000} \times \frac{3}{1000}$$

$$= 0.7236 \text{ km}^3$$

Hence, the two are not approximately equivalent.

Example 5. *An oil funnel made of tin sheet consists of a 10 cm long cylindrical portion attached to a frustum of a cone. If the total height is 22 cm, diameter of the cylindrical portion is 8 cm and the diameter of the top of the funnel is 18 cm, find the area of the tin sheet required to make the funnel (see figure).*

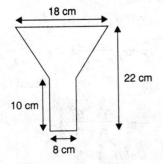

Sol. Slant height of the frustum of the cone

$$(l) = \sqrt{h^2 + (r_1 - r_2)^2}$$

$$= \sqrt{(22-10)^2 + \left(\frac{18}{2} - \frac{8}{2}\right)^2}$$

$$= \sqrt{(12)^2 + (5)^2}$$

$$= \sqrt{144 + 25}$$

$$= \sqrt{169}$$

$$= 13 \text{ cm}$$

∴ Area of the tin sheet required = Curved surface area of the cylindrical portion + Curved surface area of the frustum portion

$$= 2\pi(4)(10) + \pi(4+9)13$$

$$= 80\pi + 169\pi$$

$$= 249\pi$$

$$= 249 \times \frac{22}{7} = \frac{5478}{7}$$

$$= 782\frac{4}{7} \text{ cm}^2.$$

Example 6. *Derive the formula for the curved surface area and total surface area of the frustum of a cone, given to you in Section 13.5, using the symbols as explained.*

Example 7. *Derive the formula for the volume of the frustum of a cone, given to you in Section 13.5, using the symbols as explained.*

Sol. Let h be the height, l be the slant height and r_1 and r_2 be the radii of the bases ($r_1 > r_2$) of the frustum of a cone. We complete the conical part OCD.

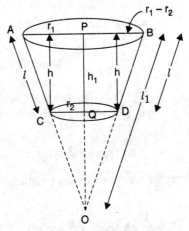

The frustum of the right circular cone can be viewed as the difference of the two right circular cones OAB and OCD. Let the height of the cone OAB be h_1 and its slant height be l_1, i.e.,

$OP = h_1$ and $OA = OB = l_1$.

Then, the height of the cone OCD
$= h_1 - h$

∴ △ OQD ~ △ OPB

| AA similarity criterion

∴ $\dfrac{h_1 - h}{h_1} = \dfrac{r_2}{r_1}$

⇒ $\dfrac{h_1}{h_1} = 1 - \dfrac{r_2}{r_1} = \dfrac{r_1 - r_2}{r_1}$

⇒ $h_1 = \dfrac{hr_1}{r_1 - r_2}$ ...(1)

Now, height of the cone OCD
$= h_1 - h$
$= \dfrac{hr_1}{r_1 - r_2} - h$
$= \dfrac{hr_2}{r_1 - r_2}$ ...(2)

∴ Volume of the frustum of cone
= Volume of the cone OAB − volume of the cone OCD

$= \dfrac{1}{3} \pi r_1^2 h_1 - \dfrac{1}{3} \pi r_2^2 (h_1 - h)$

$= \dfrac{\pi}{3} \left[r_1^2 \dfrac{hr_1}{r_1 - r_2} - r_2^2 \dfrac{hr_2}{r_1 - r_2} \right]$

| From (1) and (2)

$= \dfrac{\pi h}{3} \left(\dfrac{r_1^3 - r_2^3}{r_1 - r_2} \right)$

$= \dfrac{1}{3} \pi h (r_1^2 + r_2^2 + r_1 r_2)$

Therefore, volume of the frustum of cone is
$= \dfrac{1}{3} \pi h (r_1^2 + r_2^2 + r_1 r_2)$.

If A_1 and A_2 are the surface areas $(A_1 > A_2)$ of the two circular bases, then

$A_1 = \pi r_1^2$
and $A_2 = \pi r_2^2$

Then, volume of the frustum of the cone

$= \dfrac{h}{3} (\pi r_1^2 + \pi r_2^2 + \sqrt{\pi r_1^2} \sqrt{\pi r_2^2})$

$= \dfrac{h}{3} (A_1 + A_2 + \sqrt{A_1 A_2})$

Again, from △ DEB,
$l = \sqrt{h^2 + (r_1 - r_2)^2}$

∴ △ OQD ~ △ OPB

| AA similarity criterion

∴ $\dfrac{l_1 - l}{l_1} = \dfrac{r_2}{r_1}$

⇒ $l_1 = \dfrac{lr_1}{r_1 - r_2}$ ...(3)

∴ $l_1 - l = \dfrac{lr_1}{r_1 - r_2} - l$

$= \dfrac{lr_2}{r_1 - r_2}$ ...(4)

Hence, curved surface area of the frustum of cone

$= \pi r_1 l_1 - \pi r_2 (l_1 - l)$

$= \pi r_1 \dfrac{lr_1}{r_1 - r_2} - \pi r_2 \dfrac{lr_2}{r_1 - r_2}$

| From (3) and (4)

$= \pi l \left(\dfrac{r_1^2 - r_2^2}{r_1 - r_2} \right)$

$= \pi l (r_1 + r_2)$

Therefore, curved surface area of the frustum of cone $= \pi l (r_1 + r_2)$, where

$l = \sqrt{h^2 + (r_1 - r_2)^2}$

So, total surface area of the frustum of cone
$= \pi l (r_1 + r_2) + \pi r_1^2 + \pi r_2^2$.

MISCELLANEOUS EXERCISE

1. A solid is in the form of a right circular cylinder with hemi spherical ends. The total height of the solid is 58 cm and the diameter of the cylinder is 28 cm. Find the total surface area of the solid. $\left(\text{Use } \pi = \dfrac{22}{7}\right)$

   (AI CBSE 2006)

2. The radius of the base end the height of a solid right circular cylinder are in the ratio 2 : 3 and its volume is 1617 cu. cm. Find the total surface area of the cylinder. $\left(\text{Use } \pi = \dfrac{22}{7}\right)$.

   (CBSE 2006)

3. A right triangle, whose sides and 15 cm and 20 cm, is made to revolve about its hypotenuse. Find the volume and the surface area of the double cone so formed (Use π = 3.14).

4. A right circular cone is divided by a plane parallel to its base in two equal volumes. In what ratio will the plane divide the axis of the cone ?

5. The radius of the base of a right circular cone is r. It is cut by a plane parallel to the base at a height h from the base. The distance of the boundary of the upper surface from the centre of the base of the frustum is $\sqrt{h^2 + \dfrac{r^2}{9}}$. Show that the volume of the frustum is $\dfrac{13}{27}\pi r^2 h$.

6. A cone is 8.4 cm high and the radius of the base is 2.1 cm. It is melted and recast into a sphere. Find the radius of the sphere.

7. A spherical ball of lead, 3 cm in diameter, is melted and recast into three spherical balls. If the diameters of two of these balls are 1 cm and 1.5 cm, find the diameter of the third ball.

8. 50 circular plates, each of radius 7 cm and thickness $\dfrac{1}{2}$ cm, are placed one above another to from a solid right circular cylinder. Find the total surface area and the volume of the cylinder so formed.

9. Determine the ratio of the volume of a cube to that of a sphere which will exactly fit inside the cube.

10. Find the depth of a cylindrical tank of radius 28 m, if its capacity is equal to that of a rectangular tank of size 28 m × 16 m × 11 m.

11. How many bricks, each measuring 25 cm × 15 cm × 8 cm will be required to build a wall 10 m × 4 dm × 5 m when $\dfrac{1}{10}$ of its volume is occupied by mortar ?

12. The interior of a building is in the form of a cylinder of base radius 12 m and height 3.5 m, surmounted by a cone of equal base and slant height 12.5 m. Find the internal curved surface area and the capacity of the building.

13. A cylindrical boiler, 2 m high, is 3.5 m in diameter. It has a spherical lid. Find the remove of its interior (including the part curved by the lid). $\left(\text{Use } \pi = \dfrac{22}{7}\right)$.

14. A cone of height 15 cm and diameter 7 cm is mounted on a hemisphere of the same diameter. Determine the volume of the solid thus formed $\left(\text{Use } \pi = \dfrac{22}{7}\right)$.

15. A vessel is in the form of an invested cone. Its height is 8 m and the radius of its top which is open is 5 cm. It is filled with water upto the rim. When lead shots, each of which is a sphere of radius 0.5 cm, are dropped into the vessel, one-fourth of the water flows out. Find the number of lead shots dropped in the vessel.

Answers

1. 5104 cm²
2. 770 cm²
3. 3768 cm³, 1318.8 cm²
4. $1 : 2^{1/3} - 1$
6. 2.1 cm
7. $\dfrac{1}{2}(181)^{1/3}$ cm
8. 1408 cm², 3850 cm³
9. 6 : π
10. 2 m
11. 6000
12. 471.43 m², 2112 m³
13. 30.48 m³
14. 282.33 cm³
15. 100.

SUMMARY

1. To determine the surface area of an object formed by combining any two of the basic solids, namely, cuboid, cone, cylinder, sphere and hemisphere.
2. To find the volume of objects formed by combining any two of a cuboid, cone, cylinder, sphere and hemisphere.
3. Given a right circular cone, which is sliced through by a plane parallel to its base, when the smaller conical portion is removed, the resulting solid is called a *Frustum of a Right Circular Cone.*
4. The formulae involving the frustum of a cone are :

   (i) Volume of a frustum of a cone
   $$= \frac{1}{3} \pi h (r_1^2 + r_2^2 + r_1 r_2)$$

   (ii) Curved surface area of a frustum of a cone $= \pi l (r_1 + r_2)$ where
   $$l = \sqrt{h^2 + (r_1 - r_2)^2}$$

   (iii) Total surface area of frustum of a cone $= \pi l (r_1 + r_2) + \pi (r_1^2 + r_2^2)$ where
   h = vertical height of the frustum, l = slant height of the frustum, r_1 and r_2 are radii of the two bases (ends) of the frustum.

14
Statistics

IMPORTANT POINTS

1. Mean of Grouped Data. We know that the mean (or average) of observations, is the sum of the values of all the observations divided by the total number of observations. Also, we know that if $x_1, x_2, ..., x_n$ are observations with respective frequencies $f_1, f_2, ..., f_n$, then this means x_1 occurs f_1 times, x_2 occurs f_2 times, and so on.

Now, Sum of the values of all the observations
$$= f_1 x_1 + f_2 x_2 + ... + f_n x_n$$
and, number of observations $= f_1 + f_2 + ... + f_n$

So, the mean $\bar{x}$ of the data is given by
$$\bar{x} = \frac{f_1 x_1 + f_2 x_2 + ... + f_n x_n}{f_1 + f_2 + ... + f_n}.$$

Note that we can write this in short form by using the Greek letter Σ (capital sigma) which means summation. That is,
$$\bar{x} = \frac{\sum_{i=1}^{n} f_i x_i}{\sum_{i=1}^{n} f_i}$$

which, more briefly, is written as $\bar{x} = \frac{\Sigma f_i x_i}{\Sigma f_i}$, if it is understood that i varies from 1 to n.

2. Need of Grouped Data. In most of our real life situations, data is usually so large that to make a meaningful study, it needs to be condensed as grouped data. So, we need to convert the ungrouped data into grouped data.

3. Formation of Grouped Data from Ungrouped Data. To convert ungrouped data into grouped data, we form class-interval of a certain suitable width and allocate frequencies to each class-interval. We have a convention in our mind that the observation following in any upper class-limit would be considered in the next class.

4. Class Marks. Now, for each class-interval, we search a point which would serve as the representative of the whole class. *It is assumed that the frequency of each class-interval is centred around its* mid-point. So the mid-point (or class mark) of each class can be chosen to represent the observations falling in the class. We know that we find the mid-point of a class (or its class mark) by finding the average of its upper and lower limits. That is,

$$\text{Class Mark} = \frac{\text{Upper class limit} + \text{Lower class limit}}{2}.$$

5. Calculation of Mean (Direct Method). We find the class marks of all the class-intervals and put them in a table. These class marks serve x_i's. Now, in general, for the ith class-interval, we have the frequency f_i corresponding to the class mark x_i. Then, using the formula $\bar{x} = \dfrac{\Sigma f_i x_i}{\Sigma f_i}$, we compute the mean.

This new method of finding the mean is known as the **Direct Method**.

6. Differences of Means from Ungrouped and Grouped Data. Using the same data and employing the same formula for the calculation of the mean, the results (means) obtained from ungrouped and grouped data are different. This difference in the two values is because of the mid-point assumption in grouped data. However, the mean obtained from ungrouped data is the exact mean while the mean obtained from ungrouped data converted into grouped data is an approximate mean.

7. Assumed Mean Method for Calculation of Mean. Sometimes when the numerical values of x_i and f_i are large, finding the product of x_i and f_i, becomes tedious and time consuming. So, for such situations, we think of a method of reducing these calculations.

We can do nothing with the f_i's, but we can change each x_i to a smaller number so that our calculations become easy.

For this, we subtract a fixed number from each of these x_i's.

The first step is to choose one among the x_i's as the assumed mean, and denote it by 'a'. Also, to further reduce our calculation work, we may take 'a' to be that x_i which lies in the centre of $x_1, x_2, \ldots, x_n$. The next step is to find the difference d_i between each a and each of the x_i's, that is, the deviation of 'a' from each of the x_i's i.e.,

$$d_i = x_i - a$$

The third step is to find the product of d_i with the corresponding f_i, and take the sum of all the $f_i d_i$'s. Then, the mean of the deviations,

$$\bar{d} = \frac{\Sigma f_i d_i}{\Sigma f_i}.$$

Relation between $\bar{d}$ and $\bar{x}$

Since in obtaining d_i, we subtracted 'a' from each x_i, so, in order to get the mean $\bar{x}$, we need to add 'a' to $\bar{d}$. This can be explained mathematically as:

Mean of deviations, $\bar{d} = \dfrac{\Sigma f_i d_i}{\Sigma f_i}$.

So, $\bar{d} = \dfrac{\Sigma f_i (x_i - a)}{\Sigma f_i} = \dfrac{\Sigma f_i x_i}{\Sigma f_i} - \dfrac{\Sigma f_i a}{\Sigma f_i}$

$= \bar{x} - a \dfrac{\Sigma f_i}{\Sigma f_i} = \bar{x} - a$

So, $\bar{x} = a + \bar{d}$

STATISTICS

i.e.,
$$\bar{x} = a + \frac{\Sigma f_i d_i}{\Sigma f_i}$$

The method discussed above is called the **Assumed Mean Method.**

Note. The value of the mean does not depend on the choice of 'a', i.e., if we find the mean by taking each of x_i's as 'a', we will find that the mean determined in each case is the same.

8. Step Deviation Method. If the deviations d_i's obtained are all multiples of a certain number h (called the class size), then we divide all the values of d_i's by this number and get smaller numbers to multiply with f_i. So, let $u_i = \frac{x_i - a}{h}$, where a is the assumed mean and h is the class size.

Let
$$\bar{u} = \frac{\Sigma f_i u_i}{\Sigma f_i}$$

Relation between $\bar{u}$ and $\bar{x}$

We have, $u_i = \frac{x_i - a}{h}$

Therefore,
$$\bar{u} = \frac{\Sigma f_i \frac{(x_i - a)}{h}}{\Sigma f_i} = \frac{1}{h}\left[\frac{\Sigma f_i x_i - a\Sigma f_i}{\Sigma f_i}\right]$$

$$= \frac{1}{h}\left[\frac{\Sigma f_i x_i}{\Sigma f_i} - a\frac{\Sigma f_i}{\Sigma f_i}\right] = \frac{1}{h}[\bar{x} - a]$$

So $h\bar{u} = \bar{x} - a$ i.e., $\bar{x} = a + h\bar{u}$

So,
$$\bar{x} = a + h\left(\frac{\Sigma f_i u_i}{\Sigma f_i}\right)$$

The method discussed above is called the **Step-deviation** method.

We note that :
- The step-deviation method will be convenient to apply if all the d_i's have a common factor.
- The mean obtained by all the three methods is the same.
- The assumed mean method and step-deviation method are just simplified forms of the direct method.
- The formula $\bar{x} = a + h\bar{u}$ still holds if a and h are not as given above, but are any non-zero numbers such that $u_i = \frac{x_i - a}{h}$.

9. Appropriate choice of method. The result obtained by all the three methods is the same. So the choice of method to be used depends on the numerical values of x_i and f_i. If x_i and f_i are sufficiently small, then the direct method is an appropriate choice. If x_i and f_i are numerically large numbers, then we can go for the assumed mean method or step-deviation method. If the class sizes are unequal, and x_i, are large numerically, we can still apply the step-deviation method by taking h to be a suitable divisor of all the d_i's.

ILLUSTRATIVE EXAMPLES

[NCERT Exercise 14.1]
(Page No. 270)

Example 1. *A survey was conducted by a group of students as a part of their environment awareness programme, in which they collected the following data regarding the number of plants in 20 houses in a locality. Find the mean number of plants per house.*

Number of plants	0—2	2—4	4—6	6—8	8—10	10—12	12—14
Number of houses	1	2	1	5	6	2	3

Which method did you use for finding the mean, and why ?

Sol.

Number of plants	Number of houses (f_i)	Class mark (x_i)	$f_i x_i$
0—2	1	1	1
2—4	2	3	6
4—6	1	5	5
6—8	5	7	35
8—10	6	9	54
10—12	2	11	22
12—14	3	13	39
Total	$\Sigma f_i = 20$		$\Sigma f_i x_i = 162$

$\therefore \quad \bar{x} = \dfrac{\Sigma f_i x_i}{\Sigma f_i}$

Using direct method because numerical values of x_i and f_i are small

$= \dfrac{162}{20} = 8.1$ plants

We have used the direct method for finding the mean because numerical values of x_i and f_i are small.

Example 2. *Consider the following distribution of daily wages of 50 workers of a factory.*

Daily wages (in Rs)	100—120	120—140	140—160	160—180	180—200
Number of workers	12	14	8	6	10

Find the mean daily wages of the workers of the factory by using an appropriate method.

Sol. Take $a = 150$, $\quad h = 20$

STATISTICS

Daily Wages (in Rs.)	Number of workers (f_i)	class mark (x_i)	$d_i = x_i - 150$	$u_i = \dfrac{x_i - 150}{20}$	$f_i u_i$
100—120	12	110	−40	−2	−24
120—140	14	130	−20	−1	−14
140—160	8	150	0	0	0
160—180	6	170	20	1	6
180—200	10	190	40	2	20
Total	$\Sigma f_i = 50$				$\Sigma f_i u_i = -12$

Using the step-deviation method,

$$\bar{x} = a + \left(\dfrac{\Sigma f_i u_i}{\Sigma f_i}\right) \times h = 150 + \left(\dfrac{-12}{50}\right) \times 20$$

$$= 150 - 4.8 = 145.20.$$

Hence, the mean daily wages of the workers of the factory is Rs. 145.20

Example 3. *The following distribution shows the daily pocket allowance of children of a locality.*

The mean pocket allowance is Rs. 18. Find the missing frequency f.

Daily pocket allowance (in Rs.)	11—13	13—15	15—17	17—19	19—21	21—23	23—25
Number of children	7	6	9	13	f	5	4

Sol.

Daily pocket allowance (in Rs.)	Number of children (f_i)	class mark (x_i)	$f_i x_i$
11—13	7	12	84
13—15	6	14	84
15—17	9	16	144
17—19	13	18	234
19—21	f	20	20f
21—23	5	22	110
23—25	4	24	96
Total	$\Sigma f_i = f + 44$		$\Sigma f_i u_i = 20f + 752$

Using the direct method,

$$\bar{x} = \dfrac{\Sigma f_i x_i}{\Sigma f_i}$$

$$\Rightarrow \qquad 18 = \dfrac{20f + 752}{f + 44}$$

$$\Rightarrow \qquad 20f + 752 = 18(f + 44)$$
$$\Rightarrow \qquad 20f + 752 = 18f + 792$$
$$\Rightarrow \qquad 20f - 18f = 792 - 752$$
$$\Rightarrow \qquad 2f = 40$$
$$\Rightarrow \qquad f = \frac{40}{2} = 20$$

Hence, the missing frequency is 20.

Example 4. *Thirty women were examined in a hospital by a doctor and the number of heart beats per minute were recorded and summarised as follows. Find the mean heart beats per minute for these women, choosing a suitable method.*

Number of heart beats per minute	65—68	68—71	71—74	74—77	77—80	80—83	83—86
Number of women	2	4	3	8	7	4	2

Sol. Take $a = 75.5$, $h = 3$

Number of heart beats per minute	Number of women (f_i)	Class mark (x_i)	$d_i = x_i - 75.5$	$u_i = \dfrac{x_i - 75.5}{3}$	$f_i u_i$
65—68	2	66.5	−9	−3	−6
68—71	4	69.5	−6	−2	−8
71—74	3	72.5	−3	−1	−3
74—77	8	75.5	0	0	0
77—80	7	78.5	3	1	7
80—83	4	81.5	6	2	8
83—86	2	84.5	9	3	6
Total	$\Sigma f_i = 30$				$\Sigma f_i u_i = 4$

Using the step-deviation method,

$$\bar{x} = a + \left(\frac{\Sigma f_i u_i}{\Sigma f_i}\right) \times h = 75.5 + \left(\frac{4}{30}\right) \times 3 = 75.5 + 0.4 = 75.9$$

Hence, the mean heart beats per minute is 75.9.

Example 5. *In a retail market, fruit vendors were selling mangoes kept in packing boxes. These boxes contained varying number of mangoes. The following was the distribution of mangoes according to the number of boxes.*

Number of mangoes	50—52	53—55	56—58	59—61	62—64
Number of boxes	15	110	135	115	25

Find the mean number of mangoes kept in a packing box. Which method of finding the mean did you choose?

STATISTICS

Sol. Take $a = 57$, $h = 3$

Number of mangoes	Number of boxes (f_i)	Class mark (x_i)	$d_i = x_i - 57$	$u_i = \dfrac{x_i - 57}{3}$	$f_i u_i$
50—52	15	51	− 6	− 2	− 30
53—55	110	54	− 3	− 1	− 110
56—58	135	57	0	0	0
59—61	115	60	− 3	1	115
62—64	25	63	6	2	50
Total	$\Sigma f_i = 400$				$\Sigma f_i u_i = 25$

Using the step-deviation method,

$$\bar{x} = a + \left(\dfrac{\Sigma f_i u_i}{\Sigma f_i}\right) \times h = 57 + \left(\dfrac{25}{400}\right) \times 3$$

$$= 57 + 0.19 = 57.19.$$

Example 6. *The table below shows the daily expenditure on food of 25 households in a locality.*

Daily expenditure (in Rs.)	100—150	150—200	200—250	250—300	300—350
Number of households	4	5	12	2	2

Find the mean daily expenditure on food by a suitable method.

Sol. Take $a = 225$, $h = 50$

Daily expenditure (in Rs.)	Number of households (f_i)	Class mark (x_i)	$d_i = x_i - 225$	$u_i = \dfrac{x_i - 225}{50}$	$f_i u_i$
100—150	4	125	− 100	− 2	− 8
150—200	5	175	− 50	− 1	− 5
200—250	12	225	0	0	0
250—300	2	275	50	1	2
300—350	2	325	100	2	4
Total	$\Sigma f_i = 25$				$\Sigma f_i u_i = -7$

Using the step-deviation method,

$$\bar{x} = a + \left(\dfrac{\Sigma f_i u_i}{\Sigma f_i}\right) \times h = 225 + \left(\dfrac{-7}{25}\right) \times 50$$

$$= 225 - 14 = \text{Rs. } 211.$$

Hence, the mean daily expenditure on food is Rs. 211.

Example 7. *To find out the concentration of SO_2 in the air (in parts per million, i.e., ppm), the data was collected for 30 localities in a certain city and is presented below:*

Concentration of SO_2 (in ppm)	Frequency
0.00—0.04	4
0.04—0.08	9
0.08—0.12	9
0.12—0.16	2
0.16—0.20	4
0.20—0.24	2

Find the mean concentration of SO_2 in the air.

Sol. Take $a = 0.14$, $h = 0.04$

Concentration of SO_2 (in ppm)	Frequency (f_i)	Class mark (x_i)	$d_i = x_i - 0.14$	$u_i = \dfrac{x_i - 0.14}{0.04}$	$f_i u_i$
0.00—0.04	4	0.02	−0.12	−3	−12
0.04—0.08	9	0.06	−0.08	−2	−18
0.08—0.12	9	0.10	0.04	−1	−9
0.12—0.16	2	0.14	0	0	0
0.16—0.20	4	0.18	0.04	1	4
0.20—0.24	2	0.22	0.08	2	4
Total	$\Sigma f_i = 30$				$\Sigma f_i u_i = -31$

Using the step-deviation method,

$$\bar{x} = a + \left(\dfrac{\Sigma f_i u_i}{\Sigma f_i}\right) \times h = 0.14 + \left(\dfrac{-31}{30}\right) \times (0.04)$$

$$= 0.14 - .041 = 0.099 \text{ ppm.}$$

Therefore, the mean concentration of SO_2 in the air is 0.099 ppm.

Example 8. *A class teacher has the following absentee record of 40 students of a class for the whole term. Find the mean number of days a student was absent.*

Number of days	0—6	6—10	10—14	14—20	20—28	28—38	38—40
Number of students	11	10	7	4	4	3	1

Sol.

Number of days	Number of students (f_i)	Class mark (x_i)	$f_i x_i$
0—6	11	3	33
6—10	10	8	80
10—14	7	12	84
14—20	4	17	68
20—28	4	24	96
28—38	3	33	99
38—40	1	39	39
Total	$\Sigma f_i = 40$		$\Sigma f_i x_i = 499$

Using the direct method,

$$\bar{x} = \frac{\Sigma f_i x_i}{\Sigma f_i} = \frac{499}{40} = 12.47$$

Hence, the mean number of days a student was absent is 12.48.

Example 9. *The following table gives the literacy rate (in percentage) of 35 cities. Find the mean literacy rate.*

Literacy rate (in%)	45—55	55—65	65—75	75—85	85—95
Number of cities	3	10	11	8	3

Sol. Take $a = 70$, $h = 10$

Literacy rate (in %)	Number of cities (f_i)	Class mark (x_i)	$d_i = x_i - 70$	$u_i = \frac{x_i - 70}{10}$	$f_i u_i$
45—55	3	50	−20	−2	−6
55—65	10	60	−10	−1	−10
65—75	11	70	0	0	0
75—85	8	80	10	1	8
85—95	3	90	20	2	6
Total	$\Sigma f_i = 35$				$\Sigma f_i u_i = -2$

Using the step-deviation method,

$$\bar{x} = a + \left(\frac{\Sigma f_i u_i}{\Sigma f_i}\right) \times h = 70 + \left(\frac{-2}{35}\right) \times 10$$

$$= 70 - \frac{4}{7} = 70 - 0.57 = 69.43 \%.$$

Hence, the mean literacy rate is 69.43%

ADDITIONAL EXAMPLES

Example 10. *The Arithmetic mean of the following frequency distribution is 50. Find the value of p.*

Classes	Frequency
0—20	17
20—40	p
40—60	32
60—80	24
80—100	19

(CBSE 2006)

Sol.

Classes	Frequency f_i	Class mark x_i	$f_i x_i$
0—20	17	10	170
20—40	p	30	30p
40—60	32	50	1600
60—80	24	70	1680
80—100	19	90	1710
Total	$\Sigma f_i = p + 92$		$\Sigma f_i x_i = 30p + 5160$

Using the direct method,

$$\bar{x} = \frac{\Sigma f_i x_i}{\Sigma f_i}$$

$\Rightarrow \quad 50 = \dfrac{30p + 5160}{p + 92} \quad \Rightarrow \quad 30p + 5160 = 50(p + 92)$

$\Rightarrow \quad 30p + 5160 = 50p + 4600 \quad \Rightarrow \quad 50p - 30p = 5160 - 4600$

$\Rightarrow \quad 20p = 560$

$\Rightarrow \quad p = \dfrac{560}{20} \quad \Rightarrow \quad p = 28$

Hence, the value of p is 28.

Example 11. *Find the mean of the following distribution.*

Class	Number of students
4—8	2
8—12	12
12—16	15
16—20	25
20—24	18
24—28	12
28—32	13
32—36	3

(CBSE 2005)

Sol. Take $a = 22$, $h = 4$

Class	Number of students (f_i)	Class mark (x_i)	$d_i = x_i - 22$	$u_i = \dfrac{x_i - 22}{4}$	$f_i u_i$
4—8	2	6	−16	−4	−8
8—12	12	10	−12	−3	−36
12—16	15	14	−8	−2	−30
16—20	25	18	−4	−1	−25
20—24	18	22	0	0	0
24—28	12	26	4	1	12
28—32	13	30	8	2	26
32—36	3	34	12	3	9
Total	$\Sigma f_i = 100$				$\Sigma f_i u_i = -52$

Using the step-deviation method,

$$\bar{x} = a + \left(\dfrac{\Sigma f_i u_i}{\Sigma f_i}\right) \times h = 22 + \left(-\dfrac{52}{100}\right) \times 4$$

$$= 22 - \dfrac{52}{25} = 22 - 2.08 = 19.92$$

Hence, the mean of the given distribution is 19.92.

Example 12. *If the mean of the following data is 18.75, find the value of p :*

x_i	f_i
10	5
15	10
p	7
25	8
30	2

(A.I. CBSE 2005)

Sol.

x_i	f_i	$f_i x_i$
10	5	50
15	10	150
p	7	7p
25	8	200
30	2	60
Total	$\Sigma f_i = 32$	$\Sigma f_i x_i = 7p + 460$

Using direct method,

$$\bar{x} = \dfrac{\Sigma f_i x_i}{\Sigma f_i}$$

$$\Rightarrow \quad 18.75 = \frac{7p + 460}{32}$$
$\Rightarrow \quad 7p + 460 = 32 \times 18.75 \qquad\qquad \Rightarrow \quad 7p + 460 = 600$
$\Rightarrow \quad 7p = 600 - 460 \qquad\qquad\qquad \Rightarrow \quad 7p = 140$
$$\Rightarrow \quad p = \frac{140}{7} \qquad\qquad\qquad\qquad \Rightarrow \quad p = 20$$
Hence, the value of p is 20.

TEST YOUR KNOWLEDGE

1. The marks obtained by 30 students of Class X of a certain school in a Mathematics paper consisting of 100 marks are presented in table below. Find the mean of the marks obtained by the students.

Marks obtained (x_i)	10	20	36	40	50	56	60	70	72	80	88	92	95
Number of student (f_i)	1	1	3	4	3	2	4	4	1	1	2	3	1

2. For the following grouped frequency distribution table, calculate the mean.

Class interval	Number of students
10—25	2
25—40	3
40—55	7
55—70	6
70—85	6
85—100	6

3. The table below gives the percentage distribution of female teachers in the primary schools of rural areas of various states and union territories (U.T.) of India. Find the mean percentage of female teachers by all the three methods discussed in this section.

Percentage of female teachers	15—25	25—35	35—45	45—55	55—65	65—75	75—85
Number of States/U.T.	6	11	7	4	4	2	1

(Source : Seventh All India School Education Survey conducted by NCERT)

4. The distribution below shows the number of wickets taken by bowlers in one-day cricket matches. Find the mean number of wickets by choosing a suitable method. What does the mean signify ?

Number of wickets	20—60	60—100	100—150	150—250	300—350	350—450
Number of bowlers	7	5	16	12	2	3

STATISTICS

5. The Arithmetic Mean of the following frequency distribution is 53. Find the value of p.

Classes	Frequency
0—20	12
20—40	15
40—60	32
60—80	p
80—100	13

(CBSE 2006)

6. The Arithmetic Mean of the following frequency distribution is 52.5. Find the value of p.

Classes	Frequency
0—20	15
20—40	20
40—60	37
60—80	p
80—100	21

(CBSE 2006)

7. The Arithmetic Mean of the following distribution is 47. Determine the value of p.

Class interval	Frequency
0—20	8
20—40	15
40—60	20
60—80	p
80—100	5

(A.I. CBSE 2006)

8. The Arithmetic Mean of the following frequency distribution is 50. Determine the value of p.

Class interval	Frequency
0—20	16
20—40	p
40—60	30
60—80	32
80—100	14

(A.I. CBSE 2006)

9. The Arithmetic Mean of the following frequency distribution is 25. Determine the value of p.

Class interval	Frequency
0—10	5
10—20	18
20—30	15
30—40	p
40—50	6

(A.I. CBSE 2006)

10. The mean of the following frequency distribution is 62.8 and the sum of the frequencies is 50. Compute the missing frequency f_1 and f_2:

Class	Frequency
0—20	5
20—40	f_1
40—60	10
60—80	f_2
80—100	7
100—120	8
Total	50

(CBSE 2004)

11. The mean of the following frequency distribution is 57.6 and the sum of the observations is 50. Find the missing frequency f_1 and f_2 :

Class	Frequency
0—20	7
20—40	f_1
40—60	12
60—80	f_2
80—100	8
100—120	5

(A.I. CBSE 2004)

12. Compute the missing frequency 'f_1' and 'f_2' in the following data if the mean is $166\dfrac{9}{26}$ and the sum of observations is 52.

Classes	Frequency
140—150	5
150—160	f_1
160—170	20
170—180	f_2
180—190	6
190—200	2
Sum	52

(CBSE SP Set 1)

13. The following table shows the marks secured by 100 students in an examination:

Marks	Number of students
0 – 10	15
10 – 20	20
20 – 30	35
30 – 40	20
40 – 50	10

Find the mean marks obtained by a student.

(CBSE SP Set 2)

Answers

1. 59.3
2. 62
3. 39.71, 39.71, 39.71
4. 152.89. This tells us, on an average, the number of wickets taken by these 45 bowlers in one-day cricket is 152.89.
5. 28
6. 25
7. 12
8. 28
9. 16
10. $f_1 = 8, f_2 = 12$.
11. $f_1 = 8, f_2 = 10$,
12. $f_1 = 10, f_2 = 9$
13. 24.

MODE OF GROUPED DATA

IMPORTANT POINTS

1. Mode. Mode is that value among the observations which occurs most often, that is, the value of the observation having the maximum frequency.

2. Multimodal Data. If more than one value may have the same maximum frequency, then in such situations, the data is said to be multimodal.

3. Mode of Grouped Data. In a grouped frequency distribution, it is not possible to determine the mode by looking all the frequencies. Here, we can only locate a class with the maximum frequency, called the **modal class**. The mode is a value inside the modal class, and is given by the formula:

$$\text{Mode} = l + \left(\frac{f_1 - f_0}{2f_1 - f_0 - f_2}\right) \times h$$

where
l = lower limit of the modal class
h = size of the class interval (assuming all class sizes to be equal)
f_1 = frequency of the modal class,
f_0 = frequency of the class preceding the modal class,
f_2 = frequency of the class succeeding the modal class.

4. Remarks. (1) Mode may be less than, equal to a greater than mean.

(2) It depends upon the demand of the situation whether we are interested we finding the mean or mode.

ILLUSTRATIVE EXAMPLES

[NCERT Exercise 14.2]
(Page No. 275)

Example 1. *The following table shows the ages of the patients admitted in a hospital during a year :*

Age (in years)	5—15	15—25	25—35	35—45	45—55	55—65
Number of patients	6	11	21	23	14	5

Find the mode and the mean of the data given above. Compare and interpret the two measures of central tendency.

Sol. Mode. Here, the maximum frequency is 23 and the class corresponding to this frequency is 35–45. So, the modal class is 35–45.

Now, size $(h) = 10$ lower limit (l) of modal class = 35 frequency (f_1) of the modal class = 23 frequency (f_0) of class preceding the modal class = 21

Frequency (f_2) of class succeeding the modal class = 14

$$\therefore \quad \text{Mode} = l + \frac{f_i - f_0}{2f_i - f_0 - f_2} \times h = 35 + \frac{23 - 21}{2 \times 23 - 21 - 14} \times 10$$

$$= 35 + \frac{2}{11} \times 10 = 35 + \frac{20}{11}$$

$$= 35 + 1.8 \text{ (approx.)}$$

$$= 36.8 \text{ year's (approx.)}$$

Mean. Take $a = 40$, $h = 10$.

Age (in years)	Number of patients (f_i)	Class marks x_i	$d_i = x_i - 40$	$u_i = \dfrac{x_i - 40}{10}$	$f_i u_i$
5—15	6	10	−30	−3	−18
15—25	11	20	−20	−2	−22
25—35	21	30	−10	−1	−21
35—45	23	40	0	0	0
45—55	14	50	10	1	14
55—65	5	60	20	2	10
Total	$\Sigma f_i = 80$				$\Sigma f_i u_i = -37$

Using the step-deviation method,

$$\bar{x} = a + \left(\frac{\Sigma f_i u_i}{\Sigma f_i}\right) \times h = 40 + \left(\frac{-37}{80}\right) \times 10$$

$$= 40 - \frac{37}{8} = 40 - 4.63$$

$$= 35.37 \text{ years.}$$

Interpretation. Maximum number of patients admitted in the hospital are of the age 36.8 years (approx.), while on an average the age of a patient admitted to the hospital is 35.37 years.

Example 2. *The following data gives the information on the observed lifetimes (in hours) of 225 electrical components :*

Lifetimes (in hours)	0—20	20—40	40—60	60—80	80—100	100—120
Frequency	10	35	52	61	38	29

Determine the modal lifetimes of the components.

Sol. Here, the maximum class frequency is 61, and the class corresponding to this frequency is 60–80. So, the modal class is 60–80.

STATISTICS

Therefore,
$$h = 20$$
$$l = 60$$
$$f_1 = 61$$
$$f_0 = 52$$
$$f_2 = 38$$

$$\therefore \quad \text{Mode} = l + \frac{f_1 - f_0}{2f_1 - f_0 - f_2} \times h = 60 + \frac{61 - 52}{2 \times 61 - 52 - 38} \times 20$$

$$= 60 + \frac{9}{122 - 90} \times 20 = 60 + \frac{180}{32}$$

$$= 60 + \frac{45}{8} = 60 + 5.625 = 65.625 \text{ hours}$$

Therefore, the modal lifetime of the components is 65.625 hours.

Example 3. *The following data gives the distribution of total monthly household expenditure of 200 families of a village. Find the modal monthly expenditure of the families. Also, find the mean monthly expenditure :*

Expenditure (in Rs.)	Number of families
1000—1500	24
1500—2000	40
2000—2500	33
2500—3000	28
3000—3500	30
3500—4000	22
4000—4500	16
4500—5000	7

Sol. Mode. Since the maximum number of families have their total monthly expenditure (in Rs.) in the interval 1500–2000, the modal class is 1500–2000. Therefore,

$$l = 1500$$
$$h = 500$$
$$f_1 = 40$$
$$f_0 = 24$$
$$f_2 = 33$$

$$\therefore \quad \text{Mode} = l + \left(\frac{f_1 - f_0}{2f_1 - f_0 - f_2}\right) \times h$$

$$= 1500 + \left(\frac{40 - 24}{2 \times 40 - 24 - 33}\right) \times 500$$

$$= 1500 + \left(\frac{16}{80 - 57}\right) \times 500$$

$$= 1500 + \frac{16 \times 500}{23} = 1500 + \frac{8000}{23}$$
$$= 1500 + 347.83 = 1847.83$$

Therefore, the modal monthly expenditure of the families is Rs. 1847.83.

Mean. Take $a = 3250$, $h = 500$

Expenditure (in Rs.)	Number of families (f_i)	Class mark (x_i)	$d_i = x_i - 3250$	$u_i = \dfrac{x_i - 3250}{500}$	$f_i u_i$
1000—1500	24	1250	– 2000	– 4	– 96
1500—2000	40	1750	– 1500	– 3	– 120
2000—2500	33	2250	– 1000	– 2	– 66
2500—3000	28	2750	– 500	– 1	– 28
3000—3500	30	3250	0	0	0
3500—4000	22	3750	500	1	22
4000—4500	16	4250	1000	2	32
4500—5000	7	4750	1500	3	21
Total	$\Sigma f_i = 200$				$\Sigma f_i u_i = -235$

Using the step-deviation method,

$$\bar{x} = a + \left(\frac{\Sigma f_i u_i}{\Sigma f_i}\right) \times h = 3250 + \left(\frac{-235}{200}\right) \times 500$$

$$= 3250 - \frac{235 \times 5}{2} = 3250 - \frac{1175}{2}$$

$$= 3250 - 587.50 = 2662.50$$

Hence, the mean monthly expenditure in Rs. 2662.50.

Example 4. *The following distribution gives the state-wise teacher-student ratio in higher secondary schools of India. Find the mode and mean of this data. Interpret, the two measures.*

Number of students per teacher	Number of States/U.T.
15—20	3
20—25	8
25—30	9
30—35	10
35—40	3
40—45	0
45—50	0
50—55	2

Sol. Mode. Since the maximum number of states/U.T. have the number of students per teacher in the interval 30–35, the modal class is 30–35.

STATISTICS

Therefore,
$$l = 30$$
$$h = 5$$
$$f_1 = 10$$
$$f_0 = 9$$
$$f_2 = 3$$

$$\therefore \quad \text{Mode} = l + \left(\frac{f_1 - f_0}{2f_1 - f_0 - f_2}\right) \times h$$

$$= 30 + \left(\frac{10 - 9}{2 \times 10 - 9 - 3}\right) \times 5 = 30 + \frac{5}{8}$$

$$= 30 + 0.6 = 30.6$$

Hence, the mode of the given data is 30.6.

Mean. Take $a = 37.5$, $h = 5$.

Number of students per teacher	Number of states/U.T. (f_i)	Class mark (x_i)	$d_i = x_i - 37.5$	$u_i = \dfrac{x_i - 37.5}{5}$	$f_i u_i$
15—20	3	17.5	−20	−4	−12
20—25	8	22.5	−15	−3	−24
25—30	9	27.5	−10	−2	−18
30—35	10	32.5	−5	−1	−10
35—40	3	37.5	0	0	0
40—45	0	42.5	5	1	0
45—50	0	47.5	10	2	0
50—55	2	52.5	15	3	6
Total	$\Sigma f_i = 35$				$\Sigma f_i u_i = -58$

Using the step-deviation method,

$$\bar{x} = a + \left(\frac{\Sigma f_i u_i}{\Sigma f_i}\right) \times h$$

$$= 37.5 + \left(\frac{-58}{35}\right) \times 5$$

$$= 37.5 - 8.3$$
$$= 29.2.$$

Interpretation. Most states/U.T. have a student teacher ratio of 30.6 and on an average, this ratio is 29.2.

Example 5. *The given distribution shows the number of runs scored by some top batsmen of the world in one-day international cricket matches.*

Runs scored	Number of batsmen
3000—4000	4
4000—5000	18
5000—6000	9
6000—7000	7
7000—8000	6
8000—9000	3
9000—10000	1
10000—11000	1

Find the mode of the data.

Sol. Since the maximum number of batsmen have their runs scored in the interval 4000—5000, the modal class is 4000—5000.

Therefore, $l = 4000$
$h = 1000$
$f_1 = 18$
$f_0 = 4$
$f_2 = 9$

∴ Mode $= l + \left(\dfrac{f_1 - f_0}{2f_1 - f_0 - f_2}\right) \times h = 4000 + \left(\dfrac{18 - 4}{2 \times 18 - 4 - 9}\right) \times 1000$

$= 4000 + \dfrac{14000}{23} = 4000 + 608.7 = 4608.7$

Hence, the mode of the data is 4608.7.

Example 6. *A student noted the number of cars passing through a spot on a road for 100 periods each of 3 minutes and summarised it in the table given below. Find the mode of the data :*

Number of cars	Frequency
0—10	7
10—20	14
20—30	13
30—40	12
40—50	20
50—60	11
60—70	15
70—80	8

Sol. Here, the maximum frequency is 20, and the class corresponding to this frequency is 40—50. So, the modal class is 40—50.

Therefore, $l = 40$
$h = 10$
$f_1 = 20$
$f_0 = 12$
$f_2 = 11$

$\therefore$ $\quad$ Mode $= l + \left(\dfrac{f_1 - f_0}{2f_1 - f_0 - f_2}\right) \times h$

$= 40 + \left(\dfrac{20 - 12}{2 \times 20 - 12 - 11}\right) \times 10$

$= 40 + \dfrac{80}{17} = 40 + 4.7 = 44.7$

Hence, the mode of the data is 44.7 cars.

TEST YOUR KNOWLEDGE

1. The wickets taken by a bowler in 10 cricket matches are as follows :
   2   6   4   5   0   2   1   3   2   3
   Find the mode of the data.

2. A survey conducted on 20 households in a locality by a group of students resulted in the following frequency table for the number of family members in a household :

Family size	Number of families
1—3	7
3—5	8
5—7	2
7—9	2
9—11	1

   Find the mode of this data

3. The marks distribution of 30 students in a mathematics examination are given in the following table :

Class interval	Number of students
10—25	2
25—40	3
40—55	7
55—70	6
70—85	6
85—100	6
Total	

   Find the mode of this data. Also compare and interpret the mode and the mean.

4. Find the mode of the following grouped frequency distribution :

Class interval	Frequency
0—10	5
10—20	12
20—30	20
30—40	9
40—50	4

5. Find the mode from the following table :

Marks	Number of students
17.5—22.5	2
22.5—27.5	8
27.5—32.5	33
32.5—37.5	80
37.5—42.5	170
42.5—47.5	243
47.5—52.5	213
52.5—57.5	145
57.5—62.5	67
62.5—67.5	35
67.5—72.5	4

6. Calculate the mode from the following table :

Class interval	Frequency
0—5	20
5—10	24
10—15	32
15—20	28
20—25	20
25—30	16
30—35	34
35—40	10
40—45	8

Answers

1. 2
2. 3.286
3. Mode = 52, Mean = 62 ; the maximum number of students obtained 52 marks, while on an average a student obtained 62 marks.
4. 24.2
5. 46.04 marks
6. 13.33.

MEDIAN OF GROUPED DATA

IMPORTANT POINTS

1. Median. We know that the median is a measure of central tendency which gives the value of the middle-must observation in the data.

2. Median of ungrouped data. For finding the median of ungrouped data, we first arrange the data values of the observations in ascending order. Then, if n is odd, the median is the $\left(\dfrac{n+1}{2}\right)$ th observation. And, if n is even, then the median will be the average of the $\dfrac{n}{2}$ th and the $\left(\dfrac{n}{2}+1\right)$ th observations.

3. Median of ungrouped frequency distribution. We shall explain the method of finding median of an ungrouped frequency distribution by taking an example given below :

Suppose, we have to find the median of the following data, which gives the marks, out of 50, obtained by 100 students in a test :

Marks obtained	20	29	28	33	42	38	43	25
Number of students	6	28	24	15	2	4	1	20

First, we arrange the marks in ascending order and prepare a frequency table as follows :

Table

Marks obtained	Number of students (Frequency)
20	6
25	20
28	24
29	28
33	15
38	4
42	2
43	1
Total	100

Here, $n = 100$, which is even. The median will be the average of the $\dfrac{n}{2}$ th and the $\left(\dfrac{n}{2}+1\right)$ th observation, *i.e.*, the 50th and 51st observations. To find these observations, we proceed as follows :

Table

Marks obtained	Number of students
20	6
upto 25	6 + 20 = 26
upto 28	26 + 24 = 50
upto 29	50 + 28 = 78
upto 33	78 + 15 = 93
upto 38	93 + 4 = 97
upto 42	97 + 2 = 99
upto 43	99 + 1 = 100

Now we add another column depicting this information to the frequency table above and name it as *cumulative frequency column*.

Table

Marks obtained	Number of students	Cumulative frequency
20	6	6
25	20	26
28	24	50
29	28	78
33	15	93
38	4	97
42	2	99
43	1	100

From the table above, we see that :

50th observation is 28 as it comes under the *C.F.* 50

51st observation is 29 as it comes under the *C.F.* 78

So, Median = $\dfrac{28 + 29}{2}$ = 28.5

Remark. The part of Table consisting Column 1 and Column 3 is known as *Cumulative Frequency Table*. The median marks 28.5 conveys the information that about 50% students obtained marks less than 28.5 and another 50% students obtained marks more than 28.5.

4. Median of grouped data. We shall explain the method of finding median of a grouped data by taking an example given below :

Consider the grouped frequency distribution of marks obtained, out of 100, by 53 students, in a certain examination, as follows :

Table A

Marks	Number of students
0—10	5
10—20	3
20—30	4
30—40	3
40—50	3
50—60	4
60—70	7
70—80	9
80—90	7
90—100	8

From the above table, we find that the number of students who have scored marks less than 10 is 5. Now the number of students who have scored marks less than 20 includes the number of students who have scored marks less than 20 includes the number of students who have scored marks from 0—10 as well as the number of students who have scored marks from 10—20. So, the total number of students with marks less than 20 is 5 + 3, *i.e.*, 8.

We say that the cumulative frequency of the class 10—20 is 8.

Similarly, we can compute the cumulative frequencies of the other classes, *i.e.*, the number of students with marks less than 30, less than 40, ..., less than 100. We give them in Table given below :

Table B

Marks obtained	Number of students (Cumulative frequency)
Less than 10	5
Less than 20	5 + 3 = 8
Less than 30	8 + 4 = 12
Less than 40	12 + 3 = 15
Less than 50	15 + 3 = 18
Less than 60	18 + 4 = 22
Less than 70	22 + 7 = 29
Less than 80	29 + 9 = 38
Less than 90	38 + 7 = 45
Less than 100	45 + 8 = 53

The distribution given above is called the *cumulative frequency distribution of the less than type*. Here 10, 20, 30, ..., 100, are the upper limits of the respective class intervals.

We can similarly make the table for the number of students with scores, more than or equal to 0, more than or equal to 10, more than or equal to 20, and so on. From Table we observe that all 53 students have scored marks more than or equal to 0. Since there are 5 students scoring marks in the interval 0—10, this means that there are 53 – 5 = 48 students getting more than or equal to 10 marks. Continuing in the same manner, we get the number of students scoring 20 or above as 48 – 3 = 45, 30 or above as 45 – 4 = 41, and so on, as shown in Table.

Table

Marks obtained	Number of students (Cumulative frequency)
More than or equal 0	53
More than or equal 10	53 – 5 = 48
More than or equal 20	48 – 3 = 45
More than or equal 30	45 – 4 = 41
More than or equal 40	41 – 3 = 38
More than or equal 50	38 – 3 = 35
More than or equal 60	35 – 4 = 31
More than or equal 70	31 – 7 = 24
More than or equal 80	24 – 9 = 15
More than or equal 90	15 + 7 = 8

The table above is called a *cumulative frequency distribution of the more than type*. Here 0, 10, 20, ..., 90 give the lower limits of the respective class intervals.

Now, to find the median of grouped data, we can make use of any of these cumulative frequency distributions.

Let us combine Tables A and B to get Table given below :

Table

Mark	Number of students (f)	Cumulative frequency (cf)
0—10	5	5
10—20	3	8
20—30	4	12
30—40	3	15
40—50	3	18
50—60	4	22
60—70	7	29
70—80	9	38
80—90	7	45
90—100	8	53

Now in a grouped data, we may not be able to find the middle observation by looking at the cumulative frequencies as the middle observation will be some value in a class interval. It is, therefore, necessary to find the value inside a class that divides the whole distribution into two halves.

To find this class, we find the cumulative frequencies of all the classes and $\frac{n}{2}$. We now locate the class whose cumulative frequency is greater than (and nearest to) $\frac{n}{2}$. This is called the *median class*. In the distribution above, $n = 53$. So, $\frac{n}{2} = 26.5$. Now 60—70 is the class whose cumulative frequency 29 is greater than (and nearest to) $\frac{n}{2}$, i.e., 26.5.

Therefore, 60—70 is the median class.

After finding the median class, we use the following formula for calculating the median.

$$\text{Median} = l + \left(\frac{\frac{n}{2} - cf}{f}\right) \times h$$

where
  l = lower limit of median class,
  n = number of observation,
  cf = cumulative frequency of class preceding the median class,
  f = frequency of median class,
  h = class size (assuming class size to be equal).

Substituting the values $\frac{n}{2} = 26.5$, $l = 60$, $cf = 22$, $f = 7$, $h = 10$ in the formula above, we get

$$\text{Median} = 60 + \left(\frac{26 - 22}{7}\right) \times 10 = 60 + \frac{45}{7} = 66.4$$

So, about half the students have scored marks less than 66.4, and the other half have scored marks more 66.4.

5. Which measure would be best suited for a particular requirement

(i) **Mean.** The mean is the most frequently used measure of central tendency because it takes into account all the observations, and lies between the extremes, *i.e.*, the largest and the smallest observations of the entire data. It also enables us to compare two or more distributions. For example, by comparing the average (mean) results of students of different schools of a particular examination, we can conclude which school has a better performance.

However, extreme values in the data affect the mean. For example, the mean of classes having frequencies more or less the same is a good representative of the data. But, if one class has frequency, say 2, and the five others have frequency 20, 25, 20, 21, 18, then the mean will certainly not reflect the way the data behaves. So, in such cases, the mean is not a good representative of the data.

(ii) **Median.** In problems where individual observations are not important, and we wish to find out a 'typical' observation, the median is more appropriate, *e.g.*, finding the typical productivity rate of workers, average wage in a country, etc. These are situations where extreme values may be there. So, rather than the mean, we take the median as a better measure of central tendency.

(*iii*) **Mode.** In situations which require establishing the most frequent value or most popular item, the mode is the best choice, *e.g.*, to find the most popular T.V. programme being watched, the consumer item in greatest demand, the colour of the vehicle used by most of the people, etc.

Remark. 1. There is a empirical relationship between the three measures of central tendency :

3 Median = Mode + 2 Mean.

ILLUSTRATIVE EXAMPLES

[NCERT Exercise 14.3]
(*Page No. 287*)

Example 1. *The following frequency distribution gives the monthly consumption of electricity of 68 consumers of a locality. Find the median, mean and mode of the data and compare them.*

Monthly consumption (in units)	Number of consumers
65—85	4
85—105	5
105—125	13
125—145	20
145—165	14
168—185	8
185—205	4

Sol. Median

Monthly consumption (in units)	Number of consumers	Cumulative frequency
65—85	4	4
85—105	5	9
105—125	13	22
125—145	20	42
145—165	14	56
165—185	8	64
185—205	4	68

Now, $n = 68$

So, $\dfrac{n}{2} = \dfrac{68}{2} = 34$

This observation lies in the class 125—145. Therefore, 125—145 is the median class.

So, $l = 125$
$cf = 22$
$f = 20$
$h = 20$

$$\text{Median} = l + \left(\dfrac{\dfrac{n}{2} - cf}{f}\right) \times h = 125 + \left(\dfrac{34 - 22}{20}\right) \times 20$$

$$= 125 + 12 = 137 \text{ units}$$

Mean. Take $a = 135$, $h = 20$

Monthly consumption (in units)	Number of consumers (f_i)	Class mark (x_i)	$d_i = x_i - 135$	$u_i = \dfrac{x_i - 135}{10}$	$f_i u_i$
65—85	4	75	−60	−3	−12
85—105	5	95	−40	−2	−10
105—125	13	115	−20	−1	−13
125—145	20	135	0	0	0
145—165	14	155	20	1	14
168—185	8	175	40	2	16
185—205	4	195	60	3	12
Total	$\Sigma f_i = 68$				$\Sigma f_i u_i = 7$

Using the step-deviation method,

$$\bar{x} = a + \left(\dfrac{\Sigma f_i u_i}{\Sigma f_i}\right) \times h = 135 + \left(\dfrac{7}{68}\right) \times 20$$

$$= 135 + \dfrac{35}{17} = 135 + 2.05 = 137.05.$$

Mode. Since the maximum number of consumers have their monthly consumption (in units) in the interval 125—145, the modal class is 125—145. Therefore,

$$h = 125$$
$$h = 20$$
$$f_1 = 20$$
$$f_0 = 13$$
$$f_2 = 14$$

$$\therefore \quad \text{Mode} = l + \left(\dfrac{f_1 - f_0}{2f_1 - f_0 - f_2}\right) \times h = 125 + \left(\dfrac{20 - 13}{2 \times 20 - 13 - 14}\right) \times 20$$

$$= 125 + \dfrac{7}{13} \times 20 = 125 + \dfrac{140}{13}$$

$$= 125 + 10.76 = 135.76 \text{ units}.$$

Comparison. On comparison, we find that the three measures are approximately the same in this case.

Example 2. *If the median of the distribution given below is 28.5, find the values of x and y.*

Class interval	Frequency
0—10	5
10—20	x
20—30	20
30—40	15
40—50	y
50—60	5
Total	60

Sol.

Class interval	frequency	Cumulative frequency
0—10	5	5
10—20	x	$5 + x$
20—30	20	$25 + x$
30—40	15	$40 + x$
40—50	y	$40 + x + y$
50—60	5	$45 + x + y$
Total	60	

$n = 60$ | Given

$\Rightarrow \quad 45 + x + y = 60$

$\Rightarrow \quad x + y = 60 - 45$

$\Rightarrow \quad x + y = 15$ ...(1)

The median is 28.5, which lies in the class 20—30.
So, $\quad l = 20$
$\quad f = 20$
$\quad cf = 5 + x$
$\quad h = 10$

$\therefore \quad$ Median $= l + \left(\dfrac{\dfrac{n}{2} - cf}{f}\right) \times h$

$\Rightarrow \quad 28.5 = 20 + \left\{\dfrac{\dfrac{60}{2} - (5 + x)}{20}\right\} \times 10$

$\Rightarrow \quad 28.5 = 20 + \dfrac{25 - x}{2}$

$\Rightarrow$ $\dfrac{25-x}{2} = 28.5 - 20$

$\Rightarrow$ $\dfrac{25-x}{2} = 8.5$

$\Rightarrow$ $25 - x = 8.5 \times 2$

$\Rightarrow$ $25 - x = 17$

$\Rightarrow$ $x = 25 - 17 = 8$ ...(2)

From (1) and (2),

$8 + y = 15$

$\Rightarrow$ $y = 15 - 8 = 7$

Hence, the values of x and y are 8 and 7 respectively.

Example 3. *A life insurance agent found the following data for distribution of ages of 100 policy holders. Calculate the median age, if policies are only given to persons having age 18 years onwards but less than 60 year.*

Age (in years)	Number of policy holders
Below 20	2
Below 25	6
Below 30	24
Below 35	45
Below 40	78
Below 45	89
Below 50	92
Below 55	98
Below 60	100

Sol. To calculate the median age, we need to find the class intervals and their corresponding frequencies. It is shown below :

Class intervals	Frequency	Cumulative frequency
Below 20	2	2
20—25	4	6
25—30	18	24
30—35	21	45
35—40	33	78
40—45	11	89
45—50	3	92
50—55	6	98
55—60	2	100

Now, $n = 100$

So, $\dfrac{n}{2} = \dfrac{100}{2} = 50$

This observation lies in the class 35—40.

So, 35—40 is the median class.

Therefore,
$l = 35$
$h = 5$
$cf = 45$
$f = 33$

$$\therefore \quad \text{Median} = l + \left(\frac{\frac{n}{2} - cf}{f}\right) \times h = 35 + \left(\frac{50 - 45}{33}\right) \times 5$$

$$= 35 + \frac{25}{33} = 35 + 0.76 = 35.76 \text{ years.}$$

Hence, the median age is 35.76 years.

Example 4. *The lengths of 40 leaves of a plant are measured correct to the nearest millimetre, and the data obtained is represented in the following table :*

Length (in mm)	Number of leaves
118—126	3
127—135	5
136—144	9
145—153	12
154—162	5
163—171	4
172—180	2

Find the median length of the leaves.

*(**Hints** : The data needs to be converted to continuous classes for finding the median, since the formula assumes continuous classes. The classes then change to 117.5—126.5, 126.5—135.5, ..., 171.5—180.5).*

Sol. We shall first convert the given data to continuous classes. Then, the data become

Length (in mm)	Number of leaves	Cumulative frequency
117.5—126.5	3	3
126.5—135.5	5	8
135.5—144.5	9	17
144.5—153.5	12	29
153.5—162.5	5	34
162.5—171.5	4	38
171.5—180.5	2	40

STATISTICS

Now, $\quad n = 40$

So, $\quad \dfrac{n}{2} = \dfrac{40}{2} = 20$

This observation lies in the class 144.5—153.5. So, 144.5—153.5 is the median class.
Therefore, $\quad l = 144.5$
$\quad\quad\quad\quad h = 9$
$\quad\quad\quad\quad cf = 17$
$\quad\quad\quad\quad f = 12$

∴ $\quad$ Median $= l + \left(\dfrac{\dfrac{n}{2} - cf}{f}\right) \times h = 144.5 + \left(\dfrac{20 - 17}{12}\right) \times 9$

$\quad\quad\quad\quad\quad\quad = 144.5 + 2.25 = 146.75$ mm.

Hence, the median length of the leaves is 146.75 mm.

Example 5. *The following table gives the distribution of the life time of 400 neon lamps :*

Life time (in hours)	Number of lamps
1500—2000	14
2000—2500	56
2500—3000	60
3000—3500	86
3500—4000	74
4000—4500	62
4500—5000	48

Find the median life time of a lamp.

Sol.

Life time (in hours)	Number of lamps	Cumulative frequency
1500—2000	14	14
2000—2500	56	70
2500—3000	60	130
3000—3500	86	216
3500—4000	74	290
4000—4500	62	352
4500—5000	48	400

Now, $\quad n = 400$

So, $\quad \dfrac{n}{2} = \dfrac{400}{2} = 200$

This observation lies in the class 3000—3500.

So, 3000—3500 is the median class.
Therefore,
$l = 3000$
$h = 500$
$cf = 130$
$f = 86$

$$\therefore \quad \text{Median} = l + \left(\frac{\frac{n}{2} - cf}{f}\right) \times h = 3000 + \left(\frac{200 - 130}{86}\right) \times 500$$

$$= 3000 + 406.98 = 3406.98 \text{ hours}$$

Hence, the median life time of a lamp is 3406.98 hours.

Example 6. *100 surnames were randomly picked up from a local telephone directory and the frequency distribution of the number of letters in the English alphabets in the surnames was obtained as follows :*

Number of letters	Number of surnames
1—4	6
4—7	30
7—10	40
10—13	16
13—16	4
16—19	4

Determine the median number of letters in the surnames. Find the mean number of letters in the surnames ? Also, find the modal size of the surnames.

Sol. Median

Number of letters	Number of surnames	Cumulative frequency
1—4	6	6
4—7	30	36
7—10	40	76
10—13	16	92
13—16	4	96
16—19	4	100

Now, $n = 100$

So, $\frac{n}{2} = \frac{100}{2} = 50$

This observation lies in the class 7—10. So, 7—10 is the median class.
Therefore,
$l = 7$
$h = 3$

STATISTICS

$$f = 40$$
$$cf = 36$$

$$\therefore \quad \text{Median} = l + \left(\frac{\frac{n}{2} - cf}{f}\right) \times h = 7 + \left(\frac{50 - 36}{40}\right) \times 3$$

$$= 7 + \frac{21}{20} = 7 + 1.05 = 8.05$$

Hence, the median number of letters in the surnames is 8.05.

Mean. Take $a = 8.5$, $h = 3$

Number of letters	Number of surnames (f_i)	Class mark	$d_i = x_i - 8.5$	$u_i = \dfrac{x_i - 8.5}{3}$	$f_i u_i$
1—4	6	2.5	−6	−2	−12
4—7	30	5.5	−3	−1	−30
7—10	40	8.5	0	0	0
10—13	16	11.5	3	1	16
13—16	4	14.5	6	2	8
16—19	4	17.5	9	3	12
Total	$\Sigma f_i = 100$				$\Sigma f_i u_i = -6$

Using the step-deviation mean,

$$\bar{x} = a + \left(\frac{\Sigma f_i u_i}{\Sigma f_i}\right) \times h = 8.5 + \left(\frac{-6}{100}\right) \times 3$$

$$= 8.5 - 0.18 = 8.32$$

Hence, the mean number of letters in the surnames is 8.32.

Mode. Since the maximum number of surnames have number of letters in the interval 7—10, the modal class is 7—10.

Therefore, $\quad l = 7$

$$h = 3$$
$$f_1 = 40$$
$$f_0 = 30$$
$$f_2 = 16$$

$$\therefore \quad \text{Mode} = l + \left(\frac{f_1 - f_0}{2f_1 - f_0 - f_2}\right) \times h = 7 + \left(\frac{40 - 30}{2 \times 40 - 30 - 16}\right) \times 3$$

$$= 7 + \frac{30}{34} = 7 + 0.88 = 7.88$$

Hence, the modal size of the surnames is 7.88.

Example 7. *The distribution below gives the weights of 30 students of a class. Find the median weight of the students.*

Weight (in kg)	Number of students
40—45	2
45—50	3
50—55	8
55—60	6
60—65	6
65—70	3
70—75	2

Sol.

Weight (in kg)	Number of students	Cumulative frequency
40—45	2	2
45—50	3	5
50—55	8	13
55—60	6	19
60—65	6	25
65—70	3	28
70—75	2	30

Now, $n = 30$

So, $\dfrac{n}{2} = \dfrac{30}{2} = 15$

This observation lies in the class 55—60.
So, 55—60 is the median class.
Therefore, $l = 55$
$h = 5$
$f = 6$
$cf = 13$

$\therefore$ Median $= l + \left(\dfrac{\dfrac{n}{2} - cf}{f}\right) \times h = 55 + \left(\dfrac{15 - 13}{6}\right) \times 5$

$= 55 + \dfrac{10}{6} = 55 + \dfrac{5}{3}$

$= 55 + 1.67 = 56.67$

Hence, the median weight of the students is 56.67 kg.

STATISTICS

TEST YOUR KNOWLEDGE

1. A survey regarding the heights (in cm) of 51 girls of Class X of a school was conducted and the following data was obtained:

Height (in cm)	Number of girls
Less than 140	4
Less than 145	11
Less than 150	29
Less than 155	40
Less than 160	46
Less than 165	51

   Find the median height.

2. The median of the following data is 525. Find the values of x and y, if the total frequency is 100.

Class interval	Frequency
0—100	2
100—200	5
200—300	x
300—400	12
400—500	17
500—600	20
600—700	y
700—800	9
800—900	7
900—1000	4

3. Find the median of the following distribution:

Class	Frequency
0—10	20
10—20	36
20—30	44
30—40	33
40—50	18

4. Find the median of the following grouped frequency distribution :

Class	Frequency
5—10	5
10—15	8
15—20	16
20—25	23
25—30	10
30—35	8

5. Find the median from the following table:

Class	Frequency
30—34	1
35—39	2
40—44	5
45—49	10
50—54	17
55—59	15
60—64	9
65—69	5
70—74	3

6. Find the median of the table given below:

Class intervals	Frequency
0—5	5
5—10	17
10—15	32
15—20	31
20—25	8
25—30	5
30 and more	2

Answers

1. 149.03 cm
2. $x = 9, y = 15$
3. 24.43
4. 21.3
5. 54.06
6. 14.375.

GRAPHICAL REPRESENTATION OF CUMULATIVE FREQUENCY DISTRIBUTION

IMPORTANT POINTS

1. Importance of Graphical Representation. We know that pictures speak better than words. A graphical representation helps us in understanding given data at a glance.

2. Cumulative Frequency curve or an ogive (of the less than type). Let us consider the cumulative frequency distribution given in the following table and draw its ogive (of the less than type).

Table 1

Marks obtained	Number of students (Cumulative frequency)
Less than 10	5
Less than 20	8
Less than 30	12
Less than 40	15
Less than 50	18
Less than 60	22
Less than 70	29
Less than 80	38
Less than 90	45
Less than 100	53

Here, the values 10, 20, 30, ... 100 are the upper limits of the respective class-intervals. To represent the data in the table graphically, we mark the upper limits of the class intervals on the horizontal axis (x-axis) and their corresponding cumulative frequencies on the vertical axis (y-axis), choosing a convenient scale. The scale may not be the same on both the axis. Now we plot the points corresponding to the ordered pairs given by (upper limit, corresponding cumulative frequency), i.e., (10, 5), (20, 8), (30, 12), (40, 15), (50, 18), (60, 22), (70, 29), (80, 38), (90, 45), (100, 53) on a graph paper and join them by a free hand smooth curve. The curve we get is called a **cumulative frequency curve,** or an **ogive** (of the less than type) (see figure)

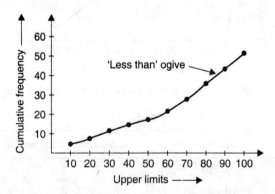

3. Cumulative frequency curve or an ogive (of the more than type). Next, again let us consider the cumulative frequency distribution given in the following Table and draw its ogive (of the more than type).

Here, the values 0, 10, 20, ..., 90 are the lower limits of the respective class intervals 0–10, 10–20, ..., 90–100. To represent 'the more than type' graphically, we plot the lower limits on the x-axis and the corresponding cumulative frequencies on the y-axis. Then we plot the points (lower limit, corresponding cumulative frequency), i.e., (0, 53), (10, 48), (20, 45), (30, 41), (40, 38), (50, 35), (60, 31) (70, 24), (80, 15), (90, 8), on a graph paper, and join them by a free hand smooth curve. The curve we get is a *cumulative frequency curve,* or an *ogive, (of the more than type)* (see figure).

Table 2

Marks obtained	Number of students (Cumulative frequency)
More than or equal to 0	53
More than or equal to 10	48
More than or equal to 20	45
More than or equal to 30	41
More than or equal to 40	38
More than or equal to 50	35
More than or equal to 60	31
More than or equal to 70	24
More than or equal to 80	15
More than or equal to 90	8

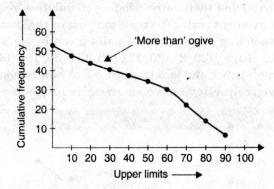

Note 1. Both the ogives above correspond to the same data, which is given in the following table:

Table 3

Marks	Number of students
0—10	5
10—20	3
20—30	4
30—40	3
40—50	3
50—60	4
60—70	7
70—80	9
80—90	7
90—100	8

Note 2. For drawing ogives, it should be ensured that the class intervals are continuous.

STATISTICS

4. Historical Facts. The term 'ogive' is pronounced as 'ojeev' and is derived from the word ogee. An ogee is a shape consisting of a concave arc flowing into a convex arc, so forming an S-shaped curve with vertical ends. In architecture, the ogee shape is one of the characteristics of the 14th and 15th century Gothic styles.

5. Ogive and median. The ogive are related to the median very well. It is possible to obtain the median from these two cumulative frequency curve corresponding to the data given in the table 3. There are two ways to obtain the median.

First way. One obvious way is to locate $\frac{n}{2} = \frac{53}{2} = 26.5$ on the y-axis (see figure). From this point, draw a line parallel to the x-axis cutting the curve at a point. From this point, draw a perpendicular to the x-axis. The point of intersection of this perpendicular with the x-axis determines the median of the data (see figure).

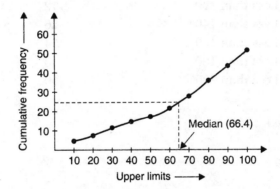

Second way. Another way of obtaining the median is the following: Draw both ogives (*i.e.*, of the less than type and of the more than type) on the same axis. The two ogives will intersect each other at a point. From this point, if we draw a perpendicular on the x-axis, the point at which it cuts the x-axis gives us the median (see figure).

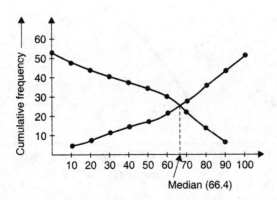

ILLUSTRATIVE EXAMPLES

[NCERT Exercise 14.4]
(Page No. 293)

Example 1. *The following distribution gives the daily income of 50 workers of a factory.*

Daily income (in Rs).	100—120	120—140	140—160	160—180	180—200
Number of workers	12	14	8	6	10

Convert the distribution above to a less than type cumulative frequency distribution, and draw its ogive.

Sol.

Less than type cumulative frequency distribution.

Daily income (in Rs.)	Number of workers
Less than 120	12
Less than 140	26
Less than 160	34
Less than 180	40
Less than 200	50

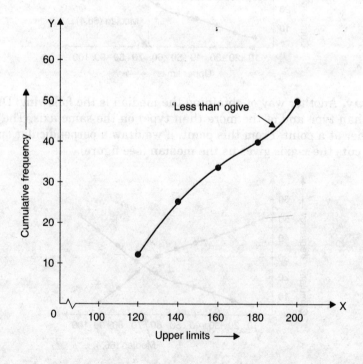

Example 2. *During the medical check up of 35 students of a class, their weights were recorded as follows:*

STATISTICS

Weight (in kg)	Number of students
Less than 38	0
Less than 40	3
Less than 42	5
Less than 44	9
Less than 46	14
Less than 48	28
Less than 50	32
Less than 52	35

Draw a less than type ogive for the given data. Hence obtain the median weight from the graph and verify the result by using the formula.

Sol. Here, $\dfrac{n}{2} = \dfrac{35}{2} = 17.5$

Locate 17.5 on the y-axis. From this point, draw a line parallel to the x-axis cutting the curve at a point. From this point, draw a perpendicular to the x-axis. The point of intersection of this perpendicular with the x-axis determines the median of the given data as **46.4 kg**.

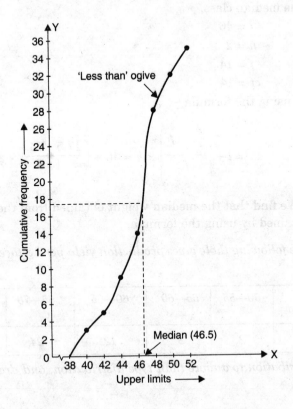

Median Weight by using the formula

Grouped Frequency Distribution

Weight (in kg)	Number of students	Cumulative frequency
0—38	0	0
38—40	3	3
40—42	2	5
42—44	4	9
44—46	5	14
46—48	14	28
48—50	4	32
50—52	3	35

Now $n = 35$

So, $\dfrac{n}{2} = \dfrac{35}{2} = 17.5$

This observation lies in the class 46 – 48.
So, 46 – 48 is the median class.
Therefore, $l = 46$
$h = 2$
$f = 14$
$cf = 14$

∴ Median (by using the formula)

$$= l + \left(\dfrac{\dfrac{n}{2} - cf}{f}\right) \times h = 46 + \left(\dfrac{17.5 - 14}{14}\right) \times 2 = 46 + \dfrac{1}{2} = 46.5 \text{ kg}.$$

Verification. We find that the median weight obtained from the graph is the same as the median weight obtained by using the formula.

Example 3. *The following table gives production yield per hectare of wheat of 100 farms of a village.*

Production yield (in kg/ha)	50—55	55—60	60—65	65—70	70—75	75 – 80
Number of farms	2	8	12	24	38	16

Change the distribution to a more than type distribution, and draw its ogive.

Sol. More than type distribution

Production yield (in kg/ha)	Number of farms
More than 50	100
More than 55	98
More than 60	90
More than 65	78
More than 70	54
More than 75	16

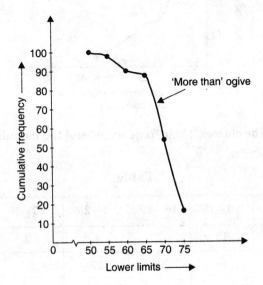

ADDITIONAL EXAMPLES

Example 4. *The annual profits earned by 30 shops of a shopping complex in a locality give rise to the following distribution:*

Profit (in lakhs in Rs.)	Number of shops (frequency)
More than or equal to 5	30
More than or equal to 10	28
More than or equal to 15	16
More than or equal to 20	14
More than or equal to 25	10
More than or equal to 30	7
More than or equal to 35	3

Draw both ogives for the data above. Hence obtain the median profit.

Sol. We first draw the co-ordinate axes, with lower limits of the profit along the horizontal axis, and the cumulative frequency along the vertical axes. Then, we plot the point (5, 30), (10, 28), (15, 16), (20, 14), (25, 10), (30, 7) and (35, 3). We join these points with a smooth curve to get the 'more than' ogive, as shown in figure.

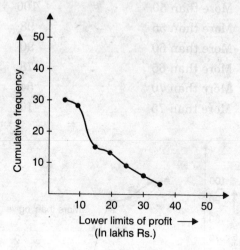

Now, let us obtain the classes, their frequencies and the cumulative frequency from the table above.

Table

Classes	5—10	10—15	15—20	20—25	25—30	30—35	35—40
No. of shops	2	12	2	4	3	4	3
Cumulative frequency	2	14	16	20	23	27	30

Using these values, we plot the points (10, 2), (15, 14), (20, 16), (25, 20), (30, 23), (35, 27), (40, 30) on the same axes as in figure to get the 'less than' ogive, as shown in figure.

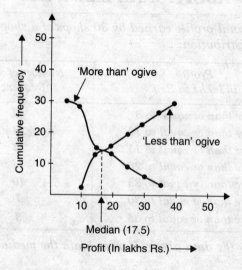

The abcissa of their point of intersection is nearly 17.5, which is the median. This can also be verified by using the formula. Hence, the median profit (in lakhs) is Rs. 17.5.

TEST YOUR KNOWLEDGE

1. The following frequency distribution table represents the weights of 36 students of a class :

Weights (in kg)	Number of students
30.5—35.5	9
35.5—40.5	6
40.5—45.5	15
45.5—50.5	3
50.5—55.5	1
55.5—60.5	2

   (i) Convert the distribution above to a less than type cumulative frequency distribution and draw its ogive.

   (ii) Convert the distribution above to a more than type cumulative frequency distribution and draw its ogive.

2. The life times of 400 new lamps were recorded as follows :

Life time (in hours)	Number of lamps
Less than 300	0
Less than 400	14
Less than 500	70
Less than 600	130
Less than 700	216
Less than 800	290
Less than 900	352
Less than 1000	400

   Draw a less than type ogive for the given data.

3. The annual profits earned by 30 shops of a shopping complex in a locality give rise to the following distribution :

Profit (in lakhs Rs.)	Number of shops (frequency)
More than or equal to 5	30
More than or equal to 10	28
More than or equal to 15	16
More than or equal to 20	14
More than or equal to 25	10
More than or equal to 30	7
More than or equal to 35	3

   Draw both ogives for the data above. Hence obtain the median profit.

MISCELLANEOUS EXERCISE

1. The following table shows marks secured by 140 students in an examination:

Marks	Number of students
0—10	20
10—20	24
20—30	40
30—40	36
40—50	20

   Calculate mean marks.

2. Calculate the mean of the following distribution:

Class interval	Frequency
0—80	22
80—160	35
160—240	44
240—320	25
320—400	24

3. Find the mean marks from the following data:

Marks	Number of students
Below 10	5
Below 20	9
Below 30	17
Below 40	29
Below 50	45
Below 60	60
Below 70	70
Below 80	78
Below 90	83
Below 100	85

4. The weekly observations on cost of living index in a certain city for the year 2005 – 2006 are given below. Compute the mean weakly cost of living index.

Cost of living Index	Number of weeks
140—150	5
150—160	10
160—170	20
170—180	9
180—190	6
190—200	2

STATISTICS

5. In the following frequency distribution, the frequency of the class-interval (40 – 50) is missing. It is known that the mean of the distribution is 52. Find the missing frequency.

Wages (in Rs.)	Number of workers
10—20	5
20—30	3
30—40	4
40—50	—
50—60	2
60—70	6
70—80	13

6. Find the mean age of 100 residents of a colony from the following data:

Age in years (Greater than or equal to)	Number of Persons
0	100
10	90
20	75
30	50
40	25
50	15
60	5
70	0

7. The mean of the following frequency table is 50 and the sum of the frequencies is 120, but the frequencies f_1 and f_2 in classes 20–40 and 60–80 respectively are not known. Find these frequencies.

Class	Frequency
0—20	17
20—40	f_1
40—60	32
60—80	f_2
80—100	19

8. Calculate median from the following table:

Income (in rupees)	Number of Persons
Less than 100	16
Less than 200	39
Less than 300	67
Less than 400	107
Less than 500	127
Less than 600	139
Less than 700	147
Less than 800	150

9. The students of two colleges secured the following marks in an examination. By finding out the median of both distributions, decide which college is better as regards teaching.

Marks secured	Number of students	
	College A	College B
0—10	2	5
10—20	8	6
20—30	12	10
30—40	10	12
40—50	8	7
50—60	4	5
60—70	5	0
70—80	2	3
80—90	1	0

10. Calculate mean, median and mode from the following distribution:

Class intervals	Frequency
0—5	3
5—10	5
10—15	7
15—20	9
20—25	10
25—30	15
30—35	13
35—40	12
40—45	8
45—50	4
50—55	2

11. Calculate the mode from the following frequency distribution:

Class intervals	Frequency
0—10	2
10—20	18
20—30	30
30—40	45
40—50	35
50—60	20
60—70	6
70—80	5

12. Find mode for the table given below:

Marks	Frequency
0—4	3
4—8	9
8—12	15
12—16	11
16—20	7
20—24	5

13. 100 surnames were randomly picked up from a local telephone directory and a frequency distribution of the number of letters in the English alphabet in the surnames was found as follows:

Number of letters	Number of surnames
1—5	6
5—9	30
9—13	44
13—17	16
17—21	4

Draw its 'more than' ogive.

14. The following table gives the distribution of students of two sections according to the marks obtained by them :

Section A		Section B	
Marks	Frequency	Marks	Frequency
0—10	3	0—10	5
10—20	9	10—20	19
20—30	17	20—30	15
30—40	12	30—40	10
40—50	9	40—50	1

Draw a 'more than' ogive for section A and a 'less than' ogive for section B.

15. If the mode and mean of a grouped frequency distributions are 30 and 30 respectively, find the median of that distribution.

Answers

1. 25.86 marks (approx.)
2. 196.8
3. 48.41 marks
4. 166.3 (approx.)
5. 7
6. 31 years
7. $f_1 = 28, f_2 = 24$
8. Rs. 321.25
9. B
10. 27.67, 28.33, 28.57
11. 36
12. 10.4
15. 30.

SUMMARY

1. The mean for grouped data can be found by :

    (i) the direct method: $\bar{x} = \dfrac{\Sigma f_i x_i}{\Sigma f_i}$

    (ii) the assumed mean method: $\bar{x} = a + \dfrac{\Sigma f_i d_i}{\Sigma f_i}$

    (iii) the step-deviation method: $\bar{x} = a + \left(\dfrac{\Sigma f_i u_i}{\Sigma f_i}\right) \times h$,

    with the assumption that the frequency of a class is centred at its mid-point, called its class mark.

2. The mode for grouped data can be found by using the formula:

    $$\text{Mode} = l + \left(\dfrac{f_1 - f_0}{2f_1 - f_0 - f_2}\right) \times h,$$

    where symbols have their usual meanings.

3. The cumulative frequency of a class is the frequency obtained by adding the frequencies of all the classes preceding the given class.

4. The median for grouped data is formed by using the formula:

    $$\text{Median} = l + \left(\dfrac{\frac{n}{2} - cf}{f}\right) \times h,$$

    where symbols have their usual meanings.

5. Representing a cumulative frequency distribution graphically as a cumulative frequency curve or an ogive of the less than type and of the more than type.

6. The median of grouped data can be obtained graphically as the x-coordinate of the point of intersection of the two ogives for this data.

15
Probability

IMPORTANT POINTS

1. Experimental (or empirical) Probabilities. Experimental probabilities of events, as the name suggests is based on the results of actual experiments and adequate recordings of the happening of the events.

An Experiment. Consider the experiment of tossing a coin 1000 times in which the frequencies of the outcomes are as follows :

   Head : 455
   Tail : 545

Based on this experiment, the empirical probability of a head is $\frac{455}{1000}$, *i.e.*, 0.455 and that of getting a tail is 0.545. Note that these probabilities are based on the results of an actual experiment of tossing a coin 1000 times. For this reason, they are called experimental or empirical probabilities. Moreover, these probabilities are only 'estimates'. If we perform the same experiment for another same number of times, we may get different data giving different probability estimates.

2. An Activity. Toss a coin many times and note the number of times it turns up heads (or tails). It may also be noted that as the number of tosses of the coin increases, the experimental probability of getting a head (or tail) comes closer and closer to the number $\frac{1}{2}$.

3. Historical Facts. Many persons from different parts of the world have done the experiment of tossing a coin and recorded the number of heads that turned up.

For example, the eighteenth century French naturalist Comte de Buffon tossed a coin 4040 times and got 2048 heads. The experimental probability of getting a head, in this case, was $\frac{2048}{4040}$, *i.e.*, 0.507. J.E. Kerrich, from Britain, recorded 5067 heads in 10000 tosses of a coin. The experimental probability of getting a head, in this case, was $\frac{5067}{10000} = 0.5067$. The statistician Karl Pearson spent some more time, making 24000 tosses of a coin. He got 12012 heads, and thus, the experimental probability of a head obtained by him was 0.5005.

4. Theoretical Probability. If the experiment of tossing a coin is carried on upto, say, one million times or 10 million times and so on, we would intuitively feel that as the number of tosses increases, the experimental probability of a head (or a tail) seems to be settling down around the number 0.5. *i.e.*, $\frac{1}{2}$, which is what we call the theoretical probability of getting a head (or getting a tail).

5. An unbiased coin. An unbiased coin is 'fair', that is, it is symmetrical so that there is no reason for it to come down more offer on one side than the other. We call this property of the coin as being 'unbiased'.

6. Random Toss. By the phrase 'random toss' we mean that the coin is allowed to fall freely without any bias or interference.

7. Equally likely outcomes

Example 1. Suppose a coin is tossed at random. We know, in advance, that the coin can only land in one of two possible ways-either head up or tail up tail up (we dismiss the possibility of its 'landing' on its edge, which may be possible, for example, if it falls on sand). We can reasonably assume that each outcome head or tail, is *as likely to occur as the other*. We refer to this by saying that *the outcomes* head and tail, *are equally likely*.

Example 2. Suppose we throw a die once. We assume a die to always mean a fair die. Then, the possible outcomes are 1, 2, 3, 4, 5, 6. Each number has the same possibility of showing up. So the equally likely outcomes of throwing a die are 1, 2, 3, 4, 5 and 6.

8. Not equally likely outcomes.

Suppose that a bag contain 4 red balls and 1 blue ball, and we draw a ball without looking into the bag. Then there are two possible outcomes—a red ball and a blue ball. Since these are 4 red balls and only one blue ball, so we are more likely to get a red ball than a blue ball. So, the outcomes (a red ball or a blue ball) are not equally likely. However, the outcome of drawing a ball of any colour from the bag is equally likely. So, all the experiments do not necessarily have equally likely outcomes.

Note : In this chapter, we will assume that all the experiments have equally likely outcomes.

9. Definition of experimental probability.

The experimental or empirical probability P(E) of an event E is defined as

$$P(E) = \frac{\text{Number of trials in which the event happened}}{\text{Total number of trials}}$$

The empirical interpretation of probability can be applied to every event associated with an experiment which can be repeated a large number of times.

10. Limitations of repeating an experiment.

The requirement of repeating an experiment has some limitations, as it may be very expensive or unfeasible in many situations. Of course, it works well in coin tossing or die throwing experiments but not in repeating the experiment of launching a satellite in order to compute the empirical probability of its failure during launching, or the repetition of the phenomenon of an earthquake to compute the empirical probability of a multistoreyed building getting destroyed in an earthquake.

11. Certain assumption lead to theoretical probability.

In experiments where we are prepared to make certain assumptions, the repetition of an experiment can be avoided, as the assumptions help in directly calculating the exact (theoretical) probability. The assumption of equally likely outcomes (which is valid in many experiments, as in the two examples above of a coin and of a die) is one such assumption that leads us to the following definition of probability of an event.

The **theoretical probability** (also called **classical probability**) of an event E, written as P(E), is defined as

$$P(E) = \frac{\text{Number of outcomes favourable to E}}{\text{Number of all possible outcomes of the experiment}},$$

Here, we assume that the outcomes of the experiment are *equally likely*. We will briefly refer to theoretical probability as probability.

12. Historical Facts.

This definition of probability was given by Pierre-Simon Laplace in 1795. Probability theory had its

Pierre-Simon Laplace
(1749—1827)

origin in the 16th century when an Italian physician and mathematician J. Cardan wrote the first book on the subject, **The Book on Games of Chance.** Since its inception, the study of probability has attracted the attention of great mathematicians. James Bernoulli

(1654-1705), A. de Moivre (1667-1754), and Pierre Simon Laplace are among those who made significant contributions to this field. Laplace's *Theorie Analytique des probabilities*, 1812, is considered to be the greatest contribution by a single person to the theory of probability. In recent years, probability has been used extensively in many areas such as biology, economics, genetics, physics, sociology etc.

13. Elementary Event. An event having only one favourable outcome of the experiment is called an elementary event.

The sum of the probabilities of all the elementary events of an experiment is 1.

14. Probability of 'not an event'. Let P(E) be the probability of the event E. Then, the probability of the event 'not E' is denoted by $P(\bar{E})$. Note that

$$P(E) + P(\text{not } E) = 1$$
$$\Rightarrow P(E) + P(\bar{E}) = 1$$
$$\Rightarrow P(\bar{E}) = 1 - P(E)$$

So, in general, it is the that for an event E, $P(\bar{E}) = 1 - P(E)$

15. Complementary Events. The event $\bar{E}$ representing 'not E', is called the complement of the event E. We also say that E and $\bar{E}$ are complementary events.

16. Impossible Event. The probability of an event which is impossible to occur is 0. Such an event is called an impossible event. For example, the event of getting a number in a single throw of a die is an impossible event.

17. Sure Event. The probability of an event which is sure (or certain) to occur is 1. Such an event is called a sure event or a certain event. For example, the event of getting number less than 7 in a single throw of a die is a sure event

Note. $0 \leq P(E) \leq 1$.

18. A Deck of Playing cards. A deck of playing cards consists of 52 cards which are divided into 4 suits of 13 cards each-spades, hearts, diamonds and clubs. Clubs and spades are of black colour, while hearts and diamonds are of red colour. The cards in each suitare arc, king queen Jack, 10, 9, 8, 7, 6, 5, 4, 3 and 2. Kings, queens and jacks are called face cards.

ILLUSTRATIVE EXAMPLES

[NCERT Exercise 15.1]
(Page No. 308)

Example 1. *Complete the following statements :*

(i) Probability of an event E + Probability of the event 'not E' = ...

(ii) The probability of an event that cannot happen is .... Such an event is called ....

(iii) The probability of an event that is certain to happen is .... Such an event is called ....

(iv) The sum of the probabilities of all the elementary events of an experiment is ....

(v) The probability of an event is greater than or equal to .... and less than or equal to ....

Sol. (*i*) Probability of an event E + Probability of the event 'not E' = **1**.

(*ii*) The probability of an event that can not happen is **0**. Such an event is called **impossible event**.

(*iii*) The probability of an event that is certain to happen is **1**. Such an event is called **sure or certain event**.

(*iv*) The sum of the probabilities of all the elementary events of an experiment is **1**.

(*v*) The probability of an event is greater than or equal to **0** and less than or equal to **1**.

Example 2. *Which of the following experiments have equally likely outcomes ? Explain.*

(i) A driver attempts to start a car. The car starts or does not start.

(ii) A player attempts to shoot a basketball. She/he shoots or misses the shot.

(iii) A trial is made to answer a true-false question. The answer is right or wrong.

(iv) A baby is born. It is a boy or a girl.

Sol. (iii) and (iv).

Example 3. *Why is tossing a coin considered to be a fair way of deciding which team should get the ball at the beginning of a football game ?*

Sol. When we toss a coin, the outcomes head and tail are equally likely. So, the result of an individual coin toss is completely unpredictable.

Example 4. *Which of the following cannot be the probability of an event ?*

(A) $\dfrac{2}{3}$ (B) -1.5

(C) 15% (D) 0.7.

Sol. (B) -1.5 cannot be the probability of an event because

$$0 \leq P(E) \leq 1.$$

Example 5. *If $P(E) = 0.05$, what is the probability of 'not E' ?*

Sol. Probability of 'not E'

$= P(\text{not } E)$

$= P(\overline{E})$

$= 1 - P(E)$

$= 1 - 0.05 = 0.95.$

Example 6. *A bag contains lemon flavoured candies only. Malini takes out one candy without looking into the bag. What is the probability that she takes out ?*

(i) *an orange flavoured candy ?*

(ii) *a lemon flavoured candy ?*

Sol. (i) Probability that she takes out an orange flavoured candy is 0 because the bag contains lemon flavoured candies only.

(ii) Probability that she takes out a lemon flavoured candy is 1 because the bag contains lemon flavoured candies only.

Example 7. *It is given that in a group of 3 students, the probability of 2 students not having the same birthday is 0.992. What is the probability that the 2 students have the same birthday ?*

Sol. Probability that the 2 students have the same birthday

$= 1 -$ probability that the 2 students have the same birthday

$= 1 - 0.992 = 0.008.$

Example 8. *A bag contains 3 red balls and 5 black balls. A ball is drawn at random from the bag. What is the probability that the ball drawn is (i) red ? (ii) not red ?*

Sol. (i) Let E be the event that 'the ball drawn is red'.

Total number of balls in the bag

$= 3 + 5 = 8$

So, the number of all possible outcomes $= 8$

There are 3 red balls in the bag.

So, the number of outcomes favourable to $E = 3$

Therefore,

$$P(E) = \dfrac{\text{Number of outcomes favourable to E}}{\text{Number of all possible outcomes}}$$

$$= \dfrac{3}{8}$$

(ii) Probability that the ball drawn is not red

$= 1 -$ probability that the ball drawn is red

$= 1 - P(E)$

$= 1 - \dfrac{3}{8} = \dfrac{5}{8}.$

Aliter. Let $\overline{E}$ be the event that 'the ball drawn is not red'. There are 5 black (not red) balls in the bag.

So, the number of favourable outcomes to the event $\overline{E} = 5.$

Therefore,

$$P(\overline{E}) = \frac{\text{Number of outcomes favourable to } \overline{E}}{\text{Number of all possible outcomes}} = \frac{5}{8}.$$

Example 9. *A box contains 5 red marbles, 8 white marbles and 4 green marbles. One marble is taken out of the box at random. What is the probability that the marble taken out will be (i) red ? (ii) white ? (iii) not green ?*

Sol. Number of red marbles in the box = 5

Number of white marbles in the box = 8

Number of green marbles in the box = 4

∴ Total number of marbles in the box
  = 5 + 8 + 4 = 17

∴ Number of all possible outcomes
  = 17

(i) Number of outcomes favourable to the event that the marble taken out will be red = 5

∴ Probability that the marble taken outcome be red

$$= \frac{\text{Number of outcomes favourable to the event that the marble taken out will be red}}{\text{Number of all possible outcomes}}$$

$$= \frac{5}{17}$$

(ii) Number of outcomes favourable to the event that the marble taken out is white = 8

∴ Probability that the marble taken out will be white

$$= \frac{\text{Number of outcomes favourable to the event that the marble taken out will be white}}{\text{Number of all possible outcomes}}$$

$$= \frac{8}{17}$$

(iii) Number of outcomes favourable to the event that the marble taken out will be green = 4

∴ Probability that the marble taken out will be green

$$= \frac{\text{Number of outcomes favourable to the event that the marble taken out will be green}}{\text{Number of all possible outcomes}}$$

$$= \frac{4}{17}$$

∴ Probability that the marble taken out will not be green = 1 – probability that the marble taken out will be green

$$= 1 - \frac{4}{17}$$

$$= \frac{13}{17}.$$

Example 10. *A piggy bank contains hundred 50 p coins, fifty Re. 1 coins, twenty Rs. 2 coins and ten Rs 5 coins. If it is equally likely that one of the coins will fall out when the bank is turned upside down, what is the probability that the coin (i) will be a 50 p coin ? (ii) will not be a Rs. 5 coin ?*

Sol. Number 50 p coins in the piggy bank = 100

Number of Re. 1 coins in the piggy bank
  = 50

Number of Rs. 2 coins in the piggy bank
  = 20

Number of Rs. 5 coins in the piggy bank
  = 10

∴ Total number of coins in the piggy bank
  = 100 + 50 + 20 + 10 = 180

∴ Number of all possible outcomes
  = 180

(i) Number of favourable outcomes to the event that the coin will be a 50 p coin
  = 100

∴ Probability that the coin will be a 50 p coin

$$= \frac{\text{Number of outcomes favourable to the event that the coin will be a 50 p coin}}{\text{Number of all possible outcomes}}$$

$$= \frac{100}{180} = \frac{5}{9}$$

(ii) Number of favourable outcomes to the event that the coin will not be a Rs. 5 coin
= 100 + 50 + 20 = 170

∴ Probability that the coin will not be Rs. 5 coin

$$= \frac{\text{Number of outcomes favourable to the event that the coin will not be a Rs. 5 coin}}{\text{Number of all possible outcomes}}$$

$$= \frac{170}{180} = \frac{17}{18}.$$

Example 11. *Gopi buys a fish from a shop for his aquarium. The shopkeeper takes out one fish at random from a tank containing 5 male fish and 8 female fish (see figure). What is the probability that the fish taken out is a male fish?*

Sol. Number of male fishes in the aquarium = 5

Number of female fishes in the aquarium = 8

∴ Total number of fishes in the aquarium = 5 + 8 = 13

∴ Number of all possible outcomes = 13

Number of favourable outcomes to the event that the fish taken out is a male fish = 5

∴ Probability that the fish taken out is a male fish

$$= \frac{\text{Number of outcomes favourable to the event that the fish taken out is a male fish}}{\text{Number of all possible outcomes}}$$

$$= \frac{5}{13}.$$

Example 12. *A game of chance consists of spinning an arrow which comes to rest pointing at one of the numbers 1, 2, 3, 4, 5, 6, 7, 8 (see figure), and these are equally likely outcomes. What is the probability that it will point at*

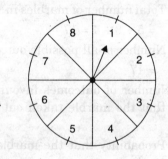

(i) 8 ?

(ii) an odd number ?

(iii) a number greater than 2 ?

(iv) a number less than 9 ?

Sol. Total numbers = 8

∴ Number of all possible outcomes = 8

(i) Number of outcomes favourable to the event that the arrow will point at 8 = 1

∴ Probability that the arrow will point at 8

$$= \frac{\text{Number of outcomes favourable to the event that the arrow will point at 8}}{\text{Number of all possible outcomes}}$$

$$= \frac{1}{8}$$

PROBABILITY

(ii) Number of outcomes favourable to the event that the arrow will point at an odd number (1, 3, 5, 7) = 4

∴ Probability that the arrow will point at an odd number

$= \dfrac{\text{Number of outcomes favourable to the event that the arrow will point at an odd number}}{\text{Number of all possible outcomes}}$

$= \dfrac{4}{8} = \dfrac{1}{2}$

(iii) Number of outcomes favourable to the event that the arrow will point at a number greater than 2

(3, 4, 5, 6, 7, 8) = 6

∴ Probability that the arrow will point at a number greater than 2

$= \dfrac{\text{Number of outcomes favourable to the event that the arrow will point at a number greater than 2}}{\text{Number of all possible outcomes}}$

$= \dfrac{6}{8} = \dfrac{3}{4}$

(iv) Number of outcomes favourable to the event that the arrow will point at a number less than 9

(1, 2, 3, 4, 5, 6, 7, 8) = 8

∴ Probability that the arrow will point at a number less than 9

$= \dfrac{\text{Number of outcomes favourable to the event that the arrow will point at a number less than 9}}{\text{Number of all possible outcomes}}$

$= \dfrac{8}{8} = 1.$

Example 13. *A die is thrown once. Find the probability of getting*

(i) *a prime number ;*
(ii) *a number lying between 2 and 6 ;*
(iii) *an odd number.*

Sol. Number of all possible outcome (1, 2, 3, 4, 5, 6) = 6

(i) Let E be the event of getting a prime number. Then, the outcomes favourable to E are 2, 3 and 5. Therefore, the number of outcomes favourable to E is 3.

So, $P(E) = \dfrac{\text{Number of outcomes favourable to E}}{\text{Number of all possible outcomes}}$

$= \dfrac{3}{6} = \dfrac{1}{2}$

(ii) Let E be the event of getting a number lying between 2 and 6.

Then, the outcomes favourable to E are 3, 4 and 5. Therefore, the number of outcomes favourable to E is 3.

So, $P(E) = \dfrac{\text{Number of outcomes favourable to E}}{\text{Number of all possible outcomes}}$

$= \dfrac{3}{6} = \dfrac{1}{2}$

(iii) Let E be the event getting an odd number.

Then, the outcomes favourable to E are 1, 3 and 5. Therefore, the number of outcomes favourable to E is 3.

So, $P(E) = \dfrac{\text{Number of outcomes favourable to E}}{\text{Number of all possible outcomes}}$

$= \dfrac{3}{6} = \dfrac{1}{2}.$

Example 14. *One card is drawn from a well-shuffled deck of 52 cards. Find the probability of getting*

(i) *a king of red colour*
(ii) *a face card*
(iii) *a red face card*

(iv) the jack of hearts
(v) a spade
(vi) the queen of diamonds.

Sol. ∵ Total number of cards = 52
∴ Number of all possible outcome
= 52

(i) Let E be the event of getting a king of red colour. Then, the outcomes favourable to E are kings of diamond and heart. Therefore, the number of outcomes favourable to E is 2.

So, $P(E) = \dfrac{\text{Number of outcomes favourable to E}}{\text{Number of all possible outcomes}}$

$= \dfrac{2}{52} = \dfrac{1}{26}$

(ii) Let E be the event of getting a face card.

Then, the outcomes favourable to E are 4 kings, 4 queens and 4 jacks.

Therefore, the number of outcomes favourable to E is 12.

So, $P(E) = \dfrac{\text{Number of outcomes favourable to E}}{\text{Number of all possible outcomes}}$

$= \dfrac{12}{52} = \dfrac{3}{13}$

(iii) Let E be the event of getting a red face card. Then, the outcomes favourable to E are 1 diamond king, 1 heart king, 1 diamond queen, 1 heart queen, 1 diamond jack and 1 heart jack. Therefore, the number of outcomes favourable to E is 6.

So, $P(E) = \dfrac{\text{Number of outcomes favourable to E}}{\text{Number of all possible outcomes}}$

$= \dfrac{6}{52} = \dfrac{3}{26}$

(iv) Let E be the event of getting the jack of hearts. Then, the number of outcomes favourable to E is 1.

So, $P(E) = \dfrac{\text{Number of outcomes favourable to E}}{\text{Number of all possible outcomes}}$

$= \dfrac{1}{52}$

(v) Let E be the event of getting a spade. Then, the outcomes favourable to E are 13 spades. Then the number of outcomes favourable to E is 13.

So, $P(E) = \dfrac{\text{Number of outcomes favourable to E}}{\text{Number of all possible outcomes}}$

$= \dfrac{13}{52} = \dfrac{1}{4}$

(vi) Let E be the event of getting the queen of diamonds. Then, the number of outcomes favourable to E is 1.

So, $P(E) = \dfrac{\text{Number of outcomes favourable to E}}{\text{Number of all possible outcomes}}$

$= \dfrac{1}{52}.$

Example 15. *Five cards – the ten, jack, queen, king and ace of diamonds, are well-shuffled with their face downwards. One card is then picked up at random.*

(i) What is the probability that the card is the queen?

(ii) If the queen is drawn and put aside, what is the probability that the second card picked up is (a) an ace? (b) a queen?

Sol. (i) Total number of cards = 5
∴ Number of all possible outcomes
= 5

Let E be the event that the card is the queen. Therefore, the number of outcomes favourable to E is 1.

So, P(E) = $\dfrac{\text{Number of outcomes favourable to E}}{\text{Number of all possible outcomes}}$

= $\dfrac{1}{5}$

(ii) Total number of cards = 5 − 1 = 4

∴ Number of all possible outcomes = 4

(a) Let E be the event that the second card picked up is an ace. Then, the number of outcomes favourable is E is 1.

So, P(E) = $\dfrac{\text{Number of outcomes favourable to E}}{\text{Number of all possible outcomes}}$

= $\dfrac{1}{4}$

(b) Let E be the event that the second card picked up is a queen. Then, the number of outcomes favourable to E is 0 (as there is no queen).

So, P(E) = $\dfrac{\text{Number of outcomes favourable to E}}{\text{Number of all possible outcomes}}$

= $\dfrac{0}{4} = 0$.

Example 16. *12 defective pens are accidentally mixed with 132 good ones. It is not possible to just look at a pen and tell whether or not it is defective. One pen is taken out at random from this lot. Determine the probability that the pen taken out is a good one.*

Sol. Number of defective pens = 12

Number of good pens = 132

∴ Total number of pens
= 12 + 132 = 144

∴ Number of all possible outcomes
= 144

Let E be the event that the pen taken out is a good one. Then, the number of outcomes favourable to E is 132.

∴ P(E) = P(pen taken out is a good one)

= $\dfrac{\text{Number of outcomes favourable to E}}{\text{Number of all possible outcomes}}$

= $\dfrac{132}{144} = \dfrac{11}{12}$.

Example 17. (i) *A lot of 20 bulbs contain 4 defective ones. One bulb is drawn at random from the lot. What is the probability that this bulb is defective ?*

(ii) *Suppose the bulb drawn in (i) is not defective and is not replaced. Now one bulb is drawn at random from the rest. What is the probability that this bulb is not defective ?*

Sol. (i) Total number of bulbs = 20

∴ Number of all possible outcomes
= 20

Let E be the event that the bulb drawn at random from the lot is defective. Then, the number of outcomes favourable to E is 4 since the number of defective bulbs is 4.

∴ P(E) = P (bulb is defective)

= $\dfrac{\text{Number of outcomes favourable to E}}{\text{Number of all possible outcomes}}$

= $\dfrac{4}{20} = \dfrac{1}{5}$

(ii) Total number of bulbs = 19

Number of defective bulbs = 4.

Let E be the event that the bulb drawn is not defective. Then, the number of outcomes favourable to E is 15 since now there are 19 − 4 = 15 not defective bulbs.

∴ P(E) = P(bulb is not defective)

= $\dfrac{\text{Number of outcomes favourable to E}}{\text{Number of all possible outcomes}}$

= $\dfrac{15}{19}$.

Example 18. *A box contains 90 discs which are numbered from 1 to 90. If one disc is drawn at random from the box, find the probability that it bears (i) a two-digit number (ii) a perfect square number (iii) a number divisible by 5.*

Sol. Total number of discs = 90

∴ Number of all possible outcomes
= 90

(i) Let E be the event that the disc bears a two-digit number. Then, the number of outcomes favourable to E is 90 – 9 = 81 as from 1 to 9, the numbers are one-digited and their number is 9.

∴ P(E) = P(a two-digit number)

$= \dfrac{\text{Number of outcomes favourable to E}}{\text{Number of all possible outcomes}}$

$= \dfrac{81}{90} = \dfrac{9}{10}$

(ii) Let E be the event that the disc bears a perfect square number. Then, the number of outcomes favourable to E (1, 4, 9, 16, 25, 36, 49, 64, 81) is 9.

∴ P(E) = P (a perfect square number)

$= \dfrac{\text{Number of outcomes favourable to E}}{\text{Number of all possible outcomes}}$

$= \dfrac{9}{90} = \dfrac{1}{10}$

(iii) Let E be the event that the disc bears a number divisible by 5. Then, the number of outcomes favourable to E (5, 10, 15, 20, 25, 30, 35, 40, 45, 50, 55, 60, 65, 70, 75, 80, 85, 90) is 18.

∴ P (E) = P (a number divisible by 5)

$= \dfrac{\text{Number of outcomes favourable to E}}{\text{Number of all possible outcomes}}$

$= \dfrac{18}{90} = \dfrac{1}{5}.$

Example 19. *A child has a die whose six faces show the letters as given below :*

| A | B | C | D | E | A |

The die is thrown once. What is the probability of getting (i) A ? (ii) D ?

Sol. Total number of letters = 6

∴ Number of all possible outcomes
= 6

(i) **A.** Let E be the event of getting A. Then, the number of outcomes favourable to E is 2 because there are 2 A's.

∴ P(E) = P(A)

$= \dfrac{\text{Number of outcomes favourable to E}}{\text{Number of all possible outcomes}}$

$= \dfrac{2}{6} = \dfrac{1}{3}$

(ii) **D.** Let E be the event of getting D. Then, the number of outcomes favourable to E is 1 since there is only one D.

∴ P(E) = P(D)

$= \dfrac{\text{Number of outcomes favourable to E}}{\text{Number of all possible outcomes}}$

$= \dfrac{1}{6}.$

Example 20*. *Suppose you drop a die at random on the rectangular region shown in figure. What is the probability that it will land inside the circle with diameter 1 m ?*

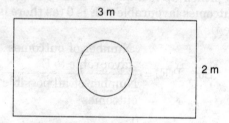

Note from the examination point of view

Sol. Area of the rectangular region
= 3 × 2 = 6 m²

Diameter of the circle = 1 m

∴ Radius of circle = $\dfrac{1}{2}$ m

∴ Area of the circle = $\pi\left(\dfrac{1}{2}\right)^2 = \dfrac{\pi}{4}$ m²

∴ Probability that the die will land inside the circle

$= \dfrac{\frac{\pi}{4}}{6} = \dfrac{\pi}{24}.$

**Note from the examination point of view

Example 21. *A lot consists of 144 ball pens of which 20 are defective and the others are good. Nuri will buy a pen if it is good, but will not buy it if it is defective. The shopkeeper draws one pen at random and gives it to her. What is the probability that*

(i) *She will buy it ?*
(ii) *She will not buy it ?*

Sol. (i) Total number of ball pens
= 144
∴ Number of all possible outcomes
= 144
Number of defective ball pens
= 20

∴ Number of good ball pens
= 144 − 20 = 124
∴ Probability that she will buy it
$= \dfrac{124}{144} = \dfrac{31}{36}$

(ii) Probability that she will not buy it
= 1 − Probability that she will buy it
$= 1 - \dfrac{31}{36}$
$= \dfrac{5}{36}.$

Example 22. *Refer to Example 13.* (i) *Complete the following table:*

Event 'Sum on 2 dice':	2	3	4	5	6	7	8	9	10	11	12
Probability	$\dfrac{1}{36}$						$\dfrac{5}{36}$				$\dfrac{1}{36}$

(ii) A student argues that 'there are 11 possible outcomes 2, 3, 4, 5, 6, 7, 8, 9, 10, 11 and 12. Therefore, each of them has a probability $\dfrac{1}{11}$'. Do you agree with this argument ? Justify your answer.

Sol. (i) Number of all possible outcomes
= 6 × 6 = 36

The outcomes favourable to the event E.

'Sum on 2 dice is 3' are : (1, 2), (2, 1)
∴ The number of outcomes favourable to E = 2
∴ $P(E) = \dfrac{2}{36}$

The outcomes favourable to the event E : 'Sum on 2 dice is 4' are :
(1, 3), (2, 2), (3, 1)
The number of outcomes favourable E = 3
∴ $P(E) = \dfrac{3}{36}$

The outcomes favourable to the event E :

'Sum on 2 dice is 5' are :
(1, 4), (2, 3), (3, 2), (4, 1)
∴ The number of outcomes favourable to E = 4
∴ $P(E) = \dfrac{4}{36}$

The outcomes favourable to the event E :

'Sum on 2 dice is 6' are :
(1, 5), (2, 4), (3, 3), (4, 2), (5, 1)
∴ The number of outcomes favourable to E = 5
∴ $P(E) = \dfrac{5}{36}$

The outcomes favourable to the event E :

'Sum on 2 dice is 7' are :
(1, 6), (2, 5), (3, 4), (4, 3), (5, 2), (6, 1)

∴ The number of outcomes favourable to E = 6

∴ $P(E) = \dfrac{6}{36}$

The outcomes favourable to the event E :

'Sum on 2 dice is 9' are :
(3, 6), (4, 5), (5, 4), (6, 3)

∴ The number of outcomes favourable to E = 4

∴ $P(E) = \dfrac{4}{36}$

The outcomes favourable to the event E :

'Sum on 2 dice is 10' are :
(4, 6), (5, 5), (6, 4)

∴ The number of outcomes favourable to E = 3

∴ $P(E) = \dfrac{3}{36}$

The outcomes favourable to the event E :

'Sum on 2 dice is 11' :
(5, 6), (6, 5)

∴ The number of outcomes favourable to E = 2

∴ $P(E) = \dfrac{2}{36}$

The outcomes favourable to the event E :

'Sum on 2 dice is 12' : (6, 6).

∴ The number of outcomes favourable to E = 1

∴ $P(E) = \dfrac{1}{36}$

Hence, the completed table is as follows :

Event : 'Sum on 2 dice'	2	3	4	5	6	7	8	9	10	11	12
Probability	$\dfrac{1}{36}$	$\dfrac{2}{36}$	$\dfrac{3}{36}$	$\dfrac{4}{36}$	$\dfrac{5}{36}$	$\dfrac{6}{36}$	$\dfrac{5}{36}$	$\dfrac{4}{36}$	$\dfrac{3}{36}$	$\dfrac{2}{36}$	$\dfrac{1}{36}$

(ii) No. We do not agree with the argument as the eleven sums are not equally likely.

Example 23. *A game consists of tossing a one rupee coin 3 times and noting its outcome each time. Hanif wins if all the tosses give the same result i.e. three heads or three tails, and loses otherwise. Calculate the probability that Hanif will lose the game.*

Sol. Let H denote the head and T denote the tail. Then, all the possible outcomes are HHH, TTT, HHT, HTH, HTT, THH, THT, TTH

∴ Number of all possible outcomes = 8

Let E be the event that Hanif will lost the game.

Then, the outcomes favourable to E are : HHT, HTH, HTT, THH, THT, TTH

∴ Number of favourable outcomes = 6

∴ P(E) = probability that Hanif will lost the game

= $\dfrac{\text{Number of outcomes favourable to E}}{\text{Number of all possible outcomes}}$

= $\dfrac{6}{8} = \dfrac{3}{4}$.

Example 24. *A die is thrown twice. What is the probability that*

(i) 5 will not come up either time ?

(ii) 5 will come up at least once ?

[**Hint :** Throwing a die twice and throwing two dice simultaneously are treated as the same experiment]

Sol. The possible outcomes are:

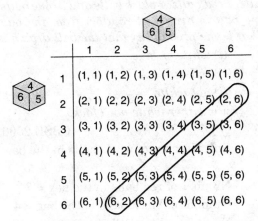

∴ Number of all possible outcomes = $6 \times 6 = 36$

(i) Let E be the event that 5 will not come up either time. Then, the favourable outcomes are :

$(1, 1), (1, 2), (1, 3), (1, 4), (1, 6),$
$(2, 1), (2, 2), (2, 3), (2, 4), (2, 6),$
$(3, 1), (3, 2), (3, 3), (3, 4), (3, 6),$
$(4, 1), (4, 2), (4, 3), (4, 4), (4, 6),$
$(6, 1), (6, 2), (6, 3), (6, 4), (6, 6)$

∴ Number of favourable outcomes
= 25

∴ P(E) = P(5 will not come up either time)

$= \dfrac{\text{Number of outcomes favourable to E}}{\text{Number of all possible outcomes}}$

$= \dfrac{25}{36}$

(ii) Let E be the event that 5 will come up at least once. Then, the favourable outcomes are :

$(1, 5), (2, 5), (3, 5), (4, 5), (5, 5), (6, 5),$
$(5, 1), (5, 2) (5, 3), (5, 4), (5, 6)$

∴ Number of favourable outcomes
= 11

∴ P(E) = P(5 will come up at least once)

$= \dfrac{\text{Number of outcomes favourable to E}}{\text{Number of all possible outcomes}}$

$= \dfrac{11}{36}.$

Example 25. *Which of the following arguments are correct and which are not correct ? Give reasons for your answer.*

(i) If two coins are tossed simultaneously there are three possible outcomes—two heads, two tails or one of each. Therefore, for each of these outcomes, the probability is $\dfrac{1}{3}$.

(ii) If a die is thrown, there are two possible outcomes—an odd number or an even number. Therefore, the probability of getting an odd number is $\dfrac{1}{2}$.

Sol. (*i*) Incorrect. We can classify the outcomes like this but they are not then 'equally likely'. Reason is that 'one of each' can result in two ways—from a head on first coin and tail on the second coin or from a tail on the first coin and head on the second coin. This makes it twicely as likely as two heads (or two tails).

(*ii*) Correct. The two outcomes considered in the question are equally likely.

ADDITIONAL EXAMPLES

Example 26. *A box contains 20 balls bearing numbers 1, 2, 3, 4, ..., 20. A ball is drawn at random from the box, what is the probability that the number on the ball is*

(i) an odd number
(ii) divisible by 2 or 3
(iii) prime number
(iv) not divisible by 10.   (CBSE 2006)

Sol. Total number of balls = 20 (bearing numbers 1, 2, 3, ..., 20).

∴ number of all possible outcomes
= 20

(i) Let E be the event of getting an odd number on the ball. Then, the outcomes favourable to E are 1, 3, 5, 7, 9, 11, 13, 15, 17, 19. Therefore, the number of outcomes favourable to E is 10.

So, P(E) = P (an odd number)

$$= \frac{\text{Number of outcomes favourable to E}}{\text{Number of all possible outcomes}}$$

$$= \frac{10}{20} = \frac{1}{2}$$

(ii) Let E be the event of getting a number divisible by 2 or 3 on the ball. Then, the outcomes favourable to E are 2, 3, 4, 6, 8, 9, 10, 12, 14, 15, 16, 18, 20. Therefore, the number of outcomes favourable to E is 13.

So, P(E) = P (a number divisible by 2 or 3)

$$= \frac{\text{Number of outcomes favourable to E}}{\text{Number of all possible outcomes}}$$

$$= \frac{13}{20}.$$

(iii) Let E be the event of getting a prime number on the ball. Then, the outcomes favourable to E are 2, 3, 5, 7, 11, 13, 17, 19. Therefore, the number of outcomes favourable to E is 8.

So, P(E) = P (prime number)

$$= \frac{\text{Number of outcomes favourable to E}}{\text{Number of all possible outcomes}}$$

$$= \frac{8}{20} = \frac{2}{5}$$

(iv) Let E be the event of getting a number not divisible by 10 on the ball. Then, the outcomes favourable to E are 1, 2, 3, 4, 5, 6, 7, 8, 9, 11, 12, 13, 14, 15, 16, 17, 18, 19. Therefore, the number of outcomes favourable to E is 18.

So, P(E) = P (not divisible by 10)

$$= \frac{\text{Number of outcomes favourable to E}}{\text{Number of all possible outcomes}}$$

$$= \frac{18}{20} = \frac{9}{10}.$$

Example 27. *A bag contains 5 white balls, 7 red balls, 4 black balls and 2 blue balls. One ball is drawn at random from the bag. What is the probability that the ball drawn is*
   (i) *white or blue*
   (ii) *red or black*
   (iii) *not white*
   (iv) *neither white nor black.*

(CBSE 2006)

Sol. Number of white balls in the bag
= 5
Number of red balls in the bag = 7
Number of black balls in the bag = 4
Number of blue balls in the bag = 2

∴ Total number of balls in the bag
= 5 + 7 + 4 + 2 = 18

∴ Number of all possible outcomes
= 18

(i) Let E be the event that the ball drawn is white or blue. Then, the number of outcomes favourable to E is 5 + 2 = 7.

So, P(E) = P (white or blue)

$$= \frac{\text{Number of outcomes favourable to E}}{\text{Number of all possible outcomes}}$$

$$= \frac{7}{18}$$

(ii) Let E be the event that the ball drawn is red or black. Then, the number of outcomes favourable to E is 7 + 4 = 11.

So, P(E) = P (red or black)

$$= \frac{\text{Number of outcomes favourable to E}}{\text{Number of all possible outcomes}}$$

$$= \frac{11}{18}.$$

(iii) Let E be the event not the ball drawn is not white. Then, the number of outcomes favourable to E is 7 + 4 + 2 = 13.

So, P(E) = P (not white)

$$= \frac{\text{Number of outcomes favourable to E}}{\text{Number of all possible outcomes}}$$

$$= \frac{13}{18}$$

PROBABILITY

(iv) Let E be the event that the ball drawn is neither white nor black. Then, the number of outcomes favourable to E is
$$7 + 2 = 9$$
So, P(E) = P (neither white nor black)
$$= \frac{\text{Number of outcomes favourable to E}}{\text{Number of all possible outcomes}}$$
$$= \frac{9}{18} = \frac{1}{2}.$$

Example 28. *A card is drawn at random from a well-shuffled deck of playing cards. Find the probability that the card drawn is*

(i) *a king or jack*

(ii) *a non-ace*

(iii) *a red card*

(iv) *neither a king nor a queen.*

(CBSE 2006)

Sol. Total number of cards in a deck of playing cards = 52

∴ Number of all possible outcomes
  = 52

(i) Let E be the event that the card drawn is a king or jackk. Then, the number of outcomes favourable to E is 4 + 4 = 8
So, P(E) = P (a king or Jack)
$$= \frac{\text{Number of outcomes favourable to E}}{\text{Number of all possible outcomes}}$$
$$= \frac{8}{52} = \frac{2}{13}.$$

(ii) Let E be the event that the card drawn is a non-ace. Then, the number of outcomes favourable to E is 52 − 4 = 48.
So, P(E) = P (a non-ace)
$$= \frac{\text{Number of outcomes favourable to E}}{\text{Number of all possible outcomes}}$$
$$= \frac{48}{52} = \frac{12}{13}.$$

(iii) Let E be the event that the card drawn is a red card. Then, the number of outcomes favourable to E is 13 + 13 = 26.

So, P(E) = P (a red-card)
$$= \frac{\text{Number of outcomes favourable to E}}{\text{Number of all possible outcomes}}$$
$$= \frac{26}{52} = \frac{1}{2}.$$

(iv) Let E be the event that the card drawn is neither a king nor a queen. Then, the number of outcomes favourable to E is 52 − (4 + 4) = 44.
So, P(E) = P (neither a king nor a queen)
$$= \frac{\text{Number of outcomes favourable to E}}{\text{Number of all possible outcomes}}$$
$$= \frac{44}{52} = \frac{11}{13}.$$

Example 29. *A card is drawn at random from a well-shuffled deck of playing cards. Find the probability that the card drawn is*

(i) *a card of spade or an ace*

(ii) *a red king*

(iii) *neither a king nor a queen*

(iv) *either a king or a queen.*

(AI CBSE 2006)

Sol. Total number of cards in a deck of playing cards = 52

∴ Number of all possible outcomes
  = 52

(i) Let E be the event that the card drawn is a card of spade or an ace. Then, the number of outcomes favourable to E is 13 + 4 − 1 = 16
So, P(E) = P (a card of spade or an ace)
$$= \frac{\text{Number of outcomes favourable to E}}{\text{Number of all possible outcomes}}$$
$$= \frac{16}{52} = \frac{4}{13}.$$

(ii) Let E be the event that the card drawn is a red king. Then, the number of outcomes favourable to E is 1 + 1 = 2
So, P(E) = P (a red king)
$$= \frac{\text{Number of outcomes favourable to E}}{\text{Number of all possible outcomes}}$$

$= \dfrac{2}{52} = \dfrac{1}{26}.$

(iii) Let E be the event that the card drawn is neither a king nor a queen. Then, the number of outcomes favourable to E is $52 - (4 + 4) = 44$.

So, P(E) = P (neither a king nor a queen)

$= \dfrac{\text{Number of outcomes favourable to E}}{\text{Number of all possible outcomes}}$

$= \dfrac{44}{52} = \dfrac{11}{13}.$

(iv) Let E be the event that the card drawn is either a king or a queen. Then, the number of outcomes favourable to E is $4 + 4 = 8$.

So, P(E) = P (either a king or a queen)

$= \dfrac{\text{Number of outcomes favourable to E}}{\text{Number of all possible outcomes}}$

$= \dfrac{8}{52} = \dfrac{2}{13}.$

Example 30. *A bag contains 4 white balls, 6 red balls, 7 black balls and 3 blue balls. One ball is drawn at random from the bag. Find the probability that the ball drawn is*

(i) *white*

(ii) *not black*

(iii) *neither white nor black*

(iv) *red or white.* (AI CBSE 2006)

Sol.

Number of white balls in the bag = 4

Number of red balls in the bag = 6

Number of black ball in the bag = 7

Number of blue balls in the bag = 3

∴ Total number of balls in the bag
$= 4 + 6 + 7 + 3 = 20$

∴ Number of all possible outcomes
$= 20$

(i) Let E be the event that the ball drawn is white. Then, the number of outcomes favourable to E is 4.

So, P(E) = P (white)

$= \dfrac{\text{Number of outcomes favourable to E}}{\text{Number of all possible outcomes}}$

$= \dfrac{4}{20} = \dfrac{1}{5}.$

(ii) Let E be the event that the ball drawn is not black. Then, the number of outcomes favourable to E is $4 + 6 + 3 = 13$.

So, P(E) = P (not black)

$= \dfrac{\text{Number of outcomes favourable to E}}{\text{Number of all possible outcomes}}$

$= \dfrac{13}{20}.$

(iii) Let E be the event that the ball drawn is neither white nor black. Then, the number of outcomes favourable to E is $6 + 3 = 9$.

So, P(E) = P (neither white nor black)

$= \dfrac{\text{Number of outcomes favourable to E}}{\text{Number of all possible outcomes}}$

$= \dfrac{9}{20}.$

(iv) Let E be the event that the ball drawn is red or white. Then, the number of outcomes favourable to E is $6 + 4 = 10$.

So, P(E) = P (red or white)

$= \dfrac{\text{Number of outcomes favourable to E}}{\text{Number of all possible outcomes}}$

$= \dfrac{10}{20} = \dfrac{1}{2}.$

Example 31. *A box contains 19 balls bearing numbers 1, 2, 3, ... 19. A ball is drawn at random from the box. Find the probability that the number on the ball is*

(i) *a prime number*

(ii) *divisible by 3 or 5*

(iii) *neither divisible by 5 nor 10*

(iv) *an even number.* (AI CBSE 2006)

Sol. Total number of balls = 19 (bearing numbers 1, 2, 3, ..., 19).

∴ Number of all possible outcomes = 19

(i) Let E be the event that the number on the ball is a prime number. Then, the outcomes favourable to E are 2, 3, 5, 7, 11, 13, 17. Therefore, the number of outcomes favourable to E is 7.

So, P(E) = P (a prime number)

$= \dfrac{\text{Number of outcomes favourable to E}}{\text{Number of all possible outcomes}}$

$= \dfrac{7}{19}.$

(ii) Let E be the event that the number on the ball is divisible by 3 or 5. Then, the outcomes favourable to E are 3, 5, 6, 9, 10, 12, 15, 18. Therefore, the number of outcomes favourable to E is 8.

So, P(E) = P (divisible by 3 or 5)

$= \dfrac{\text{Number of outcomes favourable to E}}{\text{Number of all possible outcomes}}$

$= \dfrac{8}{19}.$

(iii) Let E be the event that the number on the ball is neither divisible by 5 nor 10. Then, the outcomes favourable to E are 1, 2, 3, 4, 6, 7, 8, 9, 11, 12, 13, 14, 16, 17, 18, 19. Therefore, the number of outcomes favourable to E is 16.

So, P(E) = P (neither divisible by 5 nor by 10)

$= \dfrac{\text{Number of outcomes favourable to E}}{\text{Number of all possible outcomes}}$

$= \dfrac{16}{19}.$

(iv) Let E be the event that the number on the ball is an even number. Then, the outcomes favourable to E are 2, 4, 6, 8, 10, 12, 14, 16, 18. Therefore, the number of outcomes favourable to E is 9.

So, P(E) = P (an even number)

$= \dfrac{\text{Number of outcomes favourable to E}}{\text{Number of all possible outcomes}}$

$= \dfrac{9}{19}.$

Example 32. *A bag contains 5 red balls and some blue balls. If the probability of drawing a blue ball is double that of a red ball, find the number of blue balls in the bag.*

Sol. Let the number of blue balls in the bag be x

Then, total number of balls in the bag
$= 5 + x$

∴ Number of all possible outcomes
$= 5 + x$

Number of outcomes favourable to the event of drawing a blue ball $= x$

| ∵ there are x blue balls

∴ Probability of drawing a blue ball

$= \dfrac{\text{Number of outcomes favourable to the event of drawing a blue ball}}{\text{Number of all possible outcomes}}$

$= \dfrac{x}{5 + x}$

Similarly, probability of drawing a red ball

$= \dfrac{5}{5 + x}$

According to the answer

$\dfrac{x}{5+x} = 2\left(\dfrac{5}{5+x}\right) \Rightarrow x = 10$

Hence, the number of blue balls in the bag is 10.

TEST YOUR KNOWLEDGE

1. Find the probability of getting a head when a coin is tossed once. Also find the probability of getting a tail.
2. A bag contains a red ball, a blue ball and a yellow ball, all the balls being of the same size. Kritika takes out a ball from the bag without looking into it. What is the probability that she takes out the

    (i) yellow ball ? (ii) red ball ? (iii) blue ball?
3. Suppose we throw a die once. (i) what is the probability of getting a number greater than 4 ? (ii) what is the probability of getting a number less than or equal to 4 ?
4. One card is drawn from a well-shuffled deck of 52 cards. Calculate the probability that the card will

    (i) be an ace,   (ii) not be an ace.
5. Two players, Sangeeta and Reshma, play a tennis match. It is known that the probability of Sangeeta winning the match is 0.62. What is the probability of Reshma winning?
6. Savita and Hamida are friends. What is the probability that both will have

    (i) different birthdays ?

    (ii) the same birthdays ?

    (ignoring a leap year).
7. There are 40 students in Class X of a school of whom 25 are girls and 15 are boys. The class teacher has to select one student as a class representative. She writes the name of each student on a separate card, the cards being identical. Then she puts cards in a bag and stirs them thoroughly. She then draws one card from the bag. What is the probability that the name written on the card is the name of (i) a girl ? (ii) a boy ?
8. A box contains 3 blue, 2 white, and 4 red marbles. If a marble is drawn at *random* from the box, what is the probability that it will be

    (i) white ?   (ii) blue ?   (iii) red ?
9. Harpreet tosses two different coins simultaneously (say, one is of Re. 1 and other of Rs. 2). What is the probability that she gets *at least* one head?
10. In a musical chair game, the person playing the music has been advised to stop playing the music at any time within 2 minutes after she starts playing. What is the probability that the music will stop within the first half-minute after starting ?

11. A missing helicopter is reported to have crashed somewhere in the rectangular region shown in figure. What is the probability that it crashed inside the lake shown in the figure ?

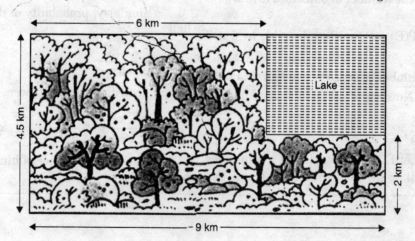

12. A carton consists of 100 shirts of which 88 are good, 8 have minor defects and 4 have major defects. Jimmy, a trader, will only accept the shirts which are good, but Sujatha, another trader, will only reject the shirts which have major defects. One shirt is drawn at random from the carton. What is the probability that

    (i) it is acceptable to Jimmy?

    (ii) it is acceptable to Sujatha?

13. Two dice, one blue and one grey, are thrown at the same time. Write down all the possible outcomes. What is the probability that the sum of the two numbers appearing on the top of the dice is

    (i) 8? (ii) 13 ?

    (iii) less than or equal to 12 ?

14. A card is drawn at random from a well-shuffled pack of 52 cards. Find the probability that the card drawn is neither a red card nor a queen. *(CBSE 2005)*

15. Find the probability that a number selected from the numbers 1 to 25 is not a prime number when each of given numbers is equally likely to be selected. *(CBSE 2005)*

16. There are 30 cards, of same size, in a bag on which numbers 1 to 30 are written. One card is taken out of the bag at random. Find the probability that the number on the selected card is not divisible by 3.

    (CBSE 2005)

17. A bag contains 8 red, 6 white and 4 black balls. A ball is drawn at random from the bag. Find the probability that the ball drawn is

    (i) red or white (ii) not black

    (iii) neither white nor black.

    (AI CBSE 2005)

18. A bag contains 5 red, 8 white and 7 black balls. A ball is drawn at random from the bag. Find the probability that the drawn ball is

    (i) red or white (ii) not black

    (iii) neither white nor black.

    (AI CBSE 2005)

19. A bag contains 7 red, 5 white and 3 black balls. A ball is drawn at random from the bag. Find the probability that the ball drawn is

    (i) red or white (ii) not black

    (iii) neither white nor black.

    (AI CBSE 2003)

20. A bag contains 5 black, 7 red and 3 white balls. A ball is drawn from the bag at random. Find the probability that the ball drawn is

    (i) red (ii) black or white

    (iii) not black. *(CBSE 2004)*

21. A bag contains 4 red, 5 black and 6 white balls. A ball is drawn from the bag at random. Find the probability that the ball drawn is

    (i) white (ii) red

    (iii) not black

    (iv) red or white. *(AI CBSE 2004)*

22. A bag contains 3 red, 5 black and 7 white balls. A ball is drawn from the bag at random. Find the probability that the ball drawn is

    (i) white (ii) red

    (iii) not black

    (iv) red or white. *(AI CBSE 2004)*

23. A bag contains 6 red, 5 black and 4 white balls. A ball is drawn from the bag at random. Find the probability that the ball drawn is

    (i) white (ii) red

    (iii) not black (iv) red or white.

24. An unbiased dice is tossed.

    (i) write the sample space of the experiment.

    (ii) find the probability of getting a number greater than 4.

    (iii) find the probability of getting a prime number. *(CBSE SP 1)*

Answers

1. $\frac{1}{2}, \frac{1}{2}$

2. (i) $\frac{1}{3}$ (ii) $\frac{1}{3}$ (iii) $\frac{1}{3}$

3. (i) $\frac{1}{3}$ (ii) $\frac{2}{3}$

4. (i) $\frac{1}{13}$ (ii) $\frac{12}{13}$ 5. 0.38

6. (i) $\frac{364}{365}$ (ii) $\frac{1}{365}$

7. (i) $\frac{5}{8}$ (ii) $\frac{3}{8}$

8. (i) $\frac{2}{9}$ (ii) $\frac{1}{3}$ (iii) $\frac{4}{9}$

9. $\frac{3}{4}$ 10. $\frac{1}{4}$ 11. $\frac{5}{27}$

12. (i) 0.88 (ii) 0.96

13. (i) $\frac{5}{36}$ (ii) 0 (iii) 1

14. $\frac{6}{13}$ 15. $\frac{16}{25}$ 16. $\frac{2}{3}$

17. (i) $\frac{7}{9}$ (ii) $\frac{7}{9}$ (iii) $\frac{4}{9}$

18. (i) $\frac{13}{20}$ (ii) $\frac{13}{20}$ (iii) $\frac{1}{4}$

19. (i) $\frac{4}{5}$ (ii) $\frac{4}{5}$ (iii) $\frac{7}{15}$

20. (i) $\frac{7}{15}$ (ii) $\frac{8}{15}$ (iii) $\frac{2}{3}$

21. (i) $\frac{2}{5}$ (ii) $\frac{4}{15}$ (iii) $\frac{2}{3}$ (iv) $\frac{2}{3}$

22. (i) $\frac{7}{15}$ (ii) $\frac{1}{5}$ (iii) $\frac{2}{3}$ (iv) $\frac{2}{3}$

23. (i) $\frac{4}{15}$ (ii) $\frac{2}{3}$ (iii) $\frac{2}{3}$ (iv) $\frac{2}{3}$

24. (i) {1, 2, 3, 4, 5, 6}, (ii) $\frac{1}{3}$ (iii) $\frac{1}{2}$.

ILLUSTRATIVE EXAMPLES

[NCERT Exercise 15.2 (Optional)]*
(Page No. 311)

Example 1. *Two customers Shyam and Ekta are visiting a particular shop in the same week (Tuesday to Saturday). Each is equally likely to visit the shop on any day as on another day. What is the probability that both will visit the shop on (i) the same day? (ii) consecutive days? (iii) different days?*

Sol. Number of all possible outcomes
= 5 × 5 = 25

(*i*) Number of outcomes favourable to the event that both will visit the shop on the same day = 5

[Tue, Tue; Wed, Wed; Th, Th; F, F; S, S]

∴ Probability that both will visit the shop on the same day

$$= \frac{5}{25} = \frac{1}{5}$$

(*ii*) Number of outcomes favourable to the event that both will visit the shop on consecutive days = 8

[T, W; W, T; W, Th; Th, W; Th, F; F, Th; F, S; S, F]

∴ Probability that both will visit the shop on consecutive days = $\frac{8}{25}$

(*iii*) Probability that both will visit the shop on different days = 1 − Probability that both will visit the shop on the same day

$$= 1 - \frac{1}{5} = \frac{4}{5}.$$

Example 2. *A die is numbered in such a way that its faces show the numbers 1, 2, 2, 3, 3, 6. It is thrown two times and the total*

---

*These exercises are not from the examination point of view.

score in two throws is noted. Complete the following table which gives a few values of the total score on the two throws:

	+	1	2	2	3	3	6
Number in second throw	1	2	3	3	4	4	7
	2	3	4	4	5	5	8
	2					5	
	3						
	3			5			9
	6	7	8	8	9	9	12

Number in first throw

What is the probability that the total score is

(i) even ? (ii) 6 ? (iii) at least 6 ?

Sol. The completed table is as follows:

	1	2	2	3	3	6
1	2	3	3	4	4	7
2	3	4	4	5	5	8
2	3	4	4	5	5	8
3	4	5	5	6	6	9
3	4	5	5	6	6	9
6	7	8	8	9	9	12

Number of all possible outcomes = 36

(i) Number of outcomes favourable to the event that the total score is even = 18

∴ Required probability = $\frac{18}{36} = \frac{1}{2}$

(ii) Number of outcomes favourable to the event that the total score is 6 = 4

∴ Required probably = $\frac{4}{36} = \frac{1}{9}$

(iii) Number of outcomes favourable to the event that the total score at last 6 = 15

∴ Required probably = $\frac{15}{36} = \frac{5}{12}$

Example 3. *A bag contains 5 red balls and some blue balls. If the probability of drawing a blue ball is double that of a red ball, determine the number of blue balls in the bag.*

Sol. Let the number of blue balls in the bag be x

Then, total number of balls in the bag
= $5 + x$

∴ Number of all possible outcomes
= $5 + x$

Number of outcomes favourable to the event of drawing a blue ball = x

[∵ there are x blue balls

∴ Probability of drawing a blue ball

$= \frac{\text{Number of outcomes favourable to the event of drawing a blue ball}}{\text{Number of all possible outcomes}}$

$= \frac{x}{5+x}$

Similarly, probability of drawing a red ball

$= \frac{5}{5+x}$

According to the question,

$\frac{x}{5+x} = 2\left(\frac{5}{5+x}\right) \Rightarrow x = 10$

Hence, the number of blue balls in the bag is 10.

Example 4. *A box contains 12 balls out of which x are black. If one ball is drawn at random from the box, what is the probability that it will be a black ball ?*

If 6 more black balls are put in the box, the probability of drawing a black ball is now double of what it was before. Find x.

Sol. Total number of balls in the box
= 12

Number of all possible outcomes = 12

Number of outcomes favourable to the event of drawing a black ball = x

∴ Required probably = $\frac{x}{12}$

Now, when 6 more black balls are put in the box, then,

Number of all possible outcomes
= 12 + 6 = 18

Number of outcomes favourable to the event of drawing a black ball = $x + 6$

∴ Probability of drawing a black ball

$$= \frac{x+6}{18}$$

According to the question,

$$\frac{x+6}{18} = 2\left(\frac{x}{12}\right)$$

$$\Rightarrow \quad x = 3.$$

Example 5. *A jar contains 24 marbles, some are green and others are blue. If a marble is drawn at random from the jar, the probability that it is green is* $\frac{2}{3}$. *Find the number of blue balls in the jar.*

Sol. Total number of marbles in the jar
= 24

∴ Number of all possible outcomes
= 24

Let the number of blue marbles in the jar be x.

Then, the number of green marbles in the jar = $24 - x$

Therefore,

Probability that the marble drawn is green

$$= \frac{24-x}{24}$$

According to the question,

$$\frac{24-x}{24} = \frac{2}{3}$$

$$\Rightarrow \quad 3(24 - x) = 48$$
$$\Rightarrow \quad 72 - 3x = 48$$
$$\Rightarrow \quad 3x = 72 - 48 = 24$$

$$\Rightarrow \quad x = \frac{24}{3} = 8.$$

Hence, the number if blue balls in the jar is 8.

SUMMARY

1. The difference between experimental probability and theoretical probability.
2. The theoretical (classical) probability of an event E, written as P(E), is defined as

$$P(E) = \frac{\text{Number of outcomes favourable to E}}{\text{Number of all possible outcomes of the experiment}}$$

where we assume that the outcomes of the experiment are equally likely.

3. The probability of a sure event (or contain event) is 1.
4. The probability of an impossible event is 0.
5. The probability of an event E is a number P(E) such that
$0 \le P(E) \le 1$
6. An event having only one outcome is called an elementary event. The sum of the probabilities of all the elementary events of an experiment is 1.
7. For any event E, $P(E) + P(\overline{E}) = 1$, where $\overline{E}$ stands for 'not E'. E and $\overline{E}$ are called complementary events.

A Note to the Reader

The experimental or empirical probability of an event is based on what has actually happened while the theoretical probability of the event attempts to predict what will happen on the basis of certain assumptions. As the number of trials in an experiment, go on increasing we may expect the experimental and theoretical probability to be nearly the same.

A1
Proofs in Mathematics

IMPORTANT POINTS

1. Use of reasoning. The ability to reason and think clearly is extremely useful in our daily life. For example, suppose a politician tells you, "If you are interested in a clean government, then you should vote for me." What he actually wants you to believe is that if you do not vote for him, then you may not get a clean government. Similarly, if an advertisement tells you, "The intelligent wear ** shoes." What the company wants you to conclude is that if you do not wear ** shoes, then you are not intelligent enough. We can ourself observe that both the above statements may mislead the general public. So, if we understand the process of reasoning correctly, we do not fall into such traps unknowingly.

2. Deductive Reasoning. The correct use of reasoning is at the core of mathematics, especially in constructing proofs. We have introduced ourselves with the idea of proofs, and have actually proved many statements, especially in geometry. We know that a proof is made up of several mathematical statements, each of which is logically deduced from a previous statement in the proof, or from a theorem proved earlier, or an axiom, or the hypotheses. The main tool, we use in constructing a proof, is the process of deductive reasoning.

3. Mathematical Statement. We know that a 'statement' is a meaningful sentence which is not an order, or an exclamation or a question. For example, 'Which two teams are playing in the World Cup Final?' is a question, not a statement. 'Go and finish your homework' is an order, not a statement. 'What a fantastic goal!' is an exclamation, not a statement.

Remember, in general, statements can be one of the following:
- *always true*
- *always false*
- *ambiguous*

We know that in mathematics, a **statement is acceptable only if it is either always true or always false.** So, ambiguous sentences are not considered as mathematical statements.

ILLUSTRATIVE EXAMPLES

[NCERT Exercise A1.1]
(Page No. 316)

Example 1. *State whether the following statements are always true, always false or ambiguous. Justify your answers.*

  (i) *All mathematics textbooks are interesting.*

  (ii) *All distance from the Earth to the Sun is approximately 1.5×10^8 km.*

  (iii) *All human beings grow old.*

  (iv) *The journey from Uttarkashi to Harsil is tiring.*

  (v) *The woman saw an elephant through a pair of binoculars.*

Sol. (i) Ambiguous because a particular textbook of mathematics may be interesting for one but not interesting for the other.

(ii) True since the astronomers have established that the distance from the Earth to the Sun is approximately 1.5×10^8 km.

(iii) True since every human being grows old.

(iv) Ambigucus because the journey from Uttarkashi to Harsil may be timing for one but not for the other.

(v) Ambiguous because it is not clear which women is being referred to.

Example 2. *State whether the following statements are true or false. Justify your answers.*

(i) *All hexagons are polygons.*

(ii) *Some polygons are pentagons.*

(iii) *Not all even numbers are divisible by 2.*

(iv) *Some real numbers are irrational.*

(v) *Not all real numbers are rational.*

Sol. (i) True because hexagon is a closed figure enclosed by 6 straight lines.

(ii) True because the polygons which are enclosed by 5 straight lines are pentagons

(iii) False because every even number is a multiple of 2 by definition.

(iv) True because the real numbers which are not rational and irrational.

(v) True because $\sqrt{2}$ is a real number but not rational. Actually it is irrational.

Example 3. *Let a and b be real numbers such that $ab \neq 0$. Then which of the following statements are true? Justify your answers.*

(i) *Both a and b must be zero.*

(ii) *Both a and b must be non-zero.*

(iii) *Either a or b must be non-zero.*

Sol. Only (ii) is true because multiplied by any number (zero or non-zero) '0' gives 0.

Example 4. *Restate the following statements with appropriate conditions, so that they become true.*

(i) *If $a^2 > b^2$, then $a > b$.*

(ii) *If $x^2 = y^2$, then $x = y$.*

(iii) *If $(x + y)^2 = x^2 + y^2$, then $x = 0$.*

(iv) *The diagonals of a quadrilateral bisect each other.*

Sol. (i) If $a > 0$ and $a^2 > b^2$, then $a > b$.

(ii) If $xy \geq 0$ and $x^2 = y^2$, then $x = y$.

(iii) If $(x + y)^2 = x^2 + y^2$ and $y \neq 0$, then $x = 0$.

(iv) The diagonals of a parallelogram bisect each other.

TEST YOUR KNOWLEDGE

1. State whether the following statements are always true, always false or ambiguous. Justify your answers.
   (i) The sun orbits the earth
   (ii) Vehicles have four wheels
   (iii) The speed of light is approximately 3×10^5 km/s.
   (iv) A road to Kolkata will be closed from November to March.
   (v) All humans are mortal.

2. State whether the following statements are true or false, and justify your answers.
   (i) All equilateral triangles are isosceles.
   (ii) Some isosceles triangles are equilateral.
   (iii) All isosceles triangles are equilateral.
   (iv) Some rational numbers are integers.
   (v) Some rational numbers are not integers.
   (vi) Not all integers are rational.
   (vii) Between any two rational numbers there is no rational number.

3. If $x < 4$, which of the following statements are true? Justify your answers.
   (i) $2x > 8$
   (ii) $2x < 6$
   (iii) $2x < 8$.

PROOFS IN MATHEMATICS

4. Restate the following statements with appropriate conditions, so that they become true statements:
   (i) If the diagonals of a quadrilateral are equal, then it is a rectangle
   (ii) A line joining two points on two sides of a triangle is parallel to the third side.
   (iii) $\sqrt{p}$ is irrational for all positive integers p.
   (iv) All quadratic equations have two real roots.

Answers

1. (i) This statement is always false, since astronomers have established that the earth orbits the sun.
   (ii) This statement is ambiguous, because we cannot decide if it is always true or always false. This depends on what the vehicle is—vehicles can have 2, 3, 4, 6, 10, etc., wheels.
   (iii) This statement is always true, as verified by physicists.
   (iv) This statement is ambiguous, because it is not clear which road is being referred to.
   (v) This statement is always true, since every human being has to die some time.

2. (i) This statement is true, because equilateral triangles have equal sides, and therefore are isosceles.
   (ii) This statement is true, because those isosceles triangles whose base angles are 60° are equilateral.
   (iii) This statement is false. Give a counter-example for it.
   (iv) This statement is true, since rational numbers of the form $\frac{p}{q}$, where p is an integer and $q = 1$, are integers (for example, $3 = \frac{3}{1}$).
   (v) This statement is true, because rational numbers of the form $\frac{p}{q}$, p, q are integers and q does not divide p, are not integers (for example $\frac{3}{2}$)
   (vi) This statement is the same as saying 'there is an integer which is not a rational number'. This is false, because all integers are rational numbers.
   (vii) This statement is false. As you know, between any two rational numbers r and s lies $\frac{r+s}{2}$, which is a rational number.

3. (i) This statement is false, because, for example, $x = 3 < 4$ does not satisfy $2x > 8$.
   (ii) This statement is false, because, for example, $x = 3.5 < 4$ does not satisfy $2x < 6$.
   (iii) This statement is true, because it is the same as $x < 4$.

4. (i) If the diagonals of a parallelogram are equal, then it is a rectangle.
   (ii) A line joining the mid-points of two sides of a triangle is parallel to the third side.
   (iii) $\sqrt{p}$ is irrational for all primes p.
   (iv) All quadratic equations have at most two real roots.

Remark : There can be other ways of restating the statements above. For instance, (iii) can also be restated as '$\sqrt{p}$ is irrational for all positive integers p which are not a perfect square'.

DEDUCTIVE REASONING

IMPORTANT POINTS

1. Deductive Reasoning. Deductive reasoning is used to deduce conclusions from given statements that we assume to be true. The given statements are called 'premises' or 'hypotheses'.

What we do is that we assume the hypotheses to be true and then apply deductive resoning to arrive at a conclusion.

What really matters is that we use the correct process of reasoning, and this process of reasoning does not depend on the truth or falsity of the hypotheses. However, it must also be noted that if we start with an incorrect premise (or hypothesis), we may arrive at a wrong conclusion.

ILLUSTRATIVE EXAMPLES

[NCERT Exercise A 1.2]
(Page No. 318)

Example 1. *Given that all women are mortal, and suppose that A is a woman, what can we conclude about A?*

Sol. Here we have two premises :
(i) All women are mortal
(ii) A is a woman
From these premises, we deduce that A is mortal.

Example 2. *Given that the product of two rational numbers is rational, and suppose a and b are rationale, what can you conclude about ab?*

Sol. Here we have two premises :
(i) The product of two rational numbers is rational.
(ii) a and b are rationals
From these premises, we deduce that ab is rational.

Example 3. *Given that the decimal expansion of irrational numbers is non-terminating, non-recurring, and $\sqrt{17}$ is irrational, what can we conclude about the decimal expansion of $\sqrt{17}$?*

Sol. Here we have two premises :
(i) The decimal expansion of irrational numbers is non-terminating, non-recurring.
(ii) $\sqrt{17}$ is irrational.
From these premises we deduce that decimal expansion of $\sqrt{17}$ is non-terminating, non-recurring.

Example 4. *Given that $y = x^2 + 6$ and $x = -1$, what can we conclude about the value of y?*

Sol. Given the two hypotheses, we get
$y = (-1)^2 + 6 = 1 + 6 = 7$.

Example 5. *Given that ABCD is a parallelogram and $\angle B = 80°$. What can you conclude about the other angles of the parallelogram?*

Sol. We are given that ABCD is a parallelogram. So, we deduce that all the properties that hold for a parallelogram hold for ABCD. Therefore, in particular the property that "the opposite angles of a Parallelogram are equal to each other", holds. So, $\angle B = \angle D$, since the know that $\angle B = 80°$, we can deduce that $\angle D = 80°$.

Again, the property that "the sum of the consecutive interior angles of a parallelogram is 180°" also holds.

Since $\angle B = 80°$
Therefore $\angle A = 180° - 80° = 100°$
Since $\angle D = 80°$
Therefore, $\angle C = 180° - 80° = 100°$.

Example 6. *Given that PQRS is a cyclic quadrilateral and also its diagonals bisect each other. What can you conclude about the quadrilateral?*

Sol. Since the diagonals of quadrilateral PQRS bisect each other. So, the quadrilateral PQRS is a parallelogram.

Since opposite angles of a parallelogram are equal and opposite angles of a cyclic quadrilateral are supplementary.

Therefore, each angle of each pair of opposite angles is 90°.

So, we deduce that the cyclic quadrilateral is a rectangle.

Example 7. *Given that $\sqrt{p}$ is irrational for all prime p and also suppose that 3721 is a prime. Can you conclude that $\sqrt{3721}$ is an irrational number ? Is your conclusion correct ? Why or why not ?*

Sol. Yes, because of the premise. No, because $\sqrt{3721}$ = 61 which is not irrational.

Since the premise that 3271 is a prime number was wrong because 3721 = 61 × 61, so the conclusion is false.

Note. The correct premises. Given that $\sqrt{p}$ is irrational for all prime p which are not perfect squares and also suppose that a particular number is a prime which is not a perfect square.

TEST YOUR KNOWLEDGE

1. Given that Bijapur is in the state of Karnataka, and suppose Shabana lives in Bijapur. In which state does Shabana live ?
2. Given that all mathematics textbooks are interesting, and suppose you are reading a mathematics textbook. What can we conclude about the textbook you are reading ?
3. Given that $y = -6x + 5$, and suppose $x = 3$. What is y ?
4. Given that ABCD is a parallelogram, and suppose AD = 5 cm, AB = 7 cm (see figure). What can you conclude about the lengths of DC and BC ?

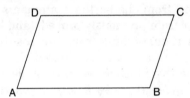

5. Given that $\sqrt{p}$ is irrational for all primes p, and suppose that 19423 is a prime. What can you conclude about $\sqrt{19423}$?

Answers

1. Shabana lives in the state of Karnataka.
2. You are reading an interesting textbook.
3. – 13
4. DC = 7 cm, BC = 5 cm
5. $\sqrt{19423}$ is irrational.

CONJECTURES, THEOREMS, PROOFS AND MATHEMATICAL REASONING

IMPORTANT POINTS

1. **Conjecture.** Consider the figure as given. The first circle has one point on it, the second two points, the third three, and so on. All possible lines connecting the points are drawn in each case.

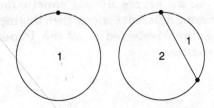
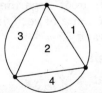

The lines divide the circle into mutually exclusive regions. We can count these and tabulate our results as shown :

Number of points	Number of regions
1	1
2	2
3	4
4	8
5	
6	
7	

We might have come up with a formula predicting the number of regions given the number of points. This intelligent guess is called a *'conjecture'*.

Suppose our conjecture is that given 'n' points on a circle, there are 2^{n-1} mutually exclusive regions, created by joining the points with all possible lines. This seems an extremely sensible guess, and we can check that if $n = 5$, we do get 16 regions. So, having verified this formula for 5 points, if we are satisfied that for any n points there are 2^{n-1} regions, then we must be able to respond for each and every value of n. For larger values of n, say $n = 25$, the verification of the formula becomes practically difficult.

To deal with such situations, we need a proof which shows beyond doubt that this result is true, or a counter-example to show that this result fails for some 'n'. Actually, if we are patient and try it out for $n = 6$, we will find that there are 31 regions, and for $n = 7$ there are 57 regions. So, $n = 6$, is a counter-example to the conjecture above. This demonstrates the power of a counter-example. Actually, to **disprove a statement, it is enough to come up with a single counter-example.** This is why we insist on a proof regarding the number of region in spite of verifying the result for $n = 1, 2, 3, 4$ and 5.

2. Need of Proof

(i) We know that

$$1 + 2 + 3 + \ldots + n = \frac{n(n+1)}{2}$$

To establish its validity, it is not enough to verify the result for $n = 1, 2, 3$, and so on, because there may be some 'n' for which this result is not true. What we need is a proof which establishats its truth beyond.

(ii) Consider, figure as given where PQ and PR are tangents to the circle drawn from P.

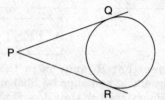

We have proved that PQ = PR (Theorem 10.2). We can not be satisfied by only drawing several such figures, measuring the lengths of the respective tangents, and verifying for ourselves that the result was true in each case.

3. What proof consists of ?

The proof consists of a sequence of statements (called **valid arguments**), each following from the earlier statements in the proof, or from previously proved (and known) results independent from the result to be proved, or from axioms, or from definitions, or from the assumptions we have made.

This is the way how any proof is constructed.

4. 'Direct' or 'Deductive' Method of Proof

In this method, we make several statements. Each is **based on previous statements**. If each statement is logically correct (*i.e*, a valid argument), it leads to a logically correct conclusion.

Note. In the sequence of steps, all are linked together. Their order is important.

5. Proof by exhaustion

If we arrive at the conclusion by eliminating different options, then this method is sometimes referred to as the **Proof by exhaustion.**

ILLUSTRATIVE EXAMPLES

[NCERT Exercise A1.3]
(Page No. 323)

In each of the following questions, we ask you to prove a statement. List all the steps in each proof, and give the reason for each step.

Example 1. *Prove that the sum of two consecutive odd numbers is divisible by 4.*

Sol.

S. No.	Statements	Analysis/comments
1.	Let two consecutive odd numbers be $(2n + 1)$ and $(2n + 3)$ for some positive integer n.	Since the result is about two consecutive odd numbers, we start with $(2n + 1)$ and $(2n + 3)$ which are odd.
2.	Their sum $= (2n + 1) + (2n + 3)$ $= 4n + 4$	The result talks about the sum of two consecutive odd numbers, so we look at $(2n + 1) + (2n + 3)$.
3.	$4n + 4 = 4(n + 1)$.	Using the known distributive property.
4.	So, $(2n + 1) + (2n + 3)$ is divisible by 4.	Since $4(n + 1)$ is divisible by 4.

Example 2. *Take two consecutive odd numbers. Find the sum of their squares, and then add 6 to the result. Prove that the new number is always divisible by 8.*

Sol.

S. No.	Statements	Analysis/comments
1.	Let two consecutive odd numbers be $(2n + 1)$ and $(2n + 3)$ for some positive integer n.	Since the result is about two consecutive odd numbers, we start with $(2n + 1)$ and $(2n + 3)$ which are odd.
2.	Adding 6 to the sum of their squares, we get $(2n + 1)^2 + (2n + 3)^2 + 6$ $= 4n^2 + 4n + 1 + 4n^2 + 12n + 9 + 6$ $= 8n^2 + 16n + 16.$	The problem directs us to do so.
3.	$8n^2 + 16n + 16 = 8(n^2 + 2n + 2).$	Using the known distributive property
4.	So, $(2n + 1)^2 + (2n + 3)^2 + 6$ is divisible by 8.	Since $8(n^2 + 2n + 2)$ is divisible by 8.

Example 3. *If $p \geq 5$ is a prime number, show that $p^2 + 2$ is divisible by 3.*
[**Hint.** *Use Example 11 page 321 of NCER Textbook*].
Sol.

S. No.	Statements	Analysis/comments
1.	Let $p \geq 5$ be a prime number.	It is given in the question.
2.	Dividing p by 6, we find that p can be of the form $6k$, $6k + 1$, $6k + 2$, $6k + 3$, $6k + 4$ or $6k + 5$ where k is a whole number.	Using Euclid's Division Lemma
3.	But $6k = 2(3k),$ $6k + 2 = 2(3k + 1)$ $6k + 3 = 3(2k + 1)$ and $6k + 4 = 2(3k + 2)$ So, they cannot be primes.	Using definition of a prime number.
4.	So, p can be of the form $6k + 1$ or $6k + 5$.	These two are the remaining possibilities.
5.	Let us first take $p = 6k + 1$ and find $p^2 + 2$. We get, $p^2 + 2 = (6k + 1)^2 + 2$ $= 36k^2 + 12k + 1 + 2$ $= 36k^2 + 12k + 3.$	The result talks about $p^2 + 2$. So, we find it.
6.	$36k^2 + 12k + 3 = 3(12k^2 + 4k + 1)$	Using the known distributive property.
7.	So, $p^2 + 2$ is divisible by 3.	Since $3(12k^2 + 4k + 1)$ is divisible by 3.
8.	So, $p^2 + 2$ is a composite number.	Since $p^2 + 2$ is divisible by 3.
9.	Then, we take $p = 6k + 5$ and find $p^2 + 2$. We get, $p^2 + 2 = (6k + 5)^2 + 2$ $= 36k^2 + 60k + 25 + 2$ $= 36k^2 + 60k + 27.$	The result talks about $p^2 + 2$. So, we find it.
10.	$36k^2 + 60k + 27 = 3(12k^2 + 20k + 9)$	Using the known distributive property.
11.	So, $p^2 + 2$ is divisible by 3.	Since, $3(12k^2 + 20k + 9)$ is divisible by 3.
12.	So, $p^2 + 2$ is a compsite number.	Since $p^2 + 2$ is divisible by 3.
13.	So, for prime number $p \geq 5$, $p^2 + 2$ is a composite number.	By S. No. 8 and 12.

Example 4. *Let x and y be rational numbers. Show that xy is a rational number.*
Sol.

S. No.	Statements	Analysis/comments
1.	Let x and y be rational numbers.	Since the result is about rationals, we start with x and y which are rationals.
2.	Let $x = \dfrac{m}{n}, n \neq 0$ and $y = \dfrac{p}{q}, q \neq 0$ where m, n, p and q are integers.	Apply the definition of rationals.
3.	So, $xy = \dfrac{m}{n} \times \dfrac{p}{q} = \dfrac{mp}{nq}$	The result talks about the product of rationals, so we look at xy.
4.	Using the properties of integers, we see that mp and nq are integers.	Using known properties of integers.
5.	Since $n \neq 0$ and $q \neq 0$, it follows that $nq \neq 0$.	Using known properties of integers.
6.	Therefore, $xy = \dfrac{mp}{nq}$ is a rational number.	Using the definition of a rational number.

Example 5. *If a and b are positive integers, then you know that $a = bq + r$, $0 \leq r < b$, where q is a whole number. Prove that HCF (a, b) = HCF (b, r).*

[**Hint :** *Let HCF (b, r) = h. So, $b = k_1 h$ and $r = k_2 h$, where k_1 and k_2 are coprime.*]
Sol.

S. No.	Statements	Analysis/comments
1.	Let HCF$(b, r) = h$, where h is a positive integer.	The result talks about HCF(b, r) and HCF (a, b). So, we took HCF $(b, r) = h$.
2.	So, $b = k_1 h$ and $r = k_2 h$ where k_1 and k_2 are coprime.	Using the definition of HCF of two numbers.
3.	We know that $a = bq + r$, $0 \leq r < b$, where q is a whole number.	Using Euclid's Division Lemma.
4.	So, $a = k_1 hq + k_2 h$ $= (k_1 q + k_2)h$	From statements 2 and 3.
5.	We have $b = k_1 h$ and $a = (k_1 q + k_2)h$	From statements 2 and 4.
6.	But $k_1 q + k_2$ and k_1 are in coprimes.	Using definition of coprime.
7.	So, HCF of $(k_1 q + k_2)h$ and $k_1 h$ is h and therefore HCF $(a, b) = h$	Using definition of HCF of two positive integers and statement 6.

Example 6. *A line parallel to side BC of a triangle ABC, intersects AB and AC at D and E respectively. Prove that* $\dfrac{AD}{DB} = \dfrac{AE}{EC}$.

Sol.

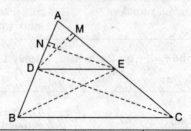

S. No.	Statements	Analysis/comments
1.	Let ABC be a triangle in which a line parallel to BC, intersects AB and AC at D and E respectively.	Since we are proving a statement about such a triangle, we begin by taking this.
2.	Join BE and CD and then draw DM ⊥ AC and EN ⊥ AB.	This is the intuitive step we have talked about that we often need to take for proving theorems.
3.	ar (ΔADE) = ½ AD × EN. ar (ΔBDE) = ½ DB × EN.	Using known formula for area of a triangle.
4.	So, $\dfrac{ar\,(\Delta ADE)}{ar\,(\Delta BDE)} = \dfrac{AD}{DB}$	From statement 3.
5.	ar (ΔADE) = ½ AE × DM ar (ΔDEC) = ½ EC × DM	Using known formula for area of a triangle.
6.	So, $\dfrac{ar\,(\Delta ADE)}{ar\,(\Delta DEC)} = \dfrac{AE}{EC}$	From statement 5.
7.	But ar(ΔBDE) = ar(ΔDEC) since ΔBDE and ΔDEC are on the same base DE and between the same parallels BC and DE.	Using known theorem.
8.	So, $\dfrac{AD}{DB} = \dfrac{AE}{EC}$	A logical deduction from statements 4, 6, and 7.

TEST YOUR KNOWLEDGE

1. Prove that the sum of two rational numbers is a rational number.
2. Prove that every prime number greater than 3 is of the form $6k + 1$ or $6k + 5$, where k is a whole number.
3. State and prove the converse of the Pythagoras theorem.

Or

If in a triangle the square of the length of one side is equal to the sum of the squares of the other two sides, then the angle opposite to the first side is a right angle. Prove.

NEGATION OF A STATEMENT

IMPORTANT POINTS

1. **Notations.** The notations make it easy for us to understand the concept of negation of a statement. We look at a statement as a single unit, and give it a name.

For example, we can denote the statement 'It rained in Delhi on September 1, 2005' by p. We can also written this by

p: It rained in Delhi on September 1, 2005.

q: All teachers are female.

r: Mike's dog has a black tail.

s: $2 + 2 = 4$.

t: Triangle ABC is equilateral.

This notation helps us to discuss properties of statments, and also to see how we can combine them.

2. **A table.** Consider the following table in which we make a new statement from each of the given statements.

Original statement	New Statement
p: It rained in Delhi on September 1, 2005.	$\sim p$: It is false that it rained in Delhi on September 1, 2005.
q: All teachers are female.	$\sim q$: It is false that all teachers are female.
r: Mike's dog has a black tail.	$\sim r$: It is false that Mike's dog has a black tail.
s: $2 + 2 = 4$.	$\sim s$: It is false that $2 + 2 = 4$.
t: Triangle ABC is equilateral.	$\sim t$: It is false that triangle ABC is equilateral.

3. **Notation $\sim p$.** Each new statement in the table given above is a *negation* of the corresponding old statement. That is, $\sim p$, $\sim q$, $\sim r$, $\sim s$ and $\sim t$ are negations of the statements p, q, r, s and t, respectively. Here, $\sim p$ is read as 'not p'. The statement $\sim p$ negates the assertion that the statement p makes. Notice that in our usual talk we would simply mean $\sim p$ as 'It did not rain in Delhi on September 1, 2005.'

4. **Precaution.** We need to be careful while writing the negation of a statement. We might think that one can obtain the negation of a statement by simply inserting the word 'not' in the given statement at a suitable place. While this works in the case of p, the difficulty comes when we have a statement that begins with 'all'. Consider, for example, the statement q: All teachers are female. We said the negation of this statement is $\sim q$: It is false that all teachers are female. This is the same as the statement 'There are some teachers who are males'. Now let us see what happens if we simply insert 'not' in q. We obtain the statement: 'All teachers are not female', or we can obtain the statement: 'Not all teachers are female'. The first statement can confuse people. It could imply (if we lay emphasis on the word 'All') that all teachers are male! This is certainly not the negation of q. However, the second statement gives the meaning of $\sim q$, *i.e.*, that there is at least one teacher who is not a female. So we have to be careful when writing the negation of a statement.

5. **Criterion.** To decide that we have the correct negation, we use the following criterion.

Let p be a statement and $\sim p$ its negation. $\sim p$ is false whenever p is true, and $\sim p$ is true whenever p is false.

For example, if it is true that Mike's dog has a black tail, then it is false that Mike's dog does not have a black tail. If it is false that 'Mike's dog has a black tail', then it is true that 'Mike's dog does not have a black tail'.

Similarly, the negations for the statements s and t are:

$s : 2 + 2 = 4$; negation, $\sim s : 2 + 2 \neq 4$.

t : Triangle ABC is equilateral; negation, $\sim t$: Triangle ABC is not equilateral.

6. Negation of Negation. We may see that $\sim(\sim s)$ would be $2 + 2 = 4$, which is s. Similarly, $\sim(\sim t)$ would be 'the triangle ABC is equilateral', *i.e.*, t. In fact, **for any statement p, $\sim(\sim p)$ is p.**

7. Working Rule. The following is the **Working Rule** for obtaining the negation of a statement :

(*i*) First write the statement with a 'not'.

(*ii*) It there is any confusion, make suitable modification, specially in the statements involving 'All' or 'Some'.

ILLUSTRATIVE EXAMPLES

[NCERT Exercise A1.4]
(Page No. 326)

Example 1. *State the negations for the following statements :*

(*i*) *Man is mortal.*

(*ii*) *Line l is parallel to line m.*

(*iii*) *This chapter has many exercises.*

(*iv*) *All integers are rational numbers.*

(*v*) *Some prime numbers are odd.*

(*vi*) *No student is lazy.*

(*vii*) *Some cats are not black.*

(*viii*) *There is no real number x, such that $\sqrt{x} = -1$.*

(*ix*) *2 divides the positive integer a.*

(*x*) *Integers a and b are coprime.*

Sol. (*i*) Man is not mortal.

(*ii*) Line *l* is not parallel to line *m*.

(*iii*) The chapter does not have many exercises.

(*iv*) Not all integers are rational numbers.

(*v*) All prime numbers are not odd.

(*vi*) Some students are lazy.

(*vii*) All cats are black.

(*viii*) There is at least one real number x, such that $\sqrt{x} = -1$.

(*ix*) 2 does not divisible the positive integer a.

(*x*) Integers a and b are not coprime.

Example 2. *In each of the following questions, there are two statements. State if the second is the negation of the first or not.*

(*i*) *Mumtaz is hungry.*
   Mumtaz is not hungry.

(*ii*) *Some cats are black.*
   Some cats are brown.

(*iii*) *All elephants are huge.*
   One elephant is not huge.

(*iv*) *All fire engines are red.*
   All fire engines are not red.

(*v*) *No man is a cow.*
   Some men are cows.

Sol. (*i*) Yes (*ii*) No (*iii*) No (*iv*) No (*v*) Yes.

TEST YOUR KNOWLEDGE

1. State the negations for the following statements:

   (*i*) Mike's dog does not have a black tail.

   (*ii*) All irrational numbers are real numbers.

   (*iii*) $\sqrt{2}$ is irrational.

   (*iv*) Some rational numbers are integers.

   (*v*) Not all teachers are males.

   (*vi*) Some horses are not brown.

   (*vii*) There is no real number x, such that $x^2 = -1$.

Answers

1. (i) It is false that Mike's dog does not have a black tail, i.e., Mike's dog has a black tail.
    (ii) It is false that all irrational numbers are real numbers, i.e., some (at least one) irrational numbers are not real numbers. One can also write this as, 'Not all irrational numbers are real numbers.'
    (iii) It is false that $\sqrt{2}$ is irrational, i.e., $\sqrt{2}$ is not irrational.
    (iv) It is false that some rational numbers are integers, i.e., no rational number is an integer.
    (v) It is false that not all teachers are males, i.e., all teachers are males.
    (vi) It is false that some horses are not brown, i.e. all horses are brown.
    (vii) It is false that there is no real number x, such that $x^2 = -1$, i.e., there is at least one real number x, such that $x^2 = -1$.

CONVERSE OF A STATEMENT

IMPORTANT POINTS

1. Notion of a 'Compound' Statement. A statement which is a combination of one or more 'simple' statements is called a 'compound' statement.

For example, the statement 'If it is raining, then it is difficult to go on a bicycle', is made up of two statements:

p: It is raining
q: It is difficult to bicycle.

Using our previous notation we can say: If p, then q. We can also say 'p implies q', and denote it by $p \Rightarrow q$.

2. Converse of the statement. Suppose we have the statement 'If the water tank is black, then it contains potable water.' This is of the form $p \Rightarrow q$, where the hypothesis is p (the water tank is black) and the conclusion is q (the tank contains potable water). Suppose we interchange the hypothesis and the conclusion. Then we get $q \Rightarrow p$, i.e., if the water in the tank is potable, then the tank must be black. This statement is called the **converse** of the statement $p \Rightarrow q$.

In general, the **converse** of the statement $p \Rightarrow q$ is $q \Rightarrow p$, where p and q are statements. Note that $p \Rightarrow q$ and $q \Rightarrow p$ are the converses of each other.

3. Note. Notice that we usually write the converse of a statement without worrying if it is true or false. For example, consider the following statement: If Ahmad is in Mumbai, then he is in India. This statement is true. Now consider the converse: If Ahmad is in India, then he is in Mumbai. This need not be true always—he could be in any other part of India.

In mathematics, especially in geometry, we come across many situations where $p \Rightarrow q$ is true, and we have to decide if the converse, i.e., $q \Rightarrow p$, is also true.

ILLUSTRATIVE EXAMPLES

[NCERT Exercise A1.5]
(Page No. 328)

Example 1. *Write the converses of the following statements.*
(i) *If it is hot in Tokyo, then Sharan sweats a lot.*
(ii) *If Shalini is hungry, then her stomach grumbles.*
(iii) *If Jaswant has a scholarship, then she can get a degree.*
(iv) *If a plant has flowers, then it is alive.*
(v) *If an animal is a cat, then it has a tail.*

Sol.
(i) If Sharan sweats a lot, then it is hot in Tokyo.
(ii) If Shalini's stomach grumbles, then she is hungry.
(iii) If Jaswant can get a degree, then she has a scholarship.
(iv) If a plant is alive, then it has flowers.
(v) If an animal has a tail, then it is a cat.

Example 2. *Write the converses of the following statements. Also, decide in each case whether the converse is true or false.*
  (i) *If triangle ABC is isosceles, then its base angles are equal.*
  (ii) *If an integer is odd, then its square is an odd integer.*
  (iii) *If $x^2 = 1$, then $x = 1$.*
  (iv) *If ABCD is a parallelogram, then AC and BD its diagonals bisect each other.*
  (v) *If a, b and c are whole numbers, then $a + (b + c) = (a + b) + c$.*
  (vi) *If x and y are two odd numbers, then $x + y$ is an even number.*
  (vii) *If vertices of a parallelogram lie on a circle, then it is rectangle.*

Sol.
(i) If the base angles of triangle ABC are equal, then it is isosceles. True.
(ii) If the square of an integer is odd, then the integer is odd. True.
(iii) If $x = 1$, then $x^2 = 1$. True.
(iv) If AC and BD bisect each other, then ABCD is a parallelogram. True.
(v) If $a + (b + c) = (a + b) + c$, then a, b and c are whole numbers. False.
(vi) If $x + y$ is an even number, then x and y are odd. False.
(vii) If a parallelogram is a rectangle, its vertices lie on a circle. True.

TEST YOUR KNOWLEDGE

1. Write the converses of the following statements.
   (i) If Jamila is riding a bicycle, then August 17th falls on a Sunday.
   (ii) If August 17th is a Sunday, then Jamila is riding a bicycle.
   (iii) If Pauline is angry, then her face turns red.
   (iv) If a person has a degree in education, then she is allowed to teach.
   (v) If a person has a viral infection, then he runs a high temperature.
   (vi) If Ahmad is in Mumbai, then he is in India.
   (vii) If triangle ABC is equilateral, then all its interior angles are equal.
   (viii) If x is an irrational number, then the decimal expansion of x is non-terminating non-recurring.
   (ix) If $x - a$ is a factor of the polynomial $p(x)$, then $p(a) = 0$.

2. State the converses of the following statements. In each case, also decide whether the converse is true or false.
   (i) If n is an even integer, then $2n + 1$ is an odd integer.
   (ii) If the decimal expansion of a real number is terminating, then the number is rational.
   (iii) If a transversal intersects two parallel lines, then each pair of corresponding angles is equal.
   (iv) If each pair of opposite sides of a quadrilateral is equal, then the quadrilateral is a parallelogram.
   (v) If two triangles are congruent, then their corresponding angles are equal.

Answers

1. (i) p: Jamila is riding a bicycle, and q, Aug. 17th falls on a Sunday. Therefore, the converse is: If August 17th falls on a Sunday, then Jamila is riding a bicycle.

(ii) This is the converse of (i). Therefore, its converse is the statement given in (i) above.

(iii) If Pauline's face turns red, then she is angry.

(iv) If a person is allowed to teach, then she has a degree in education.

(v) If a person runs a high temperature, then he has a viral infection.

(vi) If Ahmad is in India, then he is in Mumbai.

(vii) If all the interior angles of triangle ABC are equal, then it is equilateral.

(viii) If the decimal expansion of x is non-terminating non-recurring then x is an irrational number.

(ix) If $p(a) = 0$, then $x - a$ is a factor of the polynomial $p(x)$.

2. (i) The converse is 'If $2n + 1$ is an odd integer, then n is an even integer'. This is a false statement (for example, $15 = 2(7) + 1$, and 7 is odd).

(ii) 'If a real number is rational, then its decimal expansion is terminating', is the converse. This is a false statement, because a rational number can also have a non-terminating recurring decimal expansion.

(iii) The converse is 'If a transversal intersects two lines in such a way that each pair of corresponding angles are equal, then the two lines are parallel.' We have assumed, by Axiom 6.4 of Chapter 6 of Class IX textbook, that this statement is true.

(iv) 'If a quadrilateral is a parallelogram, then each pair of its opposite sides is equal' is the converse. This is true.

(v) 'If the corresponding angles in two triangles are equal, then they are congruent' is the converse. This statement is false. For example consider two equilateral triangles with the length of the sides as a units and b units ($a \neq b$). Then, the corresponding angles in two triangles are equal, but they are not congruent.

PROOF BY CONTRADICTION

IMPORTANT POINTS

1. Proof by contradiction. We know how to establish the truth of a result by using direct arguments. There is one more method to establish the truth of a result. This is known as 'proof by contradiction'. This is a very powerful tool in mathematics. It involves 'indirect' arguments. By using this, we can establish the irrationality of several numbers like $\sqrt{2}$, $\sqrt{3}$, $\sqrt{5}$, $\sqrt{7}$, etc. Moreover, we can prove some theorems.

2. What a contradiction is. In mathematics, a contradiction occurs when we get a statement p such that p is true and $\sim p$, its negation, is also true. For example,

$p: x = \dfrac{a}{b}$, where a and b are co-prime.

q: 2 divides both 'a' and 'b'.

If we assume that p is true and also manage to show that q is true, then we have arrived at a contradiction, because q implies that the negation of p is true.

3. Working of proof by contradiction. Suppose we are given that

- All women are mortal. Indira is a woman. Prove that Indira is mortal.
- Let us see how we can prove this by contradiction.
- Let us assume that we want to establish the truth of a statement p (here we want to show that p: 'Indira is mortal' is true).

- So, we begin by assuming that the statement is not true, that is, we assume that the negation of p is true (*i.e.*, Indira is not mortal).
- We then proceed to carry out a series of logical deductions based on the truth of the negation of p. (Since Indira is not mortal, we have a counter-example to the statement 'All women are mortal'. Hence, it is false that all women are mortal.)
- If this leads to a contradiction, then the contradiction arises because of our faulty assumption that p is not true. (We have a contradiction, since we have shown that the statement 'All women are mortal' are its negation, 'Not all women are mortal' is true at the same time. This contradiction arose, because we assumed that Indira is not mortal.)
- Therefore, our assumption is wrong, *i.e.*, p has to be true. (So, Indira is mortal.)

Let us write this proof down briefly.

To show: Indira is mortal.

Suppose Indira is not mortal.

Since Indira is not mortal, we have a counter-example to the statement 'All women are mortal'. Hence, it is false that all women are mortal.

We have a contradiction, since we have shown that the statement 'All women are mortal' and its negation, 'Not all women are mortal' is true at the same time. This contradiction arose because we assumed that Indira is not mortal.

Hence, Indira is mortal.

ILLUSTRATIVE EXAMPLES

[NCERT Exercise A1.6]
(*Page No. 333*)

Example 1. *Suppose $a + b = c + d$, and $a < c$. Use proof by contradiction to show $b > d$.*
Sol.

S. No.	Statements	Analysis/comments
1.	We are given that $a + b = c + d$ and $a < c$.	We begin with the given hypotheses.
2.	Let us assume that $b > d$ is not true, *i.e.*, we assume that $b < d$.	This is the starting point for 'proof by contradiction. [We assume the negation of the statement to be proved as true.]
3.	$a < c$ and $b < d \Rightarrow a + b < c + d$.	Using known properties of numbers.
4.	We reach a contradiction because we are given that $a + b = c + d$.	This is what we were looking for a contradiction.
5.	This contradiction has arisen, because we assumed that $b < d$. Therefore, $b > d$.	Logical deduction.

PROOFS IN MATHEMATICS

Example 2. *Let r be a rational number and x be an irrational number. Use proof by contradiction to show that $r + x$ is an irrational number.*

Sol.

S. No.	Statements	Analysis/comments
1.	Let r be a rational number and x be an irrational number.	We start with the given hypothesis.
2.	Let us assume that $r + x$ is not an irrational number, i.e., $r + x$ is a rational number.	We assume the negation of the statement to be proved as true.
3.	Since r is a rational number, therefore, $(-r)$ is also a rational number.	Using known property of rational numbers.
4.	Now, sum of the rational numbers $r + x$ and $(-r)$ will be a rational number, i.e., $r + x + (-r)$ is a rational number. So, x is a rational number.	Using known property (closure) of rational numbers.
5.	We reach a contradiction because we are given that x is an irrational number.	This is what we were looking for contradiction.
6.	This contradiction has arisen because of the faulty assumption that $r + x$ is a rational number. Therefore, $r + x$ is an irrational number.	Logical deduction.

Example 3. *Use proof by contradiction to prove that if for an integer a, a^2 is even, then so is a.*

[**Hint:** *Assume a is not even, that is, it is of the form $2n + 1$, for some integer n, and then proceed.*]

Sol.

S. No.	Statements	Analysis/comments
1.	Let for an integer a, a^2 is even.	We start with the given hypothesis.
2.	Let us assume that a is not even, that is, it is of the form $2n + 1$, for some integer n.	We assume the negation of the statement to be proved as true.
3.	As $a = 2n + 1$, so $a^2 = (2n + 1)^2$ $= 4n^2 + 4n + 1$	Simplifying using known identity.
4.	For any integer n, $4n^2 + 4n + 1$ is not divisible by 2 because the third term 1 is not divisible by 2.	Using known properties of numbers.
5.	So, $4n^2 + 4n + 1$, i.e., a^2 is not an even number.	Using the definition of an even number.
6.	Thus, we reach a contradiction because we are given that a^2 is even.	This is what we were looking for a contradiction.
7.	This contradiction has arisen because of our faulty assumption that a is not even. Therefore, a is even.	Logical deduction.

Example 4. *Use proof by contradiction to prove that if for an integer a, a^2 is divisible by 3, then a is divisible by 3.*

Sol.

S. No.	Statements	Analysis/comments
1.	Let for an integer a, a^2 is divisible by 3.	We start with the given hypothesis.
2.	Let us assume that a is not divisible by 3.	We assume the negation of the statement to be proved as true.
3.	As a is not divisible, it is either of the form $3n + 1$ or $3n + 2$ for a positive integer n.	Using Euclid's lemma.
4.	If $a = 3n + 1$, then $$a^2 = (3n + 1)^2$$ $$= 9n^2 + 6n + 1$$ which is not divisible by 3 because the third term 1 is not divisible by 3. Again, if $a = 3n + 2$, then $$a^2 = (3n + 2)^2$$ $$= 9n^2 + 12n + 4$$ which is not divisible by 3 because the third term 4 is not divisible by 3.	Simplifying using known identity and known properties of numbers.
5.	So, a^2 is not divisible by 3 in the cases when $a = 3n + 1$ and $a = 3n + 2$.	From statement 4.
6.	Thus, we reach a contradiction because we are given that a^2 is divisible by 3.	This is what we were looking for a contradiction.
7.	This contradiction has arisen because of our faulty assumption that a is not divisible by 3. Therefore, a is divisible by 3.	Logical deduction.

Example 5. *Use proof by contradiction to show that there is no value of n for which 6^n ends with the digit zero.*

Sol.

S. No.	Statements	Analysis/comments
1.	Let us assume that there is a value N of n for which 6^n ends with the digit zero, *i.e.*, 6^N ends with 0.	We assume the negation of the statement to be proved as true.
2.	Now, $6^N = (2 \times 3)^N = (2 \times 3) \times (2 \times 3) \times (2 \times 3) \times$ ...... N times.	Using definition of exponents and powers.
3.	The only factors occurring in 6^N are 2 and 3.	Using Fundamental theorem of Arithmetic.
4.	If a number ends with 0, then it must have 5 as one of its factors. So, 6^n will never end with 0.	Using known property of numbers.
5.	Thus, we react a contradiction because we are given that 6^N ends with 0.	This is what we were looking for a contradiction.

S. No.	Statements	Analysis/comments
6.	This contradiction has arisen because of our faulty assumption that there is a value N of n for which 6^n ends with 0. Therefore, there is no value of n for which 6^n ends with 0.	Logical deduction.

Example 6. *Prove by contradiction that two distinct lines in a plane cannot intersect in more than one point.*

Sol.

S. No.	Statements	Analysis/comments
1.	Let us assume that two lines, say l and m in a plane intersect in two (*i.e.*, more than one) points P and Q.	We assume the negation of the statement to be formed as true.
2.	So, each of the lines l and m passes through two points P and Q.	Because P and Q, both, lie on the line l as well as line m.
3.	So, we have two lines l and m passing through two points P and Q.	Because the two lines l and m are passing through P and Q.
4.	Thus, we reach a contradiction because through two distinct points, a unique line passes.	Using known axioms.
5.	This contradiction has arisen because of our faulty assumption that the two lines l and m in a plane intersect in two points P and Q. Therefore, two lines in a plane cannot intersect in more than one point.	Logical deduction.

TEST YOUR KNOWLEDGE

1. Prove that the product of a non-zero rational number and an irrational number is irrational.

2. Prove that out of all the line segments, drawn from a point to point of a line not passing through the point, the smallest is the perpendicular to the line.

SUMMARY

1. Different ingredients of a proof and other related concepts learnt in Class IX.
2. The negation of a statement.
3. The converse of a statement.
4. Proof by contradiction.

A2
Mathematical Modelling

IMPORTANT POINTS

1. Real-world Probelms and Mathematical Modelling. Consider the following results:

- An adult human body contains approximately 1,50,000 km of arteries and veins that carry blood.
- The human heart pumps 5 to 6 litres of blood in the body every 60 seconds.
- The temperature at the surface of the Sun is about 6,000° C.

It seems wonderful how our scientists and mathematicians could possibly estimate these results and get these figures. Certainly, they did not pull out veins and arteries from some adult dead bodies and measure them; they did not drain out the blood to arrive at the second result; they did not travel to the Sun with a thermometer to get the temperature of the Sun. In fact, the answer lies in **mathematical modelling.**

A mathematical model is a mathematical description of some real-life situation. Also, mathematical modelling is the process of creating a mathematical model of a problem, and using it to analyse and solve the problem.

So, in mathematical modelling, we take a real-world problem and convert it to an equivalent mathematical problem. We then solve the mathematical problem, and interpret its solution in the situation of the real-world problem. And then, it is important to see that the solution, we have obtained, 'makes sense', which is the stage of validating the model. Some examples, where mathematical modelling is of great importance, are:

(i) Finding the width and depth of a river at an unreachable place.
(ii) Estimating the mass of the Earth and other planets.
(iii) Estimating the distance between Earth and any other planet.
(iv) Predicting the arrival of the monsoon in a country.
(v) Predicting the trend of the stock market.
(vi) Estimating the volume of blood inside the body of a person.
(vii) Predicting the population of a city after 10 years.
(viii) Extimating the number of leaves in a tree.
(ix) Estimating the ppm of different pollutants in the atmosphere of a city.
(x) Estimating the effect of pollutants on the environment.
(xi) Estimating the temperature on the Sun's surface.

2. Stages in Mathematical Modelling. The main steps in mathematical modelling are as follows:

Step 1. (Understanding the Problem): Define the real problem, and if working in a team, discuss the issues that we wish to understand. Simplify by making assumptions and ignoring certain factors so that the problem is manageable.

For example, suppose our problem is to estimate the number of fishes in a lake. It is not possible to capture each of these fishes and count them. We could possibly capture a sample and from it try and estimate the total number of fishes in the lake.

Step 2. (Mathematical Description and Formulation): Describe, in mathematical terms, the different aspects of the problem. Some ways to describe the features mathematically include:

MATHEMATICAL MODELLING

- define variables
- write equations or inequalities
- gather data and organise into tables
- make graphs
- calculate probabilities

For example, having taken a sample, as stated in Step 1, to estimate the entire population, we would have to then mark the sampled fishes, allow them to mix with the remaining ones in the lake, again draw a sample from the pond, and see how many of the previously marked ones are present in the new sample. Then, using ratio and proportion, we can come up with an estimate to the total population. For instance, let us take a sample of 20 fishes from the pond and mark them, and then release them in the same pond, so as to mix with the remaining fishes. We then take another sample (say 50), from the mixed population and see how many are marked. So, we gather our data and analyse it.

One major assumption we are making is that the marked fishes mix uniformly with the marked fishes, and the sample we take is a good representative of the entire population.

Step 3. (Solving the Mathematical Problem): The simplified mathematical problem developed in Step 2 is then solved using various mathematical techniques.

For instance, suppose in the second sample in the example in Step 2, 5 fishes are marked. So, $\frac{5}{50}$, i.e., $\frac{1}{10}$, of the population is marked. If this is typical of the whole population, then $\frac{1}{10}$ th of the population = 20.

So, the whole population
$$= 20 \times 10 = 200.$$

Step 4. (Interpreting the Solution): The solution obtained in the previous step is now looked at, in the context of the real-life situation that we had started with in Step 1.

For instances, our solution in the problem in Step 3 gives us the population of fishes as 200.

Step 5. (Validating the Model): We go back to the original situation and see if the results of the mathematical work make sense. If so, we use the model until new information becomes available or assumptions change.

Sometimes, because of the simplification assumptions we make, we may lose essential aspects of the real problem while giving its mathematical description. In such cases, the solution could very often be off the mark, and not make sense in the real situation. If this happens, we reconsider the assumptions made in Step 1 and revise them to be more realistic, possibly by including some factors which were not considered earlier.

For instance, in Step 3 we had obtained an estimate of the entire population of fishes. It may not be the actual number of fishes in the pond. We next see whether this is a good estimate of the population by repeating Steps 2 and 3 a few times, and taking the mean of the results obtained. This would give a closer estimate of the population.

Another way of visualising **the process of mathematical modelling** is shown in figure.

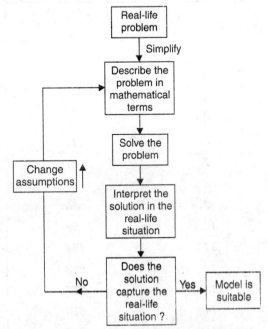

Modellers look for a balance between simplification (for ease of solution) and accuracy. They hope to approximate reality closely enough to make some progress. The best outcome is to be able to predict what will happen, or estimate an outcome, with reasonable accuracy. Remember that different assumptions we use for simplifying the problem can lead to different models. So, there are no perfect models. There are good ones and yet better ones.

ILLUSTRATIVE EXAMPLES

[NCERT Exercise A2.1]
(Page No. 338)

Example 1. *Consider the following situation.*

A problem dating back to the early 13th century, posed by Leonardo Fibonacci asks how many rabbits you would have if you started with just two and let them reproduce. Assume that a pair of rabbits produces a pair of offspring each month and that each pair of rabbits produces their first offspring at the age of 2 months. Month by month the number of pairs of rabbits is given by the sum of the rabbits in the two preceding months, except for the 0th and the 1st months.

Month	Pairs of Rabbits
0	1
1	1
2	2
3	3
4	5
5	8
6	13
7	21
8	34
9	55
10	89
11	144
12	233
13	377
14	610
15	987
16	1597

After just 16 months, you have nearly 1600 pairs of rabbits !

Clearly state the problem and the different stages of mathematical modelling in this situation.

Sol.

Statement of the Problem. A pair of rabbits produces a pair of offspring each month and that each pair of rabbits produces their first offspring at the age of 2 months. Find the number of pairs of rabbits produced after a given number of months starting with just two and let then reproduce.

Different Stages of Mathematical Modelling

Stage 1. (Understanding the Problem): First pair of rabbits will produce a rabbit after 2 months. So, the number of pairs of rabbits after 1 month is 1 and that after 2 months is $1 + 1 = 2$. After 3 months, the first pair of rabbits will produce one more pair and newly born pair will not produce any pair. So, the number of pairs of rabbits after 3 months is $2 + 1 = 3$. Now, the first two pairs will produce one pair each in the fourth month and the newly born third pair will not produce any pair. So, the number of pairs of rabbits after 4 months is $3 + 2 = 5$, and so on. Thus, we get the number of pairs of rabbits after various months as shown in the table.

Stage 2. (Mathematical Description and Formulation): The number of pairs of rabbits after different months represents a number pattern known as **Fibonacci sequence.** It is drawn below:

1, 1, 2, 3, 5, 8, 13, 21, 34, 55, 89, ...

In this sequence, each term (except for the first two months) is obtained by adding the two terms just preceding that term.

Stage 3. (Solving the Mathematical problem): Although it is not possible to write any explicit formula for the nth term of the above sequence, yet we can find the number of pairs of rabbits after any number of months as explained below:

Number of pairs after 8 months
$= 13 + 21 = 34$

Number of pairs after 9 months
$= 21 + 34 = 55$

Number of pairs after 10 months
$= 34 + 55 = 89$

Number of pairs after 11 months
$= 55 + 89 = 144$

Number of pairs after 12 months
$= 89 + 144 = 233$

and so on.

Stage 4. (Interpreting the Solution): Our solution to the problem given in stage 3 gives us the number of pairs of rabbits after a given number of months.

Stage 5. (Validating the Model): The model is valid under the given assumptions. If any assumption is changed, the model will have to be changed. For example, if each pair of rabbits produces a pair of offspring after every two months, then the above model will not be valid.

IMPORTANT POINTS

1. Instalment scheme (or plan). Not having the money you want when you need it, is a common experience for many people. Whether it is having enough money for buying essentials for daily living, or for buying comforts, we always require money. To enable the customers with limited funds to purchase goods like scooters, refrigerators, televisions, cars, etc., a scheme known as an *instalment scheme (or plan)* is introduced by traders.

Sometimes a trader introduces an *instalment* scheme as a marketing strategy to allure customers to purchase these articles. Under the instalment scheme, the customer is not required to make full payment of the article at the time of buying it. She/he is allowed to pay a part of it at the time of purchase, and the rest can be paid in instalments, which could be monthly, quarterly, half-yearly, or even yearly. Of course, the buyer will have to pay more in the instalment plan, because the seller is going to charge some interest on account of the payment made at a later date (called *deferred payment*).

2. Cash Price. The cash price of an article is the amount which a customer has to pay as full payment of the article at the time it is purchased.

3. Cash down payment. Cash down payment is the amount which a customer has to pay as part payment of the price of an article at the time of purchase.

Remark 1. If the instalment scheme is such that the remaining payment is completely made within one year of the purchase of the article, then simple interest is charged on the deferred payment.

> In the past, charging interest on borrowed money was often considered evil, and in particular, was long prohibited. One way people got around the law against paying interest was to borrow in one currency and repay in another, the interest being disguised in the exchange rate.

Remark 2. Interest rate modelling is still at its early stages and validation is still a problem of financial markets. In case different interest rates are incorporated in fixing instalments, validation becomes an important problem.

ILLUSTRATIVE EXAMPLES

[NCERT Exercise A2.2]
(Page No. 342)

In each of the problems below, show the different stages of mathematical modelling for solving the problems.

Example 1. *An ornithologist wants to estimate the number of parrots in a large field. She uses a net to catch some, and catches 32 parrots, which she rings and sets free. The following week she manages to net 40 parrots, of which 8 are ringed.*

(i) What fraction of her second catch is ringed ?

(ii) Find an estimate of the total number of parrots in the field.

Sol. Step 1. (Understanding the problem): As it is not possible to catch all the parrots, so a sample of 32 parrots has been chosen.

Step 2. (Mathematical Description and Formulation): A sample of 32 parrots has been ringed. In the second sample, there one 40 parrots, out of which 8 parrots have been found to be ringed.

Step 3. (Solving the Mathematical Problem):

(i) The fraction of her second catch that is ringed

$$= \frac{8}{40} = \frac{1}{5}$$

(ii) Estimate of the total number of parrots in the field

$$= 32 \div \frac{1}{5} = 32 \times \frac{5}{1} = 160$$

[∵ For 1 ringed parrot, total number of parrots = 5
∴ For 32 ringed parrots, total number of parrots = 5 × 32 = 160]

Steps 4. (Interpreting the solution): The estimate of the total number of parrots in the field is 160.

Steps 5. (Validating the Model): To obtain a better estimate of the number of parrots in the field, we may repeat steps 2 and 3 a few more times and take mean of the obtained results. This would give a closer estimate of the number of parrots.

Example 2. *Suppose the figure represents an aerial photograph of a forest with each dot representing a tree. Your purpose is to find the number of trees there are on this tract of land as part of an environmental census.*

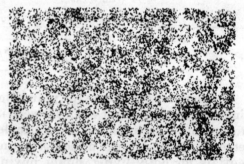

Sol. Step 1. (Understanding the Problem): It is not possible to count all dots in the photograph, so dots are counted in a small sample of area 1 cm^2.

Step 2. (Mathematical Description and Formulation): Total number of dots are counted in this sample of area 1 cm^2. Let this number be x. Then, area of the photograph is calculated. Let this area be y cm^2.

Step 3. (Solving the Mathematical Problem): Total number of dots = $x \times y$

So, total number of trees = $x \times y = xy$

[∵ 1 dot represents 1 tree in the forest.]

Steps 4. (Interpretation of the solution): The estimate of the total number of trees in the forest represented by the given photograph is xy, using the idea of unitary method. **Actual estimate will depend on the actual figure.**

Steps 5. (Validating the Model): For a better and closer estimate of the number of trees in the forest. Steps 2 and 3 can be repeated a few more times and mean may be taken of the obtained results. Again, we may also choose a sample of area 2 cm² instead of area 1 cm² and estimate the number of trees, repeating steps 2 and 3.

Example 3. *A T.V. can be purchased for Rs. 24000 cash or for Rs. 8000 cash down payment and six monthly instalments of Rs 2800 each. Ali goes to market to buy a T.V., and he has Rs. 8000 with him. He has now two options. One is to buy TV under instalment scheme or to make cash payment by taking loan from some financial society. The society charges simple interest at the rate of 18% per annum simple interest. Which option is better for Ali ?*

Sol. Step 1. (Understanding the Problem): What Ali needs to determine is whether he should by T.V. under instalment scheme or take loan from the financial society. For this, he should know the rate of interest charged under the instalment scheme and the rate of interest charged by the financial society (which is 18%).

Step 2. (Mathematical Description and Formulation): In order to decide which option is better, Ali needs to determine the rate of interest charged under the instalment scheme. He must know that in this case, simple interest shall be charged.

We know that cash price of the T.V. = Rs 24000. Also, the cash down payment under the instalment scheme = Rs 8000.

So, the balance price needs to be paid in instalments

$$= Rs\ 24000 - Rs\ 8000$$
$$= Rs\ 16000$$

Let $r\%$ per annum be the rate of interest charged under the instalment scheme.

Amount of each instalment
$$= Rs\ 2800$$

So, amount paid in instalment
$$= 2800 \times 6 = Rs\ 16800$$

So, interest paid in instalment scheme
$$= Rs\ 16800 - Rs\ 16000$$
$$= Rs\ 800 \qquad ...(1)$$

Now,
Principal for the first month
$$= Rs\ 16000$$
Principal for the second month
$$= Rs\ 16000 - Rs\ 2800$$
$$= Rs\ 13200$$
Principal for the third month
$$= Rs\ 13200 - Rs\ 2800$$
$$= Rs\ 10400$$
Principal for the fourth month
$$= Rs\ 10400 - Rs\ 2800$$
$$= Rs\ 7600$$
Principal for the fifth month
$$= Rs\ 7600 - Rs\ 2800$$
$$= Rs\ 4800$$
Principal for the sixth month
$$= Rs\ 4800 - Rs\ 2800$$
$$= Rs\ 2000$$
Total = Rs 54000.
(principal for one month)

Balance of the last principal (Rs 2000) + Interest charged (Rs 800) = Monthly instalment (Rs 2800)

[sixth instalment]

So, interest $= \dfrac{54000 \times r \times 1}{100 \times 12} = Rs\ 45r$.

...(2)

Step 3. (Solving the Mathematical Problem): From (1) and (2), we get
$$45r = 800$$
$$\Rightarrow r = \dfrac{800}{45}$$
$$\Rightarrow r = \dfrac{160}{9}$$
$$\Rightarrow r = 17.78$$

So, the rate of interest in the instalment scheme is 17.78% per annum (< 18% per annum).

Steps 4. (Interpreting the solution): The rate of interest charged in the instalment scheme is 17.78% per annum and the rate of interest charged by the financial society is 18% per annum. So, in this case, instalment scheme is a better option.

Steps 5. (Validating the model): This stage in this case is not of much importance here as the numbers are fixed. However, if the formalities for taking loan from the society such as cost of stamp paper etc. which make the efficient interest rate more than what it is the instalment scheme, then he may change his opinion.

TEST YOUR KNOWLEDGE

1. Suppose your teacher challenges you to the following guessing game. She would throw a pair of dice. Before that you need to guess then sum of the numbers that show up on the dice. For every correct answer, you get two points and for every wrong guess you lose two points. What numbers would be the best guess ?
2. Juhi wants to buy a bicycle. She goes to the market and finds that the bicycle she likes is available for Rs 1800. Juhi has Rs 600 with her. So, she tells the shopkeeper that she would not be able to buy it. The shopkeeper, after a bit of calculation, makes the following offer. He tells Juhi that she could take the bicycle by making a payment of Rs 600 cash down and the remaining money could be made in two monthly instalments of Rs 610 each. Juhi has two options one is to go for instalment scheme or to make cash payment by taking loan from a bank which is available at the rate of 10% per annum simple interest. Which option is more economical to her ?

Answers
1. Number 7.
2. To borrow the money from the bank.

WHY IS MATHEMATICAL MODELLING IMPORTANT ?

IMPORTANT POINTS

Mathematical modelling is an interdisciplinary subject. Mathematicians and specialists in other fields share their knowledge and expertise to improve existing products, develop better ones, or predict the behaviour of certain products.

There are, of course, many specific reasons for the importance of modelling, but most are related in some ways to the following:

- *To gain understanding.* If we have a mathematical model which reflects the essential behaviour of a real-world system of interest, we can understand that system better through an analysis of the model. Furthermore, in the process of building the model we find out which factors are most important in the system, and how the different aspects of the system are related.
- *To predict, or forecast, or simulate.* Very often, we wish to know what a real-world system will do in the future, but it is expensive, impractical or impossible to experiment directly with the system. For example, in weather prediction, to study drug efficacy in humans, finding an optimum design of a nuclear reactor, and so on.

Forecasting is very important in many types of organisations, since predictions of future events have to be incorporated into the decision-making process. For example:

- In marketing departments, reliable forecasts of demand help in planning of the sale strategies.
- A school board needs to able to forecast the increase in the number of school going children in various districts so as to decide where and when to start new schools.

Most often, forecasters use the past data to predict the future. They first analyse the data in order to identify a pattern that can describe it. Then this data and pattern is extended into the future in order to prepare a forecast. This basic strategy is employed in most forecasting techniques, and is based on the assumption that the pattern that has been identified will continue in the future.

- *To estimate.* Often, we need to estimate large values. You've seen examples of the trees in a forest, fish in a pond, etc. For another example, before elections, the contesting parties want to predict the probability of their party winning the elections. In particular, they want to estimate how many people in their constituency would vote for their party. Based on their predictions, they may want to decide on the campaign strategy. Exit polls have been used widely to predict the number of seats a party is expected to get in elections.

ILLUSTRATIVE EXAMPLES

[NCERT Exercise A2.3]
(Page No. 344)

Example 1. *Based upon the data of the past five years, try and forecast the average percentage of marks in Mathematics that your school would obtain in the Class X board examination at the end of the year.*

Sol. Data for different schools will be naturally different. So, on the basis of these data, students will find answers for their respective schools using the idea of experimental probability. Obviously, these answers will be different.

SUMMARY

1. A mathematical model is a mathematical description of a real-life situation. Mathematical modelling is the process of creating a mathematical model, solving it and using it to understand the real-life problem.
2. The various steps involved in modelling are : understanding the problem, formulating the mathematical model, solving it, interpreting it in the real-life situation, and, most importantly, validating the model.
3. Developed some mathematical models.
4. The importance of mathematical modelling.

ADDITIONAL MULTIPLE CHOICE QUESTIONS

Chapter 1
Real Numbers

1. Which of the following is a rational number?
   (a) $\sqrt{2}$
   (b) π
   (c) $\frac{1}{3}$
   (d) 0.1201 2001 20001 20000 ......

2. According to Euclid's Division Lemma, given positive integers a and b, there exist unique integers q and r satisfying $a = bq + r$. Here, the inequalities satisfied by r are
   (a) $0 \leq r < b$
   (b) $0 < r < b$
   (c) $0 < r \leq b$
   (d) $0 \leq r \leq b$.

3. Which of the following is the statement of Fundamental Theorem of Arithmetic?
   (a) Given positive integers a and b, there exist unique integers q and r satisfying $a = bq + r$, $0 \leq r < b$.
   (b) Every composite number can be expressed (factorised) as a product of primes, and this factorisation is unique, apart from the order in which the prime factors occur.
   (c) For any two positive integers a and b, HCF $(a, b) \times$ LCM $(a, b) = a \times b$.
   (d) HCF $(p, q, r) \times$ LCM $(p, q, r) \neq p \times q \times r$, where p, q, r are positive integers.

4. What is the HCF of the least prime number and the least composite number?
   (a) 1
   (b) 2
   (c) 3
   (d) 4.

5. What is the HCF of 10 and 3?
   (a) 3
   (b) 10
   (c) 30
   (d) 1.

6. What is the LCM of 10 and 3?
   (a) 3
   (b) 10
   (c) 30
   (d) 1.

7. What is the HCF of 4 and 19?
   (a) 4
   (b) 1
   (c) 19
   (d) 76.

8. What is the LCM of 4 and 19?
   (a) 4
   (b) 1
   (c) 19
   (d) 76.

9. What is the HCF of 81 and 3?
   (a) 3
   (b) 1
   (c) 81
   (d) 27.

10. What is the LCM of 3 and 81?
    (a) 3
    (b) 81
    (c) 1
    (d) 27.

11. What are the quotient and the remainder when 10 is divided by 3?
    (a) 3, 1
    (b) 1, 3
    (c) 1, 1
    (d) 3, 3.

12. What are the quotient and the remainder when 19 is divided by 4?
    (a) 4, 3
    (b) 3, 4
    (c) 3, 3
    (d) 4, 4.

13. What are the quotient and the remainder when 81 is divided by 3?
    (a) 4, 3 (b) 27, 0
    (c) 0, 27 (d) 27, 27.

14. If q is some integer, then a positive even integer is of the form
    (a) $2q + 1$ (b) $2q$
    (c) q (d) $q + 1$.

15. If q is some integer, then a positive odd integer is of the form
    (a) $2q$ (b) $2q + 1$
    (c) $q - 1$ (d) $q + 1$.

16. If m is some integer, then the square of any positive integer is of the form
    (a) $3m$ or $3m + 1$
    (b) $3m$ or $3m - 1$
    (c) $3m + 1$ or $3m - 1$
    (d) $3m$ or $3m + 2$.

17. The HCF of 6 and 20 is
    (a) 2 (b) 6
    (c) 20 (d) 4.

18. The LCM of 6 and 20 is
    (a) 6 (b) 20
    (c) 60 (d) 30.

19. The HCF of 6, 72 and 120 is
    (a) 2 (b) 3
    (c) 6 (d) 12.

20. The HCF of 26 and 91 is
    (a) 13 (b) 26
    (c) 2 (d) 7.

21. The HCF of 8, 9 and 25 is
    (a) 2 (b) 1
    (c) 3 (d) 5.

22. What is the HCF of two consecutive natural numbers?
    (a) 1 (b) 2
    (c) 3 (d) 4.

23. What is the HCF of any two prime numbers?
    (a) 1 (b) 2
    (c) 3 (d) 4.

24. What is the HCF of 17, 23 and 29?
    (a) 1 (b) 2
    (c) 3 (d) 4.

25. There is a circular path around a sports field. Sonia takes 18 minutes to drive one round of the field, while Ravi takes 12 minutes for the same. Suppose they both start at the same point and at the same time, and go in the same direction. After how many minutes will they meet again at the starting point?
    (a) 12 minutes (b) 18 minutes
    (c) 6 minutes (d) 36 minutes.

26. After how many digits will the decimal expansion of $\frac{3}{8}$ come to an end?
    (a) 4 (b) 3
    (c) 5 (d) 2.

27. After how many digits will the decimal expansion of $\frac{13}{125}$ come to an end?
    (a) 3 (b) 2
    (c) 4 (d) 5.

28. After how many digits will the decimal expansion of $\frac{7}{80}$ come to an end?
    (a) 2 (b) 3
    (c) 4 (d) 5.

29. Let x be a rational number whose decimal expansion terminates. Then x can be expressed in the form $\frac{p}{q}$, where p and q are coprime, and the prime factorisation of q, if n, m are non-negative integers, is of the form
    (a) $2^n \, 5^m$ (b) $2^n \, 3^m$
    (c) $3^n \, 5^m$ (d) 3^{n+m}.

30. The rational number $\frac{13}{3125}$ has a
    (a) terminating decimal expansion
    (b) non-terminating decimal expansion
    (c) non-terminating repeating decimal expansion
    (d) non-terminating non-repeating decimal expansion.

31. The rational number $\frac{29}{343}$ has a
    (a) terminating decimal expansion
    (b) non-terminating decimal expansion
    (c) non-terminating repeating decimal expansion
    (d) non-terminating non-repeating decimal expansion.

32. The decimal expansion of $\frac{17}{8}$ is
    (a) 2.125   (b) 2.25
    (c) 2.375   (d) 2.0125.

33. The decimal expansion of $\frac{6}{15}$ is
    (a) 0.04   (b) 0.004
    (c) 0.4   (d) 0.0004.

34. The prime factorisation of 96 is
    (a) $2^5 \times 3$ (b) $2^6 \times 3$
    (c) $2^4 \times 3$ (d) $2^3 \times 3$.

35. The prime factorisation of 3125 is
    (a) 5^5 (b) 5^6
    (c) 5^3 (d) 5^7.

36. The prime factorisation of 256 is
    (a) 2^6 (b) 2^7
    (c) 2^8 (d) 2^9.

37. $7 \times 11 \times 13 + 13$ is
    (a) a prime number
    (b) a composite number
    (c) an odd number
    (d) divisible by 5.

38. The number $7 \times 6 \times 5 \times 4 \times 3 \times 2 \times 1 + 5$ is
    (a) divisible by 5
    (b) an even number
    (c) a prime number
    (d) divisible by 3.

39. The HCF of 72 and 120 is 12. Find their L.C.M.
    (a) 120   (b) 240
    (c) 360   (d) 720.

40. An army contingent of 616 members is to march behind an army band of 32 members in a parade. The two groups are to march in the same number of columns. What is the maximum number of columns in which they can march?
    (a) 4   (b) 8
    (c) 7   (d) 11.

41. A sweet seller has 420 kaju barfies and 130 badam barfies. She wants to stack them in such a way that each stack has the same number, and they take up the least area of the tray. What is the number of that can be placed in each stack for this purpose?
    (a) 10   (b) 13
    (c) 14   (d) 20.

42. If a and b are any two positive integers, then, HCF $(a, b) \times$ LCM $(a, b) =$
    (a) $a + b$ (b) $a \times b$
    (c) $a^2 + b^2$ (d) $2ab$.

43. Find the missing term in the following factor tree:

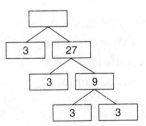

(a) 243 (b) 27
(c) 81 (d) 729.

44. Find the missing term in the following factor tree:

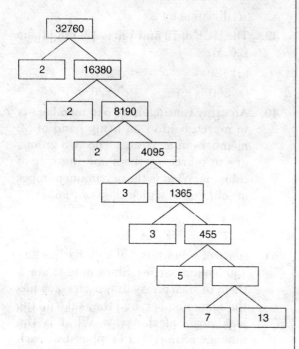

(a) 20 (b) 91
(c) 182 (d) 6.

45. $\sqrt{2}$ is
(a) a rational number
(b) an irrational number
(c) an integer
(d) a natural number.

46. $\frac{1}{\sqrt{3}}$ is
(a) an irrational number
(b) a rational number
(c) a natural number
(d) a prime number.

47. $\sqrt{9}$ is
(a) a rational number
(b) an irrational number

(c) an even integer
(d) a composite number.

48. π is
(a) an irrational number
(b) a rational number
(c) a prime number
(d) a composite number.

49. $3\sqrt{2}$ is
(a) a rational number
(b) an irrational number
(c) an odd integer
(d) an even integer.

50. $5 - \sqrt{3}$ is
(a) an irrational number
(b) a rational number
(c) a composite number
(d) a prime number.

51. $\sqrt{8}$ is
(a) a rational number
(b) an irrational number
(c) an even integer
(d) an odd integer.

52. $6 + \sqrt{2}$ is
(a) a rational number
(b) an irrational number
(c) an even integer
(d) an odd integer.

53. 0.66666 ...... is
(a) a rational number
(b) an irrational number
(c) a prime number
(d) a composite number.

54. 0.1201 2001 20001 20000 .... is
(a) an even integer
(b) a prime number
(c) an irrational number
(d) a rational number.

REAL NUMBERS

55. $43.\overline{123456789}$ is
(a) a rational number
(b) an irrational number
(c) a composite number
(d) a prime number.

56. 43.123456789 is
(a) an irrational number
(b) a rational number
(c) an odd integer
(d) an even integer.

57. $2.023\overline{1}$ is
(a) a rational number
(b) an irrational number
(c) a prime number
(d) a composite number.

58. Which of the following is not an irrational number?
(a) $\sqrt{2}$
(b) $\sqrt{3}$
(c) $\sqrt{9}$
(d) $\sqrt{7}$.

59. The solution of which equation is an irrational number?
(a) $\dfrac{1}{x} = \dfrac{3}{4}$
(b) $x^2 = \dfrac{49}{20}$
(c) $x^2 = \dfrac{9}{16}$
(d) $\dfrac{1}{2}x + \dfrac{1}{5} = \dfrac{1}{3}$.

60. The number of prime numbers between 1 and 10 is
(a) 2
(b) 3
(c) 4
(d) 5.

61. Which of the following pair of numbers is co-prime?
(a) 9 and 12
(b) 14 and 21
(c) 39 and 65
(d) 6 and 35.

62. Which of the following four digit numbers is a perfect square such that the first two digits and the last two digits considered separately also represent perfect squares?
(a) 3616
(b) 6436
(c) 8164
(d) 1681.

63. The largest number of four digits divisible by 85 is
(a) 9935
(b) 9965
(c) 9954
(d) 9945.

64. The decimal representation of $\dfrac{93}{1500}$ will be
(a) terminating
(b) non-terminating
(c) non-terminating repeating
(d) non-terminating non-repeating.

65. Find HCF (8, 9, 25) × LCM (8, 9, 25).
(a) 1
(b) 1800
(c) 900
(d) 3600.

66. An inspector of schools wishes to distribute 84 balls and 180 bats equally among a number of boys. Find the greatest number receiving the gift in this way.
(a) 6
(b) 12
(c) 18
(d) 7.

67. After how many places will the decimal expansion of rational number $\dfrac{43}{2^4 \, 5^3}$ will terminate?
(a) 1
(b) 2
(c) 3
(d) 4.

68. Complete the missing entries in the following factor tree:

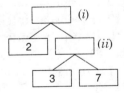

(a) 42, 21
(b) 21, 42
(c) 42, 42
(d) 21, 21.

69. In a seminar, the number of participants in English, Mathematics and Science are 36, 60 and 84 respectively. Find the minimum number of rooms required if in each

room the same number of participants are to be seated and all of them being in the same subject.
(a) 10   (b) 15
(c) 20   (d) 8.

70. If HCF (12, 40) = 40 × 1 + 4 × x, then find x
(a) – 8   (b) – 9
(c) – 10   (d) – 6.

ANSWERS

1. (c)	2. (a)	3. (b)	4. (b)	37. (b)	38. (a)	39. (d)	40. (b)
5. (d)	6. (c)	7. (b)	8. (d)	41. (a)	42. (b)	43. (c)	44. (b)
9. (a)	10. (b)	11. (a)	12. (a)	45. (b)	46. (a)	47. (a)	48. (a)
13. (b)	14. (b)	15. (b)	16. (a)	49. (b)	50. (a)	51. (b)	52. (b)
17. (a)	18. (c)	19. (c)	20. (a)	53. (a)	54. (c)	55. (a)	56. (b)
21. (b)	22. (a)	23. (a)	24. (a)	57. (a)	58. (c)	59. (b)	60. (c)
25. (d)	26. (b)	27. (a)	28. (c)	61. (d)	62. (d)	63. (d)	64. (a)
29. (a)	30. (a)	31. (c)	32. (a)	65. (b)	66. (b)	67. (d)	68. (a)
33. (c)	34. (a)	35. (a)	36. (c)	69. (b)	70. (b).		

HINTS/SOLUTIONS

1. $\frac{1}{3}$ is of the form $\frac{p}{q}$ ($q \neq 0$) where p and q are coprime integers.
2. Mentioned in Euclid's Division Lemma.
3. Statement of Fundamental Theorem of Arithmetic.
4. Least prime number = 2
   Least composite number = 4
5. 10 and 3 are coprime integers.
6. 10 and 3 are coprime integers.
7. 4 and 19 are coprime integers.
8. 4 and 19 are coprime integers.
9. 3 divides 81   10. 3 divides 81
11. 3 × 3 + 1 = 10   12. 4 × 4 + 3 = 19
13. 3 × 27 = 81
14. An even integer is divisible by 2.
15. 1, 3, 5, 7, 9, ... are odd integers.
16. Square of any positive integer is either of the form $3m$ or $3m + 1$ for some integer m.
17. 6 = 2 × 3
    20 = 2 × 2 × 5
18. 6 = 2 × 3
    20 = 2 × 2 × 5

19. 6 = 2 × 3
    72 = 2 × 2 × 2 × 3 × 3
    120 = 2 × 2 × 2 × 3 × 5
20. 26 = 2 × 13
    91 = 7 × 13
21. 8 = 2 × 2 × 2
    9 = 3 × 3
    25 = 5 × 5
22. Two consecutive natural numbers are always coprime.
23. Any two prime numbers are always coprime.
24. 17, 23 and 29 have no common factor other than 1.
25. LCM (18, 12) = 36
26. $\frac{3}{8} = \frac{3 \times 125}{8 \times 125} = \frac{375}{1000} = 0.375$
27. $\frac{13}{125} = \frac{13 \times 8}{125 \times 8} = \frac{104}{1000} = 0.104$
28. $\frac{7}{80} = \frac{7 \times 125}{80 \times 125} = \frac{875}{10000} = 0.0875$
29. This is as per theorem.

30. $3125 = 5 \times 5 \times 5 \times 5 \times 5 = 5^5$
31. $343 = 7 \times 7 \times 7 = 7^3$
32. $\dfrac{17}{8} = \dfrac{17 \times 125}{8 \times 125} = \dfrac{2125}{1000} = 2.125$
33. $\dfrac{6}{15} = \dfrac{2}{5} = \dfrac{2 \times 2}{5 \times 2} = \dfrac{4}{10} = 0.4$
34.
```
2 | 96
2 | 48
2 | 24
2 | 12
2 | 6
3 | 3
  | 1
```
35.
```
5 | 3125
5 | 625
5 | 125
5 | 25
5 | 5
  | 1
```
36.
```
2 | 256
2 | 128
2 | 64
2 | 32
2 | 16
2 | 8
2 | 4
2 | 2
  | 1
```
37. $7 \times 11 \times 13 + 13 = (7 \times 11 + 1)\,13$
    $= 78 \times 13 = 6 \times 13 \times 13$
38. $7 \times 6 \times 5 \times 4 \times 3 \times 2 \times 1 + 5$
    $= (7 \times 6 \times 4 \times 3 \times 2 \times 1 + 1)\,5$
39. $\text{LCM} = \dfrac{72 \times 120}{12} = 720$
40. HCF (32, 616) = 8
41. HCF (420, 130) = 10
42. It is an important result.
43. $3 \times 27 = 81$ 44. $7 \times 13 = 91$
47. $\sqrt{9} = 3$
53. Let $x = 0.6666 \ldots$
    Then
    $10x = 6.6666 \ldots$
    Subtracting
    $10x - x = 6$
    $\Rightarrow 9x = 6$
    $\Rightarrow x = \dfrac{6}{9} = \dfrac{2}{3}$
    Aliter non-terminating repeating
54. Non-terminating non-repeating
55. Non-terminating repeating
56. Terminating
57. Non-terminating repeating
58. $\sqrt{9} = 3$
59. $x = \sqrt{\dfrac{49}{20}} = \dfrac{7}{2\sqrt{5}}$
60. 2, 3, 5, 7
61. $6 = 2 \times 3$
    $35 = 5 \times 7$
62. $\sqrt{1681} = 41$
63. Largest number of four digits = 9999
```
      117
85)9999
    85
    ---
    149
     85
    ---
    649
    595
    ---
     54
```
∴ Required number
    = 9999 − 54 = 9945
64. $\dfrac{93}{1500} = \dfrac{31}{500}$;
    $500 = 2 \times 2 \times 5 \times 5 \times 5 = 2^2\, 5^3$
65. HCF (8, 9, 25) = 1
    LCM (8, 9, 25) = $8 \times 9 \times 25 = 1800$
66. HCF (84, 180) = 12
67. $\dfrac{43}{2^4\, 5^3} = \dfrac{43 \times 5}{2^4\, 5^4} = \dfrac{215}{10^4} = 0.0215$
68. (i) $2 \times 21 = 42$ (ii) $3 \times 7 = 21$
69. HCF (36, 60, 84) = 12
    Required number of rooms
    $= \dfrac{36}{12} + \dfrac{60}{12} + \dfrac{84}{12} = 3 + 5 + 7 = 15$
70. HCF (12, 40) = 4
    $4 = 40 \times 1 + 4 \times x$
    $\Rightarrow x = -9$

Chapter 2
Polynomials

1. The standard form of a linear polynomial is
   (a) $ax + b$
   (b) $ax^2 + bx + c$
   (c) $ax^3 + bx^2 + cx + d$
   (d) $ax^4 + bx^3 + cx^2 + dx + e$

2. The standard form of a quadratic polynomial is
   (a) $ax + b$
   (b) $ax^2 + bx + c$
   (c) $ax^3 + bx^2 + cx + d$
   (d) $ax^4 + bx^3 + cx^2 + dx + e$

3. The standard form of a cubic polynomial is
   (a) $ax + b$
   (b) $ax^2 + bx + c$
   (c) $ax^3 + bx^2 + cx + d$
   (d) $ax^4 + bx^3 + cx^2 + dx + e$

4. The standard form of a biquadratic polynomial is
   (a) $ax + b$
   (b) $ax^2 + bx + c$
   (c) $ax^3 + bx^2 + cx + d$
   (d) $ax^4 + bx^3 + cx^2 + dx + e$

5. $4x + 2$ is a
   (a) linear polynomial
   (b) quadratic polynomial
   (c) cubic polynomial
   (d) biquadratic polynomial.

6. $2y^2 - 3y + 4$ is a
   (a) linear polynomial
   (b) quadratic polynomial
   (c) cubic polynomial
   (d) biquadratic polynomial.

7. $5x^3 - 4x^2 + x - \sqrt{2}$ is a
   (a) linear polynomial
   (b) quadratic polynomial
   (c) cubic polynomial
   (d) biquadratic polynomial.

8. $2x^4 - 3x^3 - 3x^2 + 6x - 2$ is a
   (a) linear polynomial
   (b) quadratic polynomial
   (c) cubic polynomial
   (d) biquadratic polynomial.

9. A polynomial of degree 1 is called
   (a) a linear polynomial
   (b) a quadratic polynomial
   (c) a cubic polynomial
   (d) a biquadratic polynomial.

10. A polynomial of degree 2 is called
    (a) a linear polynomial
    (b) a quadratic polynomial
    (c) a cubic polynomial
    (d) a biquadratic polynomial.

11. A polynomial of degree 3 is called
    (a) a linear polynomial
    (b) a quadratic polynomial
    (c) a cubic polynomial
    (d) a biquadratic polynomial.

12. A polynomial of degree 4 is called
    (a) a linear polynomial
    (b) a quadratic polynomial
    (c) a cubic polynomial
    (d) a biquadratic polynomial.

POLYNOMIALS

13. Which of the following is not a polynomial?
    (a) $5x^2 - 6x + 3$ (b) $2 - 3y + 6y^2 - 2y^3$
    (c) $x^4 - 3x + 2$ (d) $x^{3/4} - 7x + 4$.

14. Which of the following is not a polynomial?
    (a) $2x + 5$
    (b) $\frac{3}{4}x^3 - \frac{4}{3}x^2 + 2x - 1$
    (c) $3x^2 + \sqrt{2}x - \sqrt{3}$
    (d) $x^3 + x + \frac{3}{x^2}$.

15. The degree of a constant polynomial is
    (a) 0   (b) 1
    (c) 2   (d) 3.

16. The degree of a linear polynomial is
    (a) 0   (b) 1
    (c) 2   (d) 3.

17. The degree of a zero polynomial is
    (a) not defined   (b) 1
    (c) 2   (d) 3.

18. The degree of a quadratic polynomial is
    (a) 0   (b) 1
    (c) 2   (d) 3.

19. The degree of a cubic polynomial is
    (a) 0   (b) 1
    (c) 2   (d) 3.

20. The degree of a biquadratic polynomial is
    (a) 0   (b) 2
    (c) 4   (d) 1.

21. The degree of a polynomial 2 is
    (a) 0   (b) 1
    (c) 2   (d) 3.

22. The degree of a polynomial $(x - 2)(x - 3)$ is
    (a) 1   (b) 2
    (c) 3   (d) 4.

23. The degree of the polynomial $x^5 + a^5$ is
    (a) 2   (b) 3
    (c) 5   (d) 4.

24. The degree of the polynomial $7u^6 - \frac{3}{2}u^4 + 4u^2 + u - 8$ is
    (a) 6   (b) 4
    (c) 3   (d) 0.

25. The value of $p(x) = x^2 - 3x - 4$ at $x = 0$ is
    (a) -3 (b) -4
    (c) 3   (d) 4.

26. The value of $p(x) = x^2 - 3x - 4$ at $x = -1$ is
    (a) 1   (b) -4
    (c) 0   (d) -3.

27. The zero of $p(x) = ax + b$ is
    (a) a (b) b
    (c) $-\frac{b}{a}$ (d) $-\frac{a}{b}$.

28. The number of zeroes, a linear polynomial can have at most is
    (a) 0   (b) 1
    (c) 2   (d) 3.

29. The number of zeroes, a quadratic polynomial can have at most is
    (a) 0   (b) 1
    (c) 2   (d) 2^2.

30. The number of zeroes, a quadratic polynomial can have at most is
    (a) 0   (b) 1
    (c) 2   (d) 3.

31. The number of zeroes, a biquadratic polynomial can have at most is
    (a) 1   (b) 2
    (c) 3   (d) 4.

32. The graph of the equation $y = ax^2 + bx + c$ is an upward parabola, if
    (a) $a > 0$ (b) $a < 0$
    (c) $a = 0$ (d) $a = -1$.

33. The graph of the equation $y = ax^2 + bx + c$ is a downward parabola, if
    (a) $a > 0$ (b) $a < 0$
    (c) $a = 0$ (d) $a = 1$.

34. The zeroes of the polynomial $x^2 - 3x - 4$ are

    (a) 4, −1   (b) 4, 1
    (c) −4, 1   (d) −4, −1.

35. The zeroes of the polynomial $x^3 - 4x$ are

    (a) 0, ±2   (b) 0, ±1
    (c) 0, ±3   (d) 0, 0, 0.

36. The zeroes of the polynomial $x^3 - x^2$ are

    (a) 0, 0, 1   (b) 0, 1, 1
    (c) 1, 1, 1   (d) 0, 0, 0.

37. If α and β are the roots of the quadratic polynomial $ax^2 + bx + c$, then α + β is equal to

    (a) $\dfrac{c}{a}$ (b) $-\dfrac{b}{a}$
    (c) c (d) $-b$.

38. If α and β are the roots of the quadratic polynomial $ax^2 + bx + c$, then αβ is equal to

    (a) $\dfrac{c}{a}$ (b) $-\dfrac{b}{a}$
    (c) $-\dfrac{c}{a}$ (d) $\dfrac{b}{a}$.

39. If α, β, γ are the zeroes of the cubic polynomial $ax^3 + bx^2 + cx + d$, then α + β + γ is equal to

    (a) $-\dfrac{b}{a}$ (b) $\dfrac{b}{a}$
    (c) $\dfrac{c}{a}$ (d) $\dfrac{d}{a}$.

40. If α, β, γ are the zeroes of the cubic polynomial $ax^3 + bx^2 + cx + d$, then αβ + βγ + γα is equal to

    (a) $-\dfrac{b}{a}$ (b) $\dfrac{b}{a}$
    (c) $\dfrac{c}{a}$ (d) $\dfrac{d}{a}$.

41. If α, β, γ are the zeroes of the cubic polynomial $ax^3 + bx^2 + cx + d$, then αβγ is equal to

    (a) $\dfrac{d}{a}$ (b) $-\dfrac{d}{a}$
    (c) $-\dfrac{b}{a}$ (d) $\dfrac{b}{a}$.

42. The zeroes of the polynomial $t^2 - 15$ are

    (a) $\pm\sqrt{15}$ (b) $\pm\sqrt{5}$
    (c) $\pm\sqrt{3}$ (d) ± 3.

43. The quadratic polynomial, the sum and product of whose zeroes are −3 and 2 respectively, is

    (a) $x^2 + 3x + 2$ (b) $x^2 - 3x + 2$
    (c) $x^2 + 3x - 2$ (d) $x^2 - 3x - 2$.

44. The quadratic polynomial, the sum and product of whose zeroes are 1 and 1 respectively, is

    (a) $x^2 - x + 1$ (b) $x^2 - x - 1$
    (c) $x^2 + x + 1$ (d) $x^2 + x - 1$.

45. The quadratic polynomial, the sum and product of whose zeroes are 0 and $\sqrt{5}$ respectively, is

    (a) $x^2 - \sqrt{5}$ (b) $x^2 + \sqrt{5}$
    (c) $x^2 - 5$ (d) $x^2 + 5$.

46. What is the coefficient of the first term of the quotient when $2x^2 + 3x + 1$ is divided by $x + 2$?

    (a) 1   (b) 2
    (c) 3   (d) −2.

47. What is the coefficient of the first term of the quotient when $3x^3 + x^2 + 2x + 5$ is divided by $1 + 2x + x^2$?

    (a) 1   (b) 2
    (c) 3   (d) 5.

48. What is the coefficient of the first term of the quotient when $3x^2 - x^3 - 3x + 5$ is divided by $x - 1 - x^2$?

    (a) 1   (b) −1
    (c) 2   (d) −2.

POLYNOMIALS

49. If the divisor is x^2 and quotient is x while the remainder is 1, then the dividend is
    (a) x^2 (b) x
    (c) x^3 (d) $x^3 + 1$.

50. Dividend is equal to
    (a) divisor × quotient + remainder
    (b) divisor × quotient
    (c) divisor × quotient – remainder
    (d) divisor × quotient × remainder.

51. If the zeroes of the polynomial $x^3 - 3x^2 + x + 1$ are $a - b$, a and $a + b$, then the value of a is
    (a) 1 (b) – 1
    (c) 0 (d) – 3.

52. If the zeroes of the polynomial $x^3 - 3x^2 + x + 1$ are $\dfrac{a}{r}$, a and ar, then the value of a is
    (a) 1 (b) – 1
    (c) 2 (d) – 3.

53. What can be the degree of the remainder at most when a biquadratic polynomial is divided by a quadratic polynomial?
    (a) 0 (b) 1
    (c) 2 (d) 4.

54. If the zeroes of the quadratic polynomial $ax^2 + bx + c$ are reciprocal to each other, then the value of c is
    (a) 0 (b) a
    (c) $-a$ (d) 1.

55. If the sum of the zeroes of the quadratic polynomial $ax^2 + bx + c$ is 0, then
    (a) $a = 0$ (b) $b = 0$
    (c) $c = 0$ (d) $b \neq 0$.

56. If α and β are the zeroes of the quadratic polynomial $x^2 - 2x + 1$, then $\alpha^2 + \beta^2$ is
    (a) 1 (b) 3
    (c) 2 (d) – 2.

57. For the polynomial $2x^3 - 5x^2 - 14x + 8$, find the sum of the products of the zeroes, taken two at a time.
    (a) 7 (b) 8
    (c) – 7 (d) – 8.

58. For the polynomial $3x^3 - 5x^2 - 11x - 3$, find the product of the zeroes.
    (a) 1 (b) – 1
    (c) 3 (d) – 3.

59. For the polynomial $x^3 - 3x^2 - x + 3$, find the sum of the zeroes.
    (a) 1 (b) – 1
    (c) 3 (d) – 3.

60. For the graph of $y = p(x)$ given below where $p(x)$ is the polynomial, the number zeroes of $p(x)$ is

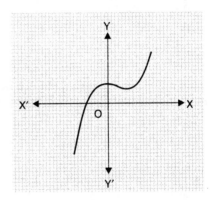

    (a) 0 (b) 1
    (c) 2 (d) 3.

61. Which of the following polynomials is in standard form?
    (a) $24x^3 - 14x^2 - 39x + 20$
    (b) $24x^3 - 39x - 14x^2 + 20$
    (c) $24x^3 + 20 - 14x^2 - 39x$
    (d) $20 - 39x - 14x^2 + 24x^3$.

62. If the graph of a polynomial does not intersect the x-axis, then the number of zeroes of the polynomial is
    (a) 0 (b) 1
    (c) – 1 (d) 3.

63. Which of the following is not a zero of the polynomial $x^3 - 6x^2 + 11x - 6$?
    (a) 1 (b) 2
    (c) 3 (d) 0.

64. Which of the following is a zero of the polynomial $x^3 - 6x^2 + 11x - 6$?
(a) 0   (b) 1
(c) 4   (d) – 2.

65. The polynomial $x^2 - 5x + 6$ has
(a) no linear factor
(b) only one linear factor
(c) two linear factors
(d) more than two linear factors.

66. A cubic polynomial with the sum, sum of the product of its zeroes taken two at a time, and the product of its zeroes are 2, – 7 and – 14 respectively, is
(a) $x^3 - 2x^2 - 7x + 14$
(b) $x^3 + 2x^2 + 7x + 14$
(c) $x^3 - 2x^2 - 7x - 14$
(d) $x^3 + 2x^2 - 7x - 14$.

67. If α and β are the zeroes of the quadratic polynomial $9x^2 - 1$, find the value of $\alpha^2 + \beta^2$.
(a) $\frac{1}{9}$ (b) $\frac{2}{9}$
(c) $\frac{1}{3}$ (d) $\frac{2}{3}$.

68. What should be subtracted from $x^3 - 2x^2 + 4x + 1$ to get?
(a) $x^3 - 2x^2 + 4x$ (b) $x^3 - 2x^2 + 4x + 1$
(c) $x^3 - 2x^2 + 4x - 1$
(d) $x^2 + 2x^2 + 4x + 1$.

69. If α, β, γ are the zeroes of the polynomial $x^3 + px^2 + qx + r$, then find $\frac{1}{\alpha\beta} + \frac{1}{\beta\gamma} + \frac{1}{\gamma\alpha}$.
(a) $\frac{p}{r}$ (b) $-\frac{p}{r}$
(c) $\frac{q}{r}$ (d) $-\frac{q}{r}$.

70. Which of the following is not a quadratic polynomial?
(a) $x^2 + 5x + 6$
(b) $x^2 - 5x + 6$
(c) $1 + (x^2 - 2x)$
(d) $(x - 2)(x + 2) - (x^2 + 5x)$.

71. If α, β, γ are the zeroes of the polynomial $ax^3 + bx^2 + cx + d$, then the value of $\frac{1}{\alpha} + \frac{1}{\beta} + \frac{1}{\gamma}$ is
(a) $\frac{c}{d}$ (b) $-\frac{c}{d}$
(c) $\frac{b}{d}$ (d) $-\frac{b}{d}$.

72. If the product of two zeroes of the polynomial $x^3 - 6x^2 + 11x - 6$ is 2, then the third zero is
(a) 1   (b) 2
(c) 3   (d) 4.

73. If the product of the zeroes of the polynomial $ax^3 - 6x^2 + 11x - 6$ is 6, then the value of a is
(a) 0   (b) – 1
(c) 2   (d) 1.

74. If the sum of the zeroes of the quadratic polynomial $kx^2 + 3x + 4k$ is equal to their product, then find the value of k.
(a) $-\frac{3}{4}$ (b) $\frac{3}{4}$
(c) $\frac{4}{3}$ (d) $-\frac{4}{3}$.

75. If α, β are the zeroes of the polynomial $x^2 - 9$, find $\alpha\beta(\alpha + \beta)$.
(a) 0   (b) 3
(c) – 3   (d) 9.

ANSWERS

1. (a)	2. (b)	3. (c)	4. (d)	13. (d)	14. (d)	15. (a)	16. (b)
5. (a)	6. (b)	7. (c)	8. (d)	17. (a)	18. (c)	19. (d)	20. (c)
9. (a)	10. (b)	11. (c)	12. (d)	21. (a)	22. (b)	23. (c)	24. (a)

POLYMIALS

25. (b)	26. (c)	27. (c)	28. (b)	53. (b)	54. (b)	55. (b)	56. (c)
29. (c)	30. (d)	31. (d)	32. (a)	57. (c)	58. (a)	59. (c)	60. (b)
33. (b)	34. (a)	35. (a)	36. (a)	61. (a)	62. (a)	63. (d)	64. (b)
37. (b)	38. (a)	39. (a)	40. (c)	65. (c)	66. (a)	67. (b)	68. (a)
41. (b)	42. (a)	43. (a)	44. (a)	69. (a)	70. (d)	71. (b)	72. (c)
45. (b)	46. (b)	47. (c)	48. (a)	73. (d)	74. (a)	75. (a).	
49. (d)	50. (a)	51. (a)	52. (b)				

HINTS/SOLUTIONS

1. Definition
2. Definition
3. Definition
4. Definition
5. Compare with $ax + b$
6. Compare with $ay^2 + by + c$
7. Compare with $ax^3 + bx^2 + cx + d$
8. Compare with $ax^4 + bx^3 + cx^2 + dx + e$
9. Definition
10. Definition
11. Definition
12. Definition
13. Power of x in $x^{3/4}$ is fractional.
14. Power of x in $\dfrac{3}{x^2}$ is negative.
15. Definition
16. Definition
17. Definition
18. Definition
19. Definition
20. Definition
21. Constant polynomial
22. $(x - 2)(x - 3) = x^2 - 5x + 6$; quadratic polynomial.
23. Highest power of x is 5
24. Highest power of u is 6
25. $p(0) = (0)^2 - 3(0) - 4 = -4$
26. $p(-1) = (-1)^2 - 3(-1) - 4 = 1 + 3 - 4 = 0$
27. $ax + b = 0 \Rightarrow x = -\dfrac{b}{a}$
28. A linear polynomial has only one real linear factor at the most
29. A quadratic polynomial can have at the most two real linear factors
30. A cubic polynomial can have at the most three real linear factors
31. A biquadratic polynomial can have at the most four real linear factors
34. $x^2 - 3x - 4 = (x - 4)(x + 1)$

35. $x^3 - 4x = 0 \Rightarrow x(x^2 - 4) = 0$
    $\Rightarrow x(x - 2)(x + 2) = 0$
36. $x^3 - x^2 = 0 \Rightarrow x^2(x - 1) = 0$
42. $t^2 - 15 = 0 \Rightarrow t^2 = 15 \Rightarrow t = \pm\sqrt{15}$
43. $x^2 - $ (sum of roots) $x + $ product of roots
46. $\dfrac{2x^2}{x} = 2x$
47. $\dfrac{3x^3}{x^2} = 3x$ 48. $\dfrac{-x^3}{-x^2} = x$
49. Dividend = divisor × quotient + remainder
50. Dividend = divisor × quotient + remainder
51. $(a - b) + a + (a + b) = 3$
    $\Rightarrow 3a = 3 \Rightarrow a = 1$
52. $\dfrac{a}{r} \cdot a \cdot ar = -1$
    $\Rightarrow a^3 = -1$
    $\Rightarrow a = -1$
54. $\alpha \cdot \dfrac{1}{\alpha} = 1 = \dfrac{c}{a} \Rightarrow c = a$
55. $\alpha + \beta = 0 \Rightarrow -\dfrac{b}{a} = 0 \Rightarrow b = 0$
56. $x^2 - 2x + 1 = 0 \Rightarrow (x - 1)^2 = 0$
    $\Rightarrow x = 1, 1$
    $\therefore \alpha^2 + \beta^2 = 1^2 + 1^2 = 2$
57. $\alpha\beta + \beta\gamma + \gamma\alpha = \dfrac{\text{coefficient of } x}{\text{coefficient of } x^3}$
58. $\alpha\beta\gamma = -\left(\dfrac{-3}{3}\right) = 1$
59. $\alpha + \beta + \gamma = -\dfrac{(-3)}{1} = 3$

60. Graph intersects the x-axis at one point only
61. Powers should be in decreasing order
62. Definition of zero
63. $(0)^3 - 6(0)^2 + 11(0) - 6 \neq 0$
64. $(1)^3 - 6(1)^2 + 11(1) - 6 = 0$
65. $x^2 - 5x + 6 = (x - 2)(x - 3)$
66. $x^3 -$ (sum of zeroes) $x^2 +$ (sum of the product of zeroes taken two at a time) $x -$ product of zeroes $= 0$
67. $9x^2 - 1 = 0 \Rightarrow x^2 = \dfrac{1}{9} \Rightarrow x = \pm \dfrac{1}{3}$

    $\therefore \quad \alpha^2 + \beta^2 = \left(\dfrac{1}{3}\right)^2 + \left(-\dfrac{1}{3}\right)^2 = \dfrac{2}{9}$
68. $(x^3 - 2x^2 + 4x + 1) - (x^3 - 2x^2 + 4x) = 1$
69. $\dfrac{1}{\alpha\beta} + \dfrac{1}{\beta\gamma} + \dfrac{1}{\gamma\alpha} = \dfrac{\alpha + \beta + \gamma}{\alpha\beta\gamma} = \dfrac{-p}{-r} = \dfrac{p}{r}$
70. $(x - 2)(x + 2) - (x^2 + 5x) = (x^2 - 4) - (x^2 + 5x) = -5x - 4$
71. $\dfrac{1}{\alpha} + \dfrac{1}{\beta} + \dfrac{1}{\gamma} = \dfrac{\alpha\beta + \beta\gamma + \gamma\alpha}{\alpha\beta\gamma} = \dfrac{\frac{c}{a}}{-\frac{d}{a}} = -\dfrac{c}{d}$
72. $\alpha\beta\gamma = 6 \Rightarrow 2\gamma = 6 \Rightarrow \gamma = 3$
73. $\alpha\beta\gamma = \dfrac{6}{a} = 6 \Rightarrow a = 1$
74. $\alpha + \beta = \alpha\beta \Rightarrow -\dfrac{3}{k} = \dfrac{4k}{k}$

    $\Rightarrow -\dfrac{3}{k} = 4 \Rightarrow k = -\dfrac{3}{4}$
75. $x^2 - 9 = 0 \Rightarrow x^2 = 9 \Rightarrow x = \pm 3$

    $\therefore \quad \alpha\beta(\alpha + \beta) = \{3(-3)\}\{3 + (-3)\} = 0.$

Chapter 3
Pair of Linear Equations in Two Variables

1. Find the value of x if
   $$y = \frac{1}{2}x \text{ and } 3x + 4y = 20$$
   (a) 1 (b) 2
   (c) 3 (d) 4.

2. Which of the following is a solution of the equation $2x + 3y = 5$?
   (a) $x = 1, y = 0$ (b) $x = 0, y = 1$
   (c) $x = 1, y = 1$ (d) $x = 0, y = 0$.

3. Which of the following is not a solution of the equation $2x + 3y = 5$?
   (a) $x = 1, y = 1$ (b) $x = -2, y = 3$
   (c) $x = 4, y = -1$ (d) $x = 1, y = 7$.

4. The solution of the pair of equations
   $$x + y = 14$$
   $$x - y = 4$$
   is
   (a) $x = 9, y = 5$ (b) $x = 5, y = 9$
   (c) $x = 9, y = 9$ (d) $x = 5, y = 5$.

5. Solve the following pair of equations:
   $$\sqrt{2}x + \sqrt{3}y = 0$$
   $$\sqrt{3}x - \sqrt{8}y = 0$$
   (a) $x = 1, y = 0$ (b) $x = 0, y = 0$
   (c) $x = 0, y = 1$ (d) $x = 1, y = 1$.

6. The difference between two numbers is 26 and one number is three times the other number. Find them.
   (a) 39, 13 (b) 30, 10
   (c) 40, 14 (d) 30, 40.

7. The larger of two supplementary angles exceeds the smaller by 18 degrees. Find them.

   (a) 98°, 80° (b) 99°, 81°
   (c) 118°, 100° (d) 100°, 80°.

8. If the pair of linear equations
   $$a_1x + b_1y + c_1 = 0$$
   $$a_2x + b_2y + c_2 = 0$$
   has a unique solution, then
   (a) $\frac{a_1}{a_2} = \frac{b_1}{b_2} \neq \frac{c_1}{c_2}$ (b) $\frac{a_1}{a_2} = \frac{b_1}{b_2} = \frac{c_1}{c_2}$
   (c) $\frac{a_1}{a_2} \neq \frac{b_1}{b_2}$ (d) $\frac{b_1}{b_2} \neq \frac{c_1}{c_2}$.

9. If the pair of linear equations
   $$a_1x + b_1y + c_1 = 0$$
   $$a_2x + b_2y + c_2 = 0$$
   has infinitely many solutions, then
   (a) $\frac{a_1}{a_2} = \frac{b_1}{b_2} = \frac{c_1}{c_2}$
   (b) $\frac{a_1}{a_2} = \frac{b_1}{b_2} \neq \frac{c_1}{c_2}$
   (c) $a_1a_2 + b_1b_2 + c_1c_2 = 0$
   (d) $a_1^2 + b_1^2 + c_1^2 = a_2^2 + b_2^2 + c_2^2$.

10. If the pair of linear equations
    $$a_1x + b_1y + c_1 = 0$$
    $$a_2x + b_2y + c_2 = 0$$
    has no solution, then
    (a) $\frac{a_1}{a_2} = \frac{b_1}{b_2} = \frac{c_1}{c_2}$ (b) $\frac{a_1}{a_2} = \frac{b_1}{b_2} \neq \frac{c_1}{c_2}$
    (c) $\frac{a_1}{a_2} \neq \frac{b_1}{b_2}$ (d) $\frac{a_1}{a_2} \neq \frac{c_1}{c_2}$.

11. The solution of the pair of linear equations
    $$a_1x + b_1y + c_1 = 0$$
    $$a_2x + b_2y + c_2 = 0$$

by cross-multiplication method is given by

(a) $x = \dfrac{b_2 c_1 - b_1 c_2}{a_1 b_2 - a_2 b_1}, y = \dfrac{a_1 c_2 - a_2 c_1}{a_1 b_2 - a_2 b_1}$

(b) $x = \dfrac{b_1 c_2 - b_2 c_1}{a_1 b_2 - a_2 b_1}, y = \dfrac{a_1 c_2 - a_2 c_1}{a_1 b_2 - a_2 b_1}$

(c) $x = \dfrac{b_2 c_1 - b_1 c_2}{a_1 b_2 - a_2 b_1}, y = \dfrac{a_2 c_1 - a_1 c_2}{a_1 b_2 - a_2 b_1}$

(d) $x = \dfrac{b_1 c_2 - b_2 c_1}{a_1 b_2 - a_2 b_1}, y = \dfrac{a_2 c_1 - a_1 c_2}{a_1 b_2 - a_2 b_1}$.

12. The solution of the pair of linear equations
$$a_1 x + b_1 y = c_1$$
$$a_2 x + b_2 y = c_2$$
by cross-multiplication method is given by

(a) $x = \dfrac{b_1 c_2 - b_2 c_1}{a_1 b_2 - a_2 b_1}, y = \dfrac{a_2 c_1 - a_1 c_2}{a_1 b_2 - a_2 b_1}$

(b) $x = -\dfrac{b_1 c_2 - b_2 c_1}{a_1 b_2 - a_2 b_1}, y = -\dfrac{a_2 c_1 - a_1 c_2}{a_1 b_2 - a_2 b_1}$

(c) $x = \dfrac{b_1 c_2 - b_2 c_1}{a_1 b_2 - a_2 b_1}, y = -\dfrac{a_2 c_1 - a_1 c_2}{a_1 b_2 - a_2 b_1}$

(d) $x = -\dfrac{b_1 c_2 - b_2 c_1}{a_1 b_2 - a_2 b_1}, y = \dfrac{a_2 c_1 - a_1 c_2}{a_1 b_2 - a_2 b_1}$.

13. For which values of p does the pair of equations given below has unique solution?
$$4x + py + 8 = 0$$
$$2x + 2y + 2 = 0$$
(a) $p = 4$ (b) $p \neq 4$
(c) $p = 2$ (d) $p \neq 2$.

14. The value of y when
$$\dfrac{x+y}{xy} = 2 \text{ and } \dfrac{x-y}{xy} = 6 \text{ is}$$
(a) $\dfrac{1}{4}$ (b) $-\dfrac{1}{2}$
(c) $-\dfrac{1}{4}$ (d) $\dfrac{1}{3}$.

15. The condition for the following system of linear equations to have a unique solution is

$$ax + by = c$$
$$lx + my = n$$
(a) $am = bl$ (b) $am \neq bl$
(c) $al = bm$ (d) $al \neq bm$.

16. If $y = 2x - 3$ and $y = 5$, then the value of x is
(a) 1 (b) 2
(c) 3 (d) 4.

17. The pair satisfying $2x + y = 6$ is
(a) (1, 2) (b) (2, 1)
(c) (2, 2) (d) (1, 1).

18. If $\dfrac{4}{x} + 5y = 7$ and $x = -\dfrac{4}{3}$, then the value of y is
(a) $\dfrac{37}{15}$ (b) 2
(c) $\dfrac{1}{2}$ (d) $\dfrac{1}{3}$.

19. If $\dfrac{3}{x} + 4y = 5$ and $y = 1$, then the value of x is
(a) 3 (b) $\dfrac{1}{3}$
(c) -3 (d) $-\dfrac{1}{3}$.

20. If $x = 1$, then the value of y in the equation $\dfrac{4}{x} + \dfrac{3}{y} = 5$ is
(a) 1 (b) $\dfrac{1}{3}$
(c) 3 (d) -3.

21. If the unit's and ten's digit of a two digit number are y and x respectively, then the number will be
(a) $10x + y$ (b) $10y + x$
(c) $x + y$ (d) xy.

22. The age of a son is one-third the age of his mother. If the present age of mother is x years, then the age of the son after 12 years is

(a) $\frac{x}{3} + 12$ (b) $\frac{x+12}{3}$

(c) $x + 4$ (d) $\frac{x}{3} - 12$.

23. Find the value of k for which the system of equations
$$3x - 4y + 7 = 0$$
$$kx + 3y - 5 = 0$$
has no solution.

(a) $\frac{9}{4}$ (b) $-\frac{9}{4}$

(c) $\frac{4}{9}$ (d) $-\frac{4}{9}$.

24. Find the value of k for which the system of equations
$$x - ky = 2$$
$$3x + 2y = -5$$
has a unique solution.

(a) $k = \frac{2}{3}$ (b) $k \neq -\frac{2}{3}$

(c) $k = \frac{3}{2}$ (d) $k \neq -\frac{3}{2}$.

25. Determine k for which the system of equations
$$4x + y = 3$$
$$8x + 2y = 5k$$
has infinite solutions.

(a) $-\frac{5}{6}$ (b) $-\frac{6}{5}$

(c) $\frac{5}{6}$ (d) $\frac{6}{5}$.

26. Determine the value of k for which the system of equations
$$2x - 3y = 1$$
$$kx + 5y = 7$$
has a unique solution.

(a) $k \neq -\frac{10}{3}$ (b) $k = \frac{10}{3}$

(c) $k = \frac{3}{10}$ (d) $k \neq \frac{3}{10}$.

27. For what value of k will the system of linear equations have infinite number of solutions?

$$kx + 4y = k - 4$$
$$16x + ky = k$$

(a) 2 (b) 4

(c) 6 (d) 8.

28. Determine the value of k for which the system of equations
$$2x + 5y = 7$$
$$3x - ky = 5$$
has a unique solution.

(a) $k \neq -\frac{15}{2}$ (b) $k = \frac{15}{2}$

(c) $k = \frac{2}{15}$ (d) $k \neq \frac{2}{15}$.

29. Determine the value of k for which the system of equations
$$kx + 2y - 1 = 0$$
$$5x - 3y + 2 = 0$$
has no solution.

(a) $k = \frac{10}{3}$ (b) $k = -\frac{10}{3}$

(c) $k = \frac{3}{10}$ (d) $k = -\frac{3}{10}$.

30. Determine the value of k for which the system of linear equations
$$(k-3)x + 3y = k$$
$$kx + ky = 12$$
has infinite number of solutions.

(a) 3 (b) 4

(c) 5 (d) 6.

31. Solve for x and y:
$$x - y = 0$$
$$2x - y = 2$$

(a) $x = 1, y = 1$ (b) $x = 2, y = 2$

(c) $x = 3, y = 3$ (d) $x = 4, y = 4$.

32. Determine the value of c for which the system of linear equations

$$cx + 3y = 3$$
$$12x + cy = 6$$
has no solution.
(a) 6 (b) −6
(c) ±6 (d) 3.

33. For what value of k, the system of equations
$$2x + ky = 1$$
$$3x - 5y = 7$$
has a unique solution?
(a) $k \neq -\dfrac{10}{3}$ (b) $k = \dfrac{10}{3}$
(c) $k = \dfrac{3}{10}$ (d) $k \neq -\dfrac{3}{10}$.

34. For what value of k, the system of equations
$$2x + ky = 1$$
$$3x - 5y = 7$$
has no solution?
(a) $k = -\dfrac{10}{3}$ (b) $k = \dfrac{10}{3}$
(c) $k = \dfrac{3}{10}$ (d) $k = -\dfrac{3}{10}$.

35. Find the value of k for which the system of equations
$$8x + 5y = 9$$
$$kx + 10y = 15$$
has no solution.
(a) 4 (b) 8
(c) 12 (d) 16.

36. The equations
$$3x - 2y = 4$$
$$6x + 2y = 4$$
have
(a) a unique solution
(b) no solution
(c) infinite solutions
(d) two solutions.

37. The equations
$$3x - 2y = 4$$
$$9x - 6y = 12$$
have
(a) no solution
(b) a unique solution
(c) infinite solutions
(d) only two solutions.

38. The equations
$$2x - 3y = 4$$
$$4x - 6y = 7$$
have
(a) a unique solution
(b) no solution
(c) infinite solutions
(d) two solutions.

39. The equations
$$2x - 3y = 4$$
$$6x - 9y = 12$$
have
(a) a unique solution
(b) no solution
(c) only two solutions
(d) infinite solutions.

40. The equations
$$4x - 5y = 3$$
$$5x - 4y = 5$$
have
(a) a unique solution
(b) no solution
(c) infinite solutions
(d) two solutions.

41. The equations
$$4x - 5y = 3$$
$$8x - 10y = 6$$
have
(a) infinite solutions
(b) no solution
(c) a unique solution
(d) only two solutions.

42. For what value of k will the system of equations
$$2x + 3y = 4$$
$$(k + 2)x + 6y = 3k + 2$$

have infinite solutions?
(a) $k = 1$ (b) $k = 2$
(c) $k = 3$ (d) $k = 4$.

43. The solution of the equations
$$\frac{x}{a} + \frac{y}{b} = 2$$
$$ax - by = a^2 - b^2$$
is
(a) $x = a, y = b$ (b) $x = -a, y = -b$
(c) $x = a, y = -b$ (d) $x = -a, y = b$.

44. The solution of the equations
$$\frac{x}{a} - \frac{y}{b} = a - b$$
$$ax + by = a^3 + b^3$$
(a) $x = a, y = b$ (b) $x = a^2, y = b^2$
(c) $x = a^3, y = b^3$ (d) $x = a^4, y = b^4$.

45. Solve for x and y:
$$\frac{bx}{a} - \frac{ay}{b} + a + b = 0$$
$$bx - ay + 2ab = 0$$
(a) $x = -a, y = b$ (b) $x = -a, y = -b$
(c) $x = a, y = -b$ (d) $x = -a, y = -b$.

46. Solve for x and y:
$$\frac{a^2}{x} - \frac{b^2}{y} = 0$$
$$\frac{a^2 b}{x} + \frac{b^2 a}{y} = a + b; x, y \neq 0$$
(a) $x = a^2, y = b^2$ (b) $x = -a^2, y = -b^2$
(c) $x = a^2, y = -b^2$ (d) $x = -a^2, y = b^2$.

47. The sum of the two digits of a two-digit number is 12. The number obtained by interchanging the two digits exceeds the given number by 18. Find the number.
(a) 75 (b) 57
(c) 66 (d) 48.

48. The sum of the numerator and the denominator of a fraction is 12. If the denominator is increased by 3, the fraction becomes $\frac{1}{2}$. Find the fraction.

(a) $\frac{5}{7}$ (b) $\frac{7}{5}$
(c) $\frac{3}{8}$ (d) $\frac{8}{3}$.

49. Which of the following is not a linear equation?
(a) $x + y = 6$ (b) $x^2 - 5x + 6 = 0$
(c) $y = 2x$ (d) $x = 0$.

50. The following pair of equations represents
$$5x - 3y + 9 = 0$$
$$10x - 6y + 18 = 0$$
(a) parallel lines
(b) coincident lines
(c) intersecting lines
(d) perpendicular lines.

51. The pair of equations
$$x = 0$$
$$y = 0$$
represents
(a) parallel lines
(b) coincident lines
(c) perpendicular lines
(d) non-intersecting lines.

52. If one linear equation is $2x - 3y + 5 = 0$, then another linear equation to have parallel line as its geometrical construction will be
(a) $4x - 6y + 10 = 0$
(b) $6x - 9y + 15 = 0$
(c) $8x - 12y + 20 = 0$
(d) $4x - 6y + 15 = 0$.

53. The point where the line $2x + y = 5$ meets the y-axis is
(a) (0, 5) (b) (0, -5)
(c) (0, 2) (d) (0, -2).

54. The point where the line $x - y - 3 = 0$ meets the y-axis is
(a) (0, -3) (b) (0, 3)
(c) (0, 2) (d) (0, 1).

55. Find the point where the line $2x + y = 8$ meets the x-axis.
(a) (0, 4) (b) (4, 0)
(c) (0, 8) (d) (8, 0).

56. Find the point where the line $3x - 2y = 12$ meets the x-axis
    (a) (0, 4)   (b) (4, 0)
    (c) (0, 8)   (d) (8, 0).

57. Find the area of the triangle formed by the coordinate axes and the line $x + y = 6$.
    (a) 6   (b) 12
    (c) 18   (d) 36.

58. Find the area of the triangle formed by the coordinate axes and the line
    $$\frac{x}{a} + \frac{y}{b} = 1.$$
    (a) ab (b) $\frac{1}{2}ab$
    (c) $a + b$ (d) $2ab$.

59. $x = 3$ is a line
    (a) parallel to x-axis
    (b) parallel to y-axis
    (c) passing through origin
    (d) passing through (0, 3).

60. $y = 5$ is a line
    (a) parallel to x-axis
    (b) parallel to y-axis
    (c) passing through origin
    (d) passing through (5, 0).

61. The point (5, 0) is a point on the line
    (a) $x = 5$ (b) $y = 5$
    (c) $x = 0$ (d) $y = 0$.

62. The point (0, 6) is a point on the line
    (a) $y = x$ (b) $y = 2x$
    (c) $y = 6$ (d) $x = 6$.

63. $y = 3x$ is a line
    (a) parallel to x-axis
    (b) parallel to y-axis
    (c) passing through origin
    (d) passing through (3, 1).

64. If the line $y = px - 2$ passes through the point (3, 2), then find the value of p.
    (a) $\frac{3}{4}$ (b) $\frac{4}{3}$
    (c) 3   (d) 4.

65. The sum of two numbers is 8 and their difference is 2. Find the numbers.
    (a) 5 and 3   (b) 6 and 4
    (c) 4 and 2   (d) 4 and 4.

66. Find the value of a so that the point $(3, a)$ lies on the line represented by
    $$2x - 3y = 5.$$
    (a) 1   (b) 3
    (c) $\frac{1}{3}$ (d) $-\frac{1}{3}$.

67. The pair of equations
    $$x + y = 14$$
    $$x - y = 4$$
    is
    (a) consistent
    (b) inconsistent
    (c) having two solutions only
    (d) representing coincident lines.

68. Solve for x and y:
    $$x - y = 0, \ 2x - y = 2$$
    (a) 2, 2   (b) 1, 1
    (c) 2, 0   (d) 0, 2.

69. The solution of the pair of equations
    $$\frac{x}{a} + \frac{y}{b} = a + b$$
    $$\frac{x}{a^2} + \frac{y}{b^2} = 2$$
    is
    (a) $x = a^2, y = b^2$ (b) $x = a, y = b$
    (c) $x = a^3, y = b^3$ (d) $x = a^4, y = b^4$.

70. If $3x + y = 10$ and $y = 4$, then find the value of x.
    (a) 0   (b) 1
    (c) 2   (d) 3.

71. What is the condition that the lines
    $$ax + by = c$$
    and $lx + my = n$ are coincident?
    (a) $\frac{a}{l} = \frac{b}{m} = \frac{c}{n}$ (b) $\frac{a}{l} = \frac{b}{m} \neq \frac{c}{n}$
    (c) $\frac{a}{l} \neq \frac{b}{m}$ (d) $\frac{b}{m} \neq \frac{c}{n}$.

72. The pair of lines
    $$5x - 4y + 8 = 0$$
    $$7x + 6y - 9 = 0$$

represents
(a) parallel lines
(b) coincident lines
(c) intersecting lines
(d) perpendicular lines.

73. The pair of lines
$$9x + 3y + 12 = 0$$
$$18x + 6y + 24 = 0$$
represents
(a) coincident lines
(b) parallel lines
(c) intersecting lines
(d) lines through origin.

74. The pair of lines
$$6x - 3y + 10 = 0$$
$$2x - y + 9 = 0$$
represents
(a) intersecting lines
(b) coincident lines
(c) parallel lines
(d) lines through origin.

75. The pair of equations
$$3x + 2y = 5,\ 2x - 3y = 7$$
(a) is consistent
(b) is inconsistent
(c) has infinite solutions
(d) represents lines through origin.

76. The pair of equations
$$2x - 3y = 8,$$
$$4x - 6y = 9$$
is
(a) consistent
(b) inconsistent

(c) has only two solutions
(d) has a unique solution.

77. For the linear equation $2x + 3y - 8 = 0$, another linear equation in two variables so that the geometrical representation of the pair so formed is parallel lines is
(a) $4x + 6y - 5 = 0$
(b) $3x + y - 5 = 0$
(c) $6x + 9y - 24 = 0$
(d) $8x + 12y - 32 = 0$.

78. The coordinates of the origin are
(a) (0, 0)   (b) (0, 1)
(c) (1, 0)   (d) (1, 1).

79. The point on x-axis is
(a) (2, 3)   (b) (2, 0)
(c) (0, 2)   (d) (2, 2).

80. The point on y-axis is
(a) (0, 2)   (b) (2, 0)
(c) (2, 2)   (d) (3, 4).

81. The point on the line $y = x$ is
(a) (1, 1)   (b) (0, 1)
(c) (1, 0)   (d) (2, 3).

82. The point on the line $y = 2x$ is
(a) (1, 2)   (b) (2, 1)
(c) (1, 1)   (d) (2, 2).

83. The point (1, 2) lies in the quadrant
(a) I   (b) II
(c) III (d) IV.

84. The point (– 1, 3) lies in the quadrant
(a) I   (b) II
(c) III (d) IV.

85. The point (– 2, – 2) lies in the quadrant
(a) I   (b) II
(c) III (d) IV.

ANSWERS

1. (d)	2. (c)	3. (d)	4. (a)		25. (d)	26. (a)	27. (d)	28. (a)
5. (b)	6. (a)	7. (b)	8. (c)		29. (b)	30. (d)	31. (b)	32. (b)
9. (a)	10. (b)	11. (d)	12. (b)		33. (a)	34. (a)	35. (d)	36. (a)
13. (a)	14. (a)	15. (b)	16. (d)		37. (c)	38. (b)	39. (d)	40. (a)
17. (c)	18. (b)	19. (a)	20. (c)		41. (a)	42. (b)	43. (a)	44. (b)
21. (a)	22. (a)	23. (b)	24. (b)		45. (a)	46. (a)	47. (b)	48. (a)

49. (b)	50. (b)	51. (c)	52. (d)	69. (a)	70. (c)	71. (a)	72. (c)
53. (a)	54. (a)	55. (b)	56. (b)	73. (a)	74. (c)	75. (a)	76. (b)
57. (c)	58. (b)	59. (b)	60. (a)	77. (a)	78. (a)	79. (b)	80. (a)
61. (a)	62. (c)	63. (c)	64. (b)	81. (a)	82. (a)	83. (a)	84. (b)
65. (a)	66. (c)	67. (a)	68. (a)	85. (c).			

HINTS/SOLUTIONS

1. $3x + 4\left(\frac{1}{2}x\right) = 20$
   $\Rightarrow \quad 5x = 20$
   $\Rightarrow \quad x = 4$

2. $2(1) + 3(1) = 2 + 3 = 5$

3. $2(1) + 3(7) = 2 + 21 = 23 \neq 5$

4. Add to get
   $2x = 18 \Rightarrow x = 9$
   Subtract to get
   $2y = 10 \Rightarrow y = 5$

5. $c_1 = 0, c_2 = 0$

6. $3x - x = 26 \Rightarrow 2x = 26$
   $\Rightarrow x = 13$
   $\therefore 3x = 39$

7. $(x + 18°) + x = 180°$
   $\Rightarrow \quad x = 81°$
   $\Rightarrow x + 18° = 99°$

11.
$$\begin{matrix} & x & y & 1 \\ b_1 & c_1 & a_1 & b_1 \\ b_2 & c_2 & a_2 & b_2 \end{matrix}$$

12.
$$\begin{matrix} & x & y & -1 \\ b_1 & c_1 & a_1 & b_1 \\ b_2 & c_2 & a_2 & b_2 \end{matrix}$$

13. $\frac{4}{2} \neq \frac{p}{2} \Rightarrow p \neq 4$

14. $\frac{1}{y} + \frac{1}{x} = 2$
   $\frac{1}{y} - \frac{1}{x} = 6$
   Add to get
   $\frac{2}{y} = 8 \Rightarrow y = \frac{1}{4}$

15. $\frac{a}{l} \neq \frac{b}{m} \Rightarrow am \neq bl$

16. $5 = 2x - 3 \Rightarrow 2x = 8 \Rightarrow x = 4$

17. $2(2) + 2 = 4 + 2 = 6$

18. $\frac{4}{\left(-\frac{4}{3}\right)} + 5y = 7 \Rightarrow -3 + 5y = 7$
   $\Rightarrow \quad 5y = 10 \Rightarrow y = 2$

19. $\frac{3}{x} + 4(1) = 5 \Rightarrow \frac{3}{x} = 1 \Rightarrow x = 3$

20. $\frac{4}{1} + \frac{3}{y} = 5 \Rightarrow \frac{3}{y} = 1 \Rightarrow y = 3$

23. $\frac{3}{k} = -\frac{4}{3} \neq \frac{7}{-5} \Rightarrow k = -\frac{9}{4}$

24. $\frac{1}{3} \neq -\frac{k}{2} \Rightarrow k \neq -\frac{2}{3}$

25. $\frac{4}{8} = \frac{1}{2} = \frac{3}{5k} \Rightarrow k = \frac{6}{5}$

26. $\frac{2}{k} \neq -\frac{3}{5} \Rightarrow k \neq = -\frac{10}{3}$

27. $\frac{k}{16} = \frac{4}{k} = \frac{k-4}{k} \Rightarrow k = 8$

28. $\frac{2}{3} \neq \frac{5}{-k} \Rightarrow k \neq -\frac{15}{2}$

29. $\frac{k}{5} = \frac{2}{-3} \neq \frac{-1}{2} \Rightarrow k = -\frac{10}{3}$

30. $\frac{k-3}{k} = \frac{3}{k} = \frac{k}{12} \Rightarrow k = 6$

31. Subtracting, we get $x = 2 \quad \therefore y = 2$

32. $\frac{c}{12} = \frac{3}{c} \neq \frac{3}{6} \Rightarrow c = -6$

33. $\frac{2}{3} \neq \frac{k}{-5} \Rightarrow k \neq -\frac{10}{3}$

34. $\frac{2}{3} = \frac{k}{-5} \neq \frac{1}{7} \Rightarrow k = -\frac{10}{3}$

35. $\frac{8}{k} = \frac{5}{10} \neq \frac{9}{15} \Rightarrow k = 16$

PAIR OF LINEAR EQUATIONS IN TWO VARIABLES

36. $\dfrac{3}{6} \neq -\dfrac{2}{2}$

37. $\dfrac{3}{9} = \dfrac{-2}{-6} = \dfrac{4}{12}$

38. $\dfrac{2}{4} = \dfrac{-3}{-6} \neq \dfrac{4}{7}$

39. $\dfrac{2}{6} = \dfrac{-3}{-9} = \dfrac{4}{12}$

40. $\dfrac{4}{5} \neq \dfrac{-5}{-4}$

41. $\dfrac{4}{8} = \dfrac{-5}{-10} = \dfrac{3}{6}$

42. $\dfrac{2}{k+2} = \dfrac{3}{6} = \dfrac{4}{3k+2} \Rightarrow k = 2$

43. $\dfrac{a}{a} + \dfrac{b}{b} = 2,\ 1+1 = 2$

44. $\dfrac{a^2}{a} - \dfrac{b^2}{b} = a - b$

45. $b(-a) - a(b) + 2ab = 0$

46. $\dfrac{a^2}{a^2} - \dfrac{b^2}{b^2} = 1 - 1 = 0$

47. $5 + 7 = 12,\ 75 - 57 = 18$

48. $5 + 7 = 12$

    $\dfrac{5}{7+3} = \dfrac{5}{10} = \dfrac{1}{2}$

49. $x^2 - 5x + 6 = 0$ is a quadratic equation

50. $\dfrac{5}{10} = \dfrac{-3}{-6} = \dfrac{9}{18}$

51. x-axis $\perp$ y-axis

52. $\dfrac{2}{4} = \dfrac{-3}{-6} \neq \dfrac{5}{15}$

53. Put $x = 0$ and find y
54. Put $x = 0$ and find y
55. Put $y = 0$ and get x
56. Put $y = 0$ and get x

57. Area = $\dfrac{6 \times 6}{2} = 18$

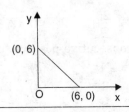

58. Area = $\dfrac{1}{2} ab$

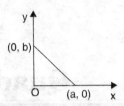

63. $(0, 0)$ satisfies $y = 3x$

64. $y = px - 2$
    $\therefore$ It passes through $(3, 2)$
    $\therefore\ 2 = 3p - 2$
    $\Rightarrow\ p = \dfrac{4}{3}$

65. $5 + 3 = 8,\ 5 - 3 = 2$

66. $2(3) - 3a = 5$
    $\Rightarrow\ 3a = 1$
    $\Rightarrow\ a = \dfrac{1}{3}$

67. Add to get $2x = 18 \Rightarrow x = 9$
    Subtract to get $2y = 10 \Rightarrow y = 5$
    $\therefore$ Solution is $(9, 5)$

68. $2 - 2 = 0,\ 2(2) - 2 = 2$

69. $\dfrac{a^2}{a} + \dfrac{b^2}{b} = a + b$

    $\dfrac{a^2}{a^2} + \dfrac{b^2}{b^2} = 1 + 1 = 2$

70. $3x + 4 = 10 \Rightarrow x = 2$

72. $\dfrac{5}{7} \neq -\dfrac{4}{6}$

73. $\dfrac{9}{18} = \dfrac{3}{6} = \dfrac{12}{24}$

74. $\dfrac{6}{2} = \dfrac{-3}{-1} \neq \dfrac{10}{9}$

75. $\dfrac{3}{2} \neq \dfrac{2}{-3}$
    $\therefore$ unique solution $\therefore$ consistent.

76. $\dfrac{2}{4} = \dfrac{-3}{-6} \neq \dfrac{8}{9}$
    $\therefore$ no solution $\therefore$ inconsistent.

77. $\dfrac{2}{4} = \dfrac{3}{6} \neq \dfrac{-8}{-5}$.

Chapter 4
Quadratic Equations

1. The standard form of a quadratic equation is
   (a) $ax + b = 0, a \neq 0$
   (b) $ax^2 + bx + c = 0, a \neq 0$
   (c) $ax^3 + bx^2 + cx + d = 0, a \neq 0$
   (d) $ax^4 + bx^3 + cx^2 + dx + e = 0, a \neq 0$.

2. Which of the following is not a quadratic equation?
   (a) $(x - 2)^2 + 1 = 2x - 3$
   (b) $x(x + 1) + 8 = (x + 2)(x - 2)$
   (c) $x(2x + 3) = x^2 + 1$
   (d) $(x + 2)^3 = x^3 - 4$.

3. Which of the following is a quadratic equation?
   (a) $(x + 1)^2 = 2(x - 3)$
   (b) $(x - 2)(x + 1) = (x - 1)(x + 3)$
   (c) $x^2 + 3x + 1 = (x - 2)^2$
   (d) $x^4 - 1 = 0$.

4. John and Jivanti together have 45 marbles. Both of them lost 5 marbles each, and the product of the number of marbles they now have is 124. Represent this situation in the form of a quadratic equation.
   (a) $(x - 5)(40 - x) = 124$
   (b) $(x - 5)(x - 40) = 124$
   (c) $(5 + x)(40 - x) = 124$
   (d) $(x + 5)(40 + x) + 124 = 0$.

5. A cottage industry produces a certain number of toys in a day. The cost of production of each toy (in rupees) was found to be 55 minus the number of toys produced in a day. On a particular day, the total cost of production was Rs. 750. Represent this situation in the form of a quadratic equation.
   (a) $x(55 - x) = 750$
   (b) $x(x - 55) = 750$
   (c) $55x = 750$
   (d) $x(x + 55) = 750$.

6. The roots of the quadratic equation $(x - 1)(2x - 3) = 0$ are
   (a) $1, \dfrac{3}{2}$ (b) $-1, -\dfrac{3}{2}$
   (c) $1, -\dfrac{3}{2}$ (d) $-1, \dfrac{3}{2}$.

7. The roots of the quadratic equation $(3x - 2)(2x + 1) = 0$ are
   (a) $-\dfrac{2}{3}, -\dfrac{1}{2}$ (b) $-\dfrac{2}{3}, \dfrac{1}{2}$
   (c) $\dfrac{2}{3}, \dfrac{1}{2}$ (d) $\dfrac{2}{3}, -\dfrac{1}{2}$.

8. The roots of the quadratic equation $(\sqrt{3}x - \sqrt{2})(\sqrt{3}x - \sqrt{2}) = 0$ are
   (a) $\dfrac{\sqrt{2}}{\sqrt{3}}, -\dfrac{\sqrt{2}}{\sqrt{3}}$ (b) $\dfrac{\sqrt{2}}{\sqrt{3}}, \dfrac{\sqrt{2}}{\sqrt{3}}$
   (c) $-\dfrac{\sqrt{2}}{\sqrt{3}}, \dfrac{\sqrt{2}}{\sqrt{3}}$ (d) $-\dfrac{\sqrt{2}}{\sqrt{3}}, -\dfrac{\sqrt{2}}{\sqrt{3}}$.

9. The roots of the quadratic equation $(x - 12)(2x + 25) = 0$ are
   (a) $-12, -\dfrac{25}{2}$ (b) $12, -\dfrac{25}{2}$
   (c) $-12, \dfrac{25}{2}$ (d) $12, \dfrac{25}{2}$.

10. Two consecutive positive integers are
    (a) $x, x + 1$ (b) $x, x + 2$
    (c) $x, x - 2$ (d) $x, 2x$.

QUADRATIC EQUATIONS

11. Two consecutive positive integers differ by
    (a) 2   (b) 1
    (c) 3   (d) 4.

12. Two consecutive even integers are
    (a) $x, x + 2$ (b) $x, x + 1$
    (c) $x, x - 1$ (d) $x, 2x$.

13. The roots of the quadratic equation $x^2 - 2x + 1 = 0$ are
    (a) 1, 1   (b) 1, –1
    (c) –1, –1   (d) 2, 2.

14. The sum of a number and its reciprocal is $\frac{5}{2}$. Represent this situation in the form of a quadratic equation.
    (a) $x + \frac{1}{x} = \frac{5}{2}$ (b) $x - \frac{1}{x} = \frac{5}{2}$
    (c) $x^2 + \frac{1}{x^2} = \frac{5}{2}$ (d) $x + \sqrt{x} = \frac{5}{2}$.

15. The sum of the squares of two consecutive natural numbers is 25. Represent this situation in the form of a quadratic equation.
    (a) $x^2 + (x + 1)^2 = 25$
    (b) $x^2 - (x + 1)^2 = 25$
    (c) $(x + 1)^2 - x^2 = 25$
    (d) $x^2 + (x + 1)^2 + 25 = 0$.

16. The quadratic equation $ax^2 + bx + c = 0$ has two distinct real roots if
    (a) $b^2 = 4ac$ (b) $b^2 > 4ac$
    (c) $b^2 < 4ac$ (d) $b^2 = ac$.

17. The quadratic equation $ax^2 + bx + c = 0$ has two equal real roots if
    (a) $b^2 = 4ac$ (b) $b^2 < 4ac$
    (c) $b^2 > 4ac$ (d) $b^2 = ac$.

18. The quadratic equation $ax^2 + bx + c = 0$ has no real roots if
    (a) $b^2 = 4ac$ (b) $b^2 > 4ac$
    (c) $b^2 < 4ac$ (d) $b^2 = ac$.

19. If $ax^2 + bx + c = 0$, then quadratic formula is

    (a) $x = \frac{-b \pm \sqrt{b^2 - 4ac}}{2a}$
    (b) $x = \frac{-b \pm \sqrt{b^2 + 4ac}}{2a}$
    (c) $x = \frac{-b \pm \sqrt{b^2 - 4ac}}{2}$
    (d) $x = \frac{-b \pm \sqrt{b^2 + 4ac}}{2}$.

20. The discriminant of the quadratic equation $2x^2 - 4x + 3 = 0$ is
    (a) 8   (b) –8
    (c) 4   (d) –4.

21. The roots of the quadratic equation $2x^2 - 4x + 3 = 0$ are
    (a) real and distinct
    (b) real and equal
    (c) no real roots
    (d) reciprocal roots.

22. The discriminant of the quadratic equation $3x^2 - 2x + \frac{1}{3} = 0$ is
    (a) 0   (b) 3
    (c) $\frac{1}{3}$ (d) –2.

23. The roots of the quadratic equation $3x^2 - 2x + \frac{1}{3} = 0$ are
    (a) real and distinct
    (b) real and equal
    (c) not real
    (d) consecutive natural numbers.

24. The value of k so that the quadratic equation $2x^2 + kx + 3 = 0$ has two equal roots is
    (a) $\pm 2\sqrt{6}$ (b) $\pm \sqrt{6}$
    (c) $\pm 2\sqrt{3}$ (d) $\pm \sqrt{3}$.

25. Find the value of k if the roots of the quadratic equation $kx(x - 2) + 6 = 0$ are equal.
    (a) 3   (b) 4
    (c) 5   (d) 6.

26. The roots of the quadratic equation $2x^2 - 8 = 0$ are
    (a) ± 3
    (b) ± 2
    (c) ± 1
    (d) ± 4.

27. The roots of the quadratic equation $\dfrac{x^2 - 8}{x^2 + 20} = \dfrac{1}{2}$ are
    (a) ± 3
    (b) ± 2
    (c) ± 6
    (d) ± 4.

28. The roots of the quadratic equation $3(x + 3)^2 = 48$ are
    (a) 1, − 7
    (b) 1, 7
    (c) − 1, 7
    (d) − 1, − 7.

29. The roots of the quadratic equation $2x^2 - 18 = 0$ are
    (a) ± 1
    (b) ± 2
    (c) ± 3
    (d) ± 4.

30. The roots of the quadratic equation $x^2 - 36 = 0$ are
    (a) ± 3
    (b) ± 4
    (c) ± 5
    (d) ± 6.

31. The roots of the quadratic equation $\dfrac{x}{a} = \dfrac{a}{x}$ are
    (a) a, a
    (b) $a, -a$
    (c) $-a, -a$
    (d) $-a^2, a^2$.

32. The roots of the quadratic equation $(2x + 3)^2 - 4 = 0$ are
    (a) $-\dfrac{1}{2}, -\dfrac{5}{2}$
    (b) $\dfrac{1}{2}, \dfrac{5}{2}$
    (c) $\dfrac{1}{2}, -\dfrac{5}{2}$
    (d) $-\dfrac{1}{2}, \dfrac{5}{2}$.

33. The roots of the quadratic equation $3x^2 = 36$ are
    (a) $\pm 2\sqrt{3}$
    (b) $\pm \sqrt{3}$
    (c) $\pm 3\sqrt{2}$
    (d) $\pm \sqrt{2}$.

34. The roots of the quadratic equation $7x^2 - 9 = 0$ are
    (a) $\pm \dfrac{3}{\sqrt{7}}$
    (b) $\pm \dfrac{3}{7}$
    (c) $\pm \dfrac{\sqrt{3}}{7}$
    (d) ± 3.

35. The roots of the quadratic equation $x - \dfrac{1}{x} = 0$ are
    (a) ± 1
    (b) ± 2
    (c) ± 3
    (d) ± 4.

36. The roots of the quadratic equation $\dfrac{x^2}{a^2 - b^2} = \dfrac{a + b}{a - b}$ are
    (a) $\pm (a + b)$
    (b) $\pm a$
    (c) $\pm b$
    (d) a, b.

37. The roots of the quadratic equation $\dfrac{11}{3 + x} = 4(3 - x)$ are
    (a) $\pm \dfrac{1}{5}$
    (b) $\pm \dfrac{5}{2}$
    (c) ± 2
    (d) ± 5.

38. The roots of the quadratic equation $\dfrac{a}{x^2 - 27} = \dfrac{25}{x^2 - 11}$ are
    (a) ± 3
    (b) ± 4
    (c) ± 6
    (d) ± 5.

39. The roots of the quadratic equation $x^2 + 7x = 0$ are
    (a) 0, − 7
    (b) 0, 7
    (c) 7, − 7
    (d) − 7, − 7.

40. The roots of the quadratic equation $\left(x - \dfrac{1}{2}\right)^2 = 4$ are
    (a) $\dfrac{5}{2}, -\dfrac{3}{2}$
    (b) $\dfrac{5}{2}, \dfrac{3}{2}$
    (c) $-\dfrac{5}{2}, -\dfrac{3}{2}$
    (d) $-\dfrac{5}{2}, \dfrac{3}{2}$.

41. The roots of the quadratic equation $x^2 + 8x + 7 = 0$ are
    (a) − 1, 7
    (b) 1, − 7
    (c) 1, 7
    (d) − 1, − 7.

QUADRATIC EQUATIONS

42. The roots of the quadratic equation $5x^2 + 8x = 0$ are

    (a) $0, -\dfrac{8}{5}$ (b) $0, \dfrac{8}{5}$

    (c) $\dfrac{8}{5}, \dfrac{8}{5}$ (d) $-\dfrac{8}{5}, -\dfrac{8}{5}$.

43. The roots of the quadratic equation $(x + 2)(5x + 6) = 0$ are

    (a) $-2, -\dfrac{6}{5}$ (b) $2, -\dfrac{6}{5}$

    (c) $-2, \dfrac{6}{5}$ (d) $2, \dfrac{6}{5}$.

44. The roots of the quadratic equation $(2x + b + a)(2x + b - a) = 0$ are

    (a) $-\dfrac{a+b}{2}, \dfrac{a-b}{2}$ (b) $\dfrac{a+b}{2}, \dfrac{a-b}{2}$

    (c) $\dfrac{a+b}{2}, \dfrac{a+b}{2}$ (d) $\dfrac{a-b}{2}, \dfrac{a-b}{2}$.

45. Form a quadratic equation whose roots are $2 + \sqrt{5}$ and $2 - \sqrt{5}$.

    (a) $x^2 + 4x + 1 = 0$ (b) $x^2 - 4x - 1 = 0$

    (c) $x^2 + 4x - 1 = 0$ (d) $x^2 - 4x + 1 = 0$.

46. Form a quadratic equation whose roots are 1 and $\dfrac{1}{2}$.

    (a) $x^2 - \dfrac{3}{2}x + \dfrac{1}{2} = 0$

    (b) $x^2 + \dfrac{3}{2}x + \dfrac{1}{2} = 0$

    (c) $x^2 - \dfrac{3}{2}x - \dfrac{1}{2} = 0$

    (d) $x^2 + \dfrac{3}{2}x - \dfrac{1}{2} = 0$.

47. Form a quadratic equation whose roots are a and $\dfrac{1}{a}$.

    (a) $x^2 - \left(a + \dfrac{1}{a}\right)x + 1 = 0$

    (b) $x^2 + \left(a + \dfrac{1}{a}\right)x + 1 = 0$

    (c) $x^2 - \left(a + \dfrac{1}{a}\right)x - 1 = 0$

    (d) $x^2 + \left(a + \dfrac{1}{a}\right) - 1 = 0$.

48. Find the discriminant of the quadratic equation $5y^2 + 12y - 9 = 0$.

    (a) 16^2 (b) 18^2
    (c) 20^2 (d) 15^2.

49. The discriminant of the quadratic equation $2x^2 + 3x + 4 = 0$ is

    (a) 23   (b) -23
    (c) 24   (d) -24.

50. The discriminant of the quadratic equation $x^2 - 4x + 4 = 0$ is

    (a) 0   (b) 4
    (c) -4 (d) -16.

51. If $(x + 4)(x - 4) = 9$, then the values of x are

    (a) ± 5 (b) $\pm \dfrac{1}{5}$

    (c) $\dfrac{1}{5}, \dfrac{1}{5}$ (d) $5, 5$.

52. The roots of the equation $(x - 3)^2 = 3$ are

    (a) $3 \pm \sqrt{3}$ (b) $-3 \pm \sqrt{3}$
    (c) $3, -3$ (d) $\sqrt{3}, -\sqrt{3}$.

53. The product of two consecutive natural numbers is 72. Find the natural numbers.

    (a) 8, 9   (b) 6, 12
    (c) 24, 3   (d) 36, 2.

54. The roots of the quadratic equation $3x^2 - 6x + 5 = 0$ are

    (a) not real
    (b) real and distinct
    (c) real and equal
    (d) real and reciprocal.

55. The roots of the quadratic equation $p^2x^2 + 4pqx + 4q^2$ are

    (a) real and equal
    (b) real and distinct
    (c) not real
    (d) real and reciprocal.

56. The roots of the quadratic equation $5y^2 + 12y - 9 = 0$ are

    (a) real and equal
    (b) real and distinct
    (c) not real
    (d) real and reciprocal.

57. The roots of the quadratic equation $6x^2 - 7x + 2 = 0$ are
    (a) not real
    (b) real and equal
    (c) real and distinct
    (d) real and perfect squares.

58. Form a quadratic equation whose roots are $1+\sqrt{5}$ and $1-\sqrt{5}$.
    (a) $x^2 - 2x - 4 = 0$ (b) $x^2 + 2x + 4 = 0$
    (c) $x^2 + 2x - 4 = 0$ (d) $x^2 - 2x + 4 = 0$.

59. Form a quadratic equation whose roots are a and $-2a$.
    (a) $x^2 + ax + 2a^2 = 0$
    (b) $x^2 + ax - 2a^2 = 0$
    (c) $x^2 - ax - 2a^2 = 0$
    (d) $x^2 - ax + 2a^2 = 0$.

60. Form a quadratic equation whose roots are $3+\sqrt{2}$ and $3-\sqrt{2}$.
    (a) $x^2 - 6x - 7 = 0$
    (b) $x^2 + 6x + 7 = 0$
    (c) $x^2 + 6x - 7 = 0$
    (d) $x^2 - 6x + 7 = 0$.

61. If the equation $x^2 + 4x + k$ has real and distinct roots, then
    (a) $k = 4$ (b) $k \geq 4$
    (c) $k < 4$ (d) $k \leq 4$.

62. If the equation $x^2 - kx + 1 = 0$ has no real roots, then
    (a) $-2 < k < 2$ (b) $-3 < k < 3$
    (c) $-4 < k < 4$ (d) $k > 2$.

63. If p and q are the roots of the equation $x^2 - qx + p = 0$, then
    (a) $p = -2, q = 1$ (b) $p = 1, q = -2$
    (c) $p = 0, q = 1$ (d) $p = 1, q = 0$.

64. If the equation $kx^2 + 2x + k$ has two real equal roots, then
    (a) $k = 0$ (b) $k = \pm 1$
    (c) $k = 2$ (d) $k = -2$.

65. If the quadratic equation $ax^2 + bx + c = 0$ has equal roots, then $c =$
    (a) $\dfrac{b}{2a}$ (b) $-\dfrac{b}{2a}$
    (c) $-\dfrac{b^2}{4a}$ (d) $\dfrac{b^2}{4a}$.

66. The roots of the equation $(a+b)x^2 - 2ax + (a-b) = 0$ are
    (a) real and distinct
    (b) real and equal
    (c) not real
    (d) real and reciprocal.

67. The roots of the equation $x^2 - x - 12 = 0$ are
    (a) rational and unequal
    (b) equal
    (c) not real
    (d) irrational.

68. If the difference of the roots of the quadratic equation $x^2 - ax + b$ is 1, then
    (a) $a^2 - 4b = 0$ (b) $a^2 - 4b = -1$
    (c) $a^2 - 4b = 1$ (d) $a^2 - 4b = 4$.

69. If the roots of the quadratic equation $ax^2 + bx + c$ are $\sin \alpha$ and $\cos \alpha$, then
    (a) $a^2 + b^2 = c^2$ (b) $a^2 - 2bc = b^2$
    (c) $b^2 + 2ac = a^2$ (d) $b^2 - 2ac = a^2$.

70. What is the condition that one root of the quadratic equation $ax^2 + bx + c$ is reciprocal of the other?
    (a) $a = c$ (b) $a = b$
    (c) $b = c$ (d) $a + b + c = 0$.

71. The condition that both the roots of the two quadratic equations $a_1 x^2 + b_1 x + c_1 = 0$ and $a_2 x^2 + b_2 x + c_2 = 0$ are common is that
    (a) $a_1 a_2 + b_1 b_2 + c_1 c_2 = 0$
    (b) $\dfrac{a_1}{a_2} = \dfrac{b_1}{b_2} = \dfrac{c_1}{c_2}$
    (c) $a_1 + b_1 + c_1 = a_2 + b_2 + c_2$
    (d) $a_1^2 + b_1^2 + c_1^2 = a_2^2 + b_2^2 + c_2^2$.

72. If one root of the two quadratic equations $x^2 + ax + b = 0$ and $x^2 + bx + a = 0$ is common, then
    (a) $a + b = 1$ (b) $a + b = -1$
    (c) $ab = 1$ (d) $ab = -1$.

QUADRATIC EQUATIONS

73. If the two roots of the two quadratic equations $x^2 + mx + 1 = 0$ and $ax^2 + bx + a = 0$ are common, then the value of m is
    (a) ab
    (b) b
    (c) $\dfrac{a}{b}$
    (d) $\dfrac{b}{a}$.

74. The quadratic equation whose one root is $3 + \sqrt{5}$ is
    (a) $x^2 - 6x + 4 = 0$
    (b) $x^2 - 6x - 4 = 0$
    (c) $x^2 + 6x + 4 = 0$
    (d) $x^2 + 6x + 5 = 0$.

75. If the sum of the roots of the quadratic equation $3x^2 + (2k + 1)x - (k + 5) = 0$ is equal to the product of the roots, then the value of k is
    (a) 2
    (b) 3
    (c) 4
    (d) 5.

ANSWERS

1. (b)	2. (b)	3. (a)	4. (a)	41. (d)	42. (a)	43. (a)	44. (a)
5. (a)	6. (a)	7. (d)	8. (b)	45. (b)	46. (a)	47. (a)	48. (b)
9. (b)	10. (a)	11. (b)	12. (a)	49. (b)	50. (a)	51. (a)	52. (a)
13. (a)	14. (a)	15. (a)	16. (b)	53. (a)	54. (a)	55. (a)	56. (b)
17. (a)	18. (c)	19. (a)	20. (b)	57. (c)	58. (a)	59. (b)	60. (d)
21. (c)	22. (a)	23. (b)	24. (a)	61. (c)	62. (a)	63. (c)	64. (b)
25. (d)	26. (b)	27. (c)	28. (a)	65. (d)	66. (a)	67. (a)	68. (c)
29. (c)	30. (d)	31. (b)	32. (a)	69. (d)	70. (a)	71. (b)	72. (b)
33. (a)	34. (a)	35. (a)	36. (a)	73. (d)	74. (a)	75. (c).	
37. (b)	38. (c)	39. (a)	40. (a)				

HINTS/SOLUTIONS

1. Definition
2. $x(x + 1) + 8 = (x + 2)(x - 2)$
   $\Rightarrow x^2 + x + 8 = x^2 - 4$
   $\Rightarrow x + 12 = 0$
3. $(x + 1)^2 = 2(x - 3)$
   $\Rightarrow x^2 + 2x + 1 = 2x - 6$
   $\Rightarrow x^2 + 7 = 0$
4. Let John have x marbles. Then, Jivanti has $(45 - x)$ marbles. When 5 marbles are lost by each of them, then,
   John has $(x - 5)$ marbles
   Jivanti has $(45 - x - 5)$ or $(40 - x)$ marbles.
5. Let the cottage industry produce x toys in a day.
   $\therefore$ Cost of production of each toy
   $= $ Rs. $(55 - x)$

6. $x - 1 = 0 \Rightarrow x = 1, 2x - 3 = 0$
   $\Rightarrow x = \dfrac{3}{2}$
7. $3x - 2 = 0 \Rightarrow x = \dfrac{2}{3}, 2x + 1 = 0$
   $\Rightarrow x = -\dfrac{1}{2}$.
8. $\sqrt{3}x - \sqrt{2} = 0 \Rightarrow x = \dfrac{\sqrt{2}}{\sqrt{3}}$.
9. $x - 12 = 0 \Rightarrow x = 12, 2x + 25 = 0$
   $\Rightarrow x = -\dfrac{25}{2}$
10. Definition   11. Definition
12. Definition
13. $x^2 - 2x + 1 = 0$
    $\Rightarrow (x - 1)^2 = 0 \Rightarrow x = 1, 1$

14. Let the number be x. Then, its reciprocal $= \dfrac{1}{x}$.

15. Let the two consecutive natural numbers be x and $x + 1$. Then, their squares are x^2 and $(x + 1)^2$ respectively.

16. Result   17. Result
18. Result   19. Formula.
20. $b^2 - 4ac = (-4)^2 - 4(2)(3) = -8$
21. $\because$ discriminant < 0.
22. $b^2 - 4ac = (-2)^2 - 4(3)\left(\dfrac{1}{3}\right) = 0$
23. $\because$ discriminant $= 0$
24. $b^2 = 4ac \Rightarrow k^2 = 4(2)(3)$
    $\Rightarrow k = \pm 2\sqrt{6}$
25. $b^2 = 4ac$
    $\Rightarrow (-2k)^2 = 4k(6)$
    $\Rightarrow 4k^2 = 24k$
    $\Rightarrow k = 6.$
26. $2x^2 - 8 = 0$
    $\Rightarrow 2x^2 = 8$
    $\Rightarrow x^2 = 4 \Rightarrow x = \pm 2.$
27. $\dfrac{x^2 - 8}{x^2 + 20} = \dfrac{1}{2}$
    $\Rightarrow 2x^2 - 16 = x^2 + 20$
    $\Rightarrow x^2 = 36 \Rightarrow x = \pm 6$
28. $3(x + 3)^2 = 48$
    $\Rightarrow (x + 3)^2 = 16$
    $\Rightarrow x + 3 = \pm 4$
    $\Rightarrow x = 1, -7$
29. $2x^2 - 18 = 0 \Rightarrow 2x^2 = 18$
    $\Rightarrow x^2 = 9 \Rightarrow x = \pm 3$
30. $x^2 - 36 = 0$
    $\Rightarrow x^2 = 36 \Rightarrow x = \pm 6$
31. $\dfrac{x}{a} = \dfrac{a}{x} \Rightarrow x^2 = a^2 \Rightarrow x = \pm a$
32. $(2x + 3)^2 - 4 = 0$
    $\Rightarrow (2x + 3)^2 = 4$
    $\Rightarrow 2x + 3 = \pm 2$
    $\Rightarrow 2x = -1, -5$
    $\Rightarrow x = -\dfrac{1}{2}, -\dfrac{5}{2}$

33. $3x^2 = 36$
    $\Rightarrow x^2 = 12 \Rightarrow x = \pm 2\sqrt{3}$
34. $7x^2 - 9 = 0$
    $\Rightarrow x^2 = \dfrac{9}{7} \Rightarrow x = \pm \dfrac{3}{\sqrt{7}}$
35. $x - \dfrac{1}{x} = 0 \Rightarrow x^2 - 1 = 0$
    $\Rightarrow x^2 = 1 \Rightarrow x = \pm 1.$
36. $\dfrac{x^2}{a^2 - b^2} = \dfrac{a + b}{a - b}$
    $\Rightarrow \dfrac{x^2}{(a - b)(a + b)} = \dfrac{a + b}{a - b}$
    $\Rightarrow \dfrac{x^2}{a + b} = a + b$
    $\Rightarrow x^2 = (a + b)^2$
    $\Rightarrow x = \pm(a + b)$
37. $\dfrac{11}{3 + x} = 4(3x)$
    $\Rightarrow 9 - x^2 = \dfrac{11}{4}$
    $\Rightarrow x^2 = 9 - \dfrac{11}{4} = \dfrac{25}{4}$
    $\Rightarrow x = \pm \dfrac{5}{2}$
38. $\dfrac{9}{x^2 - 27} = \dfrac{25}{x^2 - 11}$
    $\Rightarrow 25x^2 - 675 = ax^2 - 99$
    $\Rightarrow 16x^2 = 576$
    $\Rightarrow x^2 = 36 \Rightarrow x = \pm 6$
39. $x^2 + 7x = 0 \Rightarrow x(x + 7) = 0$
40. $\left(x - \dfrac{1}{2}\right)^2 = 4 \Rightarrow x - \dfrac{1}{2} = \pm 2$
41. $x^2 + 8x + 7 = 0 \Rightarrow (x + 1)(x + 7) = 0$
42. $5x^2 + 8x = 0 \Rightarrow x(5x + 8) = 0$
43. $x + 2 = 0 \Rightarrow x = -2, 5x + 6 = 0$
    $\Rightarrow x = -6/5$
44. $2x + b + a = 0 \Rightarrow x = -\dfrac{a + b}{2}$
    $2x + b - a = 0 \Rightarrow x = \dfrac{a - b}{2}$
45. $(2 + \sqrt{5}) + (2 - \sqrt{5}) = 4$
    $(2 + \sqrt{5})(2 - \sqrt{5}) = 4 - 5 = -1$

46. $1 + \dfrac{1}{2} = \dfrac{3}{2}$, $(1)\left(\dfrac{1}{2}\right) = \dfrac{1}{2}$

47. Sum $= a + \dfrac{1}{a}$, Product $= (a)\left(\dfrac{1}{a}\right) = 1$

48. $b^2 - 4ac = (12)^2 - 4(5)(-9)$
    $= 144 + 180 = 324 = 18^2$

49. $b^2 - 4ac = (3)^2 - 4(2)(4) = -23$

50. $b^2 - 4ac = (-4)^2 - 4(1)(4) = 0$

51. $(x+4)(x-4) = 9 \Rightarrow x^2 - 16 = 9$
    $\Rightarrow x^2 = 25$
    $\Rightarrow x = \pm 5$

52. $(x-3)^2 = 3 \Rightarrow x - 3 = \pm\sqrt{3}$
    $\Rightarrow x = 3 \pm \sqrt{3}$

53. 8 and 9 are consecutive natural numbers

54. $b^2 - 4ac = (-6)^2 - 4(3)(5)$
    $= 36 - 60 = -24 (<0)$

55. $b^2 - 4ac = (4pq)^2 - 4p^2(4q^2) = 0$

56. $b^2 - 4ac = (12)^2 - 4(5)(-9) = 144 + 180$
    $= 324 = 18^2$

57. $b^2 - 4ac = (-7)^2 - 4(6)(2) = 49 - 48 = 1 > 0$

58. Sum $= (1 + \sqrt{5}) + (1 - \sqrt{5}) = 2$
    Product $= (1 + \sqrt{5})(1 - \sqrt{5}) = 1 - 5 = -4$

59. Sum $= a + (-2a) = -a$
    Product $= (a)(-2a) = -2a^2$

60. Sum $= (3 + \sqrt{2}) + (3 - \sqrt{2}) = 6$
    Product $= (3 + \sqrt{2})(3 - \sqrt{2}) = 9 - 2 = 7$

61. $b^2 > 4ac$
    $\Rightarrow (4)^2 > 4(1)(k) \Rightarrow k < 4$

62. $b^2 < 4ac$
    $\Rightarrow (-k)^2 < 4(1)(1)$
    $\Rightarrow k^2 < 4$
    $\Rightarrow k^2 - 4 < 0$
    $\Rightarrow (k-2)(k+2) < 0$
    $\Rightarrow -2 < k < 2$

63. $p + q = q$ ...(1)
    $pq = p$ ...(2)
    Solving (1) and (2), we get $p = 0, q = 1$

64. $b^2 = 4ac \Rightarrow (2)^2 = 4(k)(k) \Rightarrow k = \pm 1$

65. $b^2 = 4ac \Rightarrow c = \dfrac{b^2}{4a}$

66. $B^2 - 4AC$
    $= (-2a)^2 - 4(a+b)(a-b)$
    $= 4b^2 > 0$

67. $b^2 - 4ac = (-1)^2 - 4(1)(-12) = 49 = 7^2$

68. $\alpha - \beta = 1 = \sqrt{(\alpha + \beta)^2 - 4\alpha\beta}$
    $\Rightarrow 1 = \sqrt{a^2 - 4b}$

69. $\sin^2 \alpha + \cos^2 \alpha = 1$
    $\Rightarrow (\sin\alpha + \cos\alpha)^2 - 2\sin\alpha\cos\alpha = 1$
    $\Rightarrow \left(-\dfrac{b}{a}\right)^2 - 2\left(\dfrac{c}{a}\right) = 1$
    $\Rightarrow b^2 - 2ac = a^2$

70. $\alpha\beta = 1 \Rightarrow \dfrac{c}{a} = 1 \Rightarrow c = a$

71. $\alpha + \beta = -\dfrac{b_1}{a_1} = -\dfrac{b_2}{a_2}$
    $\alpha\beta = \dfrac{c_1}{a_1} = \dfrac{c_2}{a_2}$

72. $\alpha^2 + a\alpha + b = 0$ ...(1)
    $\alpha^2 + b\alpha + a = 0$ ...(2)
    Subtraction gives
    $(a-b)\alpha + (b-a) = 0$
    $\Rightarrow \alpha = 1$
    Put $\alpha = 1$ in (1) to get $a + b = -1$

73. $\dfrac{1}{a} = \dfrac{m}{b} = \dfrac{1}{a} \Rightarrow m = \dfrac{b}{a}$

74. Other roots $= 3 - \sqrt{5}$
    Sum $= 6$, Product $= 4$

75. $\alpha + \beta = \alpha\beta$
    $-(2k+1) = -(k+5)$
    $\Rightarrow k = 4$

Chapter 5
Arithmetic Progressions

1. Which of the following is an A.P.?
   (a) $a, a + d, a + 2d, a + 3d, \ldots$
   (b) $a, ar, ar^2, ar^3, \ldots$
   (c) $a, a^2, a^3, a^4, \ldots$
   (d) $\dfrac{1}{a}, \dfrac{1}{a+d}, \dfrac{1}{a+2d}, \dfrac{1}{a+3d}, \ldots$

2. Which of the following list of numbers does not form an A.P.?
   (a) 1, 1, 1, 2, 2, 2, 3, 3, 3, ...
   (b) 4, 10, 16, 22, ...
   (c) 1, – 1, – 3, – 5, ...
   (d) – 5, – 1, 3, 7, ...

3. Which of the following list of numbers does not form an A.P.?
   (a) 3, 3, 3, 3, ...
   (b) – 1.0, – 1.5, – 2.0, – 2.5, ...
   (c) 1, 2, 3, 4, ...
   (d) – 2, 2, – 2, 2, – 2, ...

4. Which of the following list of numbers forms an A.P.?
   (a) 2, 4, 8, 16, ...
   (b) 0.2, 0.22, 0.222, 0.2222, ...
   (c) 1, 3, 9, 27, ...
   (d) – 10, – 6, – 2, 2, ...

5. Which of the following list of numbers forms an A.P.?
   (a) $\sqrt{3}, \sqrt{6}, \sqrt{9}, \sqrt{12}, \ldots$
   (b) $1^2, 3^2, 5^2, 7^2, \ldots$
   (c) $1^2, 5^2, 7^2, 73, \ldots$
   (d) $a, a^2, a^3, a^4, \ldots$.

6. In an A.P. $a, a + d, a + 2d, a + 3d, \ldots$, what is a called?
   (a) common difference
   (b) common ratio
   (c) base
   (d) first term.

7. In an A.P. $a, a + d, a + 2d, a + 3d, \ldots$, what is d called?
   (a) common ratio
   (b) common difference
   (c) first term
   (d) last term.

8. What is nth term of the A.P. $a, a + d, a + 2d, a + 3d, \ldots$?
   (a) $a + nd$
   (b) $a + (n + 1)d$
   (c) $a + (n – 1)d$
   (d) $a + (2n – 1)d$.

9. What is the last term of the A.P. $a, a + d, a + 2d, a + 3d, \ldots$ containing m terms?
   (a) $a + (m – 1)d$
   (b) $a + md$
   (c) $a + (m + 1)d$
   (d) $a + (2m + 1)d$.

10. What is the sum of the first n terms of the A.P. $a, a + d, a + 2d, a + 3d, \ldots$?
    (a) $\dfrac{n}{2}[2a + (n + 1)d]$
    (b) $\dfrac{n}{2}[2a + (n – 1)d]$
    (c) $\dfrac{n}{2}[a + (n – 1)d]$
    (d) $\dfrac{n}{2}[a + (n + 1)d]$.

11. What is the sum of the following A.P. if there are n terms in the A.P.?

$a, a+d, a+2d, \ldots, l$.

(a) $\frac{n}{2}(a+l)$ (b) $\frac{n}{2}(2a+l)$

(c) $n(a+l)$ (d) $n(2a+l)$.

12. If a, b, c are in A.P., then
   (a) $2b = a+c$ (b) $b = a+c$
   (c) $b = ac$ (d) $b = \sqrt{ac}$.

13. The arithmetic mean of a and b is
   (a) $a+b$ (b) $\frac{a+b}{2}$
   (c) $\sqrt{ab}$ (d) ab.

14. What is the sum of first n natural numbers?
   (a) $\frac{n(n+1)}{2}$ (b) n^2
   (c) $\frac{n(n-1)}{2}$ (d) $\left\{\frac{n(n+1)}{2}\right\}^2$.

15. What is the 100th term of the series $1, 1, 1, 1, 1, \ldots$?
   (a) 1   (b) 100
   (c) 50   (d) 200.

16. a, b, c are in A.P. The arithmetic mean of a and b is x. The arithmetic mean of b and c is y. Then, the arithmetic mean of x and y is
   (a) a (b) b
   (c) c (d) $a+c$.

17. The A.P. with first term 10 and common difference 5 is
   (a) $10, 2, \frac{2}{5}, \frac{2}{25}, \ldots$
   (b) $10, 50, 250, 1250, \ldots$
   (c) $10, 5, 0, -5, -10, \ldots$
   (d) $10, 15, 20, 25, 30, \ldots$.

18. The A.P. with first term –100 and common difference 30 is
   (a) $-100, -70, -40, -10, 20, \ldots$
   (b) $-100, -130, -160, -190, -220, \ldots$
   (c) $-100, -160, -220, -280, \ldots$
   (d) $-100, -40, 20, 80, \ldots$.

19. The A.P. with first term 1 and common difference $\frac{1}{4}$ is
   (a) $1, \frac{3}{4}, \frac{1}{2}, \frac{1}{4}, 0, \ldots$
   (b) $1, \frac{5}{4}, \frac{3}{2}, \frac{7}{4}, 2, \ldots$
   (c) $1, \frac{1}{2}, 0, -\frac{1}{2}, -1, \ldots$
   (d) $1, \frac{3}{2}, 2, \frac{5}{2}, \ldots$.

20. The A.P. with first term 2 and common difference –1 is
   (a) $2, 0, -2, -4, \ldots$
   (b) $2, 4, 6, 8, \ldots$
   (c) $2, 1, 0, -1, -2, -3, \ldots$
   (d) $2, 3, 4, 5, 6, 7, \ldots$.

21. If d is the common difference of an A.P. whose kth term is a_k, then $a_{k+1} - a_k$ is equal to
   (a) $2d$ (b) d
   (c) 2 (d) 1.

22. If the common difference of an A.P. is d, then $a_3 - a_1$ is equal to
   (a) d (b) $2d$
   (c) $3d$ (d) $4d$.

23. What is the common difference of the A.P. $7, 3, -1, -5, -9, \ldots$
   (a) -2 (b) 2
   (c) 4 (d) -4.

24. The sum of $1 + 2 + 3 + \ldots + 100$ is
   (a) 3030   (b) 4040
   (c) 5050   (d) 6060.

25. There are n terms in an A.P. Then, fourth term from the beginning is which term from the last?
   (a) $(n-3)$th   (b) $(n-2)$th
   (c) $(n-1)$th   (d) $(n-4)$th.

26. How many middle terms are there in an A.P. whose number of terms is odd?
   (a) 2   (b) 1
   (c) 3   (d) 4.

27. How many middle terms are there in an A.P. whose number of terms is even?
    (a) 1 (b) 2
    (c) 4 (d) 3.

28. In an A.P. of n terms, where n is odd, which term is the middle term?
    (a) $\left(\dfrac{n+1}{2}\right)$th (b) $\left(\dfrac{n+3}{2}\right)$th
    (c) $\left(\dfrac{n-1}{2}\right)$th (d) nth.

29. In an A.P. of n terms, where n is even, which two terms are the middle terms?
    (a) $\left(\dfrac{n}{2}\right)$th and $\left(\dfrac{n}{2}+1\right)$th
    (b) nth and $(n-1)$th
    (c) $\left(\dfrac{n}{2}\right)$th and $\left(\dfrac{n}{2}-1\right)$th
    (d) $\left(\dfrac{n}{2}\right)$th and $\left(\dfrac{n}{2}+2\right)$th.

30. Find the third term of the sequence whose nth term is given by
    $$a_n = 3 + \dfrac{2}{3}n$$
    (a) 2 (b) 3
    (c) 4 (d) 5.

31. What is the common difference of the A.P. $a-b, a, a+b, \ldots$?
    (a) a (b) b
    (c) $a-b$ (d) $a+b$.

32. Find the next term of the A.P. 51, 59, 67, 75, ...
    (a) 91 (b) 83
    (c) 85 (d) 93.

33. What is the next term of the A.P. 75, 67, 59, 51, ...?
    (a) 41 (b) 42
    (c) 43 (d) 44.

34. The 11th term of 10.0, 10.5, 11.0, 11.5, ... is
    (a) 13 (b) 14
    (c) 15 (d) 16.

35. Find k, if the given value of x is the kth term of the given A.P.
    25, 50, 75, 100, ... ; $x = 1000$
    (a) 20 (b) 40
    (c) 50 (d) 100.

36. Find k, if the given value of x is the kth term of the given A.P.
    $-1, -3, -5, -7, \ldots$; $x = -151$
    (a) 75 (b) 77
    (c) 76 (d) 78.

37. Find $a_{30} - a_{20}$ for the A.P.
    $-9, -14, -19, -24, \ldots$
    (a) 50 (b) -50
    (c) 25 (d) -25.

38. Find $a_{30} - a_{20}$ for the A.P.
    $a, a+d, a+2d, a+3d, \ldots$
    (a) $5d$ (b) $-5d$
    (c) $10d$ (d) $-10d$.

39. Two A.P.'s have the same common difference. The first term of one of these is 3, and that of the other is 8. What is the difference between their 2nd terms?
    (a) 2 (b) 3
    (c) 4 (d) 5.

40. Two A.P.'s have the same common difference. The first term of one of these is -1, and that of the other is -3. What is difference between their second term?
    (a) 2 (b) 1
    (c) -2 (d) -1.

41. Two A.P.'s have the same common difference. The difference between their 100th term is 111 222 333. What is the difference between their millionth terms?
    (a) 111 222 333 (b) 222 444 666
    (c) 333 666 999 (d) 333 222 111.

42. Two A.P.'s have the same common difference. The difference between

their millionth terms is 111 222 333. What is the difference between their 100th terms?
   (a) 111 222 333   (b) 333 222 111
   (c) 222 333 111   (d) 333 111 222.
43. Find the sum, 2 + 4 + 6 + ... + 200
   (a) 5050   (b) 10100
   (c) 15150   (d) 20200.
44. nth term of a sequence is $a + nb$. Find $a_4 - a_3$.
   (a) a (b) b
   (c) $a + b$ (d) $2b$.
45. Which A.P. does the sequence with nth term $6 - n$ form?
   (a) 5, 4, 3, 2, 1, ... (b) 6, 7, 8, 9, 10, ...
   (c) 5, 3, 1, ...   (d) 6, 8, 10, ...
46. Find the sum of first 20 terms of the A.P. 3, 3, 3, 3, ...
   (a) 30   (b) 60
   (c) 90   (d) 120.
47. Which term of the A.P. 21, 18, 15, ... is 0?
   (a) 6th   (b) 7th
   (c) 8th   (d) 9th.
48. Determine the A.P. whose 3rd term is 5 and 7th term is 9.
   (a) 1, 2, 3, 4, ... (b) 2, 3, 4, 5, ...
   (c) 3, 4, 5, 6, ... (d) 4, 5, 6, 7, ...
49. How many two-digit numbers are divisible by 3?
   (a) 10   (b) 20
   (c) 40   (d) 30.
50. Find the 11th term from the last term of the A.P.:
   10, 7, 4, ..., – 62
   (a) 32   (b) – 32
   (c) 16   (d) – 16.
51. In a flower bed, there are 23 rose plants in the first row, 21 in the second, 19 in the third, and so on. There are 5 rose plants in the last row. How many rows are there in the flower bed?
   (a) 4   (b) 6
   (c) 8   (d) 10.
52. 30th term of the A.P. 10, 7, 4, ... is
   (a) 97   (b) 77
   (c) – 77 (d) – 87.
53. 11th term of the A.P. $-3, -\frac{1}{2}, 2, ...$ is
   (a) 28   (b) 22
   (c) – 38 (d) $-48\frac{1}{2}$.
54. Find the missing term in the box
   2, ☐, 26
   (a) 12   (b) 14
   (c) 16   (d) 10.
55. Find the missing terms in the boxes:
   ☐, 13, ☐, 3
   (a) 18, 8   (b) 14, 9
   (c) 16, 11 (d) 17, 12.
56. Find the missing terms in the boxes:
   $5, ☐, ☐, 9\frac{1}{2}$
   (a) 4, 3   (b) 6, 7
   (c) $7\frac{1}{2}, 10$ (d) $6\frac{1}{2}, 8$.
57. For what value of n, are the nth terms of two A.P.'s : 63, 65, 67, ... and 3, 10, 17, ... equal?
   (a) 11   (b) 12
   (c) 13   (d) 14.
58. If $a = 7, d = 3, n = 8$, find a_n.
   (a) 25   (b) 26
   (c) 27   (d) 28.
59. If $a = -18, n = 10, a_n = 0$, find d.
   (a) 0   (b) 1
   (c) 2   (d) – 2.
60. If $d = -3, n = 18, a_n = -5$, find a.
   (a) 46   (b) 47
   (c) 45   (d) 40.
61. If $a = -18.9, d = 2.5, a_n = 3.6$, find n.
   (a) 8   (b) 9
   (c) 10   (d) 12.

62. If $a = 3.5$, $d = 0$, $n = 105$, find a_n.
(a) 2.5 (b) 3.5
(c) 4.5 (d) 5.

63. If 6 times the 6th term of an A.P. is equal 7 times the 7th term, then 13th term of the A.P. is
(a) 0 (b) 6
(c) 7 (d) 13.

64. If pth, qth and rth terms of an A.P. are a, b, c respectively, then
$a(q-r) + b(r-p) + c(p-q) =$
(a) 0 (b) $a + b + c$
(c) $p + q + r$ (d) pqr.

65. Find the number of odd numbers between 0 and 50.
(a) 24 (b) 25
(c) 23 (d) 26.

66. If $l = 28$, $s = 144$, $n = 9$, find a.
(a) 1 (b) 2
(c) 3 (d) 4.

67. If $a = 3$, $n = 8$, $s = 192$, find d.
(a) 3 (b) 4
(c) 5 (d) 6.

68. Find $2 + 7 + 12 + \ldots$ to 10 terms.
(a) 245 (b) 255
(c) 250 (d) 235.

69. Find the sum of first 15 multiples of 8.
(a) 960 (b) 1000
(c) 940 (d) 1060.

70. The A.P. whose sum of n terms is $2n^2 + 2$ is given by
(a) 3, 7, 11, 15, ... (b) 3, 6, 9, 12, ...
(c) 3, 8, 13, 18, ... (d) 3, 4, 5, 6, ...

ANSWERS

1. (a)	2. (a)	3. (d)	4. (d)	37. (b)	38. (c)	39. (d)	40. (a)
5. (c)	6. (d)	7. (b)	8. (c)	41. (a)	42. (a)	43. (b)	44. (b)
9. (a)	10. (b)	11. (a)	12. (a)	45. (a)	46. (b)	47. (c)	48. (c)
13. (b)	14. (a)	15. (a)	16. (b)	49. (d)	50. (b)	51. (d)	52. (c)
17. (d)	18. (a)	19. (b)	20. (c)	53. (b)	54. (b)	55. (a)	56. (d)
21. (b)	22. (b)	23. (d)	24. (c)	57. (c)	58. (d)	59. (c)	60. (a)
25. (a)	26. (b)	27. (b)	28. (a)	61. (c)	62. (b)	63. (a)	64. (a)
29. (a)	30. (d)	31. (b)	32. (b)	65. (b)	66. (d)	67. (d)	68. (a)
33. (c)	34. (c)	35. (b)	36. (c)	69. (a)	70. (a).		

HINTS/SOLUTIONS

1. Standard form
2. $1 - 1 = 0$, $2 - 1 = 1$
3. $2 - (-2) \neq -2 - 2$
4. $-6 - (-10) = -2 - (-6) = 2 - (-2) = 4$
5. $5^2 - 1^2 = 7^2 - 5^2 = 73 - 7^2 = 24$
6. Definition   7. Definition
8. Formula   9. Formula
10. Formula   11. Formula
12. $b - a = c - b \Rightarrow 2b = a + c$
13. Definition
14. $a = 1$, $l = n$
15. Each term is 1
16. $2b = a + c$
$2x = a + b$
$2y = b + c$
$x + y = \dfrac{2x + 2y}{2} = \dfrac{(a+b) + (b+c)}{2}$
$= \dfrac{a + c + 2b}{2} = \dfrac{2b + 2b}{2} = 2b$
$\therefore \dfrac{x+y}{2} = b$

ARITHMETIC PROGRESSIONS

17. A.P. is 10, 10 + 5, 10 + 5 + 5, 10 + 5 + 5 + 5, ...
18. A.P. is $-100, -100 + 30, -100 + 2(30), -100 + 3(30)$, ...
19. A.P. is $1, 1 + \frac{1}{4}, 1 + 2\left(\frac{1}{4}\right), 1 + 3\left(\frac{1}{4}\right)$, ...
20. A.P. is $2, 2 + (-1), 2 + 2(-1), 2 + 3(-1)$, ...
21. a_k and a_{k+1} are successive terms
22. $a_3 = a_1 + 2d$
23. $d = 3 - 7 = -4$
24. Sum $= \frac{n}{2}(a + l) = \frac{100}{2}(1 + 100) = 5050$
25. $\{(n - 4) + 1\}$th term
30. Put $n = 3$,
$$a_3 = 3 + \frac{2}{3}(3) = 5$$
31. $d = a - (a - b) = b$
32. $d = 59 - 51 = 8$
next term $= 75 + 8 = 83$
33. $d = 67 - 75 = -8$
next term $= 51 + (-8) = 43$
34. 11th term $= a + 10d$
$= 10 + 10(0.5) = 15$
35. $a_k = 1000$
$\Rightarrow \quad a + (k - 1)d = 1000$
$\Rightarrow \quad 25 + (k - 1)25 = 1000$
$\Rightarrow \quad k = 40$
36. $a_k = -151$
$\Rightarrow \quad a + (k - 1)d = -151$
$\Rightarrow \quad (-1) + (k - 1)(-2) = -151$
$\Rightarrow \quad 1 + 2(k - 1) = 151$
$\Rightarrow \quad 2(k - 1) = 150$
$\Rightarrow \quad k - 1 = 75$
$\Rightarrow \quad k = 76$
37. $d = -14 - (-9) = -5$
$30 - 20 = 10$
$a_{30} - a_{20} = 10d = 10(-5) = -50$
38. $a_{30} - a_{20} = 10d$ | $30 - 20 = 10$
39. $a_1 = 3, d_1 = d$
$a_2 = 8, d_2 = d$

$A_2 \sim A_1 = (a_1 + d_1) \sim (a_2 + d_2)$
$= a_1 \sim a_2 = 5$
40. $a_1 = -1, a_2 = -3$
$d_1 = d, d_2 = d$
Difference $= (-1) \sim (-3) = 2$
41. 100th term of I A.P. $= a_1 + 99d$
100th term of II A.P. $= a_2 + 99d$
Difference $= |a_1 - a_2| = 111222333$
∴ Difference between their millionth terms
$= |a_1 - a_2| = 111222333$
42. Difference between millionth terms
$= |a_1 - a_2| = 111222333$
Difference between 100th terms
$= |a_1 - a_2| = 111222333$
43. Sum $= \frac{n}{2}(a + l) = \frac{100}{2}(2 + 200)$
$= 10100$
44. $a_4 - a_3 = (a + 4b) - (a + 3b) = b$
45. Put $n = 1, 2, 3, ...$ in succession. The A.P. is 5, 4, 3, 2, 1, ...
46. Sum $= 3 \times 20 = 60$
$\because$ Each term is same
$\therefore d = 0$
47. $a = 21, d = -3$
$a + (n - 1)d = 0$
$\Rightarrow 21 + (n - 1)(-3) = 0 \Rightarrow n = 8$
48. $a_3 = 5 \Rightarrow a + 2d = 5$ ...(1)
$a_7 = 9 \Rightarrow a + 6d = 9$ ...(2)
Solve (1) and (2), to get
$a = 3, d = 1$
49. Two digit numbers divisible by 3 are 12, 15, 18, ..., 99
$a = 12, d = 3, l = 99$
use $l = a + (n - 1)d$
50. Write A.P. inverse order
$-62, -59, -56, -53, ...$
$a = -62$
$d = 3$
11th term $= a + 10d$
$= -62 + 10(3) = -32$

51. 23, 21, 19, ..., 5
    $a = 23$
    $d = -2$
    $l = 5$
    Use $l = a + (n-1)d$

52. $a = 10$
    $d = -3$
    $a_{30} = a + 29d = 10 - 87 = -77$

53. $a = -3$
    $d = \dfrac{5}{2}$
    $a_{11} = a + 10d = 22$

54. $\square - 2 = 26 - \square \Rightarrow \square = 14$

55. $a + d = 13$
    $a + 3d = 3$
    Solve to get $a = 18, d = -5$

56. $a = 5$
    $a + 3d = 9\dfrac{1}{2} \Rightarrow 5 + 3d = \dfrac{19}{2}$
    $\Rightarrow 3d = \dfrac{9}{2} \Rightarrow d = \dfrac{3}{2}$

57. $63 + (n-1)2 = 3 + (n-1)7 \Rightarrow n = 13$

58. $a_n = a + (n-1)d$

59. $a_n = a + (n-1)d$

60. $a_n = a + (n-1)d$

61. $a_n = a + (n-1)d$

62. $a_n = a + (n-1)d$

63. $6a_6 = 7a_7$,
    $6(a + 5d) = 7(a + 6d)$
    $\Rightarrow a + 12d = 0 \Rightarrow a_{13} = 0$

64. $\{A + (p-1)D\}(q-r) + \{A + (q-1)D\}(r-p) + \{A + (r-1)D\}(p-q) = 0$

65. 1, 3, 5, ..., 49
    $a = 1, d = 2, l = 49$
    $l = a + (n-1)d$

66. Use $s = \dfrac{n}{2}(a + l)$

67. Use $s = \dfrac{n}{2}[2a + (n-1)d]$

68. Use $s = \dfrac{n}{2}[2a + (n-1)d]$

69. $a = 8, d = 8, n = 15$
    Use $s = \dfrac{n}{2}[2a + (n-1)d]$

70. $s_1 = 3, s_2 = 10, s_3 = 21, s_4 = 36$;
    $a_1 = s_1 = 3$
    $a_2 = s_2 - s_1 = 7$
    $a_3 = s_3 - s_2 = 11$ and so on.

Chapter 6
Triangles

1. If in two triangles, corresponding angles are equal, then the two triangles are similar. This criterion is known as
   (a) SSS similarity criterion
   (b) SAS similarity criterion
   (c) AA similarity criterion
   (d) AAA similarity criterion.

2. If in two triangles, two angles of one triangle are respectively equal to the two angles of the other triangle, then the two triangles are similar. This criterion is known as
   (a) SSS similarity criterion
   (b) SAS similarity criterion
   (c) AA similarity criterion
   (d) AAA similarity criterion.

3. If in two triangles, corresponding sides are in the same ratio, then the two triangles are similar. This criterion is known as
   (a) SSS similarity criterion
   (b) SAS similarity criterion
   (c) AA similarity criterion
   (d) AAA similarity criterion.

4. If one angle of a triangle is equal to one angle of the other triangle, and the sides including these angles are proportional, then the two triangles are similar. This criterion is known as
   (a) SSS similarity criterion
   (b) SAS similarity criterion
   (c) AA similarity criterion
   (d) AAA similarity criterion.

5. If a line is drawn parallel to one side of a triangle, to intersect the other two sides in distinct points, then the other two sides are divided in the same ratio. This theorem is known as
   (a) Basic Proportionality theorem
   (b) Pythagoras theorem
   (c) Mid-point theorem
   (d) Proportional Intercepts theorem.

6. In a right triangle, the square of the hypotenuse is equal to the sum of the squares of the other two sides. This theorem is known as
   (a) Converse of Pythagoras theorem
   (b) Pythagoras theorem
   (c) Thales theorem
   (d) Converse of Thales theorem.

7. The ratio of the areas of two similar triangles is equal to
   (a) the ratio of their corresponding sides
   (b) the square of the ratio of their corresponding sides
   (c) the cube of the ratio of their corresponding sides
   (d) the ratio of their corresponding altitudes.

8. ABC and BDE are two equilateral triangles such that D is the mid-point of BC. Ratio of the areas of triangles ABC and BDE is
   (a) 2 : 1   (b) 1 : 2
   (c) 4 : 1   (d) 1 : 4.

9. Side of two similar triangles are in the ratio 4 : 9. Areas of these triangles are in the ratio
   (a) 2 : 3
   (b) 4 : 9
   (c) 81 : 16
   (d) 16 : 81.

10. In Δ ABC, AB = $6\sqrt{3}$, AC = 12 cm and BC = 6 cm. The angle B is
    (a) 120°
    (b) 60°
    (c) 90°
    (d) 45°.

11. In the following figure, DE ∥ BC, AD = 4 cm, DB = 6 cm and AE = 5 cm. Then, the value of EC is
    (a) 6.5 cm
    (b) 7.0 cm
    (c) 7.5 cm
    (d) 8.0 cm.

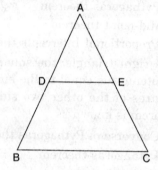

12. In the following figure, AD bisects ∠A. AB = 6 cm, BD = 8 cm, DC = 6 cm. Then, the value of AC is
    (a) 4.0 cm
    (b) 4.5 cm
    (c) 5 cm
    (d) 5.5 cm.

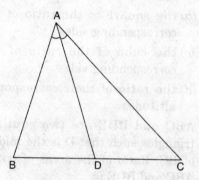

13. In the following figure, DE ∥ BC. Find the value of x.

    (a) $\sqrt{5}$
    (b) $\sqrt{6}$
    (c) $\sqrt{3}$
    (d) $\sqrt{7}$.

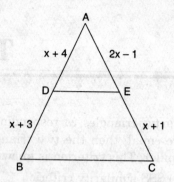

14. In the following figure, ∠BAD = ∠CAD, AB = 3.4 cm, BD = 4 cm, BC = 10 cm. Then, the value of AC is
    (a) 5.1 cm
    (b) 3.4 cm
    (c) 6 cm
    (d) 5.3 cm.

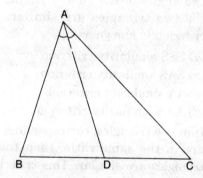

15. The areas of two similar triangles are 25 cm² and 36 cm². If the median of the smaller triangle is 10 cm, then the median of the larger triangle is
    (a) 12 cm
    (b) 15 cm
    (c) 10 cm
    (d) 18 cm.

16. In a trapezium ABCD, AB ∥ CD and its diagonals intersect at P. If AB = 6 cm and DC = 3 cm, then find the ratio of the areas of Δ AOB and Δ COD.
    (a) 4 : 1
    (b) 1 : 2
    (c) 2 : 1
    (d) 1 : 4.

TRIANGLES

17. In △ABC and △DEF, ∠A = 50°, ∠B = 70°, ∠C = 60°, ∠D = 60°, ∠E = 70°, ∠F = 50°. Then which of the following is correct?
    (a) △ABC ~ △DEF (b) △ABC ~ △EDF
    (c) △ABC ~ △DFE (d) △ABC ~ △FED.

18. If △ABC ~ △DEF and AB = 10 cm, DE = 8 cm, then ar (△ABC) : ar (△DEF) =
    (a) 25 : 16   (b) 16 : 25
    (c) 4 : 5   (d) 5 : 4.

19. In △ABC, points D and E lie on sides AB and AC such that DE ∥ BC and AD = 8 cm, AB = 12 cm, AE = 12 cm. Then, the length CE is
    (a) 6 cm   (b) 18 cm
    (c) 9 cm   (d) 15 cm.

20. The length of the shadow of a 12 cm long vertical rod is 8 cm. At the same time, the length of the shadow of a tower is 40 m. Find the height of the tower.
    (a) 60 m   (b) 60 cm
    (c) 40 cm   (d) 80 cm.

21. In a triangle ABC, AB = 3.9 cm, AC = 5.2 cm and the bisector of ∠A meets BC at D. If DC = 2.8 cm, then find BD.
    (a) 2.4 cm   (b) 2.3 cm
    (c) 2.1 cm   (d) 1.4 cm.

22. In the figure, the length of the hypotenuse of a right triangle ABC is 8 cm. Find the length of the median BM.
    (a) 2 cm   (b) 3 cm
    (c) 4 cm   (d) 5 cm.

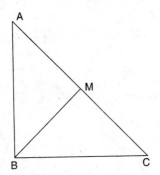

23. A man goes 15 m due east and then 20 m due north. Find his distance from the starting point.
    (a) 35 m   (b) 5 m
    (c) 25 m   (d) 15 m.

24. In the figure, the sides AB and AC of △ABC are 2.5 cm and 3.5 cm respectively. If AD is the bisector of ∠BAC, then find BD : DC.
    (a) 4 : 5   (b) 7 : 5
    (c) 6 : 7   (d) 5 : 7.

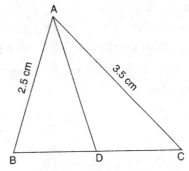

25. In figure, in △ABC, AD is a median of △ABC. P is the mid-point of AD. Then, the ratio of AE and AC is
    (a) 1 : 1   (b) 1 : 2
    (c) 1 : 3   (d) 2 : 3.

26. In figure, AD is the bisector line of ∠BAC. AB = 2 cm, AC = 4 cm, BD = 1.2 cm. Then, the length of BC is
    (a) 2.4 cm   (b) 3.6 cm
    (c) 4.8 cm   (d) 6.0 cm.

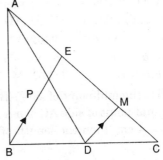

27. The perimeters of two similar triangles are 30 cm and 20 cm. If one altitude of the former triangle is 12 cm, then the length of the corresponding altitude of the latter triangle is
    (a) 8 cm   (b) 10 cm
    (c) 12 cm   (d) 15 cm.

28. The perimeters of two similar triangles are 40 cm and 50 cm. Then, the ratio of the areas of the first and second triangles is
    (a) 4 : 5   (b) 5 : 4
    (c) 25 : 16   (d) 16 : 25.

29. In two triangles ABC and DEF, $\angle A = \angle E$ and $\angle B = \angle F$. Then, $\dfrac{AB}{AC}$ is equal to
    (a) $\dfrac{DE}{DF}$ (b) $\dfrac{ED}{EF}$
    (c) $\dfrac{EF}{ED}$ (d) $\dfrac{EF}{DF}$.

30. In a right triangle ABC, $\angle C = 90°$, AC = 3 cm and BC = 4 cm. Then, the length of the median through point C is
    (a) 2.5 cm   (b) 3.0 cm
    (c) 3.5 cm   (d) 4.0 cm.

31. If $\triangle$ ABC, $\angle B$ is a right angle. BD is perpendicular to AC. If AD = a and CD = b, then AB^2 is equal to
    (a) $a(a + b)$ (b) $b(a + b)$
    (c) $b(b - a)$ (d) ab.

32. If D is the mid-point of side AB and E is the mid-point of side AC of a $\triangle$ ABC and DE = 4 cm, then the length of the side BC is
    (a) 4 cm   (b) 6 cm
    (c) 8 cm   (d) 10 cm.

33. The length of the diagonal of a square is $7\sqrt{2}$ cm. Then, the area of the square in cm^2 is
    (a) 28   (b) $14\sqrt{2}$
    (c) 21   (d) 49.

34. In figure, in $\triangle$ ABC, DE ∥ BC, AD = 1 cm, AE = 2 cm and EC = 6 cm. Find length DB.
    (a) 2 cm   (b) 4 cm
    (c) 3 cm   (d) 6 cm.

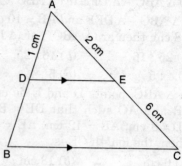

35. In the figure, the points D and E divide the side AB and AC in the ratio 1 : 3. If DE = 2.4 cm, then the length of BC is
    (a) 4.8 cm   (b) 7.2 cm
    (c) 9.6 cm   (d) 12.0 cm.

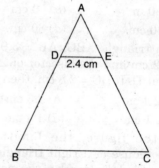

36. In the figure, PT is the bisector of $\angle$P. If PQ = 10.5 cm and PR = 7 cm, then the ratio QT : TR is
    (a) 2 : 3   (b) 2 : 1
    (c) 3 : 1   (d) 3 : 2.

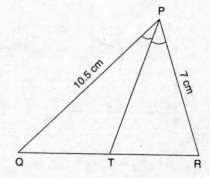

37. The sides of a triangle are 6 cm, 8 cm and 10 cm. The largest angle of the triangle is
    (a) acute angle (b) right angle
    (c) obtuse angle (d) reflex angle.

38. If the sides of a triangle are 6 cm, 8 cm and 12 cm, then the nature of the triangle will be
    (a) right angled
    (b) obtuse angled
    (c) acute angled
    (d) triangle is not possible.

39. The height of an equilateral triangle of side a is
    (a) $\dfrac{a}{2}$ (b) $a\sqrt{3}$
    (c) $\dfrac{a\sqrt{3}}{2}$ (d) $\dfrac{a\sqrt{3}}{4}$.

40. The sides AB and AC of a right triangle ABC are such that AB = 5 cm and AC = 13 cm. If $\angle B = 90°$, then the length of BC is
    (a) 10 cm (b) 12 cm
    (c) 18 cm (d) 8 cm.

41. $\triangle$ PQR is formed by joining the mid-points of the sides of $\triangle$ ABC. Then, the ratio of the areas of $\triangle$ PQR and $\triangle$ ABC is
    (a) 4 : 1 (b) 1 : 4
    (c) 2 : 1 (d) 1 : 2.

42. In $\triangle$ ABC, a line DE, drawn parallel to base BC, intersects the sides AB and AC at points D and E respectively. If AD = 2, BD = 4 and DE = 2, then base BC is equal to
    (a) 2 (b) 4
    (c) 6 (d) 8.

43. The areas of two similar triangles are 36 cm^2 and 64 cm^2. If one side of the former triangle is 6 cm, then the corresponding side of the latter triangle will be
    (a) 6 cm (b) 8 cm
    (c) 10 cm (d) 12 cm.

44. D and E are the mid-points of the sides AB and AC of $\triangle$ ABC. If DE measures 3 cm, then the side BC measures
    (a) 6 cm (b) 7 cm
    (c) 8 cm (d) 9 cm.

45. The ratio between the corresponding sides of two similar triangles is 1 : 2. Find the ratio of their areas.
    (a) 1 : 2 (b) 1 : 3
    (c) 1 : 4 (d) 2 : 3.

46. The areas of two similar triangles are 16 cm^2 and 24 cm^2. If the square of one side of the first triangle is 12 cm^2, then find the square of the corresponding side of the second triangle.
    (a) 8 cm^2 (b) 12 cm^2
    (c) 18 cm^2 (d) 24 cm^2.

47. The lengths of the diagonals of a rhombus are 8 cm and 6 cm. Find the length of each side of the rhombus.
    (a) 2 cm (b) 3 cm
    (c) 4 cm (d) 5 cm.

48. Which of the following are the sides of a right triangle?
    (a) 5, 12, 13 (b) 3, 6, 8
    (c) 7, 25, 26 (d) 50, 80, 100.

49. $\triangle$ ABC is an isosceles triangle in which $\angle C = 90°$. If BC = 4 cm, find AB.
    (a) $4\sqrt{2}$ cm (b) $2\sqrt{2}$ cm
    (c) 4 cm (d) 2 cm.

50. In a triangle, the internal bisector of an angle bisects the opposite side. Find the nature of the triangle.
    (a) right angled (b) equilateral
    (c) scalene (d) isosceles.

51. Which of the following pairs shows similar figures
    (a) △ ○ (b) △ △
    (c) □ □ (d) □ ▱

52. If $\triangle$ ABC ~ $\triangle$ DEF, $\angle C = 60°$, $\angle B = 75°$, then $\angle F =$
    (a) 45° (b) 75°
    (c) 60° (d) 90°.

ANSWERS

1. (d)	2. (c)	3. (a)	4. (b)	29. (c)	30. (a)	31. (a)	32. (c)
5. (a)	6. (b)	7. (b)	8. (c)	33. (d)	34. (c)	35. (b)	36. (d)
9. (d)	10. (c)	11. (c)	12. (b)	37. (b)	38. (b)	39. (c)	40. (b)
13. (d)	14. (a)	15. (a)	16. (a)	41. (b)	42. (c)	43. (b)	44. (a)
17. (d)	18. (a)	19. (a)	20. (a)	45. (c)	46. (c)	47. (d)	48. (a)
21. (c)	22. (c)	23. (c)	24. (d)	49. (a)	50. (d)	51. (c)	52. (c).
25. (c)	26. (b)	27. (a)	28. (d)				

HINTS/SOLUTIONS

1. Theorem
2. Theorem
3. Theorem
4. Theorem
5. Theorem
6. Theorem
7. Theorem
8. $\because$ $\triangle ABC$ and $\triangle BDE$ are equilateral
   $\therefore$ They are similar
   $\therefore$ $\dfrac{ar(\triangle ABC)}{ar(\triangle BDE)} = \left(\dfrac{AB}{BD}\right)^2$
   $= \left(\dfrac{BC}{BD}\right)^2 = \left(\dfrac{2BD}{BD}\right)^2 = 4 : 1$

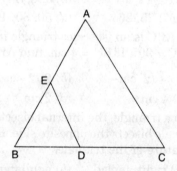

9. Ratio of areas $= \left(\dfrac{4}{9}\right)^2 = 16 : 81$

10. $AB^2 + BC^2 = (6\sqrt{3})^2 + (6)^2 = 108 + 36$
    $= 144 = (12)^2 = AC^2$
    $\therefore$ $\angle B = 90°$
    | By converse of Pythagoras theorem

11. $\because$ DE ∥ BC
    $\therefore$ $\dfrac{AD}{DB} = \dfrac{AE}{EC}$ | By Thales theorem

    $\Rightarrow$ $\dfrac{4}{6} = \dfrac{5}{EC}$ $\Rightarrow$ EC = 7.5 cm.

12. $\because$ AD bisects $\angle A$
    $\therefore$ $\dfrac{AB}{AC} = \dfrac{BD}{DC}$
    $\Rightarrow$ $\dfrac{6}{AC} = \dfrac{8}{6}$ $\Rightarrow$ AC = 4.5 cm

13. $\because$ DE ∥ BC
    $\therefore$ $\dfrac{AD}{DB} = \dfrac{AE}{EC}$
    $\Rightarrow$ $\dfrac{x+4}{x+3} = \dfrac{2x-1}{x+1}$ $\Rightarrow$ $x = \sqrt{7}$

14. $\because$ AD bisects $\angle A$
    $\therefore$ $\dfrac{AB}{AC} = \dfrac{BD}{DC}$
    $\Rightarrow$ $\dfrac{3.4}{AC} = \dfrac{4}{6}$ $\Rightarrow$ AC = 5.1 cm

15. $\because$ The ratio of the areas of two similar triangles is equal to the square of the ratio of their corresponding medians
    $\therefore$ $\dfrac{ar\ (smaller\ triangle)}{ar\ (larger\ triangle)}$
    $= \left(\dfrac{median\ of\ smaller\ triangle}{median\ of\ larger\ triangle}\right)^2$
    $\Rightarrow$ $\dfrac{25}{36} = \left(\dfrac{10}{median\ of\ larger\ triangle}\right)^2$
    $= 12$ cm

TRIANGLES

16. Δ OAB ~ Δ OCD
    | AAA criterion of similarity

    ∴ $\frac{ar(\Delta OAB)}{ar(\Delta OCD)} = \left(\frac{AB}{CD}\right)^2 = \left(\frac{6}{3}\right)^2 = 4:1$

17. Δ ABC ~ Δ FED
    | AAA criterion of similarity
    ∵ Δ ABC ~ Δ DEF

18. ∴ $\frac{ar(\Delta ABC)}{ar(\Delta DEF)} = \left(\frac{AB}{DE}\right)^2 = \left(\frac{10}{8}\right)^2 = \frac{25}{16}$
    $= 25:16$

    ∵ The ratio of the areas of two similar triangles is equal to the square of the ratio of their corresponding sides

19. ∵ DE ∥ BC
    ∴ By Thales theorem,
    $\frac{AD}{DE} = \frac{AE}{EC}$
    ⇒ $\frac{8}{12-8} = \frac{12}{CE}$ ⇒ CE = 6 cm

20. $\frac{\text{Length of rod}}{\text{Length of shadow of the rod}}$
    $= \frac{\text{Length of tower}}{\text{Length of shadow of the tower}}$
    ⇒ $\frac{12}{8} = \frac{\text{Length of tower}}{40}$
    ⇒ Length of tower = 60 m

21. ∵ AD bisects ∠A
    ∴ $\frac{AB}{AC} = \frac{BD}{DC}$ ⇒ $\frac{3.9}{5.2} = \frac{BD}{2.8}$
    ⇒ BD = 2.1 cm.

22. ∵ The mid-point of the hypotenuse of a right triangle is equidistant from its vertices
    ∴ AM = BM = CM
    ∴ BM = $\frac{1}{2}$AC = 4 cm.

23. OB = $\sqrt{OA^2 + AB^2}$
    | By Pythagoras theorem

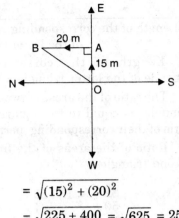

    $= \sqrt{(15)^2 + (20)^2}$
    $= \sqrt{225 + 400} = \sqrt{625} = 25.$

24. ∵ AD bisects ∠BAC
    ∴ $\frac{AB}{AC} = \frac{BD}{DC}$ ⇒ $\frac{2.5}{3.5} = \frac{BD}{DC}$
    ⇒ $\frac{5}{7} = \frac{BD}{DC}$.

25. In Δ CBE,
    ∵ D is the mid-point of BC
    and DM ∥ BE
    ∴ M is the mid-point of EC
    | By Thales theorem
    ∴ CM = ME   ...(1)
    In Δ ADM,
    ∵ P is the mid-point of AD
    and PE ∥ DM
    ∴ E is the mid-point of AM
    | By Thales theorem
    ∴ AE = EM   ...(2)
    From (1) and (2),
    AE = EM = MC
    ∴ AE = $\frac{1}{3}$ AC.

26. ∵ AD bisects ∠BAC
    ∴ $\frac{AB}{AC} = \frac{BD}{DC}$ ⇒ $\frac{2}{4} = \frac{1.2}{DC}$
    ⇒ DC = 2.4 cm ∴ AC = 3.6 cm.

27. ∵ The ratio of the perimeters of two similar triangles is equal to the ratio of the corresponding altitudes of the two triangles
    ∴ $\frac{30}{20}$

$$= \frac{12}{\text{Length of the corresponding altitude of the latter triangle}}$$

⇒ Length of the corresponding altitude of the latter triangle = 8 cm.

28. ∵ The ratio of the areas of two similar triangles is equal to the square of the ratio of their corresponding perimeters

∴ Ratio of the areas of the first and second triangles

$$= \left(\frac{40}{50}\right)^2 = 16 : 25$$

29. △ABC ~ △EFD

| AA criterion of similarity

∴ $\frac{AB}{EF} = \frac{AC}{ED}$

| ∵ Corresponding sides of two similar triangles are proportional

⇒ $\frac{AB}{AC} = \frac{EF}{ED}$

30. $AB^2 = AC^2 + BC^2$

| By Pythagoras theorem
$= 3^2 + 4^2 = 5^2$

⇒ AB = 5 cm

∴ Length of the median through the point C

$= \frac{5}{2}$ cm = 2.5 cm

| ∵ The mid-point of the hypotenuse of a right triangle is equidistant from its vertices

31. $AB^2 = AD \cdot AC = a(a + b)$

32. △ ADE ~ △ ABC

∴ $\frac{AD}{AB} = \frac{DE}{BC}$

| ∵ Corresponding sides of two similar triangles are proportional

⇒ $\frac{AD}{2AD} = \frac{4}{BC}$ ⇒ BC = 8 cm

33. Side $= \frac{\text{diagonal}}{\sqrt{2}} = 7$ cm

∴ area = 7 × 7 = 49 cm²

34. ∵ DE ∥ BC

∴ $\frac{AD}{DB} = \frac{AE}{EC}$ | By Thales theorem

⇒ $\frac{1}{DB} = \frac{2}{6}$ ⇒ DB = 3 cm

35. ∵ $\frac{AD}{DB} = \frac{AE}{EC}$ $\left(= \frac{1}{3} \text{ each}\right)$

∴ By converse of Thales theorem,
DE ∥ BC

∴ △ ADE ~ △ ABC

∴ $\frac{AD}{AB} = \frac{DE}{BC}$

| ∵ Corresponding sides of two similar triangles are proportional

⇒ $\frac{1}{3} = \frac{2.4}{BC}$ ⇒ BC = 7.2 cm

36. ∵ PT bisects ∠P

∴ $\frac{PQ}{PR} = \frac{QT}{TR}$ ⇒ $\frac{10.5}{7} = \frac{QT}{TR}$

⇒ QT : TR = 3 : 2

37. ∵ $6^2 + 8^2 = 10^2$

∴ By converse of Pythagoras theorem
The angle between sides of length 6 cm and 8 cm = 90°

which is the largest angle and a right angle.

38. Let AB = 6 cm, BC = 8 cm and CA = 12 cm. Then,
$CA^2 > AB^2 + BC^2$

∴ ∠C is obtuse

∴ △ ABC is obtuse angled.

39. $AD^2 = AB^2 - BD^2$

| By Pythagoras theorem

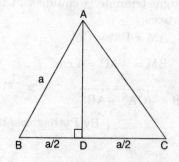

TRIANGLES 611

$= a^2 - \dfrac{a^2}{4} = \dfrac{3a^2}{4}$

$\Rightarrow \quad AD = \dfrac{a\sqrt{3}}{2}$

40. $\because \quad \angle B = 90°$

$\therefore \quad AC^2 = AB^2 + BC^2$

| By Pythagoras theorem

$\Rightarrow \quad 13^2 = 5^2 + BC^2$

$\Rightarrow \quad BC = 12$ cm

41. $\triangle ABC \sim \triangle PQR$

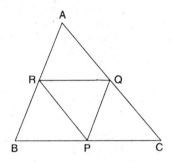

$\therefore \quad \dfrac{ar(\triangle PQR)}{ar(\triangle ABC)} = \left(\dfrac{PQ}{AB}\right)^2$

| ∵ The ratio of the areas of two similar triangles is equal to the square of the ratio of their corresponding sides

$= \left(\dfrac{PQ}{2PQ}\right)^2 = \dfrac{1}{4} = 1:4$

42. $\because \quad DE \parallel BC$

$\therefore \quad \triangle ADE \sim \triangle ABC$

$\therefore \quad \dfrac{AD}{AE} = \dfrac{DE}{BC}$

$\Rightarrow \quad \dfrac{AD}{AD+DE} = \dfrac{DE}{BC}$

$\Rightarrow \quad \dfrac{2}{2+4} = \dfrac{2}{BC} \Rightarrow BC = 6.$

43. The ratio of the areas of two similar triangles is equal to the square of the ratio of their corresponding sides

$\therefore \quad \dfrac{36}{64} = \left(\dfrac{6}{\text{Corresponding side of the latter triangle}}\right)^2$

$\Rightarrow$ Corresponding side of the latter triangle = 8 cm.

44. $\because$ D and E are the mid-points of AB and AC

$\therefore \quad DE \parallel BC$

$\therefore \quad \triangle ADE \sim \triangle ABC$

$\therefore \quad \dfrac{AD}{AB} = \dfrac{DE}{BC}$

$\Rightarrow \quad \dfrac{AD}{2AD} = \dfrac{4}{BC}$

$\Rightarrow \quad BC = 8$ cm

45. Required ratio $= \left(\dfrac{1}{2}\right)^2 = 1:4$

46. $\dfrac{16}{24} = \dfrac{12}{\text{square of the corresponding side of the second triangle}}$

$\Rightarrow$ square of the corresponding side of the second triangle = 18 cm^2

| ∵ The ratio of the areas of two similar triangles is equal to the ratio of the squares of their corresponding sides

47. We know that the diagonals of a rhombus bisect and are perpendicular to each other

$\therefore$ Length of each side $= \sqrt{\left(\dfrac{8}{2}\right)^2 + \left(\dfrac{6}{2}\right)^2}$

$= 5$ cm | By Pythagoras theorem

48. $5^2 + 12^2 = 13^2$

49. AC = BC = 4 cm

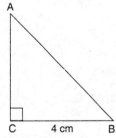

$\therefore \quad AB^2 = AC^2 + BC^2$

| By Pythagoras theorem

$= 4^2 + 4^2$

$\Rightarrow \quad AB = 4\sqrt{2}$ cm

50. ∵ AD bisects ∠A

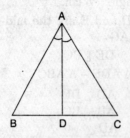

∴ $\dfrac{AB}{AC} = \dfrac{AD}{CD} = 1$

⇒ AB = AC

51. Two squares are similar.

52. ∠F = ∠C = 60°.

Chapter 7
Coordinate Geometry

1. The number coordinate axes in a plane is
   (a) 1 (b) 2
   (c) 3 (d) 4.

2. The coordinates of the point of intersection of x-axis and y-axis are
   (a) (0, 0) (b) (0, 1)
   (c) (1, 0) (d) (1, 1).

3. The coordinates of origin are
   (a) (0, 0) (b) (0, 1)
   (c) (1, 0) (d) (1, 1).

4. For each point on x-axis, y-coordinate is equal to
   (a) 1 (b) 2
   (c) 3 (d) 0.

5. For each point on y-axis, x-coordinate is equal to
   (a) 1 (b) 2
   (c) 3 (d) 0.

6. The distance of the point (3, 4) from x-axis is
   (a) 3 (b) 4
   (c) 1 (d) 7.

7. The distance of the point (3, 4) from y-axis is
   (a) 3 (b) 4
   (c) 7 (d) 1.

8. The distance of the point (3, 4) from origin is
   (a) 3 (b) 4
   (c) 5 (d) 1.

9. The distance of the point (5, – 2) from x-axis is
   (a) 1 (b) 2
   (c) 3 (d) 4.

10. The distance of the point (– 2, 5) from y-axis is
    (a) 1 (b) 4
    (c) 2 (d) 3.

11. The angle between x-axis and y-axis is
    (a) 0° (b) 45°
    (c) 60° (d) 90°.

12. The distance of the point (α, β) from origin is
    (a) $\alpha + \beta$ (b) $\alpha^2 + \beta^2$
    (c) $\sqrt{\alpha^2 - \beta^2}$ (d) $\sqrt{\alpha^2 + \beta^2}$.

13. The distance between the point (x_1, y_1) and (x_2, y_2) is
    (a) $\sqrt{(x_2 - x_1)^2 + (y_2 - y_1)^2}$
    (b) $\sqrt{(x_2 + x_1)^2 + (y_2 - y_1)^2}$
    (c) $\sqrt{(x_2 - x_1)^2 + (y_2 + y_1)^2}$
    (d) $\sqrt{(x_2 + x_1)^2 + (y_2 + y_1)^2}$.

14. Which of the following is the distance formula?
    (a) The distance between two points (x_1, y_1) and (x_2, y_2) is
    $\sqrt{(x_2 - x_1)^2 + (y_2 - y_1)^2}$
    (b) The mid-point of the line segment joining the points (x_1, y_1) and (x_2, y_2) is $\left(\dfrac{x_1 + x_2}{2}, \dfrac{y_1 + y_2}{2}\right)$
    (c) The coordinates of the centroid of the triangle ABC where A, B, C are $(x_1, y_1), (x_2, y_2), (x_3, y_3)$ respectively

are $\left(\dfrac{x_1+x_2+x_3}{3}, \dfrac{y_1+y_2+y_3}{3}\right)$

(d) The coordinates of the point dividing the line segment joining the points (x_1, y_1) and (x_2, y_2) internally in the ratio $m_1 : m_2$ are $\left(\dfrac{m_1 x_2 + m_2 x_1}{m_1 + m_2}, \dfrac{m_1 y_2 + m_2 y_1}{m_1 + m_2}\right)$.

15. The distance between the points $(0, 3)$ and $(-2, 0)$ is

    (a) $\sqrt{14}$ (b) $\sqrt{15}$
    (c) $\sqrt{13}$ (d) $\sqrt{5}$.

16. The triangle with vertices $(-2, 1)$, $(2, -2)$ and $(5, 2)$ is

    (a) scalene   (b) equilateral
    (c) isosceles   (d) right angled.

17. The point equidistant from the points $(0, 0), (2, 0)$ and $(0, 2)$ is

    (a) $(1, 2)$ (b) $(2, 1)$
    (c) $(2, 2)$ (d) $(1, 1)$.

18. If ABCD is a square where A $\rightarrow (0, 0)$, B $\rightarrow (2, 0)$, D $\rightarrow (0, 2)$, then find the coordinates of C.

    (a) $(1, 2)$ (b) $(2, 1)$
    (c) $(2, 2)$ (d) $(1, 1)$.

19. The coordinates of the point, dividing the join of the points $(5, 0)$ and $(0, 4)$ in the ratio $2 : 3$ internally, are

    (a) $\left(3, \dfrac{8}{5}\right)$ (b) $\left(1, \dfrac{4}{5}\right)$
    (c) $\left(\dfrac{5}{2}, \dfrac{3}{4}\right)$ (d) $\left(2, \dfrac{12}{5}\right)$.

20. The coordinates of the point, which divides the join of (x_1, y_1) and (x_2, y_2) in the ratio $m_1 : m_2$ internally, are

    (a) $\left(\dfrac{m_1 x_2 + m_2 x_1}{m_1 + m_2}, \dfrac{m_1 y_2 + m_2 y_1}{m_1 + m_2}\right)$

    (b) $\left(\dfrac{m_1 x_2 - m_2 x_1}{m_1 - m_2}, \dfrac{m_1 y_2 - m_2 y_1}{m_1 - m_2}\right)$

    (c) $\left(\dfrac{m_1 x_2 + m_2 x_1}{m_1 - m_2}, \dfrac{m_1 y_2 + m_2 y_1}{m_1 - m_2}\right)$

    (d) $\left(\dfrac{m_1 x_2 - m_2 x_1}{m_1 + m_2}, \dfrac{m_1 y_2 - m_2 y_1}{m_1 + m_2}\right)$.

21. The coordinates of the mid-point of the line-segment joining the points (x_1, y_1) and (x_2, y_2) are

    (a) $(x_1 + x_2, y_1 + y_2)$
    (b) $\left(\dfrac{x_1+x_2}{2}, \dfrac{y_1+y_2}{2}\right)$
    (c) $\left(\dfrac{x_1+x_2}{4}, \dfrac{y_1+y_2}{4}\right)$
    (d) $(x_1 x_2, y_1 y_2)$.

22. The coordinates of the centroid of the triangle with vertices $(0, 0), (3a, 0)$ and $(0, 3b)$ are

    (a) $(-a, -b)$ (b) $(-a, b)$
    (c) $(a, -b)$ (d) (a, b).

23. If the points $(0, 0), (a, 0)$ and $(0, b)$ are collinear, then

    (a) $a = b$ (b) $a + b = 0$
    (c) $ab = 0$ (d) $a \neq b$.

24. The area of the triangle whose vertices are $(0, 0), (a, 0)$ and $(0, b)$ is

    (a) ab (b) $\dfrac{1}{2} ab$
    (c) $a + b$ (d) $a^2 + b^2$.

25. If the points $(1, 2), (-1, x)$ and $(2, 3)$ are collinear, then the value of x is

    (a) 2   (b) 0
    (c) -1 (d) 1.

26. If the distance between the points $(3, 9)$ and $(4, 1)$ is $\sqrt{10}$, then, find the values of a.

    (a) $3, -1$ (b) $2, -2$
    (c) $4, -2$ (d) $5, -3$.

27. If the point (x, y) is equidistant from the points $(2, 1)$ and $(1, -2)$, then

    (a) $x + 3y = 0$ (b) $3x + y = 0$
    (c) $x + 2y = 0$ (d) $2y + 3x = 0$.

COORDINATE GEOMETRY

28. The closed figure with vertices (– 2, 0), (2, 0), (2, 2), (0, 4) and (– 2, 2) is a
    (a) triangle (b) quadrilateral
    (c) pentagon (d) hexagon.

29. In what ratio does the point (3, 4) divide the line segment joining the points (1, 2) and (6, 7)?
    (a) 1 : 2 (b) 2 : 3
    (c) 3 : 4 (d) 1 : 1.

30. The opposite vertices of a square are (5, – 4) and (– 3, 2). The length of its diagonal is
    (a) 6 (b) 8
    (c) 10 (d) 12.

31. One end of a line is (4, 0) and its middle point is (4, 1). Find the coordinates of the other end.
    (a) (8, 2) (b) (4, 2)
    (c) (4, 8) (d) (8, 4).

32. The distance of the mid-point of the line-segment joining the points (6, 8) and (2, 4) from the point (1, 2) is
    (a) 3 (b) 4
    (c) 5 (d) 6.

33. Find the ratio in which the point (4, 8) divide the line-segment joining the points (5, 7) and (3, 9).
    (a) 1 : 1 (b) 1 : 2
    (c) 2 : 1 (d) 1 : 3.

34. Find the ratio in which the point (4, 8) divide the line-segment joining the points (8, 6) and (0, 10).
    (a) 1 : 1 (b) 1 : 2
    (c) 2 : 1 (d) 3 : 1.

35. If OPQR is a rectangle where O is the origin and P (3, 0) and R (0, 4), then the coordinates of Q are
    (a) (3, – 4) (b) (3, 4)
    (c) (4, 3) (d) (– 3, – 4).

36. If the coordinates of P and Q are ($a \cos \theta$, $b \sin \theta$) and ($- a \sin \theta$, $b \cos \theta$), then $OP^2 + OQ^2 =$
    (a) $a^2 + b^2$ (b) $a + b$
    (c) ab (d) $2ab$.

37. Find the coordinates of the centroid of the triangle whose vertices are (8, – 5), (– 4, 7) and (11, 13).
    (a) (2, 2) (b) (3, 3)
    (c) (4, 4) (d) (5, 5).

38. The coordinates of vertices A, B and C of a triangle ABC are (0, – 1), (2, 1) and (0, 3). Find the length of the median through B.
    (a) 1 (b) 2
    (c) 3 (d) 4.

39. The vertices of a triangle are (3, 5), (5, 7) and (– 5, 9). Find the x-coordinate of the centroid of the triangle.
    (a) 0 (b) 1
    (c) 2 (d) 3.

40. The vertices of a triangle are (3, 6), (2, – 2) and (1, 2). Find the y-coordinate of the centroid of the triangle.
    (a) 0 (b) 1
    (c) 2 (d) – 2.

41. Two vertices of a triangle are (3, 5) and (– 4, – 5). If the centroid of the triangle is (4, 3), find the third vertex.
    (a) (13, 9) (b) (9, 13)
    (c) (13, – 9) (d) (– 9, – 13).

42. The vertices of a triangle are (4, y), (6, 9) and (x, 4). The coordinates of its centroid are (3, 6). Find the values of x and y.
    (a) – 1, – 5 (b) 1, – 5
    (c) 1, 5 (d) – 1, 5.

43. The coordinates of the mid-point of the line-segment joining (– 8, 13) and (x, 7) is (4, 10). Find the value of x.
    (a) 16 (b) 10
    (c) 4 (d) 8.

44. What is the area of the triangle formed by the points (0, 0), (3, 0) and (0, 4)?
    (a) 6 (b) 12
    (c) 3 (d) 24.

45. Which of the points A (1, 3), B (– 3, 2), C(3, 4) and D(4, 1) is nearest to the origin?
    (a) A   (b) B
    (c) C   (d) D

46. If a vertex of a parallelogram is (2, 3) and the diagonals cut at (3, – 2), find the opposite vertex.
    (a) (4, – 7)   (b) (4, 7)
    (c) (– 4, 7)   (d) (– 4, – 7).

47. In which quadrant does the point (– 3, – 3) lie?
    (a) I   (b) II
    (c) III   (d) IV.

48. Find the value of k if the distance between $(k, 3)$ and $(2, 3)$ is 5.
    (a) 5   (b) 6
    (c) 7   (d) 8.

49. What is the coordination that A, B, C are the successive points of a line?
    (a) AB + BC = AC
    (b) BC + CA = AB
    (c) CA + AB = BC
    (d) AB + BC = 2AC.

50. If the points $(a, 0)$, $(0, b)$ and $(1, 1)$ are collinear, then $\frac{1}{a} + \frac{1}{b} =$
    (a) 0   (b) 1
    (c) 2   (d) – 1.

51. If the centroid of the triangle (a, b), (b, c) and (c, a) is O(0, 0), then the value of $a^3 + b^3 + c^3$ is
    (a) 0   (b) abc
    (c) $3abc$ (d) $2abc$.

52. The distance between the points ($a \cos \theta + b \sin \theta$, 0) and (0, $a \sin \theta - b \cos \theta$) is
    (a) $a^2 + b^2$ (b) $\sqrt{a^2 + b^2}$
    (c) $a^2 - b^2$ (d) $a + b$.

53. The distance between the points ($a \cos \alpha$, 0) and (0, $a \sin \alpha$) is
    (a) a (b) a^2
    (c) $\sqrt{a}$ (d) 1.

54. The distance between the points $(\sin \theta, \sin \theta)$ and $(\cos \theta, - \cos \theta)$ is
    (a) $\sqrt{2}$ (b) 2
    (c) $\sqrt{3}$ (d) 1.

55. Three consecutive vertices of a parallelogram are (– 2, 1), (1, 0) and (4, 3). Find the fourth vertex.
    (a) (1, 4)   (b) (1, – 2)
    (c) (– 1, 2)   (d) (– 1, – 2).

56. If (– 2, – 1), $(a, 0)$, $(4, b)$ and (1, 2) are the vertices of a parallelogram, then the values of a and b are
    (a) 1, 3   (b) 1, 4
    (c) 2, 3   (d) 3, 1.

ANSWERS

1. (b)   2. (a)   3. (a)   4. (d)   29. (b)   30. (c)   31. (b)   32. (c)
5. (d)   6. (b)   7. (a)   8. (c)   33. (a)   34. (a)   35. (b)   36. (a)
9. (b)   10. (c)   11. (d)   12. (d)   37. (d)   38. (b)   39. (b)   40. (c)
13. (a)   14. (a)   15. (c)   16. (c)   41. (a)   42. (a)   43. (a)   44. (a)
17. (d)   18. (c)   19. (a)   20. (a)   45. (a)   46. (a)   47. (c)   48. (c)
21. (b)   22. (d)   23. (c)   24. (b)   49. (a)   50. (b)   51. (c)   52. (b)
25. (b)   26. (c)   27. (a)   28. (c)   53. (a)   54. (a)   55. (a)   56. (a).

COORDINATE GEOMETRY

HINTS/SOLUTIONS

1. x-axis and y-axis are the coordinate axes.
2. Point of intersection is O, the origin.
3. O $\to$ (0, 0).
4. We do not have to move along y-axis.
5. We do not have to more alone x-axis.
6. y-coordinate = 4.
7. x-coordinate = 3.
8. Distance = $\sqrt{(3)^2 + (4)^2} = 5$
9. $|y\text{-coordinate}| = 2$
10. $|x\text{-coordinate}| = 2$
11. y-axis is perpendicular to x-axis.
12. Distance = $\sqrt{\alpha^2 + \beta^2}$
13. Distance formula.
14. Distance formula.
15. $d = \sqrt{(-2-0)^2 + (0-3)^2} = \sqrt{13}$
16. AB = $\sqrt{(2+2)^2 + (-2-1)^2} = 5$
    BC = $\sqrt{(5-2)^2 + (2+2)^2} = 5$
    CA = $\sqrt{(-2-5)^2 + (1-2)^2} = \sqrt{10}$
    $\therefore$ AB = BC.
17.

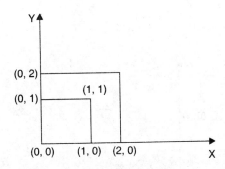

18. Clearly, C $\to$ (2, 2)

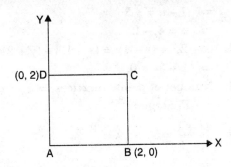

19. $x = \dfrac{2(0) + 3(5)}{2+3} = 3$

    $y = \dfrac{2(4) + 3(0)}{2+3} = \dfrac{8}{5}$

20. Internal division formula.
21. Mid-point formula.
22. $x = \dfrac{0 + 3a + 0}{3} = a$

    $y = \dfrac{0 + 0 + 3b}{3} = b$.

23. $0(0 - b) + a(b - 0) + 0(0 - 0) = 0$
    $\Rightarrow ab = 0$.
24. Area = $\dfrac{ab}{2}$

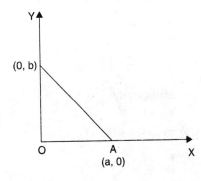

25. $1(x - 3) - 1(3 - 2) + 2(2 - x) = 0$
    $\Rightarrow x - 3 - 3 + 2 + 4 - 2x = 0$
    $\Rightarrow x = 0$

26. $(4-3)^2 + (1-a)^2 = (\sqrt{10})^2$
    $\Rightarrow (1-a)^2 = 9$
    $\Rightarrow 1-a = \pm 3$
    $\Rightarrow a = 4, -2.$

27. $(x-2)^2 + (y-1)^2 = (x-1)^2 + (y+2)^2$
    $\Rightarrow x + 3y = 0.$

28. Number of points (vertices) = 5
    $\therefore$ Pentagon.

29. $\dfrac{6\lambda + 1}{\lambda + 1} = 3 \Rightarrow \lambda = \dfrac{2}{3}$

    $\dfrac{7\lambda + 2}{\lambda + 1} = 4 \Rightarrow \lambda = \dfrac{2}{3}.$

30. Length = $\sqrt{(-3-5)^2 + (2+4)^2} = 10$

31. Let other end $\to (x, y)$. Then
    $\dfrac{x+4}{2} = 4 \Rightarrow x = 4$
    $\dfrac{y+0}{2} = 1 \Rightarrow y = 2.$

32. Mid-point $\to \left(\dfrac{6+2}{2}, \dfrac{8+4}{2}\right)$ i.e., $(4, 6).$
    Its distance from $(1, 2)$
    $= \sqrt{(4-1)^2 + (6-2)^2} = 5.$

33. $\dfrac{5+3}{2} = 4, \dfrac{7+9}{2} = 8$
    $\therefore$ mid-point
    $\therefore$ 1 : 1.

34. $\dfrac{8+0}{2} = 4, \dfrac{6+10}{2} = 8$
    $\therefore$ mid-point
    $\therefore$ 1 : 1

35. Clearly, $Q \to (3, 4)$

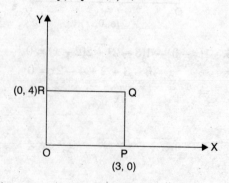

36. $OP^2 + OQ^2 = \{(a\cos\theta)^2 + (b\sin\theta)^2\}$
    $+ \{(-a\sin\theta)^2 + (b\cos\theta)^2\}$
    $= a^2(\cos^2\theta + \sin^2\theta) + b^2(\sin^2\theta + \cos^2\theta)$
    $= a^2 + b^2.$

37. $x = \dfrac{8(-4) + 11}{3} = 5$

    $y = \dfrac{-5 + 7 + 13}{3} = 5$

38. Mid-point of AC $= \left(\dfrac{0+0}{2}, \dfrac{-1+3}{2}\right)$
    $= (0, 1)$
    $\therefore$ Median through B
    $= \sqrt{(0-2)^2 + (1-1)^2} = 2$

39. $x = \dfrac{3 + 5 + (-5)}{3} = 1$

40. $y = \dfrac{6 + (-2) + 2}{3} = 2$

41. Let the third vertex be (x, y)
    $\dfrac{3 + (-4) + x}{3} = 4 \Rightarrow x = 13$
    $\dfrac{5 + (-5) + y}{3} = 3 \Rightarrow y = 9$

42. $\dfrac{4 + 6 + x}{3} = 3 \Rightarrow x = -1$
    $\dfrac{y + 9 + 4}{3} = 6 \Rightarrow y = 5$

43. $\dfrac{-8 + x}{2} = 4 \Rightarrow x = 16$

44. Area = $\dfrac{3 \times 4}{2} = 6.$

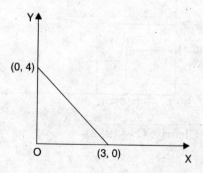

45. $OA = \sqrt{10}$
    $OB = \sqrt{13}$

COORDINATE GEOMETRY

$OC = \sqrt{25}$

$OD = \sqrt{17}$.

46. Let the opposite vertex be (x, y)

$\dfrac{x+2}{2} = 3 \Rightarrow x = 4$

$\dfrac{y+3}{2} = -2 \Rightarrow y = -7$.

47. x and y both are –ve

48. $\sqrt{(k-2)^2 + (3-3)^2} = 5$

$\Rightarrow k - 2 = 5$

$\Rightarrow k = 7$

49. A———B———C

50. $a(b-1) + 0(1-0) + 1(0-b) = 0$

$ab = a + b \Rightarrow \dfrac{a+b}{ab} = 1$

$\Rightarrow \dfrac{1}{a} + \dfrac{1}{b} = 1$

51. $= \dfrac{a+b+c}{3} = 0 \Rightarrow a+b+c = 0$

$\therefore a^3 + b^3 + c^3 - 3abc$

$= (a+b+c)(a^2 + b^2 + c^2 - ab - b(-ca)$

$= 0$

$\therefore a^3 + b^3 + c^3 = 3abc$.

52. $d = \sqrt{(a\cos\theta + b\sin\theta)^2 + \{(a\sin\theta - b\cos\theta)\}^2}$

$= \sqrt{a^2 + b^2}$

53. $d = \sqrt{a^2\cos^2\alpha + a^2\sin^2\alpha} = a$

54. $d = \sqrt{(\sin\theta - \cos\theta)^2 + (\sin\theta + \cos\theta)^2}$

$= \sqrt{2}$

55. Let the fourth vertex be (x, y). Then,

$\dfrac{x+1}{2} = \dfrac{-2+4}{2} \Rightarrow x = 1$

$\dfrac{y+0}{2} = \dfrac{1+3}{2} \Rightarrow y = 4$

| ∵ Diagonals of a prallelogram bisect each other

56. $\dfrac{-2+4}{2} = \dfrac{1+a}{2} \Rightarrow a = 1$

$\dfrac{-1+b}{2} - \dfrac{2+0}{2} \Rightarrow b = 3$

| ∵ Diagonals of a parallelogram bisect each other

Chapter 8
Introduction to Trigonometry

1. If $\tan \theta = \sqrt{3}$, then the value of $\sin \theta$ is
   (a) $\dfrac{1}{\sqrt{3}}$
   (b) $\dfrac{\sqrt{3}}{2}$
   (c) $\dfrac{2}{\sqrt{3}}$
   (d) 1.

2. If $\sin \theta = \dfrac{5}{13}$, then the value of $\tan \theta$ is
   (a) $\dfrac{5}{12}$
   (b) $\dfrac{12}{13}$
   (c) $\dfrac{13}{12}$
   (d) $\dfrac{12}{5}$.

3. If $\sqrt{3} \cos A = \sin A$, then the value of $\cot A$ is
   (a) $\sqrt{3}$
   (b) 1
   (c) $\dfrac{1}{\sqrt{3}}$
   (d) 2.

4. In the following figure, the value of $\cot A$ is

   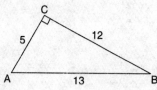

   (a) $\dfrac{12}{13}$
   (b) $\dfrac{5}{12}$
   (c) $\dfrac{5}{13}$
   (d) $\dfrac{13}{5}$.

5. In the following figure, find the value of $\tan \theta$

   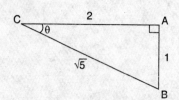

   (a) 2
   (b) $\dfrac{1}{\sqrt{5}}$
   (c) $\dfrac{2}{\sqrt{5}}$
   (d) $\dfrac{1}{2}$.

6. In the following figure, the value of $\operatorname{cosec} \alpha$ is

   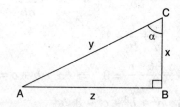

   (a) $\dfrac{y}{x}$
   (b) $\dfrac{y}{z}$
   (c) $\dfrac{x}{z}$
   (d) $\dfrac{x}{y}$.

7. The value of $\sin^2 30° + \cos^2 30°$ is
   (a) 0
   (b) 2
   (c) 3
   (d) 1.

8. The value of $\operatorname{cosec}^2 55° - \cot^2 55°$ is
   (a) 1
   (b) 2
   (c) 3
   (d) 0.

9. If $\cot \phi = \dfrac{20}{21}$, then the value of $\operatorname{cosec} \phi$ is
   (a) $\dfrac{21}{20}$
   (b) $\dfrac{20}{29}$
   (c) $\dfrac{29}{21}$
   (d) $\dfrac{21}{29}$.

INTRODUCTION TO TRIGONOMETRY

10. If in $\triangle ABC$, $\angle B = 90°$, $AB = 12$ cm and $BC = 9$ cm, then the value of cos C is
    (a) $\dfrac{3}{5}$
    (b) $\dfrac{3}{4}$
    (c) $\dfrac{5}{3}$
    (d) $\dfrac{4}{3}$.

11. The value of $(\sec 40° + \tan 40°)(\sec 40° - \tan 40°)$ is
    (a) -1
    (b) 1
    (c) $\cos 40°$
    (d) $\sin 40°$.

12. $\dfrac{1}{\sin\theta - \tan\theta}$ is equal to
    (a) $\dfrac{\cot\theta}{\cos\theta - 1}$
    (b) $\dfrac{\cot\theta}{\cot\theta - \csc\theta}$
    (c) $\csc\theta - \cot\theta$
    (d) $\cot\theta$.

13. $\dfrac{\sec A - 1}{\sec A + 1}$ is equal to
    (a) $\dfrac{1 + \cos A}{1 - \cos A}$
    (b) $\dfrac{\cos A - 1}{1 + \cos A}$
    (c) $\dfrac{1 - \cos A}{1 + \cos A}$
    (d) $\dfrac{\cos A - 1}{1 - \cos A}$.

14. $\cot^2\theta - \dfrac{1}{\sin^2\theta}$ is equal to
    (a) 2
    (b) 1
    (c) 0
    (d) -1.

15. If $\csc\theta = \dfrac{41}{40}$, then the value of $\tan\theta$ is
    (a) $\dfrac{40}{9}$
    (b) $\dfrac{9}{41}$
    (c) $\dfrac{40}{41}$
    (d) $\dfrac{9}{40}$.

16. If $\csc\theta = \dfrac{41}{40}$, then the value of $\cos\theta$ is
    (a) $\dfrac{40}{9}$
    (b) $\dfrac{9}{41}$
    (c) $\dfrac{40}{41}$
    (d) $\dfrac{9}{40}$.

17. If in $\triangle ABC$, $\angle B = 90°$, $AB = 12$ cm and $BC = 5$ cm, then the value of sin A is
    (a) $\dfrac{5}{13}$
    (b) $\dfrac{5}{12}$
    (c) $\dfrac{12}{13}$
    (d) $\dfrac{13}{5}$.

18. If in $\triangle ABC$, $\angle B = 90°$, $AB = 12$ cm and $BC = 5$ cm, then the value of tan A is
    (a) $\dfrac{5}{13}$
    (b) $\dfrac{5}{12}$
    (c) $\dfrac{13}{5}$
    (d) $\dfrac{12}{5}$.

19. If in $\triangle ABC$, $\angle B = 90°$, $AB = 12$ cm and $BC = 5$ cm, then the value of sin C is
    (a) $\dfrac{5}{12}$
    (b) $\dfrac{12}{5}$
    (c) $\dfrac{13}{12}$
    (d) $\dfrac{12}{13}$.

20. If in $\triangle ABC$, $\angle B = 90°$, $AB = 12$ cm and $BC = 5$ cm, then the value of cot C is
    (a) $\dfrac{13}{5}$
    (b) $\dfrac{5}{12}$
    (c) $\dfrac{12}{5}$
    (d) $\dfrac{5}{13}$.

21. If $\cos\theta = \dfrac{3}{5}$, then the value of $\dfrac{\sin\theta - \cot\theta}{2\tan\theta}$ is
    (a) $\dfrac{3}{160}$
    (b) $\dfrac{7}{160}$
    (c) $\dfrac{1}{160}$
    (d) $\dfrac{11}{160}$.

22. If $\cos\theta = \dfrac{21}{29}$, then the value of $\dfrac{\sec\theta}{\tan\theta - \sin\theta}$ is
    (a) $\dfrac{160}{41}$
    (b) $\dfrac{841}{160}$
    (c) $\dfrac{160}{84}$
    (d) $\dfrac{41}{160}$.

23. If $\tan\theta = \dfrac{4}{3}$, then the value of $\dfrac{3\sin\theta + 2\cos\theta}{3\sin\theta - 2\cos\theta}$ is
    (a) 1
    (b) 2
    (c) 3
    (d) 4.

24. If $\cot \theta = \dfrac{b}{a}$, then the value of $\dfrac{\cos \theta + \sin \theta}{\cos \theta - \sin \theta}$ is
    (a) $\dfrac{b-a}{b+a}$
    (b) $b - a$
    (c) $b + a$
    (d) $\dfrac{b+a}{b-a}$.

25. If $\operatorname{cosec} A = 2$, then the value of $\cot A + \dfrac{\sin A}{1 + \cos A}$ is
    (a) 1
    (b) 2
    (c) 3
    (d) 4.

26. If $\cot \theta = \dfrac{1}{\sqrt{3}}$, then the value of $\dfrac{1 - \cos^2 \theta}{2 - \sin^2 \theta}$ is
    (a) $\dfrac{1}{5}$
    (b) $\dfrac{2}{5}$
    (c) $\dfrac{3}{5}$
    (d) $\dfrac{5}{3}$.

27. $\sqrt{\sec^2 A + \operatorname{cosec}^2 A} =$
    (a) $\tan A + \cot A$
    (b) $\tan A - \cot A$
    (c) $\sin A + \cos A$
    (d) $\sin A - \cos A$.

28. $\dfrac{\tan \alpha + \tan \beta}{\cot \alpha + \cot \beta} =$
    (a) $\tan \alpha \tan \beta$
    (b) $\tan \alpha \cot \beta$
    (c) $\tan \beta \cot \alpha$
    (d) $\cot \beta \cot \alpha$.

29. $\dfrac{\tan \alpha}{\sqrt{1 + \tan^2 \alpha}} =$
    (a) $\sin \alpha$
    (b) $\cos \alpha$
    (c) $\operatorname{cosec} \alpha$
    (d) $\sec \alpha$.

30. $\dfrac{1 - \tan^2 \alpha}{\cot^2 \alpha - 1} =$
    (a) $\tan^2 \alpha$
    (b) $\cot^2 \alpha$
    (c) $-\tan^2 \alpha$
    (d) $-\cot^2 \alpha$.

31. $\cos^4 \theta - \sin^4 \theta =$
    (a) $1 - 2\sin^2 \theta$
    (b) $1 - 2\cos^2 \theta$
    (c) $1 - \sin^2 \theta$
    (d) $1 - \cos^2 \theta$.

32. $\sec^2 \theta - \operatorname{cosec}^2 \theta =$
    (a) $\tan^2 \theta - \cot^2 \theta$
    (b) $\tan^2 \theta + \cot^2 \theta$
    (c) $\cot^2 \theta - \tan^2 \theta$
    (d) $\sec^2 \theta \operatorname{cosec}^2 \theta$.

33. The value of $\dfrac{\tan \theta}{\sqrt{1 + \tan^2 \theta}}$ is
    (a) $\cos \theta$
    (b) $\sin \theta$
    (c) $\sec \theta$
    (d) $\cot \theta$.

34. $\dfrac{\sqrt{\operatorname{cosec}^2 \theta - 1}}{\operatorname{cosec} \theta} =$
    (a) $\cos \theta$
    (b) $\sec \theta$
    (c) $\sin \theta$
    (d) $\operatorname{cosec} \theta$.

35. $\sin \theta \operatorname{cosec} \theta + \cos \theta \sec \theta$ is equal to
    (a) 2
    (b) 1
    (c) $\dfrac{1}{2}$
    (d) -1.

36. $\dfrac{1}{\sin^2 \theta} - \cot^2 \theta =$
    (a) -2
    (b) -1
    (c) 2
    (d) 1.

37. $\dfrac{2 - \tan \theta}{2 \operatorname{cosec} \theta - \sec \theta} =$
    (a) $\tan \theta$
    (b) $\cos \theta$
    (c) $\sin \theta$
    (d) $\cot \theta$.

38. If $\sin \theta = \dfrac{1}{2}$, then the value of $(\tan \theta + \cot \theta)^2$ is
    (a) $\dfrac{16}{3}$
    (b) $\dfrac{8}{3}$
    (c) $\dfrac{4}{3}$
    (d) $\dfrac{10}{3}$.

39. If $\cos \theta = \dfrac{1}{2}$, then the value of $\dfrac{\cot \theta + \tan \theta}{\operatorname{cosec} \theta}$ is
    (a) 1
    (b) 3
    (c) 4
    (d) 2.

40. The value of $\cot 30°$ is
    (a) $\dfrac{1}{\sqrt{3}}$
    (b) $\sqrt{3}$
    (c) $\dfrac{\sqrt{3}}{2}$
    (d) $\sqrt{2}$.

INTRODUCTION TO TRIGONOMETRY

41. The value of sec 45° is
    (a) $\sqrt{2}$
    (b) $\frac{1}{\sqrt{2}}$
    (c) $\frac{1}{\sqrt{3}}$
    (d) $\sqrt{3}$.

42. The value of $\cos^2 45°$ is
    (a) $\frac{1}{\sqrt{2}}$
    (b) $\frac{\sqrt{3}}{2}$
    (c) $\frac{1}{2}$
    (d) $\frac{1}{\sqrt{3}}$.

43. The value of $\tan^2 60°$ is
    (a) 3
    (b) $\frac{1}{3}$
    (c) 1
    (d) ∞.

44. sin 65° is equal to
    (a) cosec 45°
    (b) cos 55°
    (c) sin 25°
    (d) cos 25°.

45. tan 81° is equal to
    (a) sin 9°
    (b) cos 19°
    (c) cot 9°
    (d) sec 81°.

46. If $\tan \theta = \frac{1}{\sqrt{3}}$, then the value of θ is
    (a) 30°
    (b) 60°
    (c) 45°
    (d) 90°.

47. If $\cosec \theta = \frac{2}{\sqrt{3}}$, then the value of θ is
    (a) $\frac{\pi}{4}$
    (b) $\frac{\pi}{3}$
    (c) $\frac{\pi}{2}$
    (d) $\frac{\pi}{6}$.

48. The value of $2 \sin^2 60° \cos 60°$ is
    (a) $\frac{4}{3}$
    (b) $\frac{5}{2}$
    (c) $\frac{3}{4}$
    (d) $\frac{1}{3}$.

49. The value of $\frac{\cosec 39°}{\sec 51°}$ is
    (a) 0
    (b) 2
    (c) 3
    (d) 1.

50. The value of $\sin^2 60° \cot^2 60°$ is
    (a) $\frac{1}{4}$
    (b) 3
    (c) $\frac{\sqrt{3}}{2}$
    (d) $\frac{3}{4}$.

51. The value of $\tan \frac{\pi}{6} \tan \frac{\pi}{3}$ is
    (a) 0
    (b) 1
    (c) –1
    (d) 2.

52. The value of $4 \cos^3 30° - 3 \cos 30°$ is
    (a) 0
    (b) 3
    (c) 2
    (d) 4.

53. The value of $\frac{\sin 45° \cos (90° - \theta)}{\cos 45° \sin (90° - \theta)}$ is
    (a) tan θ
    (b) cot θ
    (c) 1
    (d) – tan θ.

54. The value of $\{\cos 0° - \sin (90° - \theta)\}\{\cos 0° + \sin (90° - \theta)\}$ is
    (a) sin θ
    (b) $\sin^2 \theta$
    (c) cos θ
    (d) $\cos^2 \theta$.

55. The value of tan 5° tan 25° tan 45° tan 65° tan 85° is
    (a) 2
    (b) 3
    (c) 1
    (d) 4.

56. If for angle θ $\cos^2 \theta = \frac{1}{2}$, then the value of $\sin^2 \theta$ is
    (a) $\frac{1}{4}$
    (b) $\frac{\sqrt{3}}{2}$
    (c) $\frac{1}{\sqrt{2}}$
    (d) $\frac{1}{2}$.

57. $\cos \left(\frac{\pi}{2} - 30°\right)$ is equal to
    (a) $\frac{1}{\sqrt{3}}$
    (b) $\frac{1}{\sqrt{2}}$
    (c) $\frac{1}{2}$
    (d) $\frac{\sqrt{3}}{2}$.

58. The value of $\frac{\tan 35°}{\cot (90° - 35°)} + \frac{\tan 15°}{\cot (90° - 15°)} - 2 \tan 45°$ is
    (a) 4
    (b) 2
    (c) 1
    (d) 0.

59. The value of cos 25° sin 65° + sin 25° cos 65° is
    (a) 0
    (b) sin 40°
    (c) 1
    (d) cos 40°.

60. If $\sin \theta = \frac{\sqrt{3}}{2}$, then the value of θ is
    (a) 120° (b) 60°
    (c) 45° (d) 90°.

61. $(\csc^2 \theta - \cot^2 \theta)(1 - \cos^2 \theta)$ is equal to
    (a) $\csc^2 \theta$ (b) $\tan^2 \theta$
    (c) $\sec^2 \theta$ (d) $\sin^2 \theta$.

62. The value of $(1 + \cot^2 \theta)(1 + \cos \theta)(1 - \cos \theta)$ is
    (a) $\sin^2 \theta$ (b) $\csc^2 \theta$
    (c) 1 (d) $\sec^2 \theta$.

63. If $\sin \alpha = \cos \alpha$, then the value of α is
    (a) 30° (b) 45°
    (c) 60° (d) 90°.

64. The value of cos 54° cosec 54° tan 54° is
    (a) 4 (b) 1
    (c) $\sqrt{2}$ (d) $\sqrt{3}$.

65. The value of $\tan^2 \theta (\csc^2 \theta - 1)$ is
    (a) $\tan^2 \theta$ (b) $\csc^2 \theta$
    (c) $\cot^2 \theta$ (d) 1.

66. If $\sin \theta = \frac{1}{2}$, then the value of $\sin 2\theta$ is
    (a) 1 (b) $\frac{\sqrt{3}}{2}$
    (c) $\frac{1}{2}$ (d) $-\frac{\sqrt{3}}{2}$.

67. If θ is an acute angle and $\sec \theta = \frac{13}{12}$, then the value of tan θ is
    (a) $-\frac{5}{12}$ (b) $\frac{5}{12}$
    (c) $\frac{12}{13}$ (d) $-\frac{12}{13}$.

68. The value of $\frac{\sin 19°}{\cos 71°}$ is equal to
    (a) > 1 (b) 1
    (c) < 1 (d) 0.

69. If $\cos \theta = \frac{1}{2}$, then the value of $\csc^2 \theta$ is
    (a) $\frac{1}{2}$ (b) $\frac{\sqrt{3}}{2}$
    (c) $\frac{3}{4}$ (d) $\frac{4}{3}$.

70. If θ = 45°, then the value of $\frac{1 - \cos 2\theta}{\sin 2\theta}$ is

    (a) 0 (b) 1
    (c) 2 (d) ∞.

71. The value of $\sin^2 \theta + \frac{1}{1 + \tan^2 \theta}$ is
    (a) $\sin^2 \theta$ (b) $\cos^2 \theta$
    (c) $\sec^2 \theta$ (d) 1.

72. The value of cos 20° cos 70° − sin 20° sin 70° is equal to
    (a) 0 (b) 1
    (c) ∞ (d) cos 50°.

73. In any triangle ABC, the value of $\sin\left(\frac{B+C}{2}\right)$ is
    (a) $\cos \frac{A}{2}$ (b) $\sin \frac{A}{2}$
    (c) $-\sin \frac{A}{2}$ (d) $-\cos \frac{A}{2}$.

74. If sec θ + tan θ = m and sec θ − tan θ = n, then the value of mn is
    (a) 2 (b) 1
    (c) ± 1 (d) ± 2.

75. If $\sin \theta + \cos \theta = \sqrt{2} \cos(90° - \theta)$, then cot θ is equal to
    (a) $-\sqrt{2}$ (b) $\sqrt{2}$
    (c) $\sqrt{2} + 1$ (d) $\sqrt{2} - 1$.

76. If $2 \sin^2 \theta = \frac{1}{2}$, 0° < θ < 90°, then θ is equal to
    (a) 60° (b) 30°
    (c) 45° (d) 90°.

77. In the given figure, ABCD is a rectangle in which segments AP and AQ are drawn as shown. Find the length of (AP + AQ)

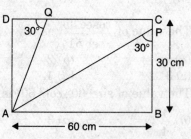

    (a) 120 cm (b) 180 cm
    (c) 150 cm (d) 100 cm.

INTRODUCTION TO TRIGONOMETRY

78. In the given figure, ABCD is a rectangle with AD = 8 cm and CD = 12 cm. Line-segment CE is drawn, making an angle of 60° with AB, intersecting AB in E. Find the lengths of CE and BE.

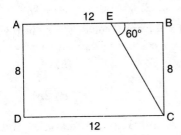

   (a) $\dfrac{16}{\sqrt{3}}$ cm, $\dfrac{8}{\sqrt{3}}$ cm
   (b) $\dfrac{8}{\sqrt{3}}$ cm, $\dfrac{16}{\sqrt{3}}$ cm
   (c) $\dfrac{16}{\sqrt{3}}$ cm, $\dfrac{16}{\sqrt{3}}$ cm
   (d) $\dfrac{8}{\sqrt{3}}$ cm, $\dfrac{8}{\sqrt{3}}$ cm.

79. If $\sqrt{3}\tan\theta = 3\sin\theta$, then the value of $\sin^2\theta - \cos^2\theta$ is
   (a) $\dfrac{1}{\sqrt{3}}$
   (b) $\dfrac{1}{\sqrt{2}}$
   (c) $\dfrac{1}{3}$
   (d) $\dfrac{1}{4}$.

80. What is the maximum value of $\dfrac{1}{\csc\theta}$?
   (a) 0
   (b) 1
   (c) 2
   (d) not defined.

81. What is the maximum value of $\tan\theta$?
   (a) 1
   (b) not defined
   (c) $\sqrt{3}$
   (d) $\dfrac{1}{\sqrt{3}}$.

82. The value of $(\sin^2 60° + \cos^2 60°)^2$ is
   (a) 1
   (b) 2
   (c) 4
   (d) 9.

83. Which of the following is not true?
   (a) $\sin^2\theta + \cos^2\theta = 1$
   (b) $\csc^2\theta = 1 + \cot^2\theta$
   (c) $\sec^2\theta = 1 - \tan^2\theta$
   (d) $\tan\theta \cot\theta = 1$.

84. Which of the following is true?
   (a) $\tan\theta = \dfrac{\sin\theta}{\cos\theta}$
   (b) $\sin\theta = \dfrac{1}{\sec\theta}$
   (c) $\cos\theta = \dfrac{1}{\csc\theta}$
   (d) $\cot\theta = \dfrac{\cos\theta}{\sec\theta}$.

85. Which of the following is not possible?
   (a) $\sin\theta = 1, \cos\theta = 0$
   (b) $\sin\theta = 0, \cos\theta = 1$
   (c) $\sin\theta = \dfrac{\sqrt{3}}{2}, \cos\theta = \dfrac{1}{2}$
   (d) $\sin\theta = 1, \cos\theta = \dfrac{1}{\sqrt{2}}$.

86. If $x = a\tan\theta$ and $y = b\sec\theta$, then
   (a) $\dfrac{y^2}{b^2} - \dfrac{x^2}{a^2} = 1$
   (b) $\dfrac{x^2}{a^2} + \dfrac{y^2}{b^2} = 1$
   (c) $\dfrac{x^2}{a^2} - \dfrac{y^2}{b^2} = 1$
   (d) $\dfrac{x^2}{a^2} - \dfrac{y^2}{b^2} = 0$.

87. If $\cos\theta = x$ and $\csc\theta = y$, then $\cot\theta$ is equal to
   (a) $\dfrac{x}{y}$
   (b) $\dfrac{y}{x}$
   (c) xy
   (d) $\dfrac{1}{xy}$.

88. The value of $\dfrac{1+\tan^2\theta}{\sec^4\theta}$ is
   (a) $\sin^2\theta$
   (b) $\cos^2\theta$
   (c) $\sec^2\theta$
   (d) $\csc^2\theta$.

89. The value of $1 + \tan 5° \cot 85°$ is equal to
   (a) $\sin^2 5°$
   (b) $\cos^2 5°$
   (c) $\sec^2 5°$
   (d) $\csc^2 5°$.

90. The value of $4\sin 30° \cos 60°$ is
   (a) 0
   (b) 1
   (c) $\dfrac{1}{2}$
   (d) 2.

91. 8 cosec² A − 8 cot² A is equal to
    (a) 8   (b) −8
    (c) 4   (d) −4.

92. The value of $\dfrac{\sin\theta}{1+\cos\theta}$ is equal to
    (a) $\dfrac{1-\cos\theta}{\sin\theta}$ (b) $\dfrac{1+\cos\theta}{\sin\theta}$
    (c) $\dfrac{\sin\theta}{1-\cos\theta}$ (d) $\dfrac{\sin\theta}{1+\cos\theta}$.

93. sec⁴ A − sec² A is equal to
    (a) tan⁴ A − tan² A
    (b) tan⁴ A + tan² A
    (c) tan² A − tan⁴ A
    (d) − tan² A − tan⁴ A.

94. $\sqrt{\dfrac{1+\sin\theta}{1-\sin\theta}}$ is equal to
    (a) sec θ + tan θ   (b) sec θ − tan θ
    (c) tan θ − sec θ   (d) sec² θ + tan² θ.

95. $\sqrt{\dfrac{1+\cos\theta}{1-\cos\theta}}$ is equal to
    (a) cosec θ + cot θ   (b) cosec θ − cot θ
    (c) cot θ − cosec θ   (d) cosec² θ + cot² θ.

96. If $a\cos\theta + b\sin\theta = m$ and $a\sin\theta - b\cos\theta = n$, then $a^2 + b^2 =$
    (a) $m^2 - n^2$ (b) $n^2 - m^2$
    (c) $m^2 + n^2$ (d) $m^2 n^2$.

97. If $x = r\sin\theta\cos\phi$, $y = r\sin\theta\sin\phi$ and $z = r\cos\theta$, then $x^2 + y^2 + z^2 =$
    (a) r (b) r^2
    (c) $\dfrac{r^2}{2}$ (d) $2r^2$.

98. If A and B are acute angles and sin A = cos B, then A + B is equal to
    (a) 60°   (b) 90°
    (c) 45°   (d) 0°.

99. If sin 3θ = cos (θ − 6°), find θ
    (a) 24°   (b) 36°
    (c) 12°   (d) 30°.

100. If tan 2θ = cot (θ + 6°), then θ =
    (a) 24°   (b) 28°
    (c) 12°   (d) 45°.

ANSWERS

1. (b)	2. (a)	3. (c)	4. (b)	53. (a)	54. (b)	55. (c)	56. (c)
5. (d)	6. (b)	7. (d)	8. (a)	57. (c)	58. (d)	59. (c)	60. (b)
9. (c)	10. (a)	11. (b)	12. (a)	61. (d)	62. (c)	63. (b)	64. (b)
13. (c)	14. (d)	15. (a)	16. (b)	65. (d)	66. (b)	67. (b)	68. (b)
17. (a)	18. (b)	19. (d)	20. (b)	69. (c)	70. (b)	71. (d)	72. (a)
21. (a)	22. (b)	23. (c)	24. (d)	73. (a)	74. (b)	75. (d)	76. (b)
25. (b)	26. (c)	27. (a)	28. (a)	77. (b)	78. (a)	79. (c)	80. (b)
29. (a)	30. (a)	31. (a)	32. (a)	81. (b)	82. (a)	83. (c)	84. (a)
33. (b)	34. (a)	35. (a)	36. (d)	85. (d)	86. (a)	87. (c)	88. (b)
37. (c)	38. (a)	39. (d)	40. (b)	89. (c)	90. (b)	91. (a)	92. (a)
41. (a)	42. (c)	43. (a)	44. (d)	93. (b)	94. (a)	95. (a)	96. (c)
45. (c)	46. (a)	47. (b)	48. (c)	97. (b)	98. (b)	99. (a)	100. (b).
49. (d)	50. (a)	51. (b)	52. (a)				

INTRODUCTION TO TRIGONOMETRY

HINTS/SOLUTIONS

1. $\tan\theta = \sqrt{3} = \tan 60° \Rightarrow \theta = 60°$

   $\therefore \sin\theta = \sin 60° = \dfrac{\sqrt{3}}{2}$

2. $\tan\theta = \dfrac{\sin\theta}{\cos\theta} = \dfrac{\sin\theta}{\sqrt{1-\sin^2\theta}}$

   $= \dfrac{\frac{5}{13}}{\sqrt{1-\left(\frac{5}{13}\right)^2}} = \dfrac{\frac{5}{13}}{\frac{12}{13}} = \dfrac{5}{12}$

3. $\sqrt{3}\cos A = \sin A \Rightarrow \tan A = \sqrt{3}$

   $\therefore \cot A = \dfrac{1}{\sqrt{3}}$

4. $\cot A = \dfrac{\text{Base}}{\text{Perpendicular}} = \dfrac{AC}{BC} = \dfrac{5}{12}$

5. $\tan\theta = \dfrac{\text{Perpendicular}}{\text{Base}} = \dfrac{AB}{AC} = \dfrac{1}{2}$

6. $\text{cosec}\,\alpha = \dfrac{\text{Hypotenuse}}{\text{Perpendicular}} = \dfrac{AC}{AB} = \dfrac{y}{z}$

7. $\sin^2\theta + \cos^2\theta = 1$

8. $\text{cosec}^2\theta - \cot^2\theta = 1$

9. $\text{cosec}\,\phi = \sqrt{1+\cot^2\phi} = \sqrt{1+\left(\dfrac{20}{21}\right)^2} = \dfrac{29}{21}$

10. $\cos C = \dfrac{BC}{AC} = \dfrac{BC}{\sqrt{AB^2+BC^2}}$

    $= \dfrac{9}{\sqrt{12^2+9^2}} = \dfrac{9}{15} = \dfrac{3}{5}$

11. $(\sec 40° + \tan 40°)(\sec 40° - \tan 40°)$
    $= \sec^2 40° - \tan^2 40° = 1$

12. $\dfrac{1}{\sin\theta - \tan\theta} = \dfrac{\frac{1}{\tan\theta}}{\frac{\sin\theta}{\tan\theta} - \frac{\tan\theta}{\tan\theta}}$

    $= \dfrac{\cot\theta}{\cos\theta - 1}$

13. $\dfrac{\sec A - 1}{\sec A + 1} = \dfrac{\frac{1}{\cos A} - 1}{\frac{1}{\cos A} + 1} = \dfrac{1 - \cos A}{1 + \cos A}$

14. $\cot^2\theta - \dfrac{1}{\sin^2\theta} = \cot^2\theta - \text{cosec}^2\theta = -1$

15. $\tan\theta = \dfrac{1}{\cot\theta} = \dfrac{1}{\sqrt{\text{cosec}^2\theta - 1}}$

    $= \dfrac{1}{\sqrt{\left(\frac{41}{40}\right)^2 - 1}} = \dfrac{40}{9}$

16. $\text{cosec}\,\theta = \dfrac{41}{40}$

    $\Rightarrow \dfrac{1}{\sin\theta} = \dfrac{41}{40} \Rightarrow \sin\theta = \dfrac{40}{41}$

    $\therefore \cos\theta = \sqrt{1-\sin^2\theta} = \sqrt{1-\left(\dfrac{40}{41}\right)^2}$

    $= \dfrac{9}{41}$

17. $\sin A = \dfrac{BC}{AC} = \dfrac{BC}{\sqrt{AB^2+BC^2}}$

    $= \dfrac{5}{\sqrt{12^2+5^2}} = \dfrac{5}{13}$

18. $\tan A = \dfrac{BC}{AB} = \dfrac{5}{12}$

19. $\sin C = \dfrac{AB}{AC} = \dfrac{AB}{\sqrt{AB^2+BC^2}}$

    $= \dfrac{12}{\sqrt{12^2+5^2}} = \dfrac{12}{13}$

20. $\cot C = \dfrac{BC}{AB} = \dfrac{5}{12}$

21. $\sin\theta = \sqrt{1-\cos^2\theta} = \sqrt{1-\left(\dfrac{3}{5}\right)^2} = \dfrac{4}{5}$

    $\therefore \tan\theta = \dfrac{\sin\theta}{\cos\theta} = \dfrac{4}{3}$

    $\cot\theta = \dfrac{1}{\tan\theta} = \dfrac{3}{4}$

$\therefore \quad \dfrac{\sin\theta - \cot\theta}{2\tan\theta} = \dfrac{\dfrac{4}{5} - \dfrac{3}{4}}{2\left(\dfrac{4}{3}\right)} = \dfrac{\dfrac{1}{20}}{\dfrac{8}{3}} = \dfrac{3}{160}$

22. $\dfrac{\sec\theta}{\tan\theta - \sin\theta} = \dfrac{1}{\cos\theta\left(\dfrac{\sin\theta}{\cos\theta} - \sin\theta\right)}$

$= \dfrac{1}{\sin\theta - \sin\theta\cos\theta}$

$= \dfrac{1}{\sin\theta(1 - \cos\theta)}$

$= \dfrac{1}{\sqrt{1 - \cos^2\theta}\,(1 - \cos\theta)}$

$= \dfrac{1}{\sqrt{1 - \left(\dfrac{21}{29}\right)^2}\left(1 - \dfrac{21}{29}\right)}$

$= \dfrac{29 \times 29}{20 \times 8} = \dfrac{841}{160}$

23. $\dfrac{3\sin\theta + 2\cos\theta}{3\sin\theta - 2\cos\theta} = \dfrac{3\dfrac{\sin\theta}{\cos\theta} + 2}{3\dfrac{\sin\theta}{\cos\theta} - 2}$

$= \dfrac{3\tan\theta + 2}{3\tan\theta - 2} = \dfrac{4 + 2}{4 - 2} = 3$

24. $\dfrac{\cos\theta + \sin\theta}{\cos\theta - \sin\theta} = \dfrac{\cot\theta + 1}{\cot\theta - 1}$

$= \dfrac{\dfrac{b}{a} + 1}{\dfrac{b}{a} - 1} = \dfrac{b + a}{b - a}$

25. $\operatorname{cosec} A = 2 = \operatorname{cosec} 30° \Rightarrow A = 30°$

$\therefore \quad \cot A + \dfrac{\sin A}{1 + \cos A}$

$= \cot 30° + \dfrac{\sin 30°}{1 + \cos 30°}$

$= \sqrt{3} + \dfrac{\dfrac{1}{2}}{1 + \dfrac{\sqrt{3}}{2}} = \sqrt{3} + \dfrac{1}{2 + \sqrt{3}}$

$= \dfrac{2\sqrt{3} + 3 + 1}{2 + \sqrt{3}} = \dfrac{2(2 + \sqrt{3})}{2 + \sqrt{3}} = 2$

26. $\cot\theta = \dfrac{1}{\sqrt{3}} = \cot 60° \Rightarrow \theta = 60°$

$\therefore \quad \dfrac{1 - \cos^2\theta}{2 - \sin^2\theta} = \dfrac{1 - \cos^2 60°}{2 - \sin^2 60°}$

$= \dfrac{1 - \dfrac{1}{4}}{2 - \dfrac{3}{4}} = \dfrac{3}{5}$

27. $\sqrt{\sec^2 A + \operatorname{cosec}^2 A}$

$= \sqrt{1 + \tan^2 A + 1 + \cot^2 A}$

$= \sqrt{\tan^2 A + \cot^2 A + 2\tan^2 A \cot^2 A}$

$= \tan A + \cot A$

28. $\dfrac{\tan\alpha + \tan\beta}{\cot\alpha + \cot\beta} = \dfrac{\tan\alpha + \tan\beta}{\dfrac{1}{\tan\alpha} + \dfrac{1}{\tan\beta}}$

$= \tan\alpha \tan\beta$

29. $\dfrac{\tan\alpha}{\sqrt{1 + \tan^2\alpha}} = \dfrac{\tan\alpha}{\sec\alpha} = \dfrac{\dfrac{\sin\alpha}{\cos\alpha}}{\dfrac{1}{\cos\alpha}} = \sin\alpha$

30. $\dfrac{1 - \tan^2\alpha}{\cot^2\alpha - 1} = \dfrac{1 - \tan^2\alpha}{\dfrac{1}{\tan^2\alpha} - 1}$

$= \dfrac{(1 - \tan^2\alpha)\tan^2\alpha}{1 - \tan^2\alpha} = \tan^2\alpha$

31. $\cos^4\theta - \sin^4\theta$

$= (\cos^2\theta - \sin^2\theta)(\cos^2\theta + \sin^2\theta)$

$= \cos^2\theta - \sin^2\theta = 1 - \sin^2\theta - \sin^2\theta$

$= 1 - 2\sin^2\theta$

32. $\sec^2\theta - \operatorname{cosec}^2\theta$

$= (1 + \tan^2\theta) - (1 + \cot^2\theta)$

$= \tan^2\theta - \cot^2\theta$

33. $\dfrac{\tan\theta}{\sqrt{1 + \tan^2\theta}} = \dfrac{\tan\theta}{\sec\theta} = \dfrac{\dfrac{\sin\theta}{\cos\theta}}{\dfrac{1}{\cos\theta}} = \sin\theta$

34. $\dfrac{\sqrt{\operatorname{cosec}^2\theta - 1}}{\operatorname{cosec}\theta} = \dfrac{\cot\theta}{\operatorname{cosec}\theta} = \dfrac{\dfrac{\cos\theta}{\sin\theta}}{\dfrac{1}{\sin\theta}} = \cos\theta$

35. $\sin\theta \operatorname{cosec}\theta + \cos\theta \sec\theta$
$$= \sin\theta\left(\frac{1}{\sin\theta}\right) + \cos\theta\left(\frac{1}{\cos\theta}\right)$$
$$= 1 + 1 = 2$$

36. $\dfrac{1}{\sin^2\theta} - \cot^2\theta = \operatorname{cosec}^2\theta - \cot^2\theta = 1$

37. $\dfrac{2 - \tan\theta}{2\operatorname{cosec}\theta - \sec\theta}$
$$= \frac{2 - \tan\theta}{\operatorname{cosec}\theta\left(2 - \dfrac{\sec\theta}{\operatorname{cosec}\theta}\right)}$$
$$= \frac{2 - \tan\theta}{\operatorname{cosec}\theta\,(2 - \tan\theta)} = \sin\theta$$

38. $\sin\theta = \dfrac{1}{2} = \sin 30° \Rightarrow \theta = 30°$
$\therefore (\tan\theta + \cot\theta)^2 = (\tan 30° + \cot 30°)^2$
$$= \left(\frac{1}{\sqrt{3}} + \sqrt{3}\right)^2 = \left(\frac{4}{\sqrt{3}}\right)^2 = \frac{16}{3}$$

39. $\cos\theta = \dfrac{1}{2} = \cos 60°$
$\Rightarrow A = 60°$
$\therefore \dfrac{\cot\theta + \tan\theta}{\operatorname{cosec}\theta} = \dfrac{\cot 60° + \tan 60°}{\operatorname{cosec} 60°}$
$$= \frac{\dfrac{1}{\sqrt{3}} + \sqrt{3}}{\dfrac{2}{\sqrt{3}}} = 2$$

40. See table
41. See table
42. $\cos^2 45° = \left(\dfrac{1}{\sqrt{2}}\right)^2 = \dfrac{1}{2}$
43. $\tan^2 60° = (\sqrt{3})^2 = 3$
44. $\sin 65° = \sin(90° - 25°) = \cos 25°$
45. $\tan 81° = \tan(90° - 9°) = \cot 9°$
46. $\tan\theta = \dfrac{1}{\sqrt{3}} = \tan 30° \Rightarrow \theta = 30°$
47. $\operatorname{cosec}\theta = \dfrac{2}{\sqrt{3}}$
$\Rightarrow \dfrac{1}{\sin\theta} = \dfrac{2}{\sqrt{3}}$

$\Rightarrow \sin\theta = \dfrac{\sqrt{3}}{2} = \sin\dfrac{\pi}{3}$
$\Rightarrow \theta = \dfrac{\pi}{3}$

48. $2\sin^2 60°\cos 60° = 2\left(\dfrac{\sqrt{3}}{2}\right)^2 \dfrac{1}{2} = \dfrac{3}{4}$

49. $\dfrac{\operatorname{cosec} 39°}{\sec 51°} = \dfrac{\operatorname{cosec} 39°}{\sec(90° - 39°)}$
$$= \frac{\operatorname{cosec} 39°}{\operatorname{cosec} 39°} = 1$$

50. $\sin^2 60°\cot^2 60° = \sin^2 60°\,\dfrac{\cos^2 60°}{\sin^2 60°}$
$$= \cos^2 60° = \left(\frac{1}{2}\right)^2 = \frac{1}{4}$$

51. $\tan\dfrac{\pi}{6}\tan\dfrac{\pi}{3} = \dfrac{1}{\sqrt{3}}\sqrt{3} = 1$

52. $4\cos^3 30° - 3\cos 30°$
$$= 4\left(\frac{\sqrt{3}}{2}\right)^3 - 3\left(\frac{\sqrt{3}}{2}\right)$$
$$= 4\,\frac{3\sqrt{3}}{8} - \frac{3\sqrt{3}}{2} = 0$$

53. $\dfrac{\sin 45°\cos(90° - \theta)}{\cos 45°\sin(90° - \theta)}$
$$= \tan 45°\cot(90° - \theta)$$
$$= 1(\tan\theta) = \tan\theta$$

54. $\{\cos 0° - \sin(90° - \theta)\}$
  $\{\cos 0° + \sin(90° - \theta)\}$
$$= (1 - \cos\theta)(1 + \cos\theta)$$
$$= 1 - \cos^2\theta = \sin^2\theta$$

55. $\tan 5°\tan 25°\tan 45°\tan 65°\tan 85°$
$= \tan 5°\tan 25°(1)$
  $\tan(90° - 25°)\tan(90° - 5°)$
$= \tan 5°\tan 25°\cot 25°\cot 5°$
$= \tan 5°\tan 25°\,\dfrac{1}{\tan 25°}\,\dfrac{1}{\tan 5°} = 1.$

56. $\sin^2\theta = 1 - \cos^2\theta = 1 - \dfrac{1}{2} = \dfrac{1}{2}$

57. $\cos\left(\dfrac{\pi}{2} - 30°\right) = \sin 30° = \dfrac{1}{2}$

58. $\dfrac{\tan 35°}{\cot (90° - 35°)} + \dfrac{\tan 15°}{\cot (90° - 15°)}$
$\qquad\qquad\qquad\qquad - 2 \tan 45°$
$= \dfrac{\tan 35°}{\tan 35°} + \dfrac{\tan 15°}{\tan 15°} - 2 (1)$
$= 1 + 1 - 2 = 0$

59. $\cos 25° \sin 65° + \sin 25° \cos 65°$
$= \cos 25° \sin (90° - 25°)$
$\qquad + \sin 25° \cos (90° - 25°)$
$= \cos 25° \cos 25° + \sin 25° \sin 25°$
$= \cos^2 25° + \sin^2 25° = 1$

60. $\sin \theta = \dfrac{\sqrt{3}}{2} = \sin 60° \Rightarrow \theta = 60°$

61. $(\csc^2 \theta - \cot^2 \theta)(1 - \cos^2 \theta)$
$= (1)(\sin^2 \theta) = \sin^2 \theta$

62. $(1 + \cot^2 \theta)(1 + \cos \theta)(1 - \cos \theta)$
$= \csc^2 \theta (1 - \cos^2 \theta)$
$= \csc^2 \theta \sin^2 \theta$
$= \dfrac{1}{\sin^2 \theta} \sin^2 \theta = 1$

63. $\sin \alpha = \cos \alpha \Rightarrow \tan \alpha = 1 = \tan 45°$
$\Rightarrow \alpha = 45°$

64. $\cos 54° \csc 54° \tan 54°$
$= \cos 54° \dfrac{1}{\sin 54°} \dfrac{\sin 54°}{\cos 54°} = 1$

65. $\tan^2 \theta (\csc^2 \theta - 1)$
$= \tan^2 \theta \cot^2 \theta$
$= \tan^2 \theta \dfrac{1}{\tan^2 \theta} = 1$

66. $\sin \theta = \dfrac{1}{2} = \sin 30° \Rightarrow \theta = 30°$
$\therefore \sin 2\theta = \sin 60° = \dfrac{\sqrt{3}}{2}$

67. $\tan \theta = \sqrt{\sec^2 \theta - 1} = \sqrt{\left(\dfrac{13}{12}\right)^2 - 1} = \dfrac{5}{12}$

68. $\dfrac{\sin 19°}{\cos 71°} = \dfrac{\sin 19°}{\cos (90° - 19°)} = \dfrac{\sin 19°}{\sin 19°} = 1$

69. $\cos \theta = \dfrac{1}{2} = \cos 60° \Rightarrow \theta = 60°$

$\therefore \csc^2 \theta = \csc^2 60°$
$= \left(\dfrac{2}{\sqrt{3}}\right)^2 = \dfrac{4}{3}$

70. $\dfrac{1 - \cos 2\theta}{\sin 2\theta} = \dfrac{1 - \cos 90°}{\sin 90°} = \dfrac{1 - 0}{1} = 1$

71. $\sin^2 \theta + \dfrac{1}{1 + \tan^2 \theta}$
$= \sin^2 \theta + \dfrac{1}{\sec^2 \theta}$
$= \sin^2 \theta + \cos^2 \theta = 1$

72. $\cos 20° \cos 70° - \sin 20° \sin 70°$
$= \cos 20° \cos (90° - 20°)$
$\qquad - \sin 20° \sin (90° - 20°)$
$= \cos 20° \sin 20° - \sin 20° \cos 20° = 0$

73. $\sin \left(\dfrac{B + C}{2}\right) = \sin \left(\dfrac{180° - A}{2}\right)$
$= \sin \left(90° - \dfrac{A}{2}\right) = \cos \dfrac{A}{2}$

74. $mn = (\sec \theta + \tan \theta)(\sec \theta - \tan \theta)$
$= \sec^2 \theta - \tan^2 \theta = 1$

75. $\sin \theta + \cos \theta = \sqrt{2} \cos (90° - \theta)$
$\Rightarrow \sin \theta + \cos \theta = \sqrt{2} \sin \theta$
$\Rightarrow \sin \theta (\sqrt{2} - 1) = \cos \theta$
$\Rightarrow \cot \theta = \sqrt{2} - 1$

76. $2 \sin^2 \theta = \dfrac{1}{2}$
$\Rightarrow \sin^2 \theta = \dfrac{1}{4}$
$\Rightarrow \sin \theta = \dfrac{1}{2} = \sin 30° \Rightarrow \theta = 30°$

77. $\sin 30° = \dfrac{AB}{AP}$
$\Rightarrow \dfrac{1}{2} = \dfrac{60}{AP} \Rightarrow AP = 120$ cm
$\sin 30° = \dfrac{AD}{AQ}$
$\Rightarrow \dfrac{1}{2} = \dfrac{30}{AQ} \Rightarrow AQ = 60$ cm
$\therefore AP + AQ = 180$ cm

INTRODUCTION TO TRIGONOMETRY

78. $\sin 60° = \dfrac{BC}{EC} \Rightarrow \dfrac{\sqrt{3}}{2} = \dfrac{8}{CE}$

$\Rightarrow CE = \dfrac{16}{\sqrt{3}}$ cm

$\cos 60° = \dfrac{BE}{CE} \Rightarrow \dfrac{1}{2} = \dfrac{BE}{\frac{16}{\sqrt{3}}}$

$\Rightarrow BE = \dfrac{8}{\sqrt{3}}$ cm

79. $\sqrt{3} \tan \theta = 3 \sin \theta$

$\Rightarrow \sqrt{3} \dfrac{\sin \theta}{\cos \theta} = 3 \sin \theta$

$\Rightarrow \cos \theta = \dfrac{1}{\sqrt{3}} \Rightarrow \cos^2 \theta = \dfrac{1}{3}$

$\therefore \sin^2 \theta - \cos^2 \theta = (1 - \cos^2 \theta) - \cos^2 \theta$

$= 1 - 2 \cos^2 \theta = 1 - \dfrac{2}{3} = \dfrac{1}{3}$

80. $\dfrac{1}{\csc \theta} = \sin \theta$ and $-1 \le \sin \theta \le 1$

81. $\tan 90° = \infty$

82. $(\sin^2 60° + \cos^2 60°)^2 = 1^2 = 1$

83. Actually, $\sec^2 \theta = 1 + \tan^2 \theta$

84. $\sin \theta = \dfrac{1}{\csc \theta}$, $\cos \theta = \dfrac{1}{\sec \theta}$,

$\cot \theta = \dfrac{\cos \theta}{\sin \theta}$

85. $\sin \theta = 1 = \sin 90° \Rightarrow \theta = 90°$

$\cos \theta = \dfrac{1}{\sqrt{2}} = \cos 45° \Rightarrow \theta = 45°$

86. $\dfrac{y^2}{b^2} - \dfrac{x^2}{a^2} = \sec^2 \theta - \tan^2 \theta = 1$

87. $\cot \theta = \dfrac{\cos \theta}{\sin \theta} = \cos \theta \csc \theta = xy$

88. $\dfrac{1 + \tan^2 \theta}{\sec^4 \theta} = \dfrac{\sec^2 \theta}{\sec^4 \theta} = \dfrac{1}{\sec^2 \theta} = \cos^2 \theta$

89. $1 + \tan 5° \cot 85°$

$= 1 + \tan 5° \cot (90° - 5°)$

$= 1 + \tan 5° \tan 5°$

$= 1 + \tan^2 5° = \sec^2 5°$

90. $4 \sin 30° \cos 60° = 4 \cdot \dfrac{1}{2} \cdot \dfrac{1}{2} = 1$

91. $8 \csc^2 A - 8 \cot^2 \theta$

$= 8 (\csc^2 A - \cot^2 A)$

$= 8 (1) = 8$

92. $\dfrac{\sin \theta}{1 + \cos \theta} = \dfrac{\sin \theta (1 - \cos \theta)}{1 - \cos^2 \theta}$

$= \dfrac{\sin \theta (1 - \cos \theta)}{\sin^2 \theta} = \dfrac{1 - \cos \theta}{\sin \theta}$

93. $\sec^4 A - \sec^2 A = \sec^2 A (\sec^2 - 1)$

$= \sec^2 A \tan^2 A$

$= (1 + \tan^2 A) \tan^2 A$

$= \tan^2 A + \tan^4 A$

94. $\sqrt{\dfrac{1 + \sin \theta}{1 - \sin \theta}} = \sqrt{\dfrac{(1 + \sin \theta)(1 + \sin \theta)}{1 - \sin^2 \theta}}$

$= \sqrt{\dfrac{(1 + \sin \theta)^2}{\cos^2 \theta}} = \dfrac{1 + \sin \theta}{\cos \theta}$

$= \sec \theta + \tan \theta$

95. $\sqrt{\dfrac{1 + \cos \theta}{1 - \cos \theta}} = \sqrt{\dfrac{(1 + \cos \theta)^2}{1 - \cos^2 \theta}}$

$= \dfrac{1 + \cos \theta}{\sin \theta} = \csc \theta + \cos \theta$

96. $m^2 + n^2 = (a^2 \cos^2 \theta + b^2 \sin^2 \theta$

$+ 2ab \cos \theta \sin \theta) + (a^2 \sin^2 \theta + b^2 \cos^2 \theta$

$- 2ab \sin \theta \cos \theta) = a^2 + b^2$

97. $x^2 + y^2 + z^2 = r^2 \sin^2 \theta (\cos^2 \phi + \sin^2 \phi)$

$+ r^2 \cos^2 \theta$

$= r^2 (\sin^2 \theta + \cos^2 \theta) = r^2$

98. $\sin A = \cos B = \sin (90° - B)$

$\Rightarrow A = 90° - B \Rightarrow A + B = 90°$

99. $\sin 3\theta = \cos (\theta - 6°)$

$\Rightarrow \cos (90° - 3\theta) = \cos (\theta - 6°)$

$\Rightarrow 90° - 3\theta = \theta - 6° \Rightarrow \theta = 24°$

100. $\tan 2\theta = \cot (\theta + 6°)$

$\Rightarrow \cot (90° - 2\theta) = \cot (\theta + 6°)$

$\Rightarrow 90° - 2\theta = \theta + 6°$

$\Rightarrow 3\theta = 84° \Rightarrow \theta = 28°.$

Chapter 9
Some Applications of Trigonometry

1. The angle of elevation of the top of a tower from a point at a distance of 100 m from the base of the tower is found to be 45°. The height of the tower is
   (a) 50 m (b) 100 m
   (c) $50\sqrt{2}$ m (d) $50\sqrt{3}$ m.

2. Find the angle of elevation of the top of a tower, whose height is 100 m, at a point whose distance from the base of the tower is 100 m.
   (a) 30° (b) 60°
   (c) 45° (d) none of these.

3. The angle of elevation of the top of a tree from a point at a distance of 200 m from its base is 60°. The height of the tree is
   (a) $50\sqrt{3}$ m (b) $100\sqrt{3}$ m
   (c) $200\sqrt{3}$ m (d) $\dfrac{200}{\sqrt{3}}$ m.

4. Find the angle of elevation of the top of a tree of height $200\sqrt{3}$ m at a point at a distance of 200 m from the base of the tree.
   (a) 30° (b) 45°
   (c) 60° (d) none of these.

5. From a bridge, 25 m high, the angle of depression of a boat is 45°. Find the horizontal distance of the boat from the bridge.
   (a) 25 m (b) $\dfrac{25}{2}$ m
   (c) 50 m (d) $\dfrac{25}{\sqrt{3}}$ m.

6. Find the angle of depression of a boat from the bridge at a horizontal distance of 25 m from the bridge if the height of the bridge is 25 m.
   (a) 45° (b) 60°
   (c) 30° (d) 15°.

7. If the shadow of 10 m high tree is $10\sqrt{3}$ m, then find the angle of elevation of the Sun.
   (a) 60° (b) 90°
   (c) 45° (d) 30°.

8. Find the length of the shadow of 10 m high tree if the angle of elevation of the Sun is 30°.
   (a) 10 m (b) $\dfrac{10}{\sqrt{3}}$ m
   (c) $10\sqrt{3}$ m (d) 20 m.

9. The length of the shadow of a tower is equal to its height. Find the angle of elevation of the Sun.
   (a) 30° (b) 45°
   (c) 60° (d) 90°.

10. If the angle of elevation of the Sun is 45°, then find the length of the shadow of a tower whose height is h m.
    (a) $\dfrac{h}{2}$ m (b) h m
    (c) $2h$ m (d) $h\sqrt{3}$ m.

11. If the shadow of a tree is $\dfrac{1}{\sqrt{3}}$ times the height of the tree, then find the angle of elevation of the Sun.
    (a) 30° (b) 90°
    (c) 45° (d) 60°.

SOME APPLICATIONS OF TRIGONOMETRY

12. If the angle of elevation of the Sun is 60°, then find the ratio of the height of a tree with its shadow.
    (a) $\sqrt{3} : 1$
    (b) $1 : \sqrt{3}$
    (c) $3 : 1$
    (d) $1 : 3$.

13. From the top of a 10 m high tower, the angle of depression of a point on the ground is found to be 30°. Find the distance of the point from the base of the tower.
    (a) 10 m
    (b) $10\sqrt{3}$ m
    (c) $\dfrac{10}{\sqrt{3}}$ m
    (d) $5\sqrt{3}$ m.

14. Find the angle of depression, of a point on the ground at a distance of $10\sqrt{3}$ from the base of tower, from the top of the tower whose height is 10 m.
    (a) 45°
    (b) 30°
    (c) 60°
    (d) 90°.

15. The slope of a hill makes an angle of 60° with the horizontal. If one has to walk 500 m to reach the top of the hill, then the height of the hill is
    (a) $500\sqrt{3}$ m
    (b) $\dfrac{500}{\sqrt{3}}$ m
    (c) $250\sqrt{3}$ m
    (d) $\dfrac{250}{\sqrt{3}}$ m.

16. The angle of depression of a point on the horizontal from the slope of a hill is 60°. If one has to walk 300 m to reach the top from this point, then the distance of this point from the base of the hill is
    (a) $300\sqrt{3}$ m
    (b) 150 m
    (c) $150\sqrt{3}$ m
    (d) $\dfrac{150}{\sqrt{3}}$ m.

17. The angle of elevation of the Sun is 45°. Then, the length of the shadow of a 12 m high tree is
    (a) $6\sqrt{3}$ m
    (b) $12\sqrt{3}$ m
    (c) $\dfrac{12}{\sqrt{3}}$ m
    (d) 12 m.

18. The upper end of a ladder reaches the top of a wall. The lower end of the ladder is at a distance of 1.5 m from the wall and makes an angle of 60° from the plane. The height of the wall is
    (a) $3\sqrt{3}$ m
    (b) $\sqrt{3}$ m
    (c) $\dfrac{\sqrt{3}}{2}$ m
    (d) $\dfrac{3\sqrt{3}}{2}$ m.

19. The length of the string of a kite is 100 m which makes an angle of 60° with the horizontal. Assuming the string to be a straight line and one end of the string on the ground, find the height of the kite from the ground.
    (a) $50\sqrt{3}$ m
    (b) 50 m
    (c) $\dfrac{50}{\sqrt{3}}$ m
    (d) 100 m.

20. From the upper end of a vertical pole situated in a horizontal plane, a tight string is tied and the other end of the string lies on the ground. A circus artist climbs the string from the ground. The height of the pole is 12 m and the string makes an angle of 30° from the ground. Find the distance traversed by the artist in reaching the upper end of the pole.
    (a) 12 m
    (b) 18 m
    (c) 24 m
    (d) 30 m.

21. The length of that portion of a bridge over a river, which is exactly above the river, is 150 m and the bridge makes an angle of 45° with the bank of the river. Find the width of the river.
    (a) $\dfrac{150}{\sqrt{2}}$ m
    (b) 75 m
    (c) $150\sqrt{3}$ m
    (d) $150\sqrt{2}$ m.

22. At a distance of 28.5 m from a tower, 30 m high, an observer of height 1.5 m is standing. Find the angle of elevation of the top of the tower from the eye of the observer.
    (a) 30°
    (b) 45°
    (c) 60°
    (d) none of these.

23. A tree breaks into two parts due to heavy wind such that the upper part makes an angle of 30° with the plane. The place where the upper part of the tree touches the ground is at a distance of 10 m from the base point of the tree. Find the height of the tree.

    (a) $10\sqrt{3}$ m (b) $10\sqrt{2}$ m
    (c) $\dfrac{10}{\sqrt{3}}$ m (d) $5\sqrt{2}$ m.

24. The angles of elevation two points at distances a and b in a horizontal line through the base of the tower, of the top of the tower are complementary to each other. Then, the height of the tower is

    (a) $a + b$ (b) ab
    (c) $\sqrt{ab}$ (d) $2ab$.

25. From the top of the tower, the angles of depression of two points at distances 4 m and 9 m from the base of the tower are complementary to each other. The height of the tower is

    (a) 3 m (b) 6 m
    (c) 8 m (d) 12 m.

26. If the shadow of a tree is $\sqrt{3}$ times its height, then find the angle of elevation of the Sun.

    (a) 30° (b) 60°
    (c) 45° (d) none of these.

27. The angle of depression of a boat from a 50 m high bridge is 30°. Find the horizontal distance of the boat from the bridge.

    (a) $50\sqrt{3}$ m (b) 50 m
    (c) $25\sqrt{3}$ m (d) 25 m.

28. The angle of elevation of an aeroplane, flying at a height of 1500 m from the ground, is found to be 60° from the airport. Find the horizontal distance of the aeroplane from the airport.

    (a) $500\sqrt{3}$ m (b) $1500\sqrt{3}$ m
    (c) $1000\sqrt{3}$ m (d) 1500 m.

29. A steel pole is 10 m high. To keep the pole upright, one end of a steel wire is tied to the top of the pole while the other end has been fixed on the ground. If the steel wire makes an angle of 45° with the horizontal through the base point of the pole, then find the length of the steel wire.

    (a) $10\sqrt{2}$ m (b) $10\sqrt{3}$ m
    (c) 5 m (d) $5\sqrt{2}$ m.

30. The angle of elevation of the top of a tower from a point situated at a distance of 100 m from the base of tower is 30°. Find the height of the tower.

    (a) $\dfrac{100}{\sqrt{3}}$ m (b) $100\sqrt{3}$ m
    (c) $\dfrac{50}{\sqrt{3}}$ m (d) $50\sqrt{3}$ m.

31. The angle of elevation of a bird sitting on the top of a tree as seen from the point at a distance of 20 m from the base of the tree is 60°. Find the height of the tree.

    (a) $20\sqrt{3}$ m (b) $10\sqrt{3}$ m
    (c) 20 m (d) 10 m.

32. The angle of elevation of the top of a pillar from a point on the ground is 15°. On walking 100 m towards the pillar, the angle of elevation becomes 30°. Find the height of the pillar.

    (a) 25 m (b) 50 m
    (c) $50\sqrt{2}$ m (d) $25\sqrt{2}$ m.

33. A tower stands on a horizontal plane. The shadow of the tower when the angle of elevation of the Sun is 30° is 45 m more than when the angle of elevation of the Sun is 60°. Find the height of the tower.

    (a) $\dfrac{45\sqrt{3}}{2}$ m (b) $45\sqrt{3}$ m
    (c) $45\sqrt{2}$ m (d) $\dfrac{45}{\sqrt{3}}$ m.

SOME APPLICATIONS OF TRIGONOMETRY

ANSWERS

1. (b)   2. (c)   3. (c)   4. (c)   21. (a)   22. (b)   23. (a)   24. (c)
5. (a)   6. (a)   7. (d)   8. (c)   25. (b)   26. (a)   27. (a)   28. (a)
9. (b)   10. (b)   11. (d)   12. (a)   29. (a)   30. (a)   31. (a)   32. (b)
13. (b)   14. (b)   15. (c)   16. (b)   33. (a).
17. (d)   18. (d)   19. (a)   20. (c)

HINTS/SOLUTIONS

1. $\tan 45° = \dfrac{AB}{BC}$

   $\Rightarrow 1 = \dfrac{AB}{100}$

   $\Rightarrow AB = 100$ m.

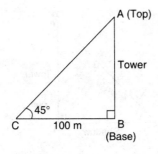

2. $\tan \theta = \dfrac{AB}{BC}$

   $\Rightarrow \tan \theta = \dfrac{100}{100} = 1 = \tan 45°$

   $\Rightarrow \theta = 45°$.

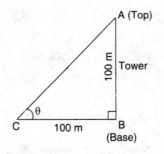

3. $\tan 60° = \dfrac{AB}{BC}$

   $\Rightarrow \sqrt{3} = \dfrac{AB}{200}$

   $\Rightarrow AB = 200\sqrt{3}$ m.

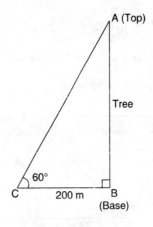

4. $\tan \theta = \dfrac{AB}{BC} = \dfrac{200\sqrt{3}}{200} = \sqrt{3} = \tan 60°$

   $\Rightarrow \theta = 60°$.

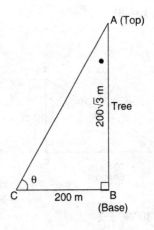

5. $\tan 45° = \dfrac{AB}{BC}$

$\Rightarrow \quad 1 = \dfrac{25}{BC} \Rightarrow BC = 25$ m.

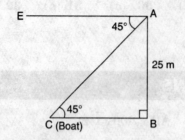

6. $\tan \theta = \dfrac{AB}{BC}$

$\Rightarrow \quad \tan \theta = \dfrac{25}{25} = 1 = \tan 45°$

$\Rightarrow \quad \theta = 45°.$

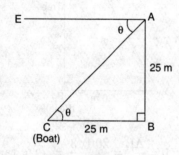

7. $\tan \theta = \dfrac{AB}{BC} = \dfrac{10}{10\sqrt{3}} = \dfrac{1}{\sqrt{3}} = \tan 30°$

$\Rightarrow \quad \theta = 30°.$

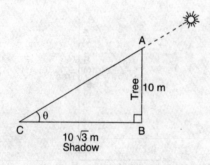

8. $\tan 30° = \dfrac{AB}{BC}$

$\Rightarrow \quad \dfrac{1}{\sqrt{3}} = \dfrac{10}{BC}$

$\Rightarrow BC = 10\sqrt{3}$ m.

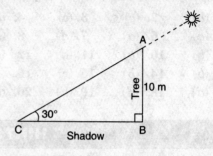

9. $\tan \theta = \dfrac{AB}{BC} = 1 = \tan 45°$

$\Rightarrow \quad \theta = 45°.$

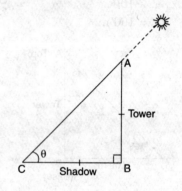

10. $\tan 45° = \dfrac{AB}{BC}$

$\Rightarrow \quad 1 = \dfrac{h}{BC}$

$\Rightarrow BC = h$ m.

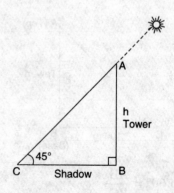

11. $\cot \theta = \dfrac{BC}{AB} = \dfrac{1}{\sqrt{3}} = \cot 60°$

$\Rightarrow \quad \theta = 60°.$

SOME APPLICATIONS OF TRIGONOMETRY

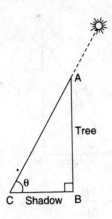

12. $\tan 60° = \dfrac{AB}{BC}$

$\Rightarrow \sqrt{3} = \dfrac{AB}{BC}$

$\Rightarrow AB : BC = \sqrt{3} : 1.$

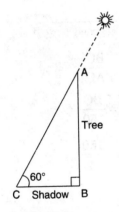

13. $\tan 30° = \dfrac{AB}{BC}$

$\Rightarrow \dfrac{1}{\sqrt{3}} = \dfrac{10}{BC}$

$\Rightarrow BC = 10\sqrt{3}$ m.

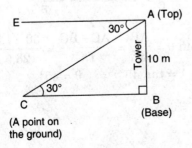

14. $\tan \theta = \dfrac{AB}{BC} = \dfrac{10}{10\sqrt{3}} = \dfrac{1}{\sqrt{3}} = \tan 30°$

$\Rightarrow \theta = 30°.$

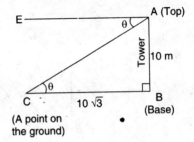

15. $\sin 60° = \dfrac{AB}{AC}$

$\Rightarrow \dfrac{\sqrt{3}}{2} = \dfrac{AB}{500}$

$\Rightarrow AB = 250\sqrt{3}$ m.

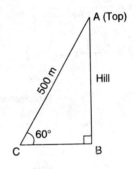

16. $\cos 60° = \dfrac{BC}{AC}$

$\Rightarrow \dfrac{1}{2} = \dfrac{BC}{300}$

$\Rightarrow BC = 150$ m.

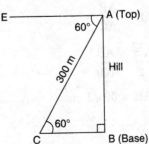

17. $\tan 45° = \dfrac{AB}{BC}$

$\Rightarrow 1 = \dfrac{12}{BC} \Rightarrow BC = 12$ m.

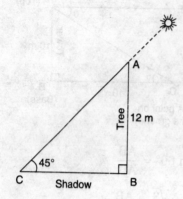

18. $\tan 60° = \dfrac{AB}{1.5}$

$\Rightarrow \sqrt{3} = \dfrac{2AB}{3}$

$\Rightarrow AB = \dfrac{3\sqrt{3}}{2}$ m.

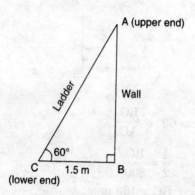

19. $\sin 60° = \dfrac{AB}{AC}$

$\Rightarrow \dfrac{\sqrt{3}}{2} = \dfrac{AB}{100}$

$\Rightarrow AB = 50\sqrt{3}$ m.

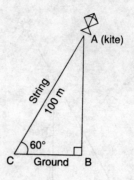

20. $\sin 30° = \dfrac{AB}{AC}$

$\Rightarrow \dfrac{1}{2} = \dfrac{12}{AC} \Rightarrow AC = 24$ m.

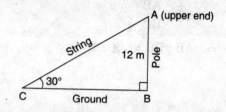

21. $\cos 45° = \dfrac{BC}{AC}$

$\Rightarrow \dfrac{1}{\sqrt{2}} = \dfrac{BC}{150}$

$\Rightarrow BC = \dfrac{150}{\sqrt{2}}$ m.

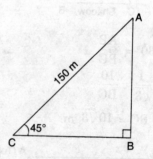

22. $\tan \theta = \dfrac{AB}{BE} = \dfrac{AC - BC}{DC} = \dfrac{30 - 1.5}{28.5}$

$= 1 = \tan 45° \Rightarrow \theta = 45°$.

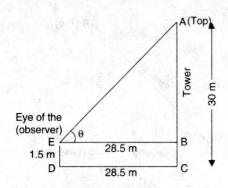

23. $\tan 30° = \dfrac{BC}{CD}$

$\Rightarrow \dfrac{1}{\sqrt{3}} = \dfrac{BC}{10} \Rightarrow BC = \dfrac{10}{\sqrt{3}}$ m

$\cos 30° = \dfrac{CD}{BD} \Rightarrow \dfrac{\sqrt{3}}{2} = \dfrac{10}{BD}$

$\Rightarrow BD = \dfrac{20}{\sqrt{3}}$

$\therefore AC = AB + BC = BD + CD$

$= \dfrac{20}{\sqrt{3}} + \dfrac{10}{\sqrt{3}} = \dfrac{30}{\sqrt{3}} = 10\sqrt{3}$ m.

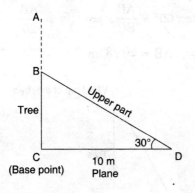

24. $\tan \theta = \dfrac{AB}{BD} = \dfrac{AB}{b}$ ...(1)

$\tan(90° - \theta) = \dfrac{AB}{BC} = \dfrac{AB}{a}$

$\Rightarrow \cot \theta = \dfrac{AB}{a}$ ...(2)

$\therefore \tan \theta \cdot \cot \theta = \dfrac{AB}{b} \cdot \dfrac{AB}{a}$

$\Rightarrow 1 = \dfrac{AB^2}{ab} \Rightarrow AB = \sqrt{ab}$.

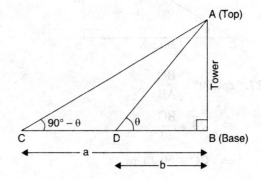

25. $\tan \theta = \dfrac{AB}{BC} = \dfrac{AB}{4}$ ...(1)

$\tan(90° - \theta) = \dfrac{AB}{BD}$

$\Rightarrow \cot \theta = \dfrac{AB}{9}$ ...(2)

$\therefore \tan \theta \cot \theta = \dfrac{AB}{4} \cdot \dfrac{AB}{9}$

$\Rightarrow 1 = \dfrac{AB^2}{36} \Rightarrow AB = 6$ m.

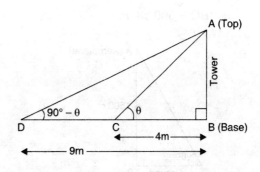

26. $\cot \theta = \dfrac{BC}{AB} = \sqrt{3} = \cot 30°$

$\Rightarrow \theta = 30°$.

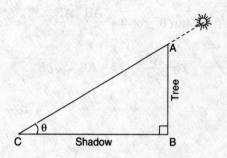

27. $\cot 30° = \dfrac{BC}{AB}$

$\Rightarrow \sqrt{3} = \dfrac{BC}{50}$

$\Rightarrow BC = 50\sqrt{3}$ m.

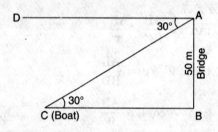

28. $\cot 60° = \dfrac{BC}{AB}$

$\Rightarrow \dfrac{1}{\sqrt{3}} = \dfrac{BC}{1500}$

$\Rightarrow BC = 500\sqrt{3}$ m.

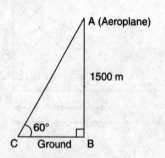

29. $\sin 45° = \dfrac{AB}{AC}$

$\Rightarrow \dfrac{1}{\sqrt{2}} = \dfrac{10}{AC}$

$\Rightarrow AC = 10\sqrt{2}$ m.

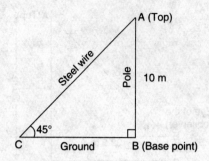

30. $\tan 30° = \dfrac{AB}{BC}$

$\Rightarrow \dfrac{1}{\sqrt{3}} = \dfrac{AB}{100} \Rightarrow AB = \dfrac{100}{\sqrt{3}}$ m.

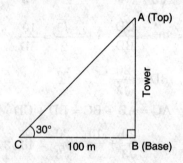

31. $\tan 60° = \dfrac{AB}{BC} \Rightarrow \sqrt{3} = \dfrac{AB}{20}$

$\Rightarrow AB = 20\sqrt{3}$ m.

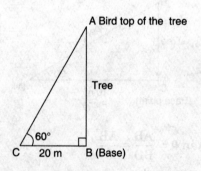

32. $\angle CAD = 30° - 15° = 15° = \angle ADC$

$\therefore \quad CA = CD = 100$ m

$\sin 30° = \dfrac{AB}{AC}$

SOME APPLICATIONS OF TRIGONOMETRY

$\Rightarrow \quad \dfrac{1}{2} = \dfrac{AB}{100} = AB = 50$ m.

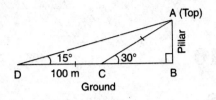

33. $\angle DAC = \angle ACD = 30°$
   $\therefore \quad AD = CD = 45$ m
   $\sin 60° = \dfrac{AB}{AD}$

$\Rightarrow \quad \dfrac{\sqrt{3}}{2} = \dfrac{AB}{45}$

$\Rightarrow \quad AB = \dfrac{45\sqrt{3}}{2}$ m.

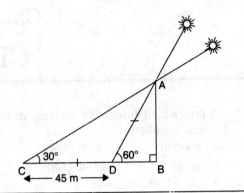

Chapter 10
Circles

1. A line which intersects a circle in two points is called
   (a) a secant (b) a chord
   (c) an arc (d) a tangent.

2. A line which intersects a circle in only one point is called
   (a) a secant (b) a tangent
   (c) a chord (d) a diameter.

3. A tangent to a circle intersects the circle in
   (a) one point only (b) two points
   (c) no point (d) three points.

4. A secant of a circle intersects the circle in
   (a) one point only (b) two points
   (c) three points (d) no point.

5. The point where a tangent line intersects a circle is called the
   (a) centre (b) point of contact
   (c) end-point (d) none of these.

6. The angle between the tangent at any point of a circle and the radius through the point of contact is
   (a) 60° (b) 90°
   (c) 45° (d) 30°.

7. How many tangents can be drawn to a circle at any point of it?
   (a) 1 (b) 2
   (c) 3 (d) none of these.

8. How many parallel tangents can a circle have at the most?
   (a) 1 (b) 2
   (c) 4 (d) 3.

9. Two circles of radii 5 cm and 3 cm touch each other externally. The distance between their centres is
   (a) 5 cm (b) 3 cm
   (c) 2 cm (d) 8 cm.

10. Two circles of radii 5 cm and 3 cm touch each other internally. The distance between their centres is
    (a) 5 cm (b) 3 cm
    (c) 2 cm (d) 8 cm.

11. A tangent PQ at a point P of a circle of radius 5 cm meets a line through the centre O at a point Q so that OQ = 12 cm. Length PQ is
    (a) 13 cm (b) 7 cm
    (c) 12 cm (d) $\sqrt{119}$ cm.

12. From a point Q, the length of the tangent to a circle is 24 cm and the distance of Q from the centre is 25 cm. The radius of the circle is
    (a) 7 cm (b) 5 cm
    (c) 12 cm (d) 1 cm.

13. If TP and TQ are the two tangents to a circle with centre O so that ∠POQ = 100°, then ∠PTQ is equal to
    (a) 60° (b) 80°
    (c) 50° (d) 120°.

14. If tangents PA and PB from a point P to a circle with centre O are inclined to each other at angle of 100°, then ∠POA is equal to
    (a) 50° (b) 20°
    (c) 30° (d) 40°.

15. The tangents at the end points of a diameter of a circle are
    (a) perpendicular (b) parallel
    (c) intersecting   (d) inclined at 60°.

16. The length of the tangent from a point A at distance 5 cm from the centre of the circle is 4 cm. The radius of the circle is
    (a) 3 cm   (b) 2 cm
    (c) 5 cm   (d) 4 cm.

17. Two concentric circles are of radii 5 cm and 3 cm. Find the length of the chord of the larger circle which touches the smaller circle.
    (a) 8 cm   (b) 6 cm
    (c) 4 cm   (d) 10 cm.

18. The angle between the two tangents drawn from an external point to a circle and the angle subtended by the line-segment joining the points of contact at the centre are
    (a) complementary (b) supplementary
    (c) equal   (d) 90° each.

19. From a point P, 10 cm away from the centre of a circle, a tangent PT of length 8 cm is drawn. Find the radius of the circle
    (a) 3 cm   (b) 4 cm
    (c) 5 cm   (d) 6 cm.

20. How many tangent lines can be drawn to a circle from a point outside the circle?
    (a) 1   (b) 2
    (c) none   (d) none of these.

21. Three circles are drawn with the vertices of a triangle as centres such that each circle touches the other two. If the sides of the triangle are 2 cm, 3 cm and 4 cm, find the diameter of the smallest circle.
    (a) 1 cm   (b) 3 cm
    (c) 5 cm   (d) 4 cm.

22. Two circles touch each other externally. The distance between the centres of the circle is 3 cm and the radius of one circle is 1 cm. Find the radius of the other circle.
    (a) 4 cm   (b) 2 cm
    (c) 1.5 cm (d) 2.5 cm.

23. Two circles touch each other internally. The distance between their centres is 1 cm and the radius of larger circle is 3.5 cm. Find the radius of the smaller circle.
    (a) 1 cm   (b) 2.5 cm
    (c) 2 cm   (d) 4.5 cm.

24. How many tangents can a circle have
    (a) 2   (b) infinitely many
    (c) one (d) no.

25. A circle is inscribed in a quadrilateral ABCD. Then,
    (a) AB + CD = BC + DA
    (b) AB + CD = AC + BD
    (c) AC + AD = BD + CD
    (d) AC + AD = BC + BD.

26. In the following figure, find ∠PBA.

    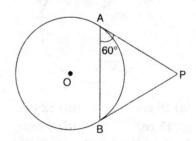

    (a) 60°   (b) 30°
    (c) 45°   (d) none of these.

27. In the following figure, find AC if AB = 5 cm

    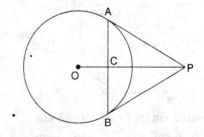

    (a) 2.5 cm   (b) 3 cm
    (c) 2 cm   (d) 1 cm.

28. In the following figure, find CD if AB = 5 cm

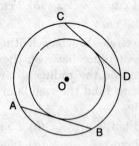

(a) 5 cm   (b) 4 cm
(c) 3 cm   (d) 2 cm.

29. In the following figure, find the perimeter of △APQ if AB = 6 cm.

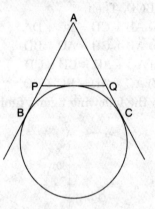

(a) 10 cm   (b) 12 cm
(c) 15 cm   (d) 6 cm.

30. In the following figure, find the length of the chord AB if PA = 6 cm and ∠PAB = 60°

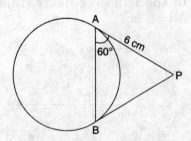

(a) 2 cm   (b) 4 cm
(c) 6 cm   (d) 5 cm.

31. In the following figure, △OQP is an isosceles triangle. Find ∠POQ.

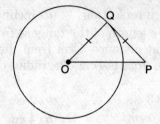

(a) 30°   (b) 60°
(c) 45°   (d) none of these.

32. In the following figure,

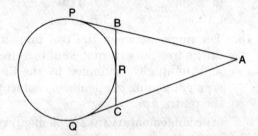

(a) AP = AB + BC + CA
(b) 3AP = AB + BC + CA
(c) 4AP = AB + BC + CA
(d) 2AP = AB + BC + CA.

33. In the following figure, OP = diameter of the circle. Then ∠APB is equal to

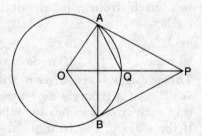

(a) 30°   (b) 60°
(c) 45°   (d) none of these.

CIRCLES

34. Two tangents BC and BD are drawn to a circle with centre O such that ∠CBD = 120°. Then OB =

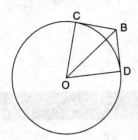

(a) 2BC (b) BC
(c) 3BC (d) $\dfrac{BC}{2}$.

35. In the following figure, find the radius of the inscribed circle.

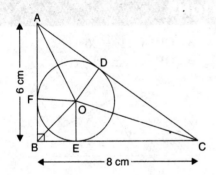

(a) 1 cm (b) 2 cm
(c) 3 cm (d) 4 cm.

36. In the following figure, OP = diameter of the circle. Then △APB is

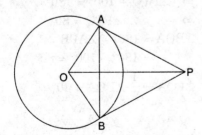

(a) isosceles (b) equilateral
(c) scalene   (d) right angled.

37. In the following figure, find the length of AB if CD = 2 cm.

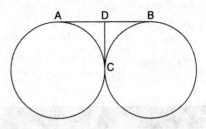

(a) 2 cm (b) 4 cm
(c) 6 cm (d) none of these.

38. In the following figure, find ∠ACB.

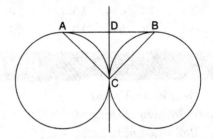

(a) 60° (b) 45°
(c) 90° (d) 30°.

39. In the following figure, CP and CQ are two tangent lines of a circle with centre O. ARB is another tangent line of the circle which touches the circle at the point R. If CP = 11 cm and BC = 7 cm, then find the length of BR.

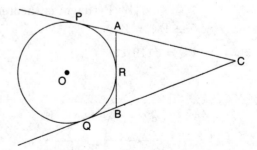

(a) 1 cm (b) 2 cm
(c) 3 cm (d) 4 cm.

40. In the following figure, ΔABC circumscribes a circle. Find the length of BC.
    (a) 6 cm
    (b) 8 cm
    (c) 10 cm
    (d) 12 cm.

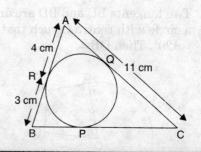

ANSWERS

1. (a)	2. (b)	3. (a)	4. (b)	21. (a)	22. (b)	23. (b)	24. (b)
5. (b)	6. (b)	7. (a)	8. (b)	25. (a)	26. (a)	27. (a)	28. (a)
9. (d)	10. (c)	11. (d)	12. (a)	29. (b)	30. (c)	31. (c)	32. (d)
13. (b)	14. (d)	15. (c)	16. (a)	33. (b)	34. (a)	35. (b)	36. (b)
17. (a)	18. (b)	19. (d)	20. (b)	37. (b)	38. (c)	39. (d)	40. (c)

HINTS/SOLUTIONS

1. Definition of secant
2. Definition of tangent
3. Definition of tangent
4. Definition of secant
5. Definition of point of contact
6. Theorem
7. Concept of tangent
8. Draw parallel tangents and visualise
9. Distance = 5 cm + 3 cm = 8 cm
10. Distance = 5 cm − 3 cm = 2 cm
11. $OP^2 + PQ^2 = OQ^2$
    | By Pythagoras theorem
    $\Rightarrow 5^2 + PQ^2 = 12^2$
    $\Rightarrow PQ = \sqrt{119}$ cm.

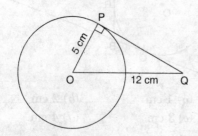

12. $OP^2 + PQ^2 = OQ^2$
    | By Pythagoras theorem

$\Rightarrow OP^2 + 24^2 = 25^2$
$\Rightarrow OP = 7$ cm

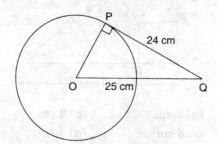

13. $\angle PTQ + \angle POQ = 180°$
    $\Rightarrow \angle PTQ + 100° = 180°$
    $\Rightarrow \angle PTQ = 80°$
14. $\angle BOA = 180° - \angle APB$
    $= 180° - 100° = 80°$
    $\angle POA = \dfrac{1}{2} \angle BOA = 40°$

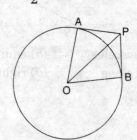

CIRCLES

15. ∠XAB + ∠SBA = 90° + 90° = 180°
∴ XY ∥ RS

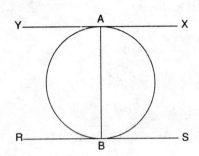

16. $OB^2 + AB^2 = OA^2$
    | By Pythagoras theorem
    ⇒ $OB^2 + 4^2 = 5^2$
    ⇒ OB = 3 cm

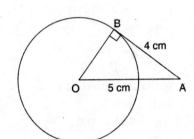

17. $OP^2 + AP^2 = OA^2$
    | By Pythagoras theorem
    ⇒ $3^2 + AP^2 = 5^2$
    ⇒ AP = 4 cm
    ∴ AB = 2AP = 8 cm

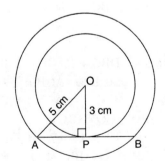

18. Theorem
19. $OT^2 + PT^2 = OP^2$
    | By Pythagoras theorem
    ⇒ $OT^2 + 8^2 = 10^2$
    ⇒ OT = 6 cm

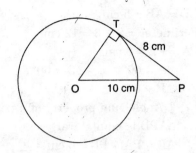

20. Visualise by drawing tangent lines
21. $r_1 + r_2 = 2$ cm
    $r_2 + r_3 = 3$ cm
    $r_3 + r_1 = 4$ cm
    ∴ $2(r_1 + r_2 + r_3) = 9$
    ∴ $r_1 + r_2 + r_3 = 4.5$
    ∴ $r_2 = 4.5 - 4 = 0.5$
    ∴ $2r_2 = 1$
22. Radius of the other circle = 3 cm − 1 cm
    = 2 cm
23. Radius of the smaller circle = 3.5 cm
    − 1 cm = 2.5 cm
24. Visualise geometrically
25. AB + CD = (AP + PB) + (CR + RD)
    = (AS + BQ) + (CQ + DS)
    = (AS + DS) + (BQ + CQ)
    = AD + BC

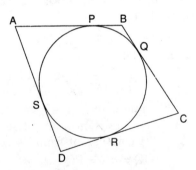

26. ∵ PA = PB
    ∴ ∠PAB = ∠PBA = 60°
    | Angles opposite to equal sides of a
    triangle are equal
27. ΔPAC ≅ ΔPBC
    | AAS congruence rule
    ∴ AC = BC = $\frac{1}{2}$ AB = $\frac{1}{2}$ (5) = 2.5 cm.

28. CD = AB = 5 cm
29. Perimeter = 2 × 6 = 12 cm
30. PA = PB = 6 cm
    ∠A = 60° ∴ ∠B = 60°
    ∴ ∠P = 60°
    | Angle sum property of a triangle
    ∴ ΔAPB is equilateral
    ∴ AB = PA = PB = 6 cm
31. ∠OQP = 90°
    ∵ The tangent at any point of a circle is perpendicular to the radius through the point of contact
    ∴ ∠OPQ + ∠POQ = 90°   ...(1)
    | Angle sum property of a triangle
    ∵ ΔOQP is isosceles
    ∴   OQ = PQ
    ∴ ∠OPQ = ∠POQ   ...(2)
    (1) and (2) give
    ∠POQ = 45°
32. AB + BC + CA
    = AB + (BR + RC) + CA
    = (AB + BR) + (RC + CA)
    = (AB + BP) + (CQ + CA)
    = AB + AC = AB + AB = 2AB
33. OP = 2OQ
    ∴ OQ = QP = r
    ∠OAP = 90°
    | Angle between tangent and radius through the point of contact
    ∵ The mid-point of the hypotenuse of a right angled triangle is equidistant from its vertices
    ∴   OQ = AQ = PQ
    ∴   OA = OQ = AQ = r
    ∴ ΔOAQ is equilateral
    ∴ ∠AQO = 60°
    ∴ ∠AQP = 120°
    ∵   AQ = QP
    ∴ ∠QAP = ∠APQ = 30°
    ∴ ∠APB = 2∠APQ = 60°

34. ∠CBO = $\frac{1}{2}$ ∠CBD = 60°
    ∠OCB = 90°
    $\cos 60° = \frac{BC}{OB}$
    ⇒ $\frac{1}{2} = \frac{BC}{OB}$
    ⇒ OB = 2BC
    AC = $\sqrt{AB^2 + BC^2}$
    | By Pythagoras theorem
    = $\sqrt{6^2 + 8^2}$ = 10 cm
35. Area of ΔABC = Area of ΔOBC + Area of ΔOCA + Area of ΔOAB
    ⇒ $\frac{6 \times 8}{2} = \frac{8 \times r}{2} + \frac{10 \times r}{2} + \frac{6 \times r}{2}$
    ⇒   r = 2 cm
36. ∠A = ∠B = ∠P = 60°
    ∴ ΔAPB is equilateral
37. AB = AD + BD
       = DC + DC
       = 2CD = 4 cm
38. ∵   DA = DC
    ∴ ∠DAC = ∠DCA   ...(1)
    ∵   DB = DC
    ∴ ∠DBC = ∠DCB   ...(2)
    Add (1) and (2) to get
    ∠DAC + ∠DBC = ∠DCA + ∠DCB
                = ∠ACB
    But,
    ∠DAC + ∠DBC + ∠ACB = 180°
    ∴   ∠ACB + ∠ACB = 180°
    ⇒   ∠ACB = 90°
39. BR = BQ = CQ − CB
       = CP − CB = 11 − 7 = 4 cm
40. BC = BP + PC = BR + CQ
       = 3 + (AC − AQ) = 3 + AC − AR
       = 3 + 11 − 4 = 10 cm.

Chapter 12*
Areas Related to Circles

1. If r is the radius of a circle, then the circumference of the circle is
   (a) πr (b) $2\pi r$
   (c) πr^2 (d) $\frac{1}{2}\pi r^2$.

2. If r is the radius of a circle, then the area of the circle is
   (a) πr (b) $2\pi r$
   (c) πr^2 (d) $\frac{1}{2}\pi r^2$.

3. If r is the radius of a circle, then the area of the semicircle is
   (a) πr (b) $2\pi r$
   (c) πr^2 (d) $\frac{1}{2}\pi r^2$.

4. If r is the radius of a circle, then the area of the quadrant of the circle is
   (a) πr^2 (b) $2\pi r$
   (c) $\frac{1}{4}\pi r^2$ (d) $\frac{1}{2}\pi r^2$.

5. What is the ratio between the circumference and diameter of a circle of radius r?
   (a) $\pi : 1$ (b) $1 : \pi$
   (c) $1 : 2$ (d) $2 : 1$.

6. The area of a sector of a circle of radius r and of angle $\theta°$ is
   (a) $\pi r^2 \frac{\theta}{360}$ (b) $2\pi r \frac{\theta}{360}$
   (c) $\pi r^2 \frac{\theta}{180}$ (d) $\pi r^2 \frac{\theta}{90}$.

7. Length of an arc of a sector of a circle of radius r and angle with degree measure θ is
   (a) $\frac{\theta}{360} \times 2\pi r$ (b) $\frac{\theta}{360} \times \pi r^2$
   (c) $\frac{\theta}{360} \times \frac{\pi r^2}{2}$ (d) $\frac{\theta}{360} \times \frac{\pi r^2}{4}$.

8. Length of the arc of a semicircle of radius r is
   (a) $2\pi r$ (b) πr
   (c) $2r$ (d) r.

9. Length of the arc of a quadrant of a circle of radius r is
   (a) $\frac{\pi r}{2}$ (b) πr
   (c) $2\pi r$ (d) $\frac{\pi r}{2} + 2r$.

10. The sum of areas of a major sector and the corresponding minor sector of a circle is equal to
    (a) area of the circle
    (b) $\frac{1}{2}$ area of the circle
    (c) $\frac{1}{4}$ area of the circle
    (d) $\frac{3}{4}$ area of the circle.

11. If the perimeter and the area of a circle are equal numerically, then the diameter of the circle is
    (a) 2 units (b) π units
    (c) 4 units (d) 7 units.

*MCQs of chapter 11 not required.

12. Which one of the following is not the approximate value of π?
    (a) $\frac{7}{22}$
    (b) $\frac{22}{7}$
    (c) 3.14
    (d) $\frac{355}{113}$.

13. The area of an annulus, whose internal and external radii are r_1 and r_2 respectively, is
    (a) $\pi(r_2^2 - r_1^2)$
    (b) $\pi(r_2 + r_1)$
    (c) $\pi(r_2 - r_1)$
    (d) $\pi r_1 r_2$.

14. The radius of a circle is 21 cm. Find its area.
    (a) 1386 cm²
    (b) 2π(21)
    (c) $\frac{\pi}{2}(21)^2$
    (d) $\frac{\pi}{4}(21)^2$.

15. The radius of a circle is 21 cm. Find its circumference.
    (a) 66 cm
    (b) 33 cm
    (c) 132 cm
    (d) 99 cm.

16. Find the radius of the circle whose circumference is 176 cm.
    (a) 7 cm
    (b) 14 cm
    (c) 21 cm
    (d) 28 cm.

17. Find the radius of the circle whose area is 154 cm².
    (a) 7 cm
    (b) 14 cm
    (c) 21 cm
    (d) 28 cm.

18. The radius of a semicircular compound is 35 m. Find its perimeter.
    (a) 45 m
    (b) 90 m
    (c) 135 m
    (d) 180 m.

19. Find the angle through which the minute hand of a clock moves from 6 p.m. to 6.35 pm.
    (a) 60°
    (b) 45°
    (c) 90°
    (d) 210°.

20. The radius of a circle is 18 cm and the angle subtended by an arc of this circle at the centre is 30°. Find the length of this arc.
    (a) π
    (b) 2π
    (c) 3π
    (d) 4π.

21. The radius of a circle is 5 cm. Find the area of the sector formed by an arc of this circle of length 9 cm.
    (a) 45 cm²
    (b) 22.5 cm²
    (c) 67.5 cm²
    (d) 2.25 cm².

22. A chord of a circle of radius 6 cm subtends an angle of measure 60° at the centre. Find the area of the minor segment formed by this chord.
    (a) $3(2\pi - 3\sqrt{3})$ cm²
    (b) $3(3\pi - \sqrt{3})$ cm²
    (c) $3(3\pi - 2\sqrt{3})$ cm²
    (d) $3(3\pi - 3\sqrt{3})$ cm².

23. The diameter of a semicircle is 8 cm. Its area is
    (a) 2π cm²
    (b) 4π cm²
    (c) 8π cm²
    (d) 16π cm².

24. The external and internal diameters of a circular path are 10 m and 6 m respectively. The area of the circular path is
    (a) 9π m²
    (b) 16π m²
    (c) 25π m²
    (d) 36π m².

25. The radii of two concentric circles are 6 cm and 4 cm. Find the ratio of their areas.
    (a) 9 : 4
    (b) 3 : 4
    (c) 9 : 2
    (d) 9 : 16.

26. The circumference of a circle inscribed in a square of side 14 cm is
    (a) 11 cm
    (b) 22 cm
    (c) 33 cm
    (d) 44 cm.

27. The area swept out by a horse tied in a rectangular grass field with a rope 8 m long is
    (a) 16π m²
    (b) 64π m²
    (c) 48π m²
    (d) 32π m².

28. The measure of the central angle of a circle is
    (a) 45°
    (b) 90°
    (c) 180°
    (d) 360°.

AREAS RELATED TO CIRCLES

29. The central angle of a quadrant of a circle measures
    (a) 90°   (b) 135°
    (c) 45°   (d) 180°.

30. What is the supplementary angle of the central angle of a semicircle?
    (a) 0°   (b) 90°
    (c) 180°   (d) 360°.

31. If the diameter of a semicircular protractor is 14 cm, then, find its perimeter.
    (a) 36 cm   (b) 18 cm
    (c) 27 cm   (d) 30 cm.

32. A square circumscribes a circle of radius 6 cm. Find the length of the diagonal of the square.
    (a) $5\sqrt{2}$ cm   (b) $12\sqrt{2}$ cm
    (c) $3\sqrt{2}$ cm   (d) $2\sqrt{2}$ cm.

33. Find the perimeter of a quadrant of a circle of radius 7 cm.
    (a) 10 cm   (b) 15 cm
    (c) 20 cm   (d) 25 cm.

34. The ratio of the areas of a circle and an equilateral triangle, whose diameter and side are equal, is
    (a) $\sqrt{3} : \pi$ (b) $\pi : \sqrt{2}$
    (c) $\pi : \sqrt{3}$ (d) $\pi : 1$.

35. A wire is in the shape of a circle of radius 42 cm. It is bent into a square. The side of the square is
    (a) 11 cm   (b) 22 cm
    (c) 33 cm   (d) 66 cm.

36. Find the circumference of a sector of a circle of radius 7 cm and central angle 45°.
    (a) 19.5 cm (b) 39 cm
    (c) 14 cm   (d) 7 cm.

37. In the following figure, O is the centre of the circle. The area of the sector $\triangle APB$ is $\dfrac{5}{18}$ part of the area of the circle. Find the value of x.

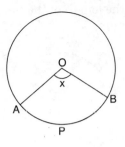

    (a) 30°   (b) 60°
    (c) 45°   (d) 100°.

38. In the following figure, find the perimeter of the figure where $\overset{\frown}{AED}$ is a semicircle and ABCD is a rectangle.

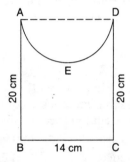

    (a) 152 cm   (b) 38 cm
    (c) 76 cm   (d) 104 cm.

39. Find the area of the major segment of a circle of the area of the minor segment is 25 cm² and the area of the circle is 100 cm².
    (a) 25 cm²   (b) 100 cm²
    (c) 75 cm²   (d) 50 cm².

40. The minute hand of a clock is of length 4 cm. Find the angle swept by the minute hand in 15 minutes.
    (a) 90°   (b) 30°
    (c) 45°   (d) 60°.

41. The length of the minute hand of a clock is $\sqrt{21}$ cm. Find the angle moved by the minute hand from 7 a.m. to 7.10 a.m.
    (a) 30°   (b) 60°
    (c) 45°   (d) 90°.

42. Find the length of the arc in the following figure.

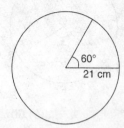

   (a) 11 cm   (b) 22 cm
   (c) 33 cm   (d) 44 cm.

43. The radii of two circles are 8 cm and 6 cm. Find the radius of the circle whose area is equal to the sum of the areas of these two circles.
   (a) 5 cm   (b) 10 cm
   (c) 8 cm   (d) 6 cm.

44. The radii of the circles are 20 cm and 10 cm. Find the radius of the circle whose circumference is equal to the sum of the circumferences of the two circles.
   (a) 10 cm   (b) 20 cm
   (c) 30 cm   (d) 15 cm.

45. A square is circumscribing a circle. The side of the square is 14 cm. Find the area of the square not included in the circle.

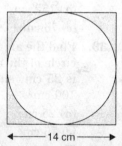

   (a) 21 cm^2 (b) 42 cm^2
   (c) 48 cm^2 (d) 196 cm^2.

46. The area of a circle whose circumference is 44 cm is
   (a) 77 cm^2 (b) 38.5 cm^2
   (c) 154 cm^2 (d) 115.5 cm^2.

47. Find the area of a right angled triangle if the radius of the semicircle is 3 cm and the altitude drawn to the hypotenuse is 4 cm.
   (a) 8 cm^2 (b) 12 cm^2
   (c) 10 cm^2 (d) 20 cm^2.

48. Find the area of a right angled triangle if the diameter of its circumcircle is 8 cm and altitude drawn to the hypotenuse is 2.5 cm.
   (a) 10 cm   (b) 15 cm
   (c) 20 cm   (d) 30 cm.

49. Find the area of a ring-shaped region enclosed between two concentric circles of radii 4 cm and 3 cm.
   (a) 22 cm^2 (b) 11 cm^2
   (c) 33 cm^2 (d) 44 cm^2.

50. A bicycle wheel makes 5000 revolutions in moving 11 km. Find the circumference of the wheel.
   (a) 55 cm   (b) 110 cm
   (c) 165 cm   (d) 220 cm.

ANSWERS

1. (b)   2. (c)   3. (d)   4. (c)   29. (a)   30. (a)   31. (a)   32. (b)
5. (a)   6. (a)   7. (a)   8. (b)   33. (d)   34. (c)   35. (d)   36. (a)
9. (a)   10. (a)   11. (c)   12. (a)   37. (d)   38. (c)   39. (c)   40. (a)
13. (a)   14. (a)   15. (c)   16. (d)   41. (b)   42. (b)   43. (b)   44. (d)
17. (a)   18. (d)   19. (d)   20. (c)   45. (b)   46. (c)   47. (b)   48. (a)
21. (b)   22. (a)   23. (c)   24. (b)   49. (a)   50. (d).
25. (a)   26. (d)   27. (b)   28. (d)

HINTS/SOLUTIONS

1. Formula
2. Formula
3. Formula
4. Formula
5. $\dfrac{\text{Circumference}}{\text{Diameter}} = \dfrac{2\pi r}{2r} = \dfrac{\pi}{1} = \pi : 1$
6. Formula
7. Formula
8. Formula
9. Formula
10. The sum of the angles subtended by the minor and major sectors of the circle at the centre of the circle is 360°.
11. $2\pi r = \pi r^2$
    $\Rightarrow r = 2$
    $\Rightarrow 2r = 4$
12. $\pi = \dfrac{22}{7}$ (approx.)
13. Formula
14. Area $= \pi r^2 = \dfrac{22}{7} \times 21 \times 21 = 1386$ cm^2
15. Circumference $= 2\pi r = 2 \times \dfrac{22}{7} \times 21$
    $= 132$ cm
16. $2 \times \pi \times r = 176$
    $\Rightarrow 2 \times \dfrac{22}{7} \times r = 176$
    $\Rightarrow r = 28$ cm
17. $\pi r^2 = 154$
    $\Rightarrow \dfrac{22}{7} \times r^2 = 154$
    $\Rightarrow r = 7$ cm
18. Perimeter $= \pi r + 2r$
    $= \dfrac{22}{7} \times 35 + 2 \times 35$
    $= 180$ m
19. Angle $= \dfrac{360°}{60°} \times 35 = 210°$
20. Length of arc $= \dfrac{30°}{360°} \times 2\pi \times 18 = 3\pi$
21. $2\pi r \dfrac{\theta}{360°} = 9$

$\Rightarrow 2 \times \pi \times 5 \times \dfrac{\theta}{360°} = 9$

$\Rightarrow \theta = \dfrac{324}{\pi}$

Area swept $= \pi r^2 \dfrac{\theta}{360°}$

$= \pi \times 5 \times 5 \times \dfrac{324}{\pi \times 360°}$

$= 22.5$ cm^2.

22. Area of minor segment
$= \pi (6)^2 \dfrac{60°}{360°} - \dfrac{\sqrt{3}}{4} (6)^2$
$= 6\pi - 9\sqrt{3} = 3(2\pi - 3\sqrt{3})$ cm^2

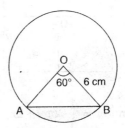

23. $r = \dfrac{8}{2} = 4$ cm

$\therefore$ Area $= \dfrac{1}{2}\pi r^2 = \dfrac{1}{2}\pi(4)^2 = 8\pi$ cm^2

24. Area of the circular path
$= \pi\left\{\left(\dfrac{10}{2}\right)^2 - \left(\dfrac{6}{2}\right)^2\right\} = 16\pi$ m^2

25. Ratio of areas $= \dfrac{\pi(6)^2}{\pi(4)^2} = \dfrac{9}{4} = 9 : 4$

26. Circumference $= 2\pi r$
$= 2\pi\left(\dfrac{14}{2}\right) = 2 \times \dfrac{22}{7} \times 7 = 44$ cm

27. Area swept out $= \pi r^2 = \pi(8)^2 = 64\pi$ m^2
28. The sum of two straight angles is 360°.
29. Central angle $= \dfrac{1}{4} \times 360° = 90°$

30. Central angle of a semicircle
$$= \frac{360°}{2} = 180°$$
∴ Its supplementary angle
$$= 180° - 180° = 0°$$

31. $r = \frac{14}{2} = 7$ cm
∴ Perimeter $= \pi r + 2r$
$$= \frac{22}{7} \times 7 + 2 \times 7 = 36 \text{ cm}$$

32. Side of square = 12 cm
∴ Length of diagonal $= 12\sqrt{2}$ cm

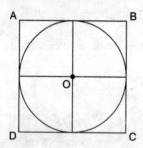

33. Perimeter $= \frac{1}{4}(2\pi r) + 2r$
$$= \frac{\pi r}{2} + 2r$$
$$= \frac{22}{7} \times \frac{1}{2} \times 7 + 2 \cdot 7 = 25 \text{ cm}$$

34. $\dfrac{\text{Area of circle}}{\text{Area of equilateral triangle}} = \dfrac{\pi \left(\frac{r}{2}\right)^2}{\frac{\sqrt{3}}{4}r^2}$

$$= \frac{\pi \frac{r^2}{4}}{\sqrt{3}\frac{r^2}{4}} = \pi : \sqrt{3}$$

35. Perimeter of square = Circumference of circle
$$= 2\pi r = 2 \times \frac{22}{7} \times 42 = 264$$

∴ Side of the square $= \frac{264}{4} = 66$ cm

36. Circumference $= r + r + 2\pi r \times \frac{45°}{360°}$
$$= 7 + 7 + 2 \cdot \frac{22}{7} \cdot 7 \cdot \frac{1}{8} = 19.5 \text{ cm}$$

37. $\pi r^2 \dfrac{x}{360°} = \dfrac{5}{18} \times \pi r^2$
$\Rightarrow x = 100°$

38. Perimeter of the figure
$$= 14 + 20 + 20 + \pi \left(\frac{14}{2}\right)$$
$$= 54 + \frac{22}{7} \times \frac{14}{2} = 76 \text{ cm}$$

39. Area of the major segment
= Area of the circle − Area of the minor segment
$$= 100 \text{ cm}^2 - 25 \text{ cm}^2 = 75 \text{ cm}^2$$

40. Angle swept $= \dfrac{360}{60} \times 15 = 90°$

41. Angle moved $= \dfrac{360}{60} \times 10 = 60°$

42. Length of arc $= 2 \times \dfrac{22}{7} \times 21 \times \dfrac{60}{360}$
$$= 22 \text{ cm}$$

43. $\pi r^2 = \pi(8)^2 + \pi(6)^2 \Rightarrow r = 10$ cm

44. $2\pi r = 2\pi(20) + 2\pi(10) \Rightarrow r = 30$

45. Required area
= Area of square − Area of circle
$$= 14 \times 14 - \pi \left(\frac{14}{2}\right)^2$$
$$= 196 - 154 = 42 \text{ cm}^2$$

46. $2\pi r = 44$
$\Rightarrow 2 \times \dfrac{22}{7} \times r = 44$
$\Rightarrow r = 7$
∴ $\pi r^2 = \dfrac{22}{7} \times 7 \times 7 = 154 \text{ cm}^2$

47. Area = $\dfrac{6 \times 4}{2}$ = 12 cm²

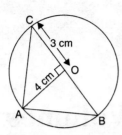

48. Area = $\dfrac{8 \times 2.5}{2}$ = 10 cm²

49. Area = $\pi \{(4)^2 - (3)^2\} = \dfrac{22}{7} \times 7$ = 22 cm²

50. Circumference = $\dfrac{11 \times 1000 \times 100}{5000}$

= 220 cm.

Chapter 13

Surface Areas and Volumes

1. A cylinder, a cone and a hemisphere are of equal base and have the same height. What is the ratio of their volumes?
   (a) 3 : 1 : 2   (b) 3 : 2 : 1
   (c) 1 : 2 : 3   (d) 1 : 3 : 2.

2. If S_1 denotes the total surface area of a sphere of radius r and S_2 denotes the total surface area of a cylinder of base radius r and height $2r$, then
   (a) $S_1 = S_2$ (b) $S_1 > S_2$
   (c) $S_1 < S_2$ (d) $S_1 = 2S_2$.

3. The volume of the greatest cylinder that can be cut from a solid wooden cube of length of edge 14 cm is
   (a) 2156 cm^3 (b) 1078 cm^3
   (c) 539 cm^3 (d) none of these.

4. Find the ratio of the volume of a cube to that of the sphere which will exactly fit inside the cube.
   (a) 6 : π (b) 4 : π
   (c) 2 : π (d) 3 : π.

5. The surface areas of two spheres are in the ratio 1 : 4. Then, the ratio of their volumes is
   (a) 1 : 4   (b) 1 : 8
   (c) 1 : 16   (d) 1 : 64.

6. The diameter of the base of a right circular cylinder is 28 cm and its height is 21 cm. Find the curved surface area.
   (a) 1848 cm^2 (b) 3080 cm^2
   (c) 12936 cm^2 (d) none of these.

7. The volume of a vessel in the form of a right circular cylinder is 448π cm^3 and its height is 7 cm. Find the radius of the base.
   (a) 4 cm   (b) 6 cm
   (c) 8 cm   (d) 2 cm.

8. The radius of the base and the height of a right circular cone are 7 cm and 24 cm respectively. Find the volume of the cone.
   (a) 1232 cm^3 (b) 616 cm^3
   (c) 308 cm^3 (d) none of these.

9. How many balls, each of radius 1 cm, can be made from a solid sphere of lead of radius 8 cm?
   (a) 64   (b) 216
   (c) 512   (d) 16.

10. A conical flask is full of water. The flask has base-radius r and height h. The water is poured into a cylindrical flask of base-radius mr. Find the height of water in the cylindrical flask.
    (a) $\dfrac{h}{3m^2}$ (b) $\dfrac{h}{3m}$
    (c) $\dfrac{3m^2}{h}$ (d) $\dfrac{3m}{h}$.

11. A cone and a hemisphere have equal bases and equal volumes. Find the ratio of their heights.
    (a) 2 : 1   (b) 3 : 1
    (c) 4 : 1   (d) 1 : 1.

12. The base radii of two right circular cones of the same height are in the ratio 3 : 5. Find the ratio of their volumes.
    (a) 3 : 5   (b) 5 : 3
    (c) 9 : 25   (d) 27 : 125.

13. The ratio of the volumes of two cones is 4 : 5 and the ratio of the radii of their bases is 2 : 3. Find the ratio of their vertical heights.
    (a) 4 : 5   (b) 9 : 5
    (c) 3 : 5   (d) 2 : 5.

14. The internal and external radii of a hemispherical bowl and r_1 and r_2 respectively. The surface area of the bowl is
    (a) $\pi(r_1^2 + r_2^2)$ (b) $2\pi(r_1^2 + r_2^2)$
    (c) $2\pi(r_2^2 - r_1^2)$ (d) $\pi(r_2^2 - r_1^2)$.

15. The ratio of the radii of two spheres is 4 : 5. Find the ratio of their total surface area is
    (a) 4 : 5   (b) $2 : \sqrt{5}$
    (c) 5 : 4   (d) 16 : 25.

16. The volume of a right circular cone of height 14 cm is 168π cm^3. The radius of the cone is
    (a) 6 cm   (b) 8 cm
    (c) 10 cm   (d) 12 cm.

17. If the diameter of a sphere is d, then its volume is
    (a) $\frac{1}{3}\pi d^3$ (b) $\frac{1}{24}\pi d^3$
    (c) $\frac{4}{3}\pi d^3$ (d) $\frac{1}{6}\pi d^3$.

18. Two solid right circular cones have the same height. The radii of their bases are r_1 and r_2. They are melted and recast into a cylinder of same height. Then, the radius of the base of the cylinder is
    (a) $\frac{r_1^2 + r_2^2}{3}$ (b) $\sqrt{\frac{r_1^2 + r_2^2}{3}}$
    (c) $\sqrt{r_1^2 + r_2^2}$ (d) $\sqrt{r_1 r_2}$.

19. Find the volume of the largest right circular cone that can be carved out of a solid hemisphere of radius r.
    (a) $\frac{4}{3}\pi r^3$ (b) $\frac{2}{3}\pi r^3$
    (c) $4\pi r^2$ (d) $\frac{1}{3}\pi r^3$.

20. If a cone is cut into two parts by a horizontal plane passing through the mid-point of the axis, the ratio of the volumes of the upper part and the cone is
    (a) 1 : 2   (b) 1 : 4
    (c) 1 : 8   (d) 1 : 6.

21. A solid is hemispherical at the bottom and conical above. If the surface areas of the two parts are equal, then the ratio of its radius and the height of its conical part is
    (a) $1 : \sqrt{3}$ (b) 1 : 1
    (c) $\sqrt{3} : 1$ (d) 1 : 3.

22. A solid sphere of radius r is melted and cast into the shape of a solid cone of height r. The radius of the base of the cone is
    (a) r (b) $2r$
    (c) $3r$ (d) $4r$.

23. A solid consists of a right circular cylinder at the top of which a right circular cone is fitted exactly. The height of the cone is H. The total volume of the solid is three times the volume of the cone. Then, the height of the right circular cylinder is
    (a) H   (b) $\frac{H}{3}$
    (c) $\frac{3H}{2}$ (d) $\frac{2H}{3}$.

24. A cylindrical rod of iron, whose height is 4 times its radius, is melted and cast into spherical balls of the same radius. The number of balls cast is
    (a) 1   (b) 2
    (c) 3   (d) 4.

25. How many lead balls of radius 2 cm can be made from a ball of radius 4 cm?
    (a) 1   (b) 2
    (c) 4   (d) 8.

26. Three shots are made by melting a solid metallic sphere of radius 6 cm. If the radii of the two shots are 3 cm and 4 cm respectively, then find the radius of the third shot.
    (a) 2 cm (b) 3 cm
    (c) 4 cm (d) 5 cm.

27. A hollow cylinder of height 3 cm is melted and cast into a solid cylinder of height 9 cm. If the internal and external radii of the hollow cylinder are 1 cm and 2 cm respectively, then find the radius of the solid cylinder.
    (a) 1 cm (b) 2 cm
    (c) 3 cm (d) 4 cm.

28. The volume and surface area of a sphere are numerically the same. Find the volume of the smallest cylinder in which the sphere is exactly kept.
    (a) 54π (b) 27π
    (c) 36π (d) 9π.

29. From a wooden cube of side a cm, a right circular cylinder of maximum volume is cut. Find the volume of this cylinder.
    (a) πa^3 cm^3 (b) $\dfrac{\pi a^3}{4}$ cm^3
    (c) $\dfrac{\pi a^2}{4}$ cm^3 (d) $\dfrac{4}{3}\pi a^3$ cm^3.

30. The volume of a frustum of a cone of height h and ends-radii r_1 and r_2 is
    (a) $\dfrac{1}{3}\pi h\,(r_1^2 + r_2^2 + r_1 r_2)$
    (b) $\dfrac{1}{3}\pi h\,(r_1^2 + r_2^2 - r_1 r_2)$
    (c) $\pi h\,(r_1^2 + r_2^2 + r_1 r_2)$
    (d) $\pi h\,(r_1^2 + r_2^2 - r_1 r_2)$.

ANSWERS

1. (a)	2. (c)	3. (a)	4. (a)	17. (d)	18. (b)	19. (d)	20. (c)
5. (b)	6. (a)	7. (c)	8. (a)	21. (a)	22. (b)	23. (d)	24. (c)
9. (c)	10. (a)	11. (a)	12. (c)	25. (d)	26. (d)	27. (a)	28. (a)
13. (b)	14. (b)	15. (d)	16. (a)	29. (b)	30. (a).		

HINTS/SOLUTIONS

1. $r_1 : r_2 : r_3 = \pi r^2 r : \dfrac{1}{3}\pi r^2 r : \dfrac{2}{3}\pi r^3$

    $= 1 : \dfrac{1}{3} : \dfrac{2}{3}$

    $= 3 : 1 : 2$

2. $S_1 = 4\pi r^2$
    $S_2 = 2\pi r^2 + 2\pi r(2r) = 6\pi r^2$
    $\therefore\ S_1 < S_2$

3. Required volume $= \pi r^2 h$

    $= \dfrac{22}{7} \times \left(\dfrac{14}{2}\right)^2 \times 14$

    $= \dfrac{22}{7} \times 7 \times 7 \times 14 = 2156$ cm^3

4. Required ratio $= \dfrac{\text{volume of cube}}{\text{volume of sphere}}$

    $= \dfrac{a^3}{\dfrac{4}{3}\pi\left(\dfrac{a}{2}\right)^3} = 6 : \pi$

5. $S_1 : S_2 = 1 : 4$
    $\Rightarrow\ 4\pi r_1^2 : 4\pi r_2^2 = 1 : 4$
    $\Rightarrow\ r_1^2 : r_2^2 = 1^2 : 2^2$
    $\Rightarrow\ \dfrac{r_1}{r_2} = \dfrac{1}{2}$

    $\therefore\ \dfrac{r_1}{r_2} = \dfrac{\dfrac{4}{3}\pi r_1^3}{\dfrac{4}{3}\pi r_2^3} = \left(\dfrac{r_1}{r_2}\right)^3 = \left(\dfrac{1}{2}\right)^3$

SURFACE AREAS AND VOLUMES

$$\therefore \frac{1}{8} = 1:8$$

6. $r = \frac{28}{2} = 14$ cm

   $h = 21$ cm

   $\therefore$ Curved surface area
   $= 2\pi r h$
   $= 2 \times \frac{22}{7} \times 14 \times 21 = 1848$ cm^2

7. $\pi r^2 h = 448\pi$
   $\Rightarrow \pi r^2 7 = 448\pi$
   $\Rightarrow r^2 = 64$
   $\Rightarrow r = 8$ cm

8. $r = 7$ cm
   $h = 24$ cm
   $\therefore v = \frac{1}{3}\pi r^2 h = \frac{1}{3} \times \frac{22}{7} \times (7)^2 \times (24)$
   $= 1232$ cm^3

9. Number of balls $= \dfrac{\frac{4}{3}\pi(8)^3}{\frac{4}{3}\pi(1)^3} = 512$

10. $\frac{1}{3}\pi r^2 h = \pi (mr)^2 H$

    $\Rightarrow H = \frac{h}{3m^2}$

11. $\frac{2}{3}\pi r^3 = \frac{1}{3}\pi r^2 h$
    $\Rightarrow 2r = h$
    $\Rightarrow \frac{h}{r} = \frac{2}{1} = 2:1$

12. $\frac{v_1}{v_2} = \dfrac{\frac{1}{3}\pi(3x)^2 h}{\frac{1}{3}\pi(5x)^2 h} = \frac{9}{25} = 9:25$

13. $\frac{v_1}{v_2} = \dfrac{\frac{1}{3}\pi r_1^2 h_1}{\frac{1}{3}\pi r_2^2 h_2} = \frac{r_1^2 h_1}{r_2^2 h_2} = \left(\frac{r_1}{r_2}\right)^2 \frac{h_1}{h_2}$

    $\Rightarrow \frac{4}{5} = \left(\frac{2}{3}\right)^2 \frac{h_1}{h_2}$

$\Rightarrow \frac{h_1}{h_2} = \frac{9}{5} = 9:5$

14. Surface area of the bowl
    = surface area of the internal surface
    + surface area of the external surface
    $= 2\pi r_1^2 + 2\pi r_2^2$
    $= 2\pi(r_1^2 + r_2^2)$

15. Ratio of total surface area
    $= \dfrac{4\pi r_1^2}{4\pi r_2^2} = \left(\frac{r_1}{r_2}\right)^2 = \left(\frac{4}{5}\right)^2 = \frac{16}{25}$

16. $\frac{1}{3}\pi r^2 h = 168\pi$

    $\Rightarrow \frac{r^2 h}{3} = 168$

    $\Rightarrow \frac{r^2}{3}(14) = 168$

    $\Rightarrow r = 6$ cm

17. Volume $= \frac{4}{3}\pi \left(\frac{d}{2}\right)^3 = \frac{1}{6}\pi d^3$

18. $\pi r^2 h = \frac{1}{3}\pi r_1^2 h + \frac{1}{3}\pi r_2^2 h$

    $\Rightarrow r = \sqrt{\frac{r_1^2 + r_2^2}{3}}$

19. Volume $= \frac{1}{3}\pi r^2 h = \frac{1}{3}\pi r^2 \cdot r$
    $= \frac{1}{3}\pi r^3$.

20. $\frac{r_2}{r_1} = \frac{h_2}{h_1} = \frac{1}{2}$

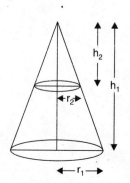

$$\therefore \quad \frac{v_2}{v_1} = \frac{\frac{1}{3}\pi r_2^2 h_2}{\frac{1}{3}\pi r_1^2 h_1} = \left(\frac{h_2}{h_1}\right)^3 = \left(\frac{1}{2}\right)^3 = \frac{1}{8}$$

$$= 1 : 8$$

21. $2\pi r^2 = \pi r \sqrt{r^2 + h^2}$

$\Rightarrow \quad 3r^2 = h^2 \quad \Rightarrow \quad \dfrac{r^2}{h^2} = \dfrac{1}{3}$

$\therefore$ Ratio $= \dfrac{r}{h} = \dfrac{1}{\sqrt{3}} = 1 : \sqrt{3}$

22. $\dfrac{4}{3}\pi r^3 = \dfrac{1}{3}\pi R^2 (r) \quad \Rightarrow \quad R = 2r$

23. Volume of solid = 3 volume of cone

$\Rightarrow \quad \dfrac{1}{3}\pi r^2 H + \pi r^2 h = 3 \cdot \left(\dfrac{1}{3}\pi r^2 H\right)$

$\Rightarrow \quad \pi r^2 h = \dfrac{2}{3}\pi r^2 H$

$\Rightarrow \quad h = \dfrac{2H}{3}$

24. Required number $= \dfrac{\pi r^2 (4r)}{\dfrac{4}{3}\pi r^3} = 3$

25. Number of balls $= \dfrac{\dfrac{4}{3}\pi(2)^3}{\dfrac{4}{3}\pi(1)^3} = 8$

26. $\dfrac{4}{3}\pi(6)^3 = \dfrac{4}{3}\pi(3)^2 + \dfrac{4}{3}\pi(4)^3 + \dfrac{4}{3}\pi(R)^3$

$\Rightarrow \quad R = 5$ cm

27. $\pi (2^2 - 1^2) 3 = \pi R^2 (9) \quad \Rightarrow \quad R = 1$ cm

28. $\dfrac{4}{3}\pi r^3 = 4\pi r^2 \quad \Rightarrow \quad r = 3$

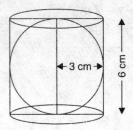

$\therefore$ Volume of cylinder $= \pi r^2 h$
$= \pi (3)^2 (3 + 3) = 54\pi$

29. Volume $= \pi \left(\dfrac{a}{2}\right)^2 \cdot a = \dfrac{\pi a^3}{4}$ cm^3

30. Formula.

Chapter 14
Statistics

1. The median class of the distribution

Class	Frequency
0–10	4
10–20	4
20–30	8
30–40	10
40–50	12
50–60	8
60–70	4

    is
    (a) 20–30 (b) 30–40
    (c) 40–50 (d) 60–70.

2. Which measure of central tendency is given by the x-coordinate of the point of intersection of the 'more than ogive' and 'less than ogive'?
    (a) Mean (b) Median
    (c) Mode (d) Weighted Mean.

3. Find the median class of the following data:

Marks obtained	Frequency
0–10	8
10–20	10
20–30	12
30–40	22
40–50	30
50–60	18

    (a) 0–10 (b) 20–30
    (c) 10–20 (d) 30–40.

4. Find the class marks of classes 10–25 and 35–55.
    (a) 10, 35 (b) 25, 55
    (c) 15, 20 (d) 17.5, 45.

5. What is the empirical relationship between the three measures of central tendency?
    (a) 3 median = mode – 2 mean
    (b) 3 median = mode + mean
    (c) 3 median = 2 mode + mean
    (d) 3 median = mode + 2 mean.

6. The most frequently used measure of central tendency is
    (a) mean (b) mode
    (c) median (d) none of these.

7. Which is the better measure of central tendency when individual observations are not important?
    (a) mode (b) median
    (c) mean (d) none of these.

8. The item value which occurs most frequently is called
    (a) mean (b) median
    (c) mode (d) none of these.

9. The class with maximum frequency is called
    (a) median class (b) modal class
    (c) mean class (d) average class.

10. The class, whose cumulative frequency is greater than (and nearest to) $\frac{n}{2}$, is called
    (a) average class (b) median class
    (c) modal class (d) mean class.

11. The lower limit of the class 0–10 is
    (a) 0   (b) 5
    (c) 10   (d) 1.

12. The upper limit of class 90–100 is
    (a) 90   (b) 95
    (c) 100   (d) none of these.

13. The width of the class interval 40–50 is
    (a) 40   (b) 50
    (c) 45   (d) 10.

14. The wickets taken over by a bowler in 10 cricket matches are as follows:
    2 6 4 5 0 2 1 3 2 3
    Find the mode of this data.
    (a) 0   (b) 1
    (c) 2   (d) 3.

15. Which of the following is correct ?
    (a) Class mark = $\dfrac{\text{upper class limit} + \text{lower class limit}}{2}$
    (b) Class mark = lower class limit
    (c) Class mark = upper class limit
    (d) Class mark = upper class limit − lower class limit.

16. A student draws a cumulative frequency curve for the marks obtained by 40 students of a class as shown below. Find the median marks obtained by the students of the class.

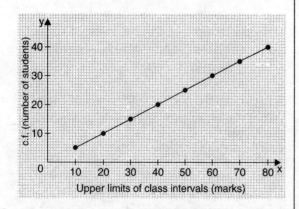

    (a) 20   (b) 30
    (c) 40   (d) 50.

17. What is the value of one median of the data using the graph given below of less than ogive and more than ogive?

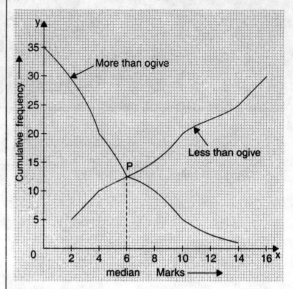

    (a) 2   (b) 4
    (c) 6   (d) 8.

18. The mean of a distribution is 8 and mode of the same distribution is 5. Find the median.
    (a) 5   (b) 6
    (c) 7   (d) 8.

19. The median and mode of a distribution are 7.5 and 6.5 respectively. Find the mean.
    (a) 4   (b) 8
    (c) 6.5   (d) 7.5.

20. The median and mean of a distribution are 3 and 3 each. Find the mode.
    (a) 1   (b) 2
    (c) 4   (d) 3.

STATISTICS

21. The cumulative frequency of the class 40–50 in the following distribution is

Class	Frequency
40–50	10
50–60	25
60–70	28
70–80	12
80–90	10
90–100	15

(a) 10   (b) 100
(c) 50   (d) 51.

22. The cumulative frequency of the class 40–50 in the following distribution is

Class	Frequency
0–10	6
10–20	10
20–30	13
30–40	7
40–50	4

(a) 6   (b) 40
(c) 36   (d) none of these.

23. The cumulative frequency of the class 30–35 in the following frequency distribution is

Expenditure on water (in rupees)	Number of houses
15–20	7
20–25	5
25–30	7
30–35	8
35–40	9
40–45	11
45–50	7
50–55	5
55–60	4
60–65	4
65–70	3

(a) 27   (b) 19
(c) 12   (d) 7.

24. What is the frequency of the class preceding the class 250–300, if the distribution is as follows?

Expenditure (in rupees)	Number of houses
100–150	24
150–200	40
200–250	33
250–300	28
300–350	30
350–400	22
400–450	16
450–500	7

(a) 24   (b) 33
(c) 40   (d) 30.

25. What is the frequency of the class succeeding the class 70–85 in the following frequency distribution?

Class	f_i
10–25	6
25–40	20
40–55	44
55–70	26
70–85	3
85–100	1

(a) 1   (b) 100
(c) 10   (d) 6.

26. The cumulative frequency of the class 55–58 is how much greater than the frequency of the class 58–61 in the following distribution?

Height (in cm)	Number of students
52–55	10
55–58	20
58–61	25
61–64	10

(a) 2   (b) 3
(c) 4   (d) 5.

27. What is the frequency of the modal class in the following distribution?

Marks	Number of students
0–10	5
10–20	12
20–30	14
30–40	10
40–50	8
50–60	6

(a) 14 (b) 8
(c) 6 (d) 5.

28. The frequency of the class preceding the modal class in the following frequency distribution is

Marks	Number of students
20–30	4
30–40	28
40–50	42
50–60	20
60–70	6

(a) 28 (b) 32
(c) 20 (d) 100.

29. The frequency of the class succeeding the modal class in the following frequency distribution is

Class	Frequency
10–15	3
15–20	7
20–25	16
25–30	12
30–35	9
35–40	5
40–45	3

(a) 3 (b) 6
(c) 9 (d) 12.

30. The cumulative frequency of the class preceding the median class in the following frequency distribution is

Class	f_i
10–25	6
25–40	20
40–55	44
55–70	26
70–85	3
85–100	1

(a) 26 (b) 32
(c) 6 (d) 100.

31. The mode of the distribution 2, 5, 7, 5, 3, 1, 5, 8, 7, 5 is
(a) 1 (b) 2
(c) 3 (d) 5.

32. The mode of the following distribution is 2, 4, 6, 2, 6, 6, 7, 8.
(a) 2 (b) 4
(c) 6 (d) 7.

33. If the mode of the following distribution is 2.8, then find the value of x.
2.5, 2.5, 2.1, 2.7, 2.8, 2.5, x, 2.8, 2.8, 2.7
(a) 2.8 (b) 2.7
(c) 2.5 (d) 2.1.

34. The number of family members of 30 families of a village is according to the following table. Find their mode.

Number of members	Number of families
2	1
3	2
4	4
5	6
6	10
7	3
8	5

(a) 2 (b) 4
(c) 3 (d) 6.

STATISTICS

35. The ages (in years) of 20 students of a class are as follows:
15 16 13 14 14 13 15 14 13 13
14 12 15 14 16 13 14 14 13 15
Their mode (in years) is
(a) 13 (b) 14
(c) 15 (d) 16.

36. The position average is
(a) mean (b) mode
(c) median (d) geometric mean.

37. The modal value of a series is
(a) middle most value
(b) value whose frequency is maximum
(c) value whose frequency is minimum
(d) limiting value.

38. The median value of the series
520, 20, 340, 190, 35, 800, 1210, 50, 80 is
(a) 1210 (b) 520
(c) 190 (d) 35.

39. The marks of four students in statistics are 53, 75, 42 and 70 respectively. The arithmetic mean of their marks is
(a) 42 (b) 64
(c) 60 (d) 56.

40. The mean of first ten odd natural numbers is
(a) 5 (b) 10
(c) 20 (d) 19.

41. The median of the following distribution is

x	f
1	8
2	10
3	11
4	16
5	20
6	25
7	15
8	9
9	6

(a) 1 (b) 2
(c) 3 (d) 5

42. The formula for finding mode in a grouped frequency distribution is
$$\text{Mode} = l + \frac{f_1 - f_0}{2f_1 - f_0 - f_2} \times h$$
What does f_1 represent here?
(a) Frequency of the modal class
(b) Frequency of the class preceding the modal class
(c) Frequency of the class succeeding the modal class
(d) Frequency of the last class.

43. The formula for finding out the median in a grouped frequency distribution is
$$\text{Median} = l + \frac{\frac{N}{2} - F}{f} \times h$$
What does F represent here?
(a) Frequency of the class preceding the median class
(b) Cumulative frequency of the class preceding the median class
(c) Frequency of the class succeeding the median class
(d) Cumulative frequency of the class succeeding the median class.

44. The mean of the frequency distribution

x	f
1	45
2	25
3	19
4	8
5	2
6	1

is
(a) 1 (b) 2
(c) 3 (d) 4.

45. Find the mode of the following distribution:

Size of the shoes	Number of shoes
4.5	1
5.0	2
5.5	4
6.0	5
6.5	15
7.0	30
7.5	60
8.0	95
8.5	82
9.0	75

(a) 6   (b) 7
(c) 8   (d) 9.

46. The curve of less than ogive in a cumulative frequency distribution is
(a) increasing
(b) decreasing
(c) circle
(d) straight line always.

47. The curve of more than ogive in a cumulative frequency distribution is
(a) increasing
(b) decreasing
(c) straight line always
(d) circle.

48. The mean of first n natural numbers is
(a) $\dfrac{n+1}{2}$ (b) $\dfrac{n(n+1)}{2}$
(c) $\dfrac{n(n-1)}{2}$ (d) n^2.

49. An ogive is a graphical representation of
(a) an ordinary frequency distribution
(b) a cumulative frequency distribution
(c) ungrouped item values
(d) none of these.

50. Which of the following formulae is true for finding mean?

(a) $\bar{x} = \dfrac{\sum_{i=1}^{n} x_i}{n}$ (b) $\bar{x} = \dfrac{\sum_{i=1}^{n} x_i^2}{n}$

(c) $\bar{x} = \dfrac{\sum_{i=1}^{n} x_i}{n^2}$ (d) $\bar{x} = \dfrac{\sum_{i=1}^{n} x_i^2}{n^2}$.

51. Which of the following is not a measure of central tendency?
(a) mean   (b) median
(c) mode   (d) range.

52. Which measure of central tendency cannot be determined graphically?
(a) mean   (b) mode
(c) median   (d) none of these.

53. The mode of a frequency distribution can be determined graphically by drawing
(a) ogive
(b) frequency curve
(c) frequency polygon
(d) histogram.

54. The algebraic sum of all the deviations of all the observations from their mean is always
(a) 0
(b) +ve
(c) –ve
(d) equal to the number of observations.

55. If mean = (3 median – mode) m, then $k =$
(a) 1   (b) $\dfrac{1}{2}$
(c) 2   (d) none of these.

56. Find the sum of the deviations of the variate values 3, 4, 6, 7, 8, 14 from their mean.
(a) 0   (b) 3
(c) 4   (d) 1.

57. The mean of n observations $x_1, x_2, x_3,$ ......, x_n is $\bar{x}$. If each observation is multiplied by p, then the mean of the new observations is
(a) $\dfrac{\bar{x}}{p}$ (b) $p\bar{x}$
(c) $\bar{x}$ (d) $p + \bar{x}$.

58. The mean of n observations $x_1, x_2, x_3, ..., x_n$ is $\bar{x}$. If each observation is divided by p_1 then the mean of the new observations is

(a) $\dfrac{\bar{x}}{p}$ (b) $p\bar{x}$

(c) $\bar{x}$ (d) $p + \bar{x}$.

59. If the mean of the following distribution is 6, find the value of p.

x	y
2	3
4	2
6	3
10	1
$p + 5$	2

(a) 4   (b) 5

(c) 6   (d) 7.

60. Arun scored 36 marks in English, 44 marks in Hindi, 75 marks in Mathematics and x marks in Science. If he has scored an average of 50 marks, find the value of x.

(a) 45   (b) 40

(c) 50   (d) 48.

ANSWERS

1. (b)   **2.** (b)   **3.** (d)   **4.** (d)   **33.** (a)   **34.** (d)   **35.** (b)   **36.** (c)
5. (d)   **6.** (a)   **7.** (b)   **8.** (c)   **37.** (b)   **38.** (c)   **39.** (c)   **40.** (b)
9. (b)   **10.** (b)   **11.** (a)   **12.** (c)   **41.** (d)   **42.** (a)   **43.** (b)   **44.** (b)
13. (c)   **14.** (c)   **15.** (a)   **16.** (c)   **45.** (c)   **43.** (a)   **47.** (b)   **48.** (a)
17. (c)   **18.** (c)   **19.** (b)   **20.** (d)   **49.** (b)   **50.** (a)   **51.** (d)   **52.** (a)
21. (a)   **22.** (b)   **23.** (a)   **24.** (b)   **53.** (d)   **54.** (a)   **55.** (b)   (a)
25. (a)   **26.** (d)   **27.** (a)   **28.** (a)   **57.** (b)   **58.** (a)   **59.** (d)   **60.** (a).
29. (d)   **30.** (b)   **31.** (d)   **32.** (c)

HINTS/SOLUTIONS

1.

Class	Frequency	Cumulative Frequency
0–10	4	4
10–20	4	8
20–30	8	16
30–40	10	26
40–50	12	38
50–60	8	46
60–70	4	50

$n = 50$

$\therefore \dfrac{n}{2} = 25$ which lies in the class 30–40

$\therefore$ 30–40 is the median class.

2. A graphical method of finding the median.

3.

Marks obtained	Frequency	Cumulative Frequency
0–10	8	8
10–20	10	18
20–30	12	30
30–40	22	52
40–50	30	82
50–60	18	100

$n = 100$

$\therefore \dfrac{n}{2} = 50$ which lies in the class 30–40

$\therefore$ 30–40 is the median class.

4. Class marks = $\frac{10+25}{2}, \frac{35+55}{2}$
   = 17.5, 45.
5. Formula.
6. As it takes into account all the observations and lies between the extremes.
7. Median is a typical observation which is not affected by external values.
8. Definition of mode.
9. Definition of modal class.
10. Definition of median class.
11. Definition of lower limit of a class.
12. Definition of the upper limit of a class.
13. Width = 50 – 40 = 10.
14. The observation 2 has the maximum frequency (3).
15. Definition of class mark.
16.

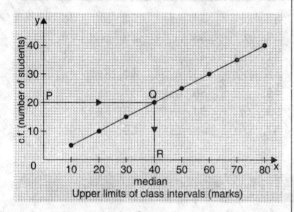

$n = 40$
$\therefore \quad \frac{n}{2} = 20$
which corresponds to P on Y-axis
From P_1 draw PQ ∥ X-axis to intersect the curve at Q. From Q, draw QR ∥ Y-axis to intersects the X-axis at R. R corresponds to 40 on X-axis.
$\therefore$ median marks = 40.

17. x-coordinate of the point of intersection of 'more than ogive' and 'less than ogive' = 6.
18. 3 Median = 2 Mean + Mode
    = 2 × 8 + 5 = 21

$\Rightarrow$ Median = 7.

19. 3 Median = 2 Mean + Mode
    $\Rightarrow$ 3 × 7.5 = 2 × Mean + 6.5
    $\Rightarrow$ Mean = 8.
20. 3 Median = 2 Mean + Mode
    $\Rightarrow$ 3 × 3 = 2 × 3 + Mode.
    $\Rightarrow$ Mode = 3.
21. Being the first class, the cumulative frequency = 10.
22. Being the last class, the cumulative frequency
    = 6 + 10 + 13 + 7 + 4 = 40.
23. Cumulative frequency
    = 7 + 5 + 7 + 8 = 27.
24. The class preceding the class 250–300 is 200–250.
25. The class exceeding the class 70–85 is 85–100.
26. Cumulative frequency of the class
    55–58 = 10 + 20 = 30
    Frequency of the class 58 – 61 = 25.
27. Modal class is 30–40.
    Its frequency = 10.
28. The class preceding the modal class (40–50) is 30–40.
29. The modal class is 20–25. The class succeeding this class is 25–30.
30.

Class	f_i	c.f.
10–25	6	6
25–40	20	26
40–55	44	70
55–70	26	96
70–85	3	99
85–100	1	100

$n = 100 \Rightarrow \frac{n}{2} = 50$

which lies in the class 40–55.
$\therefore$ 40–55 is the median class.
The class preceding the median class is 25–40.

31. The item value 5 has the maximum frequency (4).

STATISTICS

32. The item value 6 has the maximum frequency (3).
33. Frequency of 2.5 = 3 = Frequency of 2.8
    ∴ $x = 2.8$.
34. The maximum number of families have 6 members in the family.
35. Maximum number of students (6) are of age 14 years.
36. Median divides the distribution in two class such that the item values prior to this value are all lesser than the median and the item values post to this value are all greater than the median.
37. Definition of mode.
38. Write the item values in ascending order:
    20, 35, 50, 80, 190, 340, 520, 800, 1210
    $n = 9$
    ∴ median = $\left(\dfrac{n+1}{2}\right)$th item value
    = 5th item value = 190.
39. Arithmetic mean = $\dfrac{53 + 75 + 42 + 70}{4}$
    = $\dfrac{240}{4} = 60$.
40. First ten odd natural numbers are 1, 3, 5, 7, 9, 11, 13, 15, 17 and 19.
    Their mean = $\dfrac{1 + 3 + 5 + 7 + 9 + 11 + 13 + 15 + 17 + 19}{10}$
    = 10.

41.

x	f	cf
1	8	8
2	10	18
3	11	29
4	16	45
5	20	65
6	25	90
7	15	105
8	9	114
9	6	120

$n = 120$

∴ $\dfrac{n}{2} = 60$
∴ median = 5.

42. Formula.
43. Formula.
44. Mean = $\dfrac{1 \times 45 + 2 \times 25 + 3 \times 19 + 4 \times 8 + 5 \times 2 + 6 \times 1}{45 + 25 + 19 + 8 + 2 + 1}$
    = $\dfrac{45 + 50 + 57 + 32 + 10 + 6}{100}$
    = $\dfrac{200}{100} = 2$.
45. The size 8 has maximum frequency (95)
    ∴ mode = 8.
46. Can be visualised from ogive.
47. Can be visualised from ogive.
48. Mean = $\dfrac{1 + 2 + 3 + \ldots + n}{n}$
    = $\dfrac{n(n+1)}{2n} = \dfrac{n+1}{2}$.
49. Definition of ogive.
50. Formula for finding mean.
51. Range = highest item value − lowest item value.
52. Mean is simply the sum of all item values divided by the number of item values.
53. The height of the rectangles in a histogram are proportional to frequencies of the corresponding class intervals.
54. (a) result.
55. 3 median = 2 mean + mode
56. (a) result.
57. New mean = $\dfrac{px_1 + px_2 + px_3 + \ldots + px_n}{n}$
    = $p\left(\dfrac{x_1 + x_2 + x_3 + \ldots + x_n}{n}\right) = p\bar{x}$.
58. New mean = $\dfrac{\dfrac{x_1}{p} + \dfrac{x_2}{p} + \dfrac{x_3}{p} + \ldots + \dfrac{x_n}{p}}{n}$
    = $\dfrac{1}{p}\left(\dfrac{x_1 + x_2 + x_3 + \ldots + x_n}{n}\right) = \dfrac{\bar{x}}{p}$.

59.

x	y	xy
2	3	6
4	2	8
6	3	18
10	1	10
$p+5$	2	$2(p+5)$
	11	$2p+52$

$\bar{x} = \dfrac{\Sigma xy}{\Sigma y}$

$6 = \dfrac{2p+52}{11} \Rightarrow p = 7.$

60. $\dfrac{36+44+75+x}{4} = 50$

$\Rightarrow 155 + x = 200 \Rightarrow x = 45.$

Chapter 15
Probability

1. The probability of a sure event is
    (a) 0
    (b) 1
    (c) $\frac{1}{2}$
    (d) greater than 0 and less than 1.

2. The probability of an impossible event is
    (a) 0
    (b) 1
    (c) $\frac{1}{2}$
    (d) a number lying between 0 and 1.

3. If p is the probability of the occurrence of an event, then the probability of non-occurrence of that event is
    (a) $1 - p$
    (b) p
    (c) $2p$
    (d) $\frac{1}{2}p$.

4. If the probability of the non-happening of an event is q, then the probability of happening of that event is
    (a) $1 - q$
    (b) q
    (c) $\frac{q}{2}$
    (d) $2q$.

5. If p is the probability of happening of an event and q is the probability of non-happening of that event, then, $p + q$ is equal to
    (a) 0
    (b) 1
    (c) -1
    (d) 2.

6. Which of the following is true?
    (a) $0 \leq P(E) \leq 1$
    (b) $P(E) > 1$
    (c) $P(E) < 0$
    (d) $-\frac{1}{2} \leq P(E) \leq \frac{1}{2}$.

7. The sum of the probabilities of all the elementary events of an experiment is
    (a) 1
    (b) 2
    (c) 3
    (d) 4.

8. If $P(E) = 0.06$, then $P(\overline{E})$ is equal to
    (a) 0.06
    (b) 0.04
    (c) 0.01
    (d) -0.06.

9. If the probability of winning a game is 0.3, what is the probability of losing it?
    (a) -0.3
    (b) 0
    (c) 1
    (d) 0.7.

10. How many faces does a coin have?
    (a) 1
    (b) 2
    (c) 3
    (d) 4.

11. How many faces does a die have?
    (a) 2
    (b) 3
    (c) 6
    (d) 4.

12. How many cards are there in a deck of playing cards?
    (a) 13
    (b) 26
    (c) 39
    (d) 52.

13. How many colours are there of cards in a deck of playing cards?
    (a) 1
    (b) 2
    (c) 3
    (d) 4.

14. How many cards are of red colour in a deck of playing cards?

(a) 13 (b) 26
(c) 39 (d) 52.

15. How many cards are of black colour in deck of playing cards?
(a) 26 (b) 52
(c) 13 (d) 4.

16. How many face cards are there in a deck of playing cards?
(a) 4 (b) 3
(c) 12 (d) 13.

17. Which of the following is not a face card?
(a) jack (b) queen
(c) king (d) ace.

18. Which of the following terms is not related with a deck of playing cards?
(a) heart (b) diamond
(c) gold (d) club.

19. How many club cards are there in a deck of playing cards?
(a) 13 (b) 26
(c) 39 (d) 52.

20. The number of non-face cards in a deck of playing cards is
(a) 10 (b) 20
(c) 30 (d) 40.

21. What is the probability of getting a tail on throwing an unbiased coin?
(a) 1 (b) 0
(c) $\frac{1}{2}$ (d) $\frac{1}{4}$.

22. What is the probability of getting a head on throwing an unbiased coin?
(a) 0 (b) 1
(c) –1 (d) $\frac{1}{2}$.

23. Which of the following cannot be the probability of an event?
(a) $\frac{1}{2}$ (b) 20%
(c) 0.3 (d) $\frac{5}{2}$.

24. Two coins are tossed simultaneously. The probability of getting a head on only one of the two coins is
(a) 1 (b) $\frac{1}{2}$

(c) $\frac{1}{4}$ (d) $\frac{3}{4}$.

25. From a well shuffled pack of cards, a card is drawn at random. Find the probability of getting a black queen.
(a) $\frac{1}{26}$ (b) $\frac{1}{13}$
(c) $\frac{1}{52}$ (d) $\frac{1}{4}$.

26. A die is thrown once. Find the probability of getting an even prime number.
(a) $\frac{1}{2}$ (b) $\frac{1}{3}$
(c) $\frac{1}{6}$ (d) $\frac{1}{4}$.

27. A die is thrown once. Find the probability of getting a multiple of 3.
(a) $\frac{1}{2}$ (b) $\frac{1}{3}$
(c) $\frac{1}{4}$ (d) $\frac{1}{6}$.

28. A bag contains 4 red and 6 black balls. A ball is taken out of the bag at random. Find the probability of getting a black ball.
(a) $\frac{2}{5}$ (b) $\frac{3}{5}$
(c) $\frac{4}{5}$ (d) $\frac{1}{5}$.

29. A die is thrown once. Find the probability of getting a number less than 3.
(a) $\frac{1}{2}$ (b) $\frac{1}{3}$
(c) $\frac{1}{4}$ (d) $\frac{1}{6}$.

30. A bag contains 4 red, 5 black and 3 yellow balls. A ball is taken out of the bag at random. Find the probability that the ball taken out is of yellow colour.
(a) $\frac{1}{2}$ (b) $\frac{1}{3}$
(c) $\frac{1}{4}$ (d) $\frac{1}{5}$.

PROBABILITY

31. A bag contains 4 red, 5 black and 3 yellow balls. A ball is taken out of the bag at random. Find the probability that the ball taken out is not of red colour.
    (a) $\frac{2}{3}$ (b) $\frac{1}{3}$
    (c) $\frac{1}{4}$ (d) $\frac{1}{2}$.

32. A die is thrown once. Find the probability of getting a number greater than 5.
    (a) $\frac{1}{6}$ (b) $\frac{1}{2}$
    (c) $\frac{1}{3}$ (d) $\frac{1}{4}$.

33. A dice is thrown once. Find the probability of getting a number less than 5.
    (a) $\frac{1}{3}$ (b) $\frac{2}{3}$
    (c) $\frac{1}{4}$ (d) $\frac{1}{2}$.

34. A card is drawn from a well-shuffled pack of 52 cards. What is the probability that it is an ace of spade?
    (a) $\frac{1}{13}$ (b) $\frac{1}{26}$
    (c) $\frac{1}{39}$ (d) $\frac{1}{52}$.

35. In one thousand lottery tickets, there are 50 prizes to be given. Find the probability of Manish winning a prize, who bought one ticket.
    (a) $\frac{1}{10}$ (b) $\frac{1}{100}$
    (c) $\frac{1}{1000}$ (d) $\frac{1}{20}$.

36. Two coins are tossed simultaneously. What is the number of all possible outcomes?
    (a) 1 (b) 2
    (c) 3 (d) 4.

37. Three coins are tossed simultaneously. What is the number of all possible outcomes?
    (a) 2 (b) 4
    (c) 6 (d) 8.

38. A dice is thrown once. What is the number of all possible outcomes?
    (a) 2 (b) 4
    (c) 6 (d) 8.

39. Two dice are thrown at the same instant of time. What is the number of all possible outcomes?
    (a) 6 (b) 12
    (c) 30 (d) 36.

40. Cards marked with number 2, 4, 6, 8, 10, ..., 50 are placed in a bag and mixed thoroughly. One card is then drawn. What is the probability that the card is marked with a prime number?
    (a) $\frac{1}{25}$ (b) $\frac{1}{50}$
    (c) $\frac{1}{100}$ (d) $\frac{1}{10}$.

41. A box contains 30 chocolates in all out of which some are milk chocolates and the remaining are coco chocolates. If one chocolate is taken out at random from the box and the probability that it is a milk chocolate is $\frac{2}{3}$, then the number of coco chocolates in the box is
    (a) 10 (b) 20
    (c) 30 (d) 15.

42. A bag contains 10 red flags and some green flags. If the probability of drawing a green flag is thrice that of drawing a red flag, then the number of green flags in the bag is
    (a) 10 (b) 20
    (c) 30 (d) 3.

43. A purse contains 10 notes of Rs. 1000, 20 notes of Rs. 500 and 20 notes of Rs. 100. One note is drawn at random. What is the probability that the note of Rs. 500?
    (a) $\frac{1}{5}$ (b) $\frac{2}{5}$
    (c) $\frac{3}{5}$ (d) $\frac{4}{5}$

44. What is the probability of getting 53 Sundays in an ordinary year?
    (a) $\frac{1}{7}$ (b) $\frac{2}{7}$
    (c) $\frac{1}{365}$ (d) $\frac{1}{366}$.

45. The probability that it will rain tomorrow is 0.3. What is the probability that it will not rain tomorrow?
    (a) 0.3 (b) 0.2
    (c) 0.7 (d) 0.07.

46. What is the probability that a number selected from the numbers 1, 2, 3, ..., 15 is a multiple of 5.
    (a) $\frac{1}{3}$ (b) $\frac{2}{3}$
    (c) $\frac{1}{2}$ (d) $\frac{1}{5}$.

47. A card is drawn from a well shuffled deck of playing cards. Find the probability of drawing a black face card.
    (a) $\frac{1}{26}$ (b) $\frac{3}{26}$
    (c) $\frac{1}{52}$ (d) $\frac{3}{52}$.

48. A letter is chosen from the word CONGRATULATIONS. What is the probability that it is a vowel?
    (a) $\frac{2}{3}$ (b) $\frac{1}{2}$
    (c) $\frac{1}{4}$ (d) $\frac{1}{5}$.

49. A letter is chosen at random from the English alphabet. Find the probability that the letter chosen succeeds X.
    (a) $\frac{1}{13}$ (b) $\frac{1}{52}$
    (c) $\frac{1}{26}$ (d) $\frac{1}{2}$.

50. A letter is chosen at random from the English alphabet. Find the probability that the letter chosen is a vowel.
    (a) $\frac{1}{26}$ (b) $\frac{3}{26}$

    (c) $\frac{5}{26}$ (d) $\frac{1}{52}$.

51. A letter is chosen from the word APALA. Find the probability of choosing letter A.
    (a) $\frac{1}{5}$ (b) $\frac{2}{5}$
    (c) $\frac{3}{5}$ (d) $\frac{4}{5}$.

52. Cards each marked with one of the numbers 4, 5, 6, ..., 20 are placed in a box and mixed thoroughly. One card is drawn at random from the box. What is the probability of getting an even prime number?
    (a) 0 (b) 1
    (c) – 1 (d) $\frac{1}{2}$.

53. What is the probability that the first snowfall in Srinagar next winter will take place on Sunday?
    (a) $\frac{1}{7}$ (b) $\frac{2}{7}$
    (c) $\frac{3}{7}$ (d) $\frac{6}{7}$.

54. A child has a block in the shape of a cube with one letter written on each face as follows:

    The cube is thrown once. What is the probability of getting A?
    (a) $\frac{1}{3}$ (b) $\frac{1}{6}$
    (c) $\frac{1}{2}$ (d) $\frac{1}{4}$.

55. Apala and Meenu are friends. What is the probability that both will have the same birth day (ignoring a leap year)?
    (a) $\frac{1}{365}$ (b) $\frac{1}{366}$
    (c) $\frac{364}{365}$ (d) none of these.

56. A number x is chosen at random from $-4, -3, -2, -1, 0, 1, 2, 3, 4$
Find the probability that $|x| \leq 4$
(a) 0 (b) 1
(c) $\frac{1}{2}$ (d) $\frac{1}{9}$.

57. A die is in the shape of a tetrahedron as shown in the following figure. On each face, one of the numbers 1, 2, 3 or 4 is written. This die is tossed. Find the probability of getting a prime number.

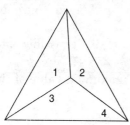

(a) $\frac{1}{2}$ (b) $\frac{1}{3}$
(c) 1 (d) 0.

58. A digit is randomly taken from a logarithmic table. Find the probability that the digit is 0 or 9.
(a) $\frac{2}{5}$ (b) $\frac{1}{5}$
(c) $\frac{3}{5}$ (d) $\frac{4}{5}$.

59. In class IX, there are 18 girls and 12 boys. The names of all the girls and boys are written on a card and are placed in a box. These cards are mixed up thoroughly and a card is drawn at random. What is the probability that the name card will be that of a girl student?
(a) $\frac{1}{5}$ (b) $\frac{2}{5}$
(c) $\frac{3}{5}$ (d) $\frac{4}{5}$.

60. Suppose you drop a die at random on the rectangular region shown in the figure given below. What is the probability that it will land inside the square of side 1 cm?

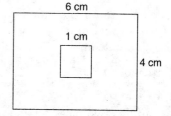

(a) $\frac{1}{6}$ (b) $\frac{1}{12}$
(c) $\frac{1}{10}$ (d) $\frac{1}{24}$.

ANSWERS

1. (b)	2. (a)	3. (a)	4. (a)	33. (b)	34. (d)	35. (d)	36. (d)
5. (b)	6. (a)	7. (a)	8. (b)	37. (d)	38. (c)	39. (d)	40. (a)
9. (d)	10. (b)	11. (c)	12. (d)	41. (a)	42. (c)	43. (b)	44. (a)
13. (d)	14. (b)	15. (a)	16. (c)	45. (c)	46. (a)	47. (b)	48. (a)
17. (d)	18. (c)	19. (a)	20. (d)	49. (a)	50. (c)	51. (c)	52. (a)
21. (c)	22. (d)	23. (d)	24. (b)	53. (a)	54. (a)	55. (a)	56. (b)
25. (a)	26. (c)	27. (b)	28. (b)	57. (a)	58. (b)	59. (b)	60. (d).
29. (b)	30. (c)	31. (a)	32. (a)				

HINTS/SOLUTIONS

1. Each outcome is a favourable outcome. So, number of favourable outcomes = number of all possible outcomes.
2. No outcome is a favourable outcome. So, number of favourable outcomes = 0.
3. $P(E) + P(\overline{E}) = 1$
4. $P(E) + P(\overline{E}) = 1$
5. $P(E) + P(\overline{E}) = 1$
6. Minimum value of number of favourable outcomes = 0
   Maximum value of number of favourable outcomes = Number of all possible outcomes.
7. Sum of the number of favourable outcomes to all the elementary events of an experiment = Number of all possible outcomes.
8. $P(E) + P(\overline{E}) = 1$
9. $P(E) + P(\overline{E}) = 1$
10. A coin has two faces – head and tail.
11. A die has six faces as it is cuboidal in shape.
12. Total number of cards in a deck of playing cards
    = 13 (diamond) + 13 (heart) + 13 (club) + 13 (spade) = 52.
13. Number of colours of cards in a deck of playing cards
    = 1 (diamond) + 1 (heart) + 1 (club) + 1 (spade) = 4
14. Number of cards of red colour in a deck of playing cards
    = 13 (diamond) + 13 (heart) = 26.
15. Number of cards of black colour in a deck of playing cards
    = 13 (club) + 13 (spade) = 26
16. Number of face cards in a deck of playing cards
    = 4 (jacks) + 4 (queens) + 4 (kings) = 12.
17. The face cards include jacks, queens and kings only.
18. Gold is a metal not a colour.
19. Number of club cards = 13.
20. Number of non-face cards
    = 52 – 12 = 40.
21. Number of favourable outcomes (Tail) = 1
    Number of all possible outcomes (Head, Tail) = 2
22. Number of favourable outcomes (Head) = 1
    Number of all possible outcomes (Head, Tail) = 2
23. The maximum value of the probability of any event is 1. $\left(\dfrac{5}{2} > 1\right)$.
24. Number of favourable outcomes (HT, TH) = 2
    Number of all possible outcomes (HH, HT, TH, TT) = 4.
25. Number of favourable outcomes (a queen of club, a queen of spade) = 2
    Number of all possible outcomes = 52
26. Number of favourable outcomes (2) = 1
    Number of all possible outcomes (1, 2, 3, 4, 5, 6) = 6
27. Number of favourable outcomes (3, 6) = 2
    Number of all possible outcomes (1, 2, 3, 4, 5, 6) = 6.
28. Number of favourable outcomes = 6 as there are 6 black balls.
    Number of all possible outcomes = 4 + 6 = 10.
29. Number of favourable outcomes (1, 2) = 2.
    Number of all possible outcomes (1, 2, 3, 4, 5, 6) = 6.
30. Number of favourable outcomes = 3 as there are 3 yellow balls.
    Number of all possible outcomes
    = 4 (red) + 5 (black) + 3 (yellow) = 12

31. Number of favourable outcomes
    = 5 (black) + 3 (yellow) = 8
    Number of all possible outcomes
    = 4 (red) + 5 (black) + 3 (yellow) = 12.
32. Number of favourable outcomes (6) = 1
    Number of all possible outcomes (1, 2, 3, 4, 5, 6) = 6.
33. Number of favourable outcomes
    (1, 2, 3, 4) = 4.
    Number of all possible outcomes
    (1, 2, 3, 4, 5, 6) = 6.
34. Number of favourable outcomes = 1 as there is only one ace of spade.
    Number of all possible outcomes = 52
35. Number of favourable outcomes = 50 as there are 50 prizes.
    Number of all possible outcomes = 1000
36. Number of all possible outcomes (HH, HT, TH, TT) = 4.
37. Number of all possible outcomes (HHH, HHT, HTH, HTT, THH, THT, TTH, TTT) = 8.
38. Number of all possible outcomes (1, 2, 3, 4, 5, 6) = 6.
39. Number of all possible outcomes
    = 6 × 6 = 36.
40. Number of favourable outcomes = 1 as only 2 is the prime number.
    Number of all possible outcomes = 25
41. Let the number of milk chocolates be x.
    Then, $\dfrac{2}{3} = \dfrac{x}{30}$ ⇒ $x = 20$
    ∴ Number of coco chocolates
    = 30 − 20 = 10.
42. Let the number of green flags be x.
    Then,
    $3\left(\dfrac{10}{10+x}\right) = \dfrac{x}{10+x}$
    ⇒ $x = 30$.
43. Number of favourable outcomes = 20
    Number of all possible outcomes
    = 10(Rs. 1000) + 20(Rs. 500)
    + 20(Rs. 100)
    = 50.
44. There are seven days in a week.
    52 × 7 = 364
    365 − 364 = 1
    This remaining 1 day may be any one of the seven days of the week.
    ∴ Required probability = $\dfrac{1}{7}$.
45. $P(E) + P(\overline{E}) = 1$.
46. Number of favourable outcomes
    (5, 10, 15) = 3
    Number of all possible outcomes = 15.
47. Number of favourable outcomes (2 black jacks, 2 black queens, 2 black kings) = 6
    Number of all possible outcomes = 52.
48. Number of favourable outcomes
    (O, A, U, A, I, O) = 6
    Number of all possible outcomes = 15 as there are 15 letters in the word.
49. Number of favourable outcomes
    (y, z) = 2
    Number of all possible outcomes = 26.
50. Number of favourable outcomes
    (a, e, i, o, u) = 5
    Number of all possible outcomes = 26.
51. Number of favourable outcomes = 3 as there are 3 A's in APALA.
    Number of all possible outcomes = 5 as there are 5 letters in the word APALA.
52. Number of favourable outcomes = 0 as the only even prime number is 2 which is not included in the given numbers.
    Number of all possible outcomes = 17.
53. There are 7 days is a week.
    Number of favourable outcomes (Sunday) = 1
    Number of all possible outcomes
    (S, M, T, W, Th, F, S) = 7.
54. Number of favourable outcomes = 2 as there are two A's.
    Number of all possible outcomes = 6.
55. Number of favourable outcomes = 1
    Number of all possible outcomes = 365

56. Number of favourable outcomes
    $(-4, -3, -2, -1, 0, 1, 2, 3, 4) = 9$
    Number of all possible outcomes
    $(-4, -3, -2, -1, 0, 1, 2, 3, 4) = 9$.
57. Number of all possible outcomes
    $(1, 2, 3, 4) = 4$
    Number of favourable outcomes
    $(2, 3) = 2$.
58. Number of favourable outcomes
    $(0, 9) = 2$

Number of all possible outcomes
$(0, 1, 2, 3, 4, 5, 6, 7, 8, 9) = 10$.
59. Number of favourable outcomes = 18 as there are 18 girls in class IX.
Number of all possible outcomes [18 (girls) + 12 (boys)] = 30.
60. Required probability
$$= \frac{\text{Area of square}}{\text{Area of rectangle}}$$
$$= \frac{1 \times 1}{6 \times 4} = \frac{1}{24}.$$

Higher Order Thinking Skills (HOTS) Questions

Chapter 1 REAL NUMBERS

1. What is an algorithm ?
2. Define a lemma.
3. What is Euclid's division lemma ?
4. Write the statement of 'Fundamental Theorem of Arithmetic'.
5. Find the prime factorisation of the number 1188.
6. Find the missing numbers in the following prime factorisation :

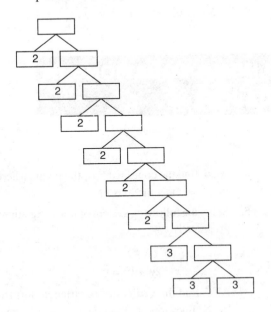

7. The HCF of two numbers is 103 and their LCM is 19261. If one of the numbers is 1133, find the other.
8. Find the LCM and HCF of the following integers by applying the prime factorisation method:

    72, 90, 120.

9. Use Euclid's division algorithm to find the HCF of 10524 and 12752.
10. Prove that one of every three consecutive positive integers is divisible by 3.
11. Show that 8^n cannot end with the digit 0 for any natural number n.
12. Prove that $\sqrt{2}$ is irrational.
13. Write the condition to be satisfied by q so that a rational number $\dfrac{p}{q}$ has a terminating decimal expansion.

Chapter 2 POLYNOMIALS

1. Define a biquadratic polynomial.
2. The graph of $y = p(x)$ is given in the figure below, for some polynomial $p(x)$. Find the number of zeroes of $p(x)$.

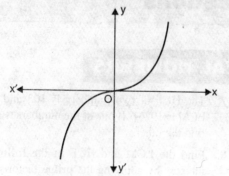

3. Find a quadratic polynomial, the sum and product of whose zeroes are 3 and 0.
4. Find a cubic polynomial with the sum, sum of the products of its zeroes taken two at a time and product of its zeroes as 1, 3, 6, respectively.
5. Find the zeroes of the polynomial $x^2 + x$ and verify and relationship between the zeroes and its coefficients.
6. Verify that 1, 2, 3 are the zeroes of the cubic polynomial $x^3 - 6x^2 + 11x - 6$, and then verify the relationship between the zeroes and its coefficients.
7. If α, β are the zeroes of the polynomial $2 - 2x + x^2$, then find the value of $\dfrac{1}{\alpha} + \dfrac{1}{\beta}$.
8. If the zeroes of the polynomial $x^3 - 14x^2 + 37x - 60$ are α, β, γ, then find
   (i) $\alpha + \beta + \gamma$
   (ii) $\alpha\beta + \beta\gamma + \gamma\alpha$
   (iii) $\alpha\beta\gamma$.
9. Prove that $2x^4 - 6x^3 + 3x^2 + 3x - 2$ is exactly divisible by $x^2 - 3x + 2$.
10. Obtain all other zeroes of $2x^4 - 2x^3 - 7x^2 + 3x + 6$ if two of its zeroes are $\pm\sqrt{\dfrac{3}{2}}$.

Chapter 3 PAIR OF LINEAR EQUATIONS IN TWO VARIABLES

1. Solve the following system of equations:
   $15x + 4y = 61$
   $4x + 15y = 72$
2. For what value of k will the system of linear equations
   $kx + 4y = k - 4$
   $16x + ky = k$
   have infinite number of solutions?
3. For what value of k, the following system of equations have
   (i) a unique solution?
   (ii) no solution?
   $2x + ky = 1, \; 3x - 5y = 7$.
4. Solve graphically the following system of linear equations:
   $2x - y = 2$
   $4x - y = 8$
   Also, find the coordinates of the points where the lines meet the axis of x.
5. Solve the following system of linear equations graphically:
   $2x + y + 6 = 0$
   $3x - 2y - 12 = 0$
   Also, find the vertices of the triangle formed by the lines representing the above equations and x-axis.
6. Solve the following system of linear equations graphically:
   $x - y = 1$
   $2x + y = 8$

Shade the area bounded by these two lines and the y-axis.

7. Solve for x and y:

$$\frac{4}{x} + 3y = 14$$

$$\frac{3}{x} - 4y = 23.$$

8. Solve the following system of equations:

$$\frac{11}{v} - \frac{7}{u} = 1$$

$$\frac{9}{v} - \frac{4}{u} = 6.$$

9. Solve the following system of linear equations:

$$(a+c)x - (a-c)y = 2ab$$
$$(a+b)x - (a-b)y = 2ab.$$

10. Solve for x and y:

$$\frac{6}{x+y} = \frac{7}{x-y} + 3$$

$$\frac{1}{2(x+y)} = \frac{1}{3(x-y)}.$$

11. Solve the following system of equations:

$$35x + 23y = 209$$
$$23x + 35y = 197.$$

12. 3 chairs and 2 tables cost Rs. 700 and 5 chairs and 3 tables cost Rs. 1100. Find the cost of 2 chairs and 2 tables.

13. If 2 is added to the numerator of a fraction, it reduces to $\frac{1}{2}$ and if 1 is subtracted from the denominator, it reduces to $\frac{1}{3}$. Find the fraction.

Chapter 4 QUADRATIC EQUATIONS

1. Solve for x:
   $$12\,ab\,x^2 - (9a^2 - 8b^2)\,x - 6ab = 0.$$

2. Using quadratic formula, solve the following quadratic equation:
   $$x^2 - 4ax + 4a^2 - b^2 = 0.$$

3. Solve for x:
   $$\frac{x+1}{x-1} + \frac{x-2}{x+2} = 3 \,(x \ne 1, -2)$$

4. Find the value of k, if the equation $x^2 - 2(k+1)x + k^2 = 0$ has equal roots.

5. Find the discriminant of the quadratic equation $2x^2 - 4x + 3 = 0$ and hence find the nature of its roots.

6. Determine k so that the equation $x^2 - 4x + k = 0$ has two real roots.

7. A train travels a distance of 300 km at a constant speed. If the speed of the train is increased by 5 km per hour, the journey would have taken 2 hours less. Find the original speed of the train.

8. 300 apples are distributed equally among a certain number of students. Had there been 10 more students, each would have received one apple less. Find the number of students.

9. The sum of two number is 18. The sum of their reciprocals is $\frac{1}{4}$. Find the numbers.

10. Two numbers differ by 3 and their product is 504. Find the numbers.

11. A number consists of two digits whose product is 18. When 27 is subtracted from the number, its digits change their places. Find the number.

Chapter 5 ARITHMETIC PROGRESSIONS

1. If three consecutive terms of an AP are $3x$, $x + 2$ and 8, then find the value of x. Also, find its 4th term.
2. Find the sum $1 + 2 + 3 + ... + 100$.
3. The sum of n terms of an AP is $3n^2 - n$. Find out the first term and the common difference.
4. If 7 times the 7th term of an AP is equal to 11 times the 11th term, show that the 18th term of the AP is zero.
5. Find the sixth term from end of the AP $17, 14, 11, ..., -40$.
6. Find the sum of first 21 terms of the AP whose 2nd term is 8 and 4th term is 14.
7. If the sum of first n terms of an AP is given by $S_n = 5n^2 + 3n$, find the nth term of the AP.
8. Find the sum of all multiples of 9 lying between 300 and 700.
9. The nth term (a_n) of an AP is given by $a_n = 5n - 3$. Find the sum of the first 20 terms of AP.
10. The 8th term of an AP is 37 and its 12th term is 57. Find the AP.
11. If the three sides of a right angled triangle are in AP, then prove that they are in the ratio $3 : 4 : 5$.
12. If the angles of a quadrilateral be in AP such that the common difference is $20°$, then find the angles.
13. How many terms of the series $24 + 20 + 16 + ...$ give the sum 72?
Give reason for the two answers.

Chapter 6 TRIANGLES

1. In the given figure, if $DE \parallel BC$ and $AD = 4x - 3$, $AE = 8x - 7$, $BD = 3x - 1$ and $CE = 5x - 3$, find x.

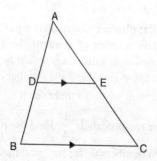

2. Determine the length of AD in terms of b and c.

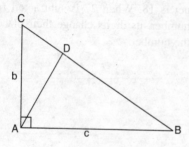

3. The area of two similar triangles are 81 cm² and 49 cm² respectively. If the altitude of the first triangle is 6.3 cm, find the corresponding altitude of the other.
4. A man goes 80 m due east and then 150 m due north. How far is he from the starting point?
5. In the given figure, O is a point in side $\triangle PQR$ such that $\angle POR = 90°$, $OP = 6$ cm and $OR = 8$ cm. If $PQ = 24$ cm and $QR = 26$ cm, prove that $\triangle PQR$ is a right angled triangle.

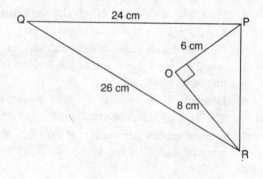

6. In the given figure, D divides AB such that AD : DB = 3 : 2 and E is a point on BC such that DE ∥ AC. Find the ratio of areas of △BAC and △BDE.

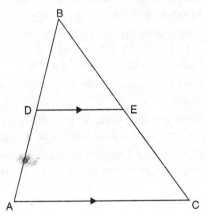

7. Prove that the altitude of an equilateral triangle of side $2a$ is $\sqrt{3}\,a$.

8. The diagonals of a quadrilateral ABCD intersect each other at the point O such that $\dfrac{AO}{OC} = \dfrac{BO}{OD}$. Show that ABCD is a trapezium.

9. In the following figure, ABCD is a quadrilateral and P, Q, R, S are the points of trisection of the sides AB, BC, CD and DA respectively. Prove that PQRS is a parallelogram.

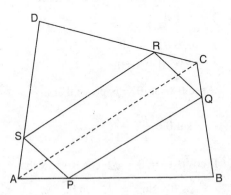

10. In △ABC, D and E are two points on AB such that AD = BE. If DP ∥ BC and EQ ∥ AC, prove that PQ ∥ AB.

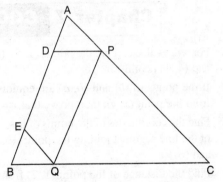

11. In the following figure, ∠BAC = 90° and AD ⊥ BC. Prove that
    $$AD^2 = BD \times DC.$$

12. The diagonal BD of a parallelogram ABCD intersects the segment AE at the point F, where E is any point on the side BC. Prove that
    $$DF \times EF = FB \times FA.$$

13. Equilateral triangles are drawn on the sides of a right triangle. Show that the area of the triangle on the hypotenuse is equal to the sum of the areas of triangles on the other two sides.

14. In the following figure, ABC is a right angled triangle right angled at B.

    AD and CE are the two medians drawn from A and C respectively.

    If AC = 5 cm and AD = $\dfrac{3\sqrt{5}}{2}$ cm, find the length of CE.

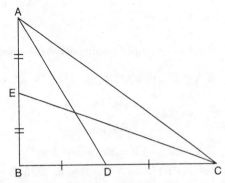

15. In an equilateral triangle ABC, the side BC is trisected at D.
    Prove that $9\,AD^2 = 7\,AB^2$.

Chapter 7 — COORDINATE GEOMETRY

1. For what value of x are the points $(1, 5)$, $(x, 1)$ and $(4, 11)$ collinear.
2. If the points (a, b) and (b, a) are equidistant from the point (x, y), then prove that $x = y$.
3. Find the coordinates of the points of trisection of the line segment joining the points $(3, -3)$ and $(6, 9)$.
4. Find the distance of the point $(1, 2)$ from the mid-point of the line segment joining the points $(6, 8)$ and $(2, 4)$.
5. Show that the points $A(1, 0)$, $B(5, 3)$, $C(2, 7)$ and $D(-2, 4)$ are the vertices of a parallelogram.
6. Find the lengths of the medians of the triangle whose vertices are $(1, -1)$, $(0, 4)$ and $(5, 3)$.
7. The vertices of a triangle are $(-4, 2)$, $(a, 7)$ and $(2, b)$ and its centroid is $(-1, 3)$. Find the values of a and b.
8. If the points $P(a, -11)$, $Q(5, b)$, $R(2, 15)$ and $S(1, 1)$ are the vertices of a parallelogram PQRS, find the values of a and b.
9. Find the distance of the point $(6, 8)$ from the origin.
10. Show that the points $A(1, 2)$, $B(5, 4)$, $C(3, 8)$ and $D(-1, 6)$ are the vertices of a square.
11. Find the coordinates of the circumcentre of a triangle whose vertices are $A(4, 6)$, $B(0, 4)$ and $C(6, 2)$. Also find its circumcentre.

Chapter 8 — INTRODUCTION TO TRIGONOMETRY

1. Solve for θ :
   $2 \sin^2 \theta = \dfrac{1}{2}$, $0° < \theta < 90°$.
2. If $5 \cot \theta = 3$, then find the value of
   $\dfrac{5 \sin \theta - 3 \cos \theta}{4 \sin \theta + 3 \cos \theta}$.
3. If $15 \cot A = 8$, find $\sin A$ and $\sec A$.
4. If $\tan \theta + \cot \theta = 2$, find the value of $\tan^2 \theta + \cot^2 \theta$.
5. If $\cos \theta = \dfrac{3}{5}$, find the value of $\cot \theta + \csc \theta$.
6. If $A = 30°$, verify that
   $\cos 2A = \dfrac{1 - \tan^2 A}{1 + \tan^2 A}$.
7. If $A = 60°$ and $B = 30°$, verify that
   $\cos(A - B) = \cos A \cos B + \sin A \sin B$.
8. Evaluate
   $\csc(65° + \theta) - \sec(25° - \theta) - \tan(55° - \theta) + \cot(35° + \theta)$.
9. Evaluate
   $\sin(50° + \theta) - \cos(40° - \theta)$
   $\qquad + \tan 1° \tan 10° \tan 20° \tan 70° \tan 80° \tan 89°$.
10. Prove that
    $\sec^2 \theta + \csc^2 \theta = \sec^2 \theta \csc^2 \theta$.
11. Show that
    $\dfrac{\sin \theta - 2 \sin^3 \theta}{2 \cos^3 \theta - \cos \theta} = \tan \theta$.
12. If $\sec \theta + \tan \theta = p$, prove that
    $\sin \theta = \dfrac{p^2 - 1}{p^2 + 1}$.
13. If $\cos \theta + \sin \theta = \sqrt{2} \cos \theta$, show that
    $\cos \theta - \sin \theta = \sqrt{2} \sin \theta$.
14. If $\tan \theta + \sin \theta = m$ and $\tan \theta - \sin \theta = n$, show that
    $m^2 - n^2 = 4\sqrt{mn}$.

HIGHER ORDER THINKING SKILLS (HOTS) QUESTIONS

Chapter 9 — SOME APPLICATIONS OF TRIGONOMETRY

1. An aeroplane flying horizontally at a height of 1.5 km above the ground is observed at a certain point on earth to subtend an angle of 60°. After 15 seconds, its angle of elevation at the same point is observed to be 30°. Calculate the speed of the aeroplane in km/hr.

2. From the top of a tower 60 m high, the angles of depression of the top and bottom of a building whose base is in the same straight line with the base of the tower are observed to be 30° and 60° respectively. Find the height of the building.

3. Determine the height of a mountain if the elevation of its top at an unknown distance from the base is 30° and at a distance 10 km further off from the mountain, along the same line, the angle of elevation is 15°.
   (Use tan 15° = 0.27).

4. The angle of elevation of the top Q of a vertical tower PQ from a point X on the ground is 60°. At a point Y, 40 m vertically above X, the angle of elevation is 45°. Find the height of the tower PQ and the distance XQ.

5. The angle of elevation of a cloud from a point 60 m above a lake is 30° and the angle of depression of the reflection of cloud in the lake is 60°. Find the height of the cloud.

6. A vertically straight tree, 15 m high, is broken by the wind in such a way that it top just touches the ground and makes an angle of 60° with the ground. At what height from the ground did the tree break ?

7. A vertical tower stands on a horizontal plane and is surmounted by a vertical flagstaff of height 5 metres. At a point on the plane, the angles of elevation of the bottom and the top of the flagstaff are respectively 30° and 60°. Find the height of the tower.

8. On the same side of a tower, two objects are located. When observed from the top of the tower, their angles of depression are 45° and 60°. If the height of the tower is 150 m, find the distance between the objects.

9. If the shadow of a tower is 100 m long, when the sun's altitude is 45°, what is the length of the shadow when the sun's altitude is 60° ?

10. The angle of elevation θ of a vertical tower from a point on the ground is such that its tangent is $\frac{5}{12}$. On walking 192 metres towards the tower in the same straight line, the tangent of the angle ϕ is found to be $\frac{3}{4}$. Find the height of the tower.

11. Two men are on opposite sides of a tower. They observe the angles of elevation of the top of the tower as 30° and 45° respectively. If the height of the tower 50 m, find the distance between the two men.

12. A tower subtends an angle α at a point A in the plane of its base and the angle of depression of the foot of the tower at a point b meter just above A is β. Prove that the height of the tower is $b \tan \alpha \cot \beta$.

Chapter 10 CIRCLES

1. What is the angle between a tangent and radius through the point of contact ?
2. In how many points does a tangent to a circle intersect it ?
3. How many parallel tangents can a circle have at the most ?
4. How many tangents can be drawn to a circle from a point lying cut side the circle ?
5. Can you draw a tangent to a circle from a point lying in side the circle ?
6. How many tangents can be drawn to a circle at a point lying on the circumference of the circle ?
7. How many tangents can a circle have ?
8. PA and PB are two tangents to a circle from a point P lying outside the circle. If PA = 5 cm, find the length of PB.
9. The radius of a circle is 3 cm. The length of a tangent from a point A to the circle is 4 cm. Find the distance of the point from the centre of the circle.
10. In the given figure, O is the centre of the circle and PT is a tangent at T. If PC = 3 cm and PT = 6 cm, calculate the radius of the circle.

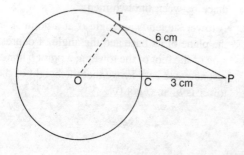

11. Find the locus of centres of circles which touch a given line at a given point.
12. Two circles touch internally at a point P and from a point T on the common tangent at P, tangent segments TQ, TR and draw to the two circles. Prove that
    TQ = TR.
13. Find the locus of centres of circles which touch two intersecting lines.
14. The in circle of a △ABC touches the sides AB, BC and CA at the points F, D and E respectively. Prove that
    AF + BD + CE = FB + DC + EA
    $= \frac{1}{2}$ (Perimeter of △ABC).
15. In the given figure, prove that
    BD = DC if AB = AC.

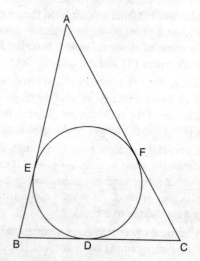

Chapter 11 CONSTRUCTIONS

1. Draw a line segment of length 4.9 cm and divide it in the ratio 3 : 4. Measure the two parts.
2. Draw a line segment of length 7 cm and divide it in the ratio 5 : 2. Measure the two parts.
3. Construct a triangle of sides 4.5 cm, 5 cm and 6 cm and then a triangle similar to it whose sides are $\frac{3}{5}$ of the corresponding sides of the first triangle.

4. Construct an isosceles triangle whose base is 6 cm and altitude 3 cm and then another triangle whose sides are $1\frac{1}{2}$ times the corresponding sides of the isosceles triangle.

5. Draw a circle of radius 5 cm from a point 8 cm away from its centre, construct the pair of tangents to the circle and measure their length.

6. Draw a pair of tangents to a circle of radius 4 cm which are inclined to each other at an angle of 60°.

Chapter 12 AREAS RELATED TO CIRCLES

1. The radius of a circle is 28 cm. An arc of this circle subtends an angle of 36° at the centre. Find the length of the arc.

2. What is the measure of the central angle of a circle ?

3. Find the area of the shaded region in the following figure :

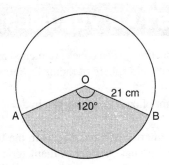

4. Find the area of a quadrant of a circle whose circumference is 44 cm.

5. If the diameter of a semi-circular protractor is 14 cm, then find its perimeter.

6. The perimeter and area of a square are numerically equal. Find the area of the square.

7. The radii of two circles are 6 cm and 8 cm. Find the diameter of the circle having area equal to the sum of the areas of the two circles.

8. The minute hand of a clock is 6 m long. Find the area of the face of the clock described by the minute hand in 35 minutes.

9. The diameters of two circles are 20 cm and 10 cm respectively. Find the diameter of the circle which has circumference equal to the sum of the circumferences of the two circles.

10. A wire is looped in the form of a circle of radius 28 cm. It is re-bent into a square form. Determine the length of the side of the square.

11. Two circles touch internally. The sum of their areas is $116\,\pi$ cm² and distance between their centres is 6 cm. Find the radii of the circles.

12. A bicycle wheel makes 5000 revolution in moving 11 km. Find the diameter of the wheel.

13. Find the area of the shaded region in the figure given below :

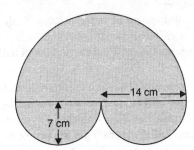

14. ABCD is a flower bed. If OA = 21 m and OC = 14 m, find the area of the bed. $\left(\text{use } \pi = \dfrac{22}{7}\right)$.

15. In the following figure, find the area of the shaded region. The four corners are circle quadrants and at the centre, there is a circle. Take $\pi = 3.14$

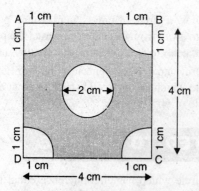
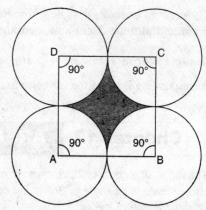

16. Four equal circles are described about the four corners of a square so that each touches two of the others, as shown in the figure. Find the area of the shaded region, if each side of the square measures 14 cm.

17. A chord of a circle of radius 14 cm subtends a right angle at the centre. Find the areas of the minor and major segments of the circle.

Chapter 13 SURFACE AREAS AND VOLUMES

1. Three cubes each of side 5 cm and joined end to end. Find the surface area of the resulting cuboid.

2. A solid sphere of radius 3 cm is melted and then cast into small spherical balls each of diameter 0.6 cm. Find the number of balls thus obtained.

3. A conical vessel whose internal radius is 5 cm and height 24 cm is full of water. The water is emptied into a cylindrical vessel with internal radius 10 cm. Find the height to which the water rises.

4. Water is flowing at the rate of 7 m/s through a circle pipe whose internal diameter is 2 cm into a cylindrical tank the radius of whose base is 40 cm. Determine the increase in the water level in $\frac{1}{2}$ hour.

5. Two solid right circular cones have the same height. The radii of their bases are r_1 and r_2. They are melted and recast into a cylinder of same height. Show that the radius of the base of the cylinder is $\sqrt{\dfrac{r_1^2 + r_2^2}{3}}$.

6. The radii of the bases of two right circular solid cones of same height are r_1 and r_2 respectively. The cones are melted and recast into a solid sphere of radius R. Show that the height of each cone is given by $h = \dfrac{4R^3}{r_1^3 + r_2^3}$.

7. The interior of a building is in the form of a right circular cylinder of diameter 4.2 m and height 4 m surmounted by a cone. The vertical height of cone is 2.1 m. Find the outer surface area and volume of the building. $\left(\text{Use } \pi = \dfrac{22}{7}\right)$

8. A solid wooden toy is in the shape of a right circular cone mounted on a hemisphere. If the radius of the hemisphere is 4.2 cm and the total height of the toy is 10.2 cm, find the volume of the wooden toy.

9. A toy is in the shape of a right circular cylinder with hemisphere on one end and a cone on the other. The height and radius of the cylindrical part are 13 cm and 5 cm respectively. The radius of the hemispherical and conical parts are the same as that of the cylindrical part. Calculate the surface area of the toy if the height of the conical point is 12 cm.

HIGHER ORDER THINKING SKILLS (HOTS) QUESTIONS

10. The perimeters of the ends of a frustum of a right circular cone are 44 cm and 33 cm respectively. If the height of the fustum be 16 cm, find its volume and the total surface.

11. A bucket is in the form of a frustum of a cone and holds 28.490 litres of water. The radii of the top and bottom are 28 cm and 21 cm respectively. Find the height of the bucket.

Chapter 14 STATISTICS

1. Name the three measures of control tendency.
2. Write the empirical relationship between the three measures of central tendency.
3. What is the most frequently used measure of central tendency?
4. Define mode.
5. What does the *x*-coordinate of the point of intersection of two ogives represent?
6. Calculate the mean for the following distribution :

Weekly wages (in rupees)	No. of workers
150	20
170	30
190	40
210	50
230	60
Total	200

7. Find the mean for the following distribution using step-deviation method.

Class Interval	Frequency
0–50	8
50–100	15
100–150	32
150–200	26
200–250	12
250–300	7

8. Compute the missing frequencies 'f_1' and 'f_2' in the following data if the mean is $166\frac{9}{26}$ and the sum of observations is 52.

Classes	Frequency
140–150	5
150–160	f_1
160–170	20
170–180	f_2
180–190	6
190–200	2

9. Find the mode of the following distribution :

Class Interval	Frequency
25–30	25
30–35	34
35–40	50
40–45	42
45–50	38
50–55	14

10. Calculate the median from the following table:

Income (in rupees)	Number of persons
Less than 100	16
Less than 200	39
Less than 300	67
Less than 400	107
Less than 500	127
Less than 600	139
Less than 700	147
Less than 800	150

11. Draw a less than type ogive for the following data :

Marks	Number of students
Marks less than 10	3
Marks less than 20	11
Marks less than 30	28
Marks less than 40	48
Marks less than 50	70

Chapter 15 PROBABILITY

1. If E is an event such that P(E) = 0.45, then find P(not E).
2. If the probability of winning a game is 0.9, what is the probability of loosing it ?
3. If the probability of happening of an event is $\frac{5}{11}$, what is the probability of not happening of the event ?
4. What is the probability of a sure event ?
5. What is the probability of an impossible event ?
6. A letter of english alphabet is chosen at random. Find the probability that the letter so chosen is a vowel.
7. Find the probability of getting 53 Fridays in a non-leap year.
8. Find the probability of getting 53 Fridays in a leap year.
9. A jar contains 32 marbles out of which some are red and the others are green. The probability of drawing a red marble from the jar is $\frac{3}{4}$. Find the number of green marbles in the jar.
10. It is known that a box of 200 electric bulbs contains 20 defective bulbs. One bulb is taken out at random from this box. What is the probability that it is a non-defective bulb ?
11. Meenu and Apala are classmates. What is the probability that both will have different birthdays ? (Ignore a leap year).
12. In an N.C.C. camp, there are 20 boys and 15 girls. The best cadet is to be chosen. What is the probability that the best cadet is a girl ?
13. A letter is selected from the letters of the word (MATHEMATICS). What is the probability that it is M ?
14. 15 cards numbered 1, 2, 3, ..., 15 are put in a box and mixed throughly. Manish drawn a card from the box. What is probability that the number on card is prime ?
15. A bag contains 5 white, 7 red, 4 black and 2 blue balls, one ball is drawn at random from the bag. What is the probability that the ball drawn is neither white nor black ?
16. A card is drawn at random from a pack of 52 playing cards. Find the probability that the card drawn neither a queen nor jack.
17. A die is thrown once. Find the probability of getting a number between 3 and 6.
18. Three coins are tossed simultaneously. Find the probability of getting at leat two heads.

Solutions to Higher Order Thinking Skills (HOTS) Questions

Chapter 1 REAL NUMBERS

1. An algorithm is a series of well defined steps which gives a procedure for solving a type of problems.

2. A lemma is a proven statement used for proving another statement.

3. The statement of Euclid's division lemma is as follows:

   "Given positive integers a and b, there exist unique integers q and r satisfying

   $a = bq + r, 0 \leq r < b$.

4. The statement of 'Fundamental Theorem of Arithmetic' is as follows:

   "Every composite number can be expressed (factorised) as a product of primes, and this factorisation is unique, apart from the order in which the prime factors occurs".

5.

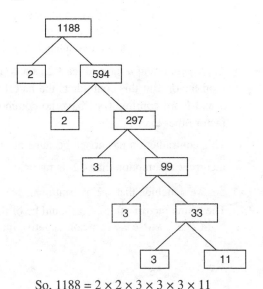

So, $1188 = 2 \times 2 \times 3 \times 3 \times 3 \times 11$
$= 2^2 \times 3^3 \times 11$.

6. $3 \times 3 = \mathbf{9}$
   $3 \times 9 = \mathbf{27}$
   $2 \times 27 = \mathbf{54}$
   $2 \times 54 = \mathbf{108}$
   $2 \times 108 = \mathbf{216}$
   $2 \times 216 = \mathbf{432}$
   $2 \times 432 = \mathbf{864}$
   $2 \times 864 = \mathbf{1728}$

   Hence, the missing numbers from the top to bottom in succession are

   1728, 864, 432, 216, 108, 54, 27 and 9.

7. Other number
   $= \dfrac{103 \times 19261}{1133} = 1751$.

8. $72 = 2 \times 2 \times 2 \times 3 \times 3 = 2^3 \times 3^2$
   $90 = 2 \times 3 \times 3 \times 5 = 2 \times 3^2 \times 5$
   $120 = 2 \times 2 \times 2 \times 3 \times 5 = 2^3 \times 3 \times 5$
   $\therefore$ HCF $= 2 \times 3 = 6$
   LCM $= 2^3 \times 3^2 \times 5 = 360$.

9. $12752 = 10524 \times 1 + 2228$
   $10524 = 2228 \times 4 + 1612$
   $2228 = 1612 \times 1 + 616$
   $1612 = 616 \times 2 + 380$
   $616 = 380 \times 1 + 236$
   $380 = 236 \times 1 + 144$
   $236 = 144 \times 1 + 92$
   $114 = 92 \times 1 + 52$
   $92 = 52 \times 1 + 40$
   $52 = 40 \times 1 + 12$

$$40 = 12 \times 3 + 4$$
$$12 = 4 \times 3 + 0$$
$\therefore$ HCF = 4.

10. Let $n, n + 1, n + 2$ be three consecutive positive integers. We know that n is of the form $3q$, $3q + 1$ or $3q + 2$ for some integer q. So, there arise three cases:

    Case I. When n = 3q

    Then, n is divisible by 3 but neither of $n + 1$ and $n + 2$ is divisible by 3.

    Case II. When n = 3q + 1

    Then, $n + 2 = 3q + 1 + 2$
    $$= 3q + 3$$
    $$= 3(q + 1)$$

    is divisible by 3 but neither of n and $n + 1$ is divisible by 3.

    Case III. When n = 3q + 2

    Then, $n + 1 = 3q + 2 + 1$
    $$= 3q + 3$$
    $$= 3(q + 1)$$

    is divisible by 3 but neither of n and $n + 2$ is divisible by 3.

    Hence, one of $n, n + 1$ and $n + 2$ is divisible by 3.

11. If the number 8^n, for any n, were to end with the digit zero, then it would be divisible by 5. That is, prime factorisation of 8^n would contain the prime 5. This is not possible because $8^n = (2)^{3n}$; so the only prime in the factorisation of 8^n is 2. So, the uniqueness of the Fundamental Theorem of Arithmetic guarantees that these are no other primes in the factorisation of 8^n. So, there is no natural number n for which 8^n ends with the digit zero.

12. Let us assume to the contrary, that $\sqrt{2}$ is rational.

    That is, we can find integers r and $s \, (\neq 0)$ such that
    $$\sqrt{2} = \frac{r}{s}$$

    Suppose r and s have a common factor other than 1. Then, we can divide by the common factor to get $\sqrt{2} = \frac{a}{b}$ where a and b are coprime.

    So, $\qquad b\sqrt{2} = a$

    Squaring on both sides, and rearranging, we get
    $$2b^2 = a^2$$
    Therefore, 2 divides a^2.

    It follows that 2 divides a.

    > Let p be a prime number. If p divides a^2, then p divides a, where a is a positive integer.

    So, we can write
    $$a = 2c \text{ for some integer } c.$$
    Substituting for a, we get
    $$2b^2 = 4c^2, \text{ that is,}$$
    $$b^2 = 2c^2$$
    This means that 2 divides b^2, and so 2 divides b.

    > Let p be a prime number. If p divides a^2, then p divides a, where a is a positive integer.

    Therefore, a and b have at least 2 as a common factor. But this contradicts the fact that a and b are coprime (*i.e.*, have no common factor other than 1).

    This contradiction has arisen because of our incurrent assumption that $\sqrt{2}$ is rational.

    So, we conclude that $\sqrt{2}$ is irrational.

13. The prime factorisation of q should be of the form $2^n 5^m$, where n, m are non-negative integers.

Chapter 2 POLYNOMIALS

1. A polynomial of degree 4 is called a biquadratic polynomial.
2. The number of zeroes is 1 as the graph intersects the x-axis at one point only.
3. The quadratic polynomial is
   $x^2 - (3)x + (0)$ or, $x^2 - 3x$.
4. The cubic polynomial is
   $x^3 - (1)x^2 + (3)x - (6)$
   or, $x^3 - x^2 + 3x - 6$.
5. $\quad x^2 + x = 0$
   $\Rightarrow x(x+1) = 0$
   $\Rightarrow \quad x = 0 \ \text{ or } \ x + 1 = 0$
   $\Rightarrow \quad x = 0 \ \text{ or } \ x = -1$
   $\Rightarrow \quad x = 0, -1$

   Sum of zeroes
   $= (0) + (-1)$
   $= -1$
   $= \dfrac{-1}{1} = \dfrac{-\text{coefficient of } x}{\text{coefficient of } x^2}$

   Product of zeroes
   $= (0)(-1)$
   $= 0$
   $= \dfrac{0}{1} = \dfrac{\text{constant term}}{\text{coefficient of } x^2}$.

6. Let $p(x) = x^3 - 6x^2 + 11x - 6$. Then,
   $p(1) = (1)^3 - 6(1)^2 + 11(1) - 6$
   $\quad = 1 - 6 + 11 - 6 = 0$
   $p(2) = (2)^3 - 6(2)^2 + 11(2) - 6$
   $\quad = 8 - 24 + 22 - 6 = 0$
   $p(3) = (3)^3 - 6(3)^2 + 11(3) - 6$
   $\quad = 27 - 54 + 33 - 6 = 0$

   Hence, 1, 2, 3 are the zeroes of $p(x)$.
   Now,
   $1 + 2 + 3 = 6 = -\dfrac{(-6)}{1}$
   $= -\dfrac{\text{coefficient of } x^2}{\text{coefficient of } x^3}$

   $(1)(2) + (2)(3) + (3)(1) = 11 = \dfrac{11}{1}$
   $= \dfrac{\text{coefficient of } x}{\text{coefficient of } x^3}$.

   $(1)(2)(3) = 6 = \dfrac{-(-6)}{1}$
   $= -\dfrac{\text{constant term}}{\text{coefficient of } x^3}$.

7. $\dfrac{1}{\alpha} + \dfrac{1}{\beta} = \dfrac{\alpha + \beta}{\alpha\beta} = \dfrac{-\dfrac{(-2)}{1}}{\dfrac{2}{1}} = 1$.

8. (i) $\alpha + \beta + \gamma = -\dfrac{(-14)}{1} = 14$

   (ii) $\alpha\beta + \beta\gamma + \gamma\alpha = \dfrac{37}{1} = 37$

   (iii) $\alpha\beta\gamma = \dfrac{-(-60)}{1} = 60$.

9. $\quad x^2 - 3x + 2 \overline{) 2x^4 - 6x^3 + 3x^2 + 3x - 2}$
   $\qquad\qquad\quad 2x^4 - 6x^3 + 4x^2$
   $\qquad\qquad\quad\ \ -\quad\ +\qquad -$
   $\qquad\qquad\qquad\qquad\qquad -x^2 + 3x - 2$
   $\qquad\qquad\qquad\qquad\qquad -x^2 + 3x - 2$
   $\qquad\qquad\qquad\qquad\quad\ \ +\quad\ -\quad\ +$
   $\qquad\qquad\qquad\qquad\qquad\qquad\qquad 0$

   Quotient $= 2x^2 - 1$
   Remainder $= 0$
   ∵ The remainder is 0
   ∴ $2x^4 - 6x^3 + 3x^2 + 3x - 2$ is exactly divisible by $x^2 - 3x + 2$.

10. $\left(x - \sqrt{\dfrac{3}{2}}\right)\left(x + \sqrt{\dfrac{3}{2}}\right) = x^2 - \dfrac{3}{2}$

    Let us divide $2x^4 - 2x^3 - 7x^2 + 3x + 6$
    by $\qquad x^2 - \dfrac{3}{2}$

$$x^2 - \frac{3}{2} \overline{)\begin{array}{l} 2x^4 - 2x^3 - 7x^2 + 3x + 6 \\ 2x^4 - 3x^2 \\ -+ \\ \hline -2x^3 - 4x^2 + 3x + 6 \\ -2x^3 + 3x \\ +- \\ \hline -4x^2 + 6 \\ -4x^2 + 6 \\ +- \\ \hline 0 \end{array}} \quad 2x^2 - 2x - 4$$

Now, $2x^2 - 2x - 4$
$= 2(x^2 - x - 2)$
$= 2(x^2 - 2x + x - 2)$
$= 2\{x(x-2) + 1(x-2)\}$
$= 2(x-2)(x+1)$

Hence, the other two zeros are 2 and -1.

Chapter 3 PAIR OF LINEAR EQUATIONS IN TWO VARIABLES

1. $15x + 4y = 61$ ...(1)
   $4x + 15y = 72$ ...(2)

   Multiplying (1) by 15 and (2) by 4, we get
   $225x + 60y = 915$ ...(3)
   $16x + 60y = 288$ ...(4)

   Subtracting (4) from (3), we get
   $209x = 627$
   $\Rightarrow \quad x = \dfrac{627}{209} = 3$

   Put $x = 3$ in (1), we get
   $15(3) + 4y = 61$
   $\Rightarrow \quad 45 + 4y = 61$
   $\Rightarrow \quad 4y = 16$
   $\Rightarrow \quad y = \dfrac{16}{4} = 4.$

2. For infinite number of solutions,
   $$\dfrac{a_1}{a_2} = \dfrac{b_1}{b_2} = \dfrac{c_1}{c_2}$$
   $\Rightarrow \quad \dfrac{k}{16} = \dfrac{4}{k} = \dfrac{k-4}{k}$

   First two members give
   $k^2 = 64$
   $\Rightarrow \quad k = \pm 8$

   Last two numbers give
   $4k = k^2 - 4k$
   $\Rightarrow \quad k^2 - 8k = 0$
   $\Rightarrow \quad k(k-8) = 0$
   $\Rightarrow \quad k = 0, 8$

   $k = 8$ is common to both. So, the required value of k is 8.

3. (i) For a unique solution
   $$\dfrac{2}{3} \neq \dfrac{k}{-5}$$
   $\Rightarrow \quad k \neq \dfrac{-10}{3}.$

   (ii) For no solution
   $$\dfrac{2}{3} = \dfrac{k}{-5} \neq \dfrac{1}{7}$$
   $\Rightarrow \quad k = \dfrac{-10}{3}.$

4. $2x - y = 2$ ...(1)
   $\Rightarrow \quad y = 2x - 2$

HIGHER ORDER THINKING SKILLS (HOTS) SOLUTIONS

Table of solutions

x	0	1
y	-2	0

Plot $(0, -2)$ and $(1, 0)$ on a graph paper and join by a line. This line is the graph of eqn. (1).

$$4x - y = 8 \qquad \ldots(2)$$
$$\Rightarrow \quad y = 4x - 8$$

Table of solutions

x	2	3
y	0	4

Plot $(2, 0)$ and $(3, 4)$ on the same graph paper and join by a line. This line is the graph of eqn. (2).

From graph, we see that the two lines intersect at the point $(3, 4)$. Hence, the required solution is given by

$$x = 3, y = 4.$$

Also line (1) meets the axis of x in the point $(1, 0)$ while line (2) meets the axis of x in the point $(2, 0)$.

Table of solutions

x	-3	-4
y	0	2

Plot $(-3, 0)$ and $(-4, 2)$ on a graph paper and join by a line. This line is the graph of equation (1).

$$3x - 2y - 12 = 0 \qquad \ldots(2)$$
$$\Rightarrow \quad 2y = 3x - 12$$
$$\Rightarrow \quad y = \frac{3x - 12}{2}$$

Table of solutions

x	4	0
y	0	-6

Plot $(4, 0)$ and $(0, -6)$ on the same graph paper and join by a line. This line is the graph of equation (2).

From graph, we see that the two lines intersect at the point $(0, -6)$. Hence, the required solution is given by

$$x = 0, y = -6.$$

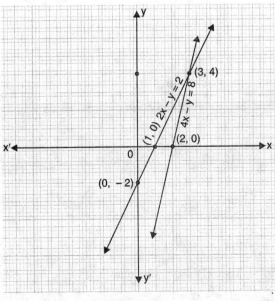

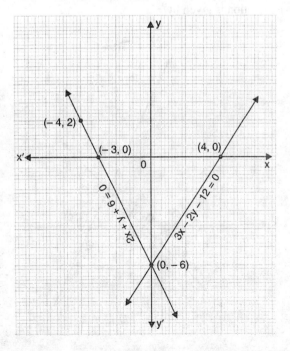

5. $2x + y + 6 = 0 \qquad \ldots(1)$
   $\Rightarrow \quad y = -2x - 6$

Also, the vertices of the triangle formed by the given lines and x-axis are (– 3, 0), (4, 0) and (0, – 6).

6. $x - y = 1$ ...(1)
$\Rightarrow y = x - 1$

Table of solutions

x	0	1
y	-1	0

Plot $(0, -1)$ and $(1, 0)$ on a graph paper and join by a line. This line is the graph of equation (1).

$2x + y = 8$ ...(2)
$\Rightarrow y = 8 - 2x$

Table of solutions

x	4	2
y	0	4

Plot $(4, 0)$ and $(2, 4)$ on the same graph paper and join by a line. This line is the graph of equation (2).

For graph, we see that the two lines intersect at the point $(3, 2)$. Hence, the required solution is given by

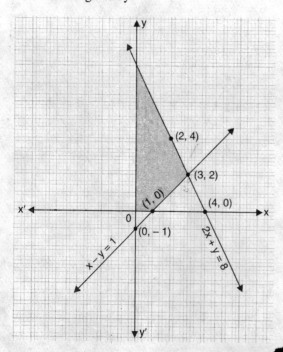

$x = 3, y = 2$

The area bounded by these two lines and the y-axis has been shaded.

7. We have $\dfrac{4}{x} + 3y = 14$ ...(1)

$\dfrac{3}{x} - 4y = 23$ ...(2)

Put $\dfrac{1}{x} = X$

Then (1) and (2) become

$4X + 3y = 14$
$3X - 4y = 23$
$\Rightarrow 4X + 3y - 14 = 0$
$3X - 4y - 23 = 0$

$$\begin{array}{ccc} X & y & 1 \\ 3 \diagdown -14 & 4 \diagdown 3 \\ -4 \diagup -23 & 3 \diagup -4 \end{array}$$

Then, $\dfrac{X}{(3)(-23) - (-4)(-14)}$

$= \dfrac{y}{(-23)(4) - (-14)(3)}$

$= \dfrac{1}{(4)(-4) - (3)(3)}$

$\Rightarrow \dfrac{X}{-125} = \dfrac{y}{-50} = \dfrac{1}{-25}$

$\Rightarrow X = 5 \Rightarrow \dfrac{1}{x} = 5 \Rightarrow x = \dfrac{1}{5}$

$y = 2$.

8. $\dfrac{11}{v} - \dfrac{7}{u} = 1$ ...(1)

$\dfrac{9}{v} - \dfrac{4}{u} = 6$ ...(2)

Put $\dfrac{1}{v} = y$ and $\dfrac{1}{u} = x$.

Then (1) and (2) become

$11y - 7x = 1$ ...(3)
$9y - 4x = 6$ ...(4)

From (3),
$$7x = 11y - 1$$
$$\Rightarrow x = \frac{11y - 1}{7}$$

Put this value of x in (4), we get
$$9y - 4\left(\frac{11y - 1}{7}\right) = 6$$
$$\Rightarrow 63y - 44y + 4 = 42$$
$$\Rightarrow 19y = 38$$
$$\Rightarrow y = \frac{38}{19} = 2$$
$$\Rightarrow \frac{1}{v} = 2$$
$$\Rightarrow v = \frac{1}{2}$$

Put $y = 2$ in (4), we get
$$9(2) - 4x = 6$$
$$\Rightarrow 4x = 12$$
$$\Rightarrow x = \frac{12}{4} = 3$$
$$\Rightarrow \frac{1}{u} = 3$$
$$\Rightarrow u = \frac{1}{3}.$$

9. $(a + c)x - (a - c)y = 2ab$
$(a + b)x - (a - b)y = 2ab$
$$\Rightarrow (a + c)x - (a - c)y - 2ab = 0$$
$$(a + b)x - (a - b)y - 2ab = 0$$

x	y	1
$-(a-c)$	$-2ab$	$a+c$	$-(a-c)$
$-(a-b)$	$-2ab$	$a+b$	$-(a-b)$

Then,
$$\frac{x}{\{-(a-c)\}(-2ab) - \{-(a-b)(-2ab)\}} = \frac{y}{(-2ab)(a+b) - (-2ab)(a+c)}$$
$$= \frac{1}{(a+c)\{-(a-b)\} - (a+b)\{-(a-c)\}}$$

$$\Rightarrow \frac{x}{2ab(a-c) - 2ab(a-b)} = \frac{y}{-2ab(a+b) + 2ab(a+c)}$$

$$= \frac{1}{-(a+c)(a-b) + (a+b)(a-c)}$$

$$\Rightarrow \frac{x}{2ab(a - c - a + b)} = \frac{y}{2ab(a + c - a - b)}$$

$$= \frac{1}{-(a^2 + ac - ab - bc) + (a^2 - ac + ab - bc)}$$

$$\Rightarrow \frac{x}{2ab(b-c)} = \frac{y}{2ab(c-b)} = \frac{1}{2a(b-c)}$$

$$\Rightarrow x = b, y = -b.$$

10. Put $\frac{1}{x+y} = X$, $\frac{1}{x-y} = Y$

Then, the given eqns. become
$$6X - 7Y = 3$$
$$\frac{1}{2}X - \frac{1}{3}Y = 0$$
$$\Rightarrow 6X - 7Y = 3 \quad ...(1)$$
$$3X - 2Y = 0 \quad ...(2)$$

From (1),
$$Y = \frac{6X - 3}{7} \quad ...(3)$$

From (2),
$$Y = \frac{3X}{2} \quad ...(4)$$

Equating the two values of Y from (3) and (4), we get
$$\frac{6X - 3}{7} = \frac{3X}{2}$$
$$\Rightarrow 12X - 6 = 21X$$
$$\Rightarrow 9X = -6$$
$$\Rightarrow X = \frac{-2}{3}$$
$$\Rightarrow \frac{1}{x+y} = -\frac{2}{3}$$
$$\Rightarrow x + y = -\frac{3}{2} \quad ...(5)$$

Put $X = -\dfrac{2}{3}$ in (4), we get

$$Y = \dfrac{3}{2}\left(-\dfrac{2}{3}\right) = -1$$

$\Rightarrow \quad \dfrac{1}{x-y} = -1$

$\Rightarrow \quad x - y = -1$ ...(6)

Adding (5) and (6), we get

$$2x = -\dfrac{5}{2}$$

$$x = -\dfrac{5}{4}$$

Subtracting (6) from (5), we get

$$2y = -\dfrac{1}{2}$$

$\Rightarrow \quad y = -\dfrac{1}{4}$.

11. We have $35x + 23y = 209$ ...(1)
$23x + 35y = 197$ ...(2)

Adding (1) and (2), we get

$58x + 58y = 406$

$\Rightarrow \quad x + y = 7$ ...(3)

Subtracting (2) from (1), we get

$12x - 12y = 12$

$= x - y = 1$ ...(4)

Solving (3) and (4), we get

$x = 4, y = 3$.

12. Let the cost of one chair be Rs. x and of one table be Rs. y. Then, according to the question,

$3x + 2y = 700$ ...(1)
$5x + 3y = 1100$ ...(2)

Multiplying (1) by 3 and (2) by 2, we get

$9x + 6y = 2100$ ...(3)
$10x + 6y = 2200$ ...(4)

Subtracting (3) from (4), we get

$x = 100$

Put $x = 100$ in (1), we get

$3(100) + 2y = 700$

$\Rightarrow \quad y = 200$

$\therefore$ Cost of 2 chairs and 2 tables

$= 2x + 2y$

$= 2 \times 100 + 2 \times 200$

$= 200 + 400 = $ Rs. 600.

13. Let the fraction be $\dfrac{x}{y}$. Then, according to the question,

$$\dfrac{x+2}{y} = \dfrac{1}{2}$$

$\Rightarrow \quad 2x - y = -4$ ...(1)

and $\dfrac{x}{y-1} = \dfrac{1}{3}$

$\Rightarrow \quad 3x - y = -1$ ...(2)

Subtracting (1) from (2), we get

$x = 3$

Put $x = 3$ in (1), we get

$2(3) - y = -4$

$\Rightarrow \quad 6 - y = -4$

$\Rightarrow \quad y = 10$

Hence, the required fraction is $\dfrac{3}{10}$.

Chapter 4 QUADRATIC EQUATIONS

1. $12abx^2 - (9a^2 - 8b^2)x - 6ab = 0$

$\Rightarrow 12abx^2 - 9a^2 x + 8b^2 x - 6ab = 0$

$\Rightarrow 3ax(4bx - 3a) + 2b(4bx - 3a) = 0$

$\Rightarrow (4bx - 3a)(3ax + 2b) = 0$

$\Rightarrow 4bx - 3a = 0$ or $3ax + 2b = 0$

$\Rightarrow x = \dfrac{3a}{4b}$ or $x = -\dfrac{2b}{3a}$

$\Rightarrow x = \dfrac{3a}{4b}, -\dfrac{2b}{3a}$.

2. $x^2 - 4ax + 4a^2 - b^2 = 0$

Comparing with $Ax^2 + Bx + C = 0$, we get

$A = 1$

$$B = -4a$$
$$C = 4a^2 - b^2$$

Using Quadratic Formula,

$$x = \frac{-B \pm \sqrt{B^2 - 4AC}}{2A}$$

$$= \frac{4a \pm \sqrt{16a^2 - 4(1)(4a^2 - b^2)}}{2(1)}$$

$$= \frac{4a \pm 2b}{2} = 2a \pm b.$$

3. $\dfrac{x+1}{x-1} + \dfrac{x-2}{x+2} = 3$

$\Rightarrow (x+1)(x+2) + (x-2)(x-1)$
$\quad = 3(x-1)(x+2)$

$\Rightarrow (x^2 + 2x + x + 2) + (x^2 - 2x - x + 2)$
$\quad = 3(x^2 + 2x - 2)$

$\Rightarrow \quad 2x^2 + 4 = 3x^2 + 3x - 6$

$\Rightarrow \quad x^2 + 3x - 10 = 0$

$\Rightarrow \quad x^2 + 5x - 2x - 10 = 0$

$\Rightarrow \quad x(x+5) - 2(x+5) = 0$

$\Rightarrow \quad (x+5)(x-2) = 0$

$\Rightarrow \quad x = -5, 2.$

4. Here, $A = 1$, $B = -2(k+1)$, $C = k^2$

For equal roots,
$$B^2 = 4AC$$
$\Rightarrow \quad 4(k+1)^2 = 4k^2$

$\Rightarrow \quad k = -\dfrac{1}{2}.$

5. Here, $a = 2$, $b = -4$, $c = 3$

∴ discriminant
$$= b^2 - 4ac$$
$$= (-4)^2 - 4(2)(3)$$
$$= -8\ (<0)$$

Hence, the quadratic equation has no real roots.

6. Here, $a = 1$, $b = -4$, $c = k$

∴ discriminant
$$= b^2 - 4ac$$
$$= (-4)^2 - 4(1)(k)$$
$$= 16 - 4k$$

for two real roots, discriminant ≥ 0

$\Rightarrow \quad 16 - 4k \geq 0$

$\Rightarrow \quad 16 \geq 4k$

$\Rightarrow \quad 4k \leq 16$

$\Rightarrow \quad k \leq 4.$

7. Let the original speed of the train be x km per hour.

Then, increased speed of the train $= (x+5)$ km per hour.

Time taken with original speed
$$= \frac{300}{x} \text{ hr.}$$

Time taken with increased speed
$$= \frac{300}{x+5} \text{ hr.}$$

According to the question,
$$\frac{300}{x} - \frac{300}{x+5} = 2$$

$\Rightarrow \quad \dfrac{1}{x} - \dfrac{1}{x+5} = \dfrac{1}{150}$

$\Rightarrow \quad \dfrac{x+5-x}{x(x+5)} = \dfrac{1}{150}$

$\Rightarrow \quad x(x+5) = 750$

$\Rightarrow \quad x^2 + 5x - 750 = 0$

$\Rightarrow \quad x^2 + 30x - 25x - 750 = 0$

$\Rightarrow \quad x(x+30) - 25(x+30) = 0$

$\Rightarrow \quad (x+30)(x-25) = 0$

$\Rightarrow \quad x = -30, 25$

$x = -30$ is inadmissible as the speed cannot be negative.

∴ $\quad x = 25$

Hence, the original speed of the train is 25 km per hour.

8. Let the number of students be x. Then,
$$\frac{300}{x} - \frac{300}{x+10} = 1$$

$\Rightarrow \quad \dfrac{1}{x} - \dfrac{1}{x+10} = \dfrac{1}{300}$

$\Rightarrow \quad \dfrac{x+10-x}{x(x+10)} = \dfrac{1}{300}$

$\Rightarrow \quad x(x+10) = 3000$

$\Rightarrow \quad x^2 + 10x - 3000 = 0$

$\Rightarrow \quad x^2 + 60x - 50x - 3000 = 0$

$\Rightarrow \quad x(x+60) - 50(x+60) = 0$

$\Rightarrow \quad (x+60)(x-50) = 0$

$\Rightarrow \quad x = -60, 50$

$x = -60$

is inadmissible as the number of students cannot be –ve.

$\therefore \quad x = 50$

Hence, the number of students is 50.

9. Let the two numbers be x and $18 - x$. Then,

$$\dfrac{1}{x} + \dfrac{1}{18-x} = \dfrac{1}{4}$$

$\Rightarrow \quad \dfrac{18-x+x}{x(18-x)} = \dfrac{1}{4}$

$\Rightarrow \quad x(18-x) = 72$

$\Rightarrow \quad x^2 - 18x + 72 = 0$

$\Rightarrow \quad x^2 - 6x - 12x + 72 = 0$

$\Rightarrow \quad x(x-6) - 12(x-6) = 0$

$\Rightarrow \quad (x-6)(x-12) = 0$

$\Rightarrow \quad x = 6, 12$

$\Rightarrow \quad 18 - x = 12, 6$

Hence the two numbers are 6 and 12.

10. Let the two numbers be x and $x - 3$. Then,

$$x(x-3) = 504$$

$\Rightarrow \quad x^2 - 3x - 504 = 0$

$\Rightarrow \quad x^2 - 24x + 21x - 504 = 0$

$\Rightarrow \quad x(x-24) + 21(x-24) = 0$

$\Rightarrow \quad (x-24)(x+21) = 0$

$\Rightarrow \quad x = 24, -21$

$\Rightarrow \quad x - 3 = 21, -24$

Hence, the two numbers are
21, 24 or –21, –24.

11. Let the unit's digit be x.

Then, the ten's digit $= \dfrac{18}{x}$

$\therefore$ Number $= x + \left(\dfrac{18}{x}\right)(10)$

$= x + \dfrac{180}{x}$

According to the question,

$x + \dfrac{180}{x} - 27 = 10x + \dfrac{18}{x}$

$\Rightarrow \quad 9x - \dfrac{162}{x} = -27$

$\Rightarrow \quad x - \dfrac{18}{x} = -3$

$\Rightarrow \quad x^2 + 3x - 18 = 0$

$\Rightarrow \quad x^2 + 6x - 3x - 18 = 0$

$\Rightarrow \quad x(x+6) - 3(x+6) = 0$

$\Rightarrow \quad (x+6)(x-3) = 0$

$\Rightarrow \quad x = -6, 3$

$x = -6$

is inadmissible as the unit's digit cannot be –ve

$\therefore \quad x = 3$

$\therefore \quad \dfrac{18}{x} = 6$

Hence, the required number is 63.

Chapter 5 ARITHMETIC PROGRESSIONS

1. $(x+2) - 3x = 8 - (x+2)$

$\Rightarrow \quad 2 - 2x = 6 - x$

$\Rightarrow \quad x = -4$

$a = 3x = -12$

$d = 2 - 2x = 10$

$\therefore \quad$ 4th term $= a + (4-1)d$

$= a + 3d$

$= (-12) + 3(10) = 18.$

2. $a = 1$

$n = 100$

$l = 100$

$\therefore \quad$ Sum $= \dfrac{n}{2}(a+l)$

$= \dfrac{100}{2}(1+100) = 5050.$

3. $S_n = 3n^2 - n$
   Put $n = 1, 2$
   $S_1 = 3(1)^2 - 1 = 2 = a_1$
   $S_2 = 3(2)^2 - 2 = 10 = a_1 + a_2$
   $\therefore \quad a_2 = 10 - 2 = 8$
   $\therefore \quad a = 2$
   $d = a_2 - a_1 = 8 - 2 = 6.$

4. Let the first term be a and the common difference be d. Then,
   $7a_7 = 11a_{11}$
   $\Rightarrow \quad 7(a + 6d) = 11(a + 10d)$
   $\Rightarrow \quad 7a + 42d = 11a + 110d$
   $\Rightarrow \quad 4a + 68d = 0$
   $\Rightarrow \quad a + 17d = 0$
   $\Rightarrow \quad a_{18} = 0.$

5. $a = 17$
   $d = 14 - 17 = -3$
   $l = -40$
   Let the number of terms of the AP be n.
   Then, $l = a + (n - 1)d$
   $\Rightarrow \quad -40 = 17 + (n - 1)(-3)$
   $\Rightarrow \quad -57 = (n - 1)(-3)$
   $\Rightarrow \quad n - 1 = 19$
   $\Rightarrow \quad n = 20$
   $\therefore$ 6th term from the end
   $= (20 - 6 + 1)$th term from the beginning
   $= 15$th term from the beginning
   $= a_{15}$
   $= a + 14d$
   $= 17 + 14(-3)$
   $= 17 - 42 = -25.$

6. Let a be the first term and d be the common difference of the AP. Then,
   $a_2 = 8$
   $\Rightarrow \quad a + d = 8$ ...(1)
   $a_4 = 14$
   $\Rightarrow \quad a + 3d = 14$ ...(2)
   Solving (1) and (2), we get
   $a = 5, d = 3$
   $\therefore \quad S_{21} = \dfrac{21}{2}[2(5) + (21 - 1)3]$
   $= \dfrac{21}{2}[10 + 60] = 735.$

7. $S_n = 5n^2 + 3n$
   $S_{n-1} = 5(n-1)^2 + 3(n-1)$
   $= 5n^2 - 7n + 2$
   $\therefore \quad a_n = S_n - S_{n-1}$
   $= 10n - 2.$

8. Multiples of 9 lying between 300 and 700 are 306, 315, 324, ..., 693
   $a = 306, d = 315 - 306 = 9, l = 693$
   Let the number of terms be n. Then,
   $l = a + (n - 1)d$
   $\Rightarrow \quad 693 = 306 + (n - 1)9$
   $\Rightarrow \quad 387 = (n - 1)9$
   $\Rightarrow \quad 43 = n - 1$
   $\Rightarrow \quad n = 44$
   $\therefore \quad S_n = \dfrac{n}{2}(a + l)$
   $= \dfrac{44}{2}(306 + 693)$
   $= (22)(999) = 21978.$

9. $a_n = 5n - 3$
   $\therefore \quad a_1 = 5(1) - 3 = 2$
   $\quad a_2 = 5(2) - 3 = 7$
   $\therefore \quad a = 2$
   $d = a_2 - a_1 = 7 - 2 = 5$
   $\therefore$ Sum of the first 20 terms of AP
   $= \dfrac{20}{2}[2(2) + (20 - 1)5]$
   $= 10[4 + 95] = 990.$

10. Let a be the first term and d be the common difference of the AP. Then,
    8th term $= 37$
    $\Rightarrow \quad a + 7d = 37$ ...(1)
    12th term $= 57$
    $\Rightarrow \quad a + 11d = 57$ ...(2)
    Solving (1) and (2), we get
    $a = 2, d = 5$
    $\therefore$ The A.P. is
    $2, 2 + 5, 2 + 5 + 5, 2 + 5 + 5 + 5, ...$
    or $2, 7, 12, 17, ...$

11. Let the three sides in AP be $a, a + d, a + 2d$. Then,
$$(a + 2d)^2 = a^2 + (a + d)^2$$
    | ∵ Triangle is right angled
$\Rightarrow a^2 + 4d^2 + 4ad = a^2 + a^2 + d^2 + 2ad$
$\Rightarrow a^2 - 3d^2 - 2ad = 0$
$\Rightarrow a^2 - 3ad + ad - 3a^2 = 0$
$\Rightarrow a(a - 3d) + d(a - 3d) = 0$
$\Rightarrow (a + d)(a - 3d) = 0$
∵ $a + d = 0$ is impossible
    | ∵ Length of a side cannot be zero
∴ $a - 3d = 0$
$\Rightarrow a = 3d$
∴ The three sides are $3d, 4d, 7d$
∴ Their ratio $= 3 : 4 : 7$.

12. Let the angles of the quadrilateral be $a, a + d, a + 2d, a + 3d$
$a + (a + d) + (a + 2d) + (a + 3d) = 360°$
    | Angle sum property of a quadrilateral
$\Rightarrow 4a + 6d = 360°$
$\Rightarrow 4a + 6(20°) = 360°$
$\Rightarrow 4a + 120° = 360°$
$\Rightarrow 4a = 240°$
$\Rightarrow a = 60°$

Hence, the four angles of the quadrilateral are 60°, 80°, 100° and 120°.

13. $a = 24, d = 20 - 24 = -4, S_n = 72$
$\Rightarrow \dfrac{n}{2}[2(24) + (n - 1)(-4)] = 72$
$\Rightarrow n[48 - 4n + 4] = 144$
$\Rightarrow n[52 - 4n] = 144$
$\Rightarrow 4n^2 - 52n + 144 = 0$
$\Rightarrow n^2 - 13n + 36 = 0$
$\Rightarrow (n - 4)(n - 9) = 0$
$\Rightarrow n = 4, 9$

Reason for two answers. Sum of terms from 5th to 9th is zero.

Chapter 6 TRIANGLES

1. ∵ DE ∥ BC
   ∴ By Thales Theorem,
   $$\dfrac{AD}{BD} = \dfrac{AE}{CE}$$
   $\Rightarrow \dfrac{4x - 3}{3x - 1} = \dfrac{8x - 7}{5x - 3}$
   $\Rightarrow (4x - 3)(5x - 3) = (8x - 7)(3x - 1)$
   $\Rightarrow 20x^2 - 27x + 9 = 24x^2 - 29x + 7$
   $\Rightarrow 4x^2 - 2x - 2 = 0$
   $\Rightarrow 2x^2 - x - 1 = 0$
   $\Rightarrow 2x^2 - 2x + x - 1 = 0$
   $\Rightarrow 2x(x - 1) + x - 1 = 0$
   $\Rightarrow (x - 1)(2x + 1) = 0$
   $\Rightarrow x = 1, -\dfrac{1}{2}$

   $x = -\dfrac{1}{2}$ is inadmissible as then AD = –5 and length cannot be negative.
   ∴ $x = 1$.

2. Area of right triangle ABC
   $$= \dfrac{AB \times AC}{2} = \dfrac{BC \times AD}{2}$$
   $\Rightarrow AB \times AC = BC \times AD$
   $\Rightarrow c \times b = \sqrt{b^2 + c^2} \times AD$
       | By Pythagoras Theorem
   $\Rightarrow AD = \dfrac{bc}{\sqrt{b^2 + c^2}}$.

3. Let the corresponding altitude of the other triangle be x cm. Then,
   $$\dfrac{81}{49} = \left(\dfrac{6.3}{x}\right)^2$$
   $\Rightarrow \left(\dfrac{9}{7}\right)^2 = \left(\dfrac{6.3}{x}\right)^2$
   $\Rightarrow \dfrac{9}{7} = \dfrac{6.3}{x}$
   $\Rightarrow x = 4.9$

   Hence, the required altitude is 4.9 cm.

4. His distance from the starting point

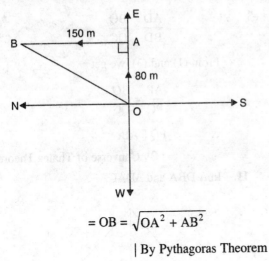

$= OB = \sqrt{OA^2 + AB^2}$

| By Pythagoras Theorem

$= \sqrt{(80)^2 + (150)^2}$

$= \sqrt{6400 + 22500}$

$= \sqrt{28900} = 170$ m

Hence, his distance from the starting point is 170 m.

5. $PR = \sqrt{OP^2 + OR^2}$

| By Pythagoras Theorem

$= \sqrt{6^2 + 8^2}$

$= \sqrt{36 + 64} = \sqrt{100}$

$= 10$ cm

Again,

$\because PQ^2 + PR^2$

$= 24^2 + 10^2$

$= 576 + 100 = 676$

$= 26^2 = QR^2$

$\therefore$ By converse of Pythagoras Theorem,

$\angle PQR = 90°$

Hence, ΔPQR is a right angled triangle.

6. $\because$ DE || AC

$\therefore \quad \Delta BAC \sim \Delta BDE$

$\therefore \quad \dfrac{ar(\Delta BAC)}{ar(\Delta BDE)} = \left(\dfrac{BA}{BD}\right)^2$ ...(1)

Now, $\dfrac{AD}{DB} = \dfrac{3}{2}$

$\Rightarrow \dfrac{AD}{DB} + 1 = \dfrac{3}{2} + 1$

$\Rightarrow \dfrac{AD + DB}{BD} = \dfrac{5}{2}$

$\Rightarrow \dfrac{BA}{BD} = \dfrac{5}{2}$ ...(2)

$\therefore \dfrac{ar(\Delta BAC)}{ar(\Delta BDE)} = \left(\dfrac{5}{2}\right)^2 = \dfrac{25}{4} = 25 : 4$.

| From (1) and (2)

7. In right triangles ADB and ADC,

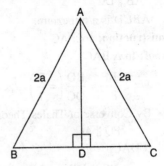

Hyp AB = Hyp AC

side AD = side AD

$\therefore \quad \Delta ADB \cong \Delta ADC$

| RHS congruence Axiom

$\therefore \quad BD = CD = \dfrac{1}{2} BC = \dfrac{1}{2}(2a) = a$

In right triangle ADB,

$AD^2 + BD^2 = AB^2$

$\Rightarrow AD^2 + a^2 = (2a)^2$

$\Rightarrow AD = \sqrt{3}\, a$.

8. **Construction:** Draw OE || CD meeting AD at E.

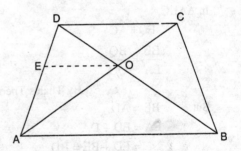

Proof: In $\triangle ACD$,

∵ $OE \parallel CD$

∴ $\dfrac{AO}{OC} = \dfrac{AE}{ED}$ ...(1)

| By Thales Theorem

But $\dfrac{AO}{OC} = \dfrac{BO}{OD}$ ...(2) | Given

∴ $\dfrac{BO}{OD} = \dfrac{AE}{ED}$

∴ $EO \parallel AB$

| By Converse of Thales Theorem

But $EO \parallel DC$

∴ $AB \parallel DC$

∴   ABCD is a trapezium.

9. Construction: Join AC.

Proof: In $\triangle BAC$,

$\dfrac{BP}{PA} = \dfrac{BQ}{QC} \left(= \dfrac{2}{1}\right)$

∴ By Converse of Thales Theorem,

    $PQ \parallel AC$ ...(1)

In $\triangle DAC$,

$\dfrac{DS}{SA} = \dfrac{DR}{RC} \left(= \dfrac{2}{1}\right)$

∴ By Converse of Thales Theorem,

    $SR \parallel AC$ ...(2)

From (1) and (2),

    $SR \parallel PQ$

Similarly, we can prove that

    $RQ \parallel SP$

Hence, PQRS is a parallelogram.

10. In $\triangle ABC$,

    $DP \parallel BC$

∴ $\dfrac{AD}{DB} = \dfrac{AP}{PC}$ ...(1)

| By Thales Theorem

In $\triangle ABC$,

    $EQ \parallel AC$

∴ $\dfrac{BE}{EA} = \dfrac{BQ}{QC}$ ...(2)

| By Thales Theorem

But, $BE = AD$

    $EA = ED + DA$

       $= ED + BE = BD$

∴ From (2),

    $\dfrac{AD}{BD} = \dfrac{BQ}{QC}$ ...(3)

From (1) and (3), we get

    $\dfrac{AP}{PC} = \dfrac{BQ}{QC}$

∴ $PQ \parallel AB$.

| By Converse of Thales Theorem

11. In $\triangle DBA$ and $\triangle DAC$,

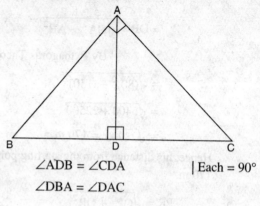

    $\angle ADB = \angle CDA$ | Each = 90°

    $\angle DBA = \angle DAC$

| $\angle DBA + \angle DAB = 90°$
| $\angle DAC + \angle DAB = \angle BAC = 90°$
| ∴ $\angle DBA = \angle DAC$

∴ $\triangle DBA \sim \triangle DAC$

| A A criterion of similarity

∴ $\dfrac{BD}{AD} = \dfrac{AD}{CD}$

$\Rightarrow$ $AD^2 = BD \times CD$

$\Rightarrow$ $AD^2 = BD \times DC$.

12. In $\triangle FDA$ and $\triangle FBE$,

    $\angle FDA = \angle FBE$ | Alt. Angles

    $\angle AFD = \angle EFB$

| Vert. Opp. Angles

∴ $\triangle FDA \sim \triangle FBE$

| A A similarity criterion

∴ $\dfrac{FD}{FB} = \dfrac{FA}{FE}$

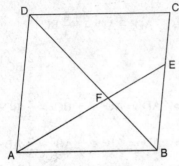

$\Rightarrow$ DF × EF = FB × FA

13. $\because$ △ PAB, △ QBC and △ RAC are equilateral

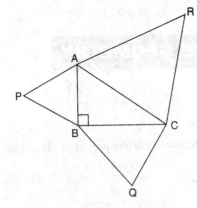

$\therefore$ They are equiangular
$\therefore$ They are similar.

$\therefore \quad \dfrac{ar(\triangle PAB)}{ar(\triangle RAC)} = \dfrac{AB^2}{AC^2}$ ...(1)

$\dfrac{ar(\triangle QBC)}{ar(\triangle RAC)} = \dfrac{BC^2}{AC^2}$ ...(2)

Adding (1) and (2), we get

$= \dfrac{ar(\triangle PAB)}{ar(\triangle RAC)} + \dfrac{ar(\triangle QBC)}{ar(\triangle RAC)}$

$= \dfrac{AB^2}{AC^2} + \dfrac{BC^2}{AC^2}$

$= \dfrac{AB^2 + BC^2}{AC^2} = \dfrac{AC^2}{AC^2} = 1$

| By Pythagoras Theorem

$\Rightarrow \dfrac{ar(\triangle PAB) + ar(\triangle QBC)}{ar(\triangle RAC)} = 1$

$\Rightarrow$ ar(△ PAB) + ar(△ QBC)
$= $ ar(△ RAC).

14. In right triangle ABD,
$AD^2 = AB^2 + BD^2$

| By Pythagoras Theorem

$= AB^2 + \left(\dfrac{BC}{2}\right)^2$

$= AB^2 + \dfrac{BC^2}{4}$ ...(1)

In right triangle BCE,
$CE^2 = BC^2 + BE^2$

$= BC^2 + \left(\dfrac{AB}{2}\right)^2$

$= BC^2 + \dfrac{AB^2}{4}$ ...(2)

Adding (1) and (2), we get

$AD^2 + CE^2 = \dfrac{5}{4}(AB^2 + BC^2)$

$= \dfrac{5}{4} AC^2$

| By Pythagoras Theorem

$\Rightarrow \left(\dfrac{3\sqrt{5}}{2}\right)^2 + CE^2 = \dfrac{5}{4}(5)^2$

$\Rightarrow \quad CE = 2\sqrt{5}$ cm.

15.

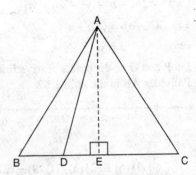

Construction: Draw AE ⊥ BC.

Proof: $\because \quad \triangle AEB \cong \triangle AEC$

| RHS Congruence criterion

∴ BE = EC
Then,
$$BD = \frac{1}{3} BC$$
$$DC = \frac{2}{3} BC$$
$$BE = EC = \frac{1}{2} BC$$
∵ ∠ACB = 60°
∴ ∠ACB is acute
∴ In Δ ADC,
$$AD^2 = AC^2 + DC^2 - 2DC \cdot EC$$

$$\Rightarrow AD^2 = AC^2 + \left(\frac{2}{3} BC\right)^2$$
$$- 2\left(\frac{2}{3} BC\right)\left(\frac{1}{2} BC\right)$$
$$\Rightarrow AD^2 = AC^2 + \frac{4}{9} BC^2 - \frac{2}{3} BC^2$$
$$\Rightarrow AD^2 = AB^2 + \frac{4}{9} AB^2 - \frac{2}{3} AB^2$$
| ∵ AB = BC = CA
$$\Rightarrow AD^2 = \frac{7}{9} AB^2$$
$$\Rightarrow 9AD^2 = 7AB^2.$$

Chapter 7 COORDINATE GEOMETRY

1. Since the given points are collinear, the area of the triangle formed by them must be zero.
∴ $\frac{1}{2}[1(1-11) + x(11-5) + 4(5-1)] = 0$
$\Rightarrow \quad -10 + 6x + 16 = 0$
$\Rightarrow \quad 6x = -6$
$\Rightarrow \quad x = -1.$

2. Let A → (a, b), B → (b, a) and P → (x, y)
Then, PA = PB
$\Rightarrow$ PA^2 = PB^2
$\Rightarrow (x-a)^2 + (y-b)^2 = (x-b)^2 + (y-a)^2$
$\Rightarrow x^2 - 2ax + a^2 + y^2 - 2by + b^2$
$= x^2 - 2bx + b^2 + y^2 - 2ay + a^2$
$\Rightarrow 2(b-a)x = 2(b-a)y$
$\Rightarrow x = y.$

3. Let P and Q be the points of trisection. Then, P divides AB in the ratio 1 : 2.

A———•———•———B
(3, −3) P   Q   (6, 9)

∴ P →
$$\left\{\frac{(1)(6) + (2)(3)}{1+2}, \frac{(1)(9) + (2)(-3)}{1+2}\right\}$$
$\Rightarrow$ P → (4, 1)
Again, ∵ Q is the mid-point of PB

∴ $Q \to \left(\frac{4+6}{2}, \frac{1+9}{2}\right)$
$\Rightarrow Q \to (5, 5).$

4. Mid-point of the join of (6, 8) and (2, 4)
$= \left(\frac{6+2}{2}, \frac{8+4}{2}\right)$ or (4, 6)
∴ Required distance
$= \sqrt{(4-1)^2 + (6-2)^2}$
$= \sqrt{9 + 16} = \sqrt{25} = 5.$

5. Mid-point of diagonal AC
$= \left(\frac{1+2}{2}, \frac{0+7}{2}\right) = \left(\frac{3}{2}, \frac{7}{2}\right)$
Mid-point of diagonal BD
$= \left(\frac{5+(-2)}{2}, \frac{3+4}{2}\right) = \left(\frac{3}{2}, \frac{7}{2}\right)$
Since the mid-points of the diagonals AC and BD are the same, therefore, the points A, B, C and D are the vertices of a parallelogram.

6. D → $\left(\frac{0+5}{2}, \frac{4+3}{2}\right)$ or $\left(\frac{5}{2}, \frac{7}{2}\right)$
E → $\left(\frac{5+1}{2}, \frac{3+(-1)}{2}\right)$ or (3, 1)
F → $\left(\frac{0+1}{2}, \frac{4+(-1)}{2}\right)$ or $\left(\frac{1}{2}, \frac{3}{2}\right)$

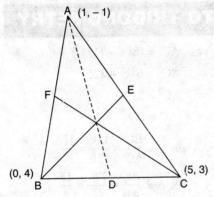

Length of median AD

$$= \sqrt{\left(\frac{5}{2}-1\right)^2 + \left(\frac{7}{2}+1\right)^2} = \sqrt{\frac{9}{4}+\frac{81}{4}}$$

$$= \sqrt{\frac{90}{4}} = \sqrt{\frac{45}{2}} = \frac{3\sqrt{5}}{\sqrt{2}}$$

Length of median BE

$$= \sqrt{(3-0)^2 + (1-4)^2} = 3\sqrt{2}$$

Length of median CF

$$= \sqrt{\left(\frac{1}{2}-5\right)^2 + \left(\frac{3}{2}-3\right)^2}$$

$$= \sqrt{\frac{81}{4}+\frac{9}{4}} = \sqrt{\frac{90}{4}} = \sqrt{\frac{45}{2}} = \frac{3\sqrt{5}}{\sqrt{2}}.$$

7. $\dfrac{-4+a+2}{3} = -1$

$\Rightarrow \quad a - 2 = -3$

$\Rightarrow \quad a = -1$

$\dfrac{2+7+b}{3} = 3$

$\Rightarrow \quad b = 0.$

8. ∵ The diagonals of a parallelogram bisect each other

∴ Mid-point of diagonal PR = Mid-point of diagonal QS

$\Rightarrow \left(\dfrac{a+2}{2}, \dfrac{-11+15}{2}\right) = \left(\dfrac{5+1}{2}, \dfrac{b+1}{2}\right)$

$\Rightarrow \left(\dfrac{a+2}{2}, 2\right) = \left(3, \dfrac{b+1}{2}\right)$

$\Rightarrow \dfrac{a+2}{2} = 3$

$2 = \dfrac{b+1}{2}$

$\Rightarrow \quad a = 4, b = 3.$

9. Required distance

$$= \sqrt{(6-0)^2 + (8-0)^2}$$

$$= \sqrt{36+64} = \sqrt{100}$$

$$= 10.$$

10. $AB = \sqrt{(5-1)^2 + (4-2)^2} = 2\sqrt{5}$

$BC = \sqrt{(3-5)^2 + (8-4)^2} = 2\sqrt{5}$

$CD = \sqrt{(-1-3)^2 + (6-8)^2} = 2\sqrt{5}$

$DA = \sqrt{(1+1)^2 + (2-6)^2} = 2\sqrt{5}$

$AC = \sqrt{(3-1)^2 + (8-2)^2} = 2\sqrt{10}$

$BD = \sqrt{(-1-5)^2 + (6-4)^2} = 2\sqrt{10}$

∵ AB = BC = CD = DA

i.e., sides are equal and, AC = BD

i.e., diagonals are equal

∴ ABCD is a square.

11. Let P → (x, y) be the circumcentre of △ ABC. Then,

PA = PB = PC

$\Rightarrow PA^2 = PB^2 = PC^2$

$\Rightarrow (x-4)^2 + (y-6)^2$

$= (x-0)^2 + (y-4)^2$

$= (x-6)^2 + (y-2)^2$

First two numbers give

$2x + y = 9$ ...(1)

Last two numbers give

$3x - y = 6$ ...(2)

Solving (1) and (2), we get $x = 3, y = 3$

∴ P → (3, 3)

and, circumradius = PA

$= \sqrt{(4-3)^2 + (6-3)^2} = \sqrt{10}.$

Chapter 8 — INTRODUCTION TO TRIGONOMETRY

1. $2\sin^2\theta = \dfrac{1}{2}$

$\Rightarrow \sin^2\theta = \dfrac{1}{4}$

$\Rightarrow \sin\theta = \dfrac{1}{2}$ $\quad | \because 0° < \theta < 90°$

$\Rightarrow \sin\theta = \sin\dfrac{\pi}{6}$

$\Rightarrow \theta = \dfrac{\pi}{6}.$

2. $\dfrac{5\sin\theta - 3\cos\theta}{4\sin\theta + 3\cos\theta}$

$= \dfrac{5 - 3\cot\theta}{4 + 3\cot\theta}$

| dividing the numerator and denominator by $\sin\theta$

$= \dfrac{5 - 3\left(\dfrac{3}{5}\right)}{4 + 3\left(\dfrac{3}{5}\right)}$ $\quad | \because 5\cot\theta = 3$

$= \dfrac{16}{29}.$

3. Draw a right $\triangle ABC$

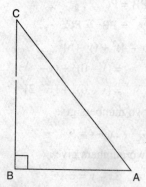

$\cot A = \dfrac{AB}{BC} = \dfrac{8}{15}$ $\quad |$ Given

If $AB = 8k$, then $BC = 15k$, where k is a positive number.

By Pythagoras Theorem,

$AC^2 = AB^2 + BC^2$

$\Rightarrow AC^2 = (8k)^2 + (15k)^2$

$\Rightarrow AC^2 = 64k^2 + 225k^2$

$\Rightarrow AC^2 = 289k^2$

$\Rightarrow AC = 17k$

$\therefore \sin A = \dfrac{BC}{AC} = \dfrac{15k}{17k} = \dfrac{15}{17}$

$\sec A = \dfrac{AC}{AB} = \dfrac{17k}{8k} = \dfrac{17}{8}.$

4. $\tan\theta + \cot\theta = 2$

$\Rightarrow (\tan\theta + \cot\theta)^2 = 4$

$\Rightarrow \tan^2\theta + \cot^2\theta + 2\tan\theta\cot\theta = 4$

$\Rightarrow \tan^2\theta + \cot^2\theta + 2\tan\theta\left(\dfrac{1}{\tan\theta}\right) = 4$

$\Rightarrow \tan^2\theta + \cot^2\theta + 2 = 4$

$\Rightarrow \tan^2\theta + \cot^2\theta = 2.$

5. $\sin\theta = \sqrt{1 - \cos^2\theta}$

$= \sqrt{1 - \left(\dfrac{3}{5}\right)^2} = \dfrac{4}{5}$

$\therefore \cot\theta = \dfrac{\cos\theta}{\sin\theta} = \dfrac{3/5}{4/5} = \dfrac{3}{4}$

$\csc\theta = \dfrac{1}{\sin\theta} = \dfrac{5}{4}$

$\therefore \cot\theta + \csc\theta$

$= \dfrac{3}{4} + \dfrac{5}{4} = 2.$

6. L.H.S. $= \cos 2A$

$= \cos 2(30°)$

$= \cos 60° = \dfrac{1}{2}$ ...(1)

R.H.S. $= \dfrac{1 - \tan^2 A}{1 + \tan^2 A}$

$= \dfrac{1 - \tan^2 30°}{1 + \tan^2 30°}$

$= \dfrac{1 - \left(\dfrac{1}{\sqrt{3}}\right)^2}{1 + \left(\dfrac{1}{\sqrt{3}}\right)^2} = \dfrac{1 - \dfrac{1}{3}}{1 + \dfrac{1}{3}}$

HIGHER ORDER THINKING SKILLS (HOTS) SOLUTIONS

$$= \frac{\frac{2}{3}}{\frac{4}{3}} = \frac{1}{2} \qquad ...(2)$$

From (1) and (2),

L.H.S. = R.H.S.

7. L.H.S. $= \cos(A - B)$
$= \cos(60° - 30°)$
$= \cos 30° = \frac{\sqrt{3}}{2} \qquad ...(1)$

R.H.S. $= \cos A \cos B + \sin A \sin B$
$= \cos 60° \cos 30° + \sin 60° \sin 30°$
$= \frac{1}{2} \cdot \frac{\sqrt{3}}{2} + \frac{\sqrt{3}}{2} \cdot \frac{1}{2}$
$= \frac{\sqrt{3}}{2} \qquad ...(2)$

From (1) and (2),

L.H.S. = R.H.S.

8. $\operatorname{cosec}(65° + \theta) - \sec(25° - \theta)$
$\qquad - \tan(55° - \theta) + \cot(35° + \theta)$
$= \operatorname{cosec}(90° - \overline{25° - \theta})$
$\qquad - \sec(25° - \theta) - \tan(90° - \overline{35° + \theta})$
$\qquad + \cot(35° + \theta)$
$= \sec(25° - \theta) - \sec(25° - \theta)$
$\qquad - \cot(35° + \theta) + \cot(35° + \theta) = 0.$

9. $\sin(50° + \theta) - \cos(40° - \theta)$
$\qquad + \tan 1° \tan 10° \tan 20° \tan 70°$
$\qquad\qquad \tan 80° \tan 89°$
$= \sin(90° - \overline{40° - \theta}) - \cos(40° - \theta)$
$\qquad + \tan 1° \tan 10° \tan 20° \tan(90° - 20°)$
$\qquad\qquad \tan(90° - 10°) \tan(90° - 1°)$
$= \cos(40° - \theta) - \cos(40° - \theta)$
$\qquad + \tan 1° \tan 10° \tan 20° \cot 20°$
$\qquad\qquad \cot 10° \cot 1°$
$= \tan 1° \tan 10° \tan 20°$
$\qquad\qquad \frac{1}{\tan 20°} \frac{1}{\tan 10°} \frac{1}{\tan 1°}$
$\Rightarrow 1.$

10. L.H.S.
$= \sec^2\theta + \operatorname{cosec}^2\theta$
$= \frac{1}{\cos^2\theta} + \frac{1}{\sin^2\theta}$
$= \frac{\sin^2\theta + \cos^2\theta}{\cos^2\theta \sin^2\theta}$
$= \frac{1}{\cos^2\theta \sin^2\theta}$
$= \frac{1}{\cos^2\theta} \cdot \frac{1}{\sin^2\theta}$
$= \sec^2\theta \operatorname{cosec}^2\theta$
$= $ R.H.S.

11. L.H.S.
$= \frac{\sin\theta - 2\sin^3\theta}{2\cos^3\theta - \cos\theta}$
$= \frac{\sin\theta(1 - 2\sin^2\theta)}{\cos\theta(2\cos^2\theta - 1)}$
$= \frac{\sin\theta(\sin^2\theta + \cos^2\theta - 2\sin^2\theta)}{\cos\theta(2\cos^2\theta - \sin^2\theta - \cos^2\theta)}$
$\qquad\qquad |\because \sin^2\theta + \cos^2\theta = 1$
$= \frac{\sin\theta(\cos^2\theta - \sin\theta)}{\cos\theta(\cos^2\theta - \sin^2\theta)}$
$= \frac{\sin\theta}{\cos\theta} = \tan\theta = $ R.H.S.

12. $\sec\theta + \tan\theta = p \qquad ...(1) \mid$ Given

We know that
$\sec^2\theta - \tan^2\theta = 1$
$\Rightarrow (\sec\theta + \tan\theta)(\sec\theta - \tan\theta) = 1$
$\Rightarrow p(\sec\theta - \tan\theta) = 1$
$\Rightarrow \sec\theta - \tan\theta = \frac{1}{p} \qquad ...(2)$

Solving eqs. (1) and (2), we get
$\sec\theta = \frac{1}{2}\left(p + \frac{1}{p}\right)$
$\tan\theta = \frac{1}{2}\left(p - \frac{1}{p}\right)$

Now, $\sin\theta = \frac{\tan\theta}{\sec\theta}$

709

$$= \frac{\frac{1}{2}\left(p - \frac{1}{p}\right)}{\frac{1}{2}\left(p + \frac{1}{p}\right)} = \frac{p^2 - 1}{p^2 + 1}.$$

13. $\cos\theta + \sin\theta = \sqrt{2}\,\cos\theta$

   $\Rightarrow (\cos\theta + \sin\theta)^2 = 2\cos^2\theta$

   $\Rightarrow \cos^2\theta + \sin^2\theta + 2\cos\theta\sin\theta = 2\cos^2\theta$

   $\Rightarrow \cos^2\theta - 2\sin\theta\cos\theta = \sin^2\theta$

   $\Rightarrow \cos^2\theta - 2\sin\theta\cos\theta + \sin^2\theta = 2\sin^2\theta$

   $\Rightarrow (\cos\theta - \sin\theta)^2 = 2\sin^2\theta$

   $\Rightarrow \cos\theta - \sin\theta = \sqrt{2}\,\sin\theta.$

14. L.H.S. $= (\tan\theta + \sin\theta)^2 - (\tan\theta - \sin\theta)^2$

   $= 4\tan\theta\sin\theta$ ...(1)

   R.H.S. $= 4\sqrt{mn}$

   $= 4\sqrt{(\tan\theta + \sin\theta)(\tan\theta - \sin\theta)}$

   $= 4\sqrt{\tan^2\theta - \sin^2\theta}$

   $= 4\sqrt{\dfrac{\sin^2\theta}{\cos^2\theta} - \sin^2\theta}$

   $= 4\sqrt{\dfrac{\sin^2\theta - \sin^2\theta\cos^2\theta}{\cos^2\theta}}$

   $= 4\sqrt{\dfrac{\sin^2\theta(1 - \cos^2\theta)}{\cos^2\theta}}$

   $= 4\sqrt{\dfrac{\sin^2\theta\sin^2\theta}{\cos^2\theta}} = 4\,\dfrac{\sin^2\theta}{\cos\theta}$

   $= 4\sin\theta\,\dfrac{\sin\theta}{\cos\theta} = 4\sin\theta\tan\theta$ ...(2)

   From eq. (1) and (2),

   L.H.S. = R.H.S.

Chapter 9   SOME APPLICATIONS OF TRIGONOMETRY

1. In right triangle ACP,

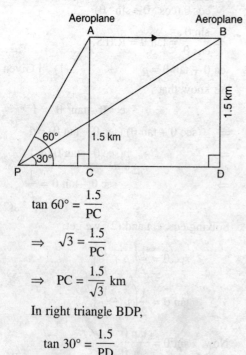

   $\tan 60° = \dfrac{1.5}{PC}$

   $\Rightarrow \sqrt{3} = \dfrac{1.5}{PC}$

   $\Rightarrow PC = \dfrac{1.5}{\sqrt{3}}$ km

   In right triangle BDP,

   $\tan 30° = \dfrac{1.5}{PD}$

   $\Rightarrow \dfrac{1}{\sqrt{3}} = \dfrac{1.5}{PC + CD}$

   $\Rightarrow \dfrac{1}{\sqrt{3}} = \dfrac{1.5}{\dfrac{1.5}{\sqrt{3}} + CD}$

   $\Rightarrow \dfrac{1}{\sqrt{3}} = \dfrac{1.5\sqrt{3}}{1.5 + \sqrt{3}\,CD}$

   $\Rightarrow 4.5 = 1.5 + \sqrt{3}\,CD$

   $\Rightarrow CD = \sqrt{3}$

   $\Rightarrow AB = \sqrt{3}$ km

   $\therefore$ Speed of the aeroplane

   $= \dfrac{\sqrt{3}}{\dfrac{15}{60 \times 60}}$

   $= \dfrac{1.732 \times 60 \times 60}{15}$ km/hr

   $= 415.68$ km/hr.

2. Let the height of the building be h m.

In right triangle ACD,

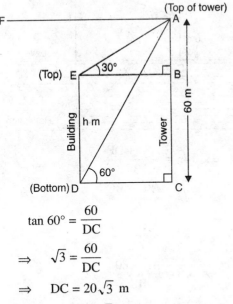

$$\tan 60° = \frac{60}{DC}$$

$$\Rightarrow \sqrt{3} = \frac{60}{DC}$$

$$\Rightarrow DC = 20\sqrt{3} \text{ m}$$

$$\Rightarrow EB = 20\sqrt{3} \text{ m}$$

In right triangle ABE,

$$\tan 30° = \frac{AB}{EB}$$

$$\Rightarrow \frac{1}{\sqrt{3}} = \frac{AB}{20\sqrt{3}}$$

$$\Rightarrow AB = 20 \text{ m}$$

Now, ED = BC
= AC − AB
= 60 − 20 = 40 m

Hence, the height of the building is **40 m**.

3. In right triangle ABC,

$$\tan 30° = \frac{AB}{BC}$$

$$\Rightarrow \frac{1}{\sqrt{3}} = \frac{AB}{BC} \quad\quad ...(1)$$

In right triangle ABD,

$$\tan 15° = \frac{AB}{BD}$$

$$\Rightarrow 0.27 = \frac{AB}{BC + DC}$$

$$\Rightarrow 0.27 = \frac{AB}{BC + 10}$$

$$\Rightarrow 0.27 = \frac{AB}{\sqrt{3}AB + 10} \quad | \text{From (1)}$$

$$\Rightarrow 2.7 + 0.27\sqrt{3}\,AB = AB$$

$$\Rightarrow AB(1 - 0.27\sqrt{3}) = 2.7$$

$$\Rightarrow AB = \frac{2.7}{1 - 0.27\sqrt{3}}$$

$$\Rightarrow AB = \frac{2.7}{1 - 0.27 \times 1.73}$$

$$= \frac{2.7}{1 - 0.46} = \frac{2.7}{0.54} = 5$$

Hence, the height of the mountain is **5 km**.

4. In right triangle QRY,

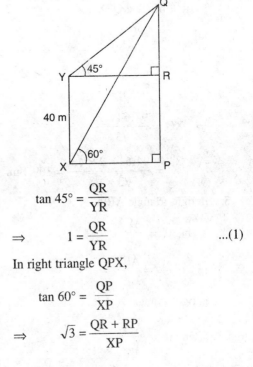

$$\tan 45° = \frac{QR}{YR}$$

$$\Rightarrow 1 = \frac{QR}{YR} \quad\quad ...(1)$$

In right triangle QPX,

$$\tan 60° = \frac{QP}{XP}$$

$$\Rightarrow \sqrt{3} = \frac{QR + RP}{XP}$$

$\Rightarrow \quad \sqrt{3} = \dfrac{QR + XY}{XP}$

$\Rightarrow \quad \sqrt{3} = \dfrac{QR + 40}{YR}$ ...(2)

Dividing (2) by (1), we get

$\dfrac{QR + 40}{QR} = \sqrt{3}$

$\Rightarrow \quad QR(\sqrt{3} - 1) = 40$

$\Rightarrow \quad QR = \dfrac{40}{\sqrt{3} - 1}$

$\qquad = \dfrac{40(\sqrt{3} + 1)}{(\sqrt{3} - 1)(\sqrt{3} + 1)}$

$\qquad = 20(\sqrt{3} + 1)$

$\qquad = 20(1.732 + 1)$

$\qquad = 20(2.732) = 54.64$

∴ Height of the tower PQ

$\qquad = PR + QR$

$\qquad = XY + QR$

$\qquad = 40 + 54.64$

$\qquad = 94.64$ m

In right triangle QPX,

$\sin 60° = \dfrac{PQ}{XQ}$

$\Rightarrow \quad \dfrac{\sqrt{3}}{2} = \dfrac{94.64}{XQ}$

$\Rightarrow \quad XQ = \dfrac{94.64 \times 2}{\sqrt{3}}$

$\qquad = \dfrac{94.64 \times 2 \times \sqrt{3}}{\sqrt{3} \times \sqrt{3}} = 109.3$ m.

5. In right triangle AEB,

$\tan 30° = \dfrac{AE}{BE}$

$\Rightarrow \quad \dfrac{1}{\sqrt{3}} = \dfrac{AE}{BE}$ ...(1)

In right triangle A′ EB,

$\tan 60° = \dfrac{A'E}{BE}$

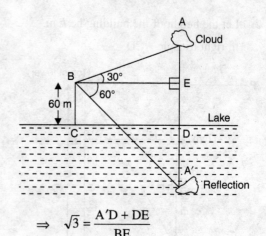

$\Rightarrow \quad \sqrt{3} = \dfrac{A'D + DE}{BE}$

$\Rightarrow \quad \sqrt{3} = \dfrac{A'D + BC}{BE}$

$\Rightarrow \quad \sqrt{3} = \dfrac{A'D + 60}{BE}$

$\Rightarrow \quad \sqrt{3} = \dfrac{AD + 60}{BE}$

$\Rightarrow \quad \sqrt{3} = \dfrac{AE + DE + 60}{BE}$

$\Rightarrow \quad \sqrt{3} = \dfrac{AE + 60 + 60}{BE}$

$\Rightarrow \quad \sqrt{3} = \dfrac{AE + 120}{BE}$ ...(2)

Dividing (2) by (1), we get

$\dfrac{AE + 120}{AE} = 3$

$\Rightarrow \quad AE = 60$ m

$\Rightarrow \quad AD = AE + ED = 60 + 60 = 120$ m

Hence, the height of the cloud is 120 m.

6. Let $\quad BC = x$

Then, $AB = 15 - x$

∴ $\quad BD = 15 - x$

In right triangle BCD,

$\sin 60° = \dfrac{BC}{BD}$

$\Rightarrow \quad \dfrac{\sqrt{3}}{2} = \dfrac{x}{15 - x}$

$\Rightarrow \quad \dfrac{1.732}{2} = \dfrac{x}{15 - x}$

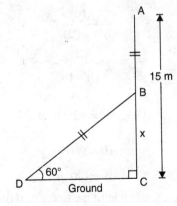

$\Rightarrow \quad 0.866 = \dfrac{x}{15-x}$

$\Rightarrow \quad 12.99 - 0.866\,x = x$

$\Rightarrow \quad 1.866x = 12.99$

$\Rightarrow \quad x = \dfrac{12.99}{1.866}$

$\Rightarrow \quad x = 6.96$

Hence, the tree broke at a height 6.96 m from the ground.

7. In right triangle BCP,

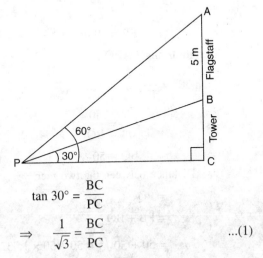

$\tan 30° = \dfrac{BC}{PC}$

$\Rightarrow \quad \dfrac{1}{\sqrt{3}} = \dfrac{BC}{PC}$ ...(1)

In right triangle ACP,

$\tan 60° = \dfrac{AC}{PC}$

$\Rightarrow \quad \sqrt{3} = \dfrac{AB + BC}{PC}$

$\Rightarrow \quad \sqrt{3} = \dfrac{5 + BC}{PC}$ ...(2)

Dividing eq. (2) by (1), we get

$\dfrac{5 + BC}{BC} = \dfrac{\sqrt{3}}{\left(\dfrac{1}{\sqrt{3}}\right)} = 3$

$\Rightarrow \quad BC = \dfrac{5}{2} = 2.5$ m

Hence, the height of the tower is 2.5 m.

8. In right triangle ABP,

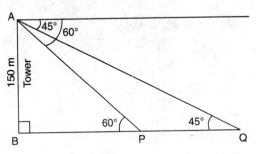

$\tan 60° = \dfrac{AB}{BP}$

$\Rightarrow \quad \sqrt{3} = \dfrac{150}{BP}$

$\Rightarrow \quad BP = 50\sqrt{3}$ m

In right triangle ABQ,

$\tan 45° = \dfrac{AB}{BQ}$

$\Rightarrow \quad 1 = \dfrac{AB}{BP + PQ}$

$\Rightarrow \quad 1 = \dfrac{150}{50\sqrt{3} + PQ}$

$\Rightarrow \quad 50\sqrt{3} + PQ = 150$

$\Rightarrow \quad 50 \times 1.732 + PQ = 150$

$\Rightarrow \quad 86.6 + PQ = 150$

$\Rightarrow \quad PQ = 150 - 86.6 = 63.4$ m

Hence, the distance between the objects is 63.4 m.

9. In right triangle ABC,

$\tan 45° = \dfrac{AB}{BC}$

$\Rightarrow \quad 1 = \dfrac{AB}{100} \quad \Rightarrow \quad AB = 100$ m

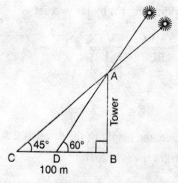

In right triangle ABD,

$\tan 60° = \dfrac{AB}{BD}$

$\Rightarrow \sqrt{3} = \dfrac{100}{BD}$

$\Rightarrow BD = \dfrac{100}{\sqrt{3}} = 57.7$

Hence, the length of the shadow is 57.7 m.

10. $\tan \theta = \dfrac{5}{12}$ ...(1)

$\tan \phi = \dfrac{3}{4}$ ...(2) Given

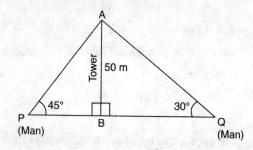

In right triangle ABD,

$\tan \phi = \dfrac{AB}{DB}$

$\Rightarrow \dfrac{3}{4} = \dfrac{AB}{DB}$

$\Rightarrow DB = \dfrac{4}{3} AB$ ...(3)

In right triangle ABC,

$\tan \theta = \dfrac{AB}{BC}$

$\Rightarrow \dfrac{5}{12} = \dfrac{AB}{BD + CD}$

$\Rightarrow \dfrac{5}{12} = \dfrac{AB}{\dfrac{4}{3} AB + 192}$

$\Rightarrow \dfrac{20}{3} AB + 960 = 12\, AB$

$\Rightarrow \dfrac{16}{3} AB = 960$

$\Rightarrow AB = 180$ m

Hence, the height of the tower is 180 m.

11. In right triangle ABP,

$\tan 45° = \dfrac{AB}{PB}$

$\Rightarrow 1 = \dfrac{50}{PB}$

$\Rightarrow PB = 50$ m ...(1)

In right triangle ABQ,

$\tan 30° = \dfrac{AB}{BQ}$

$\Rightarrow \dfrac{1}{\sqrt{3}} = \dfrac{50}{BQ}$

$\Rightarrow BQ = 50\sqrt{3}$ ...(2)

∴ Distance between the two men

= PQ

= PB + BQ

= $50 + 50\sqrt{3} = 50 + 50 \times 1.732$

= $50 + 86.6 = 136.6$ m.

12. In right triangle BAQ,

$\tan \beta = \dfrac{AB}{AQ}$

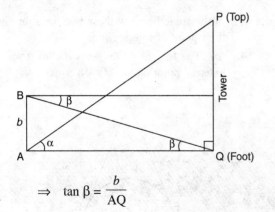

$\Rightarrow \tan \beta = \dfrac{b}{AQ}$

$\Rightarrow AQ = b \cot \beta$ ...(1)

In right triangle PQA,

$$\tan \alpha = \dfrac{PQ}{AQ}$$

$\Rightarrow \tan \alpha = \dfrac{PQ}{b \cot \beta}$ | From (1)

$\Rightarrow PQ = b \tan \alpha \cot \beta$

Hence, the height of the tower is $b \tan \alpha \cot \beta$.

Chapter 10 — CIRCLES

1. The angle between a tangent and radius through the point of contact is 90°.
2. A tangent to a circle intersects it in only one point.
3. A circle can have two parallel tangents at the most.
4. Two tangents can be drawn to a circle from a point lying outside the circle.
5. No ! we cannot draw a tangent to a circle from a point lying inside the circle.
6. Only one tangent can be drawn to a circle at a point lying on the circumference of the circle.
7. A circle can have an infinite number of tangents.
8. PB = PA = 5 cm.
9.

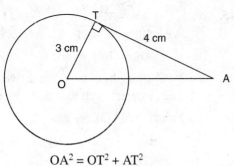

$$OA^2 = OT^2 + AT^2$$
$$= 3^2 + 4^2$$
$$= 9 + 16 = 25 = 5^2$$

$\Rightarrow OA = 5$ cm

Hence, the distance of the point A from the centre of the circle is 5 cm.

10. Join OT.

Let the radius of the circle be r cm.

Then, OT = r cm

OC = r cm

In right triangle OTP,

$$OP^2 = OT^2 + PT^2$$

$\Rightarrow (r + 3)^2 = r^2 + 6^2$

$\Rightarrow r^2 + 6r + 9 = r^2 + 36$

$\Rightarrow 6r = 27$

$\Rightarrow r = \dfrac{27}{6} = 4.5$

Hence, the radius of the circle is 4.5 cm.

11. Let AB be the given line and P be the given point on it. Let a circle with cantre O touch the line AB at P. Then,

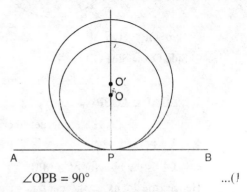

$\angle OPB = 90°$ ...(1)

Again, let a circle with centre O′ touch t line AB at P. Then,

∠O'PB = 90°   ...(2)

(1) and (2) are possible only and only when O and O' lie in the same line POO'.

Hence, the locus of the centres of circles which touch a given line AB at a given point P is a line through P perpendicular to AB.

12. Since the lengths of the tangents drawn from an external point to a circle are equal, therefore,

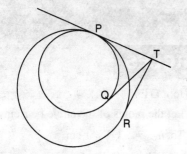

TP = TQ   ...(1)
TP = TR   ...(2)

(1) and (2) give

TQ = TR.

13. In right triangles OAP and OBP,

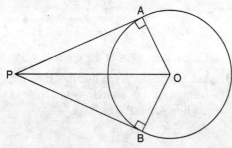

Hyp OP = Hyp OP | Common
Side OA = Side OB
          | **Radii of the same circle**
∴ ΔOAP ≅ ΔOBP
          | **R.H.S. congruence criterion**
∴ ∠APO = ∠BPO
OP is the bisector of ∠APB

ce, the locus of the centres of circles ing two intersecting lines is the bisector of the angle between these two lines through their point of intersection.

14. ∵ The lengths of tangents darwn from an external point to a circle are equal

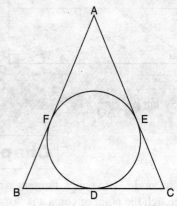

∴   AF = AE   ...(1)
      BD = BF   ...(2)
      CE = CD   ...(3)

Adding eqs. (1), (2) and (3), we get

AF + BD + CE = AE + BF + CD

$= \frac{1}{2}$(AF + BD + CE + AE + BF + CD)

$= \frac{1}{2}$(AF + BF + BD + CD + AE + CE)

$= \frac{1}{2}$(AB + BC + AC)

$= \frac{1}{2}$ (Perimeter of Δ ABC).

15.   AB = AC | Given
⇒   AE + BE = AF + CF   ...(1)
But   AE = AF   ...(2)
          | ∵ The lengths of tangents drawn from an external point to a circle are equal
∴ From (1) and (2), we get
          BE = CF
⇒   BD = CD
          | ∵ BE = BD and CF = CD

Chapter 11 CONSTRUCTIONS

1. **Given:** A line segment AB of length 4.9 cm.
   Required: To divide it in the ratio 3 : 4. Also, to measure the two parts.

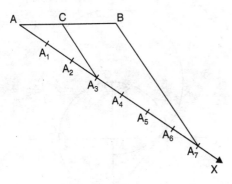

   Steps of Construction :
   1. Draw a line segment AB of length 4.9 cm.
   2. Draw any ray AX, making an acute angle with AB.
   3. Locate 7(= 3 + 4) points $A_1, A_2, A_3, A_4, A_5, A_6$ and A_7 on AX such that
      $AA_1 = A_1A_2 = A_2A_3 = A_3A_4$
      $= A_4A_5 = A_5A_6 = A_6A_7$.
   4. Join BA_7.
   5. Through the point A_3, draw a line parallel to A_7B intersecting AB at the point C.

   Then, AC : CB = 3 : 4
   On measurement,
   $\quad\quad$ AC = 2.1 cm
   and $\quad$ CB = 2.8 cm.

2. **Given:** A line segment AB of length 7 cm.
   Required: To divide it in the ratio 5 : 2. Also, to measure the two parts.
   Steps of Construction:
   1. Draw a line segment AB of length 7 cm.
   2. Draw any ray AX, making an acute angle with AB.
   3. Locate 7(= 5 + 2) points $A_1, A_2, A_3, A_4, A_5, A_6$ and A_7 on AX such that
      $AA_1 = A_1A_2 = A_2A_3 = A_3A_4 = A_4A_5$
      $= A_5A_6 = A_6A_7$.
   4. Join BA_7.
   5. Through the point A_4, draw a line parallel to A_7B intersecting AB in C.

   Then AC : CB = 5 : 2.
   On measurement,
   $\quad\quad$ AC = 5 cm, CB = 2 cm.

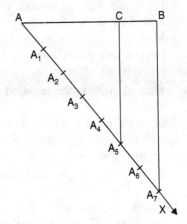

3. **Steps of Construction:**
   1. Draw a triangle ABC of sides 4.5 cm, 5 cm and 6 cm.
   2. Draw any ray BX making an acute angle with BC on the side opposite to the vertex A.
   3. Locate 5 points B_1, B_2, B_3, B_4 and B_5 on BX such that $BB_1 = B_1B_2 = B_2B_3 = B_3B_4 = B_4B_5$.

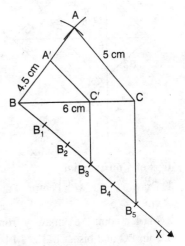

   4. Join B_5C and draw a line through B_3 parallel to B_5C intersecting BC at C'.

5. Draw a line through C' parallel to the line CA to intersect BA at A'.

Then, △ A' BC' is the required triangle.

4. **Steps of Construction:**
   1. Draw a line segment BC of length 6 cm.
   2. Draw perpendicular bisector of BC. Let it meet BC at D.
   3. Mark a point A on the perpendicular bisector such that AD = 3 cm.
   4. Join AB and AC.
      Then, △ ABC is the required isosceles triangle.
   5. Draw any ray BX making an acute angle with BC on the side opposite to the vertex A.
   6. Locate 3 points B_1, B_2 and B_3 on BX such that $BB_1 = B_1B_2 = B_2B_3$.
   7. Join B_2 to C and draw a line through B_3 parallel to B_2C, intersecting the extended line segment BC at C'.
   8. Draw a line through C' parallel to CA intersecting the extended line segment BA at A'.

   Then, A' BC' is the required triangle.

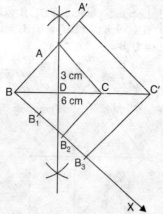

5. **Steps of Construction:**
   1. Draw a circle with centre O and radius 5 cm.
   2. Mark a point P 8 cm away from O.
   3. Join PO and bisect it. Let M be the midpoint of PO.
   4. Taking M as centre and MO as radius, draw a circle. Let it intersect the given circle at the points Q and R.
   5. Join PQ and PR.

   Then, PQ and PR are the required two tangents on measurement,

   PQ = PR = 6.9 cm.

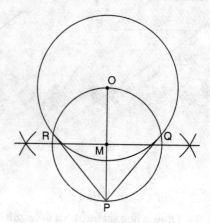

6. **Steps of Construction:**
   1. Draw a circle with centre O and radius 4 cm.
   2. Draw any diameter AOB of this circle.
   3. Construct ∠BOC = 60° such that the radius OC meets the circle at C.
   4. Draw AM ⊥ AB and CN ⊥ OC.

   Let AM and CN intersect each other at P.

   Then, PA and PC are the required tangents to the given circle inclined at an angle of 60°.

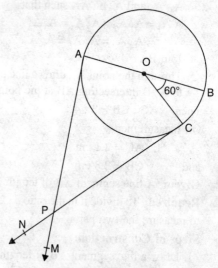

Chapter 12 AREAS RELATED TO CIRCLES

1. Length of the arc $= 2\pi r \dfrac{\theta}{360°}$

   $= 2 \times \dfrac{22}{7} \times 28 \times \dfrac{36°}{360°}$

   $= 17.6$ cm.

2. The measure of the central angle of any circle is 360°.

3. Area of the shaded region $= \pi r^2 \dfrac{\theta}{360°}$

   $= \dfrac{22}{7} \times 21 \times 21 \times \dfrac{120°}{360°}$

   $= 462$ cm^2

4. Let the radius of the circle be r cm.
   Then, $2\pi r = 44$
   $\Rightarrow 2 \times \dfrac{22}{7} \times r = 44$
   $\Rightarrow r = 7$ cm
   $\therefore$ Area of a quadrant
   $= \dfrac{1}{4} \pi r^2$
   $= \dfrac{1}{4} \times \dfrac{22}{7} \times (7 \times 7) = 38.5$ cm^2

5. $2r = 14$
   $\Rightarrow r = 7$ cm
   $\therefore$ Perimeter
   $= \pi r + 2r$
   $= \dfrac{22}{7} \times 7 + 2 \times 7$
   $= 22 + 14 = 36$ cm.

6. Let the side of the square be a cm. Then,
   Perimeter = Area
   $\Rightarrow 4a = a^2$
   $\Rightarrow a = 4$
   $\therefore$ Area of the square
   $= a^2 = 4^2 = 16$ unit^2.

7. Let the radius be r cm. Then,
   $\pi r^2 = \pi(6)^2 + \pi(8)^2$
   $\Rightarrow r = 10$ cm
   $\Rightarrow 2r = 20$ cm

   Hence, the diameter of the circle is 20 cm.

8. $\because$ Angle described by the minute hand in 60 minutes = 360°
   $\therefore$ Angle described by the minute hand in 35 minutes $= \dfrac{360°}{60°} \times 35 = 210°$

   $\therefore$ Required area $= \pi r^2 \dfrac{\theta}{360°}$
   $= \dfrac{22}{7} \times 6 \times 6 \times \dfrac{210°}{360°}$
   $= 66$ cm^2

9. Let the radius of the circle be r cm. Then,
   $2\pi r = 2\pi \left(\dfrac{20}{2}\right) + 2\pi \left(\dfrac{10}{2}\right)$
   $\Rightarrow r = 15$ cm
   $\Rightarrow 2r = 30$ cm

   Hence, the diameter of the circle is 30 cm.

10. Let the side of the square be a cm.
    Then,
    $4a = 2 \times \dfrac{22}{7} \times 28$
    $\Rightarrow a = 44$

    Hence, the length of the side of the square is 44 cm.

11. Let the radii of the two circles be R cm and r cm (R > r) respectively. Then,

    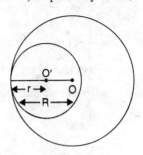

    $\pi R^2 + \pi r^2 = 116\pi$
    $\Rightarrow R^2 + r^2 = 116$
    $R - r = 6$

$(R + r)^2 + (R - r)^2 = 2(R^2 + r^2)$
$\Rightarrow (R + r)^2 + (6)^2 = 2(116)$
$\Rightarrow (R + r)^2 = 196$
$\Rightarrow R + r = 14$ ...(3)

Solving (2) and (3), we get
$R = 10$ cm
$r = 4$ cm

Hence, the radii of the two circles are 10 cm and 4 cm respectively.

12. Let the radius of the wheel be r m. Then, distance covered by the wheel in 1 revolution.

$= \dfrac{\text{Distance covered}}{\text{Revolutions made}}$

$= \dfrac{11 \times 1000}{5000} = \dfrac{11}{5}$

$\Rightarrow 2\pi r = \dfrac{11}{5}$

$\Rightarrow 2 \times \dfrac{22}{7} \times r = \dfrac{11}{5}$

$\Rightarrow r = \dfrac{7}{20}$ m

$\Rightarrow r = \dfrac{7 \times 100}{20}$ cm

$\Rightarrow r = 35$ cm

$\Rightarrow 2r = 70$ cm

Hence, the diameter of the wheel is 70 cm.

13. Area of the bigger semicircle

$= \dfrac{1}{2}\pi r^2$

$= \dfrac{1}{2}\pi (14)^2$

$= 98\pi$ cm^2

Area of one smaller semicircle

$= \dfrac{1}{2}\pi R^2$

$= \dfrac{1}{2}\pi (7)^2$

$= \dfrac{49}{2}\pi$ cm^2

∴ Area of the two smaller semicircles

$= 2 \times \dfrac{49}{2}\pi$

$= 49\pi$ cm^2

∴ Total shaded Area

$= 98\pi + 49\pi$

$= 147\pi$ cm^2

$= 147 \times \dfrac{22}{7}$ cm^2

$= 462$ cm^2.

14. Area of the bed $= \dfrac{1}{4}\pi(21)^2 - \dfrac{1}{4}\pi(14)^2$

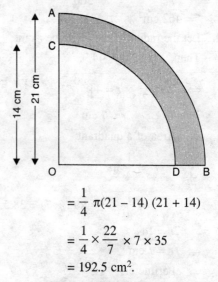

$= \dfrac{1}{4}\pi(21 - 14)(21 + 14)$

$= \dfrac{1}{4} \times \dfrac{22}{7} \times 7 \times 35$

$= 192.5$ cm^2.

15. Area of the square ABCD
$= 4 \times 4 = 16$ cm^2

Area of circle of diameter 2 cm

$= \pi \left(\dfrac{2}{2}\right)^2 = \pi$ cm^2

Area of a circle quadrant at a corner

$= \dfrac{1}{4}\pi(1)^2 = \dfrac{\pi}{4}$ cm^2

$\Rightarrow$ Area of 4 circle quadrants

$= 4\left(\dfrac{\pi}{4}\right) = \pi$ cm^2

∴ Area of the shaded region

$= 16 - (\pi + \pi)$

$= 16 - 2\pi$
$= 16 - 2 \times 3.14$
$= 16 - 6.28$
$= 9.72$ cm².

16. Area of the square $= (14)^2 = 196$ cm²

Area of one quadrant $= \dfrac{1}{4} \pi (7)^2$

∴ Area of four quadrants

$= 4 \times \dfrac{1}{4} \pi (7)^2$

$= \pi(7)^2 = \dfrac{22}{7}(7)^2 = 154$ cm²

∴ Area of the shaded region
$= 196$ cm² $- 154$ cm² $= 42$ cm².

17. Area of sector OACBO $= \pi r^2 \dfrac{\theta}{360°}$

$= \dfrac{22}{7} \times 14 \times 14 \times \dfrac{90°}{360°} = 154$ cm²

Area of Δ AOB $= \dfrac{14 \times 14}{2}$ cm² $= 98$ cm²

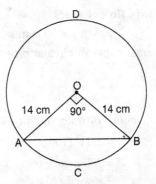

∴ Area of the minor segment ACBA
= Area of sector OACB − Area of Δ AOB
$= 154$ cm² $- 98$ cm² $= 56$ cm²

∴ Area of the major segment = BDAB
= Area of the circle
  − Area of the minor segment
$= \dfrac{22}{7} \times 14 \times 14$ cm² $- 56$ cm²
$= 616$ cm² $- 56$ cm² $= 560$ cm².

Chapter 13 SURFACE AREAS AND VOLUMES

1. For the resulting cuboid,
Length $= 5 + 5 + 5 = 15$ cm
breadth $= 5$ cm
height $= 5$ cm
∴ Surface area
$= 2(15 \times 5 + 5 \times 5 + 5 \times 15)$
$= 350$ cm².

2. Volume of sphere $= \dfrac{4}{3} \pi (3)^3$ cm³

Volume of a spherical ball

$= \dfrac{4}{3} \pi \left(\dfrac{0.6}{2}\right)^3$ cm³

$= \dfrac{4}{3} \pi (0.3)^3$ cm³

∴ Number of balls obtained

$= \dfrac{\dfrac{4}{3}\pi(3)^3}{\dfrac{4}{3}\pi(0.3)^3} = 1000$.

3. For conical vessel,
$r = 5$ cm
$h = 24$ cm

∴ Volume of conical vessel $= \dfrac{1}{3}\pi r^2 h$

$= \dfrac{1}{3}\pi(5)^2(24)$
$= 200\pi$ cm³

For cylindrical vessel
$R = 10$ cm
Let the height to which the water rises be H cm.
Then, volume of cylindrical vessel
$= \pi R^2 H$
$= \pi(10)^2 H$
$= 100\pi H$ cm³
According to the question,
$100\pi H = 200\pi$
$\Rightarrow$ $H = 2$ cm
Hence, the required height is 2 cm.

4. Rate of flow of water
   = 7 m/s = 700 cm/s
   Internal radius of circular pipe (r)
   $= \dfrac{2}{2} = 1$ cm

   ∴ Volume of water that flows in 1 second
   $= \pi(1)^2 (700)$
   $= 700 \pi$ cm³

   ∴ Volume of water that flows in $\dfrac{1}{2}$ hr.
   $= 700 \pi \times 30 \times 60$
   $= 1260000 \pi$ cm³

   Let the increase in water level be h cm.
   Then,
   $\pi(40)^2 h = 1260000 \pi$
   $\Rightarrow h = \dfrac{1260000 \pi}{\pi (40)^2}$
   $\Rightarrow h = 787.5$ cm

   Hence, the required increase in the water level is 787.5 cm.

5. Let the common height of the two cones be h. Let the radius of the base of the cylinder be R. Then,
   $\dfrac{1}{3}\pi r_1^2 h + \dfrac{1}{3}\pi r_2^2 h = \pi R^2 h$
   $\Rightarrow R = \sqrt{\dfrac{r_1^2 + r_2^2}{3}}$.

6. Let the height of each cone be h.
   Then, $\dfrac{1}{3}\pi r_1^3 h + \dfrac{1}{3}\pi r_2^3 h = \dfrac{4}{3}\pi R^3$
   $\Rightarrow h = \dfrac{4R^3}{r_1^3 + r_2^3}$.

7. **For Cylinder**
   $r = \dfrac{4.2}{2} = 2.1$ m
   $h = 4$ m
   ∴ Curved surface area
   $= 2\pi r h$

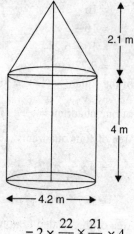

   $= 2 \times \dfrac{22}{7} \times \dfrac{21}{10} \times 4$
   $= 52.8$ m²

   Volume $= \pi r^2 h$
   $= \dfrac{22}{7} \times \dfrac{21}{10} \times \dfrac{21}{10} \times 4 = 55.44$ m³

   For cone
   R = 2.1 m
   H = 2.1 m
   $L = \sqrt{R^2 + H^2} = \sqrt{(2.1)^2 + (2.1)^2}$
   $= 2.1\sqrt{2}$

   ∴ Curved surface area
   $= \pi R L$
   $= \dfrac{22}{7} \times \dfrac{21}{10} \times 2.1 \times 1.414$
   $= 19.598$ m²

   Volume $= \dfrac{1}{3}\pi R^2 H$
   $= \dfrac{1}{3} \times \dfrac{22}{7} \times \dfrac{21}{10} \times \dfrac{21}{10} \times \dfrac{21}{10}$
   $= 9.702$ m³

   ∴ Outer surface area of the building
   $= 52.8 + 19.598$
   $= 72.398$ m²

   Volume of the building
   $= 55.44 + 9.702$
   $= 65.142$ m³.

8. For Cone

$r = 4.2$ cm

$h = 10.2 - 4.2 = 6$ cm

∴ Volume $= \dfrac{1}{3} \pi r^2 h$

$= \dfrac{1}{3} \times \dfrac{22}{7} \times \dfrac{42}{10} \times \dfrac{42}{10} \times 6$

$= 110.88$ cm²

For hemisphere

$R = 4.2$ cm

∴ Volume $= \dfrac{2}{3} \pi R^3$

$= \dfrac{2}{3} \times \dfrac{22}{7} \times \dfrac{42}{10} \times \dfrac{42}{10} \times \dfrac{42}{10}$

$= 155.23$ cm³

∴ Volume of the wooden toy

$= 110.88$ cm³ $+ 155.23$ cm³

$= 266.11$ cm³.

9. For Cone: $r = 5$ cm

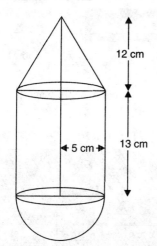

$h = 12$ cm

∴ $l = \sqrt{r^2 + h^2}$

$= \sqrt{5^2 + 12^2} = 13$ cm

∴ Curved surface area

$= \pi r l$

$= \dfrac{22}{7} \times 5 \times 13 = \dfrac{1430}{7}$ cm²

For cylinder

$r = 5$ cm

$h = 13$ cm

∴ Curved surface area

$= 2\pi r h$

$= 2 \times \dfrac{22}{7} \times 5 \times 13$

$= \dfrac{2860}{7}$ cm²

For hemisphere

$r = 5$ cm

∴ Curved surface area

$= 2\pi r^2$

$= 2 \times \dfrac{22}{7} \times (5)^2$

$= \dfrac{1100}{7}$ cm²

∴ Curved surface area of the toy

$= \dfrac{1430}{7} + \dfrac{2860}{7} + \dfrac{1100}{7}$

$= 770$ cm².

10. Let the radii of ends be r_1 cm and r_2 cm respectively. Then,

$2\pi r_1 = 44$

$\Rightarrow 2 \times \dfrac{22}{7} \times r_1 = 44$

$\Rightarrow r_1 = 7$ cm

and, $2\pi r_2 = 33$

$\Rightarrow 2 \times \dfrac{22}{7} \times r_2 = 33$

$\Rightarrow r_2 = \dfrac{21}{4}$ cm

$h = 16$ cm

∴ $l = \sqrt{h^2 + (r_1 - r_2)^2} = \sqrt{256 + \left(7 - \dfrac{21}{4}\right)^2}$

$= \sqrt{256 + \dfrac{49}{16}} = 16.09$ cm

∴ Volume $= \dfrac{1}{3} \pi h (r_1^2 + r_2^2 + r_1 r_2)$

$$= \frac{1}{3} \times \frac{22}{7} \times 16 \times \left\{ 7^2 + \left(\frac{21}{4}\right)^2 + \frac{7 \times 21}{4} \right\}$$

$$= \frac{352}{21} \left(49 + \frac{441}{16} + \frac{147}{4} \right)$$

$$= \frac{352}{21} \left(49 + \frac{1029}{16} \right)$$

$$= \frac{352}{21} \left(\frac{1813}{16} \right)$$

$$= 1899.33 \text{ cm}^3$$

Total surface $= \pi r_1^2 + \pi r_2^2 + \pi(r_1 + r_2)l$

$$= \frac{22}{7}(7)^2 + \frac{22}{7} \times \frac{21}{4} \times \frac{21}{4} + \frac{22}{7}\left(7 + \frac{21}{4}\right)(16.09)$$

$$= 154 + 86.625 + 619.465$$

$$= 860.09 \text{ cm}^2.$$

11. $r = 28.490$ L
$= 28490$ cm^3
$r_1 = 28$ cm
$r_2 = 21$ cm

Let the height of the bucket be h cm.
Then,

$$V = \frac{1}{3} \pi h (r_1^2 + r_2^2 + r_1 r_2)$$

$$\Rightarrow 28490 = \frac{1}{3} \times \frac{22}{7} \times h \times [(28)^2 + (21)^2 + (28)(21)]$$

$$\Rightarrow 28490 = \frac{22}{21} \times h \times (784 + 441 + 588)$$

$$\Rightarrow 28490 = \frac{22}{21} \times h \times 1813$$

$$\Rightarrow h = \frac{28490 \times 21}{22 \times 1813} = 15 \text{ cm}$$

Hence, the height the bucket is 15 cm.

Chapter 14 STATISTICS

1. The three measures of central tendency are mean, median and mode.
2. 3 Median = Mode + 2 Mean.
3. Mean is the most frequently used measure of central tendency.
4. Mode is that value among the observations which occurs most often.
5. The x-coordinate of the point of intersection of two ogives represents median.
6.

Weakly wages (in rupees) x_i	No. of workers f_i	$f_i x_i$
150	20	3000
170	30	5100
190	40	7600
210	50	10500
230	60	13800
Total	200	40,000

$$\therefore \text{Mean}(\bar{x}) = \frac{\Sigma f_i x_i}{\Sigma f_i} = \frac{40000}{200} = 200$$

Hence, the mean weakly wage is Rs. 200.

7. Take $a = 175$, $h = 50$

Class Interval	Frequency f_i	Class Mark x_i	$u_i = \dfrac{x_i - 175}{50}$	$f_i u_i$
0–50	8	25	–3	–24
50–100	15	75	–2	–30
100–150	32	125	–1	–32
150–200	26	175	0	0
200–250	12	225	1	12
250–300	7	275	2	14
Total	100			–60

$$\therefore \bar{x} = a + h\bar{u}$$

$$= a + h \frac{\Sigma f_i x_i}{\Sigma f_i}$$

$$= 175 + 50 \left(\frac{-60}{100} \right)$$

$$= 175 - 30$$

$$= 145.$$

8. Take $a = 175$, $h = 10$

Classes	Class Mark x_i	Frequency f_i	$u_i = \frac{x_i - 175}{10}$	$f_i u_i$
140–150	145	5	–3	–15
150–160	155	f_1	–2	$-2f_1$
160–170	165	20	–1	–20
170–180	175	f_2	0	0
180–190	185	6	1	6
190–200	195	2	2	4
Total		$f_1 + f_2 + 33$		$-2f_1 - 25$

Sum of observations = 52

$\Rightarrow f_1 + f_2 + 33 = 52 \Rightarrow f_1 + f_2 = 19$ ...(1)

$$\bar{x} = a + h \frac{\Sigma f_i u_i}{\Sigma f_i}$$

$\Rightarrow \quad 166 \frac{9}{26} = 175 + 10 \left(\frac{-2f_1 - 25}{52} \right)$

$\Rightarrow \quad \frac{4325}{26} - 175 = -10 \left(\frac{2f_1 + 25}{52} \right)$

$\Rightarrow \quad \frac{4325 - 4550}{26} = -10 \left(\frac{2f_1 + 25}{52} \right)$

$\Rightarrow \quad \frac{-225}{26} = -10 \left(\frac{2f_1 + 25}{52} \right)$

$\Rightarrow \quad 45 = 2f_1 + 25$

$\Rightarrow \quad f_1 = 10$

Put $f_1 = 10$ in (1), we get

$10 + f_2 = 19$

$\Rightarrow \quad f_2 = 9$

Hence, the missing frquency 'f_1' and 'f_2' are 10 and 9 respectively.

9.

Class Interval	Frequency
25–30	25
30–35	34
35–40	50
40–45	42
45–50	38
50–55	14

Here, the maximum class frequency is 50, and the class corresponding to this frequency is 35–40. So, the modal class is 35–40.

So,

$l = 35$

$h = 5$

$f_1 = 50$

$f_0 = 34$

$f_2 = 42$

$\therefore$ Mode $= l + \left(\frac{f_1 - f_0}{2f_1 - f_0 - f_2} \right) \times h$

$= 35 + \left(\frac{50 - 34}{2 \times 50 - 34 - 42} \right) \times 5$

$= 35 + \frac{80}{24} = 35 + 3.33 = 38.33.$

10. To calculate the median, we need to find the class intervals and their corresponding frequencies.

Class Intervals	Frequency	Cumulative Frequency
Below 100	16	16
100–200	23	39
200–300	28	67
300–400	40	107
400–500	20	127
500–600	12	139
600–700	8	147
700–800	3	150

Now, $n = 150$

$\therefore \quad \frac{n}{2} = 75$

Then observation lies in the class 300–400. So, 300–400 is the median class.

$\therefore \quad l = 300$
$f = 40$
$cf = 67$
$h = 100$

Using the formula,

$$\text{Median} = l + \left(\dfrac{\dfrac{n}{2} - cf}{f}\right) \times h$$

$\text{Median} = 300 + \left(\dfrac{75 - 67}{40}\right) \times 100$

$= 300 + 20 = 320$

Hence, the median income is Rs. 320.

11. **Less than type Cumulative Frequency Distribution**

Marks	Numbers of students
Less than 10	3
Less than 20	11
Less than 30	28
Less than 40	48
Less than 50	70

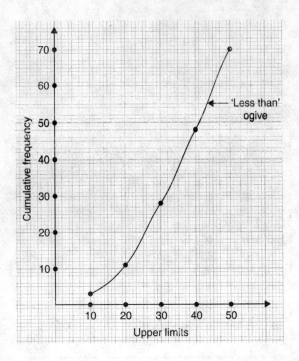

Chapter 15 PROBABILITY

1. P(not E) = 1 − P(E)
   = 1 − 0.45
   = 0.55.

2. P(loosing the game)
   = 1 − P(winning the game)
   = 1 − 0.9 = 0.1.

3. P(not happening the event)
   = 1 − P(happening the event)
   = $1 - \dfrac{5}{11} = \dfrac{6}{11}$.

4. Probability of a sure event = 1

5. Probability of an impossible event is 0.

6. Total number of letters in English alphabet = 26
   $\therefore$ Number of all possible outcomes = 26

Let E be the event that the letter so chosen is a vowel.

$\therefore$ Total number of vowels in English alphabet = 5

$\therefore$ Number of outcomes favourable to E = 5

Thus,

$P(E) = \dfrac{\text{Number of outcomes favourable to E}}{\text{Number of all possible outcomes}}$

$= \dfrac{5}{26}$.

7. A non-leap year has 365 days, *i.e.*, 52 weeks and 1 day.

This day can be any one out of the 7 days of the week.

∴ Required probability = $\frac{1}{7}$.

8. A leap year has 366 days, *i.e.*, 52 weaks and two days.

   Clearly, a leap year will have 52 Fridays essentially.

   Now, the remaining two days can be :

   (*i*) Sunday and Monday
   (*ii*) Monday and Tuesday
   (*iii*) Tuesday and Wednesday
   (*iv*) Wednesday and Thursday
   (*v*) Thursday and Friday
   (*vi*) Friday and Saturday
   (*vii*) Saturday and Sunday

   For 53 Friday, the outcomes (*v*) and (*vi*) are the only favourable outcomes out of the seven outcomes.

   ∴ Required probability = $\frac{2}{7}$.

9. Let the number of green marbles in the jar be x. Then, the number of red marbles in the jar $= 32 - x$.

   Probability of drawing a red marble

   $= \dfrac{\text{Number of favourable outcomes}}{\text{Numbers of all possible outcomes}}$

   $\Rightarrow \dfrac{3}{4} = \dfrac{32-x}{32}$

   $\Rightarrow x = 8$

   Hence, their are 8 green marbles in the jar.

10. Total number of bulbs = 200

    Number of defective bulbs = 20

    ∴ Number of non-defective bulbs
    $= 200 - 20$
    $= 180$

    ∴ Probability of getting a non-defective bulb

    $= \dfrac{\text{Number of favourable outcomes}}{\text{Number of all possible outcomes}}$

    $= \dfrac{180}{200} = \dfrac{9}{10}$.

11. Total number of days in an ordinary year = 365

    Meenu may have any day as her birthday. Also, Apala may have any day as her birthday.

    ∴ Total number of ways in which Meenu and Apala may have their birthday
    $= 365 \times 365$

    Now, Meenu and Apala may have same birthday on any one day of 365 days of the year.

    ∴ Number of ways in which Meenu and Apala have the same birthday = 365.

    ∴ P(Meenu and Apala have the same birthday)

    $= \dfrac{365}{365 \times 365} = \dfrac{1}{365}$

    ∴ P(Meenu and Apala have different birthdays)

    $= 1 - $ P

    (Meenu and Apala have the same birthday)

    $= 1 - \dfrac{1}{365} = \dfrac{364}{365}$.

12. Total number of cadets
    $= 20 + 15 = 35$

    ∴ Number of all possible outcomes
    $= 35$

    Number of girl candidates = 15

    ∴ Number of the favourable outcomes
    $= 15$

    ∴ Required probability $= \dfrac{15}{35} = \dfrac{3}{7}$.

13. Total number of letters in the word 'MATHEMATICS' = 11

    ∴ Number of all possible outcomes
    $= 11$

    Number of M,S = 2

    ∴ Number of favourable outcomes
    $= 2$

∴ Required probability = $\frac{2}{11}$.

14. Total number of cards = 15
 ∴ Number of all possible outcomes
 = 15
 Number of favourable outcomes (2, 3, 5, 7, 11, 13) = 6
 ∴ Required probability = $\frac{6}{15} = \frac{2}{5}$.

15. Total number of balls in the bag
 = 5 + 7 + 4 + 2 = 18
 ∴ Number of all possible outcomes
 = 18
 Number of favourable outcomes
 = Number of red balls
                 + Number of blue balls
 = 7 + 2 = 9

∴ Required probability = $\frac{9}{18} = \frac{1}{2}$.

16. Number of all possible outcomes = 52
 Number of favourable outcomes
 = 52 − (4 + 4) = 44
 ∴ Required probabily = $\frac{44}{52} = \frac{11}{13}$.

17. Number of all possible outcomes = 6
 Number of favourable outcomes (4, 5)
 = 2
 ∴ Required probability = $\frac{2}{6} = \frac{1}{3}$.

18. Number of all possible outcomes
 = 2 × 2 × 2 = 8
 Number of favourable outcomes (HHT, HTH, THH, HHH) = 4
 ∴ Required probability = $\frac{4}{8} = \frac{1}{2}$.

New Age Revision Test Paper-2011

MATHEMATICS
CLASS X

Time : 3 Hours   Maximum Marks : 80

General Instructions:
1. All questions are compulsory.
2. The question paper consists of thirty questions divided into 4 Section A, B, C and D. Section A comprises of ten questions of 01 mark each, Section B comprises of five questions of 02 marks each, Section C comprises of ten questions of 03 marks each and Section D comprises of five questions of 06 marks each.
3. All questions in Section A are to be answered in one word, one sentence or as per the exact requirement of the question.
4. There is no overall choice. However, internal choice has been provided in one question of 02 marks each, three questions of 03 marks each and two questions of 06 marks each. You have to attempt only one of the alternatives in all such questions.
5. In question on construction, drawings should be neat and exactly as per the given measurements.
6. Use of calculator is not permitted. However, you may ask for mathematical tables.

SECTION A

1. Given that HCF (306, 657) = 9, find LCM (306, 657).
2. What is a polynomial of degree 1 called?
3. Find a quadratic polynomial with the numbers 0, $\sqrt{5}$ as the sum and product of its zeros respectively.
4. Find the discriminant of the equation $3x^2 - 2x + \dfrac{1}{3} = 0$.
5. Find the area of a quadrant of a circle where circumference is 22 cm.
6. From a point Q, the length of the tangent to a circle is 24 cm and the distance of Q from the centre is 25 cm. Find the radius of the circle.
7. What is the empirical relationship between the three measures: 'Mean, Median and Mode' of central tendency?
8. If P(E) = 0.05, what is the probability of 'not E'?
9. Prove that the tangents drawn at the ends of a diameter of a circle are parallel.
10. The height of a tower is 20 m. The height of its shadow is 20 m. Calculate the altitude of the sun.

SECTION B

11. Find the sum of first 24 terms of the list of numbers where with term is given by
$$a_n = 3 + 2n.$$

12. If A, B and C are interior angles of a triangle ABC, then show that
$$\sin\left(\frac{B+C}{2}\right) = \cos\frac{A}{2}.$$

Or

Express $\sin 75° + \cos 67°$ in terms of trigonometric ratios of angles between 0° and 45°.

13. The diagonals of a quadrilateral ABCD intersect each other at the point O such that $\frac{AO}{BO} = \frac{CO}{DO}$. Show that ABCD is a trapezium.

14. Find the value of k if the points A(2, 3), B(4, k) and C(6, –3) one collinear.

15. A bag contains 3 red balls and 5 black balls. A ball is drawn at random from the bag. What is the probability that the ball drawn is (i) red? (ii) not red?

SECTION C

16. Find the zeroes of the quadratic polynomial $3x^2 - x - 4$ and verify the relationship between the zeroes and the coefficients.

17. Prove that $3 + 2\sqrt{5}$ is irrational.

18. For which value of k will the following pair of linear equations have no solution?
$$3x + y = 1$$
$$(2k - 1)x + (k - 1)y = 2k + 1$$

Or

Solve the following pair of equations by reducing them to a pair of linear equations:
$$\frac{5}{x-1} + \frac{1}{y-2} = 2$$
$$\frac{6}{x-1} - \frac{3}{y-2} = 1; \quad x \ne 1, \ y \ne 2.$$

19. Find the sum of first 51 terms of an AP whose second and third terms are 14 and 18 respectively.

Or

Two APs have the same common difference. The difference between their 100th terms is 100, what is the difference between their 1000th terms?

20. Prove by using the identity $\csc^2 A = 1 + \cot^2 A$ that
$$\frac{\cos A - \sin A + 1}{\cos A + \sin A - 1} = \csc A + \cot A.$$

21. Find the coordinates of the points of trisection (i.e., points dividing in three equal parts) of the line segment giving the points A(2, –2) and B(–7, 4).

Or

If Q(0, 1) is equidistant from P(5, –3) and R(x, 6), find the value of x. Also, find the distances QR and PR.

22. Find the ratio in which the y-axis divides the line segment giving the points (5, – 6) and (–1, – 4). Also find the points of intersection.

23. Draw a triangle ABC with side BC = 7 cm, $\angle$ B = 45°, $\angle$ A = 105°. Then, construct a triangle whose sides are $\frac{4}{3}$ times the corresponding sides of $\triangle$ ABC.

24. Two tangents TP and TQ are drawn to a circle with center O from an external point T. Prove that $\angle$ PTQ = 2 $\angle$ OPQ.

25. A car has two wipers which do not overlap. Each wiper has a blade of length 25 cm sweeping through an angle of 115°. Find the total area cleaned at each sweep of the blades.

SECTION D

26. Form the pair of linear equations in the following problem and find the solution graphically:

    5 pencils and 7 pens together cost Rs. 50, whereas 7 pencils and 5 pens together cost Rs. 46. Find the cost of one pencil and that of one pen.

27. Form the top of a 7 m high building, the angle of elevation of the top of a cable tower is 60° and the angle of depression of its foot is 45°. Determine the height of the tower.

28. Prove that the ratio of the areas of two similar triangles is equal to the square of the ratio of their corresponding sides.

    Use the above theorem, in the following:

    Let $\triangle$ ABC ~ $\triangle$ DEF and their areas be, respectively, 64 cm² and 121 cm². If EF = 15.4 cm, find BC.

    Or

    State and prove Pythagoras theorem.
    Use the above theorem, in the following:

    In an equilateral triangle ABC, D is a point an side BC such that BD = $\frac{1}{3}$ BC.

    Prove that $9 AD^2 = 7 AB^2$.

29. A container, opened from the top and made up of a metal sheet, is in the form of a frustum of a cone of height 16 cm with radii of its lower and upper ends as 8 cm and 20 cm, respectively. Find the cost of the milk which can completely fill the container, at the rate of Rs. 20 per litre.

    Also find the cost of metal sheet used to make the container, if it cost Rs. 8 per 100 cm². (Take π = 3.14).

    Or

    A solid consisting of a right circular cone of height 120 cm and radius 60 cm standing on a hemisphere of radius 60 cm is placed up right in a right circular cylinder full of water such that it touches the bottom. Find the volume of water left in the cylinder, if the radius of the cylinder is 60 cm and its height is 180 cm.

30. A life insurance agent found the following data for distribution of ages of 100 policy holders. Calculate the mediam age, if policies are given only to persons having age 18 years onwards but less than 60 years:

Age (in years)	Number of policy holders
Below 20	2
Below 25	6
Below 30	24
Below 35	45
Below 40	78
Below 45	89
Below 50	92
Below 55	98
Below 60	100

EXAMINATION PAPERS

MATHEMATICS—X

CBSE 2009

Set—1

Time allowed : 3 hours *Maximum marks : 80*

General Instructions :
(i) All questions are compulsory.
(ii) The question paper consists of **30** questions divided into four sections – A, B, C and D. Section A comprises **ten** questions of 1 mark each, Section B comprises of **five** questions of 2 marks each, Section C comprises of **ten** questions of 3 marks each and Section D comprises of **five** questions of **6** marks each.
(iii) All questions in Section A are to be answered in one word, one sentence or as per the exact requirement of the question.
(iv) There is no overall choice. However, an internal choice has been provided in **one** question of 2 marks each, **three** questions of 3 marks each and **two** questions of 6 marks each. You have to attempt only one of the alternatives in all such questions.
(v) In question on construction, the drawings should be neat and exactly as per the given measurements.
(vi) Use of calculators is not permitted.

SECTION A

Question number 1 to 10 carry 1 mark each.

1. The decimal expansion of the rational number $\dfrac{43}{2^4 \cdot 5^3}$, will terminate after how many places of decimals ?
2. For what value of k, (-4) is a zero of the polynomial $x^2 - x - (2k+2)$?
3. For what value of p, are $2p - 1$, 7 and $3p$ three consecutive terms of an A.P. ?
4. In Fig. 1, CP and CQ are tangents to a circle with centre O. ARB is another tangent touching the circle at R. If CP = 11 cm, and BC = 7 cm, then find the length of BR.

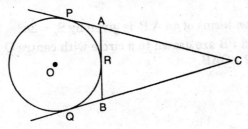

Fig. 1

5. In Fig. 2, ∠M = ∠N = 46°. Express x in terms of a, b and c where a, b and c are lengths of LM, MN and NK respectively.

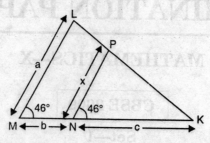

Fig. 2

6. If $\sin \theta = \dfrac{1}{3}$, then find the value of $(2 \cot^2 \theta + 2)$.

7. Find the value of a so that the point $(3, a)$ lies on the line represented by $2x - 3y = 5$.

8. A cylinder and a cone are of same base radius and of same height. Find the ratio of the volume of cylinder to that of the cone.

9. Find the distance between the points $\left(\dfrac{-8}{5}, 2\right)$ and $\left(\dfrac{2}{5}, 2\right)$.

10. Write the median class of the following distribution :

Classes	Frequency
0–10	4
10–20	4
20–30	8
30–40	10
40–50	12
50–60	8
60–70	4

SECTION B

Question number 11 to 15 carry 2 marks each.

11. If the polynomial $6x^4 + 8x^3 + 17x^2 + 21x + 7$ is divided by another polynomial $3x^2 + 4x + 1$, the remainder comes out to be $(ax + b)$, find a and b.

12. Find the value(s) of k for which the pair of linear equations $kx + 3y = k - 2$ and $12x + ky = k$ has no solution.

13. If S_n, the sum of first n terms of an A.P. is given by $S_n = 3n^2 - 4n$, then find its nth term.

14. Two tangents PA and PB are drawn to a circle with centre O from an external point P. Prove that ∠APB = 2 ∠OAB.

EXAMINATION PAPERS

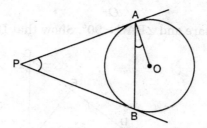

Fig. 3

Or

Prove that the parallelogram circumscribing a circle is a rhombus.

15. Simplify : $\dfrac{\sin^3 \theta + \cos^3 \theta}{\sin \theta + \cos \theta} + \sin \theta \cos \theta$

SECTION C

Question number 16 to 25 carry 3 marks each.

16. Prove that $\sqrt{5}$ is an irrational number.
17. Solve the following pair of equations :

$$\dfrac{5}{x-1} + \dfrac{1}{y-2} = 2$$

$$\dfrac{6}{x-1} - \dfrac{3}{y-2} = 1$$

18. The sum of 4th and 8th terms of an A.P. is 24 and sum of 6th and 10th terms is 44. Find A.P.
19. Construct a $\triangle ABC$ in which BC = 6.5 cm, AB = 4.5 cm and $\angle ABC = 60°$. Construct a triangle similar to this triangle whose sides are $\dfrac{3}{4}$ of the corresponding sides of the triangle ABC.
20. In Fig. 4, $\triangle ABC$ is right angled at C and DE $\perp$ AB. Prove that $\triangle ABC \sim \triangle ADE$ and hence find the lengths of AE and DE.

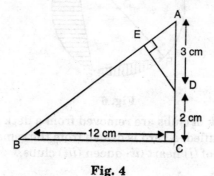

Fig. 4

Or

In Fig. 5, DEFG is a square and $\angle BAC = 90°$. Show that $DE^2 = BD \times EC$.

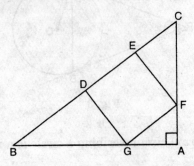

Fig. 5

21. Find the value of sin 30° geometrically.

Or

Without using trigonometrical tables, evaluate :

$$\frac{\cos 58°}{\sin 32°} + \frac{\sin 22°}{\cos 68°} - \frac{\cos 38° \operatorname{cosec} 52°}{\tan 18° \cdot \tan 35° \tan 60° \tan 72° \tan 55°}$$

22. Find the point on y-axis which is equidistant from the points $(5, -2)$ and $(-3, 2)$.

Or

The line segment joining the points A $(2, 1)$ and B $(5, -8)$ is trisected at the points P and Q such that P is nearer to A. If P also lies on the line given by $2x - y + k = 0$, find the value of k.

23. If P (x, y) is any point on the line joining the points A $(a, 0)$ and B $(0, b)$, then show that $\frac{x}{a} + \frac{y}{b} = 1$.

24. In Fig. 6 PQ = 24 cm, PR = 7 cm and O is the centre of the circle. Find the area of shaded region (take $\pi = 3.14$)

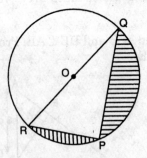

Fig. 6

25. The king, queen and jack of clubs are removed from a deck of 52 playing cards and the remaining cards are shuffled. A card is drawn from the remaining cards. Find the probability of getting a card of (*i*) heart (*ii*) queen (*iii*) clubs.

SECTION D

Question number 26 to 30 carry 6 marks each.

26. The sum of the squares of two consecutive odd numbers is 394. Find the numbers.

    Or

    Places A and B are 100 km apart on a highway. One car starts from A and another from B at the same time. If the cars travel in the same direction at different speeds, they meet in 5 hours. If they travel towards each other, they meet in 1 hour. What are the speeds of the two cars?

27. Prove that, if a line is drawn parallel to one side of a triangle to intersect the other two sides in distinct points, the other two sides are divided in the same ratio.

    Using the above result, do the following:

    In Fig. 7, DE ∥ BC and BD = CE. Prove that ΔABC is an isosceles triangle.

    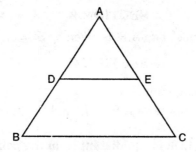

    Fig. 7

28. A straight highway leads to the foot of a tower. A man standing at the top of the tower observes a car at an angle of depression of 30°, which is approaching the foot of the tower with a uniform speed. Six seconds later the angle of depression of the car is found to be 60°. Find the time taken by the car to reach the foot of the tower from this point.

29. From a solid cylinder whose height is 8 cm and radius 6 cm, a conical cavity of height 8 cm and of base radius 6 cm, is hollowed out. Find the volume of the remaining solid correct to two places of decimals. Also find the total surface area of the remaining solid. (take π = 3.1416)

    Or

    In Fig. 8, ABC is a right triangle right angled at A. Find the area of shaded region if AB = 6 cm, BC = 10 cm and O is the centre of the incircle of ΔABC. (take π = 3.14)

    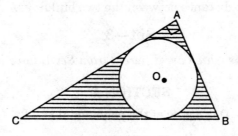

    Fig. 8

30. The following table gives the daily income of 50 workers of a factory :

Daily income (in Rs.)	100–120	120–140	140–160	160–180	180–200
Number of workers	12	14	8	6	10

Find the Mean, Mode and Median of the above data.

Set—2

[*Only those questions which are different from Set 1, have been given here.*]

SECTION A

3. For what value of p are $2p + 1$, 13, $5p - 3$, three consecutive terms of an AP ?

SECTION B

13. If S_n, the sum of first n terms of an A.P. is given by $S_n = 5n^2 + 3n$, then find its nth term.

SECTION C

16. Prove that $\sqrt{3}$ is an irrational number.

18. The sum of 5th and 9th terms of an A.P. is 72 and the sum of 7th and 12th terms is 97. Find the A.P.

22. Find the point on x-axis which is equidistant from the points $(2, -5)$ and $(-2, 9)$.

Or

The line segment joining the points $P(3, 3)$ and $Q(6, -6)$ is trisected at the points A and B such that A is nearer to P. If A also lies on the line given by $2x + y + k = 0$, find the value of k.

SECTION D

27. Prove that, in a right triangle, the square of the hypotenuse is equal to the sum of squares of the other two sides.

    Using the above, do the following :

    Prove that, in a $\triangle ABC$, if AD is perpendicular to BC, then $AB^2 + CD^2 = AC^2 + BD^2$.

28. The angles of depression of the top and bottom of an 8 m tall building from the top of a multi-storeyed building are 30° and 45°, respectively. Find the height of the multi-storeyed building and the distance between the two buildings.

Set—3

[*Only those questions which are different from Set 1, have been given here.*]

SECTION A

2. For what value of p, (-4) is a zero of the polynomial $x^2 - 2x - (7p + 3)$?

SECTION B

11. If the polynomial $x^4 + 2x^3 + 8x^2 + 12x + 18$ is divided by another polynomial $x^2 + 5$, the remainder comes out to be $px + q$. Find the values of p and q.

EXAMINATION PAPERS 739

SECTION C

16. Prove that $\sqrt{2}$ is an irrational number.
17. Solve the following pair of equations :

$$\frac{10}{x+y} + \frac{2}{x-y} = 4$$

$$\frac{15}{x+y} - \frac{5}{x-y} = -2$$

19. Construct a $\triangle ABC$ in which BC = 6.5 cm, AB = 4.5 cm and $\angle ACB = 60°$. Construct another triangle similar to $\triangle ABC$ such that each side of new triangle is $\frac{4}{5}$ of the corresponding sides of $\triangle ABC$.

SECTION D

27. Prove that the ratio of the areas of two similar triangles is equal to the square of the ratio of their corresponding sides.

    Using the above, do the following :

    In a trapezium ABCD, AC and BD are intersecting at O, AB || DC and AB = 2 CD. If area of $\triangle AOB = 84$ cm^2, find the area of $\triangle COD$.

28. A pole of height 5 m is fixed on the top of a tower. The angle of elevation of the top of the pole as observed from a point A on the ground is 60° and the angle of depression of the point A from the top of the tower is 45°. Find the height of the tower. (take $\sqrt{3} = 1.732$)

A.I. CBSE 2009

Set—1

SECTION A

Question number 1 to 10 carry 1 mark each.

1. Find the [HCF × LCM] for the numbers 100 and 190.
2. If 1 is a zero of the polynomial $p(x) = ax^2 - 3(a-1)x - 1$, then find the value of a.
3. In $\triangle$ LMN, $\angle$ L = 50° and $\angle$ N = 60°. If $\triangle$ LMN ~ $\triangle$ PQR, then find $\angle$ Q.
4. If $\sec^2 \theta (1 + \sin \theta)(1 - \sin \theta) = k$, then find the value of k.
5. If the diameter of a semicircular protractor is 14 cm, then find its perimeter.
6. Find the number of solutions of the following pair of linear equations :

   $x + 2y - 8 = 0$
   $2x + 4y = 16$

7. Find the discriminant of the quadratic equation

   $3\sqrt{3}x^2\ 10x + \sqrt{3} = 0$.

8. If $\frac{4}{5}, a, 2$ are three consecutive terms of an A.P., then find the value of a.

9. In Figure 1, △ ABC is circumscribing a circle. Find the length of BC.

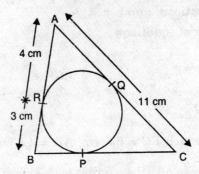

Figure 1

10. Two coins are tossed simultaneously. Find the probability of getting exactly one head.

SECTION B

Question number 11 to 15 carry 2 marks each.

11. Find all the zeroes of the polynomial $x^3 + 3x^2 - 2x - 6$, if two of its zeroes are $-\sqrt{2}$ and $\sqrt{2}$.

12. Which term of the A.P., 3, 15, 27, 39,... will be 120 more than its 21st term?

13. In Figure 2, △ ABD is a right triangle, right-angled at A and AC ⊥ BD. Prove that $AB^2 = BC \cdot BD$.

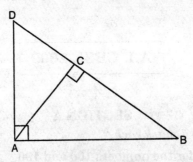

Figure 2

14. If $\cot\theta = \dfrac{15}{8}$, then evaluate $\dfrac{(2+2\sin\theta)(1-\sin\theta)}{(1+\cos\theta)(2-2\cos\theta)}$.

Or

Find the value of tan 60°, geometrically.

15. If the points A (4, 3) and B(x, 5) are on the circle with the centre O (2, 3), find the value of x.

SECTION C

Question number 16 to 25 carry 3 marks each.

16. Prove that $3 + \sqrt{2}$ is an irrational number.

17. Solve for x and y:

$$\frac{ax}{b} - \frac{by}{a} = a+b$$
$$ax - by = 2ab$$

Or

The sum of two numbers is 8. Determine the numbers if the sum of their reciprocals is $\frac{8}{15}$.

18. The sum of first six terms of an arithmetic progression is 42. The ratio of its 10th term to its 30th term is 1 : 3. Calculate the first and the thirteenth term of the A.P.

19. Evaluate :

$$\frac{2}{3} \operatorname{cosec}^2 58° - \frac{2}{3} \cot 58° \tan 32° - \frac{5}{3} \tan 13° \tan 37° \tan 45° \tan 53° \tan 77°$$

20. Draw a right triangle in which sides (other than hypotenuse) are of lengths 8 cm and 6 cm. Then construct another triangle whose sides are $\frac{3}{4}$ times the corresponding sides of the first triangle.

21. In Figure 3, AD $\perp$ BC and BD = $\frac{1}{3}$ CD. Prove that

$$2\,CA^2 = 2\,AB^2 + BC^2.$$

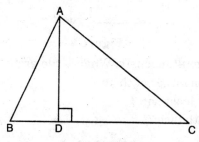

Figure 3

Or

In Figure 4, M is mid-point of side CD of a parallelogram ABCD. The line BM is drawn intersecting AC at L and AD produced at E.
Prove that EL = 2 BL.

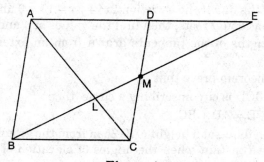

Figure 4

22. Find the ratio in which the point (2, y) divides the line segment joining the points A (– 2, 2) and B (3, 7). Also find the value of y.
23. Find the area of the quadrilateral ABCD whose vertices are A (– 4, – 2), B (– 3, – 5), C (3, – 2), and D (2, 3).
24. The area of an equilateral triangle is $49\sqrt{3}$ cm². Taking each angular point as centre, circles are drawn with radius equal to half the length of the side of the triangle. Find the area of triangle not included in the circles. [Take $\sqrt{3}$ = 1.73]

Or

Figure 5 shows a decorative block which is made of two solids – a cube and a hemisphere. The base of the block is a cube with edge 5 cm and the hemisphere, fixed on the top, has a diameter of 4.2 cm. Find the total surface area of the block. [Take $\pi = \dfrac{22}{7}$]

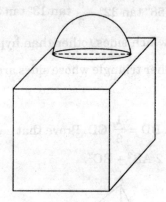

Figure 5

25. Two dice are thrown simultaneously. What is the probability that
    (i) 5 will not come up on either of them ?
    (ii) 5 will come up on at least one ?
    (iii) 5 will come up at both dice ?

SECTION D

Question number 26 to 30 carry 6 marks each.

26. Solve the following equation for x :
$$9x^2 - 9(a + b)x + (2a^2 + 5ab + 2b^2) = 0$$

Or

If (– 5) is a root of the quadratic equation $2x^2 + px - 15 = 0$ and the quadratic equation $p(x^2 + x) + k = 0$ has equal roots, then find the values of p and k.

27. Prove that the lengths of the tangents drawn from an external point to a circle are equal.

Using the above theorem prove that :

If quadrilateral ABCD is circumscribing a circle, then
    AB + CD = AD + BC.

28. An aeroplane when flying at a height of 3125 m from the ground passes vertically below another plane at an instant when the angles of elevation of the two planes from the

same point on the ground are 30° and 60° respectively. Find the distance between the two planes at that instant.

29. A juice seller serves his customers using a glass as shown in Figure 6. The inner diameter of the cylindrical glass is 5 cm, but the bottom of the glass has a hemispherical portion raised which reduces the capacity of the glass. If the height of the glass is 10 cm, find the apparent capacity of the glass and its actual capacity. (Use π = 3.14).

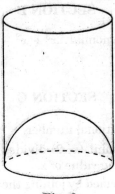

Fig. 6

Or

A cylindrical vessel with internal diameter 10 cm and height 10.5 cm is full of water. A solid cone of base diameter 7 cm and height 6 cm is completely immersed in water. Find the volume of

(i) water displaced out of the cylindrical vessel.

(ii) water left in the cylindrical vessel. [Take π = $\frac{22}{7}$]

30. During the medical check-up of 35 students of a class their weights were recorded as follows :

Weight (in kg)	Number of students
38 – 40	3
40 – 42	2
42 – 44	4
44 – 46	5
46 – 48	14
48 – 50	4
50 – 52	3

Draw a less than type and a more than type ogive from the given data. Hence obtain the median weight from the graph.

Set—2

[*Only those questions which are different from Set 1, have been given here.*]

SECTION A

1. Find the [HCF × LCM] for the numbers 105 and 120.

SECTION B

11. Find all the zeroes of the polynomial $2x^3 + x^2 - 6x - 3$, if two of its zeroes are $-\sqrt{3}$ and $\sqrt{3}$.

SECTION C

16. Prove that $(5 - 2\sqrt{3})$ is an irrational number.
22. Find the ratio in which the point $(x, 2)$ divides the line segment joining the points $(-3, -4)$ and $(3, 5)$. Also find the value of x.
23. Find the area of the triangle formed by joining the mid-points of the sides of the triangle whose vertices are $(0, -1)$, $(2, 1)$ and $(0, 3)$.

SECTION D

27. In a triangle, if the square on one side is equal to the sum of the squares on the other two sides, prove that the angle opposite to the first side is a right angle.

    Use the above theorem to find the measure of $\angle PKR$ in Figure 7.

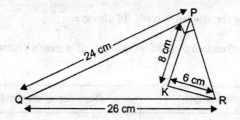

Figure 7

28. From the top of a building 60 m high, the angles of depression of the top and bottom of a vertical lamp post are observed to be 30° and 60° respectively. Find
    (*i*) The horizontal distance between the building and the lamp post.
    (*ii*) The height of the lamp post.

    [Take $\sqrt{3} = 1.732$]

Set—3

[*Only those questions which are different from Set 1, have been given here.*]

SECTION A

3. If the areas of two similar triangles are in the ratio 25 : 64, write the ratio of their corresponding sides.

SECTION B

12. Which term of the A.P. 4, 12, 20, 28, ... will be 120 more than its 21st term?

SECTION C

16. Prove that $(3 + 5\sqrt{2})$ is an irrational number.

22. Find the ratio in which the point $(x, -1)$ divides the line segment joining the points $(-3, 5)$ and $(2, -5)$. Also find the value of x.

23. Find the area of the quadrilateral ABCD whose vertices are A $(1, 0)$, B $(5, 3)$, C $(2, 7)$ and D $(-2, 4)$.

SECTION D

27. Prove that the tangent at any point of a circle is perpendicular to the radius through the point of contact.

Using the above, do the following:

In figure 7, O is the centre of the two concentric circles. AB is a chord of the larger circle touching the smaller circle at C. Prove that AC = BC.

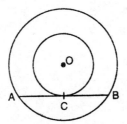

Figure 7

28. A 1.2 m tall girl spots a balloon moving with the wind in a horizontal line at a height of 88.2 m from the ground. The angle of elevation of the balloon from the eyes of the girl at that instant is 60°. After some time, the angle of elevation reduces to 30°. Find the distance travelled by the balloon during the interval.

SECTION B

12. Which been at the 2nd, 14, 20, 26... is/will the 20 more than 1st 21st term.

SECTION C

16. Prove that $7^4 \cdot 5\sqrt{2}$ is an irrational number.
21. Find the ratio in which the point $(x, -1)$ divides the line segment joining the points $(-3, 3)$ and $(6, -5)$. Also find the value of x.
22. Find the area of the quadrilateral ABCD whose vertices are $A(-3, -1)$, $B(-2, -4)$, $C(4, -1)$ and $D(3, 4)$.

SECTION D

27. Prove that the tangent at any point of a circle is perpendicular to the radius through the point of contact.

Using th. above, do the following:

In figure 7.03, PA is chord of the two concentric circles. PA is tangent to the smaller circle touching the smaller circle at C. Prove that $AC = BC$.

Figure 7

28. A 1.2 m tall girl spots a balloon moving with the wind in a horizontal line at a height of 88.2 m from the ground. The angle of elevation of the balloon from the eyes of the girl at that instant is 60°. After some time, the angle of elevation reduces to 30°. Find the distance travelled by the balloon during the interval.

EXAMINATION PAPERS

MATHEMATICS—X
A.I. CBSE 2009
Set—1

Time Allowed : 3 hours   Maximum Marks : 80

General Instructions :
  (i) All questions are compulsory.
  (ii) The question paper consists of **30** questions divided into four sections – **A, B, C** and **D**. Section **A** comprises **ten** questions of *1* mark each, Section **B** comprises of *five* questions of **2** marks each, Section **C** comprises of **ten** questions of *3* marks each and Section **D** comprises of *five* questions of **6** marks each.
  (iii) All questions in Section **A** are to be answered in one word, one sentence or as per the exact requirement of the question.
  (iv) There is no overall choice. However, an internal choice has been provided in one question of **2** marks each, three questions of **3** marks each and two questions of **6** marks each. You have to attempt only one of the alternatives in all such questions.
  (v) In question on construction, the drawings should be neat and exactly as per the given measurement.
  (vi) Use of calculators is not permitted.

SECTION A

Questions number 1 to 10 carry 1 mark each.

1. Write the prime factors of 84.
2. Write the polynomial whose zeros are – 5 and 4.
3. Is $x = 2, y = 3$ a solution of the linear equation $2x + 3y - 13 = 0$?
4. Write the next term of the A.P. $\sqrt{2}, \sqrt{8}, \sqrt{18}, \ldots$.
5. If $\cos A = \dfrac{3}{5}$, find $9 \cot^2 A - 1$.
6. If the probability of winning a game is $\dfrac{5}{11}$, what is the probability of losing it ?
7. In Figure 1, $\angle ABC = 90°$ and P is the mid-point of AC. Find the length of AP.

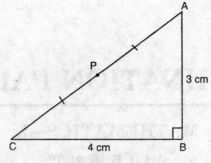

Figure 1

8. In Figure 2, find the perimeter of △ ABC if AP = 10 cm.

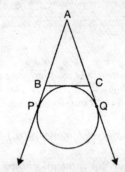

Figure 2

9. The radii of two circles are 3 cm and 4 cm. Find the radius of a circle whose area is equal to the sum of the areas of the two circles.
10. The point of intersection of the ogives (more than and less than type) is given by (20.5, 30.4). What is the median ?

SECTION B

Questions number 11 to 15 carry 2 marks each.

11. Find the zeros of the quadratic polynomial $2x^2 - 9 - 3x$ and verify the relationship between the zeros and the coefficients.
12. If $3 \cot \theta = 4$, find the value of $\dfrac{5 \sin \theta - 3 \cos \theta}{5 \sin \theta + 3 \cos \theta}$.

Or

Without using trigonometric tables, find the value of the following :

$$\left(\dfrac{\tan 20°}{\cosec 70°}\right)^2 + \left(\dfrac{\cot 20°}{\sec 70°}\right)^2 + 2 \tan 15° \tan 45° \tan 75°$$

13. If the vertices of a triangle are $(1, k)$, $(4, -3)$, $(-9, 7)$ and its area is 15 square units, find the value of k.
14. In Figure 3, DE ∥ AC and $\dfrac{BE}{EC} = \dfrac{BC}{CP}$. Prove that DC ∥ AP.

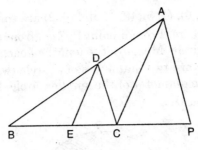

Figure 3

15. A box contains cards bearing numbers from 6 to 70. If one card is drawn at random from the box, find the probability that it bears (*i*) a one digit number, (*ii*) a number divisible by 5.

SECTION C

Questions number 16 to 25 carry 3 marks each.

16. Using prime factorisation method, find the HCF and LCM of 72, 126 and 168. Also show that HCF × LCM ≠ Product of the three numbers.

17. Represent the following system of linear equations graphically. From the graph, find the points where the lines intersect x-axis:
    $$2x - y = 2, \quad 4x - y = 8$$

18. Solve the following system of equations for x and y:
    $$(a - b)x + (a + b)y = a^2 - 2ab - b^2$$
    $$(a + b)(x + y) = a^2 + b^2$$

    Or

    For what value of 'm' will the equation $2m\,x^2 - 2(1 + 2m)x + (3 + 2m) = 0$ have real but distinct roots ? When will the roots be equal ?

19. Prove that :
    $$\frac{\cos\theta - \sin\theta + 1}{\cos\theta + \sin\theta - 1} = \csc\theta + \cot\theta$$

    Or

    Prove that :
    $$(\csc\theta - \sin\theta)(\sec\theta - \cos\theta) = \frac{1}{\tan\theta + \cot\theta}$$

20. Find the sum of all two digit natural numbers which when divided by 3 yield 1 as remainder.

21. The line joining the points (2, –1) and (5, –6) is bisected at P. If P lies on the line $2x + 4y + k = 0$, find the value of k.

22. In △ ABC, if AD is the median, show that
    $$AB^2 + AC^2 = 2(AD^2 + BD^2)$$

    Or

    In △ ABC, ∠A is acute. BD and CE are perpendicular on AC and AB respectively. Prove that
    $$AB \times AE = AC \times AD$$

23. Show that the points (5, 6), (1, 5), (2, 1) and (6, 2) are the vertices of a square.
24. Draw a circle of radius 4 cm. From a point P, 7 cm from the centre of the circle, draw a pair of tangents to the circle. Measure the length of each tangent segment.
25. In Figure 4, AB and CD are two diameters of a circle (with centre O) perpendicular to each other and OD is the diameter of the smaller circle. If OA = 14 cm, find the area of the shaded region.

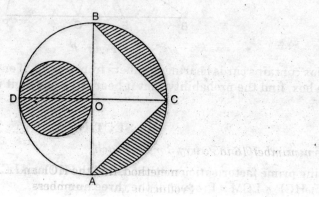

Figure 4

SECTION D

Questions number 26 to 30 carry 6 marks each.

26. Rs. 6,500 is divided equally among a certain number of persons. Had there been 15 more persons, each would have got Rs. 30 less. Find the original number of persons.

Or

A train travels 360 km at a uniform speed. If the speed of the train had been 5 km/hour more, it would have taken one hour less for the same journey. Find the original speed of the train.

27. The angle of elevation of a cloud from a point 60 m above a lake is 30° and the angle of depression of the reflection of the cloud in the lake is 60°. Find the height of the cloud.

28. Prove that the length of the tangents drawn from an external point to a circle are equal. Using the above do the following :

Prove that the angle between the two tangents drawn from an external point to a circle is supplementary to the angle subtended by the line segment joining the points of contact at the centre.

Or

Prove that the ratio of the areas of two similar triangles is equal to the ratio of the squares on their corresponding sides.

Using the above, prove the following :

If the areas of two similar triangles are equal, prove that they are congruent.

29. A bucket is in the form of a frustum of a cone whose radii of bottom and top are 7 cm and 28 cm respectively. If the capacity of the bucket is 21560 cm³, find the whole surface area of the bucket.

30. Find the mean, mode and median for the following data :

Classes	10 – 20	20 – 30	30 – 40	40 – 50	50 – 60	60 – 70	70 – 80
Frequency	4	8	10	12	10	4	2

Set—2

[Only those questions which are different from Set 1, have been given here.]

SECTION A

9. Write the prime factors of 546.

10. Write the polynomial whose zeros are $\frac{2}{3}$ and $-\frac{1}{3}$.

SECTION B

11. What is the probability of having 53 Mondays in a leap year?

14. Find the zeros of the quadratic polynomial $2x^2 - 3 + 5x$ and verify the relationship between the zeros and the coefficients.

SECTION C

23. Prove that:
$$\frac{\text{cosec A}}{\text{cosec A} - 1} + \frac{\text{cosec A}}{\text{cosec A} + 1} = 2 \sec^2 A$$

Or

Without using trigonometric tables, evaluate the following:
$$\frac{\text{cosec}^2 65° - \tan^2 25°}{\sin^2 17° + \sin^2 73°} + \frac{1}{\sqrt{3}} (\tan 10° \tan 30° \tan 80°)$$

24. Find the sum of all two-digit natural numbers which when divided by 7 yield 1 as remainder.

SECTION D

26. Find the mean, mode and median for the following data:

Classes	5 – 15	15 – 25	25 – 35	35 – 45	45 – 55	55 – 65	65 – 75
Frequency	2	3	5	7	4	2	2

29. As observed from the top of a 75 m high lighthouse (from sea-level), the angles of depression of two ships are 30° and 45°. If one ship is exactly behind the other on the same side of the lighthouse, find the distance between the two ships.

Set—3

[Only those questions which are different from Set 1, have been given here.]

SECTION A

8. Is $x = 3$, $y = 4$ a solution of the linear equation $4x + 3y - 30 = 0$?

9. Write the polynomial whose zeros are $2 + \sqrt{3}$ and $2 - \sqrt{3}$.

SECTION B

11. From a bag, containing 5 red, 8 black and 7 blue balls, a ball is selected at random. Find the probability that (i) it is not a red ball, (ii) it is not a blue ball.

12. If the vertices of a triangle are (2, 4), (5, k), (3, 10) and its area is 15 square units, find the value of k.

SECTION C

24. Prove the following :
$$\frac{1}{\sec A + \tan A} - \frac{1}{\cos A} = \frac{1}{\cos A} - \frac{1}{\sec A - \tan A}$$
Or

Without using trigonometric tables, evaluate the following :
$$\sec 41° \sin 49° + \cos 49° \cdot \csc 41° - \frac{2}{\sqrt{3}} (\tan 20° \cdot \tan 60° \tan 70°).$$

25. Find the sum of all two digit numbers greater than 50 which when divided by 7 leave a remainder of 4.

SECTION D

26. The angle of elevation of a jet-plane from a point P on the ground is 60°. After a flight of 15 seconds, the angle of elevation changes to 30°. If the jet-plane is flying at a constant height of $1500\sqrt{3}$ m, find the speed of the jet-plane in km/hour.

28. Find the mean, mode and median for the following data :

Classes	0 – 10	10 – 20	20 – 30	30 – 40	40 – 50	50 – 60	60 – 70
Frequency	5	8	15	20	14	8	5

CBSE 2008
Set—1
SECTION A

1. If $\frac{p}{q}$ is a rational number ($q \neq 0$), what is condition on q so that the decimal representation of $\frac{p}{q}$ is terminating ?

2. Write the zeroes of the polynomial $x^2 + 2x + 1$.

3. Find the value of k so that the following system of equations has no solution :
$$3x - y - 5 = 0 \;;\; 6x - 2y - k = 0$$

4. The nth term of an A.P. is $7 - 4n$. Find its common difference.

5. In Fig. 1, AD = 4 cm, BD = 3 cm and CB = 12 cm, find cot θ.

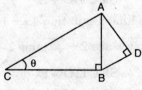

Fig. 1

6. In Fig. 2, P and Q are points on the sides AB and AC respectively of △ABC such that AP = 3.5 cm, PB = 7 cm, AQ = 3 cm and QC = 6 cm. If PQ = 4.5 cm, find BC.

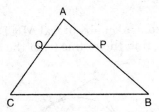

Fig. 2

7. In Fig. 3, PQ = 24 cm, QR = 26 cm, ∠PAR = 90°, PA = 6 cm and AR = 8 cm. Find ∠QPR.

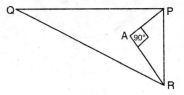

Fig. 3

8. In Fig. 4, O is the centre of a circle. The area of sector OAPB is $\frac{5}{18}$ of the area of the circle. Find x.

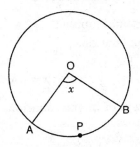

Fig. 4

9. Which measure of central tendency is given by the x-coordinate of the point of intersection of the "more than ogive" and "less than ogive"?

10. From a well shuffled pack of cards, a card is drawn at random. Find the probability of getting a black queen.

SECTION B

11. Find the zeroes of the quadratic polynomial $6x^2 - 3 - 7x$ and verify the relationship between the zeroes and the co-efficients of the polynomial.

12. Without using the trigonometric tables, evaluate the following:

$$\frac{11}{7}\frac{\sin 70°}{\cos 20°} - \frac{4}{7}\frac{\cos 53° \operatorname{cosec} 37°}{\tan 15° \tan 35° \tan 55° \tan 75°}$$

13. For what value of p, are the points (2, 1), (p, –1) and (–1, 3) collinear?

14. ABC is an isosceles triangle, in which AB = AC, circumscribed about a circle. Show that BC is bisected at the point of contact.

Or

In Fig. 5, a circle is inscribed in a quadrilateral ABCD in which $\angle B = 90°$. If AD = 23 cm, AB = 29 cm and DS = 5 cm, find the radius (r) of the circle.

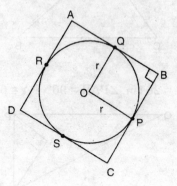

Fig. 5

15. A die is thrown once. Find the probability of getting
    (i) a prime number
    (ii) a number divisible by 2

SECTION C

16. Show that $5 - 2\sqrt{3}$ is an irrational number.

17. Find the roots of the following equation :
$$\frac{1}{x+4} - \frac{1}{x-7} = \frac{11}{30} \; ; x \neq -4, 7$$

18. Represent the following system of linear equations graphically. From the graph, find the points where the lines intersect y-axis :
$$3x + y - 5 = 0 \; ; \; 2x - y - 5 = 0$$

19. The sum of n terms of an A.P. is $5n^2 - 3n$. Find the A.P. Hence, find its 10th term.

20. Prove that :
$$\frac{\cot A - \cos A}{\cot A + \cos A} = \frac{\operatorname{cosec} A - 1}{\operatorname{cosec} A + 1}$$

Or

Prove that :
$$(1 + \cot A - \operatorname{cosec} A)(1 + \tan A + \sec A) = 2$$

21. Determine the ratio in which the line $3x + 4y - 9 = 0$ divides the line-segment joining the points (1, 3) and (2, 7).

22. Construct a $\triangle ABC$ in which AB = 6.5 cm, $\angle B = 60°$ and BC = 5.5 cm. Also construct a triangle AB'C' similar to $\triangle ABC$, whose each side is $\frac{3}{2}$ times the corresponding side of the $\triangle ABC$.

23. If the diagonals of a quadrilateral divide each other proportionally, prove that it is a trapezium.

Or

Two $\triangle$s ABC and DBC are on the same base BC and on the same side of BC in which $\angle A = \angle D = 90°$. If CA and BD meet each other at E, show that AE.EC = BE.ED.

24. If the distances of $P(x, y)$ from the points A(3, 6) and B(–3, 4) are equal, prove that $3x + y = 5$.

25. In Fig. 6, find the perimeter of shaded region where ADC, AEB and BFC are semi-circles on diameters AC, AB and BC respectively.

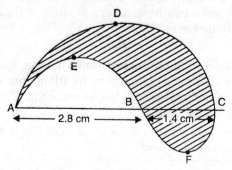

Fig. 6

Or

Find the area of the shaded region in Fig. 7, where ABCD is a square of side 14 cm.

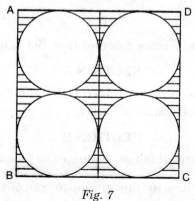

Fig. 7

SECTION D

26. In a class test, the sum of the marks obtained by P in Mathematics and Science is 28. Had he got 3 more marks in Mathematics and 4 marks less in Science, the product of marks obtained in the two subjects would have been 180. Find the marks obtained in the two subjects separately.

Or

The sum of the areas of two squares is 640 m². If the difference in their perimeters be 64 m, find the sides of the two squares.

27. A statue 1.46 m tall, stands on the top of a pedestal. From a point on the ground, the angle of elevation of the top of the statue is 60° and from the same point, the angle of elevation of the top of the pedestal is 45°. Find the height of the pedestal (use $\sqrt{3} = 1.73$).

28. Prove that the ratio of the areas of two similar triangles is equal to the ratio of squares of their corresponding sides.

    Using the above result, prove the following :

    In a $\triangle ABC$, XY is parallel to BC and it divides $\triangle ABC$ into two parts of equal area.

    Prove that $\dfrac{BX}{AB} = \dfrac{\sqrt{2}-1}{\sqrt{2}}$.

29. A gulab jamun, when ready for eating, contains sugar syrup of about 30% of its volume. Find approximately how much syrup would be found in 45 such gulab jamuns, each shaped like a cylinder with two hemispherical ends, if the complete length of each of them is 5 cm and its diameter is 2.8 cm.

    Or

    A container shaped like a right circular cylinder having diameter 12 cm and height 15 cm is full of ice-cream. This ice-cream is to be filled into cones of height 12 cm and diameter 6 cm, having a hemispherical shape on the top. Find the number of such cones which can be filled with ice-cream.

30. A survey regarding the heights (in cm) of 50 girls of Class X of a school was conducted and the following data was obtained :

Height in cm	120 – 130	130 – 140	140 – 150	150 – 160	160 – 170	Total
Number of girls	2	8	12	20	8	50

    Find the mean, median and mode of the above data.

Set—2

[Only those questions which are different from Set 1, have been given here.]

SECTION A

2. The nth term of an A.P. is $6n + 2$. Find its common difference.

4. Write the zeroes of the polynomial $x^2 - x - 6$.

SECTION B

12. Without using trigonometrical tables, evaluate the following :

    $\dfrac{\sin 18°}{\cos 72°} + \sqrt{3}\ [\tan 10° \tan 30° \tan 40° \tan 50° \tan 80°]$

14. A die is thrown once. Find the probability of getting

    (i) an even prime number

    (ii) a multiple of 3.

SECTION C

17. Find the 10th term from the end of the A.P. 8, 10, 12, ..., 126.

20. Show that $2 - \sqrt{3}$ is an irrational number.

SECTION D

28. 100 surnames were randomly picked up from a local telephone directory and the distribution of number of letters of the English alphabet in the surnames was obtained as follows :

No. of letters	1 – 4	4 – 7	7 – 10	10 – 13	13 – 16	16 – 19
Number of surnames	6	30	40	16	4	4

Determine the median and mean number of letters in the surnames. Also find the modal size of surnames.

30. A bucket made up of a metal sheet is in the form of a frustum of a cone of height 16 cm with diameters of its lower and upper ends as 16 cm and 40 cm respectively. Find the volume of the bucket. Also find the cost of the bucket if the cost of metal sheet used is Rs. 20 per 100 cm². (use $\pi = 3.14$)

Or

A farmer connects a pipe of internal diameter 20 cm from a canal into a cylindrical tank in his field which is 10 m in diameter and 2 m deep. If water flows through the pipe at the rate of 6 km/h., in how much time will the tank be filled ?

Set—3

[Only those questions which are different from Set 1, have been given here.]

SECTION A

1. Write a quadratic polynomial, the sum and product of whose zeroes are 3 and – 2 respectively.

SECTION B

11. Find the zeroes of the quadratic polynomial $5x^2 - 4 - 8x$ and verify the relationship between the zeroes and the coefficients of the polynomial.
12. For what value of p, the points (– 5, 1), (1, p) and (4, – 2) are collinear ?

SECTION C

16. Show that $5 + 3\sqrt{2}$ is an irrational number.
17. The sum of n terms of an A.P. is $3n^2 + 5n$. Find the A.P. Hence, find its 16th term.

SECTION D

26. A person standing on the bank of a river observes that the angle of elevation of the top of a tree standing on the opposite bank is 60°. When he moves 40 m away from the bank, he finds the angle of elevation to be 30°. Find the height of the tree and the width of the river. [use $\sqrt{3} = 1.732$]
27. Prove that the lengths of tangents drawn from an external point to a circle are equal. Using the above, prove the following :
A quadrilateral ABCD is drawn to circumscribe a circle (Fig. 8). Prove that
AB + CD = AD + BC.

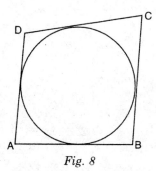

Fig. 8